Elements of Theoretical Plasma Physics

A Mathematical Description of Plasma Waves

Elements of Theoretical Plasma Physics

A Mathematical Description of Plasma Waves

Hee J. Lee

Hanyang University, South Korea

World Scientific

NEW JERSEY • LONDON • SINGAPORE • BEIJING • SHANGHAI • HONG KONG • TAIPEI • CHENNAI • TOKYO

Published by

World Scientific Publishing Co. Pte. Ltd.
5 Toh Tuck Link, Singapore 596224
USA office: 27 Warren Street, Suite 401-402, Hackensack, NJ 07601
UK office: 57 Shelton Street, Covent Garden, London WC2H 9HE

Library of Congress Control Number: 2026942693

British Library Cataloguing-in-Publication Data
A catalogue record for this book is available from the British Library.

ELEMENTS OF THEORETICAL PLASMA PHYSICS
A Mathematical Description of Plasma Waves

ISBN 978-981-98-2273-7 (hardcover)
ISBN 978-981-98-2274-4 (ebook for institutions)
ISBN 978-981-98-2275-1 (ebook for individuals)

For any available supplementary material, please visit
https://www.worldscientific.com/worldscibooks/10.1142/14568#t=suppl

Desk Editors: Eshak Nabi Akbar Ali/Muhammad Ihsan

Typeset by Stallion Press
Email: enquiries@stallionpress.com

About the Author

Hee J. Lee was born in Korea 1941. After graduating from Seoul National University (1963) with a B.S. in physics, he received a Ph.D. degree from the University of California at San Diego in plasma physics 1975. Earlier he attended the graduate school of Seoul National University majoring in fluid mechanics and Cornell University with a major in space plasma physics (M.S.). He worked at NASA Goddard Institute for Space Studies for two years in the project of general circulation of atmosphere. He was a visiting research professor at Clarkson University and University of Michigan, Ann Arbor. He has been a Professor of physics at Hanyang University (1980–2006) and a Professor Emeritus, teaching mathematical method of physics and plasma physics. He is interested in basic theoretical plasma physics, and has published more than 100 refereed papers in international journals. He served on Editorial board of *Journal of the Korean Physical Society* and the *Open Plasma Physics Journal.* His recent publications address the plasma waves predicted by relativistic kinetic equation.

Contents

Introduction

It has been 6 years since the appearance of the book *Fundamentals of Theoretical Plasma Physics, Mathematical Description of Plasma Waves* by the author. During this interval of time, I have had opportunities to teach the material therein and hear criticisms; thus being awakened to realize the necessity of refinement and improvement of the book. This new edition has incorporated the addition of new topics and omission of some, but the scope of the book is largely retained. Also, numerous non-fatal inaccuracies of the first book have been eliminated in this edition. The new inclusions are:

- Alfven surface wave,
- Entropy associated with Boltzmann equation,
- Variational principles of hydrodynamics,
- Variational principles of magnetohydrodynamic equations,
- Surface wave in cylindrical coordinates,
- Relativistic treatment of fluid theory of surface waves,
- Relativistic kinetic equation.

This book presents a coherent review of the plasma state (the fourth state of matter) in regard to its various characteristics, the modes of its description, and its interaction with electromagnetic waves. The mathematical steps leading to the desired or established conclusions are kindly exposited with mindful details, often accompanying alternative routes. The author believes that knowing two different methods that solve the same problem widens and deepens the understanding of the problem under consideration. In theoretician's view, physical reasoning is a short cut for the solution but its validity and its conformity with the governing equations should be checked against painstaking mathematical steps to make sure of the concomitance between the insight and the analysis. Some author's claim that a certain problem can be answered by means of sheer physical reasoning is often his or her physical interpretation of the result that has been acquired by clandestine effort of analytic investigation. This is the reason why we should risk the boredom of scrutinizing the algebraic steps in the relevant development. To the author's opinion, theoretical plasma physics is a discipline which requires the ability to *read and grasp* the mathematical equations without going through the detailed steps, but ironically this ability is acquired only at the expense of diligent algebraic work and necessary patience. This book aims to help the students to cultivate this ability and make their road easy, the road of trekking the rugged analytical terrain.

Putting aside the application side of plasma physics, the nature of plasma physics is theoretical; plasma physics is modeled by the Liouville equation on its foremost foundation, apart from the collaboration with the Maxwell equations. Depending upon various time and parameter scales, the Liouville equation is *evolved* into the Lenard-Balescu equation, or the Landau equation, or the Vlasov equation, or two-fluid equations, or magnetohydrodynamic equations. Therefore, the plasma physics is inherently theoretical, and its mathematical aspects cannot be downgraded. For this reason, this book starts with some salient features of the plasma kinetic theory. The simplest version of the kinetic equation is the Vlasov equation, which is the primary topic of this book.

In this book, familiarity with complex variables, Fourier transformation, and Dirac delta function is required. However, important mathematical background materials that warrant independent introduction are not simply relegated to references but has been explained to the extent that the exposition becomes self-sufficient, so that the readers can follow the relevant physical development without hindrance. For example, the author spent some pages for the mathematical theory of characteristics and the calculus of variation. The characteristic method is useful in dealing with hyperbolic partial differential equations and the calculus of variation is essential for the development of variational principles of hydrodynamics and magnetohydrodynamics. The Liouville equation, the Vlasov equation, and the Euler equation all belong to the category of hyperbolic partial differential equation. Thus the familiarity with the mathematical theory of the characteristics immensely useful in theoretical plasma physics. The usefulness of the characteristic method can be demonstrated in the study of the Liouville equation. Liouville equation and Klimontovich equation are equivalent to the dynamical equation that comprises Maxwell's equations plus Newton's law of motion. It is well known that the Liouville equation describes the manner how the ensemble density (the N-particle distribution function) of the particles composing the system in Γ-space ($6N$-dimensional space with N being the number of particles of the system) behaves. The Liouville equation states that the ensemble of the system points in the Γ-space flows as if the ensemble is an incompressible fluid, representing the conservation of the probability density:

$$\frac{\partial F^{(N)}}{\partial t} + \sum_{i=1}^{N} \left(\frac{\partial H}{\partial \mathbf{p}_i} \cdot \frac{\partial F^{(N)}}{\partial \mathbf{r}_i} - \frac{\partial F^{(N)}}{\partial \mathbf{p}_i} \cdot \frac{\partial H}{\partial \mathbf{r}_i} \right) = 0 \tag{1}$$

with

$$F^{(N)} = F^{(N)}(\mathbf{r}_1, \mathbf{r}_2, \ldots, \mathbf{r}_N, \mathbf{p}_1, \mathbf{p}_2, \ldots, \mathbf{p}_N, t), \tag{2}$$

being the N-particle distribution function; H, the Hamiltonian; $\mathbf{r}$, position vector; and $\mathbf{p}$, momentum vector. The solution of (1), which is a hyperbolic partial differential equation, satisfies the following characteristic equations

$$\frac{d\mathbf{r}_i}{dt} = \frac{\partial H}{\partial \mathbf{p}_i}, \quad \frac{d\mathbf{p}_i}{dt} = -\frac{\partial H}{\partial \mathbf{r}_i} \quad (i = 1, 2, \ldots, N) \tag{3}$$

Equation (3) is Hamilton's equations of motion for each of the N particles. Equation (1) is solved by an arbitrary function of the constants of motion that is obtained by integrating (3). Thus it is clear that $F^{(N)}$ is constant along the trajectory of the system point in the Γ-space. In fact, the trajectory represented by (3) is the curve in Γ-space that minimizes the action integral I taken along that trajectory:

$$\delta I = \delta \int_{t_0}^{t_1} L dt = 0 \tag{4}$$

with L being the Lagrangian, the sum of the kinetic energy of all the particles minus the potential energy. The variational principle expressed by (4) yields the Hamilton's equation of motion in (3). We summarize: the ensemble in Γ-space moves about in a fluid fashion, in conformity with the variational principle (4). [About variational mechanics, Reference: W. Yourgrau and S. Mandelstam, *Variational principles in dynamics and quantum theory*, Dover (1968); R. Weinstock, *Calculus of variations*, McGraw-Hill (1952).] The foregoing synopsis is introduced in this preface to highlight the power of the characteristic method which in this example laborlessly connects two physical ideas: conservation of probability and the principle of the least action.

If we go into the realm of applications of plasma physics, soon our endeavor becomes interdisciplinary. Over the last 70 years (1950–) plasma physics has expanded its application to such areas as controlled thermonuclear fusion, astrophysics, space physics (solar, magnetospheric, ionospheric physics), surface physics such as semiconductor and metal processing; all of them are closely inter-woven with human interests, immediate or remote, since our environment surrounding the earth as well as the deep cosmos consists of plasma state, and the fourth state of matter gradually becomes coexistent with the ordinary states as the human utilities require more ingenuity. This book is not directly concerned with those applications but the author hopes that solid theoretical understanding of plasma will be eventually beneficial to the students who intend to embark on these applications.

In the following, previews of some chapters are addressed to guide or motivate the readers before embarking on the details of the relevant subjects.

• Landau damping is interpreted to be an outcome of the *causality* (Chapter 7). Here the causality means the general physical notion that the response of a medium (effect) cannot be precedent to the cause. Specifically, the electric susceptibility of a medium, $\chi(t)$, is constrained by the condition $\chi(t) = 0$ for $t < 0$. In Chapter 7, Landau damping as well as the Kramers–Kronig relations is derived by applying this algebraic constraint to the susceptibility. We show that the Kramers–Kronig relations, if applied to the electrostatic plasma wave, result in the well-known Landau damping. We might say that the causal interpretation of the Landau damping is more fundamental than the mechanical interpretation that it is a consequence of a wave-particle interaction. As the famous Landau paper [L. Landau, *On the vibrations of the electronic plasma*, J. Physics (USSR) **10** (1946), p. 25] shows, Landau damping is obtained by extending the dielectric permittivity of the Vlasov–Poisson plasma with the aid of the mathematical abstraction of analytic continuation. This abstract mathematical approach somewhat obscures the implication of the causal constraint which defines the direction of time. The causal consequence is clearly and directly revealed when one employs the usual Fourier transform approach if the causal algebraic constraint is enforced.

• Landau's collision integral (Eq. (2.86)) is a type of the *Fokker–Planck equation* which can be derived from the *Smoluchowski law.* The latter applies to *Marcovian processes* and can be expressed in the form

$$f(\mathbf{v}, t + \Delta t) = \int d\Delta\mathbf{v} f(\mathbf{v} - \Delta\mathbf{v}, t) P(\mathbf{v} - \Delta\mathbf{v}, \Delta\mathbf{v}) \tag{5}$$

where $f(\mathbf{v}, t)$ is the distribution function at time t and $P(\mathbf{v} - \Delta\mathbf{v}, \Delta\mathbf{v})$ is the transition probability that the velocity changes from $(\mathbf{v} - \Delta\mathbf{v}) \to \mathbf{v}$ by a small amount $\Delta\mathbf{v}$. The time interval Δt can be assumed long enough for a large number of collisions to take place, but short enough to warrant only small $\Delta\mathbf{v}$. The Markovian change of f in the above is appropriate to describe the collisions in plasma where numerous small angle Coulomb scattering are prevalent. Generally, in a Markovian process, the rate of change of a time-dependent quantity $\partial a(t)/\partial t$ depends on the instantaneous value of the quantity $a(t)$, but not on its previous history. The change is dictated by the transition probability of the event as in the random walk. Let us expand $f(\mathbf{v}, t + \Delta t)$, $P(\mathbf{v} - \Delta\mathbf{v}, \Delta\mathbf{v})$ in Taylor series and define the mean values

$$<\Delta\mathbf{v}> = \int \Delta\mathbf{v} P(\mathbf{v}, \Delta\mathbf{v}) d\Delta\mathbf{v} \tag{6}$$

$$<\Delta v_i \Delta v_j> = \int \Delta v_i \Delta v_j P(\mathbf{v}, \Delta\mathbf{v}) d\Delta\mathbf{v} \tag{7}$$

Then we can obtain from (5)

$$\frac{\partial f}{\partial t} = -\frac{\partial}{\partial v_i} \frac{<\Delta v_i> f}{\Delta t} + \frac{1}{2} \frac{\partial^2}{\partial v_i \partial v_j} \frac{<\Delta v_i \Delta v_j> f}{\Delta t} + \cdots \tag{8}$$

which is known as the Fokker–Planck equation. Earlier, Chandrasekhar applied (5) to guiding center distribution to obtain its time rate of change in plasma [Chandrasekhar, S. *Plasma Physics*; Notes compiled by S. K. Trehan, Phoenix Books, University of Chicago Press (1960)]. In the derivation of the Landau kinetic equation (Section 2.6), which belong to the Fokker–Planck class, we shall not pursue the analysis based on the Smoluchowski law, but directly engage Taylor expansion of the Boltzmann collision integral. The result is the same.

• Magnetohydrodynamics (MHD) deals with the coupling between electromagnetic fields and motion of electrically conducting fluid in *large scale.* We find its applications in various cosmic and geomagnetic conditions. In thermonuclear research, magnetohydrodynamic equations are used as the governing equations to describe the plasma. We must emphasize the importance of the role of the magnetic field, in the magnetohydrodynamic domain, which gives rise to the Lorentz force ($\frac{e}{c}\mathbf{v} \times \mathbf{B}$ or $\mathbf{J} \times \mathbf{B}$) for the charged particles. This force changes the motion of the electrically conducting medium hydrodynamically, and we have an interplay between two disciplines: hydrodynamics and electromagnetism. One fast way to derive the magnetohydrodynamic governing equation is to include the electromagnetic stress tensor, in addition to the fluid stress tensor, to the hydrodynamic kinematics. Magnetohydrodynamics governs the realm of large scale and slow process whose time scale is much greater than the ion cyclotron period. In Chapter 3, we show that the magnetohydrodynamic linear waves can be derived from the two-fluid plasma equations, which are more intimate to Boltzmann equation, by employing the approximation $\omega \ll \omega_{\rm ci}$, the ion cyclotron frequency.

Magnetohydrodynamic equations may be regarded fundamentally as moment equations of the Liouville equation. At the same time, MHD equations can be derived from the conservation laws of Newtonian mechanics. Therefore the two approaches may be unified under the variational principle. To see this, authors demonstrated that hydrodynamic and magnetohydrodynamic equations are indeed derived from the Hamilton's principle. In actual variational calculus of this problem, the interesting provision is that subsidiary fluid equation such as the continuity equation is regarded as a constraint, thus replacing it as a Lagrange multiplier. The key variational equation to derive MHD equations is (4.25) in the text; the mathematical derivation of that equation is dealt with by allotting some pages.

• An *adiabatic invariant* remains to be an approximate constant of motion of a nearly periodic system when a parameter of the system changes slowly (Chapter 6). The word "adiabatic" appears to be coined from quantum mechanical situation where the energy states distribution is unchanged even though the Hamiltonian is changed by external means without adding heat to the system; say, by changing the volume. When we have a particle moving in a potential well (trapped particle), the *adiabatic change* of the motion is brought about by infinitesimal dislocation of turning point. If the particle encounters with a grazing turning point during its motion, the subsequent motion is subject to nonadiabatic abrupt change. The change in parameter should be slow as compared to the periods of motion and should not be related to the periods if it is an adiabatic change of motion. A well-known invariant for a particle moving in a magnetic field is its magnetic moment $\mu = W_\perp/B$ where $W_\perp$ is the kinetic energy of the perpendicular to B direction. With B changing slowly, The motion of the particle can be viewed as small perturbations from the constant field, which can be solved exactly. Here the usable small parameter is

$$\epsilon = \frac{T}{B}\frac{\partial B}{\partial t} \ll 1, \quad T: \text{ period} \tag{9}$$

in terms of which we can work on the perturbation series for the orbit. Equation (9) states that the fractional change of B over one period is $\ll 1$. We are interested in knowing the accuracy of $\mu's$ invariance. Being a constant of motion in the lowest order, i.e. in ϵ-order, is not enough. Since we are dealing with an asymptotic series, the true qualification for an invariant requires that the higher order deviation from invariance goes to zero rapidly as ϵ goes to zero. Let us designate $\Delta\mu$ the deviation of μ from the invariance. We need $\Delta\mu \ll \epsilon^N$, for all N, with N being the order. When $\epsilon = 10^{-3}$ and $N = 1$, $\Delta\mu = \epsilon^N$gives the estimation that appreciable change of μ is given rise to after 10^3 oscillations. If the experiment time is longer than 10^3 periods, the adiabatic invariant for $N = 1$ cannot be considered to be an invariant. We need a higher order invariant in cosmic or laboratory scheme where the particles bounce back and forth for a large number of periods. It is desirable to prove the existence of higher order invariants in order to explain or support the validity of the scheme utilizing magnetic invariants, such as magnetic mirror confinement or Fermi acceleration. Oscillatory motion between two turning points can be depicted in a phase plane (p, q) by a closed loop, where p and q are the generalized momentum and coordinates, respectively. Depending upon the values of the total energy, $W = mv^2/2 + V(q, \lambda)$, with V the potential energy, the trajectories represent particles with different energies. The distribution function f of particles can be expressed in terms of W and θ, where θ is the angular coordinate along the loop: $f = f(W, \theta)$. f is time-independent only if it does not depend on θ (f is constant along particle trajectories). When $V(q, \lambda(t))$ is slowly changing through a parameter λ, the orbit will not be closed completely but spiraling with tiny pitches, showing a nearly periodic trajectory. According to Liouville equation, the points in the phase space behave as an incompressible fluid: the area enclosed by the almost closed loop is conserved. That is, the phase plane area enclosed by the orbit corresponding to any one of the orbit is an approximate constant of motion:

$$\oint pdq = J \tag{10}$$

J in the above is known as the action integral and is an adiabatic invariant. Adiabatic invariance is a fundamental property of nearly periodic Hamiltonian system. In the literature, its general discussion has been done by using perturbed trapped particle motion [S. Chandrasekhar, *Plasma Physics*; Notes compiled by S. K. Trehan, Phoenix Books, University of Chicago Press (1960)] or perturbed Hamiltonian [L. D. Landau and E. M. Lifshitz, *Mechanics*, Pergamon Press Oxford (1960)].

• Intimacy between the Vlasov and Euler equations. Euler equation, the momentum equation of the cold fluid-plasma is a close kin of the Vlasov equation, closer than casual conjecture supposes. It is well-known that the fluid equation can be derived from Boltzmann equation by taking the velocity moment; but it is easily escaped that the Euler equation is contained in the Vlasov equation as a special class of the solution of the latter. This can be seen by considering the characteristics of both equations. We have the Vlasov equation for electrons written in the form

$$\frac{\partial}{\partial t}g(x,v,t) + v\frac{\partial g}{\partial x} - \frac{e}{m}E(x,t)\frac{\partial g}{\partial v} = 0 \tag{11}$$

The characteristic equations of (11) read

$$\frac{dt}{1} = \frac{dx}{v} = \frac{dv}{-eE(x,t)/m} = \frac{dg}{0} \tag{12}$$

which state that the distribution function $g(x,v,t)$ is constant along the characteristics or orbit represented by

$$\frac{dx}{dt} = v \tag{13}$$

$$\frac{dv}{dt} = -\frac{e}{m}E(x,t) \tag{14}$$

$$v\,\frac{dv}{dx} = -\frac{e}{m}E(x,t) \tag{15}$$

The solutions of (11) can be constructed with the constants of motion obtained by solving the characteristic equations, (13)–(15) [in fact, (15) is not independent]. Thus, let the implicit relation $g(x,v(x,t),t) = \mathit{const.} = c$ is an integral of (11). Then, we have per partial differentiation

$$\frac{\partial g}{\partial t} + \frac{\partial g}{\partial v}\frac{\partial v}{\partial t} = 0, \qquad \frac{\partial g}{\partial x} + \frac{\partial g}{\partial v}\frac{\partial v}{\partial x} = 0 \tag{16}$$

If the above relations are substituted into (11), we obtain

$$\frac{\partial v}{\partial t} + v\frac{\partial v}{\partial x} = -\frac{e}{m}E(x,t) \tag{17}$$

which is the Eulerian equation of motion. The characteristic equations of the Euler equation (17) reads

$$\frac{dt}{1} = \frac{dx}{v} = \frac{dv}{-eE(x,t)/m} \tag{18}$$

Therefore, (14), the particle orbit expressed in Lagrangian derivative, is the characteristics of the Vlasov equation as well as the Eulerian equation.

Conversely, we can assume that the solution of (17) is obtained in implicit form, $f(x,t,v(x,t)) = \text{const.}$ in the sense that the function f determines v

as a function of x and t. Then by partial differentiation, we have similar equations as (16) for f. Solving for $\partial v/\partial x$ and $\partial v/\partial t$, and substituting the results into (17) give

$$\frac{\partial}{\partial t}f(x,v,t) + v\frac{\partial f}{\partial x} - \frac{e}{m}E(x,t)\frac{\partial f}{\partial v} = 0$$

which is the Vlasov equation. Therefore, we may suppose that the solutions of the Euler equation belong to the subset of the solutions of the Vlasov equation. Landau damping can be derived from the cold fluid equation by regarding a plasma as collection of beams distributed with different beam velocities.

• The Van Kampen mode of plasma wave drew attention because it was discovered by employing a different mathematical approach to solve the Vlasov equation; not by using Fourier transform. The Fourier transform approach, if applied to *homogeneous* linear equations, ends up with a dispersion relation $\omega = \omega(k)$, which is the solvability condition for the algebraic set of equations. In Van Kampen's approach, the inherent singularity in the governing equation is surfaced with the form of a 'source term' and the subsequent solution need not to have a dispersion relation. Instead, the Van Kampen 'eigen functions' are complete so that they can be used for expansion of certain functions. Specifically, we should pay attention to the singularity in the form $1/(\omega - kv)$ that appears in the Vlasov or fluid equations. Historically, the singularity at $\omega = kv$ has been taken care of three different ways: (1) Vlasov simply disregarded it by taking only the principal value. (2) Landau elaborated it by analytic continuation in the Fourier domain. (3) Van Kampen paid a full account of it algebraically by introducing Dirac δ-function. Van Kampen's normal mode solutions of the Vlasov equation, which are simple harmonic waves in progressive wave form, has been proved to be capable of recovering Landau damping. Van Kampen included the δ-function term as an essential complementary solution to the Vlasov equation. In order to apply the Van Kampen eigen functions to the Vlasov initial value problem, a singular integral equation must be solved. Van Kampen's eigen functions are analogous to the eigen functions in the Sturm–Liouville theory of a differential equation.

• **Kinematics of wave.** The asymptotic state of a wave packet (multichromatic wave) can be studied by two approaches: (1) by asymptotically ($t \to \infty$) calculating the Fourier integral, (2) by assuming a slowly varying *eikonal* (phase). Section 10.1 undertakes the mathematical steps accompanying the first approach. Here, we consider the second approach for an electric field, varying slowly in space and time, which we represent in the form

$$\mathbf{E}(\mathbf{r},\mathbf{t}) = \mathbf{e}_{\mathbf{k}}\tilde{E}e^{-i\omega_{\mathbf{k}}t+i\mathbf{k}\cdot\mathbf{r}} + c.c. \tag{19}$$

where $\omega_{\mathbf{k}}$ is the eigen frequency satisfying the dispersion relation $\Delta(\omega_{\mathbf{k}},\mathbf{k}) = 0$ and $e_{\mathbf{k}}$ is the unit vector indicating the wave polarization. $e_{\mathbf{k}}$ can be obtained from the Fourier-space wave equation written in terms of ω and $\mathbf{k}$ by putting $\omega = \omega_{\mathbf{k}}$. [The solvability condition for the Fourier-space wave equation is the dispersion relation, $\triangle(\omega,\mathbf{k}) = 0$.] The vector $\mathbf{e}_{\mathbf{k}}$ may have multiplying factor, the constant phase shift $\exp(i\varphi_0)$, thus $\mathbf{e}_{\mathbf{k}}$ is complex. More generally than (19) is, we can introduce eikonal $\varphi(\mathbf{r},t)$, and write

$$\mathbf{E}(\mathbf{r},\mathbf{t}) = \mathbf{e}_{\mathbf{k}}\tilde{E}e^{-i\varphi(\mathbf{r},t)} + c.c. \tag{20}$$

Let us expand the phase φ about a local point $(\mathbf{r}_0,\ t_0)$

$$\varphi(\mathbf{r},t) = \varphi(\mathbf{r}_0,t_0) + (\mathbf{r}-\mathbf{r}_0)\cdot\nabla\varphi\big|_0 + (t-t_0)\frac{\partial\varphi}{\partial t}\bigg|_0 \tag{21}$$

where the subscript 0 means $(\mathbf{r}_0,\ t_0)$. Equations (19) and (20) should be equivalent, and we identify, as follows [for convenience, we put the reference point $(\mathbf{r}_0,\ t_0) = (0,0)$]

$$\omega = \frac{\partial\varphi}{\partial t}, \quad \mathbf{k} = -\nabla\varphi \tag{22}$$

Therefore, we have

$$\frac{\partial\mathbf{k}}{\partial t} + \nabla\omega = 0 \tag{23}$$

Equations (22) and (23) are the fundamental equations relating the eikonal with local ω and $\mathbf{k}$.

In a homogeneous medium, $\omega = \omega(\mathbf{k})$, and (23) becomes

$$\frac{\partial\mathbf{k}}{\partial t} + \mathbf{v}_g\cdot\nabla\mathbf{k} = 0 \tag{24}$$

where $\mathbf{v}_g = \partial\omega/\partial\mathbf{k}$ is the group velocity. Thus $\mathbf{k}$ is constant along the characteristics $d\mathbf{r}/dt = \mathbf{v}_g$.

In an inhomogeneous medium, $\omega = \omega(\mathbf{k},\mathbf{r})$, i.e. ω has explicit dependence on $\mathbf{r}$. In this case, (23) becomes

$$\frac{\partial\mathbf{k}}{\partial t} + \mathbf{v}_g\cdot\nabla\mathbf{k} = -\frac{\partial\omega}{\partial\mathbf{r}} = -\nabla\omega \tag{25}$$

which shows that $\mathbf{k}$ is not conserved along the curve $d\mathbf{r}/dt = \mathbf{v}_g$. Alternatively, we can write

$$\frac{\partial \omega}{\partial t} = \frac{\partial \omega}{\partial \mathbf{k}} \cdot \frac{\partial \mathbf{k}}{\partial t} \tag{26}$$

Then, use of (26) in the scalar product of (23) with $\partial\omega/\partial\mathbf{k}$ gives

$$\frac{\partial \omega}{\partial t} + \mathbf{v}_g \cdot \nabla \omega = 0 \tag{27}$$

This equation holds in inhomogeneous medium as well as in homogeneous medium. ω is conserved along the characteristics $d\mathbf{r}/dt = \mathbf{v}_g$.

Another important quantity associated with the propagation of wave packet is the energy. In plasma, the energy is the sum of the electromagnetic field energy plus the particle energy; they are all contained in the Maxwell equations. The particle energy manifests itself through the current, which also produces the electromagnetic field. So the distinction between the two is not compulsory. The usual mathematical technique in calculating the wave packet energy is two-time and two-space scale analysis: the wave packet has a slowly changing amplitude while its phase is rapidly varying, space and time-wise. The two-scale perturbation calculation is also used for deriving the *ponderomotive force* which is a potential force resulting from mixing of two waves of slightly different frequencies. See Sections 10.3 and 10.8. For the topic of wave packet energy, useful references are:

B. B. Kadomtsev, *Plasma Turbulence*, Academic Press (1965), p. 18.

B. B. Kadomtsev, *Cooperative effects in plasmas in Reviews of Plasma Physics*, Vol. 22 Edited by V. D. Shafranov, KluwerAcademic/Consultant Bureau (2001), p. 202.

G. Bekefi, *Radiation Processes on plasmas*, Wiley (1966).

Conservation of wave action (wave energy/frequency) can be derived from the argument of adiabatic invariance. It can be expressed in the form of continuity equation akin to Boltzmann equation.

Reference: Camac, M., A. R. Kantrowitz, M. M. Litvak, R. M. Patrick, and H. E. Petschek, *Shock waves in collision-free plasmas*, Nuc. Fusion Supplement **Pt. 2** (1962), pp. 423–444.

• The electric fluctuation and the Cerenkov emission (Chapter 11), being phenomena that is due to the discreteness in plasma, require a higher order description than Vlasov equation which is the lowest order crude approximation of the exact kinetic equation. A test particle approach which is known as *dressed particle* is adopted to calculate these quantities (Chapter 11). The Vlasov description of a plasma is basically akin to the fluid description which is appropriate for a continuum. The physical origin of the Cerenkov emission, which is the inverse process of the Landau damping, can be traced to the causality. This is anticipated since the fundamental reason for the Landau damping is the causality. The Cerenkov emission due to the thermal motions of plasma particles is the source of thermal noise (spontaneous emission), and its universality may be traced to causality. The fluctuations are the cause of the thermal emissions and wave scattering. The incoherent scattering of electromagnetic wave is due to the plasma density fluctuations.

• The surface waves are treated in depth by means of the fluid and kinetic equations. (Chapters 12–14 and 17) An infinite plasma (no boundary) supports various plasma normal modes whose characteristics depend upon the plasma parameters and the electromagnetic polarizations and refractive indexes. The presence of the boundary in a plasma changes the wave characteristics and excites new modes that have no corresponding modes in the infinite plasma. Mathematically speaking, the boundary imposes constraints, electromagnetic and fluid or kinetic constraints, to be satisfied on the boundary. Thus inevitably, the analysis for a bounded plasma has extra complication as compared with that in an infinite plasma. In a cold (warm) fluid plasma, the surface charges are formed (not formed) on the boundary. The existence of the surface charges on the interface yields a jump of the electric displacement across the interface, which is the electromagnetic boundary condition that must be satisfied in the surface wave problem. The absence of surface charges on the boundary of a warm plasma gives the boundary condition that the normal component of the fluid velocity on the interface is zero, which corresponds to the *specular reflection condition* in the kinetic plasma. In a Vlasov plasma, the specular reflection boundary condition is commonly used as the kinetic boundary condition for the distribution function. [Another boundary condition is the diffusive boundary condition]. In semi-bounded plasmas, the surface wave dispersion relations both in kinetic and fluid theory are well-known and thoroughly documented. In this book, we also investigate the slab geometry (the plasma is confined between two parallel infinite planes) to obtain the kinetic surface wave dispersion relations. The simultaneous specular reflection boundary conditions on the two parallel planes are successfully enforced to obtain the slab dispersion relations in closed forms. In a slab, the plasma wave has a two-mode structure, symmetric and ant-symmetric modes. Other interesting surface-wave-related subjects are: Landau damping of surface waves, plasma echoes in a bounded plasma, which are addressed in Chapter 14. The kinetic surface waves in a duct are also discussed in Chapter 14.

• Parametric instabilities (Chapter 15). The existence of various eigen modes of plasma wave is a manifestation of the collective plasma properties exhibiting medium-like continuum. For the description of such a plasma medium, a self-consistent theoretical model has been developed on the basis of the Maxwell and Vlasov equations or the Maxwell and fluid equations. The electromagnetic properties of the plasma can be investigated under the governance of these equations. Parametric instabilities are the important nonlinear interactions due to the quadratic terms in the governing plasma equations; $\mathbf{v} \cdot \nabla \mathbf{v}$ term in the fluid equation or $\mathbf{E} \cdot \frac{\partial f}{\partial \mathbf{v}}$ in the Vlasov equation. They represent the nature of plasma as a highly nonlinear dielectric medium. In particular, *nonlinear Landau damping* in unmagnetized plasma or *nonlinear cyclotron damping* in a magnetized plasma is one of the fundamental mechanisms for energy transfer from wave to particles due to wave-particle interaction, which is believed to be the cause of enhanced diffusion and anomalous resistivity of plasma. A simple picture of nonlinear Landau damping derives from the resonance condition

$$\omega - \omega' = (\mathbf{k} - \mathbf{k}') \cdot \mathbf{v}_i \tag{28}$$

where ω is the frequency of electron wave (Langmuir wave), ω' is the frequency of the scattered electron wave of slightly lower frequency, scattered by ion density perturbation, and $\mathbf{v}_i$ is the ion thermal velocity. Equation (28) is a resonance condition which indicates that the beat wave of frequency $(\omega - \omega')$ satisfies the Landau wave-particle resonance condition on ions. This ion nonlinear landau damping is responsible for the parametric decay instability. Due to this resonance, the beat wave (ion acoustic wave) gets heavily damped, discharging its wave energy to the scattered side-band wave (eventually to the ions). Because of the heavy damping, the beat wave loses its identity as a wave, and is often called a quasi mode. The ion nonlinear Landau damping transfers power from one Langmuir wave to another via an ion acoustic beat wave. Specifically, the nonlinear coupling responsible for the energy transfer is the nonlinear current of high frequency $\mathbf{J} = -en_i\mathbf{v}_e$ (subscript e denotes electron, the high frequency) and the *ponderomotive force* at the beat frequency which is derived from the nonlinear quadratic term $\mathbf{v}_e \cdot \nabla \mathbf{v}_e$ term in the fluid equation or $\mathbf{E} \cdot \frac{\partial f}{\partial \mathbf{v}}$ in the Vlasov equation.

References

Bekefi, G. *Radiation Processes on Plasmas*, Wiley (1966).

Camac, M., A. R. Kantrowitz, M. M. Litvak, R. M. Patrick, and H. E. Petschek, *Shock waves in collision-free plasmas.* Nucl. Fusion Suppl. **Pt. 2** (1962), pp. 423–444.

Chandrasekhar, S. *Plasma Physics*, Notes Compiled by S. K. Trehan, Phoenix Books, University of Chicago Press (1960).

Kadomtsev, B. B. Cooperative effects in plasmas. In Shafranov, V. D. (Ed.), *Reviews of Plasma Physics*, Vol. 22. Kluwer Academic/Consultant Bureau (2001).

Kadomtsev, B. B. *Plasma Turbulence.* Academic Press (1965).

Landau, L. *On the vibrations of the electronic plasma.* J. Phys. (USSR), **10** (1946), p. 25.

Landau, L. D. and E. M. Lifshitz, *Mechanics.* Pergamon Press, Oxford (1960).

Weinstock, R. *Calculus of Variations.* McGraw-Hill (1952).

Yourgrau, W. and S. Mandelstam, *Variational Principles in Dynamics and Quantum Theory.* Dover (1968).

Chapter 1

Liouville Equation

Liouville equation is the very origin of the various versions of kinetic equation in plasma. It comprises the fundamental concept of statistical description of plasma by describing the dynamical behavior of the *ensemble* of the system in the phase space. The ensemble is represented by the Liouville distribution function

$$F^{(N)} = F^{(N)}(\mathbf{r}_1, \mathbf{r}_2, \ldots, \mathbf{r}_N, \mathbf{p}_1, \mathbf{p}_2, \ldots, \mathbf{p}_N, t) \tag{1.1}$$

where N is the number of particles in the system, $\mathbf{r}$ and $\mathbf{p}$ denote particle position and momentum, respectively. $F^{(N)}$, the $(6N + 1)$-variable function, is governed by the Liouville equation or the Klimontovich equation, and involves a chain of multi-particle distribution functions called the BBGKY (Bogoliubov–Born–Green–Kirkwood–Yvon) hierarchy. The hierarchy stems from the potential energy of the Hamiltonian of the system. The potential energy of a particle in the system is influenced by all the rest of the particles in the system. The hierarchy cannot be unchained without introducing approximation, and the task of the kinetic theory is to invent a proper approximation to terminate the hierarchy. The approximate treatment of this hierarchy is facilitated by expanding the Liouville equation in a small parameter as referred to *plasma parameter*, the inverse of the number of particles in the Debye sphere. The development makes it possible to invent one unchained equation for the one-particle distribution function by expressing the two-particle distribution functions in terms of one-particle distribution function. We have famous kinetic equations for one particle distribution function: Lenard–Balescu equation and Boltzmann equation with Landau collision term.

1.1 Liouville Equation

Liouville distribution function of a system of N particles $F^{(N)} = F^{(N)}(\mathbf{r}_1, \mathbf{r}_2, \ldots, \mathbf{r}_N, \mathbf{p}_1, \mathbf{p}_2, \ldots, \mathbf{p}_N, t)$ specifies the distribution of systems over a statistical *ensemble*. $F^{(N)}$ may be thought of in terms of the trajectory of a phase point in $6N$-dimensional phase space ($3N$ coordinate space plus $3N$ momentum space; called Γ-space). In the course of time this point would move along a path that would be determined by the equation of motion. In stead of considering a trajectory belonging to a single phase point [the complete knowledge of the orbit of the phase point in the Γ-space is impossible to obtain], the system is represented by a collection, or an ensemble, of phase points each of which is a possible state of the system with the same macroscopic characteristics. With compatible initial conditions, $F^{(N)}$ satisfies the Liouville equation for a system of particles with Hamiltonian H:

$$\frac{\partial F^{(N)}}{\partial t} + \sum_{i=1}^{N} \left(\frac{\partial H}{\partial \mathbf{p}_i} \cdot \frac{\partial F^{(N)}}{\partial \mathbf{r}_i} - \frac{\partial F^{(N)}}{\partial \mathbf{p}_i} \cdot \frac{\partial H}{\partial \mathbf{r}_i} \right) = 0 \tag{1.2}$$

where H is assumed to be given by

$$H = \frac{1}{2m} \sum_{i=1}^{N} \left[\mathbf{p}_i - \frac{e}{c}\mathbf{A}(\mathbf{r}_i, t) \right]^2 + e \sum_{i=1}^{N} \phi(\mathbf{r}_i, t) + \frac{e^2}{2} \sum_{i=1}^{N} \sum_{j(\neq i)=1}^{N} \frac{1}{|\mathbf{r}_i - \mathbf{r}_j|} \tag{1.3}$$

where $\mathbf{A}(\mathbf{r}_i, t)$ and $\phi(\mathbf{r}_i, t)$ are vector and scalar potentials from which the *external* magnetic and electric fields acting on the ith particle are derived. Here 'external' means that its origin has nothing to do with the plasma particles composing the system. The last term represents the electrostatic inter-particle interaction, and the inter-particle magnetic interaction is neglected in H. We have for the electric and magnetic fields

$$\mathbf{B} = \nabla \times \mathbf{A} \tag{1.4}$$

$$\mathbf{E} = -\nabla\phi - \frac{1}{c}\frac{\partial \mathbf{A}}{\partial t} \tag{1.5}$$

The quantity

$$F^{(N)}(\mathbf{r}_1, \mathbf{p}_1, \mathbf{r}_2, \mathbf{p}_2, \ldots, \mathbf{r}_N, \mathbf{p}_N, t) d^3r_1 d^3r_2 \ldots d^3r_N d^3p_1 d^3p_2 \ldots d^3p_N$$

is the probability at time t that particle 1 is at the state $(\mathbf{r}_1 + d\mathbf{r}_1, \mathbf{p}_1 + d\mathbf{p}_1)$ and that particle 2 is at the state $(\mathbf{r}_2 + d\mathbf{r}_2, \mathbf{p}_2 + d\mathbf{p}_2) \ldots$, and that particle N is at the state $(\mathbf{r}_N + d\mathbf{r}_N, \mathbf{p}_N + d\mathbf{p}_N)$.

Equation (1.2) is a lengthy first order partial differential equation involving $(6N + 1)$ partial derivatives. The characteristic equation of (1.2) reads

$$
\begin{aligned}
dt &= \frac{dx_1}{\frac{\partial H}{\partial p_{1x}}} = \frac{dy_1}{\frac{\partial H}{\partial p_{1y}}} = \frac{dz_1}{\frac{\partial H}{\partial p_{1z}}} = \frac{-dp_{1x}}{\frac{\partial H}{\partial x_1}} = \frac{-dp_{1y}}{\frac{\partial H}{\partial y_1}} = \frac{-dp_{1z}}{\frac{\partial H}{\partial z_1}} \\
&= \frac{dx_2}{\frac{\partial H}{\partial p_{2x}}} = \frac{dy_2}{\frac{\partial H}{\partial p_{2y}}} = \frac{dz_2}{\frac{\partial H}{\partial p_{2z}}} = \frac{-dp_{2x}}{\frac{\partial H}{\partial x_2}} = \frac{-dp_{2y}}{\frac{\partial H}{\partial y_2}} = \frac{-dp_{2z}}{\frac{\partial H}{\partial z_2}} = \cdots \\
&= \frac{dx_N}{\frac{\partial H}{\partial p_{Nx}}} = \frac{dy_N}{\frac{\partial H}{\partial p_{Ny}}} = \frac{dz_N}{\frac{\partial H}{\partial p_{Nz}}} = \frac{-dp_{Nx}}{\frac{\partial H}{\partial x_N}} = \frac{-dp_{Ny}}{\frac{\partial H}{\partial y_N}} = \frac{-dp_{Nz}}{\frac{\partial H}{\partial z_N}}
\end{aligned}
$$

The above expression can be grouped into a compact and physically transparent form

$$
\frac{d\mathbf{r}_i}{dt} = \frac{\partial H}{\partial \mathbf{p}_i}, \quad \frac{d\mathbf{p}_i}{dt} = -\frac{\partial H}{\partial \mathbf{r}_i}, \quad (i = 1, 2, \ldots, N) \tag{1.6}
$$

which are the Hamilton's equations of motion, whose solutions are the individual particle trajectories. According to the theory of the hyperbolic partial differential equation, $F^{(N)}$ in (1.2) is constant along the characteristics—the particle trajectory. The trajectory of the phase point in the $6N$-dimensional Γ-space is determined by the vector sum of the individual path of each particle composing the N-particle system, determined by (1.6). So we conclude that the Liouville function $F^{(N)}$ is constant along the phase point trajectory in the Γ-space.

It would be instructive to show that Hamilton's equations (1.6) take the familiar form of the equation of motion. Using (1.3) in the first equation of (1.6) gives

$$
\frac{\partial H}{\partial \mathbf{p}_i} = \frac{d\mathbf{r}_i}{dt} = \mathbf{v}_i = \frac{1}{m}\left(\mathbf{p}_i - \frac{e}{c}\mathbf{A}(\mathbf{r}_i, t)\right) \tag{1.7}
$$

We calculate from (1.3) [using the identity $\frac{1}{2}\nabla F^2 = \mathbf{F} \times \nabla \times \mathbf{F} + \mathbf{F} \cdot \nabla \mathbf{F}$]

$$
\begin{aligned}
\frac{\partial H}{\partial \mathbf{r}_i} &= \frac{1}{m}\left(\mathbf{p}_i - \frac{e}{c}\mathbf{A}(\mathbf{r}_i, t)\right) \times \frac{\partial}{\partial \mathbf{r}_i} \times \left(\mathbf{p}_i - \frac{e}{c}\mathbf{A}(\mathbf{r}_i, t)\right) + \frac{1}{m}\left(\mathbf{p}_i - \frac{e}{c}\mathbf{A}(\mathbf{r}_i, t)\right) \cdot \frac{\partial}{\partial \mathbf{r}_i}\left(\mathbf{p}_i - \frac{e}{c}\mathbf{A}(\mathbf{r}_i, t)\right) \\
&\quad + e\frac{\partial}{\partial \mathbf{r}_i}\left[\phi(\mathbf{r}_i, t) + \frac{e}{2}\sum_{j(\neq i)=1}^{N} \frac{1}{|\mathbf{r}_i - \mathbf{r}_j|}\right]
\end{aligned}
$$

Noting that $\mathbf{p}$ operated by $\frac{\partial}{\partial \mathbf{r}_i}$ vanishes and using (1.7), the vector and scalar potential terms on the RHS (right hand side) simplify, and the above equation yields

$$\frac{\partial H}{\partial \mathbf{r}_i} = -\frac{e}{c}\mathbf{v}_i \times \mathbf{B}(\mathbf{r}_i, t) - \frac{e}{c}\mathbf{v}_i \cdot \nabla \mathbf{A}(\mathbf{r}_i, t) + e\frac{\partial}{\partial \mathbf{r}_i}\left[\phi(\mathbf{r}_i, t) + \frac{e}{2}\sum_{j(\neq i)=1}^{N} \frac{1}{|\mathbf{r}_i - \mathbf{r}_j|}\right] \quad (1.8)$$

On the other hand

$$\frac{d\mathbf{p}_i}{dt} = m\frac{d\mathbf{v}_i}{dt} + \frac{e}{c}\frac{d\mathbf{A}}{dt} = m\frac{d\mathbf{v}_i}{dt} + \frac{e}{c}\left(\frac{\partial \mathbf{A}}{\partial t} + \mathbf{v}_i \cdot \nabla \mathbf{A}\right) \quad (1.9)$$

Using (1.8) and (1.9) in the second equation of (1.6) yields

$$m\frac{d\mathbf{v}_i}{dt} = e\left[\mathbf{E}(\mathbf{r}_i, t) + \frac{1}{c}\mathbf{v}_i \times \mathbf{B}(\mathbf{r}_i, t)\right] \quad (1.10)$$

and

$$\mathbf{E}(\mathbf{r}_i, t) = -\frac{1}{c}\frac{\partial}{\partial t}\mathbf{A}(\mathbf{r}_i, t) - \frac{\partial}{\partial \mathbf{r}_i}\left[\phi(\mathbf{r}_i, t) + \frac{e}{2}\sum_{j(\neq i)=1}^{N} \frac{1}{|\mathbf{r}_i - \mathbf{r}_j|}\right] \equiv \mathbf{E}_i(\mathbf{r}_i, t) \quad (1.11)$$

which is the total (internal plus external) electric field acting on the particle i.

The characteristic equation of (1.10) reads

$$dt = \frac{d\mathbf{r}_i}{\mathbf{v}_i} = \frac{m d\mathbf{v}_i}{e\left(\mathbf{E}_i + \frac{1}{c}\mathbf{v}_i \times \mathbf{B}_i\right)} \qquad (i = 1, 2, \ldots, N) \quad (1.12)$$

This equation would result from the familiar form of the Liouville equation

$$\frac{\partial F^{(N)}}{\partial t} + \sum_{i=1}^{N}\left(\mathbf{v}_i \cdot \frac{\partial F^{(N)}}{\partial \mathbf{r}_i} + \mathbf{a}_i \cdot \frac{\partial F^{(N)}}{\partial \mathbf{v}_i}\right) = 0 \quad (1.13)$$

$$\text{where} \quad \mathbf{a}_i = \frac{e}{m}\left[\mathbf{E}_i(\mathbf{r}_i, t) + \frac{1}{c}\mathbf{v}_i \times \mathbf{B}_i(\mathbf{r}_i, t)\right]$$

Equation (1.13) is an alternative form of the Liouville equation which describes how the aggregate of the phase points in the $6N$-dimensional phase space move about. Defining $3N$-dimensional vectors

$$\mathbf{r} = (\mathbf{r}_1, \mathbf{r}_2, \ldots, \mathbf{r}_N)$$
$$\mathbf{v} = (\mathbf{v}_1, \mathbf{v}_2, \ldots, \mathbf{v}_N)$$
$$\mathbf{a} = (\mathbf{a}_1, \mathbf{a}_2, \ldots, \mathbf{a}_N)$$

Equation (1.13) can be written as

$$\frac{D}{Dt} f^{(N)}(\mathbf{r}, \mathbf{v}, t) = 0 \tag{1.14}$$

where

$$\frac{D}{Dt} = \frac{\partial}{\partial t} + \frac{d\mathbf{r}}{dt} \cdot \frac{\partial}{\partial \mathbf{r}} + \frac{d\mathbf{v}}{dt} \cdot \frac{\partial}{\partial \mathbf{v}} \tag{1.15}$$

$$\mathbf{v} = \frac{d\mathbf{r}}{dt}, \quad \mathbf{a} = \frac{d\mathbf{v}}{dt} \quad \text{per } (1.12) \tag{1.16}$$

In terms of the hydrodynamics analogue, the Liouville *continuum* occupying the $(\mathbf{r}, \mathbf{v})$ phase space moves like an *incompressible* fluid. Here we recall the fluid continuity equation:

$$\frac{D\rho}{dt} + \rho \nabla \cdot \mathbf{v} = 0, \quad \rho: \text{ density}$$

for *compressible* fluid. For an *incompressible* fluid, $\nabla \cdot \mathbf{v} = 0$. $\frac{D}{Dt}$ in (1.15) is called *Lagrangian derivative.*

Now we shall derive (1.13) from (1.2) rigorously. Using (1.7) and (1.8) in (1.2), the typical term (omitting the particle index) is written as

$$\frac{\partial F^{(N)}}{\partial t} + \cdots + \mathbf{v}.\frac{\partial F^{(N)}}{\partial \mathbf{r}} - \Big(-\frac{e}{c}\mathbf{v} \times \mathbf{B} - \frac{e}{c}\mathbf{v} \cdot \nabla \mathbf{A} + e\nabla \phi_T \Big) \cdot \frac{\partial F^{(N)}}{\partial \mathbf{p}} + \cdots = 0 \tag{1.17}$$

where ϕ_T is the bracketed quantity in (1.11), the total potential. We carry out the change of variables $(\mathbf{r}, \mathbf{p}, t) \to (\mathbf{r}', \mathbf{v}, t')$ in which

$$t = t', \quad \mathbf{r} = \mathbf{r}', \quad \mathbf{p} = m\mathbf{v} + \frac{e}{c}\mathbf{A}(\mathbf{r}', t')$$

So the differential operators on $F^{(N)}$ in (1.17) changes according to

$$\frac{\partial}{\partial t} = \frac{\partial}{\partial t'} + \frac{\partial}{\partial \mathbf{v}} \cdot \frac{\partial \mathbf{v}}{\partial t'} = \frac{\partial}{\partial t'} - \frac{e}{mc}\frac{\partial \mathbf{A}}{\partial t'} \cdot \frac{\partial}{\partial \mathbf{v}}$$

$$\frac{\partial}{\partial \mathbf{p}} = \frac{1}{m}\frac{\partial}{\partial \mathbf{v}}$$

$$\frac{\partial}{\partial \mathbf{r}} = \frac{\partial}{\partial \mathbf{r}'} + \frac{\partial \mathbf{v}}{\partial \mathbf{r}'} \cdot \frac{\partial}{\partial \mathbf{v}} = \frac{\partial}{\partial \mathbf{r}'} - \frac{e}{mc}\frac{\partial A_j}{\partial x_i'}\frac{\partial}{\partial v_j}$$

from which it follows that

$$\mathbf{v}\cdot\frac{\partial}{\partial\mathbf{r}}=\mathbf{v}\cdot\frac{\partial}{\partial\mathbf{r}'}-\frac{e}{mc}(\mathbf{v}\cdot\nabla')\,\mathbf{A}\cdot\frac{\partial}{\partial\mathbf{v}}$$

Substituting the above relations into (1.17) yields for a typical term

$$\frac{\partial F^{(N)}}{\partial t'}+\cdots+\mathbf{v}\cdot\frac{\partial F^{(N)}}{\partial\mathbf{r}'}+\mathbf{a}\cdot\frac{\partial F^{(N)}}{\partial\mathbf{v}}+\cdots=0 \tag{1.18}$$

In the above equation, primes can be removed; similar transformations can be carried out for all the terms to obtain (1.13). When $\frac{\partial}{\partial\mathbf{v}}\cdot\mathbf{a}=0$, (1.13) can be also written as

$$\frac{\partial F^{(N)}}{\partial t}+\sum_{i=1}^{N}\left[\frac{\partial}{\partial\mathbf{r}_i}\cdot(\mathbf{v}_iF^{(N)})+\frac{\partial}{\partial\mathbf{v}_i}\cdot(\mathbf{a}_iF^{(N)})\right]=0 \tag{1.19}$$

In the above we saw that the Liouville equation conserves the Liouville probability density in the Γ-space along the phase point trajectory, which is the characteristics of the Liouville equation. The conservation of the Liouville density along the characteristics is *Liouville theorem*. Conversely, we can derive the Liouville equation from the Liouville theorem. For notational brevity, let the representative point of a single phase point state be denoted by $\mathbf{X}(t)$. The dynamical evolution of an individual system is determined by the individual orbit equations (1.12). Let $\mathbf{X}_0$ be the initial state of the representative point: $\mathbf{X}_0=\mathbf{X}(t=0)$. Then the initial probability density is the distribution of the aggregate of the representative points, which is $F^{(N)}(\mathbf{X}_0)$. As time increases, the initial phase point $\mathbf{X}_0$ traces an orbit $\mathbf{X}(t)=\mathbf{X}(\mathbf{X}_0,t)$. The probability density at time t which is the distribution of the re-distributed representative points is $F^{(N)}(\mathbf{X}(t))$. The Liouville theorem simply states that

$$F^{(N)}(\mathbf{X}_0)=F^{(N)}(\mathbf{X}(t)) \tag{1.20}$$

The Liouville equation is deduced from the Liouville theorem in (1.20) by taking the Lagrangian derivative of $F^{(N)}(\mathbf{X}(t))$ (or differentiating along the trajectory):

$$\frac{D}{Dt}F^{(N)}(\mathbf{X}(\mathbf{t}))=0;\quad \frac{D}{Dt}=\frac{\partial}{\partial t}+\frac{d\mathbf{X}}{dt}\cdot\frac{\partial}{\partial\mathbf{X}}$$

thus obtaining (1.13).

Remark. $\mathbf{X(t)}$ in (1.20) is Lagrangian variable. A device to write $F^{(N)}$ in terms of the Eulerian coordinate is to introduce the particle orbits by means of delta functions:

$$F^{(N)}(\mathbf{X},t) = \int d\tau_0 F^{(N)}(\mathbf{X}_0) \prod_{i=1}^{N} \delta(\mathbf{r}_i - \mathbf{r}_i(\mathbf{r}_{i0},t)\ \delta(\mathbf{v}_i - \mathbf{v}_i(\mathbf{v}_{i0},t))$$

Here $\mathbf{X}$ with no argument denotes the Eulerian coordinate while Lagrangian coordinate is denoted by explicitly writing '(t)' in the parenthesis. Note that $\mathbf{r}_i, \mathbf{v}_i$ are Eulerian variables. Subscript 0 denotes the initial values. $d\tau_0$ is the volume element surrounding the point $\mathbf{X}_0$ in the Γ-space. Differentiating the above equation with respect to t, Liouville equation can be derived.

1.2 Klimontovich Equation

The Klimontovich distribution function or density for a system of species α of N number of particles expresses the microscopic particle distribution in the form

$$K_\alpha(\mathbf{r},\mathbf{v},t) = \sum_{i=1}^{N} \delta(\mathbf{r} - \mathbf{r}_{i\alpha}(t))\delta(\mathbf{v} - \mathbf{v}_{i\alpha}(t)) \tag{1.21}$$

where $\mathbf{r}_{i\alpha}(t),\ \mathbf{v}_{i\alpha}(t)$ represent the exact orbit of the ith particle of species α as determined by the forces $\mathbf{F}_{i\alpha}$ acting on it. Note that if K is integrated over the whole six dimensional phase space $(\mathbf{r},\mathbf{v})$ of the system the result is N, the total number of particles. From the equation of motion for individual particle

$$\dot{\mathbf{r}}_{i\alpha} = \mathbf{v}_{i\alpha}, \quad \dot{\mathbf{v}}_{i\alpha} = \frac{\mathbf{F}_{i\alpha}}{m_\alpha} = \frac{e_\alpha}{m_\alpha}\left(\mathbf{E}(\mathbf{r}_{i\alpha},t) + \frac{1}{c}\mathbf{v}_{i\alpha} \times \mathbf{B}(\mathbf{r}_{i\alpha},t)\right) \tag{1.22}$$

it follows that the Klimontovich density satisfies the equation

$$\frac{\partial}{\partial t}K_\alpha(\mathbf{r},\mathbf{v},t) + \mathbf{v}\cdot\frac{\partial K_\alpha}{\partial \mathbf{r}} + \frac{\mathbf{F}_\alpha}{m_\alpha}\cdot\frac{\partial K_\alpha}{\partial \mathbf{v}} = 0 \tag{1.23}$$

where $\mathbf{F}_\alpha$ is the self-consistent force that a particle of species α at $(\mathbf{r},\ \mathbf{v})$ would experience. Equation (1.23) is obtained in the following steps.

Differentiating (1.21) with respect to t gives (the index α omitted)

$$\frac{\partial K}{\partial t} = \sum_{i=1}^{N}\left[\frac{d\mathbf{r}_i}{dt}\cdot\frac{\partial}{\partial \mathbf{r}_i}\delta(\mathbf{r}-\mathbf{r}_i(t))\delta(\mathbf{v}-\mathbf{v}_i(t)) + \frac{d\mathbf{v}_i}{dt}\cdot\frac{\partial}{\partial \mathbf{v}_i}\delta(\mathbf{r}-\mathbf{r}_i(t))\delta(\mathbf{v}-\mathbf{v}_i(t))\right]$$

$$= -\sum_{i=1}^{N}\left[\mathbf{v}_i\cdot\frac{\partial}{\partial \mathbf{r}}\delta(\mathbf{r}-\mathbf{r}_i(t))\delta(\mathbf{v}-\mathbf{v}_i(t)) + \frac{\mathbf{F}_i}{m_i}\cdot\frac{\partial}{\partial \mathbf{v}}\delta(\mathbf{r}-\mathbf{r}_i(t))\delta(\mathbf{v}-\mathbf{v}_i(t))\right]$$

In the first term inside Σ, $\mathbf{v}_i$ can be replaced by $\mathbf{v}$ owing to the delta function. Also

$$\mathbf{F}_i = \mathbf{F}_i(\mathbf{r}_i, \mathbf{v}_i) = e_\alpha(\mathbf{E}(\mathbf{r}_i) + \frac{1}{c}\mathbf{v}_i \times \mathbf{B}(\mathbf{r}_i)) = e_\alpha(\mathbf{E}(\mathbf{r}, t) + \frac{1}{c}\mathbf{v} \times \mathbf{B}(\mathbf{r}, t)) \equiv \mathbf{F}_\alpha \tag{1.24}$$

due to the δ functions. In (1.24), $\mathbf{r}_i$ and $\mathbf{v}_i$ are Lagrangian variables; in the last expression of (1.24) we have put them in terms of Eulerian coordinates $\mathbf{r}$ and t $\left[\mathbf{r}_i = \mathbf{r}_i(\mathbf{r}, t)\right]$. Thus we obtain

$$\frac{\partial K_\alpha(\mathbf{r}, \mathbf{v}, t)}{\partial t} + \mathbf{v}\cdot\frac{\partial K_\alpha}{\partial \mathbf{r}} + \frac{\mathbf{F}_\alpha(\mathbf{r}, t)}{m_\alpha}\cdot\frac{\partial K_\alpha}{\partial \mathbf{v}} = 0$$

The self-consistent fields, $\mathbf{E}$ and $\mathbf{B}$, obey Maxwell's equations

$$\nabla\cdot\mathbf{E} = 4\pi\rho \tag{1.25}$$

$$\nabla\times\mathbf{B} = \frac{4\pi}{c}\mathbf{J} + \frac{1}{c}\frac{\partial \mathbf{E}}{\partial t} \tag{1.26}$$

where the source terms are generated by all the particles in the system:

$$\rho(\mathbf{x}, t) = \sum_\alpha e_\alpha \int d^3v\, K_\alpha(\mathbf{r}, \mathbf{v}, t) = \sum_\alpha e_\alpha \sum_{i=1}^{N} \delta(\mathbf{r} - \mathbf{r}_i^\alpha(t)) \tag{1.27}$$

$$\mathbf{J}(\mathbf{r}, t) = \sum_\alpha e_\alpha \int d^3v\, \mathbf{v}K_\alpha(\mathbf{r}, \mathbf{v}, t) = \sum_\alpha e_\alpha \sum_{i=1}^{N} \mathbf{v}_i^\alpha(t)\delta(\mathbf{r} - \mathbf{r}_i^\alpha(t)) \tag{1.28}$$

Equation (1.23) is entirely equivalent to (1.22). Usefulness of (1.23) requires solutions of exact orbit which is impossible to obtain. Thus it is necessary for (1.23) to be reduced to workable equation on the basis of statistical consideration.

1.3 Vlasov Equation

The most simple and crude approximation to the exact equation, Liouville equation (1.13), or Klimontovich equation (1.23), is to introduce the *one particle distribution function* by replacing the $6N$-dimensional Γ-space by ordinary 6-dimensional $(\mathbf{r}, \mathbf{v})$ space.

$$F^{(N)}(\mathbf{X}, t) \to f_\alpha(\mathbf{r}, \mathbf{v}, t)$$

$$K_\alpha(\mathbf{r}, \mathbf{v}, t) \to f_\alpha(\mathbf{r}, \mathbf{v}, t)$$

and write the *Vlasov equation*

$$\frac{\partial f_\alpha}{\partial t} + \mathbf{v} \cdot \frac{\partial f_\alpha}{\partial \mathbf{r}} + \frac{e}{m}\left(\mathbf{E} + \frac{1}{c}\mathbf{v} \times \mathbf{B}\right) \cdot \frac{\partial f_\alpha}{\partial \mathbf{v}} = 0 \tag{1.29}$$

The Vlasov description reflects the continuum aspect of plasma. In actuality, plasmas manifest discrete particle effects (besides the inter-particle collisions) which belong to the higher hierarchy of Liouville equation, or the exact orbit in Klimontovich equation. Equation (1.29) can be derived heuristically by considering the time rate of change of the number of particles enclosed by a fixed, arbitrary volume in the phase space. In the absence of collisions among the particles, this rate is determined only by the flows entering or leaving the volume across the boundary. Here the boundary is a surface in the six-dimensional phase space. The Vlasov fluid flux across the boundary is expressed by $(\frac{d\mathbf{r}}{dt}f, \frac{d\mathbf{v}}{dt}f)$, and we can write by using the mass conservation,

$$\frac{\partial}{\partial t}\int d^3r d^3v \; f = -\int d^3v d^2\mathbf{r} \cdot \frac{d\mathbf{r}}{dt} f - \int d^3r d^2\mathbf{v} \cdot \frac{d\mathbf{v}}{dt} f$$

Applying the Gauss theorem to the surface integrals on the right side, we have

$$\int d^3r d^3v \left[\frac{\partial f}{\partial t} + \frac{\partial}{\partial \mathbf{r}} \cdot \left(\frac{d\mathbf{r}}{dt} f\right) + \frac{\partial}{\partial \mathbf{v}} \cdot \left(\frac{d\mathbf{v}}{dt} f\right)\right] = 0$$

Using $m\frac{d\mathbf{v}}{dt} = e(\mathbf{E} + \frac{1}{c}\mathbf{v} \times \mathbf{B})$ in the above gives

$$\frac{\partial f}{\partial t} + \mathbf{v} \cdot \frac{\partial f}{\partial \mathbf{r}} + \frac{e}{m}\left(\mathbf{E} + \frac{1}{c}\mathbf{v} \times \mathbf{B}\right) \cdot \frac{\partial f}{\partial \mathbf{v}} = 0 \tag{1.30}$$

This derivation reflects the fluid-like behavior of the Vlasov plasma.

The dynamics of the Vlasov model of a plasma includes the charge and current densities

$$\rho = \sum_\alpha e_\alpha \int f_\alpha d^3v \tag{1.31}$$

$$\mathbf{J} = \sum_\alpha e_\alpha \int \mathbf{v}\ f_\alpha d^3v \tag{1.32}$$

1.4 BBGKY Hierarchy

We begin with the Liouville equation written in the form

$$\frac{\partial F^{(N)}}{\partial t} + \frac{1}{m}\sum_{i=1}^{N}\left(\frac{\partial H}{\partial \mathbf{v}_i}\cdot\frac{\partial F^{(N)}}{\partial \mathbf{r}_i} - \frac{\partial F^{(N)}}{\partial \mathbf{v}_i}\cdot\frac{\partial H}{\partial \mathbf{r}_i}\right) = 0$$

where the Hamiltonian H is given by

$$H(\mathbf{v}_i,\mathbf{r}_i,t) = \frac{m}{2}\sum_{i=1}^{N} v_i^2 + e\sum_{i=1}^{N}\phi_T(\mathbf{r}_i,t), \quad \phi_T(\mathbf{r}_i,t) = \phi(\mathbf{r}_i,t) + \frac{e}{2}\sum_{j(\neq i)=1}^{N}\frac{1}{|\mathbf{r}_i-\mathbf{r}_j|}$$

Using $\frac{\partial H}{\partial \mathbf{v}_i} = m\mathbf{v}_i$, we write

$$\frac{\partial F^{(N)}}{\partial t} + \sum_{i=1}^{N}\left(\mathbf{v}_i\cdot\frac{\partial F^{(N)}}{\partial \mathbf{r}_i} - \frac{e}{m}\frac{\partial \phi_T}{\partial \mathbf{r}_i}\cdot\frac{\partial F^{(N)}}{\partial \mathbf{v}_i}\right) = 0 \tag{1.33}$$

where $F^{(N)} = F^{(N)}(\mathbf{r}_1,\mathbf{v}_1,\mathbf{r}_2,\mathbf{v}_2,\ldots,\mathbf{r}_N,\mathbf{v}_N,t)$.

- **1-particle distribution function.**

$$F^{(1)}(\mathbf{r}_1,\mathbf{v}_1,t) = \int F^{(N)}(\mathbf{r}_1,\mathbf{v}_1,\mathbf{r}_2,\mathbf{v}_2,\ldots,\mathbf{r}_N,\mathbf{v}_N,t)d^3r_2d^3v_2d^3r_3d^3v_3\cdots d^3r_Nd^3v_N \tag{1.34}$$

is the probability density at time t to find the particle 1 at the state $(\mathbf{r}_1 + d\mathbf{r}_1,\ \mathbf{v}_1 + d\mathbf{v}_1)$, regardless of whereabouts of the rest of particles. Likewise

$$F^{(1)}(\mathbf{r}_2,\mathbf{v}_2,t) = \int F^{(N)}(\mathbf{r}_1,\mathbf{v}_1,\mathbf{r}_2,\mathbf{v}_2,\ldots,\mathbf{r}_N,\mathbf{v}_N,t)d^3r_1d^3v_1d^3r_3d^3v_3\cdots d^3r_Nd^3v_N$$

is the probability density at time t to find particle 2 at the state $(\mathbf{r}_2 + d\mathbf{r}_2,\ \mathbf{v}_2 + d\mathbf{v}_2)$, regardless of whereabouts of the rest of particles. Since the particles are indistinguishable, the above two probability densities are the same, and we can state that $F^{(1)}(\mathbf{r}, \mathbf{v}, t)$ which is obtained by integrating $F^{(N)}$ over all the particles except any one particle is the probability density at time t to find a particle at the state $(\mathbf{r} + d\mathbf{r},\ \mathbf{v} + d\mathbf{v})$. So $F^{(1)}(\mathbf{r}, \mathbf{v}, t)$ has the same meaning as the $f(\mathbf{r}, \mathbf{v}, t)$ in the familiar Vlasov equation.

- **2-particle distribution function.**

$$F^{(2)}(\mathbf{r}_1, \mathbf{v}_1, \mathbf{r}_2, \mathbf{v}_2, t) = \int F^{(N)}(\mathbf{r}_1, \mathbf{v}_1, \mathbf{r}_2, \mathbf{v}_2, \ldots, \mathbf{r}_N, \mathbf{v}_N, t) d^3r_3 d^3v_3 d^3r_4 d^3v_4 \cdots d^3r_N d^3v_N \tag{1.35}$$

is the joint probability density at time t to find a particle at the state $(\mathbf{r}_1 + d\mathbf{r}_1,\ \mathbf{v}_1 + d\mathbf{v}_1)$ and to find simultaneously another particle at the state $(\mathbf{r}_2 + d\mathbf{r}_2,\ \mathbf{v}_2 + d\mathbf{v}_2)$, regardless of whereabouts of the rest of particles. $F^{(2)}$ is obtained by integrating $F^{(N)}$ over all the particles except *any* two particles.

- **s-particle distribution function $F^{(s)}$.**

$$F^{(s)}(\mathbf{r}_1, \mathbf{v}_1, \mathbf{r}_2, \mathbf{v}_2, \ldots, \mathbf{r}_s, \mathbf{v}_s, t) = \int F^{(N)}(\mathbf{r}_1, \mathbf{v}_1, \mathbf{r}_2, \mathbf{v}_2, \ldots, \mathbf{r}_N,\ \mathbf{v}_N, t) d^3r_{s+1} d^3v_{s+1} \cdots d^3r_N d^3v_N \tag{1.36}$$

which is the joint probability density at time t to simultaneously find s particles at each of s phase space locations in the Γ-space.

Now we take Liouville equation (1.33) and integrate over coordinates and velocities of all the particles except those for particle 1 (say). Then the immediately integrated terms (corresponding to $i = 2, 3, \ldots, N$) vanish (no particles on the infinite surface; at the boundary) to give

$$\begin{aligned} &\frac{\partial}{\partial t} F^{(1)}(\mathbf{r}_1, \mathbf{v}_1, t) + \mathbf{v}_1 \cdot \frac{\partial F^{(1)}}{\partial \mathbf{r}_1} - \frac{e}{m} \frac{\partial \phi(\mathbf{r}_1, t)}{\partial \mathbf{r}_1} \cdot \frac{\partial F^{(1)}}{\partial \mathbf{v}_1} \\ &= \frac{e^2}{m} \int d^3r_2 d^3v_2 \ldots d^3r_N d^3v_N \left(\frac{\partial}{\partial \mathbf{r}_1} \sum_{j=2}^{N} \frac{1}{|\mathbf{r}_1 - \mathbf{r}_j|} \right) \cdot \frac{\partial F^{(N)}}{\partial \mathbf{v}_1} \end{aligned} \tag{1.37}$$

Since the particles are indistinguishable ($F^{(N)}$ is symmetric with respect to interchange of particle index), the summation on j yields $(N-1)$ identical terms (multiplied by $F^{(2)}$), and the right hand side becomes

$$\frac{(N-1)e^2}{m} \int d^3r_2 d^3v_2 \left(\frac{\partial}{\partial \mathbf{r}_1} \frac{1}{|\mathbf{r}_1 - \mathbf{r}_2|} \right) \cdot \frac{\partial}{\partial \mathbf{v}_1} F^{(2)}(\mathbf{r}_1, \mathbf{v}_1, \mathbf{r}_2, \mathbf{v}_2, t)$$

To improve visuality, (1.37) is rewritten, suppressing the subscript 1,

$$\frac{\partial}{\partial t}F^{(1)}(\mathbf{r},\mathbf{v},t)+\mathbf{v}\cdot\frac{\partial F^{(1)}}{\partial\mathbf{r}}-\frac{e}{m}\frac{\partial\phi(\mathbf{r},t)}{\partial\mathbf{r}}\cdot\frac{\partial F^{(1)}}{\partial\mathbf{v}}$$
$$=\frac{(N-1)e^2}{m}\int d^3r_2d^3v_2\left(\frac{\partial}{\partial\mathbf{r}}\frac{1}{|\mathbf{r}-\mathbf{r}_2|}\right)\cdot\frac{\partial}{\partial\mathbf{v}}F^{(2)}(\mathbf{r},\mathbf{v},\mathbf{r}_2,\mathbf{v}_2,t) \tag{1.38}$$

Equation (1.38) is not closed: it contains $F^{(2)}$. If we set up the equation for $F^{(2)}$ by integrating the Liouville equation by analogous procedure, the $F^{(2)}$-equation will contain $F^{(3)}$. Thus this chain of connection continues endlessly, and should be terminated at some stage by invoking some physical argument or approximation. As a first approximation at the earliest stage, let us put

$$F^{(2)}(\mathbf{r},\mathbf{v},\mathbf{r}_2,\mathbf{v}_2,t)=F^{(1)}(\mathbf{r},\mathbf{v},t)\ F^{(1)}(\mathbf{r}_2,\mathbf{v}_2,t) \tag{1.39}$$

Equation (1.39) means: the probability at time t to find a particle at the phase space location $(\mathbf{r},\mathbf{v})$ is independent from (uncorrelated with) the probability that another particle is simultaneously found at other phase space location $(\mathbf{r}',\mathbf{v}')$. In other words, the particles in the plasma behave as if they were not influenced by the other particles located at other phase points in the system. More refined kinetic theory requires an extra small term of joint probability on the right side of (1.39). Using (1.39) in (1.38) yields

$$\frac{\partial}{\partial t}F^{(1)}(\mathbf{r},\mathbf{v},t)+\mathbf{v}\cdot\frac{\partial F^{(1)}}{\partial\mathbf{r}}-\frac{e}{m}\frac{\partial\phi(\mathbf{r},t)}{\partial\mathbf{r}}\cdot\frac{\partial F^{(1)}}{\partial\mathbf{v}}$$
$$-\frac{(N-1)e}{m}\left[\int d^3r_2d^3v_2\left(\frac{\partial}{\partial\mathbf{r}}\frac{e}{|\mathbf{r}-\mathbf{r}_2|}\right)F^{(1)}(\mathbf{r}_2,\mathbf{v}_2,t)\right]\cdot\frac{\partial}{\partial\mathbf{v}}F^{(1)}(\mathbf{r},\mathbf{v},t)=0 \tag{1.40}$$

Here we recall that ϕ in the above equation is external. The quantity in the large square bracket is the electric field produced at $\mathbf{r}$ by the rest of electrons in the plasma. This is the ensemble-averaged force acting on the particle at $\mathbf{r}$ exerted by all the rest of the particles in the plasma. *Equation* (1.40) *is the Vlasov equation* and is obtained as a crude approximation that the 2-particle distribution function is a product of two 1-particle distribution functions.

Remark. The interaction potential which represents the force acting on the ith particle is contributed by the jth ($i\neq j$) particles. Thus it is necessary to introduce the two-particle distribution function $F^{(2)}$ in the reduction process of the Liouville equation. The kinetic equation (1.38) can be expressed as a closed form by

$$\frac{\partial f}{\partial t} + \mathbf{v} \cdot \frac{\partial f}{\partial \mathbf{r}} + \mathbf{a}_{\text{ext}} \cdot \frac{\partial f}{\partial \mathbf{v}} = I(f)$$

provided $F^{(2)}$ is expressed in terms of $F^{(1)}$. The major problem of kinetic theory is to find an approximate form for $I(f)$. To close (1.38), the interaction part contributing to the change of f should be found in terms of f itself. The Boltzmann collision term of the following form is a coarse grained equation for f which applies to diffuse gases in which strong but impulsive localized interaction occurs (See Chapter 2).

$$I_{\text{Boltzmann}}(f) = \int d\Omega \int d^3v_2 g\sigma(g,\theta)[f(v_1')f(v_2') - f(v_1)f(v_2)]$$

For an ionized gas where the inter-particle interactions are weak but long ranged, $I(f)$ is a Fokker–Planck type which takes the form of Landau collision integral.

Next we derive the equation for $F^{(2)}$. We take Liouville equation (1.33) and integrate over coordinates and velocities of all the particles except those for particles 1 and 2. Then the immediately integrated terms (corresponding to $i = 3, 4, \ldots, N$) vanish, and the resulting intermediate equation can be written as

$$\frac{\partial}{\partial t}F^{(2)}(\mathbf{r}_1,\mathbf{v}_1,\mathbf{r}_2,\mathbf{v}_2,t) + \mathbf{v}_1 \cdot \frac{\partial F^{(2)}}{\partial \mathbf{r}_1} + \mathbf{v}_2 \cdot \frac{\partial F^{(2)}}{\partial \mathbf{r}_2} - \frac{e}{m}\left[\frac{\partial \phi(\mathbf{r}_1,t)}{\partial \mathbf{r}_1} \cdot \frac{\partial F^{(2)}}{\partial \mathbf{v}_1} + \frac{\partial \phi(\mathbf{r}_2,t)}{\partial \mathbf{r}_2} \cdot \frac{\partial F^{(2)}}{\partial \mathbf{v}_2}\right]$$
$$+ \text{interaction potential term (IPT)} = 0 \tag{1.41}$$

$$\begin{aligned} IPT = &-\frac{e^2}{m}\int \frac{\partial}{\partial \mathbf{r}_1}\left(\frac{1}{|\mathbf{r}_1 - \mathbf{r}_2|} + \frac{1}{|\mathbf{r}_1 - \mathbf{r}_3|} + \frac{1}{|\mathbf{r}_1 - \mathbf{r}_4|} + \cdots\right) \cdot \frac{\partial F^{(N)}}{\partial \mathbf{v}_1} d^3r_3 d^3v_3 \cdots d^3r_N d^3v_N \\ &-\frac{e^2}{m}\int \frac{\partial}{\partial \mathbf{r}_2}\left(\frac{1}{|\mathbf{r}_2 - \mathbf{r}_1|} + \frac{1}{|\mathbf{r}_2 - \mathbf{r}_3|} + \frac{1}{|\mathbf{r}_2 - \mathbf{r}_4|} + \cdots\right) \cdot \frac{\partial F^{(N)}}{\partial \mathbf{v}_2} d^3r_3 d^3v_3 \cdots d^3r_N d^3v_N \end{aligned} \tag{1.42}$$

Here we make use of the symmetry of $F^{(N)}$ with respect to the interchange of particle index. In the integrands, the particle index 4 can be replaced by 3 (at the same time, with the interchange of 3 and 4 in $F^{(N)}$). In this way, we see that there are $(N-2)$ identical terms. Therefore we have

$$\begin{aligned} IPT = &-\frac{e^2}{m}\int \frac{\partial}{\partial \mathbf{r}_1}\left(\frac{1}{|\mathbf{r}_1 - \mathbf{r}_2|} + \frac{N-2}{|\mathbf{r}_1 - \mathbf{r}_3|}\right) \cdot \frac{\partial F^{(N)}}{\partial \mathbf{v}_1} d^3r_3 d^3v_3 \cdots d^3r_N d^3v_N \\ &-\frac{e^2}{m}\int \frac{\partial}{\partial \mathbf{r}_2}\left(\frac{1}{|\mathbf{r}_2 - \mathbf{r}_1|} + \frac{N-2}{|\mathbf{r}_2 - \mathbf{r}_3|}\right) \cdot \frac{\partial F^{(N)}}{\partial \mathbf{v}_2} d^3r_3 d^3v_3 \cdots d^3r_N d^3v_N \end{aligned}$$

$$= -\frac{e^2}{m}\left(\frac{\partial}{\partial \mathbf{r}_1}\frac{1}{|\mathbf{r}_1-\mathbf{r}_2|}\cdot\frac{\partial}{\partial \mathbf{v}_1}+\frac{\partial}{\partial \mathbf{r}_2}\frac{1}{|\mathbf{r}_2-\mathbf{r}_1|}\cdot\frac{\partial}{\partial \mathbf{v}_2}\right)F^{(2)}$$
$$-\frac{(N-2)e^2}{m}\int\left(\frac{\partial}{\partial \mathbf{r}_1}\frac{1}{|\mathbf{r}_1-\mathbf{r}_3|}\cdot\frac{\partial}{\partial \mathbf{v}_1}+\frac{\partial}{\partial \mathbf{r}_2}\frac{1}{|\mathbf{r}_2-\mathbf{r}_3|}\cdot\frac{\partial}{\partial \mathbf{v}_2}\right)F^{(3)}d^3r_3d^3v_3 \quad (1.43)$$

We rewrite the above results as a single equation (omitting the external potential term):

$$\frac{\partial F^{(2)}}{\partial t}+\mathbf{v}_1\cdot\frac{\partial F^{(2)}}{\partial \mathbf{r}_1}+\mathbf{v}_2\cdot\frac{\partial F^{(2)}}{\partial \mathbf{r}_2}-\frac{e^2}{m}\left(\frac{\partial}{\partial \mathbf{r}_1}\frac{1}{|\mathbf{r}_1-\mathbf{r}_2|}\cdot\frac{\partial}{\partial \mathbf{v}_1}+\frac{\partial}{\partial \mathbf{r}_2}\frac{1}{|\mathbf{r}_2-\mathbf{r}_1|}\cdot\frac{\partial}{\partial \mathbf{v}_2}\right)F^{(2)}$$
$$-\frac{(N-2)e^2}{m}\int\left(\frac{\partial}{\partial \mathbf{r}_1}\frac{1}{|\mathbf{r}_1-\mathbf{r}_3|}\cdot\frac{\partial}{\partial \mathbf{v}_1}+\frac{\partial}{\partial \mathbf{r}_2}\frac{1}{|\mathbf{r}_2-\mathbf{r}_3|}\cdot\frac{\partial}{\partial \mathbf{v}_2}\right)F^{(3)}d^3r_3d^3v_3=0 \quad (1.44)$$

The equation for $F^{(s)}$ is obtained from the Liouville equation (1.33) by integrating over the variables associated with (N-s) particles. The following equation called BBGKY hierarchy equation (Bogoliubov, Born, Green, Kirkwood, and Yvon) can be derived:

$$\frac{\partial F^{(s)}}{\partial t}+\sum_{i=1}^{s}\mathbf{v}_i\cdot\frac{\partial F^{(s)}}{\partial \mathbf{r}_i}+\sum_{i=1}^{s}\sum_{j=1}^{s}\mathbf{a}_{ij}\cdot\frac{\partial F^{(s)}}{\partial \mathbf{v}_i}+(N-s)\sum_{i=1}^{s}\int d^3r_{s+1}d^3v_{s+1}\mathbf{a}_{i,s+1}\cdot\frac{\partial F^{(s+1)}}{\partial \mathbf{v}_i}=0 \quad (1.45)$$

where $F^{(s)}(\mathbf{r}_1,\mathbf{v}_1,\mathbf{r}_2,\mathbf{v}_2,,\ldots,\mathbf{r}_s,\mathbf{v}_s,t)$

$$=\int F^{(N)}(\mathbf{r}_1,\mathbf{v}_1,\mathbf{r}_2,\mathbf{v}_2,\ldots,\mathbf{r}_s,\mathbf{v}_s,\ldots,\mathbf{r}_N,\mathbf{v}_N,t)d^3r_{s+1}d^3v_{s+1}\cdots d^3r_Nd^3v_N$$

$$\mathbf{a}_{ij}=\frac{e^2}{m}\frac{\mathbf{r}_i-\mathbf{r}_j}{|\mathbf{r}_i-\mathbf{r}_j|^3}=-\frac{e^2}{m}\frac{\partial}{\partial \mathbf{r}_i}\frac{1}{|\mathbf{r}_i-\mathbf{r}_j|}$$

is the interacting acceleration. We use $\mathbf{a}_{ij}=0$ when $i=j$.

Equation (1.45) can be written in terms of the dimensionless variables by expressing length in the unit of the Debye length and time in the unit of the electron plasma frequency. Then the dimensionless form of (1.45) becomes

$$\frac{\partial F^{(s)}}{\partial t}+\sum_{i=1}^{s}\mathbf{v}_i\cdot\frac{\partial F^{(s)}}{\partial \mathbf{r}_i}+\frac{1}{g}\sum_{i=1}^{s}\sum_{j=1}^{s}\mathbf{a}_{ij}\cdot\frac{\partial F^{(s)}}{\partial \mathbf{v}_i}+(N-s)$$
$$\times\sum_{i=1}^{s}\int d^3r_{s+1}d^3v_{s+1}\mathbf{a}_{i,s+1}\cdot\frac{\partial F^{(s+1)}}{\partial \mathbf{v}_i}=0$$

Notice the emergence of the factor g, which is very large ($\gg$1) in plasma: g turns out to be equal to particle number density (n_0) times the force range. Usually the latter is taken as the Debye length (λ_D) and $g=4\pi n_0\lambda_D^3$. Plasma is characterized by the property that the number of particles in the Debye sphere is very large. So the small parameter $1/g$ in the BBGKY equation allows one to approach the equation by expanding $F^{(s)}$ in perturbation series expanded in the power of $1/g$ [Reference: P. C. Clemmow and J. P. Daugherty *Electrodynamics of particles and Plasmas* (1969) Addison-Wesley London].

1.5 Klimontovich Moment Equations

We consider ensemble average of products of Klimontovich distribution functions and show that these moments can be expressed in terms of the hierarchial Liouville distribution functions $F^{(s)}$. Here we introduce short hand notation: $(\mathbf{r}, \mathbf{v}) \equiv \mathbf{x}$. Consider the ensemble average

$$
\begin{aligned}
<K(\mathbf{x},t)> &= \int F^{(N)}(\mathbf{x}_1, \mathbf{x}_2, \ldots, \mathbf{x}_N, t) K(\mathbf{x}, t) d^6x_1 d^6x_2 \ldots d^6x_N \\
&= \int F^{(N)}(\mathbf{x}_1, \mathbf{x}_2, \ldots, \mathbf{x}_N, t)\Big(\delta(\mathbf{x}-\mathbf{x}_1) + \delta(\mathbf{x}-\mathbf{x}_2) + \cdots + \delta(\mathbf{x}-\mathbf{x}_N)\Big) d^6x_1 d^6x_2 \cdots d^6x_N
\end{aligned}
\tag{1.46}
$$

Term $\delta(\mathbf{x}-\mathbf{x}_1)$ gives 1-particle distribution $V^{-1}F^{(1)}$ $(\mathbf{x}, t)$ (V is the volume of the box containing the system). Term $\delta(\mathbf{x}-\mathbf{x}_2)$ gives the same. This is the consequence of indistinguishability of particles. Thus we have

$$
<K(\mathbf{x},t)> = \frac{N}{V} F^{(1)}(\mathbf{x}, t) \tag{1.47}
$$

Next, we consider the second moment

$$
\begin{aligned}
<K(\mathbf{x},t)K(\mathbf{x}',t)> &= \int \sum_{i=1}^{N}\sum_{j=1}^{N} \delta(\mathbf{x}-\mathbf{x}_i)\delta(\mathbf{x}'-\mathbf{x}_j) F^{(N)} d^6x_1 d^6x_2 \ldots d^6x_N \\
&= \frac{N}{V}\int \delta(\mathbf{x}-\mathbf{x}_1(t))\delta(\mathbf{x}'-\mathbf{x}_1(t)) F^{(1)}(\mathbf{x}_1, t) d^6x_1 \\
&\quad + \frac{N(N-1)}{V^2}\int \delta(\mathbf{x}-\mathbf{x}_1(t))\delta(\mathbf{x}'-\mathbf{x}_2(t)) F^{(2)}(\mathbf{x}_1, \mathbf{x}_2, t) d^6x_1 d^6x_2 \\
&= \frac{N}{V} F^{(1)}(\mathbf{x}, t)\delta(\mathbf{x}-\mathbf{x}') + \frac{N(N-1)}{V^2} F^{(2)}(\mathbf{x}, \mathbf{x}', t)
\end{aligned}
\tag{1.48}
$$

where the first term is due to the terms $i = j$ and the second term stems from the cross terms. Similarly, we obtain for the third moment

$$
\begin{aligned}
&<K(\mathbf{x},t)K(\mathbf{x}',t)K(\mathbf{x}'',t)> \\
&= \int \sum_{i=1}^{N}\sum_{j=1}^{N}\sum_{k=1}^{N} \delta(\mathbf{x}-\mathbf{x}_i)\delta(\mathbf{x}'-\mathbf{x}_j)\delta(\mathbf{x}''-\mathbf{x}_k) F^{(N)} d^6x_1 d^6x_2 \ldots d^6x_N
\end{aligned}
$$

$$= \frac{N}{V}F^{(1)}(\mathbf{x},t)\delta(\mathbf{x}-\mathbf{x}')\delta(\mathbf{x}-\mathbf{x}'') + \frac{N(N-1)}{V^2}[F^{(2)}(\mathbf{x},\mathbf{x}',t)\delta(\mathbf{x}-\mathbf{x}'')$$
$$+ F^{(2)}(\mathbf{x},\mathbf{x}'',t)\delta(\mathbf{x}-\mathbf{x}') + F^{(2)}(\mathbf{x},\mathbf{x}'',t)\delta(\mathbf{x}''-\mathbf{x}')]$$
$$+ \frac{1}{V^3}N(N-1)(N-2)F^{(3)}(\mathbf{x},\mathbf{x}',\mathbf{x}'',t) \tag{1.49}$$

The ensemble averages of products of Klimontovich distribution functions or the moments of Klimontovich distribution functions can be expressed in terms of the Liouville functions $F^{(s)}$.

Next, we consider Klimontovich equation (1.23) for particles of one species (electrons) interacting through electrostatic forces:

$$\frac{\partial}{\partial t}K(\mathbf{r},\mathbf{v},t) + \mathbf{v}\cdot\frac{\partial K}{\partial \mathbf{r}} - \frac{e}{m}\left[\int \frac{\partial}{\partial \mathbf{r}}\frac{e}{|\mathbf{r}-\mathbf{r}'|}K(\mathbf{r}',\mathbf{v}',t)d^3r'd^3v'\right]\cdot\frac{\partial K(\mathbf{r},\mathbf{v},t)}{\partial \mathbf{v}} = 0 \tag{1.50}$$

Equation (1.50) is exact without any approximation, and requires to know all the details of orbits of the N particles, as expressed by the δ-functions. In order to reduce it to usable equation, statistical average is taken by multiplying it with $F^{(N)}$ and by integrating over all the $\mathbf{r}_i$ and $\mathbf{v}_i$. Noting that the differential operators in (1.50) are independent of the integration variables, the averaged equation takes the form

$$\frac{\partial\langle K\rangle}{\partial t} + \mathbf{v}\cdot\frac{\partial\langle K\rangle}{\partial \mathbf{r}} - \frac{e}{m}\int \frac{\partial}{\partial \mathbf{r}}\frac{e}{|\mathbf{r}-\mathbf{r}'|}\cdot\frac{\partial\langle KK'\rangle}{\partial \mathbf{v}}d^3r'd^3v' = 0 \tag{1.51}$$

where K' has the obvious meaning. When we substitute (1.47) and (1.48) into (1.51), we discard the $\delta(\mathbf{x}-\mathbf{x}')$ term in (1.48) because $\mathbf{r}\neq\mathbf{r}'$ in (1.50). This amounts to rejecting the self-force by putting

$$\int \delta(\mathbf{r}-\mathbf{r}')\frac{\partial}{\partial \mathbf{r}}\frac{1}{|\mathbf{r}-\mathbf{r}'|}d^3r' = 0 \tag{1.52}$$

The singular function in (1.52) repeatedly appear in the subsequent Klimontovich development and we shall set it to zero without further mention. We obtain

$$\frac{\partial}{\partial t}F^{(1)}(\mathbf{r},\mathbf{v},t) + \mathbf{v}\cdot\frac{\partial F^{(1)}}{\partial \mathbf{r}} - \frac{eN}{m}\int d^3r'd^3v'\frac{\partial}{\partial \mathbf{r}}\frac{e}{|\mathbf{r}-\mathbf{r}'|}\cdot\frac{\partial}{\partial \mathbf{v}}F^{(2)}(\mathbf{r},\mathbf{v},\mathbf{r}',\mathbf{v}',t) = 0 \tag{1.53}$$

which is the same as (1.38) with neglect of the external potential in the latter.

Next, to obtain equation for $\langle K(\mathbf{r},\mathbf{v},t)K'(\mathbf{r}',\mathbf{v}',t)\rangle$, note that

$$\frac{\partial}{\partial t}\langle KK'\rangle = \left\langle \frac{\partial}{\partial t}(KK')\right\rangle = \left\langle K'\frac{\partial K}{\partial t}\right\rangle + \left\langle K\frac{\partial K'}{\partial t}\right\rangle$$

Using (1.50), we construct two equations:

$$K'\left(\frac{\partial}{\partial t}+\mathbf{v}\cdot\frac{\partial}{\partial \mathbf{r}}\right)K = \frac{e}{m}K'\left[\int \frac{\partial}{\partial \mathbf{r}}\frac{e}{|\mathbf{r}-\mathbf{r}''|}K''d^3r''d^3v''\right]\cdot\frac{\partial K}{\partial \mathbf{v}} = 0$$

$$K\left(\frac{\partial}{\partial t}+\mathbf{v}'\cdot\frac{\partial}{\partial \mathbf{r}'}\right)K' = \frac{e}{m}K\left[\int \frac{\partial}{\partial \mathbf{r}'}\frac{e}{|\mathbf{r}'-\mathbf{r}''|}K''d^3r''d^3v''\right]\cdot\frac{\partial K'}{\partial \mathbf{v}'} = 0 .$$

Adding the above two equations and taking the average give

$$\frac{\partial}{\partial t}\langle KK'\rangle + \left(\mathbf{v}\cdot\frac{\partial}{\partial \mathbf{r}}+\mathbf{v}'\cdot\frac{\partial}{\partial \mathbf{r}'}\right)\langle KK'\rangle$$
$$= \int d^3r''d^3v''\left[\frac{\partial}{\partial \mathbf{r}}\frac{e}{|\mathbf{r}-\mathbf{r}''|}\cdot\frac{\partial}{\partial \mathbf{v}}+\frac{\partial}{\partial \mathbf{r}'}\frac{e}{|\mathbf{r}'-\mathbf{r}''|}\cdot\frac{\partial}{\partial \mathbf{v}'}\right] < KK'K'' > \tag{1.54}$$

Substituting (1.48) and (1.49) into (1.54) yields

$$\left(\frac{\partial}{\partial t}+\mathbf{v}\cdot\frac{\partial}{\partial \mathbf{r}}+\mathbf{v}'\cdot\frac{\partial}{\partial \mathbf{r}'}\right)F^{(2)}(\mathbf{x},\mathbf{x}',t)$$
$$= \frac{N}{m}\left[\frac{\partial}{\partial \mathbf{r}}\frac{e^2}{|\mathbf{r}-\mathbf{r}'|}\cdot\frac{\partial}{\partial \mathbf{v}}+\frac{\partial}{\partial \mathbf{r}'}\frac{e^2}{|\mathbf{r}'-\mathbf{r}|}\cdot\frac{\partial}{\partial \mathbf{v}'}\right]F^{(2)}(\mathbf{x},\mathbf{x}',t)$$
$$+\frac{N}{m}\int d^3r''d^3v''\left[\frac{\partial}{\partial \mathbf{r}}\frac{e^2}{|\mathbf{r}-\mathbf{r}''|}\cdot\frac{\partial}{\partial \mathbf{v}}+\frac{\partial}{\partial \mathbf{r}'}\frac{e^2}{|\mathbf{r}'-\mathbf{r}''|}\cdot\frac{\partial}{\partial \mathbf{v}'}\right]F^{(3)}(\mathbf{x},\mathbf{x}',\mathbf{x}'',t) \tag{1.55}$$

where $\mathbf{x} = (\mathbf{r},\mathbf{v})$. Equation (1.55) is the same as (1.44). (1.53), (1.55), and the higher Klimontovich moment equations are in complete agreement with the BBGKY hierarchy equation (1.45).

1.6 Lenard–Balescu Equation

In this section, we consider the Klimontovich moment equation in the lowest order, and derive the Lenard–Balescu equation. The key element of analysis is to separate the Klimontovich distribution function into two parts: the time-wise slowly varying part and the fluctuating part. The formal procedure of the analysis is very similar to that for the quasilinear theory. An alternative approach is to solve for the two Liouville distributions $F^{(1)}$ and $F^{(2)}$ in the BBGKY hierarchy by employing the operational method. The latter approach is not dealt with here. Readers who are not familiar with the mathematical details of this section are advised to read first the later section about the mathematical theory of Vlasov equation, and get back to this section.

References: D. C. Montgomery *Theory of unmagnetized plasma* Gordon and Beach New York (1971); E. M. Lifshitz and L. P. Pitaevskii *Physical kinetics* Landau and Lifshitz Course of theoretical physics Volume 10 Butterworth–Heinemann (1981).

We begin with the Klimontovich equation (1.50) and the Poisson equation:

$$\frac{\partial K(\mathbf{r},\mathbf{v},t)}{\partial t}+\mathbf{v}\cdot\frac{\partial K}{\partial \mathbf{r}}-\frac{e}{m}\frac{\partial \phi}{\partial \mathbf{r}}\cdot\frac{\partial K}{\partial \mathbf{v}}=0 \tag{1.56}$$

$$\phi(\mathbf{r},t)=\int\frac{e}{|\mathbf{r}-\mathbf{r}'|}K(\mathbf{r}',\mathbf{v}',t)d^3r'd^3v' \tag{1.57}$$

$$\nabla^2\phi=4\pi e\int d^3v\ K \tag{1.58}$$

Separate the functions in the manner

$$K=\overline{K}+\delta K,\quad \phi=\overline{\phi}+\delta\phi$$

where the overbar and δ denote, respectively, ensemble averaged (by means of $F^{(N)}$) and fluctuating parts. If a δ-quantity is averaged, the result is zero. $\overline{K}=<K>$ is given by (1.47). We have

$$\begin{aligned}&\frac{\partial}{\partial t}(\overline{K}+\delta K)+\mathbf{v}\cdot\frac{\partial}{\partial \mathbf{r}}(\overline{K}+\delta K)\\&=\frac{e}{m}\frac{\partial\overline{\phi}}{\partial \mathbf{r}}\cdot\frac{\partial}{\partial \mathbf{v}}\overline{K}+\frac{e}{m}\frac{\partial\overline{\phi}}{\partial \mathbf{r}}\cdot\frac{\partial}{\partial \mathbf{v}}\delta K+\frac{e}{m}\frac{\partial\delta\phi}{\partial \mathbf{r}}\cdot\frac{\partial}{\partial \mathbf{v}}\overline{K}+\frac{e}{m}\frac{\partial\delta\phi}{\partial \mathbf{r}}\cdot\frac{\partial}{\partial \mathbf{v}}\delta K\end{aligned} \tag{1.59}$$

The first two terms on the RHS vanish because the averaged electric field in a homogeneous plasma is zero. The homogeneous plasma is characterized by $\frac{\partial}{\partial \mathbf{r}}\overline{K}=\frac{\partial}{\partial \mathbf{r}}F^{(1)}(\mathbf{r},\mathbf{v},t)=0$. If (1.59) is averaged with the remaining terms, we are left with

$$\frac{\partial}{\partial t}\overline{K}=\frac{e}{m}\overline{\frac{\partial}{\partial \mathbf{v}}\ \delta K\cdot\frac{\partial}{\partial \mathbf{r}}\ \delta\phi} \tag{1.60}$$

So we should collect from (1.59) to write

$$\left(\frac{\partial}{\partial t}+\mathbf{v}\cdot\frac{\partial}{\partial \mathbf{r}}\right)\delta K=\frac{e}{m}\frac{\partial}{\partial \mathbf{v}}\overline{K}\cdot\frac{\partial}{\partial \mathbf{r}}\ \delta\phi \tag{1.61}$$

We shall seek the solution of the two coupled equations, (1.60) and (1.61). Equation (1.61) is identical in form to the linearized Vlasov equation. If $\overline{K}$ is time-independent, the mathematical machinery that we use for solving the Vlasov equation can be employed here. We make a compromise in this respect: while the time dependence of $\overline{K}$ is retained in (1.60), we hold $\overline{K}$ constant in (1.61). This approximate scheme turns out to be acceptable when the plasma parameter is small ($\ll$1), as justified by the multiple time perturbation analysis of (1.60) and (1.61). We first solve (1.61) which we rewrite by changing into more comfortable notations ($\delta K \to \delta f,\ \overline{K} \to F$)

$$\left(\frac{\partial}{\partial t} + \mathbf{v}\cdot\frac{\partial}{\partial \mathbf{r}}\right)\delta f(\mathbf{r},\mathbf{v},t) = \frac{1}{m}\frac{\partial F(\mathbf{v})}{\partial \mathbf{v}}\cdot\int d^3r'd^3v'\delta f(\mathbf{r}',\mathbf{v}',t)\frac{\partial\phi(\mathbf{r}-\mathbf{r}')}{\partial \mathbf{r}} \tag{1.62}$$

where $\phi(\mathbf{r}-\mathbf{r}') = \frac{e^2}{|\mathbf{r}-\mathbf{r}'|}$. In the new notations, (1.60) is written as

$$\frac{\partial}{\partial t}F(\mathbf{v},t) = \frac{e}{m}\frac{\partial}{\partial \mathbf{v}}\cdot <\delta f\delta\mathbf{E}> \tag{1.62$'$}$$

We solve the above equations by Laplace–Fourier transform, which is defined by

$$\delta f(\omega,\mathbf{k},\mathbf{v}) = \int_0^\infty dt e^{i\omega t}\int d^3r\ e^{-i\mathbf{k}\cdot\mathbf{r}}\delta f(\mathbf{r},\mathbf{v},t)$$

$$\delta f(\mathbf{r},\mathbf{v},t) = \int_{-\infty+ic}^{\infty+ic}\frac{d\omega}{2\pi}e^{-i\omega t}\int d^3k\ e^{i\mathbf{k}\cdot\mathbf{r}}\delta f(\omega,\mathbf{k},\mathbf{v})$$

Recall that in $\int d\omega$ integral above, the integration path runs above all the singularities of $\delta f(\omega,\mathbf{k},\mathbf{v})$. Laplace transforming (1.62) gives

$$\begin{aligned}&-i\omega\ \delta f(\omega,\mathbf{r},\mathbf{v}) - \delta f(0,\mathbf{r},\mathbf{v}) + \mathbf{v}\cdot\frac{\partial}{\partial \mathbf{r}}\delta f(\omega,\mathbf{r},\mathbf{v})\\&\quad= \frac{1}{m}\frac{\partial F}{\partial \mathbf{v}}\cdot\int d^3r'd^3v'\ \delta f(\mathbf{r}',\mathbf{v}',\omega)\frac{\partial\phi(\mathbf{r}-\mathbf{r}')}{\partial \mathbf{r}}\end{aligned} \tag{1.63}$$

where $\delta f(0,\mathbf{r},\mathbf{v})$ is the initial value. The Fourier transform of the integral part is a convolution integral:

$$\begin{aligned}&\int d^3r\ e^{-i\mathbf{k}\cdot\mathbf{r}}\int d^3r'd^3v'\ \delta f(\mathbf{r}',\mathbf{v}',\omega)\frac{\partial\phi(\mathbf{r}-\mathbf{r}')}{\partial \mathbf{r}}\\&\quad= i\mathbf{k}\int d^3r\ e^{-i\mathbf{k}\cdot\mathbf{r}}\int d^3r'd^3v'\delta f(\mathbf{r}',\mathbf{v}',\omega)\phi(\mathbf{r}-\mathbf{r}')\\&\quad= i\mathbf{k}\ \phi(k)\int \delta f(\mathbf{k},\mathbf{v}',\omega)d^3v'\end{aligned}$$

Using this in (1.63) gives

$$\delta f(\omega, \mathbf{k}, \mathbf{v}) = i\,\frac{\delta f(0, \mathbf{k}, \mathbf{v})}{\omega - \mathbf{k}\cdot\mathbf{v}} - \frac{\phi(k)}{m}\frac{\mathbf{k}\cdot\frac{\partial F}{\partial \mathbf{v}}}{\omega - \mathbf{k}\cdot\mathbf{v}}\int \delta f(\mathbf{k}, \mathbf{v}', \omega) d^3v'$$

Integrating both sides with $\int d^3v(\cdots)$ and substituting the result yield

$$\delta f(\omega, \mathbf{k}, \mathbf{v}) = \frac{i}{\omega - \mathbf{k}\cdot\mathbf{v}}\left[\delta f(0, \mathbf{k}, \mathbf{v}) - \frac{\omega_{pe}^2}{k^2\varepsilon(\mathbf{k}, \omega)}\,\mathbf{k}\cdot\frac{\partial F}{\partial \mathbf{v}}\int\frac{\delta f(0, \mathbf{k}, \mathbf{v})}{\omega - \mathbf{k}\cdot\mathbf{v}}d^3v\right] \tag{1.64}$$

$$\text{with}\quad \varepsilon(\mathbf{k}, \omega) = 1 + \frac{\omega_{pe}^2}{k^2}\int d^3v\,\frac{\mathbf{k}\cdot\frac{\partial F}{\partial \mathbf{v}}}{\omega - \mathbf{k}\cdot\mathbf{v}} \tag{1.65}$$

where we used

$$\phi(k) = \int \frac{e^2}{r}e^{-i\mathbf{k}\cdot\mathbf{r}}d^3r = \frac{4\pi e^2}{k^2}$$

So far we obtained $\delta f(\omega, \mathbf{k}, \mathbf{v})$ in terms of $\delta f(0, \mathbf{k}, \mathbf{v})$ and $F(\mathbf{v})$ by performing the analogous analysis as we do for Vlasov equation. The functions $\delta f(\omega, \mathbf{k}, \mathbf{v})$ and $\varepsilon(\mathbf{k}, \omega)$ are defined in the regions $\omega_i>0$, but for un-pathological functions F and $\delta f(0)$, the functions can be analytically continued into the regions $\omega_i<0$.

Next, we calculate the electric field $\delta\mathbf{E}(\mathbf{k}, \omega) = -i\mathbf{k}\delta\phi(\mathbf{k}, \omega)$ associated with the fluctuating distribution function $\delta f(\omega, \mathbf{k}, \mathbf{v})$. We have the Poisson equation

$$k^2\delta\phi(\mathbf{k}, \omega) = 4\pi e\int d^3v\;\;\delta f(\omega, \mathbf{k}, \mathbf{v})$$

Integrating (1.63), we obtain

$$\int d^3v\;\;\delta f(\omega, \mathbf{k}, \mathbf{v}) = \frac{i}{\varepsilon(\mathbf{k}, \omega)}\int d^3v\,\frac{\delta f(0, \mathbf{k}, \mathbf{v})}{\omega - \mathbf{k}\cdot\mathbf{v}}$$

$$\text{Therefore,}\quad \delta\mathbf{E}(\mathbf{k}, \omega) = \frac{\mathbf{k}}{k^2}\frac{4\pi e}{\varepsilon(\mathbf{k}, \omega)}\int d^3v\,\frac{\delta f(0, \mathbf{k}, \mathbf{v})}{\omega - \mathbf{k}\cdot\mathbf{v}} \tag{1.66}$$

So far the development is parallel to the analysis of the initial value problem of the Vlasov equation. Now we pay attention to the initial value of

δf : $\delta f(0, \mathbf{k}, \mathbf{v})$. At $t = 0$, the initial Klimontovich distribution function is

$$K(0, \mathbf{r}, \mathbf{v}) = \sum_{i=1}^{N} \delta(\mathbf{r} - \mathbf{r}_{i0})\delta(\mathbf{v} - \mathbf{v}_{i0}) \tag{1.67}$$

$$\text{Therefore,} \quad \delta f(0, \mathbf{r}, \mathbf{v}) = \sum_{i=1}^{N} \delta(\mathbf{r} - \mathbf{r}_{i0})\delta(\mathbf{v} - \mathbf{v}_{i0}) - F(\mathbf{v}) \tag{1.68}$$

where $F(\mathbf{v})$ is the ensemble average of the function in (1.67) over the Liouville function $F^{(N)}(\mathbf{x}_{10}, \mathbf{x}_{20}, \ldots, \mathbf{x}_{N0})$. Fourier transform of (1.68) is

$$\delta f(0, \mathbf{k}, \mathbf{v}) = \sum_{i=1}^{N} e^{-i\mathbf{k}\cdot\mathbf{r}_{i0}}\delta(\mathbf{v} - \mathbf{v}_{i0}) - (2\pi)^3\delta(\mathbf{k})F(\mathbf{v}) \tag{1.69}$$

Using (1.69) in (1.64) and (1.66) gives

$$\delta \mathbf{E}(\mathbf{k}, \omega) = \frac{\mathbf{k}}{k^2} \frac{4\pi e}{\varepsilon(\mathbf{k}, \omega)} \sum_{i=1}^{N} \frac{e^{-i\mathbf{k}\cdot\mathbf{r}_{io}}}{\omega - \mathbf{k}\cdot\mathbf{v}_{i0}} \tag{1.70}$$

$$\begin{aligned}\delta f(\omega, \mathbf{k}, \mathbf{v}) = {} & \frac{i}{\omega - \mathbf{k}\cdot\mathbf{v}} \left[\sum_{i=1}^{N} e^{-i\mathbf{k}\cdot\mathbf{r}_{i0}}\delta(\mathbf{v} - \mathbf{v}_{i0}) - (2\pi)^3\delta(\mathbf{k})F(\mathbf{v})\right] \\ & - \frac{i}{\omega - \mathbf{k}\cdot\mathbf{v}} \left[\frac{4\pi e^2}{mk^2\varepsilon(\mathbf{k}, \omega)}\mathbf{k}\cdot\frac{\partial F}{\partial \mathbf{v}} \sum_{i=1}^{N} \frac{e^{-i\mathbf{k}\cdot\mathbf{r}_{io}}}{\omega - \mathbf{k}\cdot\mathbf{v}_{i0}}\right]\end{aligned} \tag{1.71}$$

If we Laplace-invert the above equations, we obtain $\delta\mathbf{E}(\mathbf{k}, t)$ and $\delta f(\mathbf{k}, \mathbf{v}, t)$. We consider the time-asymptotic $(t \to \infty)$ forms of $\delta\mathbf{E}$ and δf. It is clear that the time-asymptotic forms of these functions are determined by the *upper most* singularities of the integrands. There are two kinds of singularities: the first kind associated with $\varepsilon(\mathbf{k}, \omega) = 0$ and the second, those on the real ω-axis. We assume that the plasma is stable (only Landau-damped), then the first kind makes the perturbation damp away since the singular points have negative imaginary parts. The asymptotic behavior is determined by the singularities on the real ω axis: $\omega = \mathbf{k}\cdot\mathbf{v}$ and $\omega = \mathbf{k}\cdot\mathbf{v}_{i0}$. Thus we have as $t \to \infty$:

$$\delta\mathbf{E}(\mathbf{k}, t) \longrightarrow -i\,\frac{4\pi e\mathbf{k}}{k^2} \sum_{i=1}^{N} \frac{e^{-i\mathbf{k}\cdot(\mathbf{r}_{i0}+\mathbf{v}_{i0}t)}}{\varepsilon(\mathbf{k},\ \mathbf{k}\cdot\mathbf{v}_{i0})} \tag{1.72}$$

$$\delta f(\mathbf{k},\mathbf{v},t)\longrightarrow\sum_{i=1}^{N}e^{-i\mathbf{k}\cdot(\mathbf{r}_{i0}+\mathbf{v}_{i0}t)}\delta(\mathbf{v}-\mathbf{v}_{i0})-(2\pi)^3F(\mathbf{v})\delta(\mathbf{k})$$
$$-\frac{4\pi e^2}{mk^2}\,\mathbf{k}\cdot\frac{\partial F}{\partial\mathbf{v}}\sum_{i=1}^{N}\frac{e^{-i\mathbf{k}\cdot\mathbf{r}_{i0}}}{\mathbf{k}\cdot(\mathbf{v}_{io}-\mathbf{v})}\left[\frac{e^{-i\mathbf{k}\cdot\mathbf{v}_{i0}}}{\varepsilon(\mathbf{k},\ \mathbf{k}\cdot\mathbf{v}_{i0}t)}-\frac{e^{-i\mathbf{k}\cdot\mathbf{v}t}}{\varepsilon(\mathbf{k},\ \mathbf{k}\cdot\mathbf{v})}\right]\tag{1.73}$$

• **Ensemble average $<\delta f\delta\mathbf{E}>$ and Lenard–Balescu equation.** The Lenard–Balescu equation is derived by calculating the right side of (1.60), or (1.62′).

$$\frac{\partial}{\partial t}F(\mathbf{v},t)=\frac{e}{m}\frac{\partial}{\partial\mathbf{v}}\cdot<\delta f\delta\mathbf{E}>\tag{1.74}$$

where the ensemble average of the RHS is to be calculated by using (1.72) and (1.73). We write

$$\delta f(\mathbf{k},\mathbf{v},t)\delta\mathbf{E}(\mathbf{k}',t)=-i\,\frac{4\pi e\mathbf{k}'}{k'^2}\sum_{j=1}^{N}\frac{e^{-i\mathbf{k}'\cdot(\mathbf{r}_{j0}+\mathbf{v}_{j0}t)}\ \copyright}{\varepsilon(\mathbf{k}',\ \mathbf{k}'\cdot\mathbf{v}_{j0})}$$
$$\times\left[\sum_{i=1}^{N}e^{-i\mathbf{k}\cdot(\mathbf{r}_{i0}+\mathbf{v}_{i0}t)}\delta(\mathbf{v}-\mathbf{v}_{i0})^{\copyright'}-\frac{4\pi e^2}{mk^2}\,\mathbf{k}\cdot\frac{\partial F}{\partial\mathbf{v}}\right.$$
$$\left.\times\sum_{i=1}^{N}\frac{e^{-i\mathbf{k}\cdot\mathbf{r}_{i0}}}{\mathbf{k}\cdot(\mathbf{v}_{io}-\mathbf{v})}\left(\frac{e^{-i\mathbf{k}\cdot\mathbf{v}_{i0}t}}{\varepsilon(\mathbf{k},\ \mathbf{k}\cdot\mathbf{v}_{i0})}\oplus-\frac{e^{-i\mathbf{k}\cdot\mathbf{v}t}}{\varepsilon(\mathbf{k},\ \mathbf{k}\cdot\mathbf{v})}\otimes\right)\right]\tag{1.75}$$

Taking the ensemble average of the above quantity operationally means multiplying

$$F^{(N)}(\mathbf{x}_{10},\mathbf{x}_{20},\ldots,\mathbf{x}_{N0})=F^{(1)}(\mathbf{v}_{10})F^{(1)}(\mathbf{v}_{20})\cdots F^{(1)}(\mathbf{v}_{N0})$$

and performing integration with respect to $d\mathbf{r}_{i0}d\mathbf{v}_{i0}$ and $d\mathbf{r}_{j0}d\mathbf{v}_{j0}$. Let us consider the average of the product $<\copyright\copyright'>$. Only the $i=j$ terms survive the spatial integration, and the sum $\sum_1^N$ gives N identical terms. We have two δ functions; $\delta(\mathbf{k}+\mathbf{k}')$ and $\delta(\mathbf{v}-\mathbf{v}_{i0})$. Thus we obtain

$$<\copyright\copyright'>=-4\pi ien_0\frac{\mathbf{k}'}{k^2}\frac{\delta(\mathbf{k}+\mathbf{k}')}{\varepsilon(\mathbf{k}',\mathbf{k}'\cdot\mathbf{v})}F(\mathbf{v})$$

Next we consider the product $<\copyright\oplus>$. Here again only the terms $i=j$ survive the spatial integration and the summation gives N identical terms, and $\delta(\mathbf{k}+\mathbf{k}')$ is produced; the velocity integral cannot be completed.

$$<\copyright\bigoplus> = \frac{-4\pi i e n_0}{k^2}\,\frac{4\pi e^2}{mk^2}\left(\mathbf{k}\cdot\frac{\partial F}{\partial \mathbf{v}}\right)\delta(\mathbf{k}+\mathbf{k}')\,\mathbf{k}\int d^3v'\frac{1}{\mathbf{k}\cdot(\mathbf{v}'-\mathbf{v})}\,\frac{F(\mathbf{v}')}{|\varepsilon(\mathbf{k},\mathbf{k}\cdot\mathbf{v}')|^2}$$

Finally the product $\copyright\bigotimes$ is calculated likewise:

$$<\copyright\bigotimes> = \frac{-4\pi i e n_0}{k^2}\,\frac{4\pi e^2}{mk^2}\left(\mathbf{k}\cdot\frac{\partial F}{\partial \mathbf{v}}\right)\delta(\mathbf{k}+\mathbf{k}')\,\mathbf{k}\int d^3v'\frac{-e^{-i\mathbf{k}\cdot(\mathbf{v}-\mathbf{v}')t}}{\mathbf{k}\cdot(\mathbf{v}'-\mathbf{v})}\,\frac{F(\mathbf{v}')}{|\varepsilon(\mathbf{k},\mathbf{k}\cdot\mathbf{v}')|^2}$$

where we put $\varepsilon(-\mathbf{k},-\mathbf{k}\cdot\mathbf{v}')=\varepsilon(-\mathbf{k},-\mathbf{k}\cdot\mathbf{v})$ because the velocity integral is predominantly contributed from the region around $\mathbf{v}'=\mathbf{v}$. Collecting the above results, we write

$$\begin{aligned}<\delta f(\mathbf{k},\mathbf{v},t)\delta\mathbf{E}(\mathbf{k}',t)> &= 4\pi i e n_0\delta(\mathbf{k}+\mathbf{k}')\frac{\mathbf{k}}{k^2}\\ &\times\left[\frac{F(\mathbf{v})}{\varepsilon(-\mathbf{k},-\mathbf{k}\cdot\mathbf{v})}+\frac{4\pi e^2}{mk^2}\mathbf{k}\cdot\frac{\partial F}{\partial \mathbf{v}}\int d^3v'\,\frac{1-e^{-i\mathbf{k}\cdot(\mathbf{v}-\mathbf{v}')t}}{\mathbf{k}\cdot(\mathbf{v}-\mathbf{v}')}\,\frac{F(\mathbf{v}')}{|\varepsilon(\mathbf{k},\mathbf{k}\cdot\mathbf{v})|^2}\right]\end{aligned}\tag{1.76}$$

where time t is understood to be approaching ∞. Using the formula

$$\lim_{t\to\infty}\frac{1-e^{-i\mathbf{k}\cdot(\mathbf{v}-\mathbf{v}')t}}{\mathbf{k}\cdot(\mathbf{v}-\mathbf{v}')}=i\pi\delta(\mathbf{k}\cdot(\mathbf{v}-\mathbf{v}'))$$

Equation (1.76) becomes

$$\begin{aligned}\lim_{t\to\infty}<\delta f(\mathbf{k},\mathbf{v},t)\delta\mathbf{E}(\mathbf{k}',t)> &= 4\pi i e n_0\delta(\mathbf{k}+\mathbf{k}')\frac{\mathbf{k}}{k^2}\\ &\times\left[\frac{F(\mathbf{v})}{\varepsilon(-\mathbf{k},-\mathbf{k}\cdot\mathbf{v})}+\frac{\omega_p^2}{k^2}\,\mathbf{k}\cdot\frac{\partial F}{\partial \mathbf{v}}i\pi\int d^3v'\;\delta(\mathbf{k}\cdot\mathbf{v}-\mathbf{k}\cdot\mathbf{v}')\,\frac{F(\mathbf{v}')}{|\varepsilon(\mathbf{k},\mathbf{k}\cdot\mathbf{v})|^2}\right]\end{aligned}\tag{1.77}$$

where the first term inside the square bracket is

$$\frac{\varepsilon(\mathbf{k},\mathbf{k}\cdot\mathbf{v})}{|\varepsilon(\mathbf{k},\mathbf{k}\cdot\mathbf{v})|^2}F(\mathbf{v})\tag{1.78}$$

whose imaginary part only is needed. Using the formula

$$\mathrm{Im}\ \varepsilon(\mathbf{k},\omega) = -i\ \pi \frac{\omega_p^2}{k^2} \int d^3v'\mathbf{k}\cdot \frac{\partial F(\mathbf{v}')}{\partial \mathbf{v}'} \delta(\omega - \mathbf{k}\cdot\mathbf{v}')$$

the imaginary part of (1.78) is

$$-\frac{\omega_p^2}{k^2}\, i\pi \frac{F(\mathbf{v})}{|\varepsilon(\mathbf{k},\mathbf{k}\cdot\mathbf{v})|^2} \int d^3v'\ \delta(\mathbf{k}\cdot\mathbf{v} - \mathbf{k}\cdot\mathbf{v}')\ \mathbf{k}\cdot \frac{\partial F(\mathbf{v}')}{\partial \mathbf{v}'} \tag{1.79}$$

Using (1.79) in (1.76) gives

$$\begin{aligned}\lim_{t\to\infty} <\delta f(\mathbf{k},\mathbf{v},t)\delta \mathbf{E}(\mathbf{k}',t)> &= 4\pi i e n_0 \delta(\mathbf{k}+\mathbf{k}')\frac{\mathbf{k}}{k^2}\\ \times\, i\pi\ \frac{\omega_p^2}{k^2}\ \frac{\mathbf{k}\cdot}{|\varepsilon(\mathbf{k},\mathbf{k}\cdot\mathbf{v})|^2} &\int d^3v'\ \delta(\mathbf{k}\cdot\mathbf{v} - \mathbf{k}\cdot\mathbf{v}')\left(F(\mathbf{v}')\frac{\partial F(\mathbf{v})}{\partial \mathbf{v}} - F(\mathbf{v})\frac{\partial F(\mathbf{v}')}{\partial \mathbf{v}'}\right)\end{aligned} \tag{1.80}$$

After Fourier inversion of (1.80) by performing $\int d^3k d^3k'(\cdots)e^{i\mathbf{k}\cdot\mathbf{r}}e^{i\mathbf{k}'\cdot\mathbf{r}}$ (1.74) gives the Lenard–Balescu equation in the form

$$\begin{aligned}\frac{\partial}{\partial t}F(\mathbf{v},t) = -N\ \frac{\partial}{\partial \mathbf{v}}\cdot \int d^3k\ \mathbf{k}\ &\frac{\omega_p^2}{k^2}\frac{\omega_p^2}{k^2}\frac{\pi(2\pi)^3}{|\varepsilon(\mathbf{k},\mathbf{k}\cdot\mathbf{v})|^2}\\ &\times \int d^3v'\delta[\mathbf{k}\cdot(\mathbf{v}-\mathbf{v}')]\ \mathbf{k}\cdot\left(F(\mathbf{v})\frac{\partial F(\mathbf{v}')}{\partial \mathbf{v}'} - F(\mathbf{v}')\frac{\partial F(\mathbf{v})}{\partial \mathbf{v}}\right)\end{aligned} \tag{1.81}$$

The Lenard–Balescu equation gives the time rate of change of the one-particle distribution function $F^{(1)}$ (recall (1.47)) in a self-consistent manner, obtained by taking the steps of the ensemble average over the Liouville density. The right side corresponds to the *collision term* in the Boltzmann equation. $F(\mathbf{v})$ in the right side is the same function $F(\mathbf{v},t)$ on the left.

1.7 Liouville Theorem Revisited

Let us consider a Liouville fluid consisting of N particles in $6N$-dimension Γ-space. The Liouville fluid flux across the boundary, in the course of motion, is expressed by

$$\left(\frac{d\mathbf{r}}{dt}F^{(N)}, \frac{d\mathbf{p}}{dt}F^{(N)}\right)$$

$$\mathbf{r} = (\mathbf{r}_1, \mathbf{r}_2, \ldots, \mathbf{r}_N)$$

$$\mathbf{p} = (\mathbf{p}_1, \mathbf{p}_2, \ldots, \mathbf{p}_N)$$

In accordance with the mass conservation, we have

$$\frac{\partial}{\partial t}\int d^3r d^3p\ F^{(N)} = -\int d^3v d^2\mathbf{r}\cdot\frac{d\mathbf{r}}{dt}\ F^{(N)} - \int d^3r d^2\mathbf{p}\cdot\frac{d\mathbf{p}}{dt}\ F^{(N)}$$

Applying the Gauss theorem to the surface integrals on the right side, we have

$$\int d^3r d^3p\left[\frac{\partial F^{(N)}}{\partial t} + \frac{\partial}{\partial \mathbf{r}}\cdot\left(\frac{d\mathbf{r}}{dt}F^{(N)}\right) + \frac{\partial}{\partial \mathbf{p}}\cdot\left(\frac{d\mathbf{p}}{dt}F^{(N)}\right)\right] = 0$$

Using Hamilton's equations

$$\frac{d\mathbf{r}}{dt} = \frac{\partial H}{\partial \mathbf{p}}, \quad \frac{d\mathbf{p}}{dt} = -\frac{\partial H}{\partial \mathbf{r}}$$

in the above gives the Liouville equation

$$\frac{\partial F^{(N)}}{\partial t} + \frac{\partial H}{\partial \mathbf{p}}\cdot\frac{\partial F^{(N)}}{\partial \mathbf{r}} - \frac{\partial F^{(N)}}{\partial \mathbf{p}}\cdot\frac{\partial H}{\partial \mathbf{r}} = 0$$

$$\text{or}\quad \frac{\partial F^{(N)}}{\partial t} + \sum_{i=1}^{N}\left(\frac{\partial H}{\partial \mathbf{p}_i}\cdot\frac{\partial F^{(N)}}{\partial \mathbf{r}_i} - \frac{\partial F^{(N)}}{\partial \mathbf{p}_i}\cdot\frac{\partial H}{\partial \mathbf{r}_i}\right) = 0$$

The Liouville equation amounts to $\frac{DF^{(N)}}{Dt}=0$, where D/Dt is the Lagrangian derivative along the trajectory of the representative point $(\mathbf{r}, \mathbf{p})$ in the Γ-space. The Liouville theorem is simply the mass conservation applied to the Liouville fluid as elaborated above. As is well-known, Hamilton's equations are the consequence of the variational principle, $\delta\int Ldt = 0$ (L: Lagrangian).

> It is a general feature of equations of classical physics that they can be derived from the variational principle.
>
> —Max Born

For a Liouville fluid, the incompressibility states for the fluid velocity $\mathbf{V}_\Gamma$:

$$\nabla \cdot \mathbf{V}_\Gamma = \sum_{i=1}^{N} \left(\frac{\partial \dot{\mathbf{r}}_i}{\partial \mathbf{r}_i} + \frac{\partial \dot{\mathbf{p}}_i}{\partial \mathbf{p}_i} \right) = \sum_{i=1}^{N} \left(\frac{\partial}{\partial \mathbf{r}_i} \cdot \frac{\partial H}{\partial \mathbf{p}_i} - \frac{\partial}{\partial \mathbf{p}_i} \cdot \frac{\partial H}{\partial \mathbf{r}_i} \right) = 0$$

which clearly vanishes.

1.8 Quasilinear Theory

Consider a plasma which consists of N_α ($\alpha = e,\ i$) particles and has average field $\mathbf{E} = 0$ (the plasma is homogeneous). The distribution of particles of each species is described by the Klimontovich distribution function

$$F_\alpha(\mathbf{r}, \mathbf{v}, t) = \sum_{i=1}^{N_\alpha} \delta(\mathbf{r} - \mathbf{r}_i(t))\delta(\mathbf{v} - \mathbf{v}_i(t)) \quad (\alpha = e,\ i) \tag{1.82}$$

with the initial distribution function

$$F_\alpha(\mathbf{r}, \mathbf{v}, t = 0) = \sum_{i=1}^{N_\alpha} \delta(\mathbf{r} - \mathbf{r}_{i0})\delta(\mathbf{v} - \mathbf{v}_{i0}) \tag{1.83}$$

The distribution function satisfies the Liouville equation

$$\frac{\partial F_\alpha}{\partial t} + \mathbf{v} \cdot \frac{\partial F_\alpha}{\partial \mathbf{r}} + \frac{e_\alpha}{m_\alpha}\mathbf{E} \cdot \frac{\partial F_\alpha}{\partial \mathbf{v}} = 0 \tag{1.84}$$

and the Poisson equation

$$\nabla \cdot \mathbf{E} = 4\pi \sum_\alpha n_{0\alpha} e_\alpha \int F_\alpha(\mathbf{r}, \mathbf{v}, t) d^3v \quad (n_{0\alpha} = N_\alpha/V,\ V:\ \text{volume}) \tag{1.85}$$

Fourier transform the above equations with respect to $\mathbf{r}$ gives

$$F_\alpha(\mathbf{k}, \mathbf{v}, t) = \sum_{i=1}^{N_\alpha} \delta(\mathbf{v} - \mathbf{v}_i(t)) e^{-i\mathbf{k}\cdot\mathbf{r}_i(t)} \tag{1.86}$$

$$F_\alpha(\mathbf{k}, \mathbf{v}, t = 0) = \sum_{i=1}^{N_\alpha} \delta(\mathbf{v} - \mathbf{v}_{i0}) e^{-i\mathbf{k}\cdot\mathbf{r}_{i0}} \tag{1.87}$$

$$i\mathbf{k} \cdot \mathbf{E}(\mathbf{k}, t) = 4\pi \sum_\alpha n_{0\alpha} e_\alpha \int F_\alpha(\mathbf{k}, \mathbf{v}, t) d^3v$$

To write Fourier transform of (1.84), let us use the convolution formula:

$$\int d^3 r e^{-i\mathbf{k}\cdot\mathbf{r}}\mathbf{E}(\mathbf{r})F(\mathbf{r}) = \left(\frac{1}{2\pi}\right)^3 \int d^3 k' \mathbf{E}(\mathbf{k}')F(\mathbf{k}-\mathbf{k}')$$

If one wishes to express the above in Fourier series, put $\left(\frac{1}{2\pi}\right)^3 \int d^3k' \to \sum_{\mathbf{k}'}$. Then (1.84) becomes

$$\frac{\partial}{\partial t}F_\alpha(\mathbf{k},\mathbf{v},t) + i\mathbf{k}\cdot\mathbf{v}F_\alpha + \frac{e_\alpha}{m_\alpha}\mathbf{E}(\mathbf{k},t)\cdot\frac{\partial F_\alpha(\mathbf{k}=0,\mathbf{v},t)}{\partial \mathbf{v}}$$
$$= -\frac{e_\alpha}{m_\alpha}\sum_{\mathbf{k}'}{}' \mathbf{E}(\mathbf{k}',t)\cdot\frac{\partial}{\partial \mathbf{v}}F_\alpha(\mathbf{k}-\mathbf{k}',\mathbf{v},t) \tag{1.88}$$

where we separated the $k=0$ term from the product term and the prime on $\sum$ denotes that the $\mathbf{k}'=\mathbf{k}$ term is excluded in the summation. Let us assume in (1.88) that $F_\alpha(\mathbf{k}=0,\mathbf{v},t) \gg F_\alpha(\mathbf{k},\mathbf{v},t)$ and neglect the RHS:

$$\frac{\partial}{\partial t}F_\alpha(\mathbf{k},\mathbf{v},t) + i\mathbf{k}\cdot\mathbf{v}F_\alpha + \frac{e_\alpha}{m_\alpha}\mathbf{E}(\mathbf{k},t)\cdot\frac{\partial F_\alpha(\mathbf{k}=0,\mathbf{v},t)}{\partial \mathbf{v}} = 0 \tag{1.89}$$

The crucial assumption here is that the spatially uniform part of the distribution is much greater than the spatially fluctuating part ($F_\alpha(\mathbf{k}=0,\mathbf{v},t) \gg F_\alpha(\mathbf{k},\mathbf{v},t)$). The neglect of the RHS of (1.88), the 'collision term', is the approximation that is used for the Vlasov equation. More refinement over the Vlasov-like equation, (1.89), is to take account of the time dependence of $F_\alpha(\mathbf{k}=0,\mathbf{v},t)$. The equation for $F_\alpha(\mathbf{k}=0,\mathbf{v},t)$ can be set up by extracting the terms corresponding to $\mathbf{k}=0$ in (1.88):

$$\frac{\partial}{\partial t}F_\alpha(\mathbf{k}=0,\mathbf{v},t) = -\frac{e_\alpha}{m_\alpha}\sum_{\mathbf{k}}\mathbf{E}(-\mathbf{k},t)\cdot\frac{\partial}{\partial \mathbf{v}}F_\alpha(\mathbf{k},\mathbf{v},t) \tag{1.90}$$

In addition to (1.89) and (1.90), we have two more equations:

$$i\mathbf{k}\cdot\mathbf{E}(\mathbf{k},t) = 4\pi\sum_\alpha n_{0\alpha}e_\alpha\int F_\alpha(\mathbf{k},\mathbf{v},t)d^3v \tag{1.91}$$

$$F_\alpha(\mathbf{k},\mathbf{v},t=0) = \sum_{i=1}^{N_\alpha}\delta(\mathbf{v}-\mathbf{v}_{i0})e^{-i\mathbf{k}\cdot\mathbf{r}_{i0}} \tag{1.92}$$

where $\mathbf{r}_{i0}$ and $\mathbf{v}_{i0}$ are the initial position and velocity of the ith particle of species α (the α-index omitted).

Equations (1.89)–(1.92) are the basic equations of the quasilinear theory, and further development of these equations leads to the Lenard–Balescu equation.

We solve (1.89)–(1.91) by Laplace transform with initial condition (1.92). The slow time variation of $F_\alpha(\mathbf{k} = 0, \mathbf{v}, t)$ in (1.89) will be neglected by holding that quantity constant [that quantity will be expressed as $F_\alpha(\mathbf{k} = 0, \mathbf{v}, \epsilon t)$ with the understanding that 'ϵt' represents the slow time]. Laplace transform of (1.89) yields

$$F_\alpha(\mathbf{k}, \mathbf{v}, \omega) = \frac{-i}{\omega - \mathbf{k}\cdot\mathbf{v}}\left[\frac{e_\alpha}{m_\alpha}\mathbf{E}(\mathbf{k}, \omega)\cdot\frac{\partial}{\partial\mathbf{v}}F_\alpha(\mathbf{k} = 0, \mathbf{v}, \epsilon t) - F_\alpha(\mathbf{k}, \mathbf{v}, t = 0)\right] \tag{1.93}$$

Laplace transform of (1.91) and use of (1.93) give

$$\begin{aligned}\mathbf{k}\cdot\mathbf{E}(\mathbf{k}, \omega) &= \frac{4\pi}{\varepsilon(\mathbf{k}, \omega)}\sum_\alpha n_{0\alpha}e_\alpha\int d^3v\frac{F_\alpha(\mathbf{k}, \mathbf{v}, t = 0)}{\omega - \mathbf{k}\cdot\mathbf{v}} \\ &= \frac{4\pi}{\varepsilon(\mathbf{k}, \omega)}\sum_\alpha n_{0\alpha}e_\alpha\sum_{i=1}^{N_\alpha}\frac{e^{-i\mathbf{k}\cdot\mathbf{r}_{i0}}}{\omega - \mathbf{k}\cdot\mathbf{v}_{i0}} \quad [cf.\ (1.70)]\end{aligned} \tag{1.94}$$

$$\text{where}\quad \varepsilon(\mathbf{k}, \omega) = 1 + \sum_\alpha\frac{\omega_{p\alpha}^2}{k^2}\int d^3v\frac{\mathbf{k}\cdot\frac{\partial}{\partial\mathbf{v}}F_\alpha(\mathbf{k} = 0, \mathbf{v},\ \epsilon t)}{\omega - \mathbf{k}\cdot\mathbf{v}} \tag{1.95}$$

Using (1.92) and (1.94) in (1.93) gives

$$\begin{aligned}F_\alpha(\mathbf{k}, \mathbf{v}, \omega) = &\frac{-i}{\omega - \mathbf{k}\cdot\mathbf{v}}\frac{1}{k^2\varepsilon(\mathbf{k}, \omega)}\sum_\alpha\omega_{p\alpha}^2\sum_{i=1}^{N_\alpha}\frac{e^{-i\mathbf{k}\cdot\mathbf{r}_{i0}}}{\omega - \mathbf{k}\cdot\mathbf{v}_{i0}}\mathbf{k}\cdot\frac{\partial F_\alpha(\mathbf{k} = 0, \mathbf{v},\ \epsilon t)}{\partial\mathbf{v}} \\ &+ \frac{i}{\omega - \mathbf{k}\cdot\mathbf{v}}\sum_{i=1}^{N_\alpha}\delta(\mathbf{v} - \mathbf{v}_{i0})e^{-i\mathbf{k}\cdot\mathbf{r}_{i0}}\end{aligned} \tag{1.96}$$

Inverse Laplace transform of (1.94) and (1.96) yields

$$\mathbf{k}\cdot\mathbf{E}(\mathbf{k}, t) = 4\pi i\sum_\alpha n_{0\alpha}e_\alpha\sum_{i=1}^{N_\alpha}\frac{e^{-i\mathbf{k}\cdot(\mathbf{r}_{i0}+\mathbf{v}_{i0}t)}}{\varepsilon(\mathbf{k}, \mathbf{k}\cdot\mathbf{v}_{i0})} \tag{1.97}$$

$$\begin{aligned}F_\alpha(\mathbf{k}, \mathbf{v}, t) = &\sum_{i=1}^{N_\alpha}\delta(\mathbf{v} - \mathbf{v}_{i0})e^{-i\mathbf{k}\cdot(\mathbf{r}_{i0}+\mathbf{v}t)} \\ &- \frac{1}{k^2}\sum_\alpha\omega_{p\alpha}^2\sum_{i=1}^{N_\alpha}\frac{\mathbf{k}\cdot\frac{\partial}{\partial\mathbf{v}}F_\alpha(\mathbf{k} = 0, \mathbf{v},\ \epsilon t)}{\mathbf{k}\cdot\mathbf{v} - \mathbf{k}\cdot\mathbf{v}_{i0}}\left(\frac{e^{-i\mathbf{k}\cdot(\mathbf{r}_{i0}+\mathbf{v}t)}}{\varepsilon(\mathbf{k}, \mathbf{k}\cdot\mathbf{v})} - \frac{e^{-i\mathbf{k}\cdot(\mathbf{r}_{i0}+\mathbf{v}_{i0}t)}}{\varepsilon(\mathbf{k}, \mathbf{k}\cdot\mathbf{v}_{i0})}\right)\end{aligned} \tag{1.98}$$

which is identical with (1.73) [$\delta(\mathbf{k})$-term in the latter disappears eventually due to $\mathbf{k}\delta(\mathbf{k}) = 0$]. Equation (1.97) gives upon using the electrostatic nature of $\mathbf{E}$

$$\mathbf{E}(-\mathbf{k}, t) = -\frac{4\pi i\mathbf{k}}{k^2}\sum_{\alpha} n_{0\alpha}e_{\alpha}\sum_{i=1}^{N_\alpha}\frac{e^{i\mathbf{k}\cdot(\mathbf{r}_{i0}+\mathbf{v}_{i0}t)}}{\varepsilon(-\mathbf{k}, -\mathbf{k}\cdot\mathbf{v}_{i0})} \tag{1.99}$$

Putting $\mathbf{k} \to -\mathbf{k}$, (1.99) is identical with (1.72) for one-species plasma (electrons). Substitute (1.98) and (1.99) into (1.90):

$$\frac{\partial}{\partial t}F_{\alpha}(\mathbf{k} = 0, \mathbf{v}, \epsilon t) = -\frac{\partial}{\partial \mathbf{v}}\cdot\sum_{\mathbf{k}}\frac{e_\alpha}{m_\alpha}<\mathbf{E}(-\mathbf{k}, t)F_{\alpha}(\mathbf{k}, \mathbf{v}, t)> \tag{1.100}$$

where the ensemble average should be taken as is indicated by the angular bracket. Equations (1.98)–(1.100) are entirely identical with (1.72)–(1.74), and the Lenard–Balescu equation follows by repeating the same algebra as in Section 1.6.

Problems

1.1 The Hamiltonian for a system of charged particles in (1.3) has excluded the inter-particle magnetic interaction. If the latter is included, we have the Hamiltonian (Gartenhaus, 1964; Darwin, 1920)

$$H = \frac{1}{2m}\sum_{i=1}^{N}\left[\mathbf{p}_i - \frac{e}{c}\mathbf{A}(\mathbf{r}_i, t)\right]^2 + e\sum_{i=1}^{N}\phi(\mathbf{r}_i, t) + \frac{e^2}{2}\sum_{i=1}^{N}\sum_{j(\neq i)=1}^{N}\frac{1}{r_{ij}}$$
$$-\frac{e^2}{4m^2c^2}\sum_{i=1}^{N}\sum_{j(\neq i)=1}^{N}\left(\frac{\mathbf{p}_i\cdot\mathbf{p}_j}{r_{ij}} + \frac{(\mathbf{p}_i\cdot\mathbf{r}_{ij})(\mathbf{p}_j\cdot\mathbf{r}_{ij})}{r_{ij}^3}\right)$$

where $\mathbf{r}_{ij} = \mathbf{r}_i - \mathbf{r}_j$.
Derive the equation of motion analogous to (1.10).

1.2 The Lorentz force acting on a particle of mass m and charge e that moves in an electric field $\mathbf{E}$ and a magnetic field $\mathbf{B}$ is

$$\mathbf{F} = e\left[\mathbf{E} + \frac{1}{c}(\mathbf{v}\times\mathbf{B})\right]$$

Introduce the potentials ϕ and $\mathbf{A}$ as given by (1.4) and (1.5) and use the Lorentz gauge

$$\nabla\cdot\mathbf{A} + \frac{1}{c}\frac{\partial\phi}{\partial t} = 0$$

(a) Show that the generalized potential for the charged particle is

$$U = e\left(\phi - \frac{1}{c}\mathbf{v}\cdot\mathbf{A}\right)$$

(b) Show that the momentum of the particle is

$$\mathbf{p} = m\mathbf{v} + \frac{e}{c}\mathbf{A}$$

(c) Show that the Hamiltonian for the particle is

$$H = \frac{1}{2m}\left(\mathbf{p} - \frac{e}{c}\mathbf{A}\right)^2 + e\phi$$

(d) Show that the relativistic Hamiltonian is

$$H = \sqrt{p^2c^2 + m^2c^4} + \text{ potential } V$$

with $\mathbf{p} = \gamma m\mathbf{v}$, $\gamma = \frac{1}{\sqrt{1-\frac{v^2}{c^2}}}$.

1.3 By differentiating the following expression for the Liouville density, derive the Liouville equation (Montgomery, 1971)

$$F^{(N)}(\mathbf{X}, t) = \int d\tau_0 F^{(N)}(\mathbf{X}_0) \prod_{i=1}^{N} \delta(\mathbf{r}_i - \mathbf{r}_i(\mathbf{r}_{i0}, t)\ \delta(\mathbf{v}_i - \mathbf{v}_i(\mathbf{v}_{i0}, t)).$$

Answer. Differentiating the above equation with respect to t and using the orbit equation

$$\frac{d\mathbf{x}_i(t)}{dt} = \mathbf{v}_i(t)$$
$$\frac{d\mathbf{v}_i(t)}{dt} = -\frac{e}{m}\mathbf{E}(\mathbf{x}_i(t), t)$$

give

$$\frac{\partial}{\partial t}F^{(N)}(\mathbf{X}, t) = -\int d\tau_0 F^{(N)}(\mathbf{X}_0) \sum_{i=1}^{N} \left(\mathbf{v}_i \cdot \frac{\partial}{\partial \mathbf{x}_i} - \frac{e}{m}\mathbf{E}(\mathbf{x}_i, t) \cdot \frac{\partial}{\partial \mathbf{v}_i}\right)$$
$$\times \prod_{i=1}^{N} \delta(\mathbf{r}_i - \mathbf{r}_i(\mathbf{r}_{i0}, t))\ \delta(\mathbf{v}_i - \mathbf{v}_i(\mathbf{v}_{i0}, t)).$$

Interchanging differentiation and integration yields

$$\frac{\partial}{\partial t}F^{(N)}(\mathbf{X}, t) + \sum_{i=1}^{N} \left(\mathbf{v}_i \cdot \frac{\partial}{\partial \mathbf{x}_i} - \frac{e}{m}\mathbf{E}(\mathbf{x}_i, t) \cdot \frac{\partial}{\partial \mathbf{v}_i}\right) F^{(N)} = 0.$$

1.4 In the above Problem, we integrated the Liouville theorem. the Liouville distribution function in $6N$ dimension phase space can be represented by δ functions as

$$F^{(N)}(\mathbf{r}_1, \mathbf{v}_1, \mathbf{r}_2, \mathbf{v}_2, \ldots, \mathbf{r}_N, \mathbf{v}_N, t) = \prod_{i=1}^{N} \delta(\mathbf{r}_i - \mathbf{r}_i(t))\delta(\mathbf{v}_i - \mathbf{v}_i(t))$$

This contrasts the Klimontovich distribution function in 6 dimension phase space

$$K(\mathbf{r}, \mathbf{v}, t) = \sum_{i=1}^{N} \delta(\mathbf{r} - \mathbf{r}_i(t))\delta(\mathbf{v} - \mathbf{v}_i(t)),$$

Show that the Liouville distribution function in the above satisfies the Liouville equation (1.13). Note that $\mathbf{r}_i(t)$, $\mathbf{v}_i(t)$ are Lagrangian coordinates.

1.5 Derive Eq. (1.45).

1.6 The Vlasov equation written in the form of (1.40)

$$\frac{\partial}{\partial t} F^{(1)}(\mathbf{r}, \mathbf{v}, t) + \mathbf{v} \cdot \frac{\partial F^{(1)}}{\partial \mathbf{r}} - \frac{Ne}{m}\left[\int d^3r_2 d^3v_2 \left(\frac{\partial}{\partial \mathbf{r}} \frac{e}{|\mathbf{r} - \mathbf{r}_2|}\right) F^{(1)}(\mathbf{r}_2, \mathbf{v}_2, t)\right] \cdot \frac{\partial}{\partial \mathbf{v}} F^{(1)}(\mathbf{r}, \mathbf{v}, t) = 0$$

can be written as

$$\left(\frac{\partial}{\partial t} + \mathbf{v} \cdot \frac{\partial}{\partial \mathbf{r}} - \frac{e}{m}\mathbf{E}(\mathbf{r}, t) \cdot \frac{\partial}{\partial \mathbf{v}}\right) F^{(1)}(\mathbf{r}, \mathbf{v}, t) = 0$$

What is $\mathbf{E}$?

Answer.

$$\frac{\partial F^{(1)}(\mathbf{r}, \mathbf{v}, t)}{\partial t} + \mathbf{v} \cdot \frac{\partial F^{(1)}}{\partial \mathbf{r}} + \frac{e^2}{m} \frac{\partial F^{(1)}}{\partial \mathbf{v}} \cdot \int d^3r' \frac{\mathbf{r} - \mathbf{r}'}{|\mathbf{r} - \mathbf{r}'|^3} \int d^3v' F^{(1)}(\mathbf{r}', \mathbf{v}', t) = 0$$

$$\mathbf{E}(\mathbf{r}, t) = -e \int d^3r' \frac{\mathbf{r} - \mathbf{r}'}{|\mathbf{r} - \mathbf{r}'|^3} \int d^3v' F^{(1)}(\mathbf{r}', \mathbf{v}', t)$$

The electric field in the above is the *self-consistent field* of the Vlasov equation, which is generated by the plasma particles and at the same time influence the particle distribution through the Vlasov equation.

1.7 Carry out the algebra leading to (1.55).

Answer.

The following steps will be useful. Write

$$\left(\frac{\partial}{\partial t}+\mathbf{v}\cdot\frac{\partial}{\partial \mathbf{r}}+\mathbf{v}'\cdot\frac{\partial}{\partial \mathbf{r}'}\right)\left[\frac{N}{V}F^{(1)}(\mathbf{x},t)\delta(\mathbf{x}-\mathbf{x}')+\frac{N^2}{V^2}F^{(2)}(\mathbf{x},\mathbf{x}',t)\right]$$

$$=\frac{1}{m}\int d^3r''d^3v''\left[\frac{\partial}{\partial \mathbf{r}}\frac{e^2}{|\mathbf{r}-\mathbf{r}''|}\cdot\frac{\partial}{\partial \mathbf{v}}^{(a)}+\frac{\partial}{\partial \mathbf{r}'}\frac{e^2}{|\mathbf{r}'-\mathbf{r}''|}\cdot\frac{\partial}{\partial \mathbf{v}'}^{(b)}\right]$$

$$\times\left(\frac{N}{V}F^{(1)}(\mathbf{x},t)\delta(\mathbf{x}-\mathbf{x}')\delta(\mathbf{x}-\mathbf{x}'')^{\copyright}\right.$$

$$+\frac{N^2}{V^2}F^{(2)}(\mathbf{x},\mathbf{x}',t)\delta(\mathbf{x}-\mathbf{x}'')+\frac{N^2}{V^2}F^{(2)}(\mathbf{x},\mathbf{x}'',t)\delta(\mathbf{x}-\mathbf{x}')$$

$$\left.+\frac{N^2}{V^2}F^{(2)}(\mathbf{x},\mathbf{x}'',t)\delta(\mathbf{x}''-\mathbf{x}')+\frac{N^3}{V^3}F^{(3)}(\mathbf{x},\mathbf{x}',\mathbf{x}'',t)\right)$$

The product terms $(a)\copyright=(b)\copyright=0$, per $\delta(\mathbf{x}-\mathbf{x}')=\delta(\mathbf{r}-\mathbf{r}')\delta(\mathbf{v}-\mathbf{v}')$. Also,

$$\left(\mathbf{v}\cdot\frac{\partial}{\partial \mathbf{r}}+\mathbf{v}'\cdot\frac{\partial}{\partial \mathbf{r}'}\right)[F^{(1)}(\mathbf{x},t)\delta(\mathbf{x}-\mathbf{x}')]=\delta(\mathbf{x}-\mathbf{x}')\mathbf{v}\cdot\frac{\partial}{\partial \mathbf{r}}F^{(1)}(\mathbf{x},t)$$

Omitting the $F^{(3)}$-term, we have, using (1.53),

$$\left(\frac{\partial}{\partial t}+\mathbf{v}\cdot\frac{\partial}{\partial \mathbf{r}}+\mathbf{v}'\cdot\frac{\partial}{\partial \mathbf{r}'}\right)\frac{N^2}{V^2}F^{(2)}(\mathbf{x},\mathbf{x}',t)+\frac{N^2}{V^2}\delta(\mathbf{x}-\mathbf{x}')$$

$$\times\frac{1}{m}\int d^3r''d^3v''\frac{\partial}{\partial \mathbf{r}}\frac{e^2}{|\mathbf{r}-\mathbf{r}''|}\cdot\frac{\partial}{\partial \mathbf{v}}F^{(2)}(\mathbf{x},\mathbf{x}'',t)\ ^{(d)}$$

$$=\frac{1}{m}\int d^3r''d^3v''\left[\frac{\partial}{\partial \mathbf{r}}\frac{e^2}{|\mathbf{r}-\mathbf{r}''|}\cdot\frac{\partial}{\partial \mathbf{v}}^{(a)}+\frac{\partial}{\partial \mathbf{r}'}\frac{e^2}{|\mathbf{r}'-\mathbf{r}''|}\cdot\frac{\partial}{\partial \mathbf{v}'}^{(b)}\right]$$

$$\times\frac{N^2}{V^2}\left(F^{(2)}(\mathbf{x},\mathbf{x}',t)\delta(\mathbf{x}-\mathbf{x}'')+F^{(2)}(\mathbf{x},\mathbf{x}'',t)\delta(\mathbf{x}-\mathbf{x}')^{\copyright}\right.$$

$$\left.+F^{(2)}(\mathbf{x},\mathbf{x}'',t)\delta(\mathbf{x}''-\mathbf{x}')\right)$$

Product term $(a)\copyright$ produces two terms one of which cancels term (d), the other term cancels $(b)\copyright$. Also we have vanishing terms due to (1.52), and the remaining terms give (1.55).

1.8 Prove (1.52).

1.9 Calculate directly the Fourier transform of $\frac{1}{r}$. The answer can be written down by taking the limit $\lambda_D \to \infty$ in (11.19).

1.10 Relativistic Lagrangian of N particle system (Liboff). Derive

$$
\begin{aligned}
L(\mathbf{x}, \mathbf{v}, t) &= \sum_{i=1}^{N} \left[-mc^2 \sqrt{1 - \frac{v_i^2}{c^2}} \right] \\
&\quad + \sum_{i<j} \left[-\frac{q_i q_j}{r_{ij}} + \frac{q_i q_j}{2c^2} \left(\frac{\mathbf{v}_i \cdot \mathbf{v}_j}{r_{ij}} + \frac{(\mathbf{v}_i \cdot \mathbf{r}_{ij})(\mathbf{v}_j \cdot \mathbf{r}_{ij})}{r_{ij}^3} \right) \right]
\end{aligned}
$$

where q is charge, $\mathbf{x}$ and $\mathbf{v}$ are Eulerian coordinate and velocity, respectively, and $\mathbf{x}_i$ and $\mathbf{v}_i$ are the Lagrangian coordinate and velocity of ith particle, $\mathbf{r}_j = \mathbf{x} - \mathbf{x}_j(t)$.

Hint: Use charge and current densities for the jth particle

$$
\begin{aligned}
\rho^{(j)}(\mathbf{x}, t) &= q_j \delta[\mathbf{x} - \mathbf{x}_j(t)] \\
\mathbf{J}^{(j)}(\mathbf{x}, t) &= q_j \mathbf{v}_j(t) \delta[\mathbf{x} - \mathbf{x}_j(t)]
\end{aligned}
$$

and scalar and vector potentials

$$
\begin{aligned}
\phi^{(j)}(\mathbf{x}, t) &= \int d^3x' \frac{\phi^{(j)}(\mathbf{x}', t)}{|\mathbf{x} - \mathbf{x}'|} \\
\mathbf{A}^{(j)}(\mathbf{x}, t) &= \frac{1}{c} \int d^3x' dt' \frac{\delta(t - t' - |\mathbf{x} - \mathbf{x}'|/c)}{|\mathbf{x} - \mathbf{x}'|} \\
&\quad \times \left[\mathbf{J}^{(j)}(\mathbf{x}', t) - \frac{1}{4\pi} \frac{\partial}{\partial t'} \nabla' \phi^{(j)}(\mathbf{x}', t') \right]
\end{aligned}
$$

Use the Coulomb gauge $\nabla \cdot \mathbf{A} = 0$.

1.11 The charge density and the current density of a plasma can be obtained from Klimontovich distribution

$$
\begin{aligned}
\rho(\mathbf{x}, t) &= \sum_j q_j \delta[\mathbf{x} - \mathbf{x}_j(t)] \\
\mathbf{J}(\mathbf{x}, t) &= \sum_j q_j \mathbf{v}_j(t) \delta[\mathbf{x} - \mathbf{x}_j(t)]
\end{aligned}
$$

Show that the charge and current densities satisfy the continuity equation

$$\frac{\partial \rho}{\partial t} + \nabla \cdot \mathbf{J} = 0$$

1.12 We have the equation of motion of a charged particle

$$m\frac{d\mathbf{v}}{dt} = e\left[\mathbf{E}(\mathbf{r}, t) + \frac{1}{c}\mathbf{v}(t) \times \mathbf{B}(\mathbf{r}, t)\right]$$

In terms of the canonical variables $\mathbf{P}$ and $\mathbf{Q}$, we have the relations

$$\mathbf{P}(t) = m\mathbf{v}(t) + \frac{e}{c}\mathbf{A}(\mathbf{r}(t), t)$$

$$\mathbf{Q}(t) = \mathbf{r}(t)$$

Show that the Hamiltonian H is given by

$$H(\mathbf{P}, \mathbf{Q}, t) = \frac{1}{2m}\left(\mathbf{P} - \frac{e}{c}\mathbf{A}\right) + e\phi$$

where ϕ and $\mathbf{A}$ are the scalar and vector potentials:

$$\nabla \times \mathbf{A} = \mathbf{B}$$

$$\nabla \phi = -\mathbf{E} + \frac{1}{c}\frac{\partial \mathbf{A}}{\partial t}$$

Show that

$$\frac{\partial H}{\partial P_j} = \dot{Q}_j \quad \frac{\partial H}{\partial Q_j} = -\dot{P}_j$$

Note: $\dot{\mathbf{A}} = \frac{\partial \mathbf{A}}{\partial t} + \mathbf{v} \cdot \nabla \mathbf{A}$.

References

Clemmow, P. C. and J. P. Daugherty, *Electrodynamics of Particles and Plasmas*, Addison-Wesley, London (1969).

Darwin, C. G. Proc. Cambridge Phil. Soc. **20**, 56 (1920).

Gartenhaus, S. *Elements of Plasma Physics*, Holt, Rinehart, and Winston, Inc. (1964).

Klimontovich, Y. L. *The Statistical Theory of non-Equilibrium Processes in a Plasma*, MIT Press (1967).

Lee, H. J. *Fluctuation of the Electric Field in a Plasma*, J. Korean Phys. Soc. **66** No. 8 (2015) pp. 1167–1185.

Lee, H. J. *Fundamentals of Theoretical Plasma Physics*, World Scientific (2019).

Liboff, R. L. *Kinetic Theory Classical, Quantum, and Relativistic Descriptions*, Third Edition Springer (2003).

Lifshitz, E. M. and L. P. Pitaevskii, *Physical Kinetics*, Landau and Lifshitz Course of theoretical physics, Volume 10, Butterworth-Heinemann (1981).

Montgomery, D. C. *Theory of the Unmagnetized Plasma*, Gordon and Beach, New York (1971).

Montgomery, D. C. and D. A. Tidman, *Plasma Kinetic Theory*, McGraw-Hill (1964).

Nicholson, D. R. *Introduction to Plasma Theory*, Wiley (1983).

Chapter 2

Boltzmann Equation

2.1 Heuristic Derivation of Boltzmann Equation

The Boltzmann equation is of fundamental importance in the kinetic theory of gases and plasmas. It expresses a mathematical description of the distribution function $f(\mathbf{r}, \mathbf{v})$ of the particles in the 6 dimensional phase space. It takes the form of (2.14), and gives the rate of change of f along the characteristics, the particle trajectory, due to the inter-particle collisions. The Boltzmann equation becomes the Vlasov equation when we regard the change of f due to particle collisions is zero. The *Vlasov fluid* behaves like an incompressible fluid in the 6-dimensional phase space. Landau's collision term can be derived from Boltzmann's collision term in the long range Coulomb interaction. Boltzmann's invention of the collision term is his great accomplishment. The following assumptions are made as a preliminary to the development of Boltzmann's equation:

(a) We assume that the state of the gas is described by a one-body distribution function $f(\mathbf{r}, \mathbf{v}, t)$.
(b) The density of particles is low enough that only two body interactions need to be considered, i.e. the range of inter-particle forces (r_0) are much smaller than the mean inter-particle distance.
(c) The duration of an encounter between two particles (interaction time) is much smaller than the duration of free motion of the particles, i.e. $r_0/v_{\text{av}} \equiv t_{\text{int}} \ll \lambda/v_{\text{av}}$ where λ is the mean free path and v_{av} is the mean speed of the particles.
(d) The particles are assumed to be point centers of spherically symmetric fields, so that the one-body distribution function depends only on the position $\mathbf{r}$, velocity $\mathbf{v}$ of the particles and time t. In case of exceptional models for the particles, other variables, such as the angular velocity, may be introduced.

Let $f(\mathbf{r},\mathbf{v},t)\ d^3rd^3v$ be the expected number of particles to be found in a volume element d^3rd^3v of phase space about $\mathbf{r}$ and $\mathbf{v}$ at time t. The volume element $\Delta\mu \equiv d^3rd^3v$ must be large enough to contain a sufficient number of particles in order that probability concepts can be applied at all. Thus

$$f(\mathbf{r},\mathbf{v},t) = \frac{1}{\Delta\mu}\int_{\Delta\mu} F(\mathbf{r},\mathbf{v},t)d^3r_1...d^3r_Nd^3v_1...d^3v_N \tag{2.1}$$

$$F(\mathbf{r},\mathbf{v},t) = \sum_{i=1}^{N}\delta(\mathbf{r}-\mathbf{r}_i(t))\delta(\mathbf{v}-\mathbf{v}_i(t)) \tag{2.2}$$

where i is a particle index. Moreover, the changes in f will be observed over a time Δt which is much larger than the interaction time $t_{\text{int}} = r_o/v_{\text{av}}$.

We are concerned with developing an equation which determines the temporal evolution of f given its value at some initial time t_0 for all $\mathbf{r}$ and $\mathbf{v}$. By definition the total number of particles

$$N = \int f(\mathbf{r},\mathbf{v},t)d^3rd^3v \tag{2.3}$$

where the integration is carried over the volume V in configuration space occupied by the particles and over the accessible region of velocity space. Furthermore, we define a density n

$$n(\mathbf{r},t) = \frac{1}{V}\int_{all\ v} f(\mathbf{r},\mathbf{v},t)d^3v \tag{2.4}$$

with the configuration volume V vanishingly small but large enough to contain many enough particles.

If we assume that those particles contained in d^3rd^3v do not interact with each other, then at a time $t+dt$ ($dt \gg t_{\text{int}}$) we expect these particles to be in the volume element $d^3r'd^3v'$ about $\mathbf{r}'$ and $\mathbf{v}'$ where

$$\mathbf{r}' = \mathbf{r} + \mathbf{v}dt + O(dt)^2 \tag{2.5}$$

$$\mathbf{v}' = \mathbf{v} + \mathbf{a}dt + O(dt)^2 \tag{2.6}$$

where $\mathbf{a}$ is the acceleration acquired by the particles due to the fields that may be applied by external means or those generated by the collective action of all the particles excluding those whose trajectories (internal) are under examination. Thus

$$\mathbf{a} = \mathbf{a}_e + \mathbf{a}_i \tag{2.7}$$

where the subscripts e and i correspond to 'external' and 'internal'. The new volume element $d^3r'd^3v'$ is related to the old volume element d^3rd^3v by the relation

$$d^3r'd^3v' = J\left(\frac{\mathbf{r}', \mathbf{v}'}{\mathbf{r}, \mathbf{v}}\right) d^3rd^3v \tag{2.8}$$

where J is the Jacobian determinant of the transformation written out in full as:

$$J = \begin{vmatrix} \frac{\partial r_1'}{\partial r_1} & \frac{\partial r_2'}{\partial r_1} & \frac{\partial r_3'}{\partial r_1} & 0 & 0 & 0 \\ \frac{\partial r_1'}{\partial r_2} & \frac{\partial r_2'}{\partial r_2} & \frac{\partial r_3'}{\partial r_2} & 0 & 0 & 0 \\ \frac{\partial r_1'}{\partial r_3} & \frac{\partial r_2'}{\partial r_3} & \frac{\partial r_3'}{\partial r_3} & 0 & 0 & 0 \\ 0 & 0 & 0 & \frac{\partial v_1'}{\partial v_1} & \frac{\partial v_2'}{\partial v_1} & \frac{\partial v_3'}{\partial v_1} \\ 0 & 0 & 0 & \frac{\partial v_1'}{\partial v_2} & \frac{\partial v_2'}{\partial v_2} & \frac{\partial v_3'}{\partial v_2} \\ 0 & 0 & 0 & \frac{\partial v_1'}{\partial v_3} & \frac{\partial v_2'}{\partial v_3} & \frac{\partial v_3'}{\partial v_3} \end{vmatrix}$$

This expression for the Jacobian assumes that the after-collision position of a molecule is uncorrelated with the before-collision velocity of the particle, which is the assumption of *molecular chaos*. Owing to this assumption, the chain of hierarchy of distribution functions can be terminated at the one-particle distribution function, and the Boltzmann equation becomes self-closed with the collision term that is here derived and is expressed by the Boltzmann distribution function itself. Making use of (2.5) and (2.6),

$$J = 1 + \frac{\partial a_i}{\partial v_i} dt + O(dt)^2 \tag{2.9}$$

where we note that the determinant J is equal to the product of the two 3 by 3 subdeterminants located at the upper-left and lower-right corners. The upper-left determinant is unity. The lower-right determinant is diagonal with elements

$$1 + \frac{\partial a_1}{\partial v_1} dt,$$

and the like. Here the repeated indexes are summed over.

$$d^3r'd^3v' = \left(1 + \frac{\partial a_i}{\partial v_i}dt + O(dt)^2\right)d^3rd^3v \tag{2.10}$$

and further we have

$$f(\mathbf{r}', \mathbf{v}', t + dt)d^3r'd^3v' = f(\mathbf{r}, \mathbf{v}, t)d^3rd^3v + \left(\frac{\delta f}{\delta t}\right)_c d^3rd^3vdt \tag{2.11}$$

which means that the same number of particles are in the new volume element as in the old element except for those gained or lost by interaction among the particles themselves denoted by the last term of (2.11). Expanding the left hand side (LHS) of (2.11) in a Taylor series about $(\mathbf{r}, \mathbf{v}, t)$, we have

$$\text{LHS} = \left[f(\mathbf{r}, \mathbf{v}, t) + \left(\frac{\partial f}{\partial t} + \frac{\partial f}{\partial r_i}\frac{dr_i}{dt} + \frac{\partial f}{\partial v_i}\frac{dv_i}{dt}\right)dt\right] d^3rd^3vdt \left(1 + \frac{\partial a_i}{\partial v_i}dt\right) + O(dt)^2 \tag{2.12}$$

Thus

$$\frac{\partial f}{\partial t} + \frac{\partial f}{\partial r_i}\frac{dr_i}{dt} + \frac{\partial f}{\partial v_i}\frac{dv_i}{dt} + f\,\frac{\partial a_i}{\partial v_i} = \left(\frac{\delta f}{\delta t}\right)_c \tag{2.13}$$

As we restrict ourselves to forces that are either independent of particle velocity or the Lorentz force $e\ \mathbf{v} \times \mathbf{B}$, if they depend on velocity, $\frac{\partial a_i}{\partial v_i} = 0$ always, and we finally have

$$\frac{\partial f}{\partial t} + v_i\,\frac{\partial f}{\partial r_i} + a_i\,\frac{\partial f}{\partial v_i} = \left(\frac{\delta f}{\delta t}\right)_c \tag{2.14}$$

2.2 Collision Term

Consider a two-body elastic collision. Denote the particle velocities before and after collision as $(\mathbf{v}, \mathbf{v}_1)$ and $(\mathbf{v}', \mathbf{v}_1')$, respectively, where the primed are after-collision velocities, which can be expressed in terms of the pre-collision velocities if the force law governing the interaction is known. The collision event is more conveniently described in the center-of-mass frame of the two particles, in which the collision is equivalent to a deflection of a fictitious particle of reduced mass (moving with the relative velocity) from the scattering center. Figure 2.1 shows the geometry of the scattering process. The incident particle approaches the scattering center with an *impact parameter* b and is deflected by a *scattering angle* θ. The orbit of the particle involves two angles (φ, θ) where φ is the azimuthal angle and

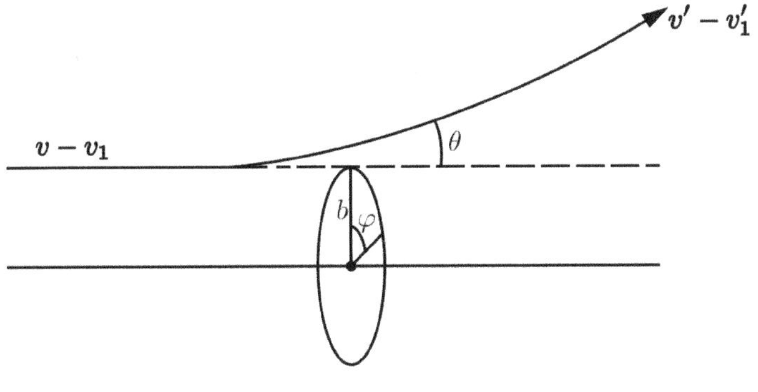

Fig. 2.1. Collision in the center-of-mass frame.

the polar angle θ corresponds to the scattering angle. The flux of particles passing through an elemental area $b\ db\ d\varphi$ is

$$b\ db\ d\varphi\,|\mathbf{v}-\mathbf{v}_1|\ f(\mathbf{r},\mathbf{v}_1,t)d^3v_1\quad(\mathrm{sec}^{-1}) \tag{2.15}$$

Then the number of collisions (per unit time) of the particles in the above flux with the particles of velocity $\mathbf{v}$ is (2.15) $\times f(\mathbf{v},\mathbf{r},t)d^3v$:

$$b\ db\ d\varphi|\mathbf{v}-\mathbf{v}_1|f(\mathbf{r},\mathbf{v}_1,t)d^3v_1\ f(\mathbf{v},\mathbf{r},t)d^3v$$

Then integrating the above equation over all $\mathbf{v}_1$ and dividing by d^3v gives the total rate of change of $f(\mathbf{v},\mathbf{r},t)$ due to collisions which scatter $\mathbf{v}$-particles out of the range $(\mathbf{v},\mathbf{v}+d\mathbf{v})$:

$$\left(\frac{\delta f}{\delta t}\right)_{\mathrm{out}}=\int d^3v_1\int_0^\infty db\ b\int_0^{2\pi}d\varphi\,|\mathbf{v}-\mathbf{v}_1|\ f(\mathbf{v},\mathbf{r},t)f(\mathbf{r},\mathbf{v}_1,t) \tag{2.16}$$

Note that $|\mathbf{v}-\mathbf{v}_1|\equiv v_R$, the relative speed, is a function of the scattering angle θ, which in turn is determined by b and φ if the force law is known. It is customary to introduce the differential scattering cross section

$$d\sigma(v_R,\theta)=b\ db\ d\varphi;\quad \theta=\theta(v_R,b)$$

For more about the scattering angle θ, Section 2.6 and Fig. 2.2 should be consulted. If we go into the center of mass frame, θ is the angle between the pre and post relative velocities, and is a function of b and the initial relative velocity. Then (2.16) is written as

$$\left(\frac{\delta f}{\delta t}\right)_{\mathrm{out}}=\int d^3v_1\int d\Omega\,\frac{d\sigma}{d\Omega}|\mathbf{v}-\mathbf{v}_1|f(\mathbf{v},\mathbf{r},t)f(\mathbf{r},\mathbf{v}_1,t),\quad d\Omega=\sin\theta\ d\theta\ d\varphi \tag{2.17}$$

The rate of scattering *into* the range $(\mathbf{v}, \mathbf{v}+d\mathbf{v})$ can be calculated in the same manner, giving

$$\left(\frac{\partial f}{\partial t}\right)_{\text{in}} = \int d^3v_1' \int \left(d\Omega \frac{d\sigma}{d\Omega}\right)' |\mathbf{v}' - \mathbf{v}_1'| f(\mathbf{r}, \mathbf{v}', t) f(\mathbf{r}, \mathbf{v}_1', t) \tag{2.18}$$

In the above equation, $\mathbf{v}', \mathbf{v}_1'$ are functions of $\mathbf{v}, \mathbf{v}_1$ because these are the velocities of the interacting particles before and after the collision, respectively. Since we have assumed an elastic collision the following conservation relations hold,

$$m\mathbf{v}' + m\mathbf{v}_1' = m\mathbf{v} + m\mathbf{v}_1 \tag{2.19}$$

$$\frac{m}{2}\mathbf{v}'^2 + \frac{m}{2}\mathbf{v}_1'^2 = \frac{m}{2}\mathbf{v}^2 + \frac{m}{2}\mathbf{v}_1^2 \tag{2.20}$$

The collision may be thought of as a linear orthogonal transformation from $(\mathbf{v}, \mathbf{v}_1)$ to $(\mathbf{v}', \mathbf{v}_1')$, and we have

$$|\mathbf{v} - \mathbf{v}_1| = |\mathbf{v}' - \mathbf{v}_1'|$$

Furthermore, for this transformation, the Jacobian is unity and therefore

$$d^3v_1' d^3v' = d^3v_1 d^3v$$

Because of the symmetry of the interaction, we also have

$$\left(\frac{d\sigma}{d\Omega} d\Omega\right)' = \left(\frac{d\sigma}{d\Omega} d\Omega\right)$$

Collecting the above equations for the collision integral, the Boltzmann equation becomes

$$\frac{\partial f}{\partial t} + v_i \frac{\partial f}{\partial r_i} + a_i \frac{\partial f}{\partial v_i} = \left(\frac{\delta f}{\delta t}\right)_c \tag{2.21}$$

where

$$\left(\frac{\delta f}{\delta t}\right)_c = \int d^3v_1 \int d\Omega \frac{d\sigma}{d\Omega} |\mathbf{v} - \mathbf{v}_1| [f(\mathbf{r}, \mathbf{v}', t) f(\mathbf{r}, \mathbf{v}_1', t) - f(\mathbf{r}, \mathbf{v}, t) f(\mathbf{r}, \mathbf{v}_1, t)]$$

The actual evaluation of this integral for small angle scattering will be performed in Section 2.6. If the fluid contains several kinds of particles, then each kind of particle has its own distribution function and we readily generalize Boltzmann's equation to

$$\frac{\partial f_\alpha}{\partial t} + v_i \frac{\partial f_\alpha}{\partial r_i} + a_i \frac{\partial f_\alpha}{\partial v_i} = \sum_\beta C_{\alpha\beta}(f_\alpha f_\beta) \tag{2.22}$$

where α and β refer to the particle species, and the summation includes $\beta = \alpha$, and $C_{\alpha\beta}(f_\alpha f_\beta)$ denotes the collision operator:

$$C_{\alpha\beta}(f_\alpha f_\beta) = \int d^3 v_\beta \int d\Omega \frac{d\sigma}{d\Omega} \, |\mathbf{v}_\alpha - \mathbf{v}_\beta| \big[f_\alpha(\mathbf{v}'_\alpha) f_\beta(\mathbf{v}'_\beta) - f_\alpha(\mathbf{v}_\alpha) f_\beta(\mathbf{v}_\beta) \big]$$

2.3 Chapman–Enskog Solution of Boltzmann Equation

In this section, we shall discuss the solution of the Boltzmann equation for a simple gas (consisting of one kind of spherical molecules having only translational energy) which is not subject to any external forces. The Chapman and Enskog treatment of the Boltzmann equation is known as the *method of moments.*

Summation invariant. Let Q be any particle property such that the sum of the $Q's$ of two particles involved in collision remains unchanged in the collision, i.e. a summation invariant. Important examples of Q are 1, $m\mathbf{v}$ (momentum), and $mv^2/2$ (kinetic energy) (see Eqs. (2.19) and (2.20)).

$$Q(\mathbf{v}) + Q(\mathbf{v}_1) = Q(\mathbf{v}') + Q(\mathbf{v}'_1) \tag{2.23}$$

Then ΔQ defined below vanishes. This is the useful property of a summation invariant.

$$\Delta Q = \int Q(\mathbf{v}) \left(\frac{\delta f}{\delta t} \right)_c d^3 v = \int Q(\mathbf{v}) [f' f'_1 - f f_1] \, g \, \frac{d\sigma}{d\Omega} \, d\Omega \; d^3 v_1 \; d^3 v \tag{2.24}$$

In the above equation, we abbreviated

$$\mathbf{g} = \mathbf{v} - \mathbf{v}_1$$
$$f = f(\mathbf{v}, \mathbf{r}, t), \quad f_1 = f(\mathbf{v}_1, \mathbf{r}, t), \quad f' = f(\mathbf{v}', \mathbf{r}, t), \quad f'_1 = f(\mathbf{v}'_1, \mathbf{r}, t) \tag{2.25}$$

The proof of $\Delta Q = 0$ goes as follows. Owing to the symmetry of the integrand, we have

$$\Delta Q = \int Q(\mathbf{v}_1) [f' f'_1 - f f_1] \, g \, \frac{d\sigma}{d\Omega} d\Omega d^3 v_1 d^3 v \tag{2.26}$$

Adding (2.24) and (2.26) gives

$$\Delta Q = \frac{1}{2} \int [Q(\mathbf{v}) + Q(\mathbf{v}_1)] [f' f'_1 - f f_1] \, g \, \frac{d\sigma}{d\Omega} d\Omega d^3 v_1 d^3 v \tag{2.27}$$

Due to the symmetry of the integrand again, (2.27) can be written as

$$\Delta Q = \frac{1}{2}\int [Q(\mathbf{v}') + Q(\mathbf{v}_1') - Q(\mathbf{v}) - Q(\mathbf{v}_1)] f f_1\, g\, \frac{d\sigma}{d\Omega} d\Omega d^3 v_1 d^3 v \tag{2.28}$$

Hence $\Delta Q = 0$.

Equilibrium. Let us consider a gas in an equilibrium (steady state in time and uniform in space). Then LHS of (2.14) vanishes and the distribution function should satisfy

$$\left(\frac{\delta f}{\delta t}\right)_c = \int d^3 v_1 d\sigma g (f' f_1' - f f_1) = 0 \tag{2.29}$$

which is satisfied if and only if $f' f_1' = f f_1$. Or

$$\log f(\mathbf{v}') + \log f(\mathbf{v}_1') = \log f(\mathbf{v}) + \log f(\mathbf{v}_1)$$

Hence the quantity $\log f(\mathbf{v})$ is a summation invariant and it should be expressible as a linear combination of the three summation invariants, 1, $m\mathbf{v}$, and $mv^2/2$:

$$\log f(\mathbf{v}) = m a_1 + \mathbf{a}_2 \cdot m\mathbf{v} - a_3 \frac{m}{2} v^2$$

where a_1, $\mathbf{a}_2$, a_3 are arbitrary constants. By completing the squares, one obtains

$$\log f(\mathbf{v}) = m\left(a_1 + \frac{a_2^2}{2a_3}\right) - \frac{m}{2} a_3 \left(\mathbf{v} - \frac{\mathbf{a}_2}{a_3}\right)^2$$

By a simple transformation, $f(\mathbf{v})$ can be put into the form

$$f(\mathbf{v}) = N\left(\frac{m}{2\pi T}\right)^{\frac{3}{2}} \exp -\frac{m(\mathbf{v} - \mathbf{V})^2}{2T} \tag{2.30}$$

where N, $\mathbf{V}$, T are related with a_1, $\mathbf{a}_2$, a_3 and defined by

$$N = \int f d^3 v, \quad \mathbf{V} = \frac{1}{N}\int \mathbf{v} f d^3 v, \quad \frac{3}{2} T = \frac{1}{N}\int \frac{m}{2} v^2 f d^3 v - \frac{m}{2} V^2 \tag{2.31}$$

Equation (2.30) is a moving Maxwell–Boltzmann distribution function. If the gas has no directional preference ($\mathbf{V} = 0$) and is distributed isotropically,

$$f(\mathbf{v}) = N\left(\frac{m}{2\pi T}\right)^{\frac{3}{2}} \exp -\frac{m\mathbf{v}^2}{2T} \tag{2.32}$$

[In the above and following, temperature T should be read as kT where k is the Boltzmann constant.] Thus the particles in equilibrium state are Maxwell–Boltzmann distributed. Particles in non-equilibrium state approaches the Maxwell–Boltzmann distribution in the course of time to establish equilibrium state (*relaxation*).

Let us consider a special case where an electrostatic force derivable from a potential,

$$\mathbf{F}(\mathbf{r}) = -e\nabla\phi(\mathbf{r})$$

acts on a plasma consisting of immobile ions and warm electrons. We also assume that a steady state is maintained ($\frac{\partial f}{\partial t} = 0$). In this case the distribution

$$f(\mathbf{r}, \mathbf{v}, t) = N_0\ e^{-e\phi(\mathbf{r})/T} \left(\frac{m}{2\pi T}\right)^{\frac{3}{2}} e^{-\frac{m\mathbf{v}^2}{2T}} \tag{2.33}$$

solves the Boltzmann equation

$$v_i \frac{\partial f}{\partial r_i} + \frac{F_i}{m}\frac{\partial f}{\partial v_i} = \left(\frac{\delta f}{\delta t}\right)_c \tag{2.34}$$

It is readily seen that LHS of (2.34) vanishes per (2.33)and also $(\frac{\delta f}{\delta t})_c = 0$ since the *Boltzmann factor* $e^{-e\phi(\mathbf{r})/T}$ can be taken out of the velocity integral in (2.29). Now the particle number density of the plasma which maintains an equilibrium under the action of an electrostatic force is calculated as

$$N(\mathbf{r}) = \int (2.33)\ d^3v = N_0 e^{-e\phi(\mathbf{r})/T} \tag{2.35}$$

The inhomogeneous distribution (2.35) is the Boltzmann distribution when the particles are in an electrostatic force field.

Boltzmann's H-theorem. Boltzmann introduced into his collision integral the famous quantity called H:

$$H(t) = \int d^3v f(\mathbf{v}, t)\ \log f(\mathbf{v}, t)$$

This is the kinetic theory version of the entropy, which is infact the negative entropy, $-S(t)$. We have

$$\frac{dH}{dt} = \int d^3v\ \frac{\partial f}{\partial t}\ \Big(1 + \log\ f(\mathbf{v}, t)\Big)$$

For a spatially homogeneous gas, using (2.21) in the above equation gives

$$\frac{dH}{dt} = \int d^3v_1 \int d^3v_2 \int bdb \int d\varphi\, |\mathbf{v_1} - \mathbf{v_2}|\ (f_2' f_1' - f_2 f_1)\,(1 + \log f_1)$$

The above integral is unchanged upon the interchange of the subscript 1 and 2, since $\int bdb \int d\varphi$ is an invariant. Thus we write:

$$\frac{dH}{dt} = \int d^3v_1 \int d^3v_2 \int bdb \int d\varphi\, |\mathbf{v_1} - \mathbf{v_2}|\, (f_2'f_1' - f_2 f_1)(1 + \log f_2)$$

Adding the preceding two equations gives

$$\frac{dH}{dt} = \frac{1}{2}\int d^3v_1 \int d^3v_2 \int bdb \int d\varphi\, |\mathbf{v_1} - \mathbf{v_2}|\, (f_2'f_1' - f_2 f_1)(2 + \log(f_1 f_2)) \tag{A}$$

The above integral is invariant upon the interchange $\mathbf{v}_1 \leftrightarrow \mathbf{v}_1'$, $\mathbf{v}_2 \leftrightarrow \mathbf{v}_2'$, since $\int bdb \int d\varphi$ is unchanged in the transformation $(\mathbf{v}_1, \mathbf{v}_2) \leftrightarrow (\mathbf{v}_1', \mathbf{v}_2')$. Also the integral is symmetric with respect to the interchange of the prime and no prime. Note that $\int bdb \int d\varphi$ is invariant in the reverse event. Also we have $d^3v_1' d^3v_2' = d^3v_1 d_2^3$. Thus we write

$$\frac{dH}{dt} = \frac{1}{2}\int d^3v_1' \int d^3v_2' \int bdb \int d\varphi\, |\mathbf{v_2'} - \mathbf{v_1'}|\, (f_2 f_1 - f_2'f_1')(2 + \log(f_1'f_2')) \tag{B}$$

Add Eqs. (A) and (B) and divide by 2 to obtain

$$\frac{dH}{dt} = \frac{1}{4}\int d^3v_1 \int d^3v_2 \int bdb \int d\varphi\, |\mathbf{v_1} - \mathbf{v_2}|\, (f_2'f_1' - f_2 f_1)\left[\log(f_1 f_2) - \log(f_1'f_2')\right]$$

where we note that $(f_2'f_1' - f_2 f_1)\left[\log\ (f_1 f_2) - \log(f_1'f_2')\right] \leq 0$ always. Therefore, we conclude that $\frac{dH}{dt} \leq 0$ (the equality holds for equilibrium state) always, which is the Boltzmann's H-theorem.

2.4 Simple Relaxation Model of the Collision Term

In a weakly ionized plasma whose major population is neutrals, the following model equation called BGK model (after Bhatnagar, Gross, and Krook, in *Phys. Rev.* **94**, 1954, p. 511) is often made use of for the Boltzmann collision term:

$$\left(\frac{\delta f}{\delta t}\right)_c = -\nu\left[f(\mathbf{r}, \mathbf{v}, t) - f_0(v)\right] \tag{2.36}$$

Here $f_0(v)$ is a suitable Maxwell–Boltzmann distribution, and $1/\nu \equiv \tau$ has the dimension of time and is termed as the relaxation time. A *Maxwellian* $f_0(v)$ makes the collision integral vanish, and the negative sign signifies that the collisions reduce the deviation of the distribution function from the equilibrium Maxwell–Boltzmann distribution. Assuming that the external force is absent from the uniform plasma under consideration, $f(\mathbf{v}, t)$ is governed by

$$\frac{\partial}{\partial t} f(\mathbf{v}, t) = -\nu[f(\mathbf{r}, \mathbf{v}, t) - f_0(v)]$$

Integrating this equation gives

$$f(\mathbf{v}, t) - f_0(v) = \Big(f(\mathbf{r}, \mathbf{v}, t) - f_0(v) \Big) \, e^{-\nu t}$$

As time t increases, $f(\mathbf{v}, t)$ approaches to the Maxwell–Boltzmann distribution $f_0(v)$.

2.5 Moment Equations

Reference: Chapman, S. and T. G. Cowling, *The mathematical theory of non-uniform gases*, 2nd Ed. Cambridge University Press (1952).

Instead of solving the Boltzmann equation as an initial-value problem, Chapman and Enskog reduced the description of $f(\mathbf{r}, \mathbf{v}, t)$ in the six-dimensional phase space to the description in terms of the moments of $n(\mathbf{r}, t), \mathbf{V}(\mathbf{r}, t), T(\mathbf{r}, t)$ in the ordinary coordinate space. In this reduction, there involves a *coarse graining* of the Boltzmann equation by introducing two time scales. The relaxation time t_0, in this case, the time for $f(\mathbf{r}, \mathbf{v}, t)$ to approach the local Maxwellian $f^{(0)}$ in (2.50), is much shorter than the time scale for the change of $n(\mathbf{r}, t), \mathbf{V}(\mathbf{r}, t), T(\mathbf{r}, t)$. Therefore the fluid equations describe slow process by suppressing the fast elements involved in Boltzmann's f. In terms of the three moments, the Boltzmann equation is transformed to a series of inhomogeneous integral equations, which we shall not step into for detailed discussion. But we will summarize the salient features of Chapman and Enskog solution.

Let us first define the following macroscopic variables which are the first three moments of the distribution function:

$$n(\mathbf{r}, t) = \int f(\mathbf{r}, t, \mathbf{v}) d^3 v \quad \text{(density)} \tag{2.37}$$

$$\mathbf{V}(\mathbf{r}, t) = \frac{1}{n} \int \mathbf{v} \; f(\mathbf{r}, t, \mathbf{v}) d^3 v \quad \text{(fluid velocity)} \tag{2.38}$$

$$T(\mathbf{r}, t) = \frac{m}{3n} \int \left(\mathbf{v} - \mathbf{V}(\mathbf{r}, t)\right)^2 f(\mathbf{r}, t, \mathbf{v}) d^3 v \quad \text{(temperature)} \tag{2.39}$$

In order to obtain the transport equations for these quantities of a simple gas (a gas consisting of one species), we multiply (2.21) by 1, $m\mathbf{v}$, $mv^2/2$, respectively, and integrate over velocity. The integrals of RHS of (2.21) vanish since these quantities are summation invariants. The integrals of LHS give

Continuity equation.

$$\frac{\partial n}{\partial t} + \nabla \cdot (n\mathbf{V}) = 0 \tag{2.40}$$

Momentum equation.

$$\frac{\partial}{\partial t}(mnV_i) + \frac{\partial}{\partial x_j}(mn<v_i v_j>) - nF_i = 0 \quad (F_i = ma_i) \tag{2.41}$$

Energy equation.

$$\frac{\partial}{\partial t}\left(\frac{mn}{2}<v^2>\right) + \frac{\partial}{\partial x_j}\left(\frac{mn}{2}<v^2 v_j>\right) - nF_i V_i = 0 \tag{2.42}$$

where the bracket $< \cdots >$ denotes the average value:

$$< \cdots > = \frac{1}{n}\int (\cdots) f d^3 v$$

It is convenient to transform (2.41) and (2.42) as follows. Let $\mathbf{w}$ be the random velocity (the '*peculiar velocity*' in Chapman and Cowling's book) defined by

$$\mathbf{w} = \mathbf{v} - \mathbf{V}(\mathbf{r}, t)$$

Then evidently, $<\mathbf{w}> = 0$ and $<v_i v_j> = V_i V_j + <w_i w_j>$. Using (2.40) in (2.41) yields

$$mn\frac{dV_i}{dt} = -\frac{\partial P_{ij}}{\partial x_j} + nF_i \tag{2.43}$$

where

$$\frac{d}{dt} = \frac{\partial}{\partial t} + V_j \frac{\partial}{\partial x_j}$$

$$P_{ij} = mn<w_i w_j> = m\int w_i w_j f d^3 v \tag{2.44}$$

The pressure tensor P_{ij} is usually split into two parts:

$$P_{ij} = p\delta_{ij} + \pi_{ij} \tag{2.45}$$

where p is the scalar pressure, and π_{ij} is the *viscous stress tensor* with the following definitions

$$p = \frac{nm}{3} {<}w^2{>} = nT(\mathbf{r}, t) = \frac{1}{3}\, P_{ii} \tag{2.46}$$

$$\pi_{ij} = nm \left({<}w_i w_j{>} - \frac{{<}w^2{>}}{3} \delta_{ij} \right) \quad (\text{trace,} \quad \pi_{ii} = 0) \tag{2.47}$$

If the velocity distribution function is isotropic, then ${<}w_x^2{>} = {<}w_y^2{>} = {<}w_z^2{>} = \frac{1}{3}{<}w^2{>}$, ${<}w_x w_y{>} = {<}w_x w_z{>} = {<}w_y w_z{>} = 0$, so $P_{ij} = p\, \delta_{ij}$. Also note that π_{ij} is traceless ($\pi_{ii} = 0$). The stress tensor π_{ij} represents the part of P_{ij} that arises as a result of the deviation of the distribution function from spherical symmetry. Carrying out similar transformations for (2.42), we can reduce it to the form

$$\frac{d}{dt}\left(\frac{3}{2} nT\right) + \frac{3}{2} nT \nabla \cdot \mathbf{V} + \nabla \cdot \mathbf{q} + P_{ij} \frac{\partial V_i}{\partial x_j} = 0 \tag{2.48}$$

$$\mathbf{q} = nm {<}\frac{w^2}{2} \mathbf{w}{>} \quad (\text{heat flux vector}) \tag{2.49}$$

In obtaining (2.48), we used (2.40) and (2.43).

So far we have obtained the first three moment equations, (2.40), (2.43), and (2.48). Our purpose is to obtain hydrodynamic equations to govern the hydrodynamic variables $n, \mathbf{V}, T$. The three moment equations contain π_{ij} and $\mathbf{q}$, which must be expressed in terms of $n, \mathbf{V}, T$. Since the quantities π_{ij} and $\mathbf{q}$ involve the integrations of the distribution function f, an approximate solution for f must be found. If f is obtained approximately in terms of $(n, \mathbf{V}, T)$, π_{ij} and $\mathbf{q}$ can be expressed in terms of $(n, \mathbf{V}, T)$, and a closed set of equations for $(n, \mathbf{V}, T)$ is obtained. The Chapman–Enskog solution for the Boltzmann equation is a successive approximation for f. At each stage of approximation f is found in terms of the three hydrodynamic variables, thereby the three moment equations form a closed set of equations for the three hydrodynamic variables. The solutions for the three moment equations then in turn serve to determine the next approximation for f, and so on. Here we shall not delve into this successive steps, but will write down the results of primary importance. The lowest order solution of the distribution function is the local Maxwellian

$$f^{(0)}(\mathbf{v}, \mathbf{r}, t) = n(\mathbf{r}, t) \left(\frac{m}{2\pi T(\mathbf{r}, t)} \right)^{\frac{3}{2}} \exp -\frac{m(\mathbf{v} - \mathbf{V}(\mathbf{r}, t))^2}{2T(\mathbf{r}, t)} \tag{2.50}$$

The moment equations for the three hydrodynamic variables are as follows.

$$\frac{\partial n}{\partial t} + \nabla \cdot (n\mathbf{V}) = 0 \tag{2.51}$$

$$mn\left(\frac{\partial \mathbf{V}}{\partial t} + (\mathbf{V}\cdot\nabla)\mathbf{V}\right) = -\nabla p + n\mathbf{F} \tag{2.52}$$

$$p = nT \tag{2.53}$$

$$\left(\frac{\partial}{\partial t} + \mathbf{V}\cdot\nabla\right)T = -\frac{2p}{3n}\,\nabla\cdot\mathbf{V} \quad \text{or} \quad \frac{d}{dt}(nT^{-3/2}) = 0 \tag{2.54}$$

Note that the energy equation (2.54) implies the adiabatic law of a perfect gas, $pn^{-\gamma} = \text{const.}$ with $\gamma = 5/3$. Recall that the specific entropy $s = \frac{3k}{2m}\ln(pn^{-\gamma})$, and (2.54) gives $ds/dt = 0$. A quick way of getting (2.54) is to put $\mathbf{q} = 0$ and $P_{ij} = p\delta_{ij}$ in (2.48).

The next higher order solutions are as follows.

$$\frac{dV_i}{dt} = \frac{F_i}{m} - \frac{1}{mn}\frac{\partial P_{ij}}{\partial x_j} \tag{2.55}$$

where

$$P_{ij} = p\delta_{ij} + \pi_{ij}; \quad \pi_{ij} = -\mu\left(\frac{\partial V_i}{\partial x_j} + \frac{\partial V_j}{\partial x_i} - \frac{2}{3}\delta_{ij}\nabla\cdot\mathbf{V}\right) \tag{2.56}$$

and μ is the *coefficient of shear viscosity.* The quantity in the parenthesis in (2.56) is called the *rate of strain tensor.* Using (2.56) in (2.55) gives the *Navier–Stokes equation* for a viscous fluid:

$$\frac{dV_i}{dt} = \frac{F_i}{m} - \frac{1}{mn}\left[\frac{\partial p}{\partial x_i} - \mu\nabla^2 V_i - \frac{\mu}{3}\frac{\partial}{\partial x_i}\nabla\cdot\mathbf{V}\right] \tag{2.57}$$

The energy equation takes the form

$$\frac{3}{2}n\frac{dT}{dt} = -\lambda\nabla^2 T - p\nabla\cdot\mathbf{V} - \frac{1}{2}\pi_{ij}\left(\frac{\partial v_i}{\partial x_j} + \frac{\partial v_j}{\partial x_i}\right) \tag{2.58}$$

where λ is *the coefficient of thermal conduction*: $\mathbf{q} = -\lambda\nabla T$.

The viscous stress tensor π_{ij} represents the internal friction due to shear and gives rise to the irreversible viscous transfer of momentum between the particles in the fluid, i.e. the exchange of momentum between the particles due to collisions. When such momentum transfer is absent from a fluid, the fluid is called *ideal* (or perfect).

Ideal two-fluid equations. The lowest order moment equations (2.51)–(2.54) are obtained when the collisional interactions between the constituent particles of a plasma are ignored. They should be compatible with the collisionless Boltzmann equation. In this context, we often use *the ideal two-fluid equations* which hold independently for each species, electron fluid and ion fluid, in the following form.

$$\frac{\partial n_\alpha}{\partial t} + \nabla \cdot (n_\alpha \mathbf{V}_\alpha) = 0 \tag{2.59}$$

$$m_\alpha n_\alpha \left(\frac{\partial \mathbf{V}_\alpha}{\partial t} + \mathbf{V}_\alpha \cdot \nabla \mathbf{V}_\alpha \right) = -\nabla p_\alpha + n_\alpha \mathbf{F} \tag{2.60}$$

$$p_\alpha = n_\alpha T_\alpha \tag{2.61}$$

$$\left(\frac{\partial}{\partial t} + \mathbf{V}_\alpha \cdot \nabla \right) (n_\alpha T_\alpha^{-3/2}) = 0 \tag{2.62}$$

where the subscript α takes e(i) for electron (ion) fluid.

2.6 Evaluation of Boltzmann's Collision Integral and Landau Collision Integral

We assume that the collisions involved in the Boltzmann's collision integral are elastic binary collisions. We evaluate the collision integral for small angle scattering. In a fully ionized plasma consisting of electrons and ions, the meaningful collisional change occurs by accumulation of small angle scattering, and the collision term can be appropriately approximated by Fokker–Planck type equation.

Kinematics of elastic binary collision. We write the collision term as

$$C_{ab}(f_a f_b) = \int d^3 v_b \int d\Omega \frac{d\sigma(g,\theta)}{d\Omega} \,|\mathbf{v}_a - \mathbf{v}_b| \left[f_a(\mathbf{v}_a') f_b(\mathbf{v}_b') - f_a(\mathbf{v}_a) f_b(\mathbf{v}_b) \right] \tag{2.63}$$

Introduce the following notations for the relative velocities,

$$\mathbf{g} = \mathbf{v}_a - \mathbf{v}_b, \quad \mathbf{g}' = \mathbf{v}_a' - \mathbf{v}_b' \tag{2.64}$$

where unprimed (primed) quantities are pre-(post-)collision velocities. Remember that the scattering angle θ is the angle between $\mathbf{g}$ and $\mathbf{g}'$. Let $\mathbf{G}$ be the center of mass velocity of two particles a and b. By the law of elastic collisions, we have

$$(m_a + m_b)\mathbf{G} = m_a \mathbf{v}_a + m_b \mathbf{v}_b = m_a \mathbf{v}_a' + m_b \mathbf{v}_b' \tag{2.65}$$

The above two equations yield

$$\mathbf{v}_a = \frac{m_b}{m_a + m_b}\mathbf{g} + \mathbf{G}, \quad \mathbf{v}_b = \frac{-m_a}{m_a + m_b}\mathbf{g} + \mathbf{G} \tag{2.66}$$

$$\mathbf{v}_a' = \frac{m_b}{m_a + m_b}\mathbf{g}' + \mathbf{G}, \quad \mathbf{v}_b' = \frac{-m_a}{m_a + m_b}\mathbf{g}' + \mathbf{G} \tag{2.67}$$

Hence the pre-collision velocities are determined by $\mathbf{G}$ and $\mathbf{g}$; the post-collision velocities by $\mathbf{G}$ and $\mathbf{g}'$. It is readily shown that

$$\frac{1}{2}(m_a v_a^2 + m_b v_b^2) = \frac{1}{2}[(m_a + m_b)^2 G^2 + \mu g^2]$$

$$\frac{1}{2}(m_a v_a'^2 + m_b v_b'^2) = \frac{1}{2}[(m_a + m_b)^2 G^2 + \mu g'^2]$$

where $\mu = \frac{m_a m_b}{m_a + m_b}$ is the reduced mass. Since the kinetic energy is conserved, the above relations give $g = g'$. The relative velocity $\mathbf{g}$ is deflected through the scattering angle θ by collision. The scattering angle θ depends in general on the magnitude of the initial relative velocity and the impact parameter b: $\theta = \theta(g, b)$. Let us introduce the unit vector $\hat{\mathbf{k}}$ as defined by (see Fig. 2.2)

$$\hat{\mathbf{k}} = \frac{\mathbf{g}' - \mathbf{g}}{|\mathbf{g}' - \mathbf{g}|}$$

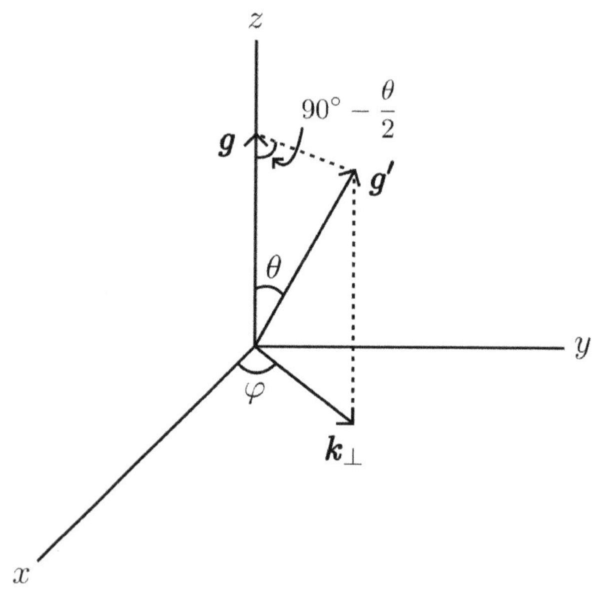

Fig. 2.2. The unit vector $\hat{\mathbf{k}} = \frac{\mathbf{g}' - \mathbf{g}}{|\mathbf{g}' - \mathbf{g}|}$.

Also introduce spherical coordinates, taking the z-direction parallel to $\mathbf{g}$: $\mathbf{g} = (0, 0, g)$, $\mathbf{g}' = (g' \sin\theta\cos\varphi,\ g'\sin\theta\sin\varphi,\ g'\cos\theta)$. Then $\hat{\mathbf{k}}$ is resolved as

$$\hat{\mathbf{k}} = \hat{\mathbf{x}}\cos\frac{\theta}{2}\cos\varphi + \hat{\mathbf{y}}\cos\frac{\theta}{2}\sin\varphi - \hat{\mathbf{z}}\sin\frac{\theta}{2}$$

Equations (2.66) and (2.67) can be written in the form

$$\mathbf{v}'_a = \mathbf{v}_a - \frac{2m_b}{m_a + m_b}(\mathbf{g}\cdot\hat{\mathbf{k}})\hat{\mathbf{k}}, \quad \mathbf{v}'_b = \mathbf{v}_b + \frac{2m_a}{m_a + m_b}(\mathbf{g}\cdot\hat{\mathbf{k}})\hat{\mathbf{k}} \tag{2.68}$$

$$\mathbf{v}_a = \mathbf{v}'_a - \frac{2m_b}{m_a + m_b}(\mathbf{g}'\cdot\hat{\mathbf{k}})\hat{\mathbf{k}}, \quad \mathbf{v}_b = \mathbf{v}'_b + \frac{2m_a}{m_a + m_b}(\mathbf{g}'\cdot\hat{\mathbf{k}})\hat{\mathbf{k}} \tag{2.69}$$

Thus when $\mathbf{v}_a$, $\mathbf{v}_b$, $\hat{\mathbf{k}}$ are given, the post-collision velocities are determined and $\hat{\mathbf{k}}$ is determined by the impact parameter b and the azimuthal angle φ. Equations (2.68) and (2.69) can be used to derive a theorem concerning the Jacobian determinant needed in transformation of the velocity integral from the pre-collision variables to post-collision variables. Let us consider the Jacobian determinants J and J':

$$J = \frac{\partial(\mathbf{v}'_a, \mathbf{v}'_b)}{\partial(\mathbf{v}_a, \mathbf{v}_b)}, \qquad J' = \frac{\partial(\mathbf{v}_a, \mathbf{v}_b)}{\partial(\mathbf{v}'_a, \mathbf{v}'_b)}$$

The partial differentiations involved in J are performed regarding $\hat{\mathbf{k}}$ as a constant. Since (2.68) which gives $\mathbf{v}'_a, \mathbf{v}'_b$ in terms of $\mathbf{v}_a, \mathbf{v}_b$ is linear, J depends only on $\hat{\mathbf{k}}$, m_a, m_b. But $\mathbf{v}_a, \mathbf{v}_b$ are given in terms of $\mathbf{v}'_a, \mathbf{v}'_b$ by (2.69) which can be obtained from (2.68) by interchanging primed and unprimed letters. Thus the Jacobian J' is equal to the same function of ($\hat{\mathbf{k}}$, m_a, m_b) as J, and therefore $J = J'$. Since $JJ' = 1$, we have $J = \pm 1$, where (-1) is irrelevant.

For small angle scattering, the change of velocity

$$\mathbf{v}'_a - \mathbf{v}_a \equiv \mathbf{u} = \frac{-2m_b}{m_a + m_b}(\mathbf{g}\cdot\hat{\mathbf{k}})\hat{\mathbf{k}} \tag{2.70}$$

will be small for most collisions, so that we may expand (2.63) in powers of $\mathbf{u}$ and keep only the first and second powers. This will be carried out. We multiply both sides of (2.63) by an arbitrary function $\Phi(\mathbf{v}_a)$ and integrate over $\mathbf{v}_a$. In the final result, Φ can be eliminated owing to the fact that it is an entirely arbitrary function. We have

$$\int \Phi(\mathbf{v}_a)\ C_{ab}(\mathbf{v}_a) d^3v_a = \int \Phi(\mathbf{v}_a)\Big[f_a(\mathbf{v}'_a)f_b(\mathbf{v}'_b) - f_a(\mathbf{v}_a)f_b(\mathbf{v}_b)\Big] g\ b\ db\ d\varphi d^3v_b d^3v_a \tag{2.71}$$

In the first integral in the right side, we change the integration variables from $(\mathbf{v}_a, \mathbf{v}_b) \to (\mathbf{v}'_a, \mathbf{v}'_b)$. In doing this, $\mathbf{v}_a$ in Φ should be expressed as a function of $\mathbf{v}'_a$ and $\mathbf{v}'_b$:

$$\int \Phi(\mathbf{v}_a) f_a(\mathbf{v}'_a) f_b(\mathbf{v}'_b) g\ b\ db\ d\varphi d^3 v_b d^3 v_a = \int \Phi(\mathbf{v}'_a - \mathbf{u}') f_a(\mathbf{v}'_a) f_b(\mathbf{v}'_b) g\ b\ db\ d\varphi d^3 v'_b d^3 v'_a$$

where $\mathbf{u}' = \frac{2m_b}{m_a+m_b}(\mathbf{g}' \cdot \hat{\mathbf{k}})\hat{\mathbf{k}}$ (see (2.69)) and we used $d^3 v_a d^3 v_b = d^3 v'_a d^3 v'_b$. In the RHS integral, we rename the integration variables by dropping the primes (then $\mathbf{g}'$ goes to $\mathbf{g}$) to write

$$\int \Phi(\mathbf{v}_a) f_a(\mathbf{v}'_a) f_b(\mathbf{v}'_b) g\ b\ db\ d\varphi d^3 v_b d^3 v_a = \int \Phi(\mathbf{v}'_a) f_a(\mathbf{v}_a) f_b(\mathbf{v}_b) g\ b\ db\ d\varphi d^3 v_b d^3 v_a \tag{2.72}$$

Using (2.72) in (2.71) gives

$$\int \Phi(\mathbf{v}_a)\ C_{ab}(\mathbf{v}_a) d^3 v_a = \int [\Phi(\mathbf{v}'_a) - \Phi(\mathbf{v}_a)] f_a(\mathbf{v}_a) f_b(\mathbf{v}_b) g\ b\ db\ d\varphi d^3 v_b d^3 v_a \tag{2.73}$$

Expanding $\Phi(\mathbf{v}'_a)$ in the powers of $\mathbf{u}$, RHS of (2.73) becomes

$$\int \left(\{u_j\} \frac{\partial \Phi(\mathbf{v}_a)}{\partial v_{aj}} + \frac{1}{2} \{u_i u_j\} \frac{\partial^2 \Phi(\mathbf{v}_a)}{\partial v_{ai} \partial v_{aj}} \right) f_a(\mathbf{v}_a) f_b(\mathbf{v}_b)\ g\ d^3 v_a d^3 v_b \tag{2.74}$$

where

$$\{\cdots\} = \int (\cdots) b\ db\ d\varphi$$

By integration by parts, (2.74) can be cast into the form

$$\int \Phi(\mathbf{v}_a) \left[-\frac{\partial}{\partial v_{ai}} (\{u_i\} f_a(\mathbf{v}_a) f_b(\mathbf{v}_b) g) + \frac{1}{2} \frac{\partial^2}{\partial v_{ai} \partial v_{aj}} (\{u_i u_j\} f_a(\mathbf{v}_a) f_b(\mathbf{v}_b) g) \right] d^3 v_a d^3 v_b$$

This is equal to LHS of (2.73) for arbitrary choice of the function Φ, and thus we conclude

$$C_{ab}(f_a f_b) = -\frac{\partial}{\partial v_{ai}}\left[f_a(\mathbf{v}_a)\int \{u_i\} f_b(\mathbf{v}_b) g d^3 v_b\right] + \frac{1}{2}\frac{\partial^2}{\partial v_{ai}\partial v_{aj}}\left[f_a(\mathbf{v}_a)\int \{u_i u_j\} f_b(\mathbf{v}_b) g d^3 v_b\right] \tag{2.75}$$

This is the *Fokker–Planck collision term.* Now the Boltzmann equation with the Fokker–Planck collision term, or the Fokker–Planck equation, is not an integro-differential equation but a differential equation.

Next, we evaluate $\{u_i\}$ and $\{u_i u_j\}$. Equation (2.70) is rewritten

$$\mathbf{u} = \frac{2m_b}{m_a + m_b} g \, \sin\frac{\theta}{2}\left(\hat{\mathbf{x}}\cos\frac{\theta}{2}\cos\varphi + \hat{\mathbf{y}}\cos\frac{\theta}{2}\sin\varphi - \hat{\mathbf{z}}\sin\frac{\theta}{2}\right) \tag{2.76}$$

Performing the φ-integration gives

$$\{u_i\} = -4\pi\frac{m_b}{m_a + m_b} g_i \int \sin^2\frac{\theta}{2}\, b\, db \tag{2.77}$$

The tensor $\{u_i u_j\}$ will be diagonal because terms like $u_x u_y$ vanish upon performing the φ-integration. We have

$$\{\mathbf{uu}\} = 2\pi\left(\frac{2m_b}{m_a + m_b}\right)^2 \int \left[\hat{\mathbf{z}}\hat{\mathbf{z}}\, g^2 \, \sin^4\frac{\theta}{2} + (\hat{\mathbf{x}}\hat{\mathbf{x}} + \hat{\mathbf{y}}\hat{\mathbf{y}})\frac{g^2}{2}\sin^2\frac{\theta}{2}\cos^2\frac{\theta}{2}\right] b db \tag{2.78}$$

To change the dyad above into tensor notation, note that

$$\hat{\mathbf{x}}\hat{\mathbf{x}} + \hat{\mathbf{y}}\hat{\mathbf{y}} + \hat{\mathbf{z}}\hat{\mathbf{z}} = \delta_{ij}, \quad \hat{\mathbf{x}}\hat{\mathbf{x}} + \hat{\mathbf{y}}\hat{\mathbf{y}} = \delta_{ij} - \frac{g_i g_j}{g^2}$$

Thus we obtain

$$\{u_i u_j\} = 4\pi g^2\left(\frac{m_b}{m_a + m_b}\right)^2 \int b\, db\left[2\frac{g_i g_j}{g^2}\sin^4\frac{\theta}{2} + \left(\delta_{ij} - \frac{g_i g_j}{g^2}\right)\sin^2\frac{\theta}{2}\cos^2\frac{\theta}{2}\right] \tag{2.79}$$

In the above equation, we neglect the terms of $\sin^4\frac{\theta}{2}$ as compared to the term of $\sin^2\frac{\theta}{2}$, which is permissible for small angle scattering. Then we finally obtain

$$\{u_i u_j\} = 4\pi g^2\left(\frac{m_b}{m_a + m_b}\right)^2\left(\delta_{ij} - \frac{g_i g_j}{g^2}\right)\int \sin^2\frac{\theta}{2}\, b\, db \tag{2.80}$$

Using the Rutherford scattering formula for coulomb collisions,

$$\tan\frac{\theta}{2} = \frac{e_a e_b}{b\,\mu\, g^2}, \quad \left(e:\ \text{charge},\quad \mu = \frac{m_a m_b}{m_a + m_b}\right)$$

the integral in (2.77) is obtained

$$\int_0^\infty \sin^2\frac{\theta}{2}\, b\, db = \frac{e_a e_b}{\mu g^2}\, J \tag{2.81}$$

$$J = \int_0^\infty \frac{\lambda d\lambda}{1+\lambda^2} \quad \left(\lambda = \frac{b\,\mu g^2}{e_a e_b}\right) \tag{2.82}$$

Using the above equations in (2.77) and (2.79) gives

$$\{u_i\} = -4\pi e_a^2 e_b^2 J\, \frac{g_i}{m_a g^4}\left(\frac{1}{m_b} + \frac{1}{m_a}\right) \tag{2.83}$$

$$\{u_i u_j\} = \frac{4\pi e_a^2 e_b^2}{g^4 m_a^2}\, J\, (g^2\delta_{ij} - g_i g_j) \tag{2.84}$$

Now the collision term (2.75) becomes upon using the above equations

$$\begin{aligned} C_{ab}(f_a f_b) &= 4\pi e_a^2 e_b^2 \frac{J}{m_a}\left(\frac{1}{m_a} + \frac{1}{m_b}\right)\frac{\partial}{\partial v_{ai}}\left[f_a \int \frac{g_i}{g^3} f_b\, d^3 v_b\right] \\ &\quad + 2\pi \frac{e_a^2 e_b^2}{m_a^2} J \frac{\partial^2}{\partial v_{ai}\partial v_{aj}}\left[f_a \int \frac{g^2\delta_{ij} - g_i g_j}{g^3} f_b d^3 v_b\right] \end{aligned} \tag{2.85}$$

Use the following identity

$$\frac{g_i}{g^3} = \frac{1}{2}\frac{\partial}{\partial v_{bj}}\left(\frac{g^2\delta_{ij} - g_i g_j}{g^3}\right) = -\frac{1}{2}\frac{\partial}{\partial v_{aj}}\left(\frac{g^2\delta_{ij} - g_i g_j}{g^3}\right)$$

in (2.85) and integrate by parts the result to finally obtain the **Landau collision integral**

$$C_{ab}(f_a f_b) = 2\pi\, \frac{e_a^2 e_b^2}{m_a}\, J\, \frac{\partial}{\partial v_{ai}} \int d^3 v_b\, U_{ij}\left[\frac{f_b}{m_a}\frac{\partial f_a}{\partial v_{aj}} - \frac{f_a}{m_b}\frac{\partial f_b}{\partial v_{bj}}\right] \tag{2.86}$$

where

$$U_{ij} = \frac{g^2\delta_{ij} - g_i g_j}{g^3} \tag{2.87}$$

The integral J diverges as $\lambda \to \infty$ (λ should run from 0 to ∞ as b does so). This is a consequence of the long range of the Coulomb potential. However, the actual interaction does not extend far beyond a Debye length since the

Coulomb potential is screened off. It is therefore reasonable to cut off the integral at $b_{\max}$, taken to be the Debye length,

$$\lambda_{\max} = \frac{\mu g^2}{e_a e_b}\sqrt{\frac{T}{4\pi n_0 e^2}}$$

With this cut-off, J in (2.82) yields $J = \log\sqrt{1+\lambda_{\max}^2}$. As will become clear shortly, 1 in the square root can be neglected as compared to $\lambda_{\max}^2$. Since the logarithmic term changes slowly, we take the mean value of g^2 by replacing it with $3T/\mu$, We then obtain

$$J \simeq \log\Lambda, \quad \Lambda = 12\pi n_0 \frac{e^2}{e_a e_b}\left[\frac{T}{4\pi n_0 e^2}\right]^{3/2} \tag{2.88}$$

Apart from a numerical factor of order 10, Λ is the number of particles in a Debye sphere, and hence much greater than 1.

Landau (1965) did not make use of the exact formula for the Rutherford scattering, but used the *impulse approximation.* One replaces the hyperbolic paths of the particles by rectilinear motion with constant velocity, and computes the force which they exert on each other during their rectilinear motion. The time integral of this force is then taken as the total momentum transfer:

$$\sin\theta = \frac{1}{g}\int F_\perp dt = \frac{1}{g^2}\int F_\perp dx = \frac{e_a e_b}{\mu g^2}\int_{-\infty}^{\infty}\frac{b\,dx}{(x^2+b^2)^{3/2}} = \frac{2e_a e_b}{b\mu g^2} \simeq \theta$$

Using the above relation, the integral in (2.81) becomes

$$\int_0^\infty \sin^2\frac{\theta}{2}\, b\, db \simeq \left(\frac{e_a e_b}{\mu g^2}\right)^2 \int \frac{db}{b} = \left(\frac{e_a e_b}{\mu g^2}\right)^2 \log\left(\frac{b_{\max}}{b_{\min}}\right)$$

where we have taken a second cut-off: $b_{\min} = \frac{e_a e_b}{3T}$, which corresponds to a deflection through an angle of $\theta = \pi/2$ (in Rutherford scattering formula, g^2 is replaced by $3T/\mu$). Taking $b_{\max}$ again to be the Debye length, one arrives at exactly the same Landau equation in the impulse approximation.

Fokker–Planck equation. One can easily verify the following relations by direct calculation.

$$\frac{\partial}{\partial v_i}\frac{1}{g} = -\frac{g_i}{g^3}$$

$$\frac{g^2\delta_{ij} - g_i g_j}{g^3} = \frac{\partial}{\partial v_i}\left(\frac{g_j}{g}\right) = \frac{\partial^2 g}{\partial v_i \partial v_j}$$

Then (2.85) can be put into the *Fokker–Planck form* of the collision integral:

$$C_{ab}(f_a f_b) = A\,\frac{\partial}{\partial v_{ai}}\left[-f_a\,\frac{\partial}{\partial v_{ai}}\int \frac{1}{g}\, f_b\; d^3v_b\right] + B\frac{\partial^2}{\partial v_{ai}\partial v_{aj}}\left[f_a\,\frac{\partial^2}{\partial v_{ai}\partial v_{aj}}\int g\; f_b\; d^3v_b\right]$$

where A and B are constants. The integrals in the above equation, $\int(\frac{1}{g}$ or $g)f_b\; d^3v_b$, are called *Rosenbluth potential.* The Landau collision integral in (2.86) is a variation of the above form obtained after integration by parts. The above expression applies to the event when two particles a and b collide. Let us change the notations: $\mathbf{v}_a \to \mathbf{v},\;\; \mathbf{v}_b \to \mathbf{v}'$. Then $\mathbf{g} = \mathbf{v} - \mathbf{v}'$, and the above equation can be written in the form

$$\begin{aligned}\left.\frac{\delta f(\mathbf{v},t)}{\delta t}\right|_c &= -A\,\frac{\partial}{\partial v_i}\left[f(\mathbf{v})\,\frac{\partial}{\partial v_i}\int \frac{f(\mathbf{v}')}{|\mathbf{v}-\mathbf{v}'|}\; d^3v'\right] \\ &\quad + B\,\frac{\partial^2}{\partial v_i\partial v_j}\left[f(\mathbf{v})\,\frac{\partial^2}{\partial v_i\partial v_j}\int |\mathbf{v}-\mathbf{v}'|\; f(\mathbf{v}')\; d^3v'\right]\end{aligned}$$

This equation gives the change of the distribution function due to the collisions between the composing particles. The effect of prime and no prime particle collisions are summed over the distribution. The first term is termed the *dynamic friction* and the second the *diffusion term.* This equation, although derived from the Boltzmann collision integral here, is valid for *Markovian process* of weak scattering and can be derived independently. For this, see References: Krall, N. A. and Trivelpiece, A. W., *Principles of Plasma Physics*, San Francisco Press, 1986; Sturrock, P. A. *Plasma Physics*, Cambridge University Press (1994). Also the Fokker–Planck equation above can be derived from Lenard-Balescu equation by taking an appropriate approximation [see Reference: Schram, P. P. J. M., *Kinetic Theory of Gases and Plasmas*, Kluwer Academic, Dordrecht (1991)].

2.7 Entropy

Boltzmann expressed *entropy* by his famous equation (inscribed on his tombstone)

$$S = k_B\; \ln\Omega \tag{2.89}$$

where $k_B = 1.3805 \times 10^{-23}$ Joule/Kelvin, the Boltzmann constant. In this book, the temperature T stands for k_BT. Ω represents the *multiplicity* of the state, which corresponds to the number of different microscopic states having the common macroscopic properties of the system. In Boltzmann statistics, the multiplicity is evaluated by the equation

$$\Omega = \frac{N!}{\eta_1!\eta_2!\cdots\eta_q!} \tag{2.90}$$

Here η_n $(n = 1, 2, \ldots q)$ is the number of particles allocated to '*room*' n with the constraint $\sum_{n=1}^{q} \eta_n = N$. Then Ω represents the number of ways that N particles (indistinguishable) are allocated to q rooms with each room being filled by number η_n. Taking logarithm gives

$$\ln \Omega = \ln N! - \sum_{n=1}^{q} \ln \eta_n! \tag{2.91}$$

Using *Sterling's formula* for large N

$$\ln N! = N \ \ln N - N \tag{2.92}$$

in the above gives

$$\ln \Omega = N \ \ln N - N - \sum_{n=1}^{q} \left(\eta_n \ \ln \eta_n - \eta_n\right) \tag{2.93}$$

Using the constraint $\sum_{n=1}^{q} \eta_n = N$ in the above yields

$$\ln \Omega = -N \sum_{n} \left[\frac{\eta_n}{N} \ \ln \frac{\eta_n}{N}\right] \tag{2.94}$$

Therefore (2.89) can be written as

$$S = -k_B \ N \sum_{n} P_n \ \ln P_n \tag{2.95}$$

where $P_n = \eta_n / N$ is the probability of realizing the η_n state.

In the kinetic theory, the entropy can be written by making correspondence $P_n \to f(\mathbf{r}, \mathbf{v})$:

$$S = -k_B \int f(\mathbf{r}, \mathbf{v}) \ \ln f \ d^3 v \tag{2.96}$$

In fact Boltzmann's H in his *H-theorem* is $H = -S$.

Variational principle for Maxwell–Boltzmann distribution. At equilibrium, the variational principle for the distribution function requires the variation of entropy to be zero:

$$\delta S = \delta \int dt \int f(t) \ \ln f(t) \ d^3v = 0 \tag{2.97}$$

where the time integral is carried over a certain interval of time. Effecting the variational operator δ gives

$$\delta S = \int dt \int (1 + \ln f) \ \delta f \ d^3v = 0 \tag{2.98}$$

While we perform the variation, the three conservation relations should be observed: the integral conservation of mass, momentum, and energy:

$$\delta \int dt \ m \int f d^3v = 0 \tag{2.99}$$

$$\delta \int dt \ m \int \mathbf{v} \ f \ d^3v = 0 \tag{2.100}$$

$$\delta \int dt \ \frac{m}{2} \int v^2 \ f \ d^3v = 0 \tag{2.101}$$

The way of incorporating these three *constraints* in the variation is to use the *the Lagrange's multiplier method.* We introduce the *Lagrange's multipliers* λ_1, $\overrightarrow{\lambda_2}$, λ_3, and form (2.98) $+\lambda_1 \times$ (2.99) $+ \overrightarrow{\lambda_2} \cdot$ (2.100) $+ \lambda_3 \times$ (2.101):

$$m \int dt \int d^3v \ \delta \ f \ \left(1 + \ln f + \lambda_1 + \overrightarrow{\lambda_2} \cdot \mathbf{v} + \lambda_3 \frac{v^2}{2}\right) = 0 \tag{2.102}$$

Thus the condition for maximizing the entropy S is

$$1 + \ln f + \lambda_1 + \overrightarrow{\lambda_2} \cdot \mathbf{v} + \lambda_3 \frac{v^2}{2} = 0 \tag{2.103}$$

Here we recall the derivation of the Maxwell–Boltzmann distribution in (2.30) via making use of the summation invariant. The Lagrange multipliers λ_1, $\overrightarrow{\lambda_2}$, λ_3, are essentially equivalent to the coefficients a_1, $\overrightarrow{a_2}$, a_3 of the summation invariant introduced in (2.30). Consequently, f ends up with the Boltzmann distribution (2.30).

Entropy conservation in collisionless plasma. Let us consider the Vlasov equation for electrons

$$\frac{\partial f(\mathbf{r}, \mathbf{v}, t)}{\partial t} + \mathbf{v} \cdot \frac{\partial f}{\partial \mathbf{r}} - \frac{e}{m} \mathbf{E} \cdot \frac{\partial f}{\partial \mathbf{v}} = 0 \tag{2.104}$$

The total entropy is

$$S(t) = -k_B N \int f \ \ln f \ d^3 r d^3 v \tag{2.105}$$

where the integral is carried over the whole space enclosed by infinite surface.

$$\frac{dS}{dt} = -k_B N \frac{\partial}{\partial t}\int f \ \ln f d\tau - k_B N \int d\tau \mathbf{v}\cdot\frac{\partial}{\partial \mathbf{r}}(f \ \ln f) + k_B N \int d\tau \frac{e}{m}\mathbf{E}\cdot\frac{\partial}{\partial \mathbf{v}}(f \ \ln f) \tag{2.106}$$

where $d\tau = d^3 r d^3 v$. The second and the third terms above vanish since the integrated terms are evaluated at the infinite surface. Thus we have

$$\frac{dS}{dt} = -k_B N \frac{\partial}{\partial t}\int f \ \ln f d\tau \tag{2.107}$$

We multiply (2.104) by $-k_B N \ \ln f$ and integrate over the whole $\mathbf{v}$, $\mathbf{r}$ space. Note

$$\begin{aligned}\int \ln f \ \frac{\partial f}{\partial t} d\tau &= \frac{\partial}{\partial t}\int f \ \ln f \ d\tau - \int \frac{\partial f}{\partial t} d\tau \\ &= \frac{\partial}{\partial t}\int f \ \ln f \ d\tau - \frac{\partial}{\partial t}\int f \ d\tau \\ &= \frac{\partial}{\partial t}\int f \ \ln f \ d\tau \end{aligned} \tag{2.108}$$

since $\int f \ d\tau = N,$ const.

$$\begin{aligned}\int \ln f \ \mathbf{v}\cdot\frac{\partial f}{\partial \mathbf{r}} d\tau &= -\int f \ \frac{\partial}{\partial r_i}(v_i \ \ln f) d\tau \\ &= -\int f \ v_i \ \frac{\partial}{\partial r_i} \ \ln f \ d\tau = -\int v_i \ \frac{\partial f}{\partial r_i} d\tau = -v_i \frac{\partial N}{\partial r_i} = 0\end{aligned} \tag{2.109}$$

Likewise we get

$$\int \ln f \ \mathbf{E}\cdot\frac{\partial f}{\partial \mathbf{v}} d\tau = 0 \tag{2.110}$$

Therefore we are led to

$$\frac{\partial}{\partial t}\int f \ \ln f \ d\tau = 0 \tag{2.111}$$

Hence, $\frac{dS}{dt} = 0$ per (2.107).

If we consider a collisional plasma, the Boltzmann equation takes the form

$$\frac{\partial f(\mathbf{r},\mathbf{v},t)}{\partial t}+\mathbf{v}\cdot\frac{\partial f}{\partial \mathbf{r}}-\frac{e}{m}\mathbf{E}\cdot\frac{\partial f}{\partial \mathbf{v}}=C(f) \tag{2.112}$$

where $C(f)$ symbolically represents the self-consistent collisional term. Equations (2.107) and (2.108) give

$$\begin{aligned}\frac{dS}{dt}&=-k_BN\int \ln f\,\frac{\partial f}{\partial t}\,d\tau\\&=-k_BN\int \ln f\;C(f)d\tau\end{aligned} \tag{2.113}$$

where we used (2.112) and integrated the scalar product terms as before. Equation (2.113) can be used to prove the H-theorem. In the Boltzmann equation, we can add the magnetic force term, $\mathbf{v}\times\mathbf{B}$ term without changing the expressions for dS/dt.

2.8 Collisional Two-Fluid Equations–Braginskii Equations

Reference: S. I. Braginskii, *Transport process in a plasma* in Reviews of plasma physics Volume 1: Consultant's Bureau, New York (1965).

The fluid equations derived in Section 2.5 apply for a single species gas or two-fluid plasma when we neglect the cross collision term in (2.22). Now we consider the electron-ion collision in (2.22) following Braginskii. Braginskii derived two-fluid equations for a plasma consisting of electrons and ions using the Landau collision term (2.86):

$$\frac{\partial f_e}{\partial t}+\mathbf{v}\cdot\frac{\partial f_e}{\partial \mathbf{r}}-\frac{e}{m_e}\left(\mathbf{E}+\frac{1}{c}\mathbf{v}\times\mathbf{B}\right)\cdot\frac{\partial f_e}{\partial \mathbf{v}}=C_{\mathrm{ee}}+C_{\mathrm{ei}} \tag{2.114}$$

$$\frac{\partial f_i}{\partial t}+\mathbf{v}\cdot\frac{\partial f_i}{\partial \mathbf{r}}+\frac{e}{m_i}\left(\mathbf{E}+\frac{1}{c}\mathbf{v}\times\mathbf{B}\right)\cdot\frac{\partial f_i}{\partial \mathbf{v}}=C_{\mathrm{ii}}+C_{\mathrm{ie}} \tag{2.115}$$

Equations (2.114) and (2.115) are two coupled equations through the cross-collision terms C_{ei} and C_{ie}. Since the Landau collision terms are appropriate for plasma of frequent Coulomb collisions, Braginskii's two-fluid equation is widely used for collisional plasma investigation.

In contrast with a simple gas, there are no summation invariants. But we have the following relations:

$$\int C_{ab}d^3v=0 \quad (a=e,\ i;\quad b=e,\ i) \tag{2.116}$$

which implies that the total number of particles of each species is not changed due to collisions between like particles as well as between unlike particles. We have for like particles

$$\int m_a \mathbf{v} C_{aa} d^3 v = 0 \tag{2.117}$$

$$\int m_a v^2 C_{aa} d^3 v = 0 \tag{2.118}$$

The above two equations denote, respectively, the conservation of momentum and energy in collisions between like particles. For unlike particles, we have

$$\int (m_i C_{\text{ie}} + m_e C_{\text{ei}}) \mathbf{v} d^3 v = 0 \tag{2.119}$$

$$\int (m_i C_{\text{ie}} + m_e C_{\text{ei}}) v^2 d^3 v = 0 \tag{2.120}$$

The above two equations express the conservation of momentum and energy for the total pair of collisions involving unlike particles.

In order to derive the moment equations, multiply both sides of (2.114) with 1, $m_e \mathbf{v}$, $\frac{m_e}{2} v^2$, respectively, and integrate over the velocity. Doing the same operation with the ion equation (2.115), we obtain the three moment equations:

$$\frac{\partial n_a}{\partial t} + \nabla \cdot (n_a \mathbf{V}_a) = 0 \tag{2.121}$$

$$m_a n_a \left(\frac{\partial}{\partial t} + \mathbf{V}_a \cdot \nabla \right) \mathbf{V}_a = -\nabla p_a - \nabla \cdot \pi_a^{ij} + e_a n_a \left(\mathbf{E} + \frac{\mathbf{V}_a}{c} \times \mathbf{B} \right) + \mathbf{R}_a \tag{2.122}$$

$$\frac{3}{2} \frac{\partial}{\partial t} (n_a T_a) + \frac{3}{2} \nabla \cdot (n_a T_a \mathbf{V}_a) + n_a T_a \nabla \cdot \mathbf{V}_a + \pi_a^{ij} \frac{\partial V_{ai}}{\partial x_j} + \nabla \cdot \mathbf{q}_a = Q_a \tag{2.123}$$

where $a = e,\ i$, and the hydrodynamic variables $n_a, \mathbf{V}_a, T_a$ are defined for each species of particles similarly to (2.37)–(2.39). $\mathbf{R}$ and Q represent, respectively, the frictional force and heat generated by collision.

$$n_a(\mathbf{r}, t) = \int f_a(\mathbf{r}, \mathbf{v}, t) d^3 v \tag{2.124}$$

$$\mathbf{V}_a(\mathbf{r}, t) = \frac{1}{n_a} \int \mathbf{v} f_a d^3 v \tag{2.125}$$

$$P_a^{ij} = m_a \int (v_i - V_{ai})(v_j - V_{aj}) f_a d^3v = p_a \delta_{ij} + \pi_a^{ij} \quad \text{(pressure tensor)} \tag{2.126}$$

$$T_a = \frac{m_a}{3n_a} \int (\mathbf{v} - \mathbf{V}_a)^2 f_a d^3v \tag{2.127}$$

The heat flux q_α is defined in (2.49):

$$\mathbf{q}_a = \frac{n_a m_a}{2} \int (\mathbf{v} - \mathbf{V}_a)(\mathbf{v} - \mathbf{V}_a)^2 f_a d^3v \tag{2-127$'$}$$

When $Q = 0$, (2.123) can be arranged into the form of (2.58).

The appearance of the quantities $\mathbf{R}_a$ and Q_a in (2.122) and (2.123) is due to the cross-collisional terms in (2.114) and (2.115); the corresponding terms in a simple gas do not exist. The quantity $\mathbf{R}_a$ represents the mean change of the momentum of particles of species 'a' due to collisions with particles of other species:

$$\mathbf{R}_e = \int C_{\text{ei}} m_e (\mathbf{v} - \mathbf{V}_e) d^3v \tag{2.128}$$

$$\mathbf{R}_i = \int C_{\text{ie}} m_i (\mathbf{v} - \mathbf{V}_i) d^3v \tag{2.129}$$

Due to the relation (2.119) we have

$$\mathbf{R}_e = -\mathbf{R}_i \tag{2.130}$$

The quantity Q_a in (2.123) is the heat generated in the particles of species 'a' as a consequence of collisions with particles of other species.

$$Q_e = \int C_{\text{ei}} \frac{m_e}{2} (\mathbf{v} - \mathbf{V}_e)^2 d^3v \tag{2.131}$$

$$Q_i = \int C_{\text{ie}} \frac{m_i}{2} (\mathbf{v} - \mathbf{V}_i)^2 d^3v \tag{2.132}$$

Adding the above two equations give

$$Q_e + Q_i = -\mathbf{R}_e \cdot (\mathbf{V}_e - \mathbf{V}_i) \tag{2.133}$$

where we used (2.118)–(2.120). The foregoing formulas have been obtained by invoking general conservation relations. Braginskii decoupled (2.114) and (2.115) by making use of the smallness of the mass ratio of electron and ion. Then the problem is reduced to that of solving the kinetic equation of single species, for which the successive approximation method of Chapman and Enskog can be applied. The key step to separate the electron and ion kinetic equations is to expand the relative velocity $\mathbf{g}$ in the collision term of (2.86) in powers of ion velocity. We shall not go into the details of the Braginskii solution, but will summarize the important results.

In the lowest order, the electron and ion distributions are the local Maxwellians

$$f_a(\mathbf{v}, \mathbf{r}, t) = n_a(\mathbf{r}, t)\left(\frac{m_a}{2\pi T_a(\mathbf{r}, t)}\right)^{\frac{3}{2}} \exp -\frac{m_a(\mathbf{v} - \mathbf{V}_a(\mathbf{r}, t))^2}{2T_a(\mathbf{r}, t)} \quad (a = e,\ i) \tag{2.134}$$

which yield the following two-fluid equations:

$$\frac{\partial n_a}{\partial t} + \nabla \cdot (n_a \mathbf{V}_a) = 0 \tag{2.135}$$

$$m_a n_a \left(\frac{\partial}{\partial t} + \mathbf{V}_a \cdot \nabla\right) \mathbf{V}_a = -\nabla p_a + e_a n_a \left[\mathbf{E} + \frac{\mathbf{V}_a}{c} \times \mathbf{B}\right] + \mathbf{R}_a^{(0)} \tag{2.136}$$

$$\frac{3}{2}\left(\frac{\partial}{\partial t} + \mathbf{V}_a \cdot \nabla\right) T_a + T_a \nabla \cdot \mathbf{V}_a = 0 \tag{2.137}$$

In the momentum equation (2.136),

$$\mathbf{R}_e^{(0)} = -\frac{m_e n_e}{\tau_e}\mathbf{u} = -\mathbf{R}_i^{(0)} \ :\ \text{frictional force} \tag{2.138}$$

$$\mathbf{u} = \mathbf{V}_e - \mathbf{V}_i, \quad \tau_e = \frac{3\sqrt{m_e} T_e^{3/2}}{4\sqrt{2\pi}\ n_i\ J\ Z^2 e^4} \ :\ \text{collision time}$$

where Z is ionic charge number and J is given by (2.82). The energy equation (2.137) is the adiabaticity: the constancy of entropy, $ds_a/dt = 0$.

In the next higher order, the momentum and energy equations take the forms of (2.122) and (2.123), respectively. But the frictional force is reduced by a factor of 0.51 and, in addition, a thermal force $\mathbf{R}_T$, which arises owing to temperature gradients, appears. So we write in (2.122)

$$\mathbf{R}_e = \mathbf{R}_u + \mathbf{R}_T \tag{2.139}$$

$$\mathbf{R}_u = -\frac{m_e n_e}{\tau_e}(0.51\mathbf{u}_{\parallel} + \mathbf{u}_{\perp}) \tag{2.140}$$

$$\mathbf{R}_T = -0.71 n_e \nabla_{\parallel} T_e - \frac{3}{2}\frac{n_e}{\omega_{ce}\tau_e}\frac{\mathbf{B}_0}{B_0} \times \nabla T_e \tag{2.141}$$

The symbols $\|$ and $\perp$ are referred to the direction of the ambient magnetic field ($\mathbf{B}_0$). If the plasma is unmagnetized, the $\perp$-components are discarded. The thermal forces are given rise to by the temperature gradients. The heat Q_a is a plasma-proper term which represents the heat generated by collisions between electrons and ions:

$$Q_i = \frac{3m_e}{m_i}\frac{n_e}{\tau_e}(T_e - T_i) \equiv Q_\Delta \tag{2.142}$$

The heat acquired by the electrons is obtained by using (2.142) in (2.133):

$$Q_e = -\mathbf{R}_e \cdot (\mathbf{V}_e - \mathbf{V}_i) - Q_\Delta \tag{2.143}$$

Another point worthy of attention in Braginskii's solution is that a heat flux ($\mathbf{q}_u^e$) is given rise to by the relative motion between electrons and ions, in addition to the usual heat flux ($\mathbf{q}_T^e$) generated by temperature gradients: $\mathbf{q}_e = \mathbf{q}_u^e + \mathbf{q}_T^e$.

$$\mathbf{q}_u^e = 0.71 n_e T_e \mathbf{u}_\| + \frac{3}{2}\frac{n_e T_e}{\omega_{ce}\tau_e}\frac{\mathbf{B}_0}{B_0} \times \mathbf{u} \tag{2.144}$$

The ion heat flux $\mathbf{q}_i$ takes the familiar form $\propto \nabla T_i$. Viscosity π_{ij}, heat flux $\mathbf{q}$, heat Q, and frictional and thermal force $\mathbf{R}$ all contribute to the production of entropy in the plasma. Define the entropy per electron as

$$s_e = \frac{3}{2}\ln T_e - \ln n_e + \text{const.} \tag{2.145}$$

The electron entropy equation is written in the form

$$\frac{\partial}{\partial t}(n_e s_e) + \nabla \cdot \left(s_e n_e \mathbf{V}_e + \frac{\mathbf{q}_e}{T_e}\right) + \frac{Q_\Delta}{T_e} = \vartheta_e \tag{2.146}$$

where ϑ_e is the entropy production per unit volume:

$$T_e \vartheta_e = -\mathbf{q}_e \cdot \nabla \ln T_e - \mathbf{R} \cdot \mathbf{u} - \frac{1}{2}\pi_{ij}^e \left(\frac{\partial V_{ei}}{\partial x_j} + \frac{\partial V_{ej}}{\partial x_i} - \frac{2}{3}\delta_{ij}\nabla \cdot \mathbf{V}_e\right) \tag{2.147}$$

Adding the analogous ion entropy equation, the entropy equation for the entire plasma can be written in the form (see Braginskii, 1965)

$$\frac{\partial}{\partial t}(s_e n_e + s_i n_i) + \nabla \cdot \left(s_e n_e \mathbf{V}_e + s_i n_i \mathbf{V}_i + \frac{\mathbf{q}_e}{T_e} + \frac{\mathbf{q}_i}{T_i}\right) = \Theta : \text{positive definite} \tag{2.148}$$

• **Use of BGK collision term.** In the Boltzmann equation for the electron distribution function, we use the BGK type collision term: $(\frac{\delta f_e}{\delta t})_c = -\nu(f_e - f_i)$. This model implies that the electron and ion distributions become the same after infinite time. We shall derive the momentum equation for the fluid velocity $\mathbf{V}_e(\mathbf{r}, t)$ by taking the moment of the equation

$$\frac{\partial f_e}{\partial t} + \mathbf{v} \cdot \frac{\partial f_e}{\partial \mathbf{r}} - \frac{e}{m_e}\left(\mathbf{E} + \frac{1}{c}\mathbf{v} \times \mathbf{B}\right) \cdot \frac{\partial f_e}{\partial \mathbf{v}} = -\nu(f_e - f_i) \tag{2.149}$$

Multiply (2.149) by $\mathbf{v}$ and integrate over d^3v.

$$m_e \int d^3v \mathbf{v} \frac{\partial f_e}{\partial t} = m_e \frac{\partial}{\partial t} \int \mathbf{v} f_e d^3v = \frac{\partial}{\partial t}(m_e n_e \mathbf{V}_e(\mathbf{r}, t))$$

$$m_e \int d^3v \mathbf{v}\, \mathbf{v} \cdot \frac{\partial f_e}{\partial \mathbf{r}} = m_e \int d^3v v_i v_j \frac{\partial f_e}{\partial x_j} = m_e \frac{\partial}{\partial x_j} \int d^3v\ v_i v_j f_e = m_e \frac{\partial}{\partial x_j}(n_e \langle v_i v_j \rangle) \tag{2.150}$$

where $\mathbf{v} = v_i$ (in the following 'i' is the free index indicating the vector component). We define the random velocity

$$w_i = v_i - V_i(\mathbf{r}, t) \tag{2.151}$$

Then we have for (2.150)

$$\begin{aligned}
&= \frac{\partial}{\partial x_j}(m_e n_e V_i V_j + m_e n_e \langle w_i w_j \rangle) \\
&= \frac{\partial}{\partial x_j}(m_e n_e V_i V_j + P_{ij}) - e \int d^3v \mathbf{v}(\mathbf{E} + \frac{1}{c}\mathbf{v} \times \mathbf{B}) \cdot \frac{\partial f_e}{\partial \mathbf{v}} d^3v \\
&= -e \int d^3v v_i \left(E_j + \frac{1}{c} e_{jlm} v_l B_m\right) \frac{\partial f_e}{\partial v_j}
\end{aligned}$$

Integrating by parts and using $\partial v_i / \partial v_j = \delta_{ij}$ give

$$\begin{aligned}
&= +e \int d^3v \left(E_j \delta_{ij} + \frac{B_m}{c} e_{jlm}(v_i \delta_{lj} + v_l \delta_{ij})\right) \quad (e_{jlm}\delta_{lj} = 0) \\
&= e n_e \left(\mathbf{E} + \frac{1}{c}\mathbf{V}_e \times \mathbf{B}\right)
\end{aligned} \tag{2.152}$$

The collision term gives

$$-\nu m_e n_e (\mathbf{V}_e - \mathbf{V}_i)$$

Collecting the above results, we write

$$\frac{\partial}{\partial t}(m_e n_e \mathbf{V}_e) + \frac{\partial}{\partial x_j}(m_e n_e V_i V_j + P_{ij}) + en_e\left(\mathbf{E} + \frac{\mathbf{V}_e}{c} \times \mathbf{B}\right) = -\nu m_e n_e (\mathbf{V}_e - \mathbf{V}_i)$$

Using the continuity equation in the above yields (see problem)

$$m_e n_e \left(\frac{\partial}{\partial t} + \mathbf{V}_e \cdot \nabla\right)\mathbf{V}_e + \nabla \cdot P_{ij} + en_e\left(\mathbf{E} + \frac{\mathbf{V}_e}{c} \times \mathbf{B}\right) = -\nu m_e n_e (\mathbf{V}_e - \mathbf{V}_i) \tag{2.153}$$

Note that $\nabla \cdot P_{ij} = \frac{\partial P_{ij}}{\partial x_j} = \frac{\partial P_{ji}}{\partial x_j}$.

Energy equation. Taking the moments of (2.114) and (2.115) with $\frac{1}{2}m_\alpha v_\alpha^2$ gives the energy equation. The first two terms yield

$$\frac{\partial}{\partial t}\left(\frac{1}{2}m_\alpha n_\alpha {<}v^2{>}_\alpha\right) + \frac{\partial}{\partial x_j}\left(\frac{1}{2}m_\alpha n_\alpha {<}v_j v^2{>}_\alpha\right)$$

The electromagnetic force term yields

$$\begin{aligned}
&\frac{e_\alpha}{2}\int d^3v v^2\left(E_j + \frac{1}{c}e_{jlm}v_l B_m\right)\frac{\partial f_\alpha}{\partial v_j} \\
&\quad = -\frac{e_\alpha}{2}\int d^3v\left(2v\frac{\partial v}{\partial v_j}\left(E_j + \frac{1}{c}e_{jlm}v_l B_m\right) + \frac{v^2}{c}e_{jlm}\delta_{jl}B_m\right) \\
&\quad = -e_\alpha n_\alpha \mathbf{E}\cdot\mathbf{V}_\alpha
\end{aligned}$$

Collecting the above results gives for the energy equation in the form

$$\frac{\partial}{\partial t}\left(\frac{1}{2}m_\alpha n_\alpha {<}v^2{>}_\alpha\right) + \frac{\partial}{\partial x_j}\left(\frac{1}{2}m_\alpha n_\alpha {<}v_j v^2{>}_\alpha\right) - e_\alpha n_\alpha \mathbf{E}\cdot\mathbf{V}_\alpha = Q_\alpha \tag{2.154}$$

Proceeding further with (2.154) leads to (2.123).

Cold plasma. When the spread of velocity vector is zero, i.e. $\mathbf{w} = 0$, the plasma is *cold*. Then $P_{ij} = 0$, and the above momentum equation becomes

$$m_e\left(\frac{\partial}{\partial t} + \mathbf{V}_e \cdot \nabla\right)\mathbf{V}_e + e\left(\mathbf{E} + \frac{\mathbf{V}_e}{c} \times \mathbf{B}\right) = -\nu m_e (\mathbf{V}_e - \mathbf{V}_i) \tag{2.155}$$

This result can be obtained by assuming the cold plasma distribution function $f(\mathbf{r}, \mathbf{v}, t) \sim \delta[\mathbf{v} - \mathbf{V}(\mathbf{r}, t)]$.

Problems

2.1 In writing (2.28), we used specifically

$$\int \Phi(\mathbf{v}_a) f_a(\mathbf{v}_a') f_b(\mathbf{v}_b') g \frac{d\sigma}{d\Omega} d\Omega d^3 v_a d^3 v_b = \int \Phi(\mathbf{v}_a') f_a(\mathbf{v}_a) f_b(\mathbf{v}_b) g \frac{d\sigma}{d\Omega} d\Omega d^3 v_a d^3 v_b$$

where Φ is a summation invariant. Prove this by using (2.19) and (2.20).

2.2 Prove Jacobian

$$J\left(\frac{\mathbf{r}', \mathbf{v}'}{\mathbf{r}, \mathbf{v}}\right) = 1$$

if the transformation $\mathbf{r}, \mathbf{v} \to \mathbf{r}', \mathbf{v}'$ is canonical. "Canonical" means each set satisfies its own Hamilton's equations.

2.3 Using (2.51) and the first equation of (2.54), derive the relation

$$\frac{D}{Dt} \ln\left(\frac{p}{n^{5/3}}\right) = 0$$

Hence the following is valid:

$$\frac{D}{Dt}\left(\frac{p}{n^{5/3}}\right) = 0$$

Hint: $\nabla \cdot \mathbf{V} = -\frac{D}{Dt} \ln n$. Using $p = nT$ in the above equation gives

$$\frac{D}{Dt}\left(n^{-2/3} T\right) = 0, \quad \text{or} \quad \frac{D}{Dt}\left(T^{-3/2} n\right) = 0$$

2.4 Prove (2.68) and (2.69).

2.5 Prove the resolution of $\hat{\mathbf{k}}$ in the text. $|\mathbf{g}' - \mathbf{g}| = 2g \sin \frac{\theta}{2}$ may be useful.

2.6 Show that a spherically symmetric velocity distribution of the Maxwell–Boltzmann distribution can be expressed as

$$f(v) = 4\pi \left(\frac{m}{2\pi kT}\right)^{3/2} v^2 \exp \, - \frac{mv^2}{2kT}$$

2.7 Derive the ideal gas law from (2.44) by assuming Maxwell–Boltzmann distribution for $\mathbf{w}$ (2.50).

Answer. $P = nk_B T$.

2.8 Find P_q maximizing entropy S in (2.95)

$$S = -k_B\, N \sum_q P_q \ \ln P_q$$

under the constraints

$$\sum_q P_q = 1, \quad N \sum_q P_q E_q = U : \ \text{const.}$$

Answer.

$$P_q = \frac{e^{-\beta E_q}}{\sum_q e^{-\beta E_q}}$$

where β is a Lagrange multiplier.

2.9 Prove H-theorem by using Landau collision integral (2.86) (Klimontovich)

2.10 For a Maxwellian distribution, show that the entropy satisfies the following thermodynamic relations

$$\left(\frac{\partial S}{\partial E}\right)_V = \frac{1}{T}, \quad \left(\frac{\partial S}{\partial V}\right)_E = \frac{p}{T} \quad \text{(Use (2.96))} \tag{A}$$

where V is the total volume, and $E = \frac{3}{2} N k_B T$ is the total energy. On the other hand we have the energy conservation

$$dE = TdS - pdV \tag{B}$$

per $dS = \frac{\delta Q}{T}$ where δQ is heat. From (B) we immediately obtain the relations in (A). Recall that (A) is derived from the Boltzmann's entropy in (2.89) and (B) contains Clausius-Kelvin entropy. This exercise suggests that the two entropies are equivalent.

2.11 Review the following properties of the local Maxwellian distribution of two fluid plasma ($\alpha = e,\ i$):

$$Q_\alpha = 0, \quad P_{ij}^\alpha = p_\alpha\ \delta_{ij}, \quad p_\alpha = n_\alpha k_B T_\alpha, \quad \frac{D}{Dt}(p_\alpha\ n_\alpha^{-\gamma}) = 0$$

where $\gamma = 5/3$ is the ratio of specific heats for particles with no internal degrees of freedom.

2.12 Expand $U_{\alpha\beta} = \left(g^2 \delta_{\alpha\beta} - g_\alpha g_\beta\right)/g^3$ in powers of ion velocity.
Hint:

$$\psi(\mathbf{r}+\mathbf{a}) = \sum_{n=0}^{\infty} \frac{1}{n!}(\mathbf{a}\cdot\nabla)^n \psi(\mathbf{r}) = \psi(\mathbf{r}) + a_i \frac{\partial}{\partial x_i}\psi(\mathbf{r}) + \frac{1}{2} a_i a_j \frac{\partial}{\partial x_i}\frac{\partial}{\partial x_j}\psi(\mathbf{r}) + \cdots$$

Note $g_\alpha = v_\alpha - v'_\alpha$ where α is the cartesian index, no prime denotes electron velocity, prime denotes ion velocity. In the above formula, put $a_\alpha = -v'_\alpha$. Then we obtain

$$U_{\alpha\beta} = V_{\alpha\beta} - \frac{\partial V_{\alpha\beta}}{\partial v_\gamma} v'_\gamma + \frac{1}{2} \frac{\partial^2 V_{\alpha\beta}}{\partial v_\gamma \partial v_\kappa} v'_\gamma v'_\kappa + \cdots$$

$$\text{where} \quad V_{\alpha\beta} = \frac{1}{v^3}\left(v^2 \delta_{\alpha\beta} - v_\alpha v_\beta\right)$$

2.13 Derive (2.153).

2.14 Equation (2.102) is an example of the Lagrange multiplier method to minimize or maximize a certain integral under given constraints. Generally the constraints are given either in integral form or in algebraic form. Review the process to obtain the minimizing condition for variational integrals.

References

Bhatnagar, P. L., E. P. Gross, and M. Krook, Phys. Rev. **94**, 511 (1954).

Braginskii, S. I. *Transport Process in a Plasma*, in Reviews of plasma physics Vol. 1: Consultant's Bureau, New York (1965).

Braginskii, S. I. *Transport phenomena in a completely ionized Two-temperature plasma,* Soviet Phys. JETP, **6** 358 (1958).

Chapman, S. and T. G. Cowling, *The Mathematical Theory of Nonuniform Gases*, Cambridge University Press (1964).

Krall, N. A. and A. W. Trivelpiece, *Principles of Plasma Physics*, San Francisco Press, 1986.

Landau, L. D. *The Transport Equation in the Case of Coulomb Interactions,* in Collected papers of L. D. Landau, Ed. D. Ter Haar Pergamon Oxford (1965) p. 163

Reif, F. *Fundamentals of Statistical and Thermal Physics*, McGraw-Hill International (1985), Chapter 13.

Sturrock, P. A. *Plasma Physics*, Cambridge University Press (1994).

Schram, P. P. J. M. *Kinetic Theory of Gases and Plasmas*, Kluwer Academic, Dordrecht (1991).

Van Kampen, N. G. and B. U. Felderhof, *Theoretical Methods in Plasma Physics*, John Wiley (1967), Chapter XV.

Wu, T.-Y. *Kinetic Equations of Gases and Plasmas*, Addison-Wesley (1966), Chapter 2.

Chapter 3

Magnetohydrodynamics

Magnetohydrodynamics (MHD) is fluid dynamics of an electrically conducting fluid in the presence of magnetic fields. A popular view of MHD medium is that it is a conducting fluid moving in a magnetic field. If a conductor moves in a magnetic field, electric field is induced as is called a *motional electric field*, and electric currents arise. The magnetic field exerts forces on the currents, and the conducting fluid changes its motion under the force. Also, the currents modify the magnetic field. Thus, we deal with a complicated system of fluid and magnetic fields. This interplay of conducting fluid and magnetic field can best be illustrated by a device known as z-pinch (Bennet pinch). A cylindrical column of conducting fluid with an axial current density $\mathbf{J}$ is depicted in Fig. 3.1. The axial current in the conducting fluid is generated by an appropriately applied electric field. The current produces azimuthal induction $\mathbf{B} = \hat{\varphi} B$ which in turn gives rise to $\mathbf{J} \times \mathbf{B}$ force. The $\mathbf{J} \times \mathbf{B}$ force acting on the fluid forces the column to be pushed inwardly in all directions, to the effect of confinement.

The MHD equations describe the dynamical change of the fluid velocity, the magnetic field, and the electric field in the plasma.

• **Distinction between plasma and magnetohydrodynamic fluid.** Roughly speaking, plasma is a collection of suitably dense charged particles which interact through electromagnetic forces generated by currents. This definition equally well applies to electrically conducting fluid that is dealt with by MHD. But a clear distinction exists between them: Magnetohydrodynamic fluid exhibits only a single fluid characteristics, always preserving the charge neutrality, while plasma is characterized by a high-frequency collective oscillations known as plasma oscillation which owes to deviation from charge neutrality. What makes a medium containing both positive and negative charged particles behave sometimes as a single charge-neutral fluid and other times as a collective medium exhibiting different inertial effects of the constituent charged particles? It depends on the inter-particle collision

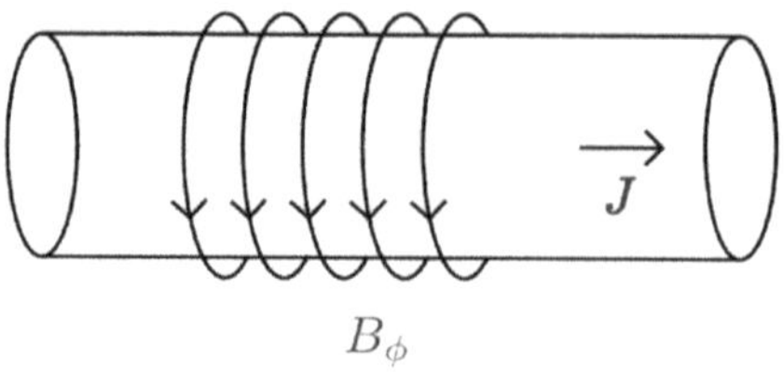

Fig. 3.1. Bennet pinch.

frequency. If the collision frequency is sufficiently high, then the electrons and the ions move in such a way that there is no separation of charge, and the medium behaves like a single fluid. If the collision frequency is low, the high-frequency electronic motion gives rise to charge separation and the medium exhibits the characteristic collective behavior. However, it may be more appropriate to accept that these two different behaviors coexist in ionized gases. Then, it depends on what part of the spectrum of the motion we are looking at. The charge separation effect is important in the timescale of inverse plasma frequency $\left(\omega_{\text{pe}}^{-1}, \omega_{\text{pe}} = \sqrt{\frac{4\pi Ne^2}{m_e}}\right)$ and in the length scale of $v_{\text{th}}/\omega_{\text{pe}} = \lambda_e$ (v_{th} is the electron thermal speed and λ_e is the electron Debye length). If one is not interested in the fine structure of motion but in the low-frequency ($\ll\omega_{\text{pe}}$) and large-scale ($\gg\lambda_e$) motion, the MHD equations will suffice.

3.1 MHD Medium as a Lumped Single Fluid

In the low-frequency motion of plasma, we can treat it, by suitable combinations of the electron and ion quantities, as a single fluid instead of regarding it as two interacting fluids via collisions and Coulomb force. Under frequent collisions, the lumped volume elements of combined electrons and ions may be considered to form single fluid particles. In this low-frequency region of plasma dynamics, the charge neutrality is often assumed: In a small volume, small but containing many electrons and ions, the net charge is approximately nil ($(n_i - n_e)/n_i \ll 1$). The magnetic field that is generated by current plays the major role in the magnetohydrodynamic plasma.

The MHD equations will be derived by two different ways. The first method now defines the lumped hydrodynamic variables as follows:

$$\varrho_m = m_i n_i + m_e n_e \quad \text{(mass density)} \tag{3.1}$$

$$\varrho_c = e(n_i - n_e) \quad \text{(charge density)} \tag{3.2}$$

$$\mathbf{V} = \frac{m_i n_i \mathbf{V}_i + m_e n_e \mathbf{V}_e}{\varrho_m} \quad \text{(center of mass velocity)} \tag{3.3}$$

$$\mathbf{J} = e(n_i \mathbf{V}_i - n_e \mathbf{V}_e) \quad \text{(current)} \tag{3.4}$$

Multiply (2.135) by m_a and sum over a to obtain continuity equation

$$\frac{\partial \varrho_m}{\partial t} + \nabla \cdot (\varrho_m \mathbf{V}) = 0 \tag{3.5}$$

Multiply (2.135) by charge 'e' and subtract electron equation from ion equation to obtain charge conservation equation

$$\frac{\partial \varrho_c}{\partial t} + \nabla \cdot \mathbf{J} = 0 \tag{3.6}$$

Taking the first moment of (2.114) (see (2.150)) gives

$$\frac{\partial}{\partial t}(m_\alpha n_\alpha \mathbf{V}_\alpha(\mathbf{r}, t)) + m_\alpha \frac{\partial}{\partial x_\mu}(n_\alpha < v_\mu v_\nu >_\alpha) - e_\alpha n_\alpha \left(\mathbf{E} + \frac{\mathbf{V}_\alpha}{c} \times \mathbf{B} \right) = \mathbf{R}_\alpha \tag{3.7}$$

where $\alpha = e,\ i (e_e = -e = e_i)$, indicating electrons and ions, μ and ν are cartesian indexes, ν is the free index indicating the vector components, and $\mathbf{R}_\alpha$ is the momentum exchange rate due to collisions (see (2.136)). Here, we define the random velocity as, in contrast with (2.151),

$$\mathbf{w} = \mathbf{v} - \mathbf{V} \quad \text{or} \quad <\mathbf{w}>_\alpha = \mathbf{V}_\alpha - \mathbf{V} \tag{3.8}$$

Then, we have

$$<v_\mu v_\nu>_\alpha = -V_\mu V_\nu + V_\mu V_{\alpha\nu} + V_\nu V_{\alpha\mu} + <w_\mu w_\nu>_\alpha \tag{3.9}$$

Substituting (3.9) into (3.7) and adding the electron and ion equations, one obtains

$$\frac{\partial}{\partial t}(\varrho_m V_\nu) + \frac{\partial}{\partial x_\mu}(\varrho_m V_\mu V_\nu + P_{\mu\nu}) - \varrho_c E_\nu - \frac{1}{c}(\mathbf{J} \times \mathbf{B})_\nu = 0 \tag{3.10}$$

where $P_{\mu\nu} = P^e_{\mu\nu} + P^i_{\mu\nu}$ is the total pressure tensor. Note that the collision terms add to zero if there are no neutral particles. Using (3.5) in the above equation, it can be written in the form

$$\varrho_m \left(\frac{\partial}{\partial t} + \mathbf{V} \cdot \nabla \right) \mathbf{V} + \nabla \cdot P_{ij} - \varrho_c \mathbf{E} - \frac{1}{c} \mathbf{J} \times \mathbf{B} = 0 \tag{3.11}$$

The energy equation (2.154) summed over α takes the form

$$\frac{\partial}{\partial t}\left(\frac{3}{2}p + \frac{1}{2}\varrho_m V^2\right) + \nabla \cdot \left(\frac{5}{2}p\mathbf{V} + +\frac{1}{2}\varrho_m V^2 \mathbf{V}\right) - \mathbf{J} \cdot \mathbf{E} + \nabla \cdot (\mathbf{q} + \pi_{ij} \cdot \mathbf{V}) = 0 \tag{3.12}$$

where

$$\mathbf{q} = \frac{1}{2} \sum_\alpha m_\alpha n_\alpha {<} w^2 \mathbf{w} {>}_\alpha$$

and the random velocity is defined by (3.8).

The energy equation (3.12) is too complex to use. We often use the equations of state, in lieu of the energy equation:

$$\frac{d}{dt}(p\varrho_m^{-\gamma}) = 0 \quad \text{(adiabatic)} \tag{3.13}$$

$$\frac{d}{dt}\frac{p}{\varrho_m} = 0 \quad \text{(isothermal)} \tag{3.14}$$

3.2 Generalized Ohm's Law

The lumped variables defined in (3.1)–(3.4) can be used to determine back the two fluid variables. In low-frequency phenomena, the charge neutrality holds. Then, we have, neglecting m_e/m_i as compared to 1,

$$n_e \simeq n_i = \frac{\varrho_m}{m_i} \tag{3.15}$$

$$\mathbf{V}_i \simeq \mathbf{V} + \frac{m_e}{m_i}\frac{\mathbf{J}}{n_i e}, \quad \mathbf{V}_e \simeq \mathbf{V} - \frac{\mathbf{J}}{n_e e} \tag{3.16}$$

Equation (2.122) can be recast into the form

$$\frac{\partial}{\partial t}(m_a n_a \mathbf{V}_a) + \frac{\partial}{\partial x_\mu}(m_a n_a V_{a\mu} V_{a\nu}) = -\nabla p_a - \frac{\partial}{\partial x_\mu}\pi_a^{\mu\nu} + e_a n_a \left[\mathbf{E} + \frac{\mathbf{V}_a}{c} \times \mathbf{B}\right] + \mathbf{R}_a \tag{3.17}$$

Multiply (3.17) by e_a/m_a, use (3.15) and (3.16), and sum over 'a' to get

$$\frac{m_i m_e}{\varrho_m e^2}\frac{\partial \mathbf{J}}{\partial t} = \mathbf{E} + \frac{\mathbf{V}}{c} \times \mathbf{B} - \frac{1}{e n_e c}\mathbf{J} \times \mathbf{B} - \frac{1}{e m_i n_i}(m_e \nabla p_i - m_i \nabla p_e) - \frac{\mathbf{R}_e}{e n_e} \tag{3.18}$$

where we omitted $\pi_{\mu\nu}$ and the quadratic terms, and neglected m_e/m_i as compared to 1. In the last term, we put

$$\frac{\mathbf{R}_e}{en_e} = \eta\mathbf{J}, \qquad \eta = \frac{m_e}{e^2 n_e \tau_e} \quad \text{(electrical resistivity)} \tag{3.19}$$

where we used (2.138) and (3.4). This equation indicates that the momentum exchange through collisions between electrons and ions is proportional to the relative fluid velocity of electrons and ions. Equation (3.18) is referred to as the *generalized Ohm's law*. In the usual expression of MHD equations, (3.18) is further simplified in the form

$$\mathbf{J} = \sigma\left(\mathbf{E} + \frac{\mathbf{V}}{c} \times \mathbf{B}\right) \tag{3.20}$$

where $\sigma = 1/\eta$ is the conductivity.

Adding the electron and ion terms of (3.17) yields another form of momentum equation in terms of the lumped variables (use (3.15) and (3.16) and $n_e \simeq n_i$):

$$\frac{\partial}{\partial t}(\varrho_m \mathbf{V}) + \frac{\partial}{\partial x_\mu}\left(\varrho_m V_\mu V_\nu + \frac{m_i m_e J_\mu J_\nu}{e^2 \varrho_m}\right) = -\nabla p - \frac{\partial}{\partial x_\mu}\pi_{\mu\nu} + \frac{1}{c}\mathbf{J} \times \mathbf{B} + \varrho_c \mathbf{E} \tag{3.21}$$

The last term in (3.21) can be neglected due to the charge neutrality, $\varrho_c \simeq 0$. Also, we have from (3.6)

$$\nabla \cdot \mathbf{J} = 0 \tag{3.22}$$

The MHD medium is inherently current-carrying (in the zero order), and the current is divergence-free.

3.3 MHD Equations for Media with Infinite Conductivity

For an infinitely conducting medium ($\sigma = \infty$), we have from (3.20)

$$\mathbf{E} + \frac{\mathbf{V}}{c} \times \mathbf{B} = 0 \tag{3.23}$$

since $\mathbf{J}$ must be finite. We have the Maxwell equations

$$\nabla \times \mathbf{E} = -\frac{1}{c}\frac{\partial \mathbf{B}}{\partial t} \tag{3.24}$$

$$\nabla \times \mathbf{B} = \frac{1}{c}\frac{\partial \mathbf{E}}{\partial t} + \frac{4\pi}{c}\mathbf{J} \tag{3.25}$$

$$\nabla \cdot \mathbf{E} = 4\pi\rho_c \tag{3.26}$$

$$\nabla \cdot \mathbf{B} = 0 \tag{3.27}$$

In MHD, $\partial/\partial t \ll \sigma$, and one neglects the displacement current and writes in place of (3.25)

$$\mathbf{J} = \frac{c}{4\pi}\nabla \times \mathbf{B} \tag{3.28}$$

Using (3.23) in (3.26) gives

$$\rho_c = -\frac{1}{4\pi c}\nabla \cdot (\mathbf{V} \times \mathbf{B}) \tag{3.29}$$

Therefore, in ordinary circumstance ($V/c \ll 1$), ρ_c is small. Hence, in momentum equation (3.21), the electric force is neglected as compared with the $\mathbf{J} \times \mathbf{B}$ force.

Using (3.23) in (3.24) gives

$$\frac{\partial \mathbf{B}}{\partial t} = \nabla \times (\mathbf{V} \times \mathbf{B}) \tag{3.30}$$

In MHD, two critical approximations are involved in the electrodynamics: neglect of the displacement current (low-frequency phenomena) and charge neutrality. The latter leads to the neglect of the electric force as compared with the $\mathbf{J} \times \mathbf{B}$ force. The idea behind the neglect of the electric body force $\varrho_c\mathbf{E}$ is that, owing to the high conductivity, small charges cause large currents so that the magnetic force $\mathbf{J} \times \mathbf{B}$ is much larger than the electric force. Also, the Maxwell displacement current in (3.25) can be neglected as compared to the conduction current $\mathbf{J} = \mathbf{E}/\eta$ (resistivity η has a dimension of time, and so conductivity $\sigma = 1/\eta$ has a dimension of frequency) owing to the high σ and the slow temporal change ($\sigma \gg \partial/\partial t$).

The Joule heat $dQ = \frac{1}{\rho_m}\mathbf{J} \cdot \mathbf{E}dt$ vanishes for media with infinite conductivity. Hence, we can assume an adiabatic process, then the pressure p and mass density ρ_m are related by the formula

$$p = C\rho_m^\gamma \tag{3.31}$$

The continuity equation (3.5), the momentum equation (3.11), the electrodynamic equations (3.28)–(3.30), and the thermodynamic equation (3.31) constitute the closed set of governing equations for magnetohydrodynamics.

3.4 Kinematics of Continuum Media

In this section, we study the kinematics of continuum media to apply it to MHD flow. Fluid can be defined as a continuum whose motion or state is determined by the properties such as density, fluid velocity, stress tensor, and thermodynamic quantities; temperature, internal energy (ε), entropy, specific heats, energy flux vector, etc. In a magnetohydrodynamic medium, the stress tensor includes not only the inter-molecular mechanical stress tensor but also the electromagnetic (Maxwell) stress tensor. Various physical quantities (such as temperature, momentum, or kinetic energy) are associated with each fluid particle; the *particle* is meant by macroscopically small volume but containing many fluid molecules. In a sufficiently dense medium, the particles constitute a continuum whose state is determined by the motion of each individual fluid particle. In the *Eulerian* description of fluid, a property of the fluid is represented by a field $\phi(\mathbf{x}, t)$. The time rate of change of $\phi(\mathbf{x}, t)$ is therefore written as

$$d\phi = \frac{\partial \phi}{\partial t} dt + \nabla \phi \cdot d\mathbf{x}$$

where $d\mathbf{x}$ is the displacement of the particle position that took place during the time interval dt. Recognizing that $d\mathbf{x}/dt = \mathbf{v}$, the particle velocity, we have

$$\frac{d\phi}{dt} = \left(\frac{\partial}{\partial t} + \mathbf{v} \cdot \nabla \right) \phi \tag{3.32}$$

d/dt on the left side is the Lagrangian derivative, taken along the particle path.

Kinematics of fluid motion begins with consideration of the rate of change of any arbitrary *material line element* during the course of fluid motion. The material line element is by definition a line element composed of fluid particles that lie on it. Imagine that a line element $d\mathbf{x}(t)$ has been stretched to $d\mathbf{x}(t + \delta t)$ (see Fig. 3.2) by the corresponding movement of $a \to a'$ and $b \to b'$.

We have for the vectors

$$\overrightarrow{aa'} = \mathbf{v}(\mathbf{x}, t)\delta t, \quad \overrightarrow{bb'} = \mathbf{v}(\mathbf{x} + d\mathbf{x}, t)\delta t$$

Thus, we write

$$\mathbf{v}(\mathbf{x}, t)\delta t + d\mathbf{x}(t + \delta t) = d\mathbf{x}(t) + \mathbf{v}(\mathbf{x} + d\mathbf{x}, t)\delta t$$

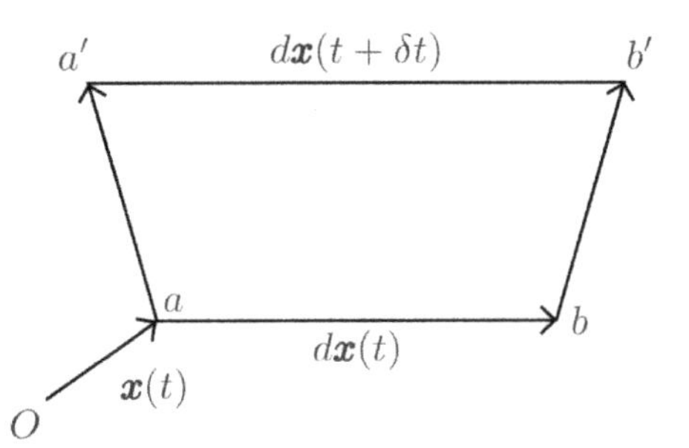

Fig. 3.2. Convective change of material line element.

Expanding the last term in Taylor series and neglecting the higher order terms, one obtains

$$\frac{d}{dt}d\mathbf{x} = \lim_{\delta t \to 0} \frac{d\mathbf{x}(t+\delta t) - d\mathbf{x}(t)}{\delta t} = (d\mathbf{x} \cdot \nabla)\mathbf{v} \tag{3.33}$$

The above equation is an important kinematic relation which describes how a material line element changes in the course of fluid motion. Next, we consider the total rate of change of a line integral associated with a fluid quantity θ:

$$\frac{d}{dt}\int_P^Q \theta(\mathbf{x},t)d\mathbf{x} \tag{3.34}$$

where P and Q are two arbitrary points in the fluid. By definition, the convective derivative of the line integral is

$$\lim_{\delta t \to 0} \frac{1}{\delta t}\left[\int_{P'}^{Q'} \theta(\mathbf{x} + \mathbf{v}(\mathbf{x})\delta t, t + \delta t)d\mathbf{x}(t+\delta t) - \int_P^Q \theta(\mathbf{x},t)d\mathbf{x}\right]$$

where P' and Q' are infinitesimally displaced position of P and Q, respectively. Expanding as

$$\theta(\mathbf{x} + \mathbf{v}(\mathbf{x})\delta t, t + \delta t) = \theta(\mathbf{x},t) + \frac{\partial \theta}{\partial t}\delta t + \nabla\theta \cdot \mathbf{v}\delta t + \cdots$$

$$d\mathbf{x}(t+\delta t) = d\mathbf{x}(t) + \left(\frac{d}{dt}d\mathbf{x}\right)\delta t + \cdots$$

Substituting from (3.33) and neglecting the higher-order terms, the line integral (3.34) becomes

$$\frac{d}{dt}\int_P^Q \theta(\mathbf{x},t)d\mathbf{x} = \int_P^Q \frac{d\theta}{dt}\, d\mathbf{x} + \int_P^Q \theta\ (d\mathbf{x} \cdot \nabla)\mathbf{v} \tag{3.35}$$

Operationally, the total derivative and the integral commute, but it is important to have the line element $d\mathbf{x}$ differentiated. According to the above formula, it goes without proof that the total differentiation can be taken into the integral sign even in a surface or a volume integral. A line integral associated with a vector field convectively changes in a like fashion:

$$\frac{d}{dt}\int \mathbf{A}(\mathbf{x},t) \cdot d\mathbf{x} = \int \frac{dA_i}{dt}d\mathbf{x}_i + \int A_i(d\mathbf{x} \cdot \nabla)v_i \tag{3.36}$$

Next, we consider a fixed arbitrary surface S surrounding a portion of fluid of volume V. Through the surface S, fluid particles move in and out, and we attempt to calculate the time rate of change of mass occupying the volume V. We assume that no mass is created or destroyed inside the volume V.

Rate of increase of mass in $V = \frac{\partial}{\partial t}\int \varrho d\tau$ = rate of net in-flow of mass through

$$S = -\int \varrho\mathbf{v}\cdot d\mathbf{S} = -\int \nabla\cdot(\varrho\mathbf{V})d\tau$$

where ϱ is the mass density and $d\tau$ is the volume element. The above equality comes from the divergence theorem, and the negative sign is due to the convention that the surface element takes positive sign when it is directed outwardly. Since the volume is fixed, we arrive at

$$\frac{\partial\varrho}{\partial t} + \nabla\cdot(\varrho\mathbf{v}) = 0 \quad \text{or} \quad \frac{d\varrho}{dt} + \varrho\nabla\cdot\mathbf{v} = 0 \tag{3.37}$$

This relation, representing the principle of mass conservation, is called continuity equation. When there is a source or sink in the fluid medium, that is, there exists any mechanism of production or annihilation of fluid mass, RHS of (3.37) should include the appropriate source or sink term.

Now, we specialize (3.37) to an infinitesimal volume element $d\tau$ containing a mass δm. The volume element is envisaged to be enlarged or shrunk as the fluid particles occupying the volume move around. But no particles should cross the boundary of the volume (as the classical law of mechanics testifies). So, δm is constant even though the size of the volume element is subject to change during the course of fluid motion. We then have from (3.37)

$$\frac{d}{dt}\left(\frac{1}{\delta\tau}\right) + \frac{1}{\delta\tau}\nabla\cdot\mathbf{v} = 0, \quad \text{or} \quad \frac{d}{dt}\delta\tau = \nabla\cdot\mathbf{v}\ \delta\tau \tag{3.38}$$

This is another kinematic law, along with (3.33), relating how a moving volume element should change as the fluid particles composing the volume element move about. It is easy to show that the continuity equation (3.37) comes out from

$$\frac{d}{dt}\int_V \varrho\ d\tau = \int_V \frac{d\varrho}{dt}d\tau + \int_V \varrho\ \frac{d}{dt}d\tau = \int_V \left(\frac{d\varrho}{dt} + \varrho\nabla\cdot\mathbf{v}\right)d\tau = 0 \tag{3.39}$$

where the material volume V now is not fixed but changes along the fluid movement.

Next, consider a tiny parallelepiped with sides δx_1, δx_2, and δx_3. Its volume $\delta\tau = \delta x_1\delta S_1 = \delta x_2\delta S_2 = \delta x_3\delta S_3 = \delta x_i\delta S_i$ (no sum on 'i'), where δS_i is the lateral elemental surface whose normal is directed to 'i'-direction

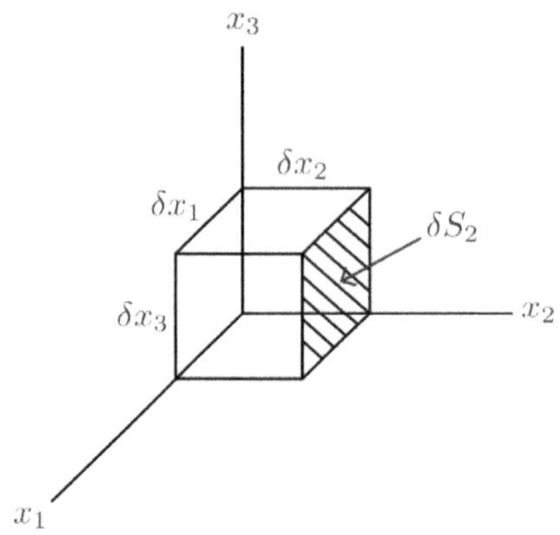

Fig. 3.3. Surface element of parallelepiped.

(see Fig. 3.3). We have

$$\frac{d}{dt}d\tau = \frac{d}{dt}(\delta x_i\ \delta S_i) = \delta S_i \frac{d}{dt}\delta x_i + \delta x_i \frac{d}{dt}\ \delta S_i$$

Substituting from (3.33) for $(d/dt)\delta x_i$ and making use of (3.38), one can easily obtain

$$\frac{d}{dt}\delta S_i = \delta S_i\ \nabla \cdot \mathbf{v} - \delta S_j\ \frac{\partial v_j}{\partial x_i} \tag{3.40}$$

This equation relates the way how an arbitrary material surface element convectively changes in the course of fluid motion.

3.5 Magnetohydrodynamic Momentum Equation Revisited

The momentum of a conducting fluid occupying a volume element $d\tau$ is $\varrho_m \mathbf{v} d\tau$. Now, we wish to calculate the total time rate of change of momentum of the fluid enclosed by a material surface S:

$$\frac{d}{dt}\int_V \varrho_m v_i\ d\tau$$

with V designating the volume enclosed by the surface S. According to Newton's law, this momentum change is brought about by the action of forces exerted on each fluid particles within and on S. The forces are divided into two types: the body force and the surface force. The body force acting on each individual elemental volume $d\tau$ is $\varrho_m F_i \tau$. The surface force acting on the elemental surface dS_j is $(P_{ij} + M_{ij})$, where P_{ij} is the mechanical pressure tensor $(=p\delta_{ij} + \pi_{ij})$ and M_{ij} denotes the Maxwell stress tensor.

Thus, we write

$$\frac{d}{dt}\int_V \varrho_m v_i d\tau = \int_V \varrho_m F_i d\tau + \int_S (P_{ij} + M_{ij}) dS_j \tag{3.41}$$

The expression for the Maxwell stress tensor is

$$M_{ij} = \frac{1}{4\pi}\left(E_i E_j - \frac{1}{2}\delta_{ij}E^2 + B_i B_j - \frac{1}{2}\delta_{ij}B^2\right) \tag{3.42}$$

Now, we calculate the divergence of M_{ij}:

$$\frac{\partial M_{ij}}{\partial x_j} = \frac{1}{4\pi}\left(E_j\frac{\partial E_i}{\partial x_j} - \frac{1}{2}\nabla E^2 + B_j\frac{\partial B_i}{\partial x_j} - \frac{1}{2}\nabla B^2\right) + \varrho_c\, E_i \tag{3.43}$$

where we used

$$\nabla \cdot \mathbf{E} = 4\pi\varrho_c \tag{3.44}$$

$$\nabla \cdot \mathbf{B} = 0 \tag{3.45}$$

Using the vector identity

$$\frac{1}{2}\nabla A^2 = \mathbf{A} \times \nabla \times \mathbf{A} + (\mathbf{A}\cdot\nabla)\mathbf{A} \tag{3.46}$$

Equation (3.43) becomes

$$\frac{\partial M_{ij}}{\partial x_j} = -\frac{1}{4\pi}(\mathbf{E} \times \nabla \times \mathbf{E} + \mathbf{B} \times \nabla \times \mathbf{B}) + \varrho_c\mathbf{E} \tag{3.47}$$

We substitute into the above equation the remaining Maxwell equations

$$\nabla \times \mathbf{E} = -\frac{1}{c}\frac{\partial \mathbf{B}}{\partial t} \tag{3.48}$$

$$\nabla \times \mathbf{B} = \frac{1}{c}\frac{\partial \mathbf{E}}{\partial t} + \frac{4\pi}{c}\mathbf{J} \tag{3.49}$$

Then, we have

$$\frac{\partial M_{ij}}{\partial x_j} = \frac{1}{4\pi c}\frac{\partial}{\partial t}(\mathbf{E} \times \mathbf{B}) + \frac{1}{c}\,\mathbf{J} \times \mathbf{B} + \varrho_c\mathbf{E} \tag{3.50}$$

Hence, (3.41) becomes

$$\varrho_m\frac{d\mathbf{v}}{dt} = \varrho_m\mathbf{F} + \nabla \cdot P_{ij} + \frac{1}{c}\,\mathbf{J} \times \mathbf{B} + \varrho_c\mathbf{E} + \frac{\partial}{\partial t}\frac{\mathbf{E} \times \mathbf{B}}{4\pi c} \tag{3.51}$$

The last term of (3.51) is the rate of change of the electromagnetic momentum density. It is customary in MHD to neglect the last two terms involving the electric field, which amounts to neglect of $\mathbf{E}$ terms in M_{ij} and the displacement current in (3.49). Thus, MHD momentum equation reads

$$\varrho_m \frac{d\mathbf{v}}{dt} = \varrho_m \mathbf{F} + \nabla \cdot P_{ij} + \frac{1}{c}\,\mathbf{J} \times \mathbf{B} \tag{3.52}$$

Some remarks on the pressure tensor P_{ij} of MHD flow are in order. The pressure tensor is the sum of electron and ion terms. Recall that the viscous stress tensor π_{ij} represents the irreversible (entropy-producing) momentum transfer between fluid particles due to inter-particle collisions. Since the ion mass is much greater than the electron mass, the dominant momentum transfer owes to the ion–ion collisions. Thus, the ion term of the viscous stress tensor dominates the electron term. When the magnetic field is sufficiently weak (ion–ion collision frequency $\gg$ ion cyclotron frequency), the MHD medium can be regarded as isotropic. Under this condition, the pressure tensor can be written as

$$P_{kl} = (p_e + p_i)\delta_{kl} + \eta_i \left(\frac{\partial v_k}{\partial x_l} + \frac{\partial v_l}{\partial x_k} - \frac{2}{3}\delta_{kl} \nabla \cdot \mathbf{v} \right)$$

where $\eta_i = 1.365 n_i T_i / \nu_i$ is the coefficient of viscosity due to ion–ion collision, with ν_i the ion–ion collision frequency as given in Shkarofsky *et al.* (1966):

$$\nu_i = \frac{4}{3}\sqrt{2\pi} n_i \left(\frac{Z^2 e^2}{T_i} \right)^2 \sqrt{\frac{T_i}{m_i}}\, A \quad (A: \text{ Coulomb logarithm})$$

When the collision frequency is much less than the ion cyclotron frequency, *the double adiabatic theory* can be used.

Ohm's law for finite conductivity

Recall (3.20):

$$\mathbf{J} = \sigma \left(\mathbf{E} + \frac{\mathbf{v}}{c} \times \mathbf{B} \right) \tag{3.53}$$

It is an empirical law that in a metal the current density $\mathbf{J}$ is linearly proportional to the electric field $\mathbf{E}$. The current and the electric field in a conducting fluid have the analogous relation with an additional complication that a conducting fluid is not stationary but moving. If a conductor is moving in a magnetic field $\mathbf{B}$, then an equivalent electric field $\mathbf{v} \times \mathbf{B}/c$ is induced in the conductor, which is known as the *motional electric field*, hence (3.53). In summary, electric currents in a conducting fluid are generated by a small amount of charge density and by the convective motion of the conducting fluid across the magnetic field. From (3.28) and (3.43), $\mathbf{E}$ can be expressed in terms of **B**:

$$\mathbf{E} = \frac{c}{4\pi\sigma}\nabla \times \mathbf{B} - \frac{\mathbf{v}}{c} \times \mathbf{B} \tag{3.54}$$

Using (3.24) in (3.54) gives the equation for **B** in a conducting fluid of finite conductivity:

$$\frac{\partial \mathbf{B}}{\partial t} = \frac{c^2}{4\pi\sigma}\nabla^2\mathbf{B} + \nabla \times (\mathbf{v} \times \mathbf{B}) \tag{3.55}$$

3.6 Magnetic Pressure and Magnetic Tension

A surface element $d\mathbf{S}$ exerted by the magnetic stress M_{ij} is subject to the force

$$\mathbf{M}_{ij} \cdot d\mathbf{S} = -\frac{B^2}{8\pi}d\mathbf{S} + \frac{1}{4\pi}\mathbf{B}\ (\mathbf{B} \cdot d\mathbf{S})$$

The first term $\frac{B^2}{8\pi}$ is the force directed normal to the surface and is naturally called magnetic pressure. The second term which is directed along the magnetic field can be interpreted as a magnetic tension. If converted into body force by divergence theorem, we have

$$\nabla \cdot \mathbf{M}_{ij} = \frac{1}{c}\mathbf{J} \times \mathbf{B} = -\nabla\left(\frac{B^2}{8\pi}\right) + \frac{1}{4\pi}(\mathbf{B} \cdot \nabla)\mathbf{B}$$

Introducing the unit vector along the field line, $\hat{\mathbf{b}} = \mathbf{B}/B$, the above expression becomes

$$-\nabla\left(\frac{B^2}{8\pi}\right) + \frac{1}{4\pi}B\frac{\partial}{\partial\xi}(\hat{\mathbf{b}}B)$$

where $d\xi$ is the infinitesimal length along the field line. Using

$$\frac{\partial\hat{\mathbf{b}}}{\partial\xi} = \frac{\hat{\mathbf{n}}}{R}$$

where $\hat{\mathbf{n}}$ is the unit normal vector and R is the local radius of curvature, we have

$$\frac{1}{c}\mathbf{J} \times \mathbf{B} = -\nabla_{\perp}\left(\frac{B^2}{8\pi}\right) + \frac{B^2}{4\pi R}\,\hat{\mathbf{n}} : \quad \text{body force} \tag{3.56}$$

where $\nabla_{\perp}$ denotes the gradient in the direction perpendicular to the magnetic field. The second term on the RHS of (3.56) operates only when the field line is curved in much the same way a tension acts on a curved string (the resultant is always toward the center of curvature). The magnetic pressure force is directed away from the region of high field strength. Bending the field line produces tension and compressing it produces a restoring force.

3.7 Energy Equation Revisited

The energy equation of fluid is basically the first law of thermodynamics which states that the increase of energy possessed by a certain portion V of the fluid is due to the work done on V by external forces plus the heat absorbed by the portion V through heat conduction, and the heat generated in V by molecular interactions. In addition, we shall see that the energy equation implies the second law of thermodynamics: The entropy of the fluid should never decrease. The fluid energy contained in a volume V of MHD flow is the sum of the kinetic energy, internal energy, and the magnetic field energy. Denoting by ε the internal energy per unit mass, we write (here we shall drop the subscript m at the mass density) for a *fixed volume* of V

$$\frac{\partial}{\partial t}\int_V \left(\frac{\varrho}{2}v^2 + \varrho\,\varepsilon + \frac{B^2}{8\pi}\right) d\tau$$

is the sum of the terms:

(1) The outgoing energy flux (non-electromagnetic) through $S+$ (2) rate of working of the body force $\mathbf{F}$ in $V+$ (3) rate of working of the surface stresses on $S+$ (4) electromagnetic energy flux (Poynting vector) through $S+$ (5) heat acquired by conduction through S:

$$= -\int_S \varrho\left(\frac{1}{2}v^2 + \varepsilon\right)\mathbf{v}\cdot d\mathbf{S} + \int_V \varrho F_i v_i d\tau + \int_S P_{ij}v_i dS_j - \int_s \mathbf{N}\cdot d\mathbf{S} + \int_S \lambda\nabla T\cdot d\mathbf{S}$$

where the Poynting vector $\mathbf{N} = \frac{c}{4\pi}\mathbf{E}\times\mathbf{B}$. Converting the surface integrals into volume integrals and using the continuity equation (3.37) and the Poynting theorem (see problem) yield

$$\varrho\frac{d}{dt}\left(\frac{v^2}{2} + \varepsilon\right) - \frac{\partial}{\partial x_j}(v_i P_{ij}) = \mathbf{J}\cdot\mathbf{E} + \nabla\cdot(\lambda\nabla T) + \mathbf{F}\cdot\mathbf{v}$$

Using the momentum equation (3.52) in the above equation gives

$$\varrho\,\frac{d\varepsilon}{dt} - \frac{p}{\varrho}\frac{d\varrho}{dt} = \varrho T\frac{ds}{dt} = \pi_{ij}\frac{\partial v_i}{\partial x_j} + \frac{c}{16\pi^2\sigma}(\nabla\times\mathbf{B})^2 + \nabla\cdot(\lambda\nabla T) \quad (3.57)$$

where s is the entropy per unit mass. The RHS of (3.57) can be shown to be positive definite.

The energy equation can be cast into the form of energy continuity equation:

$$\frac{\partial}{\partial t}\left(\frac{1}{2}\varrho v^2+\varrho\,\varepsilon+\frac{B^2}{8\pi}\right)+\nabla\cdot\left[\mathbf{v}\left(\frac{1}{2}\varrho v^2+\varrho\varepsilon\right)+\mathbf{N}\right]=\nabla\cdot(\mathbf{P}_{ij}\cdot\mathbf{v}+\lambda\nabla T) \tag{3.58}$$

The algebra leading to the above equation is left as an exercise.

• **Summary of MHD equations**

$$\frac{\partial\rho}{\partial t}+\nabla\cdot(\rho\mathbf{v})=0\text{: mass continuity equation} \tag{3.59}$$

$$\rho\frac{d\mathbf{v}}{dt}=\nabla\cdot P_{ij}+\frac{1}{c}\mathbf{J}\times\mathbf{B}+\rho\mathbf{F}\text{: momentum equation} \tag{3.60}$$

$$\mathbf{J}=\sigma\left(\mathbf{E}+\frac{1}{c}\mathbf{v}\times\mathbf{B}\right)\text{ : Ohm's law} \tag{3.61}$$

$$\nabla\times\mathbf{B}=\frac{4\pi}{c}\mathbf{J}\text{: Ampere's law} \tag{3.62}$$

$$\nabla\times\mathbf{E}=-\frac{1}{c}\frac{\partial\mathbf{B}}{\partial t}\text{: Faraday's law} \tag{3.63}$$

$$\frac{\partial\mathbf{B}}{\partial t}=\frac{c^2}{4\pi\sigma}\nabla^2\mathbf{B}+\nabla\times(\mathbf{v}\times\mathbf{B})\text{: equation of magnetic field} \tag{3.64}$$

Energy equation is often replaced by entropy conservation:

$$\frac{ds(p,\rho)}{dt}=0:\quad\text{isentropicity}$$

If the isentropicity is assumed, it would be consistent to discard all the dissipative terms in (3.60)–(3.64), in which case the pressure tensor term in (3.60) is replaced by $\nabla\cdot P_{ij}\to-\nabla p$, and $\sigma\to\infty$. Then, the resulting equations constitute the *ideal MHD equations.*

3.8 Kinematics of Magnetic Field in a Conducting Fluid

In a conducting MHD fluid of a very high conductivity ($\sigma\to\infty$), (3.55) reduces to

$$\frac{\partial\mathbf{B}}{\partial t}=\nabla\times(\mathbf{v}\times\mathbf{B})=(\mathbf{B}\cdot\nabla)\mathbf{v}-(\mathbf{v}\cdot\nabla)\mathbf{B}-\mathbf{B}\nabla\cdot\mathbf{v} \tag{3.65}$$

Substituting from the mass conservation equation (3.37) for the last term above, it can be written as

$$\frac{\partial \mathbf{B}}{\partial t} = (\mathbf{B}\cdot\nabla)\mathbf{v} - (\mathbf{v}\cdot\nabla)\mathbf{B} + \frac{\mathbf{B}}{\rho}\frac{d\rho}{dt}$$

$$\text{or} \quad \frac{d}{dt}\frac{\mathbf{B}}{\rho} = \frac{\mathbf{B}}{\rho}\cdot\nabla\,\mathbf{v} \tag{3.66}$$

In view of (3.33) which describes the way of convective change of material line element $d\mathbf{x}$, we see that $\frac{\mathbf{B}}{\rho}$ changes in the identical manner as $d\mathbf{x}$ changes. Therefore, if at $t = 0$ two vectors, $d\mathbf{x}$ and $\mathbf{B}/\rho$, are parallel, they will remain parallel, and their length will remain in the same ratio at later times. In other words, if at $t = 0$ two fluid particles belong to the same field line, then they will always belong to the same field line and the value of B/ρ will be proportional to the distance between the particles. Alfven described this by saying that in a conducting fluid of infinite conductivity the lines of magnetic field are *frozen* in the fluid and move with it (*frozen theorem*). An interesting consequence of the frozen behavior of a conducting fluid is that in a fluid of constant density, the magnetic field intensity is increased by the motion of *frozen particles* which lengthen the field lines.

Now, consider a portion of a tube of field lines (flux tube) in Fig. 3.4.

While it is carried along by the motion of the fluid, it will remain a portion of the flux tube at later instant. Since the two portions have boundaries composing of the same fluid particles, the mass contained in it remains the same always. So,

$$\rho_1 S_1 L_1 = \rho_2 S_2 L_2$$

Since the length L is proportional to B/ρ, we have

$$S_1 B_1 = S_2 B_2 \tag{3.67}$$

Thus, the magnetic flux over a material surface remains constant during the fluid motion. This is another way of expressing *Alfven's frozen theorem.*

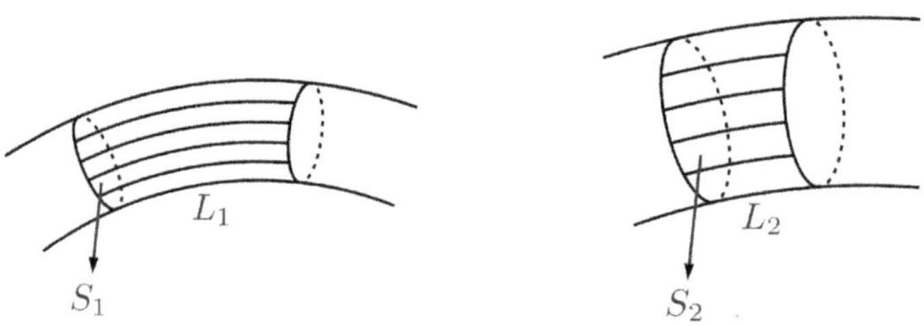

Fig. 3.4. Flux tubes.

- **Alternative way to prove the flux conservation law.**

$$\frac{d}{dt}\int_S \mathbf{B}\cdot d\mathbf{S} = \int_S \frac{d\mathbf{B}}{dt}\cdot d\mathbf{S} + \int_S \mathbf{B}\cdot\frac{d\mathbf{S}}{dt} = \int_S \left[\frac{\partial \mathbf{B}}{\partial t} - \nabla\times(\mathbf{v}\times\mathbf{B})\right]\cdot d\mathbf{S}$$

where we used (3.40) and S is an arbitrary material surface. Using (3.55) in the above gives

$$\frac{d}{dt}\int_S \mathbf{B}\cdot d\mathbf{S} = \frac{c^2}{4\pi\sigma}\int_S \nabla^2\mathbf{B}\cdot d\mathbf{S} \tag{3.68}$$

In the limit of $\sigma\to\infty$, the right-hand side vanishes, which means that the number of lines of forces passing through the material surface remains unchanged during the convective motion of fluid. In other words, any closed material contour cuts no field line in the course of motion. Any motion of the fluid perpendicular to the field lines carries them without breaking the field lines or having them slipped out, that is, fluid motion across $\mathbf{B}$ is prohibited. The motion of a low β plasma is carried along the magnetic field lines where

$$\beta = \frac{p}{B^2/8\pi} = \frac{\text{thermal kinetic energy density}}{\text{magnetic energy density}}$$

In high β plasma, the magnetic field line is convected by the fluid motion. Putting it another way, in high β plasma the disturbances are carried by the sound wave; in low β plasma the disturbances are carried by *Alfven wave.* Since any vector $\mathbf{A}$ can be resolved along two perpendicular directions as

$$\mathbf{A} = \mathbf{A}_{\parallel} + \mathbf{A}_{\perp} = (\hat{\mathbf{e}}\cdot\mathbf{A})\hat{\mathbf{e}} + \hat{\mathbf{e}}\times(\mathbf{A}\times\hat{\mathbf{e}})$$

the velocity component $\mathbf{v}_\perp$ perpendicular to the magnetic field is written as

$$\mathbf{v}_\perp = \frac{\mathbf{B}\times(\mathbf{v}\times\mathbf{B})}{B^2}$$

Substituting

$$\mathbf{E} = -\frac{1}{c}\mathbf{v}\times\mathbf{B} \tag{3.69}$$

gives

$$\mathbf{v}_\perp = \frac{c\mathbf{E}\times\mathbf{B}}{B^2} \equiv \mathbf{v}_D \tag{3.70}$$

which is called $\mathbf{E}\times\mathbf{B}$ *drift velocity.*

• **Alfven's frozen theorem** has a striking similarity with *Kelvin–Helmholtz theorem* of vorticity flux conservation in inviscid fluid, which states that the circulation or the vorticity flux is constant:

$$K = \oint_C \mathbf{v} \cdot d\mathbf{x} = \int_S \vec{\omega} \cdot d\mathbf{S} = \text{const.}$$

where the subscript C indicates *circulation* and $\vec{\omega} = \nabla \times \mathbf{v}$ is the vorticity.

When the electrical conductivity is finite, the magnetic flux passing through a material surface is not conserved, but slippage of field lines out of the contour takes place. There is a dimensionless number, analogous to Reynolds number in viscous fluid, which approximately determines whether the behavior of magnetic field line in the fluid is frozen or slipping. The fractional change of magnetic flux (Φ) in a time interval τ is

$$\frac{\frac{d\Phi}{dt}\tau}{\Phi} = O\left(\frac{c^2}{4\pi\sigma V L}\right)$$

with characteristic velocity V and length scale L. The magnetic Reynolds number R_m is defined by

$$R_m = \frac{4\pi\sigma V L}{c^2} \equiv \frac{VL}{\eta}$$

where $\eta = c^2/4\pi\sigma$ is called the magnetic viscosity (note the similarity to the kinematic viscosity in Reynolds number of a viscous fluid). When $R_m \gg 1$, the gross behavior of magnetic field lines can be viewed as frozen. In the opposite limit of $R_m \ll 1$, the second term of (3.62) can be neglected, so we have a diffusion equation for the magnetic field

$$\frac{\partial \mathbf{B}}{\partial t} = \eta \nabla^2 \mathbf{B}$$

whose quick solution can be written as

$$\mathbf{B} \sim e^{-\eta t/L^2}$$

which indicates that the magnetic field in a medium of finite conductivity suffers from dissipation with a timescale of L^2/η.

3.9 Fluid Description of Collisionless Plasma in a Strong Magnetic Field

Reference: G. F. Chew, M. L. Goldberger, and F. E. Low, *The Boltzmann equation and the one-fluid hydrodynamic equations in the absence of particle collisions*, Proc. Royal Soc. London **236A** 112.

A plasma can be described by MHD equations when the collision time is much shorter than any other characteristic times involved. If inter-particle collisions are frequent, the plasma particles tend to stick together and fluid description as a single fluid is justified. In fact, fluid equations can be derived from the Boltzmann equation assuming that the distribution function is nearly Maxwellian, which in turn requires frequent collisions in order for the fluid to be established as a local Maxwellian. In many important plasmas, the collision time is so long that collisions should be ignored. It would appear that for such collisionless plasma a fluid theory is not appropriate. However, in a plasma with a strong magnetic field such that the ion cyclotron period is much shorter than any other timescale of variation, the required localization can be provided by the magnetic field rather than collisions. The dominant perpendicular motion in a strong magnetic field is the $\mathbf{E} \times \mathbf{B}$ drift which is the same for both ions and electrons. Now, the pressure is anisotropic (the mean thermal kinetic energy per volume will be different for parallel and perpendicular directions). We start from the collisionless Boltzmann equation

$$\frac{\partial f}{\partial t} + \mathbf{v} \cdot \frac{\partial f}{\partial \mathbf{r}} + \frac{e}{m}\left(\mathbf{E} + \frac{1}{c}\mathbf{v} \times \mathbf{B}\right) \cdot \frac{\partial f}{\partial \mathbf{v}} = 0$$

for a particle species. The above Boltzmann equation can be expanded in powers of e/m. In the zero order (in a strong magnetic field, the cyclotron frequency eB/mc-term is the largest in the above), one has

$$\left(\mathbf{E} + \frac{1}{c}\mathbf{v} \times \mathbf{B}\right) \cdot \frac{\partial f_0}{\partial \mathbf{v}} = 0 \tag{3.71}$$

Now, introduce the new variable $\mathbf{u}$ such that

$$\mathbf{u} = \mathbf{v} - \mathbf{v}_d = \mathbf{v} - \frac{c\,\mathbf{E} \times \mathbf{B}}{B^2}$$

Then, (3.71) reduces to

$$(\mathbf{u} \times \mathbf{B}) \cdot \frac{\partial f_0}{\partial \mathbf{v}} = 0, \quad or \quad (\mathbf{u}_\perp \times \mathbf{B}) \cdot \frac{\partial f_0}{\partial \mathbf{v}} = 0 \tag{3.72}$$

where we assumed $\mathbf{E}_\parallel = 0$, and the symbols $\parallel$ and $\perp$ are referred to the direction of $\mathbf{B}$. Taking $\mathbf{B} = \hat{\mathbf{z}}B$, the above equation reads

$$u_y \frac{\partial f_0}{\partial u_x} - u_x \frac{\partial f_0}{\partial u_y} = 0$$

which is solved by $f_0 = f_0(u_x^2 + u_y^2)$. Then, the general solution of (3.71) is

$$f_0(\mathbf{r}, \mathbf{v}, t) = f_0(u_\perp^2, v_\parallel, \mathbf{r}, t) \tag{3.73}$$

One can show that the distribution function (3.73) gives for the average fluid velocity in (2.38)

$$\mathbf{V}(\mathbf{r}, t) = \mathbf{v}_d \tag{3.74}$$

Equation (3.74) can be easily obtained by using the anisotropic Maxwellian of the form $f_0 \sim e^{-\alpha u_\perp^2} e^{-\beta v_\parallel}$.

We can show that (3.73) makes the pressure tensor in (2.44) anisotropic. Upon introducing the unit vector $\hat{\mathbf{b}} = \frac{\mathbf{B}}{B} = (0, 0, 1)$, the pressure tensor takes the form (problem)

$$P_{ij} = p_\perp \, \delta_{ij} + (p_\parallel - p_\perp) b_i b_j \tag{3.75}$$

$$\mathbf{P_{ij}} = \begin{pmatrix} p_\perp & 0 & 0 \\ 0 & p_\perp & 0 \\ 0 & 0 & p_\parallel \end{pmatrix}$$

Therefore, one obtains

$$\frac{\partial P_{ij}}{\partial x_j} = \frac{\partial p_\perp}{\partial x_i} + \frac{\partial}{\partial x_j} \left[(p_\parallel - p_\perp) \frac{B_i B_j}{B^2} \right] = \nabla p_\perp + (\mathbf{B} \cdot \nabla) \left[\frac{(p_\parallel - p_\perp)\mathbf{B}}{B^2} \right]$$

Thus, the momentum equation in (3.52)) takes the form

$$\rho \frac{d\mathbf{V}}{dt} = -\nabla \left(p_\perp + \frac{B^2}{8\pi} \right) + (\mathbf{B} \cdot \nabla) \left[\frac{\mathbf{B}}{4\pi} - \frac{(p_\parallel - p_\perp)\mathbf{B}}{B^2} \right] \tag{3.76}$$

We need the equations for $p_\parallel$ and $p_\perp$. In the paper by Chew *et al.* (1956), the higher-order equations for the distribution function f were solved by equating the two heat fluxes $q_\parallel$ and $q_\perp$ to zero simultaneously (hence the name *double adiabatic theory*). Here, we use a little heuristic approach. The pressure p which is the trace of the pressure tensor is the average kinetic energy per unit volume. We use (2.48) and (3.75) with $\mathbf{q} = 0$ and

$$p = nT = p_\parallel + 2p_\perp$$

Then, we have

$$\frac{d}{dt} \left(p_\perp + \frac{p_\parallel}{2} \right) + \left(p_\perp + \frac{p_\parallel}{2} \right) \nabla \cdot \mathbf{V} = -p_\perp \nabla \cdot \mathbf{V} - (p_\parallel - p_\perp) b_i b_j \frac{\partial V_i}{\partial x_j}$$

or

$$\frac{d}{dt}(2p_\perp + p_\parallel) + (4p_\perp + p_\parallel) \nabla \cdot \mathbf{V} + 2(p_\parallel - p_\perp) b_i b_j \frac{\partial V_i}{\partial x_j} = 0$$

Since $p_\parallel$ and $p_\perp$ should be independent of each other, we can decompose the above equation as

$$\frac{dp_\parallel}{dt} + p_\parallel \nabla \cdot \mathbf{V} + 2p_\parallel b_i b_j \frac{\partial V_i}{\partial x_j} = 0 \tag{3.77}$$

$$\frac{dp_\perp}{dt} + 2p_\perp \nabla \cdot \mathbf{V} - p_\perp b_i b_j \frac{\partial V_i}{\partial x_j} = 0 \tag{3.78}$$

Using

$$\frac{d\mathbf{B}}{dt} = (\mathbf{B} \cdot \nabla)\mathbf{V} - \mathbf{B}\nabla \cdot \mathbf{V} \quad \text{or} \quad \frac{d\mathbf{B}}{dt} = (\mathbf{B} \cdot \nabla)\mathbf{V} + \frac{\mathbf{B}}{\rho}\frac{d\rho}{dt}$$

it follows that

$$b_i b_j \frac{\partial V_i}{\partial x_j} = \frac{\mathbf{B}}{B^2} \cdot (\mathbf{B} \cdot \nabla)\mathbf{V} = \frac{\mathbf{B}}{B^2} \cdot \left(\frac{d\mathbf{B}}{dt} - \frac{\mathbf{B}}{\rho}\frac{d\rho}{dt} \right)$$

Therefore, (3.77) and (3.78) become

$$\frac{dp_\parallel}{dt} - \frac{3p_\parallel}{\rho}\frac{d\rho}{dt} + \frac{2p_\parallel}{B}\frac{dB}{dt} = 0$$

$$\frac{dp_\perp}{dt} - \frac{p_\perp}{\rho}\frac{d\rho}{dt} - \frac{p_\perp}{B}\frac{dB}{dt} = 0$$

Or equivalently,

$$\frac{d}{dt}\left(\frac{p_\parallel B^2}{\rho^3} \right) = 0 \tag{3.79}$$

$$\frac{d}{dt}\left(\frac{p_\perp}{\rho B} \right) = 0 \tag{3.80}$$

Equations (3.79) and (3.80) are known as the double adiabatic equations for a conducting anisotropic fluid in a strong magnetic field.

3.10 Waves in Ideal MHD Flow

We now seek solutions of the *ideal* MHD equations for small amplitude linear waves. The *ideal* means that we neglect the dissipation, so we put the electrical conductivity σ to infinity in Ohm's law. The governing MHD equations are summarized in Section 3.7, which we rewrite as

$$\frac{\partial \rho}{\partial t} + \nabla \cdot (\rho \mathbf{v}) = 0 \tag{3.81}$$

$$\rho \frac{d\mathbf{v}}{dt} = -\nabla p + \frac{1}{c}\mathbf{J} \times \mathbf{B} \tag{3.82}$$

$$\mathbf{E} + \frac{1}{c}\mathbf{v} \times \mathbf{B} = 0 \tag{3.83}$$

$$\nabla \times \mathbf{E} = -\frac{1}{c}\frac{\partial \mathbf{B}}{\partial t} \tag{3.84}$$

$$\nabla \times \mathbf{B} = \frac{4\pi}{c}\mathbf{J} \tag{3.85}$$

$$\frac{d}{dt}\left(p\rho^{-\gamma}\right) = 0 \tag{3.86}$$

$$\frac{\partial \mathbf{B}}{\partial t} = \nabla \times (\mathbf{v} \times \mathbf{B}) \tag{3.87}$$

In the equilibrium state, the fluid is assumed at rest. We also assume the plasma is magnetized by a static magnetic field $\mathbf{B}_0$. Denoting perturbations associated with the wave by subscript 1, the zero-order equilibrium quantities by the subscript 0, we write

$$\mathbf{B} = \mathbf{B}_0 + \mathbf{B}_1(\mathbf{r}, t), \quad \rho = \rho_0 + \rho_1(\mathbf{r}, t), \quad \mathbf{v} = \mathbf{v}_1(\mathbf{r}, t), \quad p = p_0 + p_1(\mathbf{r}, t)$$

Linearizing the governing equations with respect to the perturbed quantities gives

$$\frac{\partial \rho_1}{\partial t} + \rho_0 \nabla \cdot \mathbf{v}_1 = 0 \tag{3.88}$$

$$\rho_0 \frac{\partial \mathbf{v}_1}{\partial t} + v_s^2 \nabla \rho_1 + \frac{\mathbf{B}_0}{4\pi} \times (\nabla \times \mathbf{B}_1) = 0 \tag{3.89}$$

$$\frac{\partial \mathbf{B}_1}{\partial t} - \nabla \times (\mathbf{v}_1 \times \mathbf{B}_0) = 0 \tag{3.90}$$

where v_s is the speed of sound:

$$v_s = \sqrt{\frac{\gamma p_0}{\rho_0}} \tag{3.91}$$

By eliminating ρ_1 and $\mathbf{B}_1$, these equations can be arranged in the form

$$\frac{\partial^2 \mathbf{v}_1}{\partial t^2} - v_s^2 \nabla\nabla \cdot \mathbf{v}_1 + \mathbf{v}_A \times \nabla \times [\nabla \times (\mathbf{v}_1 \times \mathbf{v}_A)] = 0 \tag{3.92}$$

where we introduced the Alfven velocity

$$\mathbf{v}_A = \frac{\mathbf{B}_0}{\sqrt{4\pi\rho_0}} \tag{3.93}$$

Assuming a plane wave solution for $\mathbf{v}_1$ with wave vector $\mathbf{k}$ and frequency ω, $\mathbf{v}_1(\mathbf{r}, t) = \mathbf{v}_1\, e^{i\mathbf{k}\cdot\mathbf{r} - i\omega t}$, (3.92) becomes

$$\mathbf{v}_1[-\omega^2 + (\mathbf{k} \cdot \mathbf{v}_A)^2] + \mathbf{k}[(v_s^2 + v_A^2)\mathbf{k} \cdot \mathbf{v}_1 - (\mathbf{k} \cdot \mathbf{v}_A)(\mathbf{v}_1 \cdot \mathbf{v}_A)] - \mathbf{v}_A(\mathbf{k} \cdot \mathbf{v}_1)(\mathbf{k} \cdot \mathbf{v}_A) = 0 \tag{3.94}$$

Let the coordinate axes be such that

$$\mathbf{B}_0 = \hat{\mathbf{z}} B_0, \quad \mathbf{k} = (k_x, 0, k_z)$$

Then, (3.94) is decomposed into the form

$$v_{1x}(-\omega^2 + k^2 v_A^2 + k_x^2 v_s^2) + v_{1z} k_x k_z v_s^2 = 0 \tag{3.95}$$

$$v_{1y}(-\omega^2 + k_z^2 v_A^2) = 0 \tag{3.96}$$

$$v_{1x} k_x k_z v_s^2 + v_{1z}(-\omega^2 + k_z^2 v_s^2) = 0 \tag{3.97}$$

Putting the determinant of the coefficients to zero, we obtain the normal modes of MHD wave:

$$(\omega^2 - k_z^2 v_A^2)\left[(\omega^2 - k_z^2 v_s^2)(-\omega^2 + k^2 v_A^2 + k_x^2 v_s^2) + v_s^4 k_x^2 k_z^2\right] = 0 \tag{3.98}$$

One immediate root gives the *Alfven wave*

$$\omega^2 = k_z^2 v_A^2 \tag{3.99}$$

The second factor in (3.98) is arranged in the form

$$\omega^4 - \omega^2 k^2 (v_s^2 + v_A^2) + k^2 k_z^2 v_s^2 v_A^2 = 0 \tag{3.100}$$

from which we find the *magnetosonic wave*

$$\omega_\pm^2 = \frac{k^2}{2}\left[v_s^2 + v_A^2 \pm \sqrt{(v_s^2 + v_A^2)^2 - 4 v_s^2 v_A^2 \cos^2\theta}\right] \tag{3.101}$$

where θ is the angle between the z-axis and $\mathbf{k}$ vector.

• **Conductivity tensor.** It sounds strange that we talk about *conductivity tensor* for media of infinite conductivity. The conductivity tensor σ_{ij} defines the proportionality between the current and the electric field:

$$J_i = \sigma_{ij} E_j \tag{3.102}$$

This relation is obtained from Maxwell equations even in collisionless plasmas such as Vlasov plasma or ideal MHD media. In the foregoing discussion, we eliminated $\mathbf{E}$ in the governing equations to analyze $\mathbf{B}$. Here, we do otherwise, in vice-versa way. We write the linearized ideal MHD equations again:

$$\frac{\partial \rho}{\partial t} + \rho_0 \nabla \cdot \mathbf{v} = 0 \tag{3.103}$$

$$\rho_0 \frac{\partial \mathbf{v}}{\partial t} = -\nabla p + \frac{1}{c} \mathbf{J} \times \mathbf{B}_0 \tag{3.104}$$

$$\mathbf{E} + \frac{1}{c} \mathbf{v} \times \mathbf{B}_0 = 0 \tag{3.105}$$

$$\nabla \times \mathbf{E} = -\frac{1}{c} \frac{\partial \mathbf{B}}{\partial t} \tag{3.106}$$

$$\nabla \times \mathbf{B} = \frac{4\pi}{c} \mathbf{J} \tag{3.107}$$

$$p = \frac{\gamma p_0}{\rho_0} \varrho \tag{3.108}$$

We eliminate $\mathbf{B}$ in (3.106) and (3.107) and write in terms of Fourier variables:

$$k^2 \mathbf{E} - \mathbf{k}(\mathbf{k} \cdot \mathbf{E}) = \frac{4\pi i \omega}{c^2} \mathbf{J} \tag{3.109}$$

Equations (3.103), (3.104), and (3.108) can be arranged in the form

$$\mathbf{v} = \frac{v_s^2}{\omega^2} \, \mathbf{k} \, \mathbf{k} \cdot \mathbf{v} + i \, \frac{\mathbf{J} \times \mathbf{B}_0}{c \, \omega \rho_0} \tag{3.110}$$

which yields in turn

$$\mathbf{v} = i \, \frac{\mathbf{J} \times \mathbf{B}_0}{c \, \omega \rho_0} + \frac{i}{c \, \omega \rho_0} \, \frac{v_s^2}{(\omega^2 - k^2 v_s^2)} \, \mathbf{k} \, \mathbf{J} \cdot (\mathbf{B}_0 \times \mathbf{k}) \tag{3.111}$$

We set up the coordinate axis in such a way that $\mathbf{B}_0 = \hat{\mathbf{z}} B_0$, $\mathbf{k} = \hat{\mathbf{x}} k_x + \hat{\mathbf{z}} k_z$. The latter designation of the wave vector $\mathbf{k}$ is not a loss of generality. Using (3.111) in (3.105) gives

$$\mathbf{E} = \frac{B_0^2}{i c^2 \rho_0 \omega} \left[-\mathbf{J} + \hat{\mathbf{z}} \mathbf{J} \cdot \hat{\mathbf{z}} \; - \frac{v_s^2}{\omega^2 - k^2 v_s^2} \, \mathbf{k} \times \hat{\mathbf{z}} \, \mathbf{J} \cdot (\mathbf{k} \times \hat{\mathbf{z}}) \right] \tag{3.112}$$

or

$$\mathbf{E} = \frac{4\pi i v_A^2}{c^2 \omega} \left[\hat{\mathbf{x}} J_x + \hat{\mathbf{y}} J_y \left(1 + \frac{k_x^2 v_s^2}{\omega^2 - k^2 v_s^2} \right) \right] \tag{3.113}$$

We have $E_z = 0$ as ascertained from (3.105) by dotting the equation by $\hat{\mathbf{z}}$.

- **Dispersion relation.** Equation (3.109) gives in components

$$\frac{4\pi i\omega}{c^2}J_x = k_z^2 E_x \tag{3.114}$$

$$\frac{4\pi i\omega}{c^2}J_y = k^2 E_y \tag{3.115}$$

$$\frac{4\pi i\omega}{c^2}J_z = -k_x k_z E_x \tag{3.116}$$

Substituting (3.114) and (3.115) into (3.113) yields the dispersion relations:

$$\omega^2 = v_A^2 k_z^2 \tag{3.117}$$

$$\omega^4 - \omega^2 k^2 (v_s^2 + v_A^2) + k^2 k_z^2 v_s^2 v_A^2 = 0 \tag{3.118}$$

The latter yields two magnetosonic waves as obtained in (3.101):

$$\omega_\pm^2 = \frac{k^2}{2}\left[v_s^2 + v_A^2 \pm \sqrt{(v_s^2+v_A^2)^2 - 4v_s^2 v_A^2 cos^2\theta}\right] \tag{3.119}$$

- Alternative analysis for conductivity tensor σ_{ij}:

$$J_i = \sigma_{ij}\ E_j \quad (i, j = x, y) \tag{3.120}$$

Here, we derive the conductivity tensor from the basic MHD equations (3.103)–(3.108). In components, (3.104) and (3.105) are written as

$$\omega\rho_0 v_x = k_x p + i\frac{B_0}{c}J_y \tag{3.121}$$

$$\omega\rho_0 v_y = -i\frac{B_0}{c}J_x \tag{3.122}$$

$$v_y = -\frac{c}{B_0}E_x \tag{3.123}$$

$$v_x = \frac{c}{B_0}E_y \tag{3.124}$$

Eliminating $v_{x,y}$ gives

$$\frac{c}{B_0}\omega\rho_0 E_y = k_x p + i\frac{B_0}{c}J_y \tag{3.125}$$

$$\frac{c}{B_0}\omega\rho_0 E_x = i\frac{B_0}{c}J_x \tag{3.126}$$

$$E_z = 0 \tag{3.127}$$

We need p expressed in terms of E_x and E_y. Taking divergence of (3.104) gives

$$-i\omega\rho_0\ \nabla\cdot\mathbf{v} = -\nabla^2 p - \frac{B_0}{4\pi}\nabla^2 B_z \tag{3.128}$$

On the other hand, (3.103) and (3.108) give

$$\nabla \cdot \mathbf{v} = \frac{i\omega}{c}\rho = \frac{i\omega}{\rho_0}\frac{p}{v_s^2} \tag{3.129}$$

Equations (3.128) and (3.129) give

$$p = \frac{cB_0}{4\pi\omega}\frac{v_s^2 k^2 k_x}{\omega^2 - k^2 v_s^2}E_y \tag{3.130}$$

where we used (3.106) to eliminate B_z. Using (3.130) in (3.125) yields

$$J_y = \frac{\omega}{4\pi i}\frac{c^2}{v_A^2}\left(1 - \frac{v_s^2 v_A^2 k^2 k_x^2}{\omega^2(\omega^2 - v_s^2 k^2)}\right)E_y \tag{3.131}$$

Equation (3.126) gives

$$J_x = \frac{\omega}{4\pi i}\frac{c^2}{v_A^2}E_x \tag{3.132}$$

The dielectric tensor elements of ideal MHD plasma are

$$\sigma_{xx} = \frac{\omega}{4\pi i}\frac{c^2}{v_A^2} \tag{3.133}$$

$$\sigma_{yy} = \frac{\omega}{4\pi i}\frac{c^2}{v_A^2}\left(1 - \frac{v_s^2 v_A^2 k^2 k_x^2}{\omega^2(\omega^2 - v_s^2 k^2)}\right) \tag{3.134}$$

and all other elements are zero. It is left as an exercise to show that (3.133) and (3.134) yield the dispersion relation in (3.100). The conductivity tensor elements in (3.133) and (3.134) can be obtained by inverting (3.113).

3.11 Derivation of MHD Waves from Two-Fluid Equations

In this section, we derive the MHD waves from two-fluid equations, taking the low-frequency approximation, neglecting ω/ω_{ci} as compared with unity, and neglecting also m_e/m_i, the mass ratio, as compared with unity. Thus, we indirectly show that the MHD waves are a subset of more wide spectrum of plasma waves predicted by two-fluid equations. We start from linearized two-fluid equations for electron (e) and ion (i) species:

$$\mathbf{v}_i \times \vec{\omega}_{ci} = -i\omega\mathbf{v}_i - \frac{e}{m_i}\mathbf{E} \tag{3.135}$$

$$\mathbf{v}_e \times \vec{\omega}_{ce} = -i\omega\mathbf{v}_e + \frac{e}{m_e}\mathbf{E} + i\mathbf{k}\beta^2\frac{n_e}{N} \tag{3.136}$$

where

$$\vec{\omega}_{ci} = \frac{e\mathbf{B}_0}{m_i c}, \quad \vec{\omega}_{ce} = -\frac{e\mathbf{B}_0}{m_e c}, \quad \beta^2 = \frac{\gamma_e T_e}{m_e} \tag{3.137}$$

where we assumed that ions are cold. The essential features of our development are not impaired by the assumption of cold ions. The current is

$$\mathbf{J} = Ne(\mathbf{v}_i - \mathbf{v}_e) \tag{3.138}$$

where N is the equilibrium number density of the plasma. The above equations will be combined to derive the expressions that are valid in the MHD region by taking the low-frequency approximation ($\frac{\omega}{\omega_{ci}} \ll 1$) and by filtering out the high mobility terms ($\frac{m_e}{m_i} \ll 1$).

Let us multiply (3.136) by m_e/m_i and add to (3.135) to obtain

$$\mathbf{J} \times \overrightarrow{\omega_{ci}} = -ieN\omega\mathbf{w} + ie\mathbf{k}v_s^2 n_e \tag{3.139}$$

$$v_s^2 = \frac{m_e}{m_i}\beta^2 = \frac{\gamma_e T_e}{m_i}$$

$$\mathbf{w} = \mathbf{v}_i + \frac{m_e}{m_i}\mathbf{v}_e \tag{3.140}$$

where we used $\frac{m_e}{m_i}\overrightarrow{\omega_{ce}} = -\overrightarrow{\omega_{ci}}$. The first approximation to be enforced for the low-frequency slow waves in the MHD region is to take $\frac{m_e}{m_i} \to 0$, and we can set $\mathbf{w} \approx \mathbf{v}_i$. Therefore, (3.140) is written as

$$\mathbf{J} \times \overrightarrow{\omega_{ci}} = -ieN\omega\mathbf{v}_i + ie\mathbf{k}v_s^2 n_e \tag{3.141}$$

We look at the ion equation, (3.135). Since $\omega \ll \omega_{ci}$, the inertial term can be neglected as compared with the magnetic field term, and we have

$$\mathbf{E} + \frac{1}{c}\mathbf{v}_i \times \mathbf{B}_0 = 0 \tag{3.142}$$

which is Ohm's law. An immediate consequence of the above equation is $E_z = 0$. This approximate relation greatly simplifies the analysis.

Solving for $\mathbf{v_i}$, we have

$$\mathbf{v}_i = \frac{e}{m_i}\frac{1}{\omega_{ci}^2}\mathbf{E} \times \overrightarrow{\omega_{ci}} \tag{3.143}$$

Now, we use the charge neutrality $n_e \simeq n_i$, which is not only physically acceptable but also mathematically legitimate, because (3.109) gives $\mathbf{k} \cdot \mathbf{J} = 0$, and thus, $\mathbf{k} \cdot \mathbf{v}_i = \mathbf{k} \cdot \mathbf{v}_e$. Therefore, we have in place of (3.140)

$$\mathbf{J} \times \overrightarrow{\omega_{ci}} = -iNe\omega\mathbf{v}_i + ie\mathbf{k}v_s^2\frac{N}{\omega}\mathbf{k} \cdot \mathbf{v}_i \tag{3.144}$$

Equations (3.109) and (3.143) and the ion equation (3.142) constitute the complete set for the low-frequency two-fluid plasma. If we rewrite (3.143) to reveal $\mathbf{B}_0$, it takes the form

$$\mathbf{v}_i = \frac{i\mathbf{J} \times \mathbf{B}_0}{c\omega N m_i} + \mathbf{k}\frac{v_s^2}{\omega^2}\mathbf{k} \cdot \mathbf{v}_i \tag{3.145}$$

This equation corresponds to (3.110) for ideal MHD plasma. Solving (3.144) for $\mathbf{v}_i$ yields

$$\mathbf{v}_i = \frac{i}{c\omega N m_i}\left(\mathbf{J} \times \mathbf{B}_0 + \frac{v_s^2}{\omega^2 - k^2 v_s^2}\mathbf{k}\mathbf{J} \cdot \mathbf{B}_0 \times \mathbf{k}\right) \tag{3.146}$$

Using Equations (3.141) and (3.145) yields $\mathbf{E}$ in terms of $\mathbf{J}$:

$$\mathbf{E} = \frac{iB_0^2}{c^2\omega N m_i}\left[\hat{\mathbf{x}}J_x + \hat{\mathbf{y}}J_y\left(1 + \frac{k_x^2 v_s^2}{\omega^2 - k^2 v_s^2}\right)\right] \tag{3.147}$$

Equation (3.109) with $E_z = 0$ and (3.146) gives the dispersion relations

$$\omega^2 = v_A^2 k_z^2 \qquad \left(v_A^2 = \frac{B_0^2}{4\pi N m_i}\right) \tag{3.148}$$

$$\omega^4 - \omega^2 k^2 (v_s^2 + v_A^2) + k^2 k_z^2 v_s^2 v_A^2 = 0 \tag{3.149}$$

Note that (3.146) is equivalent to (3.111) and (3.147) is equivalent to (3.112).

Dispersion relation (3.148) indicates that the Alfven wave is a wave which propagates along the direction of the zero-order magnetic field $\mathbf{B_0}$: $\mathbf{k}_A = \hat{\mathbf{z}}k_z$ (subscript A denotes Alfven wave). We have as per (3.113) $\mathbf{E} \perp \mathbf{k}_A$. Also, $\mathbf{B} = \frac{c}{\omega}\mathbf{k}_A \times \mathbf{E}$ as per (3.106). Therefore, the Alfven wave is a low-frequency electromagnetic wave which propagates along the zero-order magnetic field. The magnetosonic waves are of mixed mode; partly electromagnetic, partly electrostatic. *Electromagnetic wave* is often called *transverse electromagnetic wave* to emphasize the fact that the electric and magnetic fields are mainly or entirely perpendicular to the the direction of the propagation vector. When the electric field is primarily along the direction of wave propagation, the wave is called *electrostatic* or *longitudinal.*

Problems

3.1 Show that (3.57) can be cast into the form of energy continuity equation (3.58).

3.2 **Conservation of angular momentum.** Show that the principle of conservation of angular momentum leads to the symmetry of the pressure tensor.

Proof. Equating the rate of convective change of angular momentum possessed by the fluid enclosed inside a material surface S to the torque exerted on the same volume of the fluid, we write

$$\frac{d}{dt}\int_V \mathbf{x}\times\mathbf{v}\varrho_m\, d\tau = \int_V \varrho_m\mathbf{x}\times\mathbf{F}d\tau + \int_S \mathbf{x}\times\mathbf{P}_{ij}\cdot d\mathbf{S} \qquad \text{(A)}$$

where the pressure tensor $\mathbf{P}_{ij}$ is understood as the mechanical pressure tensor in ordinary fluid or mechanical plus Maxwell stress tensor in MHD flow. The last term is the torque exerted on the surface of the system through the surface force. Carrying out the convective differentiation and using the continuity equation and $\frac{d\mathbf{x}}{dt}\times\mathbf{v}=0$, the LHS of Eq. (A) becomes

$$\int_V \varrho_m\mathbf{x}\times\frac{d\mathbf{v}}{dt}d\tau = \int_V \mathbf{x}\times(\varrho_m\mathbf{F}+\nabla\cdot\mathbf{P})d\tau$$

Thus, we are left with

$$\int_V \mathbf{x}\times\nabla\cdot\mathbf{P}_{ij}d\tau = \int_S \mathbf{x}\times\mathbf{P}_{ij}\cdot d\mathbf{S}$$

which gives

$$\int_V e_{ijk}x_j\frac{\partial}{\partial x_l}P_{kl}d\tau = \int_S e_{ijk}x_jP_{kl}dS_l = \int_V e_{ijk}\frac{\partial}{\partial x_l}(x_jP_{kl})d\tau = \int_V e_{ijk}\left(P_{kj}+x_j\frac{P_{kl}}{\partial x_l}\right)d\tau$$

Thus, we have

$$e_{ijk}P_{kj}=0 \quad \text{i.e. } P_{kj}=P_{jk}$$

3.3 **Poynting theorem.** Prove that the Poynting vector

$$\mathbf{N}=\frac{c}{4\pi}\mathbf{E}\times\mathbf{B}$$

satisfies

$$\frac{\partial}{\partial t}\frac{B^2}{8\pi}+\nabla\cdot\mathbf{N}=-\mathbf{J}\cdot\mathbf{E}$$

3.4 Verify Eq. (3.86).

References

Alfven, H. and C. -G. Fälthammar, *Cosmical Electrodynamics*, Clarendon, Oxford (1963).

Braginskii, S. I. *Transport phenomena in a completely ionized Two-temperature plasma,* Soviet Phys. JETP, **6** 358 (1958).

Braginskii, S. I. *Transport Process in a Plasma,* in Reviews of plasma physics Vol. 1: Consultant's Bureau, New York (1965).

Chew, G. F., M. L. Goldberger, and F. E. Low, *The Boltzmann equation and the one-fluid hydromagnetic equations in the absence of particle collisions,* Proc. Royal Soc. London, **236 A**,112 (1956).

Landau, L. D. and E. M. Lifshitz, *Electrodynamics of Continuous Media,* Pergamon, London (1966).

Lee, H-.J., J. F. McKenzie, and W. I. Axford *Discontinuous transitions in a current carrying plasma,* Astrophys. and Space Sci. **51** pp. 3-32 (1977).

Shkarofsky, I. P., T. W. Johnston, and M. P. Bachynski, *The Particle Kinetics of Plasmas,* Addison-Wesley, London (1966).

Chapter 4

Variational Principle of Hydrodynamics

4.1 Mathematical Preliminary

Before we embark on the application of the variational principle to continuum media like electromagnetism and hydrodynamics, it may be necessary to introduce some mathematical background on variational calculus.

4.1.1 *Variational calculus*

To begin with, we consider an integral expression of the simplest type of *functional*

$$I = \int_{x_1}^{x_2} F(x, y, y')dx \quad \left(y' = \frac{dy}{dx}\right) \tag{4.1}$$

We assume that F has continuous second partial derivatives with respect to the three variables x, y, and y'. Also $y(x)$ is a continuously differentiable function. These assumptions will not be reminded again in the continuation. We call F a *functional* since it contains a function $y(x)$. We pose the following question: for what function $y(x)$ is the integral I *stationary*? Here the word stationary means that I is maximal, or minimal, or inflective. In terms of the variational symbol δ, that I is stationary means $\delta I = 0$ whose mathematical or operational meaning will be elaborated in the following. Let $y_0(x)$ denote the function that corresponds to the stationary value of I, I_0:

$$I_0 = \int_{x_1}^{x_2} F(x, y_0, y_0')dx \tag{4.2}$$

Now we choose another path $y(x)$ which is only infinitesimally different from $y_0(x)$ over the whole range of x:

$$y(x) = y_0(x) + \varepsilon\eta(x) \tag{4.3}$$

where ε is an infinitesimal constant, and we impose the condition $\eta(x_1) = \eta(x_2) = 0$ so that the *varied path* $y(x)$ coincides with the stationary path $y_0(x)$ at the end points of the integral. Then the variation symbol δ means

$$\begin{aligned}\delta I = I - I_0 &= \int_{x_1}^{x_2} F(x, y, y')dx - \int_{x_1}^{x_2} F(x, y_0, y_0')dx \\ &= \int_{x_1}^{x_2} F(x, y_0 + \varepsilon\eta(x), y_0' + \varepsilon\eta'(x))dx - \int_{x_1}^{x_2} F(x, y_0, y_0')dx\end{aligned} \tag{4.4}$$

where $\varepsilon \to 0$ is implied. The integrand in the first term is expanded in the powers of the small quantity ε to get

$$F(x, y_0 + \varepsilon\eta(x), y_0' + \varepsilon\eta'(x)) = F(x, y_0, y_0') + \frac{\partial F}{\partial y_0}\varepsilon\eta(x) + \frac{\partial F}{\partial y_0'}\varepsilon\eta'(x) \tag{4.5}$$

Hence

$$\delta I = \varepsilon \int_{x_1}^{x_2} \left[\frac{\partial F}{\partial y_0}\eta(x) + \frac{\partial F}{\partial y_0'}\eta'(x)\right] dx \tag{4.6}$$

The necessary condition for I to be stationary is $\delta I = 0$, which takes the form

$$\int_{x_1}^{x_2} \left[\frac{\partial F}{\partial y_0}\eta(x) + \frac{\partial F}{\partial y_0'}\eta'(x)\right] dx = 0 \tag{4.7}$$

The second term in the above equation is integrated by parts to yield (omitting the subscript 0)

$$\int_{x_1}^{x_2} \left[\frac{\partial F}{\partial y} - \frac{d}{dx}\frac{\partial F}{\partial y'}\right]\eta(x)dx + \left[\frac{\partial F}{\partial y'}\eta(x)\right]_{x_1}^{x_2} = 0 \tag{4.8}$$

Since $\eta(x)$ vanishes at the end points, by assumption, the integrated term vanishes. Finally, since $\eta(x)$ is arbitrary, the coefficient of η in the integrand must vanish identically over (x_1, x_2). [This fact, which is intuitively plausible, can be proved analytically. See Weinstock.] Therefore, if $y(x)$ minimizes (or maximizes) the integral (4.1), it must satisfy the *Euler equation*

$$\frac{d}{dx}\left(\frac{\partial F}{\partial y'}\right) - \frac{\partial F}{\partial y} = 0 \tag{4.9}$$

It should be kept in mind that d/dx is a *total derivative*, and should be expanded in the following forms:

$$\frac{dF}{dx} = \frac{\partial F}{\partial x} + F_y y' + F_{y'} y'' \tag{4.10}$$

$$\frac{d}{dx}\left(\frac{\partial F}{\partial y'}\right) = \frac{\partial}{\partial x}\left(\frac{\partial F}{\partial y'}\right) + \frac{\partial}{\partial y}\left(\frac{\partial F}{\partial y'}\right)\frac{dy}{dx} + \frac{\partial}{\partial y'}\left(\frac{\partial F}{\partial y'}\right)\frac{dy'}{dx} \tag{4.11}$$

where the subscript denotes the differentiation. We can easily prove, with the aid of the above relations, (4.9) is equivalent to

$$\frac{1}{y'}\left[\frac{d}{dx}\left(F - \frac{\partial F}{\partial y'}\frac{dy}{dx}\right) - \frac{\partial F}{\partial x}\right] = 0$$

by working backward. From the above, it follows that if F does not involve x explicitly the first integral of the Euler equation is

$$F - y'\frac{\partial F}{\partial y'} = C \quad \text{if } \frac{\partial F}{\partial x} = 0 \tag{4.12}$$

Equation (4.9) shows that if F does not involve y explicitly the first integral is

$$\frac{\partial F}{\partial y'} = C \quad \text{if } \frac{\partial F}{\partial y} = 0 \tag{4.13}$$

- **δ-notation.** For a functional expressed as a definite integral

$$I = \int_{x_1}^{x_2} F(x, y, y')dx \tag{4.14}$$

where x is the independent variable in F, the variation symbol δ is used as

$$\delta I = \delta \int_{x_1}^{x_2} F(x, y, y')dx = \int_{x_1}^{x_2} \delta F(x, y, y')dx \tag{4.15}$$

by interchanging the integral and δ. In view of (4.6), we agree with putting $\delta x = 0$ for the variation of the independent variable x. And we have the operational equality:

$$\delta F = \frac{\partial F}{\partial y}\delta y + \frac{\partial F}{\partial y'}\delta y' = \frac{\partial F}{\partial y}\varepsilon\eta + \frac{\partial F}{\partial y'}\varepsilon\eta' \tag{4.16}$$

where the second equality is due to (4.6). Thus we put

$$\delta x = 0, \quad \delta y = \varepsilon\eta(x), \quad \delta y' = \varepsilon\eta'(x) \tag{4.17}$$

It is immediately seen that δ and d/dx commute:

$$\frac{d}{dx}\delta y = \varepsilon\frac{d\eta}{dx} = \delta y' = \delta\frac{dy}{dx} \tag{4.18}$$

The differential of a function is a first-order approximation to the change in that function along a particular curve, while the variation of a functional is a first-order approximation to the change from curve to curve. Continuing from (4.15) and (4.16), we write in terms of the δ-symbol

$$\delta I = \int_{x_1}^{x_2}\left[\frac{\partial F}{\partial y}\delta y + \frac{\partial F}{\partial y'}\delta y'\right]dx \tag{4.19}$$

We replace $\delta y'$ by $d(\delta y)/dx$, and integrate by parts to obtain

$$\delta I = \int_{x_1}^{x_2}\left[\frac{\partial F}{\partial y} - \frac{d}{dx}\left(\frac{\partial F}{\partial y'}\right)\right]\delta y dx + \left[\frac{\partial F}{\partial y'}\delta y\right]_{x_1}^{x_2} \tag{4.20}$$

Equations (4.8) and (4.20) are essentially identical.

The δ notation renders the variational calculus easy and straight if one remembers a few things. The symbol δ applies to the change from curve to curve, so we have δy or $\delta y'$, not δx. The usual differential symbol d applies to *a curve*, thus to the independent variable x, so we have only dx. The operator $d\delta(\cdots) = \delta d(\cdots)$, commutative. The following algebraic steps demonstrate the easy use of δ notation in variational calculus.

$$\begin{aligned}
I &= \int_{x_1}^{x_2} F(x, y, y')dx \\
\delta I &= \int_{x_1}^{x_2} \delta F(x, y, y')dx \\
&= \int_{x_1}^{x_2}\left[\frac{\partial F}{\partial y}\delta y + \frac{\partial F}{\partial y'}\delta y'\right]dx \\
&= \int_{x_1}^{x_2}\left[\frac{\partial F}{\partial y}\delta y + \frac{\partial F}{\partial y'}\delta\frac{dy}{dx}\right]dx \\
&= \int_{x_1}^{x_2}\left[\frac{\partial F}{\partial y}\delta y + \frac{\partial F}{\partial y'}\frac{d}{dx}\delta y\right]dx \\
&= \int_{x_1}^{x_2}\left[\frac{\partial F}{\partial y} - \frac{d}{dx}\frac{\partial F}{\partial y'}\right]\delta y dx + \left[\frac{\partial F}{\partial y'}\delta y\right]_{x_1}^{x_2}
\end{aligned}$$

4.1.2 *Euler–Lagrange equation*

In this subsection, we only summarize the variational integral for a functional of several functions and several independent variables. Readers who are interested in the variational calculus for the extended variational integral should consult the references furnished at the end of this chapter.

Suppose we are given the variational principle

$$\delta \int F(\mathbf{r}, t, a(\mathbf{r}, t), a_x, a_y, a_z, a_t, b(\mathbf{r}, t), b_x, b_y, b_z, b_t, c(\mathbf{r}, t), c_x, c_y, c_z, c_t \cdots) dxdydzdt = 0 \tag{4.21}$$

where $\mathbf{r} = (x, y, z)$, and the subscripts x, y, z, t denote differentiation. We also assume that the variations vanish at the boundaries of the region of integration. Then the Euler–Lagrange equation reads

$$\begin{aligned}
&\frac{\partial}{\partial x}\frac{\partial F}{\partial a_x} + \frac{\partial}{\partial y}\frac{\partial F}{\partial a_y} + \frac{\partial}{\partial z}\frac{\partial F}{\partial a_z} + \frac{\partial}{\partial t}\frac{\partial F}{\partial a_t} - \frac{\partial F}{\partial a} = 0\\
&\frac{\partial}{\partial x}\frac{\partial F}{\partial b_x} + \frac{\partial}{\partial y}\frac{\partial F}{\partial b_y} + \frac{\partial}{\partial z}\frac{\partial F}{\partial b_z} + \frac{\partial}{\partial t}\frac{\partial F}{\partial b_t} - \frac{\partial F}{\partial b} = 0\\
&\frac{\partial}{\partial x}\frac{\partial F}{\partial c_x} + \frac{\partial}{\partial y}\frac{\partial F}{\partial c_y} + \frac{\partial}{\partial z}\frac{\partial F}{\partial c_z} + \frac{\partial}{\partial t}\frac{\partial F}{\partial c_t} - \frac{\partial F}{\partial c} = 0\\
&\qquad\qquad\cdots\cdots\cdots\cdots
\end{aligned} \tag{4.22}$$

• **Constraints and Lagrange undetermined multiplier.** The constraints are given in the form

$$\begin{aligned}
&\phi_1(\mathbf{r}, t, a(\mathbf{r}, t), a_x, a_y, a_z, a_t, b(\mathbf{r}, t), b_x, b_y, b_z, b_t, c(\mathbf{r}, t), c_x, c_y, c_z, c_t \cdots) = 0\\
&\phi_2(\mathbf{r}, t, a(\mathbf{r}, t), a_x, a_y, a_z, a_t, b(\mathbf{r}, t), b_x, b_y, b_z, b_t, c(\mathbf{r}, t), c_x, c_y, c_z, c_t \cdots) = 0\\
&\qquad\qquad\cdots\cdots\cdots\\
&\phi_m(\mathbf{r}, t, a(\mathbf{r}, t), a_x, a_y, a_z, a_t, b(\mathbf{r}, t), b_x, b_y, b_z, b_t, c(\mathbf{r}, t), c_x, c_y, c_z, c_t \cdots) = 0
\end{aligned} \tag{4.23}$$

Construct the new function G with Lagrange multiplier $\lambda_j(\mathbf{r}, t)$

$$G = F + \sum_{j=1}^{m} \lambda_j \phi_j \tag{4.24}$$

Then the Euler–Lagrange equations with the constraints incorporated are

$$\frac{\partial}{\partial x}\frac{\partial G}{\partial a_x}+\frac{\partial}{\partial y}\frac{\partial G}{\partial a_y}+\frac{\partial}{\partial z}\frac{\partial G}{\partial a_z}+\frac{\partial}{\partial t}\frac{\partial G}{\partial a_t}-\frac{\partial G}{\partial a}=0$$
$$\frac{\partial}{\partial x}\frac{\partial G}{\partial b_x}+\frac{\partial}{\partial y}\frac{\partial G}{\partial b_y}+\frac{\partial}{\partial z}\frac{\partial G}{\partial b_z}+\frac{\partial}{\partial t}\frac{\partial G}{\partial b_t}-\frac{\partial G}{\partial b}=0$$
$$\frac{\partial}{\partial x}\frac{\partial G}{\partial c_x}+\frac{\partial}{\partial y}\frac{\partial G}{\partial c_y}+\frac{\partial}{\partial z}\frac{\partial G}{\partial c_z}+\frac{\partial}{\partial t}\frac{\partial G}{\partial c_t}-\frac{\partial G}{\partial c}=0$$
$$\cdots\cdots\cdots\cdots \tag{4.25}$$

Equations (4.25) are the *necessary conditions* for the variational principle in (4.21) to be satisfied with the constraints specified in (4.23).

Example. We shall consider the variational principle of entropy at equilibrium in (2.97).

$$\delta S=\delta\int f(t)\ln f(t)d^3vdt=0 \tag{4.26}$$

with the constraints:

$$\phi_1=m=\text{const.} \tag{4.27}$$

$$\phi_2=m\mathbf{v}=\text{const.} \tag{4.28}$$

$$\phi_3=\frac{m}{2}v^2=\text{const.} \tag{4.29}$$

In the above formulas (4.22) and (4.25), we identify

$$F=f(t)\ln f(t),\quad a(t)=f(t) \tag{4.30}$$

Let us construct

$$G=f(t)\ln f(t)+\lambda_1 mf+\vec{\lambda}_2\cdot m\mathbf{v}f+\lambda_3\frac{m}{2}v^2f \tag{4.31}$$

In the formula (4.25), we have $\partial G/\partial a=\partial G/\partial f=0$, which gives

$$\ln f+1+m\lambda_1+\vec{\lambda}_2\cdot m\mathbf{v}+\lambda_3\frac{m}{2}v^2=0 \tag{4.32}$$

from which f can be put into the Maxwellian in (2.30).

4.2 Lagrangian vs. Eulerian Description of Hydrodynamics

The laws of hydrodynamics can be formulated by variational principle similarly as classical mechanics is formulated by variational principle. For an ideal fluid, variational formulation of hydrodynamic equations is always possible. There are two descriptions of fluid motion in hydrodynamics. Lagrangian form is more closely related to classical mechanics. Eulerian form is more similar to field theory. A moving fluid is described by five quantities, $\mathbf{v}(\mathbf{x},t)$, $\rho(\mathbf{x},t)$, $p(\mathbf{x},t)$: velocity, density, and pressure. From p and ρ, other thermodynamic quantities can be determined, if the equation of state is known. The Lagrangian derivative of fluid velocity reads in terms of the Eulerian derivative as follows.

$$\frac{d\mathbf{v}}{dt} = \frac{\partial \mathbf{v}}{\partial t} + \mathbf{v} \cdot \nabla \mathbf{v} \tag{4.33}$$

In the Lagrangian description, the motion of a particular fluid particle is followed in time. Let $\mathbf{x}_0$ be the initial position: $\mathbf{x}_0 = \mathbf{x}(t)|_{t=0}$. Then the fluid configuration can be represented by the vectors

$$\mathbf{x} = \mathbf{x}(\mathbf{x}_0, t) \tag{4.34}$$

For a *laminar* fluid (non-turbulent), the continuum approximation is valid.

Velocity:

$$\mathbf{v}(\mathbf{x}_0, t) = \left(\frac{\partial \mathbf{x}}{\partial t}\right)_{\mathbf{x}_0} \tag{4.35}$$

Acceleration:

$$\mathbf{a}(\mathbf{x}_0, t) = \left(\frac{\partial^2 \mathbf{x}}{\partial t^2}\right)_{\mathbf{x}_0} \tag{4.36}$$

where $\left(\frac{\partial}{\partial t}\right)_{\mathbf{x}_0} = \frac{d}{dt}$ in (4.33).

The Lagrangian coordinate, the initial position vector $\mathbf{x}_0$, can be regarded as individualization parameter of each particle. But other means of individualization can be employed depending upon the problems at hand. The function $\mathbf{x}$ in (4.34) is a single-valued function according to the law of mechanics. So the Jacobian determinant

$$\mathbf{J} = \left|\frac{\partial \mathbf{x_i}}{\partial \mathbf{x_{0j}}}\right| = \begin{vmatrix} \dfrac{\partial x_1}{\partial x_{01}} & \dfrac{\partial x_1}{\partial x_{02}} & \dfrac{\partial x_1}{\partial x_{03}} \\ \dfrac{\partial x_2}{\partial x_{01}} & \dfrac{\partial x_2}{\partial x_{02}} & \dfrac{\partial x_2}{\partial x_{03}} \\ \dfrac{\partial x_3}{\partial x_{01}} & \dfrac{\partial x_3}{\partial x_{02}} & \dfrac{\partial x_3}{\partial x_{03}} \end{vmatrix}$$

is not zero, i.e. (4.34) can be solved for $\mathbf{x}_0$ in the form of single-valued continuous functions:

$$\mathbf{x}_0 = \mathbf{x}_0(\mathbf{x}, t) \tag{4.37}$$

Mathematically, the transformation between the Eulerian coordinates $(\mathbf{x}, t)$ and the Lagrangian coordinates $(\mathbf{x}_0, \tau)$ is effected by the relations

$$t = t(\mathbf{x}_0, \tau), \quad \mathbf{x} = \mathbf{x}(\mathbf{x}_0, \tau) \tag{4.38}$$

$$\tau = \tau(\mathbf{x}, t), \quad \mathbf{x}_0 = \mathbf{x}_0(\mathbf{x}, t) \tag{4.39}$$

Specifically, the appropriate changes of variables are

$$t = \tau \tag{4.40}$$

$$\left(\frac{\partial \mathbf{x}}{\partial \tau}\right)_{\mathbf{x}_0} = \mathbf{v}(\mathbf{x}_0, \tau) \tag{4.41}$$

$$\mathbf{x}(\mathbf{x}_0, \tau = 0) = \mathbf{x}_0 \tag{4.42}$$

The time transformation is identity transformation, but it is more convenient to introduce the Lagrangian time τ to distinguish the Lagrangian derivative $\partial/\partial\tau$ from $\partial/\partial t$. The spatial transformation is defined by differential equation (4.41) with (4.42) serving as the initial condition. As a matter of fact, the functional relations between the Eulerian and the Lagrangian coordinates as expressed in (4.38) and (4.39) are determined only after solving the law of mechanics.

Integrating (4.41) gives

$$\mathbf{x}(\mathbf{x}_0, \tau) = \mathbf{x}_0 + \int_0^{\tau} \mathbf{v}(\mathbf{x}_0, \tau')d\tau' \tag{4.43}$$

By the chain rule of differentiation, we have

$$\frac{\partial}{\partial \tau} = \frac{\partial t}{\partial \tau}\frac{\partial}{\partial t} + \frac{\partial x_i}{\partial \tau}\frac{\partial}{\partial x_i} = \frac{\partial}{\partial t} + \mathbf{v} \cdot \nabla \tag{4.44}$$

$$\frac{\partial}{\partial t} = \frac{\partial \tau}{\partial t}\frac{\partial}{\partial \tau} + \frac{\partial x_{0i}}{\partial t}\frac{\partial}{\partial x_{0i}} = \frac{\partial}{\partial \tau} + \frac{\partial x_{0i}}{\partial t}\frac{\partial}{\partial x_{0i}} \tag{4.45}$$

The Lagrangian time derivative is the convective derivative in Eulerian description. A simple application of (4.44) is the particle acceleration,

$$\mathbf{a} = \left(\frac{\partial \mathbf{v}}{\partial \tau}\right)_{\mathbf{x}_0} = \frac{\partial \mathbf{v}}{\partial t} + \mathbf{v} \cdot \nabla \mathbf{v} \tag{4.46}$$

For spatial coordinates, we have

$$\frac{\partial}{\partial x_{0i}} = \frac{\partial t}{\partial x_{0i}}\frac{\partial}{\partial t} + \frac{\partial x_j}{\partial x_{0i}}\frac{\partial}{\partial x_j} = \frac{\partial x_j}{\partial x_{0i}}\frac{\partial}{\partial x_j} \tag{4.47}$$

$$\frac{\partial}{\partial x_i} = \frac{\partial \tau}{\partial x_i}\frac{\partial}{\partial \tau} + \frac{\partial x_{0j}}{\partial x_i}\frac{\partial}{\partial x_{0j}} = \frac{\partial x_{0j}}{\partial x_i}\frac{\partial}{\partial x_{0j}} \tag{4.48}$$

Let us consider the case of one-dimensional Eulerian equations

$$\frac{\partial \rho}{\partial t} + \frac{\partial}{\partial x}(\rho v) = 0 \tag{4.49}$$

$$\frac{\partial v}{\partial t} + v\frac{\partial v}{\partial x} = -\frac{1}{\rho}\frac{\partial p}{\partial x} + F \tag{4.50}$$

Our task is to write the above two equations in terms of Lagrangian variables $(\tau, \mathbf{x}_0)$. Equation (4.49) is written as

$$\frac{\partial \rho}{\partial t} + v\frac{\partial \rho}{\partial x} + \rho\frac{\partial v}{\partial x} = 0 \tag{4.51}$$

Owing to (4.44), the first two terms add to $\partial\rho/\partial\tau$. Now

$$\frac{\partial v}{\partial x} = \frac{\partial v}{\partial x_0}\frac{\partial x_0}{\partial x} = \frac{1}{J}\frac{\partial v}{\partial x_0} = \frac{1}{J}\frac{\partial}{\partial x_0}\frac{\partial x}{\partial \tau} = \frac{1}{J}\frac{\partial}{\partial \tau}\frac{\partial x}{\partial x_0} = \frac{1}{J}\frac{\partial J}{\partial \tau}$$

where we used (4.41), $J = \frac{\partial x}{\partial x_0}$, $\frac{1}{J} = \frac{\partial x_0}{\partial x}$ (this is true only in one-dimensional case). Using the above equation in (4.51) gives

$$\frac{\partial \rho}{\partial \tau} + \frac{\rho}{J}\frac{\partial J}{\partial \tau} = 0 \tag{4.52}$$

This can be immediately integrated to yield

$$\rho(\mathbf{x}, t)J = \text{const.} = \rho(\mathbf{x}_0, 0) \equiv \rho_0 \tag{4.53}$$

Equation (4.53) is physically obvious. Consider a fluid mass contained in a small volume of d^3x centered about $\mathbf{x}(\mathbf{x}_0, t)$. Let $\rho(\mathbf{x}, t)$ be the fluid density. Since the mass was contained initially in a volume d^3x_0 centered about the position $\mathbf{x}_0$, we should have $\rho(\mathbf{x}, t)d^3x = \rho(\mathbf{x}_0, 0)d^3x_0$. Since the Jacobian is the ratio of the two corresponding volume elements, $J = \frac{d^3x}{d^3x_0}$, we have (4.53). In particular, $J = 1$ for incompressible fluid.

The momentum equation (4.50) can be converted in terms of Lagrangian variables as follows. The left hand side is simply $\partial v/\partial\tau$. Now

$$\frac{1}{\rho}\frac{\partial p}{\partial x} = \frac{1}{\rho}\frac{\partial p}{\partial x_0}\frac{\partial x_0}{\partial x} = \frac{1}{J\rho}\frac{\partial p}{\partial x_0} = \frac{1}{\rho_0}\frac{\partial p}{\partial x_0}$$

Therefore the momentum equation (4.50) is written in Lagrangian coordinates as

$$\frac{\partial v}{\partial \tau} = -\frac{1}{\rho_0}\frac{\partial p}{\partial x_0} + F \tag{4.54}$$

• **Three-dimensional equations.** We have

$$\nabla \cdot \mathbf{v} = \frac{\partial x_{0j}}{\partial x_i}\frac{\partial v_i}{\partial x_{0j}}$$

Using (4.41) in the above gives

$$\nabla \cdot \mathbf{v} = \frac{\partial x_{0j}}{\partial x_i}\frac{\partial}{\partial x_{0j}}\frac{\partial x_i}{\partial \tau} = \frac{\partial x_{0j}}{\partial x_i}\frac{\partial}{\partial \tau}\frac{\partial x_i}{\partial x_{0j}} \tag{4.55}$$

The matrices $(\partial x_{0j}/\partial x_i)$ and $(\partial x_i/\partial x_{0j})$ are inverse to each other since

$$\frac{\partial x_{0j}}{\partial x_i}\frac{\partial x_i}{\partial x_{0l}} = \frac{\partial x_{oj}}{\partial x_{0l}} = \delta_{jl} \tag{4.56}$$

When a matrix $A_{ij}(\lambda)$ is an $n \times n$ matrix whose elements are functions of λ, we have the following identity (See Tensor Analysis by Sokolnikoff)

$$\frac{\partial A_{ij}}{\partial \lambda}A^{-1}_{ji} = \frac{1}{|A|}\frac{\partial |A|}{\partial \lambda}$$

Then (4.55) immediately yields

$$\nabla \cdot \mathbf{v} = \frac{1}{J}\frac{\partial J}{\partial \tau} \tag{4.57}$$

where J is the Jacobian determinant. Equation (4.57) is an important relation for the divergence of velocity field which gives, together with (4.43), fundamental connection between the Lagrangian and Eulerian descriptions. The Eulerian form of continuity equation

$$\frac{d\rho}{dt} + \rho\nabla \cdot \mathbf{v} = 0 \tag{4.58}$$

immediately changes into (4.52) upon using (4.57) and $\frac{d}{dt} \to \frac{\partial}{\partial \tau}$.

• **Entropy conservation in ideal fluid.** In an ideal fluid, energy dissipation due to viscosity, heat conduction, etc is absent. Hence no irreversible process takes place. Let $s(\mathbf{x}, t)$ be the entropy per unit mass of the fluid at $(\mathbf{x}, t)$. Then the entropy contained in a volume element at $\mathbf{x}$ satisfies the conservation relation

$$\rho(\mathbf{x},t)s(\mathbf{x},t)J = \rho(\mathbf{x}_0,0)s(\mathbf{x}_0,0) \tag{4.59}$$

Using (4.53) in the above gives

$$s(\mathbf{x},t) = s(\mathbf{x}_0,0) = s_0 \tag{4.60}$$

which means that the entropy is constant along the stream line. The Eulerian form corresponding to (4.59) is

$$\frac{\partial}{\partial t}(\rho s) + \nabla\cdot(\rho s\mathbf{v}) = 0 \tag{4.61}$$

Using the continuity equation (4.58) in the above gives

$$\frac{\partial s}{\partial t} + \mathbf{v}\cdot\nabla s = 0 \tag{4.62}$$

This is the entropy conservation equation for the *isentropic* fluid.

• **Liouville fluid.** Liouville fluid may be regarded as the aggregate of the representative points composing the ensemble in the $6N$-dimensional Γ-space. Let the single vector $\mathbf{X}$ represents the '*position*' of the '*particle*' in the Γ-space:

$$\mathbf{X} = (\mathbf{r}_1,\mathbf{r}_2,\ldots,\mathbf{r}_N,\mathbf{p}_1,\mathbf{p}_2,\ldots,\mathbf{p}_N) \tag{4.63}$$

Let $\rho(\mathbf{X},t)$ represent the Liouville density in the Γ-space. Then the Liouville density obeys the conservation equation in (4.53):

$$\rho(\mathbf{X},t)J = \text{const.} \tag{4.64}$$

Convectively differentiating this equation gives

$$\frac{d\rho}{dt} + \rho\frac{1}{J}\frac{dJ}{dt} = 0 \tag{4.65}$$

where we can put per (4.57)

$$\frac{1}{J}\frac{dJ}{dt} = \nabla_\Gamma\cdot\mathbf{V}_\Gamma \tag{4.66}$$

with

$$\mathbf{V}_\Gamma = (\dot{\mathbf{r}}_1,\dot{\mathbf{r}}_2,\ldots,\dot{\mathbf{r}}_N,\dot{\mathbf{p}}_1,\dot{\mathbf{p}}_2,\ldots,\dot{\mathbf{p}}_N) \tag{4.67}$$

$$\nabla_\Gamma = \left(\frac{\partial}{\partial\mathbf{r}_1},\frac{\partial}{\partial\mathbf{r}_2},\ldots,\frac{\partial}{\partial\mathbf{r}_N},\frac{\partial}{\partial\mathbf{p}_1},\frac{\partial}{\partial\mathbf{p}_2},\ldots,\frac{\partial}{\partial\mathbf{p}_N}\right) \tag{4.68}$$

Hence

$$\nabla_\Gamma\cdot\mathbf{V}_\Gamma = \sum_{i=1}^{N}\left(\frac{\partial\dot{\mathbf{r}}_i}{\partial\mathbf{r}_i} + \frac{\partial\dot{\mathbf{p}}_i}{\partial\mathbf{p}_i}\right) = 0 \tag{4.69}$$

where we used Hamilton's equations

$$\dot{\mathbf{r}}_i = \frac{\partial H}{\partial \mathbf{p}_i}, \quad \dot{\mathbf{p}}_i = -\frac{\partial H}{\partial \mathbf{r}_i} \tag{4.70}$$

Hence, per (4.65), we get $\frac{d\rho}{dt} = 0$, or $\frac{\partial \rho}{\partial t} + \frac{\partial \rho}{\partial \mathbf{X}} \cdot \dot{\mathbf{X}} = 0$, the Liouville equation

$$\frac{\partial \rho}{\partial t} + \sum_{i=1}^{N} \left(\frac{\partial H}{\partial \mathbf{p}_i} \cdot \frac{\partial \rho}{\partial \mathbf{r}_i} - \frac{\partial \rho}{\partial \mathbf{p}_i} \cdot \frac{\partial H}{\partial \mathbf{r}_i} \right) = 0 \tag{4.71}$$

4.3 Variational Principle of Ideal Fluid

Reference: W. Yourgrau and S. Mandelstam, *Variational principles in dynamics and quantum theory* (Dover 1968).

4.3.1 *Lagrangian variational principle*

Hamilton's principle for a conservative system is

$$\delta I = \delta \int_{t_0}^{t_1} dt(K - V) = 0 \tag{4.72}$$

where K and V are the kinetic and potential energy of the system, respectively:

$$K = \int \frac{1}{2}\rho \left(\frac{\partial x_i}{dt} \right)^2 d^3x = \int \frac{1}{2}\rho_0 \left(\frac{\partial x_i}{\partial t} \right)^2 d^3x_0 \tag{4.73}$$

$$V = \int \rho(e + U) d^3x = \int \rho_0 (e + U) d^3x_0 \tag{4.74}$$

where ρ is the density, $e(\rho, s)$ is the internal energy, U is the external potential from which the external forces are derivable. x_i is the Eulerian position vector ($i = x, y, z$), x_0 is the Lagrangian position vector. In (4.73), we used $d^3x = Jd^3x_0$ and $\rho J = \rho_0$ (see (4.53)).

All of U, e, and s are measured per unit mass. We have the thermodynamic relations for p, pressure and T, temperature.

$$p = \rho^2 \left(\frac{\partial e}{\partial \rho} \right)_s, \quad T = \left(\frac{\partial e}{\partial s} \right)_\rho \tag{4.75}$$

We regard (4.53) and (4.60) are the constraints in the formulation of the variational principle:

$$\phi_1 = \rho J - \rho_0 = 0 \tag{4.76}$$

$$\phi_2 = s - s_0 = 0 \tag{4.77}$$

Then, we can write the variational integral in the form (4.25) by introducing the Lagrange multipliers λ_1 and λ_2

$$\delta \int_{t_0}^{t_1} dt \int d^3x_0 \left[\frac{1}{2}\rho_0 \left(\frac{\partial x_i}{\partial t} \right)^2 - \rho_0(e+U) + \lambda_1(\rho J - \rho_0) + \lambda_2(s - s_0) \right] = 0 \tag{4.78}$$

where $e = e(\rho, s)$, $U = U(x_i)$. The functional, the quantity in the big bracket $[\cdots] \equiv G$, has functions x_i, ρ, s. In view of (4.21), we have

$$\frac{\partial G}{\partial \rho} = 0, \qquad \frac{\partial G}{\partial s} = 0 \tag{4.79}$$

where

$$G = \frac{1}{2}\rho_0 \left(\frac{\partial x_i}{\partial t} \right)^2 - \rho_0(e(\rho) + U) + \lambda_1(\rho J - \rho_0) + \lambda_2(s - s_0)$$

The first equation in the above gives

$$-\rho_0 \frac{\partial e}{\partial \rho} + \lambda_1 J = 0 \tag{4.80}$$

Using (4.75) and (4.53) in the above, we obtain

$$\lambda_1 \rho = p \tag{4.81}$$

The second equation in (4.79) and (4.75) give

$$\lambda_2 = \rho_0 T \tag{4.82}$$

Next, let us consider the variation with respect to the function x_i in G. The independent variables are $\mathbf{x}_0$ and t: $x_i = x_i(x_{oj}, t)$—recall (4.34). Thus we write from (4.25):

$$\frac{\partial}{\partial x_{0j}} \frac{\partial G}{\partial(\frac{\partial x_i}{\partial x_{0j}})} + \frac{\partial}{\partial t} \frac{\partial G}{\partial(\frac{\partial x_i}{\partial t})} - \frac{\partial G}{\partial x_i} = 0 \tag{4.83}$$

The first term is

$$\begin{aligned} \text{1st term} &= \frac{\partial}{\partial x_{0j}} \frac{\partial}{\partial(\frac{\partial x_i}{\partial x_{0j}})} (\lambda_1 \rho J) \\ &= \frac{\partial}{\partial x_{0j}} \left(\lambda_1 \rho \frac{\partial J}{\partial(\frac{\partial x_i}{\partial x_{0j}})} \right) \end{aligned} \tag{4.84}$$

By Laplace expansion of a determinant, we have

$$J = \frac{\partial x_i}{\partial x_{0j}} J_{ij} \tag{4.85}$$

where J_{ij} is the cofactor of the matrix $\{\frac{\partial x_i}{\partial x_{0j}}\}$—see the expression in Section 4.2. Hence

$$\text{1st term} = \frac{\partial}{\partial x_{0j}}(pJ_{ij}) = J_{ij}\frac{\partial p}{\partial x_{0j}} \tag{4.86}$$

where we used $\frac{\partial}{\partial x_{0j}}J_{ij} = 0$. The pressure gradient term in the above is desirable to be expressed in Eulerian derivative.

$$\text{1st term} = \frac{\partial p}{\partial x_k}\frac{\partial x_k}{\partial x_{0j}} \quad J_{ij} = \frac{\partial p}{\partial x_k} \quad J\delta_{ik} = J \quad \frac{\partial p}{\partial x_i} = \frac{\rho_0}{\rho}\frac{\partial p}{\partial x_i} \tag{4.87}$$

The remaining terms in (4.83) are calculated straightly to obtain

$$\left(\frac{\partial^2 x_i}{\partial t^2}\right)_{\mathbf{x}_0} + \frac{1}{\rho}\frac{\partial p}{\partial x_i} = -\frac{\partial U}{\partial x_i} \tag{4.88}$$

This is the momentum equation of an ideal fluid moving in a conservative external potential U. The first term in (4.88) is the Lagrangian acceleration as given by (4.46). Hence, putting it into more familiar form, we have Euler's equation

$$\frac{\partial \mathbf{v}}{\partial t} + \mathbf{v}\cdot\nabla\mathbf{v} + \frac{1}{\rho}\nabla p = -\nabla U \tag{4.89}$$

4.3.2 *Eulerian variational principle*

In this subsection, the Euler equation (4.89) will be derived *independently* by formulating the variational principle in terms of the Eulerian variables. We use the first expressions of (4.73) and (4.74) for the kinetic and potential energies. In the foregoing subsection, we have learned that the constancy of entropy as a constraint had no effect on the final equation. So the internal energy e will be assumed to be as a function of ρ alone: $e = e(\rho)$. Furthermore, we impose the continuity equation to be the only constraint in (4.23):

$$\phi = \frac{\partial \rho}{\partial t} + \nabla\cdot(\rho\mathbf{v}) = 0 \tag{4.90}$$

$$\delta\int_{t_0}^{t_1} dt \int d^3x \left[\frac{1}{2}\rho v^2 - \rho(e+U) - \lambda\left(\frac{\partial \rho}{\partial t} + \nabla\cdot(\rho\mathbf{v})\right)\right] = 0 \tag{4.91}$$

We restrict (4.91) to variations that vanish at the boundaries of the region of integration. Variation with respect to ρ yields

$$\frac{\partial \lambda}{\partial t} + \mathbf{v}\cdot\nabla\lambda + \frac{v^2}{2} - \frac{\partial}{\partial \rho}(e\rho) - U = 0 \tag{4.92}$$

Variation with respect to v_i yields

$$v_i + \frac{\partial \lambda}{\partial x_i} = 0, \quad \text{or} \quad \nabla\lambda = -\mathbf{v} \tag{4.93}$$

Taking gradient of (4.92) and using (4.93) yield

$$\frac{\partial \mathbf{v}}{\partial t} + \nabla\frac{v^2}{2} + \frac{1}{\rho}\nabla p = -\nabla U \tag{4.94}$$

where we used (4.75) and

$$\frac{\partial}{\partial x_i}\frac{\partial}{\partial \rho}(e\rho) = \frac{1}{\rho}\frac{\partial p}{\partial x_i} \tag{4.95}$$

Equation (4.94) is seen to be an incomplete Euler equation, deficient of the vorticity term. See (4.89). This is seen by recalling the vector identity

$$\nabla\frac{v^2}{2} - \mathbf{v}\times\nabla\times\mathbf{v} = \mathbf{v}\cdot\nabla\mathbf{v} \tag{4.96}$$

Lin's remedy. Lin (1963) pointed out that the deficiency can be remedied by introducing the Lagrangian coordinates as an extra constraint.

The relation between the Eulerian and Lagrangian coordinates

$$\mathbf{x} = \mathbf{x}(\mathbf{x}_0, t) \tag{4.97}$$

can be inverted in the form

$$\mathbf{x}_0 = \mathbf{x}_0(\mathbf{x}, t) \tag{4.98}$$

However, since $\mathbf{x}(\mathbf{x}_0, 0) = \mathbf{x}_0$, $\mathbf{x}_0$ is independent of t. Hence we have

$$\frac{dx_{0i}}{dt} = \left(\frac{\partial x_{0i}}{\partial t}\right)_{\mathbf{x}} + \frac{\partial x_{0i}}{\partial x_j}\;\frac{\partial x_j}{\partial t} = \left(\frac{\partial x_{0i}}{\partial t}\right)_{\mathbf{x}} + v_j\frac{\partial x_{0i}}{\partial x_j} = 0 \tag{4.99}$$

The last equality signifies the individuality of the stream lines, and should be imposed as the extra constraints:

$$\phi_i = \left(\frac{\partial x_{0i}}{\partial t}\right)_{\mathbf{x}} + v_j\frac{\partial x_{0i}}{\partial x_j} = 0 \quad (i = 1, 2, 3) \tag{4.100}$$

Thus the variational integral takes the form

$$\delta\int_{t_0}^{t_1} dt \int d^3x \left[\frac{1}{2}\rho v^2 - \rho(e+U) - \lambda\left(\frac{\partial \rho}{\partial t} + \nabla\cdot(\rho\mathbf{v})\right) + \rho\gamma_i\left(\frac{\partial x_{0i}}{\partial t} + v_j\frac{\partial x_{0i}}{\partial x_j}\right)\right] = 0 \tag{4.101}$$

where γ_i is the three Lagrange multipliers, and the summation convention is reminded of.

The quantity in the large bracket $[\cdots]$ above is the functional G in (4.25). Now the functions contained in G are ρ, v_i, x_{0i}, while t, x_i are independent variables. In the variation with respect to ρ, the new comer due to $\partial G/\partial \rho$ is $\left(\frac{\partial x_{0i}}{\partial t} + v_j \frac{\partial x_{0i}}{\partial x_j}\right)$, but it is zero per (4.100).

Thus, variation with respect to ρ gives the same equation as (4.92)

$$\frac{\partial \lambda}{\partial t} + \mathbf{v} \cdot \nabla \lambda + \frac{v^2}{2} - \frac{\partial}{\partial \rho}(e\rho) - U = 0 \tag{4.102}$$

Variation with respect to v_i gives

$$v_i + \frac{\partial \lambda}{\partial x_i} + \gamma_j \frac{\partial x_{0j}}{\partial x_i} = 0 \tag{4.103}$$

Variation with respect to x_{0i} gives

$$\frac{\partial}{\partial t}(\rho \gamma_i) + \frac{\partial}{\partial x_j}(\rho \gamma_i v_j) = 0 \tag{4.104}$$

Upon using (4.90), (4.104) reduces to

$$\frac{\partial \gamma_i}{\partial t} + \mathbf{v} \cdot \nabla \gamma_i = 0 \tag{4.105}$$

In addition, we have the equation for the Lagrangian coordinate (4.100):

$$\frac{\partial x_{0i}}{\partial t} = -v_j \frac{\partial x_{0i}}{\partial x_j} \tag{4.106}$$

Equations (4.102)–(4.106) boil down to the familiar Euler momentum equation after eliminating λ and γ_i. First, λ can be eliminated between (4.102) and (4.103):

$$\frac{\partial v_i}{\partial t} + \frac{\partial}{\partial x_i}\left(\frac{v^2}{2} + \frac{\partial}{\partial \rho}(e\rho) + U\right) + \frac{\partial}{\partial t}\left(\gamma_j \frac{\partial x_{0j}}{\partial x_i}\right) + \frac{\partial}{\partial x_i}\left(\gamma_j \mathbf{v} \cdot \nabla x_{0j}\right) = 0 \tag{4.107}$$

Using (4.105) and (4.106) in the above, the γ_j-containing terms add to

$$-(\mathbf{v} \cdot \nabla \gamma_j)\frac{\partial x_{0j}}{\partial x_i} + (\mathbf{v} \cdot \nabla x_{0j})\frac{\partial \gamma_j}{\partial x_i} = v_k\left(-\frac{\partial \gamma_j}{\partial x_k}\frac{\partial x_{0j}}{\partial x_i} + \frac{\partial \gamma_j}{\partial x_i}\frac{\partial x_{0j}}{\partial x_k}\right) \equiv R_i \tag{4.108}$$

where R_i is ith component of vector $\mathbf{R}$. On the other hand, taking curl of (4.103) in terms of Levi-Civita symbol gives

$$(\nabla \times \mathbf{v})_l = -e_{lpq} \frac{\partial}{\partial x_p} \left(\gamma_j \frac{\partial x_{0j}}{\partial x_q} \right) \tag{4.109}$$

Hence

$$-(\mathbf{v} \times \nabla \times \mathbf{v})_i = -e_{ikl} v_k (\nabla \times \mathbf{v})_l = v_k e_{ikl} e_{lpq} \frac{\partial}{\partial x_p} \left(\gamma_j \frac{\partial x_{0j}}{\partial x_q} \right) \tag{4.110}$$

After further algebra, the last expression becomes R_i. Hence

$$\frac{\partial \mathbf{v}}{\partial t} + \nabla \left(\frac{v^2}{2} + \frac{\partial}{\partial \rho}(e\rho) + U \right) - \mathbf{v} \times \nabla \times \mathbf{v} = 0 \tag{4.111}$$

Using the identity in (4.96) in the above gives the familiar Euler equation

$$\frac{\partial \mathbf{v}}{\partial t} + \mathbf{v} \cdot \nabla \mathbf{v} + \frac{1}{\rho} \nabla p = -\nabla U \tag{4.112}$$

Example. Show that the Lagrangian density

$$\mathcal{L} = \frac{E^2 - B^2}{8\pi} - \rho_c \phi + \frac{\mathbf{J} \cdot \mathbf{A}}{c} \tag{4.113}$$

results in the Maxwell equations

$$\nabla \cdot \mathbf{E} = 4\pi \rho_c, \quad \nabla \times \mathbf{B} = \frac{1}{c} \frac{\partial \mathbf{E}}{\partial t} + \frac{4\pi \mathbf{J}}{c} \tag{4.114}$$

where ϕ and $\mathbf{A}$ are the scalar and vector potentials, respectively,

$$\mathbf{E} = -\nabla \phi - \frac{1}{c} \frac{\partial \mathbf{A}}{\partial t}, \quad \mathbf{B} = \nabla \times \mathbf{A} \tag{4.115}$$

which yield in turn

$$\nabla \cdot \mathbf{B} = 0, \quad \nabla \times \mathbf{E} + \frac{1}{c} \frac{\partial \mathbf{B}}{\partial t} = 0 \tag{4.116}$$

Variation with respect to ϕ:

$$\frac{1}{8\pi} \left[\frac{\partial}{\partial x} \frac{\partial E^2}{\partial \phi_x} + \frac{\partial}{\partial y} \frac{\partial E^2}{\partial \phi_y} + \frac{\partial}{\partial z} \frac{\partial E^2}{\partial \phi_z} \right] = \frac{\partial \mathcal{L}}{\partial \phi} \tag{4.117}$$

where subscripts x, y, z at ϕ denote differentiation, and

$$E^2 = \left(\phi_x + \frac{1}{c} \frac{\partial A_x}{\partial t} \right)^2 + \left(\phi_y + \frac{1}{c} \frac{\partial A_y}{\partial t} \right)^2 + \left(\phi_z + \frac{1}{c} \frac{\partial A_z}{\partial t} \right)^2$$

$$\frac{\partial \mathcal{L}}{\partial \phi} = -\rho_c$$

It is readily seen that $\frac{\partial E^2}{\partial \phi_x} = -2E_x$. Therefore (4.117) yields

$$\nabla \cdot \mathbf{E} = 4\pi \rho_c \tag{4.118}$$

To calculate variation with respect to $\mathbf{A}$, write

$$\begin{aligned}\mathcal{L} = &\frac{1}{8\pi}\left[\left(\phi_x + \frac{1}{c}\frac{\partial A_x}{\partial t}\right)^2 + \left(\phi_y + \frac{1}{c}\frac{\partial A_y}{\partial t}\right)^2 + \left(\phi_z + \frac{1}{c}\frac{\partial A_z}{\partial t}\right)^2\right] \\ &- \frac{1}{8\pi}\left[\left(\frac{\partial A_z}{\partial y} - \frac{\partial A_y}{\partial z}\right)^2 + \left(\frac{\partial A_x}{\partial z} - \frac{\partial A_z}{\partial x}\right)^2 + \left(\frac{\partial A_y}{\partial x} - \frac{\partial A_x}{\partial y}\right)^2\right] \\ &- \rho_c \phi + \frac{1}{c}(J_x A_x + J_y A_y + J_z A_z)\end{aligned} \tag{4.119}$$

Variation with respect to A_x gives

$$\frac{1}{4\pi c}\frac{\partial}{\partial t}\left(\phi_x + \frac{1}{c}\frac{\partial A_x}{\partial t}\right) - \frac{1}{4\pi}\frac{\partial}{\partial z}\left(\frac{\partial A_x}{\partial z} - \frac{\partial A_z}{\partial x}\right) + \frac{1}{4\pi}\frac{\partial}{\partial y}\left(\frac{\partial A_y}{\partial x} - \frac{\partial A_x}{\partial y}\right) - \frac{J_x}{c} = 0$$

which turns out to be

$$(\nabla \times \mathbf{B})_x = \frac{1}{c}\frac{\partial E_x}{\partial t} + \frac{4\pi}{c} J_x \tag{4.120}$$

Variations with respect to $A_{y,z}$ yields the y, z component of (4.114).

• **Variational principle of Lorentz force.** For a single particle of charge q moving in an electromagnetic field, we consider Lagrangian

$$\mathcal{L}_q = \frac{m}{2}v^2 - q\phi + \frac{q}{c}\mathbf{A} \cdot \mathbf{v} \tag{4.121}$$

Compare this with (4.113). We write the *action integral*

$$\delta \int_{t_1}^{t_2} \mathcal{L}_q dt = \delta \int_{t_1}^{t_2} \left[\frac{m}{2}\left(\frac{d\mathbf{r}}{dt}\right)^2 - q\phi(\mathbf{r}) + \frac{q}{c}\mathbf{A}(\mathbf{r}) \cdot \frac{d\mathbf{r}}{dt}\right] dt = 0 \tag{4.122}$$

where the dependent variable is $\mathbf{r}(\mathbf{t})$. The Euler–Lagrange equation reads

$$\frac{d}{dt}\frac{\partial \mathcal{L}_q}{\partial \dot{\mathbf{r}}} - \frac{\partial \mathcal{L}_q}{\partial \mathbf{r}} = 0, \quad \dot{\mathbf{r}} = \frac{d\mathbf{r}}{dt} = \mathbf{v} \tag{4.123}$$

or

$$\frac{d}{dt}\left(m\dot{\mathbf{r}} + \frac{q}{c}\mathbf{A}\right) + q\nabla\phi - \frac{q}{c}\nabla(\mathbf{A} \cdot \mathbf{v}) = 0 \tag{4.124}$$

It is important to note d/dt being the total derivative; recall (4.10). Then rewriting (4.124),

$$m\frac{d\mathbf{v}}{dt} = -q\left(\nabla\phi + \frac{1}{c}\frac{\partial \mathbf{A}}{\partial t}\right) - \frac{q}{c}\left[(\mathbf{v}\cdot\nabla)\mathbf{A} - \nabla(\mathbf{A}\cdot\mathbf{v})\right]$$

One can show that $\mathbf{v}\times(\nabla\times\mathbf{A}) = \nabla(\mathbf{A}\cdot\mathbf{v}) - (\mathbf{v}\cdot\nabla)\mathbf{A} = v_j(\frac{\partial A_j}{\partial x_i} - \frac{\partial A_i}{\partial x_j})$. Hence, per (4.115)

$$m\frac{d\mathbf{v}}{dt} = q\left(\mathbf{E} + \frac{1}{c}\mathbf{v}\times\mathbf{B}\right) \tag{4.125}$$

• **Lagrangian of an assembly of charged particles.** First, the charge and current densities in (4.113) should be expressed appropriately to represent those of the assembly of the charged particles. By means of Dirac delta function,

$$\rho_c(\mathbf{r}) = \sum_{n=1}^{N} q_n\delta(\mathbf{r} - \mathbf{r}_n) \tag{4.126}$$

where q_n is the charge of the nth particle and $\mathbf{r}_n$ its position vector. Since $\mathbf{J} = \rho_c\mathbf{v}$, the last two terms in (4.113) give

$$\int d^3r\rho_c(\mathbf{r})\left[\phi(\mathbf{r}) - \frac{\mathbf{v}\cdot\mathbf{A}(\mathbf{r})}{c}\right] = \sum_{n=1}^{N} q_n\left(\phi(\mathbf{r}_n) - \frac{\mathbf{v}_n\cdot\mathbf{A}(\mathbf{r}_n)}{c}\right) \tag{4.127}$$

Then the Lagrangian associated with Maxwell equations is obtained by integrating (4.113)

$$\mathcal{L} = \int d^3r\frac{E^2 - B^2}{8\pi} - \sum_{n=1}^{N} q_n\left(\phi(\mathbf{r}_n) - \frac{\mathbf{v}_n\cdot\mathbf{A}(\mathbf{r}_n)}{c}\right) \tag{4.128}$$

The total Lagrangian L is obtained by the sum of $\mathcal{L}$ plus the Lagrangian associated with particle dynamics:

$$\mathrm{L} = \int d^3r\frac{E^2 - B^2}{8\pi} - \sum_{n=1}^{N} q_n\left(\phi(\mathbf{r}_n) - \frac{\mathbf{v}_n\cdot\mathbf{A}(\mathbf{r}_n)}{c}\right) + \frac{1}{2}\sum_{n}^{N} m_n v_n^2$$

This Lagrangian is the total Lagrangian which produces both the electromagnetic field and the Lorentz force contributed by the assembly of N charged particles. If we are dealing with charged bodies, not with point charges, the Lagrangian is

$$L = \int d^3r\frac{E^2 - B^2}{8\pi} - \int \rho_c\left(\phi - \frac{\mathbf{v}\cdot\mathbf{A}}{c}\right)d^3r + \frac{1}{2}\int \rho v^2 d^3r \tag{4.129}$$

where ρ is the charge density, ρ_c is the mass density, $\mathbf{v}$ is the velocity of the charge at any point. The $\mathbf{v}$ and ρ are treated as functions of the coordinates of the system.

4.3.3 *Lagrangian of ideal fluid equation*

In Section 4.3.1, we started from the Lagrangian $L = K - V$ in (4.73) and (4.74) and derived the equation of motion (4.88). In this subsection, we construct the Lagrangian via sheer transformation of the given equation of motion. Let us begin with the equation of motion in Lagrangian form in (4.88)

$$\frac{\partial^2 x_i}{\partial t^2} = -\frac{1}{\rho}\frac{\partial p}{\partial x_i} - \frac{\partial U}{\partial x_i} \tag{4.130}$$

where $x_i = x_i(\xi_j, t)$ or $\mathbf{x} = \mathbf{x}(\vec{\xi}, t)$ with $\vec{\xi}$ the Lagrangian coordinate. [Previous notation $\mathbf{x}_0$ is changed to $\vec{\xi}$.] Equation (4.130) can be put into the form

$$\frac{\partial^2 x_i}{\partial t^2} = -\frac{\partial V}{\partial x_i} \tag{4.131}$$

where the potential $V(x_i)$ is

$$V(x_i) = \int^{p(x_i)} \frac{dp}{\rho} + U \tag{4.132}$$

Equation (4.131) is entirely analogous to the equation of motion of a particle in classical mechanics, and the corresponding *Hamilton's principle* can be easily derived. Note that the Lagrangian coordinate ξ_j plays the role of the generalized coordinates.

First, multiply (4.131) by $\partial x_i/\partial t$ and sum over $i = 1, 2, 3$ to get the energy integral

$$\frac{1}{2}\left(\frac{\partial x_i}{\partial t}\right)^2 + V(x_i) = \text{const.} \tag{4.133}$$

Therefore, for a stream line, we define the Lagrangian L as kinetic energy minus potential energy,

$$L = \frac{1}{2}\left(\frac{\partial x_i}{\partial t}\right)^2 - V(x_i)$$

For an assembly of particles, we have the action integral in (4.72).

On the other hand, we digress a little, and multiply (4.131) by $\partial x_i/\partial\xi_j$ and sum over $i = 1, 2, 3$:

$$\frac{\partial^2 x_i}{\partial t^2}\frac{\partial x_i}{\partial \xi_j} = -\frac{\partial V}{\partial \xi_j} \quad (j = 1, 2, 3) \tag{4.134}$$

Integrate (4.134) from $t = 0$ to t:

$$\int_0^t dt\left[\frac{\partial^2 x_i}{\partial t^2}\frac{\partial x_i}{\partial \xi_j} + \frac{\partial V}{\partial \xi_j}\right] = 0 \tag{4.135}$$

Note that

$$\begin{aligned}\int_0^t \frac{\partial^2 x_i}{\partial t^2}\frac{\partial x_i}{\partial \xi_j}dt &= \left[\frac{\partial x_i}{\partial t}\frac{\partial x_i}{\partial \xi_j}\right]_0^t - \int_0^t \frac{\partial x_i}{\partial t}\frac{\partial^2 x_i}{\partial \xi_j \partial t}dt \\ &= \frac{\partial x_i}{\partial t}\frac{\partial x_i}{\partial \xi_j} - v_i(0)\delta_{ij} - \frac{\partial}{\partial \xi_j}\int_0^t \frac{1}{2}\left(\frac{\partial x_i}{\partial t}\right)^2 dt\end{aligned} \tag{4.136}$$

Hence, (4.134) takes the form

$$\frac{\partial x_i}{\partial t}\frac{\partial x_i}{\partial \xi_j} - v_i(0)\delta_{ij} = \frac{\partial \chi}{\partial \xi_j} \tag{4.137}$$

where the action integral χ is

$$\chi = \int_0^t \left[\frac{1}{2}\left(\frac{\partial x_i}{\partial t}\right)^2 - \int^p \frac{dp}{\rho} - U\right]dt \tag{4.138}$$

which gives

$$\frac{\partial \chi}{\partial t} = \frac{1}{2}\left(\frac{\partial x_i}{\partial t}\right)^2 - \int^p \frac{dp}{\rho} - U \tag{4.139}$$

The three equations in (4.137), together with (4.139), and the continuity equation are the partial differential equations to be satisfied by the five unknowns x_i, ρ, χ. The initial conditions are $x_i(0) = \xi_i$ and $\chi = 0$ at $t = 0$. The foregoing development that results in the equation for the action integral χ is known as Weber's transformation (H. Lamb, *Hydrodynamics*, Cambridge 1952). This development appears to be of only limited theoretical value.

• **Euler–Lagrange equation of Lagrangian fluid equation.** Equation (4.131) can be transformed into the form of the Euler–Lagrange equation by coordinate transformation. The transformation involves the Weber's transformation as in (4.134), which is rewritten here

$$\frac{\partial^2 x_i}{\partial t^2}\frac{\partial x_i}{\partial \xi_j} = -\frac{\partial V}{\partial \xi_j} \tag{4.140}$$

Let us use the notation $\dot{x} = \frac{\partial x}{\partial t}$, and remember $\partial/\partial t$ is the *total derivative.* Then

$$\begin{aligned}
\text{LHS} &= \frac{\partial \dot{x}_i}{\partial t}\frac{\partial x_i}{\partial \xi_j} \\
&= \frac{\partial}{\partial t}\left(\dot{x}_i\frac{\partial x_i}{\partial \xi_j}\right) - \dot{x}_i\frac{\partial}{\partial t}\left(\frac{\partial x_i}{\partial \xi_j}\right) \\
&= \frac{\partial}{\partial t}\left(\dot{x}_i\frac{\partial x_i}{\partial \xi_j}\right) - \dot{x}_i\frac{\partial}{\partial \xi_s}\left(\frac{\partial x_i}{\partial \xi_j}\right)\dot{\xi}_s \\
&= \frac{\partial}{\partial t}\left(\dot{x}_i\frac{\partial x_i}{\partial \xi_j}\right) - \dot{x}_i\frac{\partial}{\partial \xi_j}\left(\frac{\partial x_i}{\partial \xi_s}\right)\dot{\xi}_s \\
&= \frac{\partial}{\partial t}\left(\dot{x}_i\frac{\partial x_i}{\partial \xi_j}\right) - \dot{x}_i\frac{\partial}{\partial \xi_j}\left(\frac{\partial x_i}{\partial \xi_s}\dot{\xi}_s\right)
\end{aligned}$$

The last equality is due to $\partial\dot{\xi}_s/\partial\xi_j = 0$. Hence

$$\text{LHS} = \frac{\partial}{\partial t}\left(\dot{x}_i\frac{\partial x_i}{\partial \xi_j}\right) - \dot{x}_i\frac{\partial \dot{x}_i}{\partial \xi_j} \tag{4.141}$$

On the other hand

$$\frac{\partial \dot{x}_i}{\partial \dot{\xi}_j} = \frac{\partial}{\partial \dot{\xi}_j}\left(\frac{\partial x_i}{\partial \xi_s}\dot{\xi}_s\right)$$

Here x_i is independent of $\dot{\xi}_j$. Hence

$$\frac{\partial \dot{x}_i}{\partial \dot{\xi}_j} = \frac{\partial x_i}{\partial \xi_s}\frac{\partial \dot{\xi}_s}{\partial \dot{\xi}_j} = \frac{\partial x_i}{\partial \xi_s}\delta_{sj} = \frac{\partial x_i}{\partial \xi_j}$$

Therefore we obtain

$$\text{LHS} = \frac{\partial}{\partial t}\frac{\partial}{\partial \dot{\xi}_j}\left(\frac{1}{2}\dot{x}_i^2\right) - \frac{\partial}{\partial \xi_j}\left(\frac{1}{2}\dot{x}_i^2\right) \tag{4.142}$$

where $\left(\frac{1}{2}\dot{x}_i^2\right) = \frac{1}{2}\dot{x}_i\dot{x}_i = \frac{1}{2}(\dot{x}_1^2 + \dot{x}_2^2 + \dot{x}_3^2) \equiv T(\xi_i, \dot{\xi}_i, t)$ is the kinetic energy per unit mass. Thus (4.140) takes the form

$$\frac{\partial}{\partial t}\frac{\partial}{\partial \dot{\xi}_j}L(\xi_i, \dot{\xi}_i, t) - \frac{\partial}{\partial \xi_j}L = 0 \tag{4.143}$$

where

$$L = T(\xi_i, \dot{\xi}_i, t) - V(\xi_i, t) \tag{4.144}$$

is the Lagrangian of the ideal fluid per unit mass expressed in terms of the Lagrangian coordinates, and (4.143) is the corresponding Euler–Lagrange equation.

• **Eulerian variational principle** in terms of *Clebsch velocity potentials.* Equation (4.103) is the outcome of the Eulerian variation principle formulated in (4.101). It expresses the fluid velocity in terms of the scalar potential λ, the vector potential $\overrightarrow{\gamma}$, and the Lagrangian coordinates $\overrightarrow{x_0}$. The latter two represent the vorticity. Following Clebsch, the fluid velocity can be expressed by means of two potentials in the form

$$\mathbf{v} = -\nabla\phi - \lambda\nabla\mu \tag{4.145}$$

where the potentials ϕ, λ, μ are all functions of $(t, \mathbf{r})$. Both λ and μ satisfy the same equation as $\overrightarrow{x_0}$ and $\overrightarrow{\gamma}$ (see (4.105) and (4.106), viz.,

$$\frac{\partial\lambda}{\partial t} + \mathbf{v}\cdot\nabla\lambda = 0 \quad \text{and} \quad \frac{\partial\mu}{\partial t} + \mathbf{v}\cdot\nabla\mu = 0 \tag{4.146}$$

We carry out the following variational integral to obtain Euler equation including vorticity.

$$\delta\int_{t_0}^{t_1} dt\int d^3x\left[\frac{1}{2}\rho v^2 - \rho(e+U) - \phi\left(\frac{\partial\rho}{\partial t} + \nabla\cdot(\rho\mathbf{v})\right) + \rho\lambda\left(\frac{\partial\mu}{\partial t} + \mathbf{v}\cdot\nabla\mu\right)\right] = 0 \tag{4.147}$$

The algebra goes as follows:

Variation with respect to ρ yields

$$\frac{v^2}{2} - \frac{\partial}{\partial\rho}(e\rho) - U - \phi\nabla\cdot\mathbf{v} + \lambda\left(\frac{\partial\mu}{\partial t} + \mathbf{v}\cdot\nabla\mu\right) + \frac{\partial\phi}{\partial t} + \nabla\cdot(\phi\mathbf{v}) = 0 \tag{4.148}$$

After using (4.146) and cancellation, this becomes

$$\frac{v^2}{2} - \frac{\partial}{\partial\rho}(e\rho) - U + \frac{\partial\phi}{\partial t} + \mathbf{v}\cdot\nabla\phi = 0 \tag{4.149}$$

Variation with respect to $\mathbf{v}$ gives $\rho\mathbf{v} - \phi\nabla\rho + \rho\lambda\nabla\mu + \nabla(\phi\rho) = 0$, which reproduces (4.145).

Variation with respect to μ gives $\frac{\partial}{\partial t}(\rho\lambda) + \nabla \cdot (\rho\lambda\mathbf{v}) = 0$, which is an identity of no utility. Let us use (4.145)–(4.149) to eliminate the potentials. Eliminating ϕ between (4.145) and (4.149) gives

$$\frac{\partial \mathbf{v}}{\partial t} + \nabla\frac{v^2}{2} + \nabla\left(\frac{\partial}{\partial \rho}(e\rho) + U\right) + \frac{\partial}{\partial t}(\lambda\nabla\mu) + \nabla(\lambda\mathbf{v}\cdot\nabla\mu) = 0 \quad (4.150)$$

Using (4.146) in (4.150) yields

$$\frac{\partial \mathbf{v}}{\partial t} + \nabla\frac{v^2}{2} + \nabla\left(\frac{\partial}{\partial \rho}(e\rho) + U\right) + (\mathbf{v}\cdot\nabla\mu)\nabla\lambda - (\mathbf{v}\cdot\nabla\lambda)\nabla\mu = 0 \quad (4.151)$$

Taking curl of (4.145), we find

$$\nabla\times\mathbf{v} = -\nabla\lambda\times\nabla\mu, -\mathbf{v}\times\nabla\times\mathbf{v} = (\mathbf{v}\cdot\nabla\mu)\nabla\lambda - (\mathbf{v}\cdot\nabla\lambda)\nabla\mu \quad (4.152)$$

Hence, (4.151) becomes (4.111). Clebsch transformation makes the Eulerian variational calculus easier.

Problems

4.1 Derive the thermodynamic relations in (4.75).
Hint: $de = Tds - pd(\text{volume})$.

4.2 Derive (4.92) and (4.93).

4.3 Derive (4.95).

4.4 Show that Eq. (2.54) in Chapter 2 can be put into the form

$$\frac{\partial p}{\partial t} + \nabla\cdot(p\mathbf{v}) = -\frac{2}{3}p\nabla\cdot\mathbf{v}$$

Hint: $\frac{1}{n}\frac{dn}{dt} = -\nabla\cdot\mathbf{v}$.

4.5 Prove *Euler theorem*:

$$2T = \dot{x}_i^2 = \frac{\partial L}{\partial \dot{\xi}_r}\dot{\xi}_r$$

where L is the Lagrangian in (4.143) and (4.144). Using Euler theorem, show that

$$\frac{\partial}{\partial t}(T + V) = \frac{\partial}{\partial t}(2T - L) = 0$$

4.6 **Continuity equation in Lagrangian coordinates.**
Consider the element of fluid which originally occupied a rectangular parallelepiped having its center at the point (ξ_1, ξ_2, ξ_3) and its edges $\delta\xi_1, \delta\xi_2, \delta\xi_3$, parallel to the axes. At the time t, the same element forms

a deformed parallelepiped. Show that the transfigured parallelepiped volume is related with the original volume by the equation $|\frac{\partial x_i}{\partial \xi_j}|\delta\xi_1\delta\xi_2\delta\xi_3$, where $|\frac{\partial x_i}{\partial \xi_j}| = J$ is the Jacobian determinant. Recall that the volume formed by linearly independent three vectors $\mathbf{a}, \mathbf{b}, \mathbf{c}$ is $\mathbf{a} \times \mathbf{b} \cdot \mathbf{c}$. Since the mass of the element is unchanged, we have $\rho J = \rho_0$ where ρ_0 is the initial density at $\vec{\xi}$.

4.7 Show that the internal energy term in Eq. (4.91) $e\rho$ is equal to

$$e\rho = \frac{p}{\gamma - 1}$$

The first law $de = Tds - pd\rho^{-1}$ gives $e = -\int pd(\rho^{-1})$ in adiabatic fluid and we have $p\rho^{-\gamma} = C$, const. Hence $\rho = C^{-\frac{1}{\gamma}}p^{\frac{1}{\gamma}}$,

$$d(\rho^{-1}) = -\frac{1}{\gamma}C^{\frac{1}{\gamma}}p^{-\frac{1}{\gamma}-1}dp$$

Hence integrating gives

$$e = C^{\frac{1}{\gamma}}\frac{p^{-\frac{1}{\gamma}+1}}{\gamma - 1}, \quad e\rho = \frac{p}{\gamma - 1}.$$

References

Goldstein, H. *Classical Mechanics*, Addison-Wesley (1980).
Hildebrand, F. B. *Methods of Applied Mathematics*, Prentice-Hall (1965).
Lamb, H. *Hydrodynamics*, Cambridge (1952).
Lanczos, C. *The Variational Principles of Mechanics*, Dover (1986).
Lin, C. C. *Proceedings of the International School of Physics,* Varrenna New York (1963).
Weinstock, R. *Calculus of Variations*, McGraw-Hill (1952).
Yourgrau, W. and S. Mandelstam, *Variational Principles in Dynamics and Quantum Theory*, Dover (1968).

Chapter 5

Variational Principle of Magnetohydrodynamics

5.1 Mathematical Preliminary

We saw that Lagrangian formulation of fluid dynamics is advantageous for establishing variational principle in continuous media. *Lagrange expansion* is a useful tool to write the equation of motion in a continuous medium in terms of the Lagrangian *displacement vector*. Fluid quantities written in terms of Eulerian variable can be expressed in terms of the Lagrangian variable with the assistance of Lagrange expansion.

• **Lagrange expansion.** Let us consider the transformation

$$\mathbf{x}'(\mathbf{x}, \lambda) = \mathbf{x} + \lambda \overrightarrow{\xi}(\mathbf{x}) \tag{5.1}$$

where λ is a parameter. Essentially we consider the transformation

$$\mathbf{x}'(\mathbf{x}) = \mathbf{x} + \overrightarrow{\xi}(\mathbf{x}) \tag{5.2}$$

A particle at $\mathbf{x}$ is displaced by the amount of vector $\overrightarrow{\xi}$. The particle is now located at the new position $\mathbf{x}'$. As a result of the overall relocation, the original density distribution $\rho(\mathbf{x})$ will be changed to a new distribution that we write as $\rho'(\mathbf{x}) = \rho'(\mathbf{x}'(\mathbf{x}))$. We wish to calculate $\rho'(\mathbf{x})$ in terms of $\rho(\mathbf{x})$ and $\overrightarrow{\xi}(\mathbf{x})$ and λ. Our goal can be approached by using the Fourier transform technique.

$$\rho(\mathbf{x}) = \int d^3k \, e^{i\mathbf{k}\cdot\mathbf{x}} \rho(\mathbf{k}) \tag{5.3}$$

whose inversion is

$$\rho(\mathbf{k}) = (2\pi)^{-3} \int d^3x \, e^{-i\mathbf{k}\cdot\mathbf{x}} \rho(\mathbf{x}) \tag{5.4}$$

For the new distribution, the Fourier transform pair is written as

$$\rho'(\mathbf{x}') = \int d^3k\, e^{i\mathbf{k}\cdot\mathbf{x}'} \rho'(\mathbf{k}) \tag{5.5}$$

$$\begin{aligned}\rho'(\mathbf{k}) &= (2\pi)^{-3} \int d^3x'\, e^{-i\mathbf{k}\cdot\mathbf{x}'} \rho'(\mathbf{x}') \\ &= (2\pi)^{-3} \int d^3x \rho(\mathbf{x})\, e^{-i\mathbf{k}\cdot[\mathbf{x}+\lambda\vec{\xi}(\mathbf{x})]} \end{aligned} \tag{5.6}$$

where we used (5.1) and

$$d^3x'\rho'(\mathbf{x}') = d^3x\rho(\mathbf{x}) \tag{5.7}$$

Then the inversion of (5.6) is written as

$$\rho'(\mathbf{x}) = \int d^3k\, e^{i\mathbf{k}\cdot\mathbf{x}} (2\pi)^{-3} \int d^3w \rho(\mathbf{w})\, e^{-i\mathbf{k}\cdot[\mathbf{w}+\lambda\vec{\xi}(\mathbf{w})]} \tag{5.8}$$

Note that $\mathbf{w}$ is a dummy variable. We change the order of integration and expand in powers of λ to write

$$\rho'(\mathbf{x}) = (2\pi)^{-3} \int d^3w \rho(\mathbf{w}) \int d^3k e^{i\mathbf{k}\cdot(\mathbf{x}-\mathbf{w})} \left[1 - i\lambda\mathbf{k}\cdot\vec{\xi}(\mathbf{w}) - \frac{1}{2}\lambda^2(\mathbf{k}\cdot\vec{\xi}(\mathbf{w}))^2 + \cdots\right] \tag{5.9}$$

Using the delta function formula

$$(2\pi)^{-3} \int d^3k e^{i\mathbf{k}\cdot(\mathbf{x}-\mathbf{w})} = \delta(\mathbf{x}-\mathbf{w})$$

the first term in the large bracket yields $\rho(\mathbf{x})$. The second term in the bracket is calculated by using $-i\frac{\partial}{\partial x_i}e^{ik_jx_j} = k_i e^{i\mathbf{k}\cdot\mathbf{x}}$ as

$$\begin{aligned}\text{2nd term} &= -i\lambda(2\pi)^{-3} \int d^3w \rho(\mathbf{w})(-i)\frac{\partial}{\partial x_i} \int d^3k \xi_i(\mathbf{w}) e^{i\mathbf{k}\cdot(\mathbf{x}-\mathbf{w})} \\ &= -\lambda\frac{\partial}{\partial x_i}(\rho(\mathbf{x})\xi_i(\mathbf{x}))\end{aligned}$$

To calculate the λ^2 term we use

$$(\mathbf{k}\cdot\vec{\xi})^2 e^{i\mathbf{k}\cdot\mathbf{x}} = -\xi_p\xi_q \frac{\partial}{\partial x_p}\frac{\partial}{\partial x_q} e^{i\mathbf{k}\cdot\mathbf{x}}$$

Then we finally obtain the Lagrange expansion

$$\rho'(\mathbf{x}) = \rho(\mathbf{x}) - \lambda\frac{\partial}{\partial x_i}(\rho(\mathbf{x})\xi_i(\mathbf{x})) + \frac{1}{2}\lambda^2\frac{\partial^2}{\partial x_i \partial x_j}(\rho(\mathbf{x})\xi_i(\mathbf{x})\xi_j(\mathbf{x})) - \cdots \tag{5.10}$$

This equation gives the *variation of density* due to the change of configuration of the particles as expressed by (5.2):

$$\delta\rho = -\nabla \cdot (\rho\,\overrightarrow{\xi}) \tag{5.11}$$

where we neglected the quadratic term of ξ_i. Equation (5.11) must be consistent with the continuity equation

$$\frac{\partial \rho}{\partial t} + \nabla \cdot (\rho \mathbf{v}) = 0 \tag{5.12}$$

Dividing (5.11) by time interval δt and equating $\overrightarrow{\xi}/\delta t = \mathbf{v}$, we recover (5.12). Or in reverse order, we make correspondence $\frac{\partial}{\partial t} \to \frac{\delta}{\delta t}$ and $\mathbf{v} \to \overrightarrow{\xi}/\delta t$ in (5.12). After canceling δt, we have a more direct correspondence:

$$\frac{\partial}{\partial t} \to \delta, \quad \mathbf{v} \to \overrightarrow{\xi} \tag{5.13}$$

Remark. Lagrange expansion is similar to Taylor expansion but there is a big difference. Let us take a look at the expression for a Taylor expansion

$$\begin{aligned}\psi(\mathbf{r} + \delta\mathbf{r}) &= \sum_{n=0}^{\infty}(\delta\mathbf{r} \cdot \nabla)^n \psi(\mathbf{r}) \\ &= \psi(\mathbf{r}) + \delta x_i \frac{\partial \psi}{\partial x_i} + \delta x_i \delta x_j \frac{\partial}{\partial x_i}\frac{\partial \psi}{\partial x_j} + \cdots\end{aligned}$$

In ordinary Taylor expansion the displacement δr is arbitrary; no dynamical constraint is imposed on it. In Lagrange expansion (5.10) the displacement $\overrightarrow{\xi}$ should remain in the parentheses, being subject to differentiation. It is because $\overrightarrow{\xi}$ is associated with particle movement. See

$$\overrightarrow{\xi}(\mathbf{x}, t) = \int_0^t dt' \mathbf{v}(\mathbf{x}, t'), \; \frac{\partial \overrightarrow{\xi}}{\partial t} = \mathbf{v}(\mathbf{x}, t)$$

Introducing $\overrightarrow{\xi}$ in (5.12) gives

$$\frac{\partial \rho}{\partial t} + \nabla \cdot \left(\rho \frac{\partial \overrightarrow{\xi}}{\partial t}\right) = 0$$

Integrating this equation over a certain time interval, we write

$$\delta\rho = -\nabla\cdot\left(\int_0^t dt'\rho'\frac{\partial\vec{\xi}}{\partial t'}\right)$$
$$= -\nabla\cdot\left(\rho\vec{\xi} - \int_0^t dt'\frac{\partial\rho'}{\partial t'}\vec{\xi}\right)$$

where we integrated by parts. Clearly the last integral is of second order in $\vec{\xi}$. Neglecting the term gives (5.11). Lagrange expansion is useful for deriving Fokker Planck equation.

5.2 Linear Stability Analysis

The governing MHD equations will be formulated in the form of Newtonian equation of motion to investigate the stability of the plasma disturbed by the displacement vector $\vec{\xi}(\mathbf{x},t)$ introduced in Section 5.1—$\vec{\xi}(\mathbf{x},t)$ is the first order perturbation. The governing equations of ideal MHD plasma developed in Chapter 3 are written here:

$$\frac{\partial\rho}{\partial t} + \nabla\cdot(\rho\mathbf{v}) = 0 \tag{5.14}$$

$$\rho\frac{d\mathbf{v}}{dt} = -\nabla p + \frac{1}{c}\mathbf{J}\times\mathbf{B} \tag{5.15}$$

$$\nabla\times\mathbf{B} = \frac{4\pi}{c}\mathbf{J} \tag{5.16}$$

$$\left(\frac{\partial}{\partial t} + \mathbf{v}\cdot\nabla\right)(p\rho^{-\gamma}) = 0 \tag{5.17}$$

$$\frac{\partial\mathbf{B}}{\partial t} = \nabla\times(\mathbf{v}\times\mathbf{B}) \tag{5.18}$$

The above equations are linearized by setting

$$\mathbf{v}\to 0+\delta\mathbf{v},\quad \mathbf{B}\to\mathbf{B}+\delta\mathbf{B},\quad p\to p+\delta p,\quad \rho\to\rho+\delta\rho \tag{5.19}$$

where δ-quantities are perturbations, and other field variables represent the equilibrium state of zero order magnitude; they are functions of $\mathbf{x}$ only (static). In the process of linearization, we use the correspondence in (5.13).

Equation (5.14) yields

$$\delta\rho + \nabla\cdot(\rho\vec{\xi}) = 0 \tag{5.20}$$

Equation (5.15) is linearized in the form

$$\rho\frac{\partial}{\partial t}\delta\mathbf{v} = -\nabla\delta p + \frac{1}{c}\left(\delta\mathbf{J}\times\mathbf{B} + \mathbf{J}\times\delta\mathbf{B}\right) \tag{5.21}$$

We have

$$\nabla\times\delta\mathbf{B} = \frac{4\pi}{c}\delta\mathbf{J} \tag{5.22}$$

$$(\delta + \overrightarrow{\xi}\cdot\nabla)(p\rho^{-\gamma}) = 0 \tag{5.23}$$

$$\delta\mathbf{B} = \nabla\times(\overrightarrow{\xi}\times\mathbf{B}) \tag{5.24}$$

After performing the variation, δ-algebra, in (5.23) and using (5.20) therein, we obtain

$$\delta p = -\gamma p\nabla\cdot\overrightarrow{\xi} - \overrightarrow{\xi}\cdot\nabla p \tag{5.25}$$

Equations (5.22) and (5.24) give

$$\delta\mathbf{J} = \frac{c}{4\pi}\nabla\times\nabla\times(\overrightarrow{\xi}\times\mathbf{B}) \tag{5.26}$$

Substitute (5.25), (5.26), and (5.16) into (5.21) to write

$$\rho\frac{\partial^2\overrightarrow{\xi}}{\partial t^2} = \mathbf{F}(\overrightarrow{\xi}) \tag{5.27}$$

where

$$\begin{aligned}\mathbf{F} &= \nabla\left[\gamma p\nabla\cdot\overrightarrow{\xi} + (\overrightarrow{\xi}\cdot\nabla)p\right] + \frac{1}{c}\left[\delta\mathbf{J}\times\mathbf{B} + \mathbf{J}\times\delta\mathbf{B}\right] + (\nabla U)\nabla\cdot(\rho\overrightarrow{\xi})\\ &= \nabla\left[\gamma p\nabla\cdot\overrightarrow{\xi} + (\overrightarrow{\xi}\cdot\nabla)p\right] + \frac{1}{4\pi}\left[\left(\nabla\times\nabla\times(\overrightarrow{\xi}\times\mathbf{B})\right)\times\mathbf{B}\right.\\ &\quad\left. + (\nabla\times\mathbf{B})\times\left(\nabla\times(\overrightarrow{\xi}\times\mathbf{B})\right)\right] + (\nabla U)\nabla\cdot(\rho\overrightarrow{\xi})\end{aligned}$$

where U is the external potential. Note that we used the correspondence in (5.13) to write down (5.27).

Equation (5.27) is a Newtonian equation of motion. We have succeeded in transforming the set of Eulerian fluid equations in (5.14)–(5.18) to a single equation of motion of the Newtonian type. Equation (5.27) is often the starting point for linear stability analysis. It can be used to investigate the stability of the system by looking for unstable modes of oscillation by applying Fourier analysis, as is well-known in the small oscillation theory of classical mechanics. It also can be used to derive the corresponding Lagrangian and Hamiltonian in analogous way as in classical mechanics.

Assuming a time dependence $\overrightarrow{\xi} \propto \exp(-i\omega t)$, (5.27) takes the form

$$\mathbf{F}(\overrightarrow{\xi}) + \rho\omega^2 \overrightarrow{\xi} = 0 \tag{5.28}$$

This equation should be solved under certain boundary conditions, including the periodic boundary condition, for $\overrightarrow{\xi}$, hence, there is in general no solution for arbitrary values of ω^2. Equation (5.28) constitute an eigenvalue problem for the eigenvalues ω^2 and the eigenfunctions $\overrightarrow{\xi}$ of the operator $\mathbf{F}$. Then the initial value problem can be solved by expanding the initial perturbations in terms of the eigenfunctions.

The operator $\mathbf{F}$ can be shown to be *self-adjoint*:

$$\int d^3x \overrightarrow{\eta} \cdot \mathbf{F}(\overrightarrow{\xi}) = \int d^3x \overrightarrow{\xi} \cdot \mathbf{F}(\overrightarrow{\eta})$$

for any choice of $\overrightarrow{\xi}$ and $\overrightarrow{\eta}$ subject to the boundary condition. The proof is referred to the original paper by Bernstein *et al.*, or Chandrasekhar.

In the earlier works, the potential energy is defined by

$$\begin{aligned}\delta W &= -\frac{1}{2}\int d^3x \overrightarrow{\xi} \cdot \mathbf{F}(\overrightarrow{\xi}) \\ &= -\frac{1}{2}\int d^3x \overrightarrow{\xi} \cdot \left[\nabla(\gamma p \nabla \cdot \overrightarrow{\xi} + (\overrightarrow{\xi} \cdot \nabla)p) + \frac{1}{c}(\delta\mathbf{J} \times \mathbf{B} + \mathbf{J} \times \delta\mathbf{B}) + (\nabla U)\nabla \cdot (\rho\overrightarrow{\xi})\right]\end{aligned} \tag{5.29}$$

The pressure terms are the perturbations associated with magnetoacoustic wave and the acoustic wave. The $\delta\mathbf{J}$ term contains the perturbation of the Alfven wave. The $\delta\mathbf{B}$ term gives rise to *kink* and *interchange instability*. The condition for instability is that δW should be negative for some $\overrightarrow{\xi}$. Numerous cases of instability have been identified by this means, especially in thermonuclear research.

By integrating by parts, (5.29) can be separated into volume integrals and surface integrals. This can be done by using Green theorem (Schmidt). Here we directly integrate by parts to show the detail of the necessary algebra. We rewrite (5.29) by using (5.22) in the form

$$\begin{aligned}\delta W = -\frac{1}{2}\int d^3x \overrightarrow{\xi} \cdot \Big[&\nabla f(\xi, p) + \frac{1}{4\pi}(\nabla \times \delta\mathbf{B}) \times \mathbf{B} + \frac{1}{4\pi}(\nabla \times \mathbf{B}) \\ &\times \delta\mathbf{B} + (\nabla U)\nabla \cdot (\rho\overrightarrow{\xi})\Big]\end{aligned} \tag{5.30}$$

where

$$f(\xi, p) = \gamma p \nabla \cdot \vec{\xi} + (\vec{\xi} \cdot \nabla) p$$

$$\int d^3x \vec{\xi} \cdot \nabla f = \int d^3x \xi_i \frac{\partial f}{\partial x_i} = -\int f \frac{\partial \xi_i}{\partial x_i} d^3x + \int dS_i \xi_i f = -\int d^3x f \nabla \cdot \vec{\xi} + \int f \vec{\xi} \cdot d\mathbf{S}$$

where dS_i is the surface element vector directed i direction.

$$\begin{aligned} \int d^3x \vec{\xi} \cdot (\nabla \times \delta\mathbf{B}) \times \mathbf{B} &= \int d^3x \xi_i e_{ijk} e_{jlm} \frac{\partial}{\partial x_l} \delta B_m B_k \\ &= -\int d^3x \xi_i B_k \left(\frac{\partial}{\partial x_i} \delta B_k - \frac{\partial}{\partial x_k} \delta B_i \right) \\ &= \int d^3x \delta B_k \left(\frac{\partial}{\partial x_i} (B_k \xi_i) - \frac{\partial}{\partial x_i} (B_i \xi_k) \right) \\ &\quad - \int dS_i (\xi_i B_k - \xi_k B_i) \delta B_k \\ &= -\int d^3x \delta B_k \delta B_k + \int d\mathbf{S} \cdot (\vec{\xi} \times \mathbf{B}) \times \delta\mathbf{B} \end{aligned} \tag{5.31}$$

where we used (5.24). Using (5.31) in (5.30) gives

$$\begin{aligned} \delta W = \frac{1}{2} \int d^3x \Big[\frac{(\delta B)^2}{4\pi} + \frac{1}{4\pi} (\nabla \times \mathbf{B}) \cdot (\vec{\xi} \times \delta\mathbf{B}) + f(\xi, p)(\nabla \cdot \vec{\xi}) - (\vec{\xi} \cdot \nabla U) \nabla \cdot (\rho \vec{\xi}) \Big] \\ -\frac{1}{2} \int \left[\frac{1}{4\pi} (\vec{\xi} \times \mathbf{B}) \times \delta\mathbf{B} + f(\xi, p) \vec{\xi} \right] \cdot d\mathbf{S} \end{aligned} \tag{5.32}$$

This equation is useful when we study instability in a bounded MHD plasma.

The volume integral in (5.32) ($\equiv \delta W_p$) represents the plasma energy perturbation. In the absence of the surface currents flowing over the boundary surface, the surface integral in (5.32) can be converted to a volume integral over the vacuum region in the form

$$\delta W_v = \frac{1}{8\pi} \int_{\text{vacuum}} (\delta B_v)^2 d^3x$$

which is the vacuum field energy perturbation. Therefore the total potential energy perturbation δW can be written as

$$\delta W = \delta W_p + \delta W_v.$$

• **Boundary conditions.** The conditions to be satisfied at a fluid-fluid boundary or fluid-vacuum boundary can be obtained by an infinitesimal integration of the relevant equations across a thin layer coinciding with the surface. Let us introduce local cartesian coordinate system in which $x = 0$ is the tangential plane to the interface, hence, $\hat{\mathbf{x}}$ is the normal unit vector to the interface. In the infinitesimal integration

$$\lim_{\epsilon \to 0} \int_{-\epsilon}^{\epsilon} (\cdots) dx \equiv [\cdots], \tag{5.33}$$

only the perfect differential terms survive in the limit $\epsilon \to 0$. We denote by $[\,]$ the difference of the integrated terms between the two sides 1 and 2. The continuity equation (5.14) yields after the infinitesimal integration

$$[\rho v_x] = 0 \tag{5.34}$$

which implies that the mass flux is continuous. At the fluid-vacuum boundary, we simply have

$$v_x = 0 \quad (\text{at inter face}) \tag{5.35}$$

$\nabla \cdot \mathbf{B} = 0$ gives

$$[B_x] = 0 \tag{5.36}$$

Equation (5.16) yields

$$[\hat{\mathbf{x}} \times \mathbf{B}] = \frac{4\pi}{c} \mathbf{J}^* = \frac{4\pi}{c} \lim_{\epsilon \to 0} \int_{-\epsilon}^{\epsilon} \mathbf{J} dx \tag{5.37}$$

where $\mathbf{J}^*$ denotes surface current. This equation implies that the tangential component of the magnetic field is discontinuous by the amount of the surface current.

The infinitesimal integration upon (5.15) gives

$$
\begin{aligned}
[p] &= \frac{1}{c}[\mathbf{J} \times \mathbf{B}] \\
&= \frac{1}{4\pi}\int_{-\epsilon}^{\epsilon} (\nabla \times \mathbf{B}) \times \mathbf{B} dx \\
&= \frac{1}{4\pi}\int_{-\epsilon}^{\epsilon} \left(-\nabla \frac{B^2}{2} + \mathbf{B} \cdot \nabla \mathbf{B} \right) dx \\
&= -\frac{1}{8\pi}[B_y^2 + B_z^2]
\end{aligned}
$$

Adding $[B_x^2] = 0$ to the above, we can write

$$\left[p + \frac{B^2}{8\pi} \right] = 0 \tag{5.38}$$

The above formulas can be expressed in general configuration in terms of $\hat{\mathbf{n}}$, the normal to the surface vector as follows:

$$\hat{\mathbf{n}} \cdot [\rho \mathbf{v}] = 0 \tag{5.39}$$

$$\hat{\mathbf{n}} \cdot [\mathbf{B}] = 0 \tag{5.40}$$

$$\hat{\mathbf{n}} \times [\mathbf{B}] = \frac{4\pi}{c}\mathbf{J}^* \tag{5.41}$$

$$\hat{\mathbf{n}} \cdot \overrightarrow{\xi} = 0 \quad (\text{at the surface}) \tag{5.42}$$

Example. Normal modes in an unbounded gravitating adiabatic MHD fluid.

As an application of $\overrightarrow{\xi}$-equation, we consider propagation of linear waves in an infinite MHD plasma in the gravitational field and investigate the gravitational stability. The unperturbed state is of uniform density ρ_0 and is permeated with constant magnetic field $\mathbf{B}_0 = (0, 0, B_0)$ in cartesian coordinates. The zero order equilibrium state is denoted by subscript '0' and the perturbation by '1'. The momentum equation in (5.15) should be augmented by adding the gravitational potential term $-\rho_0 \nabla U$.

$$\rho_0 \frac{\partial^2 \overrightarrow{\xi}}{\partial t^2} = -\nabla p_1 - \rho_0 \nabla U + \frac{1}{4\pi}[\nabla \times \nabla \times (\overrightarrow{\xi} \times \mathbf{B}_0)] \times \mathbf{B}_0 \tag{5.43}$$

The adiabatic gas law $p\rho^{-\gamma} = p_0 \rho_0^{-\gamma}$ gives $\delta p = (\gamma p_0/\rho_0)\delta\rho$, viz.,

$$p_1 = c_0^2 \rho_1 \tag{5.44}$$

where $c_0 = \sqrt{\gamma p_0/\rho_0}$ is the speed of sound. Equation (5.20) is written in present notation

$$\rho_1 = -\rho_0 \nabla \cdot \overrightarrow{\xi} \tag{5.45}$$

Now taking divergence of (5.43) yields

$$\left(\frac{\partial^2}{\partial t^2} - c_0^2\nabla^2 - V_A^2\nabla^2 - 4\pi G\rho_0\right)\nabla \cdot \overrightarrow{\xi} = -V_A^2\nabla^2\frac{\partial \xi_z}{\partial z} \tag{5.46}$$

where V_A is the Alfven velocity, $V_A^2 = B_0^2/4\pi\rho_0$, and we used

$$\nabla^2 U = 4\pi G\rho_1 \tag{5.47}$$

where G is the gravitation constant.

The z-component of (5.43) is

$$\frac{\partial^2 \xi_z}{\partial t^2} = c_0^2\frac{\partial}{\partial z}(\nabla \cdot \overrightarrow{\xi}) - \frac{\partial U}{\partial z} \tag{5.48}$$

Eliminating U in (5.48) with the aid of (5.45) and (5.47) yields

$$\nabla^2\frac{\partial^2 \xi_z}{\partial t^2} = (c_0^2\nabla^2 + 4\pi G\rho_0)\frac{\partial}{\partial z}\nabla \cdot \overrightarrow{\xi} \tag{5.49}$$

Equations (5.46) and (5.49) are the coupled equations for $\nabla \cdot \overrightarrow{\xi}$ and ξ_z. Assuming a plane wave solution of the form

$$\nabla \cdot \overrightarrow{\xi} = A\exp(i\mathbf{k}\cdot\mathbf{r} - i\omega t)$$
$$\xi_z = C\exp(i\mathbf{k}\cdot\mathbf{r} - i\omega t)$$

we obtain a set of simultaneous homogeneous equations for the amplitudes A and C. The dispersion relation is obtained by setting to zero the determinant composed by the coefficients of unknowns A and C.

$$\omega^4 - \omega^2[k^2(c_0^2 + V_A^2) - 4\pi G\rho_0] + k_z^2V_A^2(c_0^2k^2 - 4\pi G\rho_0) = 0 \tag{5.50}$$

which is solved by quadrature

$$\omega^2 = \frac{1}{2}[k^2V_A^2 + \alpha \pm \sqrt{(k^2V_A^2 + \alpha)^2 - 4k_z^2V_A^2\alpha}] \text{ (Cf.(3.119))} \tag{5.51}$$

where

$$\alpha = c_0^2k^2 - 4\pi G\rho_0 \tag{5.52}$$

When $\mathbf{k} = \hat{\mathbf{z}}k_z$, we obtain two modes:

$$\omega_+^2 = k_z^2 c_0^2 - 4\pi G\rho_0 \tag{5.53}$$

$$\omega_-^2 = k_z^2 V_A^2 \quad \text{(Alfven wave)} \tag{5.54}$$

The mode ω_+ represents the sound wave modified by gravity.

The square root is real and $\omega^2 > 0$ in (5.51), if $\alpha > 0$, viz.,

$$k^2 c_0^2 > 4\pi G\rho_0 \tag{5.55}$$

This is the criterion for gravitational stability first obtained by Jeans—*Jeans' criterion.* If the magnetic field vanishes, (5.51) gives only one mode: $\omega^2 = \alpha$. The magnetic field splits the otherwise single mode.

5.3 Variational Principle of Magnetohydrodynamics

In this section, we establish the variational principle of the MHD equations by using the variations derived in Section 5.2. Let us remember that the displacement vector $\vec{\xi}$ is the first order perturbed quantity. Consulting (5.2) and Fig. 3.2 in Chapter 3, we can make the following correspondence:

$$\overrightarrow{ab} = \delta\mathbf{v}\delta t, \quad \overrightarrow{aa'} = \frac{\partial\vec{\xi}}{\partial t}\delta t, \quad \overrightarrow{a'b'} = \mathbf{v}\cdot\nabla\vec{\xi}\,\delta t, \quad \overrightarrow{bb'} = \vec{\xi}\cdot\nabla\mathbf{v}\delta t$$

Therefore we have the equality

$$\delta\mathbf{v} = \frac{\partial\vec{\xi}}{\partial t} + \mathbf{v}\cdot\nabla\vec{\xi} - \vec{\xi}\cdot\nabla\mathbf{v} \tag{5.56}$$

For other Eulerian fluid variables, we have the equations that we have already obtained in the preceding section

$$\delta\rho = -\nabla\cdot(\rho\vec{\xi}) \tag{5.57}$$

$$\delta p = -\gamma p\nabla\cdot\vec{\xi} - \vec{\xi}\cdot\nabla p \tag{5.58}$$

$$\delta\mathbf{B} = \nabla\times(\vec{\xi}\times\mathbf{B}) \tag{5.59}$$

Equations (5.56)–(5.59) give the Eulerian perturbations due to the displacement vector $\vec{\xi}$.

We shall show that the Lagrangian for the ideal MHD flow

$$L = \frac{\rho v^2}{2} - \frac{p}{\gamma - 1} - \frac{B^2}{8\pi} \tag{5.60}$$

satisfies the Hamilton's principle. Thus the variational integral

$$\delta \int_{t_0}^{t_1} dt \int d^3x L = \delta \int_{t_0}^{t_1} dt \int d^3x \left(\frac{\rho v^2}{2} - \frac{p}{\gamma - 1} - \frac{B^2}{8\pi} \right) = 0 \tag{5.61}$$

satisfies the MHD equations (5.14)–(5.18), and vice versa. Taking the variation in (5.60) and using (5.56)–(5.59) give

$$\delta L = \int_{t_0}^{t_1} dt \int d^3x \Bigg[\rho v_i \left(\frac{\partial \xi_i}{\partial t} + v_j \frac{\partial}{\partial x_j} \xi_i - \xi_j \frac{\partial v_i}{\partial x_j} \right) - \frac{\partial}{\partial x_i} (\rho \xi_i) \frac{v^2}{2} + \frac{\gamma p \nabla \cdot \vec{\xi}}{\gamma - 1} + \frac{\vec{\xi} \cdot \nabla p}{\gamma - 1} - \frac{1}{4\pi} \mathbf{B} \cdot \nabla \times (\vec{\xi} \times \mathbf{B}) \Bigg] = 0 \tag{5.62}$$

Integrating by parts yields

$$\delta L = \int_{t_0}^{t_1} dt \int d^3x \, \vec{\xi} \cdot \left[\rho \frac{d\mathbf{v}}{dt} + \nabla p - \frac{1}{c} \mathbf{J} \times \mathbf{B} \right] = 0 \tag{5.63}$$

Note that the third and fourth terms in (5.62) cancel. Hence, $\delta L = 0$ for all $\vec{\xi}$ vanishing at t_0, t_1, and spatial boundaries, if and only if (5.60) is satisfied. Thus we have established the *Hamilton's principle* for the MHD equations in (5.14)–(5.18).

We identify in (5.60) $kinetic\, energy = \frac{\rho v^2}{2}$ and $potential\, energy = \frac{p}{\gamma - 1} + \frac{B^2}{8\pi}$. This identification is based on the view that each fluid element is regarded as a 'particle'. Thus we write down the Hamiltonian of the ideal MHD equations

$$H = \frac{\rho v^2}{2} + \frac{p}{\gamma - 1} + \frac{B^2}{8\pi} \tag{5.64}$$

This result can be established more rigorously directly from the variational principle.

Example. Obtain the Euler–Lagrange equations from the following Lagrangian.

$$\mathcal{L} = \frac{E^2 - B^2}{8\pi} - \rho_c \phi + \frac{\mathbf{J} \cdot \mathbf{A}}{c} + \frac{1}{2} \rho v^2 - \rho (e + U) - \psi \left(\frac{\partial \rho}{\partial t} + \nabla \cdot (\rho \mathbf{v}) \right) + \lambda \rho \left(\frac{\partial \mu}{\partial t} + \mathbf{v} \cdot \nabla \mu \right) \tag{5.65}$$

where

$$\mathbf{v} = -\nabla\psi - \lambda\nabla\mu \tag{5.66}$$

$$\frac{\partial\lambda}{\partial t} + \mathbf{v}\nabla\lambda = 0 \tag{5.67}$$

$$\frac{\partial\mu}{\partial t} + \mathbf{v}\nabla\mu = 0 \tag{5.68}$$

$$\mathbf{E} = -\nabla\phi - \frac{1}{c}\frac{\partial\mathbf{A}}{\partial t} \tag{5.69}$$

$$\mathbf{B} = \nabla\times\mathbf{A} \tag{5.70}$$

Lagrangian $\mathcal{L}$ in (5.65) is a collection of the Lagrangians introduced earlier in Chapters 4 and 5. In the Lagrangian $\mathcal{L}$, the independent variables are $\mathbf{r}, t$. The functions in the functional $\mathcal{L}$ are $\phi, \mathbf{A}, \rho, \mathbf{v}$. The Euler–Lagrange equations issuing from $\mathcal{L}$ can be obtained by applying the formula (4.22). The electromagnetic part of the Lagrangian

$$\mathcal{L}_{em} = \frac{E^2 - B^2}{8\pi} - \rho_c\phi + \frac{\mathbf{J}\cdot\mathbf{A}}{c} \tag{5.71}$$

yields the Poisson equation and the Maxwell equation in (4.114)—See Example in (4.113).

In MHD context, the following Lagrangian is useful:

$$\mathcal{L}_b = -\frac{B^2}{8\pi} + \frac{\mathbf{J}\cdot\mathbf{A}}{c} \tag{5.72}$$

which yields, via variation with respect to $\mathbf{A}$,

$$\nabla\times\mathbf{B} = \frac{4\pi}{c}\mathbf{J} \quad (\text{Cf.}(4.119)) \tag{5.73}$$

Finally, the fluid part of the Lagrangian, written in terms of the Clebsch potentials ψ and μ

$$\mathcal{L}_f = \frac{1}{2}\rho v^2 - \rho(e+U) - \psi\left(\frac{\partial\rho}{\partial t} + \nabla\cdot(\rho\mathbf{v})\right) + \lambda\rho\left(\frac{\partial\mu}{\partial t} + \mathbf{v}\cdot\nabla\mu\right) \tag{5.74}$$

yields the Euler momentum equation in (4.112), as has been dealt with in Problem (4.7).

Problems

5.1 Show that the adiabatic law $p\rho^{-\gamma} = \text{const.}$ can be replaced by

$$\frac{\partial s}{\partial t} = -\mathbf{v}\cdot\nabla s$$

or

$$\frac{\partial p}{\partial t} + \nabla \cdot (p\mathbf{v}) = -(\gamma - 1)p\nabla \cdot \mathbf{v}$$

5.2 Write down the Lagrangian in the *double adiabatic fluid* (see Eq. (3.75)). *Hint*: $p/(\gamma - 1) \to p_\perp + p_\parallel/2, 3p \to 2p_\perp + p_\parallel$.

5.3 Construct energy conservation equation of ideal MHD equations (Hasegawa).

Answer. We begin with (5.15) scalar dotting with $\mathbf{v}$:

$$(1)\ \rho\mathbf{v} \cdot \frac{d\mathbf{v}}{dt} = -\mathbf{v} \cdot \nabla p + \frac{1}{c}\mathbf{v} \cdot \mathbf{J} \times \mathbf{B}$$

The left hand side is

$$\begin{aligned}\rho\mathbf{v} \cdot \frac{d\mathbf{v}}{dt} &= \rho\mathbf{v} \cdot \left(\frac{\partial \mathbf{v}}{\partial t} + \mathbf{v} \cdot \nabla\mathbf{v}\right) \\ &= \rho\left[\frac{\partial}{\partial t}\frac{v^2}{2} + \mathbf{v} \cdot \left(\nabla\frac{v^2}{2} - \mathbf{v} \times (\nabla \times \mathbf{v})\right)\right] \\ &= \frac{\partial}{\partial t}\left(\frac{\rho v^2}{2}\right) - \frac{v^2}{2}\frac{\partial \rho}{\partial t} + \nabla \cdot \left(\frac{\rho v^2\mathbf{v}}{2}\right) - \frac{v^2}{2}\nabla \cdot (\rho\mathbf{v}) \\ &= \frac{\partial}{\partial t}\left(\frac{\rho v^2}{2}\right) + \nabla \cdot \left(\frac{\rho v^2\mathbf{v}}{2}\right)\end{aligned}$$

The last term in (1):

$$\begin{aligned}\frac{1}{c}\mathbf{v} \cdot \mathbf{J} \times \mathbf{B} &= \frac{1}{c}\mathbf{J} \cdot (\mathbf{B} \times \mathbf{v}) = \mathbf{J} \cdot \mathbf{E}\ \text{(Ohm's law)} \\ &= \frac{c}{4\pi}\nabla \times \mathbf{B} \cdot \mathbf{E} \\ &= \frac{c}{4\pi}\left[\mathbf{B} \cdot \nabla \times \mathbf{E} - \nabla \cdot (\mathbf{E} \times \mathbf{B})\right] \\ &= -\frac{1}{8\pi}\frac{\partial B^2}{\partial t} - \frac{c}{4\pi}\nabla \cdot (\mathbf{E} \times \mathbf{B})\end{aligned}$$

The pressure term in (1):

$$\begin{aligned}\mathbf{v} \cdot \nabla p &= \nabla \cdot (\mathbf{v}p) - p\nabla \cdot \mathbf{v} \\ &= \nabla \cdot (\mathbf{v}p) + p\frac{d}{dt}\log \rho \\ &= \nabla \cdot (\mathbf{v}p) + \frac{p}{\gamma}\frac{d}{dt}\log p \\ &= \nabla \cdot (\mathbf{v}p) + \frac{1}{\gamma}\left(\frac{\partial p}{\partial t} + \mathbf{v} \cdot \nabla p\right)\end{aligned}$$

Therefore

$$\mathbf{v}\cdot\nabla p=\nabla\cdot\left(\frac{\gamma p\mathbf{v}}{\gamma-1}\right)+\frac{\partial}{\partial t}\left(\frac{p}{\gamma-1}\right)$$

Collecting the above results gives the energy conservation equation of ideal MHD equation in the form

$$\frac{\partial}{\partial t}\left[\frac{\rho v^2}{2}+\frac{B^2}{8\pi}+\frac{p}{\gamma-1}\right]+\nabla\cdot\left[\frac{\rho v^2\mathbf{v}}{2}+\frac{c}{4\pi}\mathbf{E}\times\mathbf{B}+\frac{\gamma p\mathbf{v}}{\gamma-1}\right]=0$$

5.4 Show that the potential in (5.29) takes the form [Van Kampen p. 74; Sturrock (16.4.6)]

$$\delta W=\frac{1}{2}\int\left[\gamma p(\nabla\cdot\vec{\xi})^2+(\nabla\cdot\vec{\xi})(\vec{\xi}\cdot\nabla)p+\frac{1}{4\pi}(\nabla\times(\vec{\xi}\times\mathbf{B}))^2\right.$$
$$\left.-\frac{1}{4\pi}(\vec{\xi}\times\nabla\times\mathbf{B})\cdot\nabla\times(\vec{\xi}\times\mathbf{B})\right]d^3x$$

5.5 Equation (5.27) is the important equation upon which linear stability analysis of MHD flow is based. In the text, it has been derived by using Lagrange expansion. In fact we can derive the equation through the familiar procedure of juggling the linearized equations. Use the following equations to derive (5.27). Subscripts '0' and '1' denote the equilibrium (static) and perturbed quantities, respectively. $\rho=\rho_0+\rho_1, p=p_0+p_1, \mathbf{v}=0+\mathbf{v}_1, \mathbf{B}=\mathbf{B}_0+\mathbf{B}_1, \mathbf{E}=0+\mathbf{E}_1$.

$$\frac{\partial\rho_1}{\partial t}+\nabla\cdot(\rho_0\mathbf{v}_1)=0$$

$$\rho_0\frac{\partial\mathbf{v_1}}{\partial t}=-\nabla p_1+\frac{1}{4\pi}(\nabla\times\mathbf{B_1})\times\mathbf{B}_0+\frac{1}{4\pi}(\nabla\times\mathbf{B_0})\times\mathbf{B}_1$$

$$\frac{\partial\mathbf{B}_1}{\partial t}=\nabla\times(\mathbf{v}_1\times\mathbf{B}_0)$$

$$\frac{\partial p}{\partial t}+\mathbf{v}\cdot\nabla p+\gamma p\nabla\cdot\mathbf{v}=0 \text{ (exact equation)}$$

Obtain (5.27) upon putting $\mathbf{v}_1=\frac{\partial\vec{\xi}}{\partial t}$.

5.6 Show that the surface integral in (5.32) can be converted to a volume integral over the vacuum region in the form (Schmidt, 1966; Biscamp, 1993)

$$\delta W_v=\frac{1}{8\pi}\int_{\text{vacuum}}(\delta B_v)^2d^3x$$

[Reference: Schmidt, Biscamp]

5.7 Integrate (5.29) by Green theorem to obtain (5.32).
Hint: use the following identities.

$$\vec{\xi} \cdot \nabla f = \nabla \cdot (\vec{\xi} f) - f \nabla \cdot \vec{\xi}$$

$$\vec{\xi} \cdot [(\nabla \times \delta \mathbf{B}) \times \mathbf{B}] = \nabla \cdot [(\vec{\xi} \times \mathbf{B}) \times \delta \mathbf{B}] - (\delta B)^2$$

$$\vec{\xi} \cdot [(\nabla \times \mathbf{B}) \times \delta \mathbf{B}] = (\nabla \times \mathbf{B}) \cdot (\delta \mathbf{B} \times \vec{\xi})$$

References

Bateman, G. *MHD Instabilities*, MIT Press (1978).

Bernstein, I. B. *The Variational Principle for Problems of Ideal Magnetohydrodynamic Stability*, Basic Plasma Physics I North-Holland (1983).

Bernstein, I. B., E. A. Frieman, M. D. Kruskal, and R. M. Kulsrud, *An Energy Principle for Hydromagnetic Stability Problems*, Proc. R. Soc. A**244**, 17.

Biscamp, D. *Nonlinear Magnetohydrodynamics*, Cambridge University Press (1993).

Chandrasekhar, S. *Hydrodynamic and Hydromagnetic Stability*, Dover (1961).

Cowling, T. G. *Magnetohydrodynamics*, Adam Hilger (1976).

Ferraro, V. C. A. and C. Plumpton, *An Introduction to Magneto-fluid Mechanics*, Clarendon Press (1966).

Kulsrud, R. M. *MHD Description of Plasma*, Basic Plasma Physics I North-Holland (1983).

Polovin, R. V. and V. P. Deumutskii, *Fundamentals of Magnetohydrodynamics*, Consultants Bureau (1990).

Schmidt, G. *Physics of High Temperature Plasmas An Introduction*, Academic Press (1966).

Sturrock, P. A. *Plasma Physics* Cambridge (1994).

Sturrock, P. A. *Generalization of the Lagrange expansion with application to physical problems*, J. Mathematical Phys. Vol. 1 Number 5 (1960).

Van Kampen, N. G. and B. U. Felderhof, *Theoretical Methods in Plasma Physics*, North-Holland (1967).

Yourgrau, W. and S. Mandelstam, *Variational Principles in Dynamics and Quantum Theory*, Dover (1968).

Chapter 6

Single Particle Motion in Electric and Magnetic Fields

If the density of a plasma is sufficiently rare, the collective effect of the medium can be neglected in the first approximation; the individual particle motion in given electric and magnetic fields need to be considered. The *Van Allen belt* formed around the earth is a collection of gyrating charged particles that move along the earth magnetic field and also drift in circumferential direction. The particles are primarily subject to the magnetic force due to the earth magnetic field. In this chapter, the electric and magnetic fields are assumed to be prescribed; the currents generated by the charged particles are ignored. Thus, the problem is purely mechanical but the necessary analysis requires a good deal of mathematical ingenuity. As a result of the study of particle orbits in a given electric and magnetic fields, scholars discovered three adiabatic invariants. A similar mathematical endeavor has been pursued in celestial mechanics. Also familiarity of the characteristics of particle motion in electric and magnetic fields is helpful to gain physical insight for rarefied collective plasma medium. Some electromagnetic properties of plasma can be derived from superposition of individual particle motion in prescribed electric and magnetic fields. The characteristic equation of the Vlasov equation is particle orbit. The equivalence of the orbit equation and the collisionless Boltzmann equation is referred to as *Jean's theorem* in astronomical literature.

6.1 Radius of Curvature Vector

Consider a positive charge moving on (x, y)-plane in a space with a static constant magnetic field $\mathbf{B} = \hat{\mathbf{z}}B$ imbedded. We have the equation of motion for single particle,

$$m\frac{d\mathbf{v}}{dt} = \frac{e}{c}\mathbf{v} \times \mathbf{B} \tag{6.1}$$

Since the force acting on the particle in (3.1) is always perpendicular to the velocity, the particle draws a circle. The sense of rotation is clockwise about the magnetic field (negative θ-direction in polar coordinate). The kinetic energy is constant, and the radius of the circle (ρ) is obtained from $\frac{mv^2}{\rho} = \frac{e}{c}vB$, giving $\rho = \frac{mcv}{eB}$, or in vector form

$$\vec{\rho} = \frac{mc}{eB^2}\mathbf{v} \times \mathbf{B} \tag{6.2}$$

The radius of curvature vector defined above is the vector drawn from the particle position to the center of *gyration* for (+) charge (for a negative charge it is the vector drawn from the gyration center to the particle). In a varying magnetic field, the center of the gyration, or the *guiding center*, drifts. In this case, we define the instantaneous radius curvature vector in analogous way.

6.2 $\vec{E} \times \vec{B}$ Drift

• **E,B = constant**. Consider an electron in constant fields:

$$m_e \frac{d\mathbf{v}}{dt} = -e\left(\mathbf{E} + \frac{1}{c}\mathbf{v} \times \mathbf{B}\right) \tag{6.3}$$

where we prescribe the fields as

$$\mathbf{E} = \hat{\mathbf{y}}E_y + \hat{\mathbf{z}}E_z, \quad \mathbf{B} = \hat{\mathbf{z}}B_0$$

In components, we have

$$\frac{dv_x}{dt} = -\omega_c v_y \qquad \left(\omega_c = \frac{eB_0}{m_e c} > 0, \text{cyclotron frequency}\right)$$

$$\frac{dv_y}{dt} = \omega_c v_x - \frac{e}{m_e}E_y, \quad \frac{dv_z}{dt} = -\frac{e}{m_e}E_z, \tag{6.4}$$

$$\frac{d^2 v_x}{dt^2} = \omega_c^2\left(-v_x + c\frac{E_y}{B_0}\right), \quad \frac{d^2 v_y}{dt^2} = -\omega_c^2 v_y \tag{6.5}$$

which are solved by

$$v_x = v_\perp \cos(\omega_c t + \beta) + v_d, \quad v_y = v_\perp \sin(\omega_c t + \beta), \tag{6.6}$$

where $v_\perp$ and β are constants of integration, and $v_d = c\frac{E_y}{B_0}$ is the *drift velocity*. Integrating once more gives

$$x = v_d t + \frac{v_\perp}{\omega_c}\sin(\omega_c t + \beta), \quad y = -\frac{v_\perp}{\omega_c}\cos(\omega_c t + \beta) \tag{6.7}$$

The orbit is visualized as follows. Its projection onto the (x, y)-plane is a circle of radius $\frac{v_\perp}{\omega_c}$ (*Larmor radius* or *gyro-radius*) whose center (guiding center) moves along the x-direction with the drift velocity v_d. Along the z-direction, the particle is accelerated according to (6.4). If $\mathbf{E}\|\mathbf{B}$ ($E_y = 0$), $v_d = 0$.

Alternatively, (6.3) can be solved by transforming as

$$\mathbf{v} = \frac{c}{B^2}\mathbf{E} \times \mathbf{B} + \mathbf{v}', \tag{6.8}$$

$$m_e \frac{d\mathbf{v}'}{dt} = \frac{-e\mathbf{E} \cdot \mathbf{B}}{B^2}\mathbf{B} - \frac{e}{c}\mathbf{v}' \times \mathbf{B} \tag{6.9}$$

The perpendicular motion to $\mathbf{B}$ is the circular motion as given by (6.1). The parallel motion along $\mathbf{B}$ is a motion of constant acceleration of $-\frac{e}{m}\frac{\mathbf{E}\cdot\mathbf{B}}{B}$. The orbit on the perpendicular plane is the circular motion represented by $\mathbf{v}'$ plus the $E \times B$ drift of the gyration center given by

$$\mathbf{v}_d = \frac{c}{B^2}\mathbf{E} \times \mathbf{B} \tag{6.10}$$

Note that the $E \times B$ drift is independent of the sign of charge.

The radius of curvature vector in constant $\mathbf{E}$ and $\mathbf{B}$ field is obtained as

$$\vec{\rho} = \frac{cm}{eB^2}\left(\mathbf{v} - \frac{c}{B^2}\mathbf{E} \times \mathbf{B}\right) \times \mathbf{B}$$

- **Constant force (F) and Constant magnetic field.**

A positive charge in this case is governed by

$$m\frac{d\mathbf{v}}{dt} = \mathbf{F} + \frac{e}{c}\mathbf{v} \times \mathbf{B}$$

The parallel motion is trivial: $m\frac{dv_\parallel}{dt} = F_\parallel$, so we assume $\mathbf{F}\perp\mathbf{B}$. One can easily prove that the motion is the gyration around $\mathbf{B}$ and the drift with the drift velocity

$$\mathbf{v}_d = \frac{c}{eB^2}\mathbf{F} \times \mathbf{B}$$

6.3 Equation for Guiding Center

Orbits in a space with a time-varying or inhomogeneous magnetic field imbedded in plasma can be approximated as circular motion about the zero order magnetic field plus the drift of the guiding center. This picture is a perturbational view that is valid when spatial and temporal variation of the magnetic field is small over the Larmor radius or the cyclotron period,

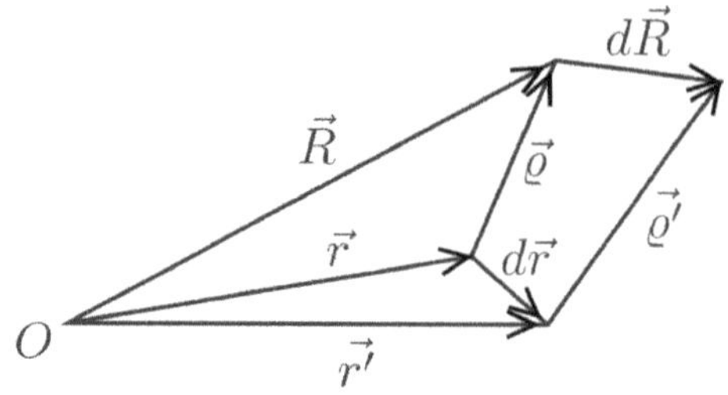

Fig. 6.1. $\vec{\rho} + d\mathbf{R} = d\mathbf{r} + \vec{\rho}'$.

respectively. The curvature of the magnetic field line can also give rise to the drift motion of the guiding center. For simplicity we assume that the magnetic field is only inhomogeneous, not time varying.

$$m\frac{d\mathbf{v}}{dt} = \frac{e}{c}\mathbf{v} \times \mathbf{B}(\mathbf{r}) \tag{6.11}$$

Consulting Fig. 6.1, where $\mathbf{r}$ is the particle position and $\mathbf{R}$ is the center of gyration (guiding center), we have $\mathbf{R}(t) = \mathbf{r}(t) + \vec{\rho}(t)$, and

$$\vec{\rho} = \frac{mc}{e}\frac{\mathbf{v} \times \mathbf{B}}{B^2} \quad \text{(for + charge)} \tag{6.12}$$

which is the radius of curvature vector. After an infinitesimal time dt, the corresponding change of the particle position is $d\mathbf{r}$, and the radius of curvature vector changes to $\vec{\rho}'$, giving the relation

$$d\mathbf{R} = d\mathbf{r} + \vec{\rho}' - \vec{\rho} \tag{6.13}$$

From (6.12), we obtain

$$\vec{\rho}' - \vec{\rho} = d\vec{\rho} = \frac{mc}{e}\left(\frac{d\mathbf{v} \times \mathbf{B} + \mathbf{v} \times d\mathbf{B}}{B^2} - 2\frac{dB}{B^3}\mathbf{v} \times \mathbf{B}\right) \tag{6.14}$$

Using (6.11) for $d\mathbf{v}$, we obtain

$$d\mathbf{R} = \frac{\mathbf{B}}{B^2}\mathbf{v} \cdot \mathbf{B}dt + \frac{mc}{e}\left(\frac{\mathbf{v} \times d\mathbf{B}}{B^2} - 2\frac{dB}{B^3}\mathbf{v} \times \mathbf{B}\right) \tag{6.15}$$

Here $\mathbf{B}$ takes the value at the particle position, so we write $\mathbf{B} = \mathbf{B}(\mathbf{r}(t))$; but not explicitly depending upon t. Then

$$d\mathbf{B} = dB_i = \frac{\partial B_i}{\partial r_j}\frac{dr_j}{dt}dt = v_j\frac{\partial B_i}{\partial r_j}dt = \mathbf{v} \cdot \nabla\mathbf{B}dt$$

Thus (6.15) gives the drift velocity of the guiding center,

$$\mathbf{v}_d = \frac{d\mathbf{R}}{dt} = \mathbf{v}_\parallel + \frac{mc}{e}\left(\frac{\mathbf{v}\times(\mathbf{v}\cdot\nabla)\mathbf{B}}{B^2} - 2\frac{\mathbf{v}\times\mathbf{B}}{B^3}(\mathbf{v}\cdot\nabla)B\right) \tag{6.16}$$

where $v_\parallel$ is the velocity component parallel to the magnetic field. (6.16) is the equation for the drift motion of the guiding center. We shall consider a few special cases of **B**.

(i) $\mathbf{B} = \hat{\mathbf{z}}B(x)$ (**B** has a gradient in x-direction). In this case we have

$$\mathbf{v}_d = \mathbf{v}_\parallel - \frac{mc}{e}\frac{v_x}{B^2}\frac{dB}{dx}(\hat{\mathbf{x}}v_y - \hat{\mathbf{y}}v_x)$$

Here the perpendicular velocity components $v_{x,y}$ are time-oscillating functions with the gyrofrequency (cyclotron frequency) ω_c. Therefore the drift velocity has a meaning when it is averaged over the period (T). We have $\langle v_x v_y\rangle = 0$ and $\langle v_x^2\rangle = \frac{1}{2}v_\perp^2$. Here $\langle\cdots\rangle = \frac{1}{T}\int_0^T(\cdots)dt$. So the above equation gives

$$\mathbf{v}_d = \mathbf{v}_\parallel + \hat{\mathbf{y}}\frac{mc}{eB^2}\frac{v_\perp^2}{2}\frac{dB}{dx} \tag{6.17}$$

(6.17) indicates that the guiding center drifts in the direction of $\mathbf{B}\times\nabla B$, and is called ∇B-drift.

(ii) **B**-field line slightly curved. As shown in Fig. 6.2, **B** vector has a small B_x component in addition to the main component $B_z (B_z \gg B_x)$.

We write

$$\mathbf{B} = \hat{\mathbf{x}}B_x(z) + \hat{\mathbf{z}}B_z(x, y)$$

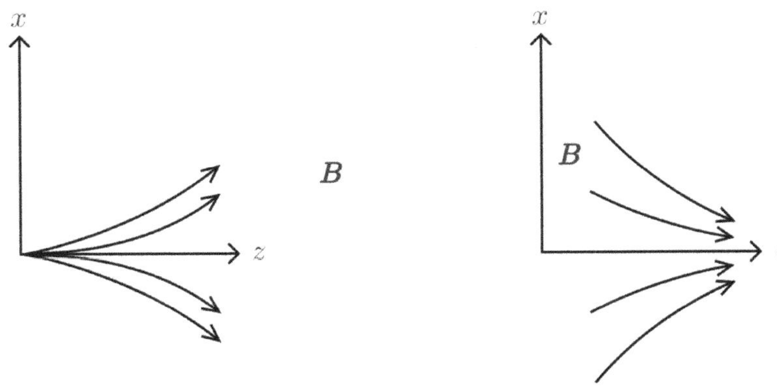

Fig. 6.2. Converging or diverging magnetic field.

Note that $\nabla \cdot \mathbf{B} = 0$. Introduce the differential operator $\mathbf{Q} = \mathbf{v}(\mathbf{v} \cdot \nabla) = v_i v_j \frac{\partial}{\partial x_j}$. Then ∇B-drift term takes a convenient form amenable to easy averaging

$$\mathbf{v}_d = \frac{mc}{eB^2}\left[\mathbf{Q} \times \mathbf{B} + \frac{2}{B}(\mathbf{B} \times \mathbf{Q})B\right] \tag{6.18}$$

Here we have $\nabla = \hat{\mathbf{z}}\frac{\partial}{\partial z} + \nabla_\perp$, with $\nabla_\perp = \hat{\mathbf{x}}\frac{\partial}{\partial x} + \hat{\mathbf{y}}\frac{\partial}{\partial y}$, and

$$\mathbf{Q} = (\hat{\mathbf{z}}v_\parallel + \mathbf{v}_\perp)\left(v_\parallel \frac{\partial}{\partial z} + \mathbf{v}_\perp \cdot \nabla_\perp\right)$$

which consists of quadratic terms of particle velocity components. $\mathbf{v}_\perp$ is a time-oscillating function expressed by sine or cosine functions with frequency ω_c. Thus we take an average over the period, by putting $\mathbf{v}_\perp = v_{\perp 0}(\hat{\mathbf{x}}\cos\omega_c t + \hat{\mathbf{y}}\sin\omega_c t)$, we obtain

$$\begin{aligned}\langle \mathbf{Q}\rangle &= \hat{\mathbf{z}}v_\parallel^2\frac{\partial}{\partial z} + \frac{v_{\perp 0}^2}{2}\left(\hat{\mathbf{x}}\frac{\partial}{\partial x} + \hat{\mathbf{y}}\frac{\partial}{\partial y}\right)\\ \langle \mathbf{Q}\rangle \times \mathbf{B} &= \hat{\mathbf{x}}\left(\frac{v_{\perp 0}^2}{2}\frac{\partial B_z}{\partial y} - v_\parallel^2\frac{\partial B_y}{\partial z}\right) - \hat{\mathbf{y}}\left(\frac{v_{\perp 0}^2}{2}\frac{\partial B_z}{\partial x} - v_\parallel^2\frac{\partial B_x}{\partial z}\right) + \hat{\mathbf{z}}\frac{v_{\perp 0}^2}{2}\left(\frac{\partial B_y}{\partial x} - \frac{\partial B_x}{\partial y}\right)\end{aligned} \tag{6.19}$$

$$\mathbf{B} \times \langle \mathbf{Q}\rangle B = \frac{v_{\perp 0}^2}{2}B_z\left(-\hat{\mathbf{x}}\frac{\partial B_z}{\partial y} + \hat{\mathbf{y}}\frac{\partial B_z}{\partial x}\right) \tag{6.20}$$

In obtaining (6.20), we put $B \simeq B_z$. Also we included the component B_y which disappears at the end. Using (6.19) and (6.20) in (6.18) gives

$$\mathbf{v}_d = \frac{cm}{eB^2}\left[-\hat{\mathbf{x}}\left(\frac{v_{\perp 0}^2}{2}\frac{\partial B_z}{\partial y} + v_\parallel^2\frac{\partial B_y}{\partial z}\right) + \hat{\mathbf{y}}\left(\frac{v_{\perp 0}^2}{2}\frac{\partial B_z}{\partial x} + v_\parallel^2\frac{\partial B_x}{\partial z}\right) + \hat{\mathbf{z}}\frac{v_{\perp 0}^2}{2}\left(\frac{\partial B_y}{\partial x} - \frac{\partial B_x}{\partial y}\right)\right] \tag{6.21}$$

Since $\mathbf{B}$ satisfies $\nabla \times \mathbf{B} = 0$, we have the relations, $\frac{\partial B_y}{\partial x} = \frac{\partial B_x}{\partial y}$, $\frac{\partial B_y}{\partial z} = \frac{\partial B_z}{\partial y}$, and $\frac{\partial B_x}{\partial z} = \frac{\partial B_z}{\partial x}$. Putting these relations into the above equation gives

$$\begin{aligned}\mathbf{v}_d &= \frac{cm}{eB^3}B_z\left(\frac{v_{\perp 0}^2}{2} + v_\parallel^2\right)\left[-\hat{\mathbf{x}}\frac{\partial B_z}{\partial y} + \hat{\mathbf{y}}\frac{\partial B_z}{\partial x}\right] = \left(\frac{v_{\perp 0}^2}{2} + v_\parallel^2\right)\frac{\hat{\mathbf{z}}B_z \times \nabla B_z}{B^2\omega_c}\\ &= \left(\frac{v_{\perp 0}^2}{2} + v_\parallel^2\right)\frac{\mathbf{B} \times \nabla B_z}{B^2\omega_c}\end{aligned} \tag{6.22}$$

The first term in (6.22) reproduces the previous result ∇B-drift. We give in the following an alternative physical interpretation of (6.22).

6.4 Magnetic Moment and ∇B Drift

Magnetic moment (μ) is defined by the product of the current and the area enclosed by the loop along which the current is flowing,

$$\mu = \frac{1}{c}\frac{e\omega_c}{2\pi}\pi r_L^2 = \frac{mv_\perp^2/2}{B}, \quad r_L = \frac{v_\perp}{\omega_c}, \text{ Larmor radius} \tag{6.23}$$

where the factor $\frac{1}{c}$ enters because we use the electrostatic unit system. In (6.22) the ∇B_z-drift reads

$$\mathbf{V}_{\nabla B_z} = \frac{v_{\perp 0}^2/2}{B^2\omega_c}\mathbf{B}\times\nabla \quad B_z = \frac{c}{eB^2}\mathbf{B}\times\mu\nabla B_z \tag{6.24}$$

Suppose a constant force $\mathbf{F}\perp\mathbf{B}$ acts upon a positive charge, in addition to the centripetal magnetic force $\frac{e}{c}\mathbf{v}\times\mathbf{B}$. Then the drift due to the force is

$$\mathbf{V}_F = \frac{c}{eB^2}\mathbf{F}\times\mathbf{B} \tag{6.25}$$

Comparison of (6.24) and (6.25) indicates that ∇B_z gives rise to an equivalent force

$$\mathbf{F} = -\mu\nabla B_z \tag{6.26}$$

acting on a charge with a magnetic moment μ.

6.5 Force Acting on a Charge Having a Magnetic Moment

In this section, we show that an electron with a magnetic moment μ is exerted by a force (6.26) due to a magnetic field whose line of force diverges as shown in Fig. 6.2. Write $\nabla\cdot\mathbf{B} = 0$ in cylindrical coordinate, assuming the azimuthal symmetry,

$$\frac{1}{r}\frac{\partial}{\partial r}(rB_r) + \frac{\partial B_z}{\partial z} = 0 \tag{6.27}$$

Integrate this equation with respect to r from 0 to ρ, the Larmor radius:

$$\rho B_r(\rho) = -\int_0^\rho r\frac{\partial B_z}{\partial z}dr$$

Since the main component of $\mathbf{B}$ is B_z and the line of force is slowly diverging, we can put $\frac{\partial B_z}{\partial z} \simeq \frac{\partial B}{\partial z}\big|_{r=0}$, which can be taken out of the integral, to obtain

$$B_r(\rho) = -\frac{\rho}{2}\frac{\partial B}{\partial z}$$

This radial component of the magnetic field together with the azimuthal velocity component of the gyrating motion $v_\theta = v_\perp$ gives rise to the Lorentz force $\mathbf{v}_\theta \times \mathbf{B}_r$ in the z-direction,

$$F_z = \frac{e}{c}v_\perp \quad B_r = -\mu\frac{\partial B}{\partial z} \quad (\text{per}\,\rho = v_\perp/\omega_c) \tag{6.28}$$

Defining magnetic moment vector,

$$\vec{\mu} = \frac{\mathbf{B}}{B^2}\frac{mv_\perp^2}{2} \tag{6.29}$$

we can write (6.28) in the form

$$F_z = -\vec{\mu}\cdot\nabla B \tag{6.30}$$

6.6 Curvature Drift

The second term in (6.22) is called the curvature drift, because it is given rise to by the centripetal force sliding along the curved line of force:

$$v_\parallel^2\frac{\mathbf{B}\times\nabla B_z}{B^2\omega_c} \;:\text{curvature drift} \tag{6.31}$$

In Fig. 6.3, the magnetic line of force is expressed by polar coordinate as $\mathbf{B} = \hat{\theta}B(r)$. $\nabla\times\mathbf{B} = 0$ gives $\frac{1}{r}\frac{\partial}{\partial r}(rB) = 0$. So we have $B \propto 1/r$, and $\nabla B/B = -\mathbf{r}/r^2$. Then the second term of (6.22) is

$$\mathbf{V}_{\text{curv}} = \frac{c}{eB^2}\frac{mv_\parallel^2}{r}\hat{\mathbf{r}}\times\mathbf{B} \tag{6.32}$$

Comparison of this equation with (6.25) indicates that the force responsible for the curvature drift is the force $\mathbf{F} = \frac{mv_\parallel^2}{r}\hat{\mathbf{r}}$, i.e. the centripetal force experienced by the particle moving along the curved line of force.

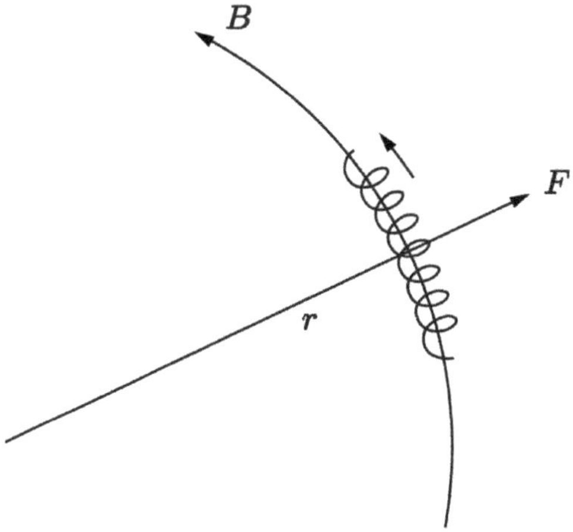

Fig. 6.3. Centrifugal force felt by a particle moving along a curved field line.

• **Magnetic moment of a charge circling around a slowly time-varying magnetic field.** This quantity is an approximate constant of motion as shown in the following. Since $\mathbf{B}(t)$ produces $\mathbf{E}(t)$ per Faraday's law, the electric force should be included in the equation of motion.

$$m\frac{d\mathbf{v}}{dt} = e\left(\mathbf{E}(t) + \frac{1}{c}\mathbf{v} \times \mathbf{B}(t)\right) \tag{6.33}$$

Assume $\mathbf{B}(t) = \hat{\mathbf{z}}B(t)$. The electric field satisfies $\nabla \cdot \mathbf{E} = 0$ and $\nabla \times \mathbf{E} = -\frac{1}{c}\frac{\partial \mathbf{B}}{\partial t}$. Hence we find

$$\mathbf{E} = -\frac{1}{2c}\hat{\mathbf{z}} \times \mathbf{r}\frac{dB}{dt} \tag{6.34}$$

$$\text{Then,} \quad \frac{d}{dt}\frac{mv^2}{2} = -\frac{e}{2c}\frac{dB}{dt}\hat{\mathbf{z}} \times \mathbf{r} \cdot \frac{d\mathbf{r}}{dt} \tag{6.35}$$

We shall consider that the particle motion takes place on the perpendicular plane, assuming $v_{\parallel} = 0$, and integrate (6.35) over one cycle of gyro-motion:

$$\delta\left(\frac{mv_{\perp}^2}{2}\right) = -\frac{e}{2c}\oint \frac{dB}{dt}\hat{\mathbf{z}} \cdot \mathbf{r} \times d\mathbf{r} \tag{6.36}$$

Here we assume that B(t) is such *a slowly varying function of time that* $\frac{dB}{dt}$ *remains almost constant over one gyro-motion*, and it can be taken out of the integral. Then the integral is twice the area of Larmor circle, $2\pi\rho^2$, giving

$$\delta\left(\frac{mv_\perp^2}{2}\right) = \frac{e}{c}\frac{dB}{dt}\pi\rho^2 = \frac{mv_\perp^2}{2B}\frac{dB/dt}{\omega_c/2\pi} \tag{6.37}$$

Note that

$$\delta(B) = \oint \frac{dB}{dt}dt \simeq \frac{dB}{dt}\frac{2\pi}{\omega_c}$$

Therefore (6.37) is the relation $\delta(\mu B) = \mu\delta B$ which yields

$$\delta\mu = 0 \quad \text{or} \quad \mu = \text{const.}$$

This proves that μ is an approximate constant of motion in a slowly changing magnetic field—*an adiabatic invariant.* The assumption that dB/dt remains constant over several gyroperiods is the *adiabatic assumption.* Putting it precisely, $B(t)$ changes slowly so that

$$\frac{dB}{dt} = \frac{\delta(B)}{\delta(t)} = \frac{\text{change of } B}{\text{corresponding time interval}} = \text{const.} \tag{6.38}$$

6.7 Adiabatic Invariant

An *adiabatic invariant* remains to be an approximate constant of motion of a nearly periodic system when a parameter of the system changes slowly. The word "adiabatic" appears to be coined from quantum mechanical situation where the energy states distribution is unchanged even though the Hamiltonian is changed by external means without adding heat to the system. When we have a particle moving in a potential well (trapped particle), the *adiabatic change* of the motion is brought about by infinitesimal dislocation of turning point. [In contrast, if the particle encounters with a grazing turning point, the subsequent motion is subject to nonadiabatic abrupt change.] The change in parameter should be slow as compared to the periods of motion and should not be related to the periods if it is an adiabatic change of motion. A well-known invariant for a particle moving in a magnetic field is its magnetic moment $\mu = W_\perp/B$ where $W_\perp$ is the kinetic energy of the perpendicular to B direction. With B changing slowly, the motion of the particle can be viewed as small perturbations from the constant field, which can be solved exactly. Here the usable small parameter is

$$\epsilon = \frac{T}{B}\frac{\partial B}{\partial t} \ll 1, \quad T: \text{period}$$

in terms of which we can work on the perturbation series for the orbit. The above equation states that the fractional change of B over one period is $\ll 1$. We are interested in knowing the accuracy of $\mu = \text{const}$. Being a constant of motion in the lowest order, i.e. in ϵ-order, is not enough. Since we are dealing with an asymptotic series, the true qualification for an invariant requires that the higher order deviation from invariance goes to zero rapidly as ϵ goes to zero. Let us designate $\Delta\mu$ the deviation of μ from the invariance. We need $\Delta\mu \ll \epsilon^N$, for all N, with N being the order. When $\epsilon = 10^{-3}$ and $N = 1$, $\Delta\mu = \epsilon^N$ gives the estimation that appreciable change of μ is given rise to after 10^3 oscillations. If the experiment time is longer than 10^3 periods, the adiabatic invariant for $N = 1$ cannot be considered to be an invariant. We need a higher order invariant in cosmic or laboratory scheme where the particles bounce back and forth for a large number of periods. It is desirable to prove the existence of higher order invariants in order to explain or support the validity of the scheme utilizing magnetic invariants, such as magnetic mirror confinement or Fermi acceleration. Oscillatory motion between two turning points can be depicted in a phase plane (p, q) by a closed loop, where p and q are the generalized momentum and coordinates, respectively. Depending upon the values of the total energy, $W = mv^2/2 + V(q, \lambda)$, with V the potential energy, the trajectories represent particles with different energies. The distribution function f of particles can be expressed in terms of W and θ, where θ is the angular coordinate along the loop: $f = f(W, \theta)$. f is time-independent only if it does not depend on θ (f is constant along particle trajectories). When $V(q, \lambda(t))$ is slowly changing through a parameter λ, the orbit will not be closed completely but spiralling with tiny pitches, showing a nearly periodic trajectory. According to Liouville equation, the points in the phase space behave as an incompressible fluid: the area enclosed by the almost closed loop is conserved. That is, the phase plane area enclosed by the orbit corresponding to any one of the orbit is an approximate constant of motion:

$$\oint pdq = J$$

J in the above is known as the action integral and is an adiabatic invariant. Adiabatic invariant is a fundamental property of nearly periodic Hamiltonian system. In the literature, its general discussion has been done by using perturbed trapped particle motion [S. Chandrasekhar, *Plasma Physics*, Notes compiled by S. K. Trehan, Phoenix Books, University of Chicago Press (1960)] or perturbed Hamiltonian [L. D. Landau and E. M. Lifshitz, *Mechanics*, Pergamon Press, Oxford (1960)].

Consider a one-dimensional motion in a potential $V(x,\lambda)$,

$$m\frac{d^2x}{dt^2} = -\frac{\partial V(x,\lambda)}{\partial x} \tag{6.39}$$

where λ is a parameter through which V can be varied, such as the length of an oscillating pendulum, subject to gravity. When λ is constant, we have an exact constant of integration,

$$\frac{m}{2}\left(\frac{dx}{dt}\right)^2 + V(x) = E : \text{const.} \tag{6.40}$$

Also the *action integral* defined below is also a constant of motion:

$$J = \oint m\frac{dx}{dt}dx \tag{6.41}$$

The integral $\oint$ means a round trip integral along the path between two turning points.

For a simple harmonic motion with $V(x) = \frac{k}{2}x^2$ (k is the force constant),

$$J = \oint m\left(\frac{dx}{dt}\right)^2 dt = \frac{k}{2}x_0^2 T = ET = 2\pi\frac{E}{\omega}, \quad (x_0 = \text{amplitude}, \omega = \sqrt{k/m}) \tag{6.42}$$

However, when λ is not constant but changes with time, (6.39) does not admit a constant of motion, hence the action integral is not a constant of motion; it is not even defined because the motion loses periodicity in the strict sense. But if $\lambda(t)$ is a *slow* enough function of t, the action integral is still an approximate constant of motion. The slow function $\lambda(t)$ changes appreciably only after many oscillations. That is, $T\frac{d\lambda}{dt} \ll \lambda$, where T is the period of oscillation defined by

$$T = \oint \frac{dx}{\dot{x}} \quad \left(\cdot = \frac{d}{dt}\right)$$

The turning points are not fixed in the long run, but they can be assumed to be fixed in a time interval of a period. We define the average

$$\langle E\rangle = \frac{1}{T}\oint dt\left[\frac{m}{2}\dot{x}^2 + V(x,\lambda)\right] \tag{6.43}$$

The turning points are obtained from

$$\frac{m}{2}\dot{x}^2 + V(x,\lambda) = \langle E\rangle \tag{6.44}$$

The corresponding action integral is

$$J = \sqrt{2m} \oint dx \sqrt{\langle E \rangle - V(x, \lambda)} \tag{6.45}$$

The proof of constancy of J goes as follows.

$$\frac{dJ}{dt} = \frac{\partial J}{\partial \langle E \rangle} \frac{d\langle E \rangle}{dt} + \frac{\partial J}{\partial \lambda} \frac{d\lambda}{dt} \tag{6.46}$$

$$\frac{\partial J}{\partial \langle E \rangle} = \sqrt{\frac{m}{2}} \oint \frac{dx}{\sqrt{\langle E \rangle - V(x, \lambda)}} = \oint \frac{dx}{\dot{x}} = T \tag{6.47}$$

$$\frac{d\langle E \rangle}{dt} = \frac{1}{T} \oint dt \left(m\dot{x}\ddot{x} + \dot{x}\frac{\partial V}{\partial x} + \dot{\lambda}\frac{\partial V}{\partial \lambda} \right) \tag{6.48}$$

The first two terms in the integral cancel. The crucial point here is that $\dot{\lambda}$ is almost constant (because λ is a slowly varying function), and can be taken out of the integral, giving

$$\frac{d\langle E \rangle}{dt} = \frac{\dot{\lambda}}{T} \oint dt \frac{\partial V}{\partial \lambda} \tag{6.49}$$

$$\frac{\partial J}{\partial \lambda} = -\sqrt{\frac{m}{2}} \oint dx \frac{\partial V / \partial \lambda}{\sqrt{\langle E \rangle - V(x, \lambda)}} = -\oint \frac{\partial V}{\partial \lambda} \frac{dx}{\dot{x}} \tag{6.50}$$

Combining the foregoing equations yields $\frac{dJ}{dt}=0$.

- **Harmonic oscillator with slowly changing frequency.**

$$\ddot{x} + \omega^2(t)x = 0 \tag{6.51}$$

We imagine that the potential changes slowly, and so does the force constant $k = m\omega^2$. The time interval between two turning points (=T), can be defined to be the approximate period, and we define $\omega = 2\pi/T$ as the frequency of oscillation. Then, the solution of (6.51) can be written as

$$x(t) = A(t)\sin(\omega t + \alpha(t)) \tag{6.52}$$

where $A(t)$ and $\alpha(t)$ are slowly varying functions, $\frac{1}{A}\frac{dA}{dt} \ll \omega$, $\frac{1}{\alpha}\frac{d\alpha}{dt} \ll \omega$. The average energy taken over the fast time scale is

$$\langle E \rangle = \frac{m}{2}\langle \dot{x}^2 \rangle + \frac{1}{2}k\langle x^2 \rangle = \frac{1}{2}m\omega^2 A^2 \tag{6.53}$$

The invariant quantity is the action in (6.45):

$$J = \sqrt{2m} \oint dx \sqrt{\frac{m\omega^2 A^2}{2} - \frac{kx^2}{2}} = m\omega \oint dx \sqrt{A^2 - x^2}$$

where we regarded ω is constant and have taken out of the integral. Hence,

$$J = 2m\omega \int_{-A}^{A} dx\sqrt{A^2 - x^2} = \pi m\omega A^2 = 2\pi\frac{\langle E\rangle}{\omega} \tag{6.54}$$

where we used $\int \sqrt{A^2 - x^2}dx = \frac{x}{2}\sqrt{A^2 - x^2} + \frac{A^2}{2}\sin^{-1}\frac{x}{A}$.

Equation (6.54) identifies the adiabatic invariant in a slowly changing simple harmonic motion.

Now we recapitulate the physical picture of the invariance of the magnetic moment in a slowly time-varying magnetic field $B(t)$. The electric field $E(t)$ given rise to by $B(t)$ changes the perpendicular energy while the particle executes the cyclotron motion with a slowly changing Larmor frequency ω_c. Analogously to the simple harmonic motion considered above, we expect $\frac{m}{2}v_\perp^2\frac{2\pi}{\omega_c}$ to be an adiabatic invariant, which is proportional to μ.

• **Charged particle motion in a time-varying magnetic field.** Here we show that the particle motion in a time-varying magnetic field can be represented by an equation of simple harmonic motion with time-varying gyro-frequency. Substituting (6.34) into (6.33) gives

$$\frac{d^2\mathbf{r}}{dt^2} + \frac{1}{2}\hat{\mathbf{z}} \times \mathbf{r}\dot{\omega}_c + \omega_c\hat{\mathbf{z}} \times \frac{d\mathbf{r}}{dt} = 0, \tag{6.55}$$

or in components

$$\frac{d^2x}{dt^2} - \frac{\dot{\omega}_c}{2}y - \omega_c\frac{dy}{dt} = 0 \tag{6.56}$$

$$\frac{d^2y}{dt^2} + \frac{\dot{\omega}_c}{2}x + \omega_c\frac{dx}{dt} = 0 \tag{6.57}$$

The above two equations can be written as a single equation by introducing $\xi = x + iy$:

$$\frac{d^2\xi}{dt^2} + \frac{i}{2}\dot{\omega}_c\xi + i\omega_c\frac{d\xi}{dt} = 0 \tag{6.58}$$

A change of variable

$$\xi = w(t)e^{-\frac{i}{2}\int\omega_c(t)dt} \quad \text{gives}$$

$$\frac{d^2w}{dt^2} + \frac{\omega_c^2(t)}{4}w = 0 \tag{6.59}$$

which is a simple harmonic equation with a varying frequency. Note that ξ in the above is a WKB type solution of (6.59).

• **Kurlsrud's proof of adiabatic invariance of the action of a slowly varying harmonic oscillator.** Reference: Kurlsrud, R. M. *Adiabatic invariant of the harmonic oscillator*, Proceedings of the International School of Physics Enrico Fermi Course XXV Advanced Plasma Theory, Academic Press (1964). Kurlsrud considered the harmonic oscillator equation

$$\frac{d^2x}{dt^2} + \omega^2(t)x = 0$$

where ω is a slowly changing function of t. By introducing a slowly varying amplitude $W(t)$ and a phase $S(t)$ of WKB type

$$x(t) = W(t)\sin\left(\int_0^t S(t)dt + \delta\right)$$

He expanded S and W in perturbation series in terms of the small parameter

$$\epsilon = \frac{1}{\omega^2}\frac{d\omega}{dt}$$

and worked out the double series and derived the following relation for the action $J = E/\omega$:

$$\frac{\Delta J}{J} \approx \exp\left(-\frac{1}{\epsilon}f(t)\right)$$

where Δ signifies the change in a period. The important point of the above 'solution' is that the exponential function in the above expression goes to zero as $\epsilon \to 0$ when $\Delta J \sim \epsilon^n$ for any n, which means that the perturbation series with coefficients ϵ^n goes to zero for *all* n. Thus, the adiabatic invariant is invariant to all orders of ϵ^n. Note that

$$\lim_{\epsilon\to 0}\frac{1}{\epsilon^n}\exp\left(-\frac{1}{\epsilon}\right) = \lim_{\lambda\to\infty}\frac{\lambda^n}{e^\lambda} \to 0, \quad \text{for any } n$$

• **Mathieu equation.** As an example, let us consider a time-varying magnetic field,

$$\mathbf{B}(t) = \hat{\mathbf{z}}B_0(1 + \epsilon\sin\Omega t) \quad (\epsilon \ll 1)$$

The induced electric field can be calculated according to (6.34). Therefore, the particle motion satisfies (6.59) with the time-varying cyclotron frequency in the form

$$\frac{d^2w}{dt^2} + \frac{\omega_{c0}^2}{4}w(1 + 2\epsilon\sin\Omega t) = 0 \quad \omega_{c0} = \frac{eB_0}{mc}$$

where we neglected the ϵ^2 term. The above equation is the *Mathieu equation.* If the natural frequency ω_{c0} and the modulation frequency Ω satisfy a certain resonance condition, the particle is accelerated and a parametric instability can be given rise to. In fact, the Mathieu equation is used as a model equation to study *parametric instability.*

6.8 Adiabatic Invariant Derived from Hamiltonian

Hamiltonian H is the expression for the total energy of system, expressed in terms of the generalized coordinates p and q, and time t. We have for the total energy

$$E = H(p, q, t)$$

If H explicitly depends on t, total energy E is not a constant of motion. When H depends on t only implicitly through $p(t)$ and $q(t)$, then E is constant. This is easily seen from

$$\frac{dE}{dt} = \frac{\partial H}{\partial p}\frac{dp}{dt} + \frac{\partial H}{\partial q}\frac{dq}{dt} + \frac{\partial H}{\partial t}$$

The first two terms on the RHS cancel owing to Hamilton's equations $\frac{\partial H}{\partial p} = \frac{dq}{dt}, \frac{\partial H}{\partial q} = -\frac{dp}{dt}$. Therefore we have $\frac{dE}{dt} = \frac{\partial H}{\partial t}$. Let $H(p, q, \lambda)$ be the Hamiltonian of the system, which depends on the parameter $\lambda(t)$. For this system, E changes according to

$$\frac{dE}{dt} = \frac{\partial H}{\partial t} = \frac{\partial H}{\partial \lambda}\frac{d\lambda}{dt} \tag{6.60}$$

The above equation is averaged over the period (T) of the motion. In doing this, we can treat $d\lambda/dt$ as constant since λ changes only very slowly, and $d\lambda/dt$ does so.

$$\left\langle \frac{dE}{dt} \right\rangle = \frac{d\lambda}{dt}\frac{1}{T}\int_0^T \frac{\partial H}{\partial \lambda} dt \tag{6.61}$$

Over one period, the Hamiltonian has a constant value E; $H(p, q, \lambda) = E$, constant. Thus, $p = p(q, \lambda)$ or $q = q(p, \lambda)$. Hence,

$$\frac{\partial H}{\partial \lambda} + \frac{\partial H}{\partial p}\frac{\partial p}{\partial \lambda} = 0 \tag{6.62}$$

We use Hamilton's equation $dq/dt = \partial H/\partial p$ to change dt to dq:

$$dt = \frac{dq}{\partial H/\partial p}, \quad T = \int_0^T dt = \oint \frac{dq}{\partial H/\partial p} \tag{6.63}$$

Using (6.62) and (6.63) in (6.61) yields

$$\left\langle \frac{dE}{dt} \right\rangle = -\frac{d\lambda}{dt} \frac{\oint \frac{\partial p}{\partial \lambda} dq}{\oint \frac{dq}{\partial H/\partial p}} \tag{6.64}$$

Also we have

$$\frac{1}{\partial H/\partial p} = \frac{1}{\partial E/\partial p} = \frac{\partial p}{\partial E} \tag{6.65}$$

Using (6.65) in (6.64) yields

$$\oint \left(\frac{\partial p}{\partial E} \left\langle \frac{dE}{dt} \right\rangle + \frac{\partial p}{\partial \lambda} \frac{d\lambda}{dt} \right) dq = 0 \tag{6.66}$$

On the other hand, let us consider the action integral

$$I = \oint p(E, \lambda, q) dq \tag{6.67}$$

We can show that the above action integral is a constant of motion by calculating dI/dt. For a loop integral, the integrated term vanishes [recall Leibnitz's rule for differentiation], and we have

$$\frac{dI}{dt} = \oint \left(\frac{\partial p}{\partial t} \right)_q dq = \oint \left(\frac{\partial p}{\partial E} \frac{dE}{dt} + \frac{\partial p}{\partial \lambda} \frac{d\lambda}{dt} \right) dq \tag{6.68}$$

In view of (6.66), the RHS is zero, and thus we proved that the action integral I is constant.

- **Three adiabatic invariants.**

(1) *μ-invariant.* magnetic moment is invariant in aperiodic magnetic field which varies slowly in time as compared with the cyclotron period. Magnetic moment is invariant in inhomogeneous magnetic field which varies little over the Larmor radius.

(2) *Longitudinal invariant.* $\int_a^b v_{\|} ds = invariant$ which is an approximate constant in a mirror type magnetic field when the magnetic field changes slowly temporally or the mirror field is not axisymmetric. In the integral, a and b are two turning points on the magnetic field line where the spiraling particle reverses its moving direction along the magnetic field line. The integral is performed along the magnetic field line, with $v_{\|}$ being the particle velocity parallel to the field line. The time scale of the change of magnetic field strength, in this case, should be long as compared with the bounce period.

(3) Φ-*invariant.* This is the magnetic flux invariant. For example, in case of a mirror field, a particle spirals along the field line while it drifts across it due to the curvature of the field line. In Van Allen belt, particles bounce back and forth between the turning points and simultaneously drift circumferentially, across the field lines—roughly speaking east-west direction. The circumferential motion completes an approximate closed contour when the particle returns to the vicinity of the starting point. In this case the line integral along the circumferential orbit, $\oint$ driftvelocity $\times\, dl$, is an adiabatic invariant. According to Stokes theorem, the line integral is the same as the surface integral, which is the flux, magnetic field $\times$ the area of the equatorial plane (say). Hence, the magnetic flux Φ in the Van Allen Belt is an adiabatic invariant.

6.9 Ponderomotive Force

The ponderomotive force is the average force acting on the electrons in an external *high frequency* electric field with an inhomogeneous amplitude. It is a potential force. We have the external electric field in the form

$$\mathbf{E}(\mathbf{r},t) = \mathbf{E}_0(\mathbf{r}) \sin \omega_0 t \tag{6.69}$$

Maxwell equation, $\nabla \times \mathbf{E} = -\frac{1}{c}\frac{\partial \mathbf{B}}{\partial t}$, gives

$$\mathbf{B}(\mathbf{r},t) = \frac{c}{\omega_0} \nabla \times \mathbf{E}_0(\mathbf{r}) \cos \omega_0 t \tag{6.70}$$

We use the Euler momentum equation for the electrons

$$\frac{\partial \mathbf{v}}{\partial t} + (\mathbf{v}\cdot\nabla)\mathbf{v} = \frac{e}{m}\mathbf{E}(\mathbf{r},t) + \frac{e}{mc}\mathbf{v}\times(\mathbf{B}_0 + \mathbf{B}(\mathbf{r},t)) \tag{6.71}$$

where $\mathbf{B}_0$ is the external static magnetic field. In the zero order, we have

$$\frac{\partial \mathbf{v}_0}{\partial t} = \frac{e}{m}\mathbf{E}_0(\mathbf{r}) \sin \omega_0 t + \frac{e}{mc}\mathbf{v}_0 \times \mathbf{B}_0 \tag{6.72}$$

For simplicity, we shall consider the case of $\mathbf{B}_0 = 0$. Then we obtain

$$\mathbf{v}_0(\mathbf{r},t) = -\frac{e}{m\omega_0}\mathbf{E}_0(\mathbf{r}) \cos \omega_0 t \tag{6.73}$$

Clearly $\langle \mathbf{v}_0 \rangle = 0$, where $\langle \cdots \rangle$ is the time average over a period. Let us put $\mathbf{v} = \mathbf{v}_0 + \langle \mathbf{v} \rangle$ in (6.71) and take time average to obtain

$$\frac{\partial}{\partial t}\langle \mathbf{v} \rangle + \langle \mathbf{v}_0 \cdot \nabla \mathbf{v}_0 \rangle = \frac{e}{m\omega_0}\langle \mathbf{v}_0 \times \nabla \times \mathbf{E}_0 \cos \omega_0 t \rangle \tag{6.74}$$

In writing (6.74) we assumed that $\langle \mathbf{v} \rangle$ has a slow time dependence, much slower than $1/\omega_0$. Using (6.73) in (6.74) gives

$$\frac{\partial}{\partial t}\langle \mathbf{v} \rangle = -\frac{1}{2}\left(\frac{e}{m\omega_0}\right)^2 [\mathbf{E}_0(\mathbf{r}) \cdot \nabla \mathbf{E}_0(\mathbf{r}) + \mathbf{E}_0(\mathbf{r}) \times \nabla \times \mathbf{E}_0(\mathbf{r})] \quad (6.75)$$

Thus the average force acting on the electrons in inhomogeneous electric field, $m\frac{\partial}{\partial t}\langle \mathbf{v} \rangle$, the *ponderomotive force*, is expressed as

$$\mathbf{F}_e = -\frac{e^2}{4m\omega_0^2} \nabla E_0^2(\mathbf{r}) \quad (6.76)$$

Time average of (6.69) gives the relation, $\langle E^2 \rangle = E_0^2/2$. Then, the force density of the ponderomotive force is obtained by multiplying number density N,

$$\mathbf{F} = -\frac{1}{8\pi}\frac{\omega_p^2}{\omega_0^2}\langle E^2 \rangle \quad (6.77)$$

where ω_p is the plasma frequency. The ponderomotive force is the *radiation pressure* arising in inhomogeneous plasma.

6.10 Polarization Drift

Let us consider a plasma under the influence of a constant magnetic field and a *slowly time-varying* uniform electric field. The electric field is quasi-static; the time-rate of change of the electric field is small compared with the gyrofrequency, and no time-varying magnetic field arises. We have the equation of motion for a charge q:

$$\frac{d\mathbf{v}}{dt} = \frac{q}{m}\left(\mathbf{E}_0(t) + \frac{1}{c}\mathbf{v} \times \mathbf{B}_0\right) \quad (6.78)$$

Perform $\times \mathbf{B}_0$ on the above equation to write

$$\frac{d}{dt}(\mathbf{v} \times \mathbf{B}_0) = \frac{q}{m}\left[\mathbf{E}_0 \times \mathbf{B}_0 + \frac{1}{c}(\mathbf{v} \cdot \mathbf{B}_0\mathbf{B}_0 - B_0^2\mathbf{v})\right] \quad (6.79)$$

The above two equations give

$$\left(\frac{d^2}{dt^2} + \omega_c^2\right)\mathbf{v} = \frac{q}{m}\frac{d\mathbf{E}_0}{dt} + \frac{q^2}{cm^2}\mathbf{E}_0 \times \mathbf{B}_0 + \omega_c^2\hat{b}\hat{b} \cdot \mathbf{v} \quad (6.80)$$

where $\omega_c = qB_0/mc, \hat{b} = \mathbf{B}_0/B_0$. Let us decompose:

$$\mathbf{v} = \mathbf{v}_\perp + \mathbf{v}_\parallel, \quad \mathbf{v}_\parallel = \hat{b}(\hat{b} \cdot \mathbf{v}) \quad (6.81)$$

Perform $\cdot\hat{b}$ on (6.80) to get

$$\frac{d^2 v_{\|}}{dt^2} = \frac{q}{m}\frac{dE_{0\|}}{dt} \tag{6.82}$$

which gives

$$v_{\|} = \frac{q}{m}E_{0\|}t + c_0, \quad (c_0; \text{constant}) \tag{6.83}$$

Subtracting (6.82) from (6.80) gives

$$\left(\frac{d^2}{dt^2} + \omega_c^2\right)\mathbf{v}_\perp = \frac{q}{m}\frac{d\mathbf{E}_{0\perp}}{dt} + \frac{q^2}{cm^2}\mathbf{E}_0 \times \mathbf{B}_0 \tag{6.84}$$

In the above equation, $\frac{d\mathbf{E}_{0\perp}}{dt}$ is regarded to be a constant vector, since E_0 is a slowly varying in time. Equation (6.84) is solved by

$$\mathbf{v}_\perp = \mathbf{v}_{\perp 0}(\cos\omega_c t + \alpha) + \mathbf{v}_d \tag{6.85}$$

where $\alpha, \mathbf{v}_{\perp 0}$ are constants, and

$$\mathbf{v}_d = \frac{\text{RHS}}{\omega_c^2} = \frac{c}{B_0^2}\mathbf{E}_0 \times \mathbf{B}_0 + \frac{mc^2}{qB_0^2}\frac{d\mathbf{E}_{0\perp}}{dt} \tag{6.86}$$

where RHS is the right hand side of (6.84). The first term on the right side of (6.86) is the $E \times B$ drift and the second term is the *polarization drift*,

$$\mathbf{v}_p = \frac{mc^2}{qB_0^2}\frac{d\mathbf{E}_{0\perp}}{dt} \tag{6.87}$$

The polarization drift gives rise to the current $\mathbf{J} = qN\mathbf{v}_p = \frac{\rho c^2}{B_0^2}\frac{d\mathbf{E}_{0\perp}}{dt}$ with ρ the mass density. We use this current in Maxwell equation

$$\nabla \times \mathbf{B} = \frac{1}{c}\frac{\partial \mathbf{E}}{\partial t} + \frac{4\pi}{c}\mathbf{J} \tag{6.88}$$

Then we write

$$\nabla \times \mathbf{B} = \frac{1}{c}\frac{\partial \mathbf{E}}{\partial t}\left(1 + \frac{4\pi\rho c^2}{B_0^2}\right) = \frac{1}{c}\frac{\partial \mathbf{D}}{\partial t} \tag{6.89}$$

where the displacement vector $\mathbf{D} = \varepsilon\mathbf{E}$, with ε dielectric constant

$$\varepsilon = 1 + \frac{4\pi\rho c^2}{B_0^2} \tag{6.90}$$

From the relation

$$\mathbf{D} = \mathbf{E} + 4\pi\mathbf{P} \tag{6.91}$$

we obtain the expression for the polarization vector

$$\mathbf{P} = \frac{\varepsilon - 1}{4\pi}\mathbf{E} = \frac{\rho c^2}{B_0^2}\mathbf{E} \tag{6.92}$$

Now we consider electromagnetic wave propagating in magnetized plasma in the light of the polarization drift. We have the Maxwell equation

$$\nabla \times \mathbf{E} = -\frac{1}{c}\frac{\partial \mathbf{B}}{\partial t} \tag{6.93}$$

Fourier transforming (6.89) and (6.93) with phasor $\exp(i\mathbf{k}\cdot\mathbf{r} - i\omega t)$ gives

$$\mathbf{k} \times \mathbf{B} = -\frac{\omega}{c}\varepsilon\mathbf{E} \tag{6.94}$$

$$\mathbf{k} \times \mathbf{E} = \frac{\omega}{c}\mathbf{B} \tag{6.95}$$

The above two equations yield the dispersion relation in the form

$$\omega^2 = \frac{c^2k^2}{\varepsilon} = \frac{c^2k^2}{1 + c^2/V_A^2} \tag{6.96}$$

where $V_A = \sqrt{4\pi\rho/B_0^2}$, Alfven velocity, and we used $\nabla\cdot\mathbf{E} = \nabla\cdot\mathbf{B} = 0$. If $V_A \ll c$, we have $\omega^2 = k^2V_A^2$.

6.11 ∇B-Drift Revisited

Equation for guiding center is derived through a perturbational analysis by expanding the inhomogeneous magnetic field in the form

$$\mathbf{B}(\mathbf{r}) = \mathbf{B}_0 + \epsilon\mathbf{r}\cdot\nabla\mathbf{B}_0 + \cdots \tag{6.97}$$

where $\mathbf{B}_0$ is the magnetic field at $\mathbf{r} = 0$; the guiding center. $\mathbf{r}$ is the particle position. In the zero order the particle executes the circular motion as given by (6.2), so we put $\mathbf{r} = \vec{\rho}$. The gradient is taken at the guiding center, and ϵ is the indicator of smallness of the term. Then we write the equation of motion

$$\frac{d\mathbf{v}}{dt} = \frac{q}{cm}(\mathbf{v} \times \mathbf{B}_0 + \epsilon\mathbf{v} \times (\vec{\rho}\cdot\nabla\mathbf{B}_0)) \tag{6.98}$$

We also expand: $\mathbf{v} = \mathbf{v}_0 + \epsilon\mathbf{v}_1 + \cdots$. In what follows we shall drop the subscript '0' at $\mathbf{B}$ with the understanding that $\mathbf{B}$ is referred to the value at the guiding center. We break down the equation in series,

$$\frac{d\mathbf{v}_0}{dt} = \frac{q}{cm}\mathbf{v}_0 \times \mathbf{B} \tag{6.99}$$

$$\frac{d\mathbf{v}_1}{dt} = \frac{q}{cm}\mathbf{v}_1 \times \mathbf{B} + \frac{q}{cm}\mathbf{v}_0 \times (\vec{\rho} \cdot \nabla \mathbf{B}) \tag{6.100}$$

where

$$\vec{\rho} = \frac{1}{\omega_c}\mathbf{v}_0 \times \hat{b} \quad \hat{b} = \mathbf{B}/B \tag{6.101}$$

The zero order equation (6.99) yields the solution in the form

$$\mathbf{v}_0 = \mathbf{v}_{\perp 0}\cos(\omega_c t + \alpha) + \hat{\mathbf{b}} v_{\parallel 0} \tag{6.102}$$

where $\mathbf{v}_{\perp 0}$ is a constant vector perpendicular to $\hat{\mathbf{b}}$, and $\alpha, v_{\parallel 0}$ are constants.

Operate $\times\mathbf{B}$ on (6.100) to write

$$\frac{d}{dt}(\mathbf{v}_1 \times \mathbf{B}) = \frac{q}{cm}[(\mathbf{v}_1 \cdot \mathbf{B})\mathbf{B} - B^2\mathbf{v}_1 + (\mathbf{v}_0 \times (\vec{\rho} \cdot \nabla \mathbf{B})) \times \mathbf{B}] \tag{6.103}$$

Differentiating (6.100) and eliminating $\frac{d}{dt}(\mathbf{v}_1 \times \mathbf{B})$ give

$$\frac{d^2\mathbf{v}_1}{dt^2} = \left(\frac{q}{cm}\right)^2 [(\mathbf{v}_1 \cdot \mathbf{B})\mathbf{B} - B^2\mathbf{v}_1 + (\mathbf{v}_0 \times (\vec{\rho} \cdot \nabla \mathbf{B})) \times \mathbf{B}] + \frac{q}{cm}\frac{d}{dt}(\mathbf{v}_0 \times (\vec{\rho} \cdot \nabla \mathbf{B})) \tag{6.104}$$

which is further arranged in the form

$$\frac{d^2\mathbf{v}}{dt^2} + \omega_c^2\mathbf{v}_\perp = \left(\frac{q}{cm}\right)^2 (\mathbf{v}_0 \times (\vec{\rho} \cdot \nabla \mathbf{B})) \times \mathbf{B} + \frac{q}{cm}\frac{d}{dt}(\mathbf{v}_0 \times (\vec{\rho} \cdot \nabla \mathbf{B})) \tag{6.105}$$

where $\mathbf{v}_\perp = \mathbf{v} - \hat{b}v_\parallel$ and we dropped the subscript '1' at $\mathbf{v}$. The $\hat{b}$ direction component of (6.105) is

$$\frac{d^2 v_\parallel}{dt^2} = \hat{b} \cdot \frac{q}{cm}\frac{d}{dt}(\mathbf{v}_0 \times (\vec{\rho} \cdot \nabla \mathbf{B})) \tag{6.106}$$

Subtracting (6.106) from (6.105) gives

$$\frac{d^2\mathbf{v}_\perp}{dt^2} + \omega_c^2\mathbf{v}_\perp = \left(\frac{q}{cm}\right)^2 (\mathbf{v}_0 \times (\vec{\rho} \cdot \nabla \mathbf{B}))\mathbf{B} + \frac{q}{cm}(1 - \hat{b}\hat{b}\cdot)\frac{d}{dt}(\mathbf{v}_0 \times (\vec{\rho} \cdot \nabla \mathbf{B})) \tag{6.107}$$

We can easily show that the time average of $\langle \frac{d}{dt}(\mathbf{v}_0 \times (\vec{\rho} \cdot \nabla \mathbf{B}))\rangle = 0$. Therefore, time-averaging (6.107) gives the drift velocity in the perpendicular direction in the form

$$\mathbf{V}_d = \frac{1}{\omega_c^2}\left(\frac{q}{cm}\right)^2 \langle(\mathbf{v}_0 \times (\vec{\rho} \cdot \nabla \mathbf{B})) \times \mathbf{B}\rangle \tag{6.108}$$

Expanding the triple vector product, we have

$$\mathbf{V}_d = \frac{1}{\omega_c^2}\left(\frac{q}{cm}\right)^2 \langle(\vec{\rho} \cdot \nabla \mathbf{B})\mathbf{v}_0 \cdot \mathbf{B} - \mathbf{v}_0 \mathbf{B} \cdot (\vec{\rho} \cdot \nabla \mathbf{B})\rangle \tag{6.109}$$

The first term vanishes upon averaging, hence

$$\mathbf{V}_d = -\frac{1}{\omega_c^2}\left(\frac{q}{cm}\right)^2 \langle \mathbf{v}_0 \mathbf{B} \cdot (\vec{\rho} \cdot \nabla \mathbf{B})\rangle \tag{6.110}$$

When $\mathbf{B} = \hat{\mathbf{z}} B(x)$ (case (i) Section 6.3), (6.110) yields the drift velocity that agrees with (6.17). The case (ii) Section 6.3 is left as an exercise.

Problems

6.1 Carry out the integral in (6.42) by showing

$$J = \sqrt{2m} \oint dx \sqrt{E - \frac{k}{2}x^2} = 2\pi \frac{E}{\omega}$$

6.2 Show that the magnetic moment is given by

$$\mu = \frac{e}{2c} \frac{|\dot{\xi}|^2}{\omega_c}.$$

where ξ satisfies (6.58).

6.3 Obtain the line integral along the loop of gyroradius $\rho = v_\perp / \omega_c$ in (6.36)

$$\oint \hat{\mathbf{z}} \cdot \mathbf{r} \times d\mathbf{r}$$

Solution. The integral is equal to, per Stokes theorem,

$$\begin{aligned}\oint \hat{\mathbf{z}} \times \mathbf{r} \cdot d\mathbf{r} &= \int d\mathbf{S} \cdot \nabla \times (\hat{\mathbf{z}} \times \mathbf{r}) \\ &= 2 \int d\mathbf{S} \cdot \hat{\mathbf{z}} = 2\pi \rho^2\end{aligned}$$

6.4 Retain the $\mathbf{v}_0 \times \mathbf{B}_0$ term in (6.72), and find the ponderomotive force.
Hint: Solve

$$\left(\frac{\partial^2}{\partial t^2} + \omega_c^2\right)\left(\mathbf{v}_0 + \frac{e}{m\omega_0}\mathbf{E}_0 \cos\omega_0 t\right)$$
$$= \frac{e^2}{cm^2}\left(\mathbf{E}_0 \times \mathbf{B}_0 \sin\omega_0 t + \frac{e}{mc\omega_0}\mathbf{B}_0 \times (\mathbf{E}_0 \times \mathbf{B}_0)\cos\omega_0 t\right)$$

where $\omega_c = eB_0/mc$, cyclotron frequency.

6.5 The Hamiltonian of a single particle of charge q moving in electric field $\mathbf{E}$ and magnetic field $\mathbf{B}$ is

$$H(\mathbf{P}, \mathbf{Q}, t) = \frac{1}{2m}\left(\mathbf{P} - \frac{q}{c}\mathbf{A}\right)^2 + q\phi(\mathbf{r})$$

where $\mathbf{A}(\mathbf{r}, t)$ is the vector potential

$$\nabla \times \mathbf{A} = \mathbf{B}$$

$\phi(\mathbf{r})$ is the scalar potential

$$\nabla\phi = -\mathbf{E} - \frac{1}{c}\frac{\partial \mathbf{A}}{\partial t}$$

The generalized momentum $\mathbf{P}$ and the generalized coordinate $\mathbf{Q}$ are identified to be

$$\mathbf{P}(t) = m\mathbf{v}(t) + \frac{q}{c}\mathbf{A}(\mathbf{r}(t), t), \quad \mathbf{Q}(t) = \mathbf{r}(t)$$

Show that the Hamilton's equations are satisfied:

$$\frac{\partial H}{\partial P_j} = \dot{Q}_j, \quad \frac{\partial H}{\partial Q_j} = -\dot{P}_j, \quad dot = \frac{\partial}{\partial t} + \mathbf{v}\cdot\nabla$$

Proof.

$$\frac{\partial H}{\partial P_j} = \frac{1}{m}\left(P_j - \frac{q}{c}A_j\right) = v_j$$

$\dot{Q}_j = \dot{r}_j = v_j$. Hence $\frac{\partial H}{\partial P_j} = \dot{Q}_j$ is checked. Next,

$$\dot{\mathbf{P}} = m\frac{d\mathbf{v}}{dt} + \frac{q}{c}\left(\frac{\partial \mathbf{A}}{\partial t} + \mathbf{v}\cdot\nabla\mathbf{A}\right)$$

$$\begin{aligned}\frac{\partial H}{\partial \mathbf{Q}} &= \frac{1}{2m}\nabla\left(\mathbf{P} - \frac{q}{c}\mathbf{A}\right)^2 + q\nabla\phi \\ &= \frac{1}{2m}\left[2\left(\mathbf{P} - \frac{q}{c}\mathbf{A}\right)\times\nabla\times\left(\mathbf{P} - \frac{q}{c}\mathbf{A}\right)\right. \\ &\quad \left. +2\left(\mathbf{P} - \frac{q}{c}\mathbf{A}\right)\cdot\nabla\left(\mathbf{P} - \frac{q}{c}\mathbf{A}\right)\right] + q\nabla\phi \\ &= -q\mathbf{v}\times\mathbf{B} - \frac{q}{c}\mathbf{v}\cdot\nabla\mathbf{A} + q\nabla\phi\end{aligned}$$

where we put $\nabla\times\mathbf{P} = \nabla\mathbf{P} = 0$. Using the above two equations in the second Hamilton's equation gives the equation of motion

$$m\frac{d\mathbf{v}}{dt} = \frac{q}{c}\mathbf{v}\times\mathbf{B} + q\mathbf{E}$$

6.6 Show that (6.1) can be transformed to

$$\left(\frac{d^2}{dt^2} + \omega_c^2\right)\mathbf{v} = \omega_c^2\hat{\mathbf{b}}\mathbf{v}\cdot\hat{\mathbf{b}} \quad \hat{\mathbf{b}} = \mathbf{B}/B$$

Solve the above equation in the form

$$\mathbf{v} = \mathbf{v}_{\perp 0}\cos(\omega_c t + \alpha) + \hat{\mathbf{b}}v_{\|}$$

where $\mathbf{v}_{\perp 0}, \alpha, v_{\|}$ are constants of motion, and ω_c is the gyrofrequency.

6.7 Use Eq. (10) Section 6.1 to show that the radius of curvature vector in the circular motion governed by (6.1) can be written as

$$\vec{\rho} = \frac{\mathbf{v}\times\hat{b}}{\omega_c} \quad (\hat{b} = \mathbf{B}/B)$$

6.8 Obtain the drift velocity from (6.110) when $\mathbf{B} = \hat{\mathbf{x}}B_x(z) + \hat{\mathbf{z}}B_z(x, y)$.

6.9 The relativistic forms of the fluid equations for electrons are

$$\left(\frac{\partial}{\partial t} + \mathbf{v} \cdot \frac{\partial}{\partial \mathbf{x}}\right)\mathbf{p} = -e\left(\mathbf{E} + \frac{1}{c}\mathbf{v} \times \mathbf{B}\right)$$

$$\frac{\partial n}{\partial t} + \nabla \cdot (n\mathbf{v}) = 0$$

with

$$\mathbf{p} = m\gamma\mathbf{v}, \quad \gamma = (1 - v^2/c^2)^{-\frac{1}{2}}$$

Eliminate $\mathbf{p}$ in favor of $\mathbf{v}$ to write

$$\left(\frac{\partial}{\partial t} + \mathbf{v} \cdot \frac{\partial}{\partial \mathbf{x}}\right)\mathbf{v} = -\frac{e}{m\gamma}\left(\mathbf{E} - \frac{\mathbf{v}\mathbf{v} \cdot \mathbf{E}}{c^2} + \frac{1}{c}\mathbf{v} \times \mathbf{B}\right)$$

References

Chandrasekhar, S. *Plasma Physics*, Notes compiled by S. K. Trehan, Phoenix Books, University of Chicago Press (1960).

Kurlsrud, R. M. *Adiabatic invariant of the harmonic oscillator*, Proceedings of the International School of Physics Enrico Fermi Course XXV Advanced Plasma Theory Academic Press (1964).

Landau, L. D. *Mechanics*, Pergamon, Oxford (1960).

Longmire, C. L. *Elementary Plasma Physics*, Interscience Publishers (1963).

Chapter 7

Plasma Waves with Dispersion

In a plasma, electromagnetic fields are induced because of the motion of charged particles; also the induced fields influence the motion of charged particles, resulting in the *self-consistent* mutual interaction of charged particles and fields. The inter-relations are described by Maxwell equations in the form

$$\nabla \times \mathbf{E} = -\frac{1}{c}\frac{\partial \mathbf{B}}{\partial t} \tag{7.1}$$

$$\nabla \times \mathbf{B} = \frac{1}{c}\frac{\partial \mathbf{E}}{\partial t} + \frac{4\pi}{c}(\mathbf{J} + \mathbf{J}_0) \tag{7.2}$$

$$\nabla \cdot \mathbf{E} = 4\pi(\rho + \rho_0) \tag{7.3}$$

$$\nabla \cdot \mathbf{B} = 0 \tag{7.4}$$

where $\mathbf{B}$ is the magnetic induction and $\mathbf{E}$ the electric field, $\mathbf{J}$ and ρ are the current density and the charge density, respectively, induced in the plasma. $\mathbf{J}_0$ and ρ_0 are the current and charge densities, respectively, due to the external sources.

Taking divergence of (7.2) and using (7.3) give the charge conservation relations

$$\frac{\partial \rho}{\partial t} + \nabla \cdot \mathbf{J} = 0 \tag{7.5}$$

$$\frac{\partial \rho_0}{\partial t} + \nabla \cdot \mathbf{J}_0 = 0 \tag{7.6}$$

The fields $\mathbf{E}$ and $\mathbf{B}$ in turn determine the particle motion through Lorentz force

$$\mathbf{F}_\alpha = e_\alpha\left(\mathbf{E} + \frac{1}{c}\mathbf{v}_\alpha \times \mathbf{B}\right) \tag{7.7}$$

where $\alpha = e, i$; electron and ion.

As we already have discussed, the current density $\mathbf{J}$ and the charge density ρ are determined in terms of $\mathbf{E}$ and $\mathbf{B}$ by employing either fluid equations or kinetic equation. Putting this result back into (7.1)–(7.4), we have a complete and consistent set of equations for $\mathbf{E}$ and $\mathbf{B}$. Together with boundary conditions, the set of equations for $\mathbf{E}$ and $\mathbf{B}$ constitute an eigenvalue problem. If periodic boundary conditions are imposed, we can Fourier transform the fields and the set of equations yield the *dispersion relation* $\omega = \omega(\mathbf{k})$ that characterize the wave under investigation.

The response of the medium against the electric and magnetic fields is conveniently reflected in the dynamics by the physical quantities such as *conductivity tensor* σ_{ij}, *dielectric tensor* ε_{ij}, *susceptibility tensor* χ_{ij}, and *polarization vector* $\mathbf{P}$.

7.1 Material Relation

The vector quantity called the *electric displacement* ($\mathbf{D}$) is introduced in the Maxwell equation in the manner

$$\nabla \times \mathbf{B} = \frac{1}{c}\frac{\partial \mathbf{E}}{\partial t} + \frac{4\pi}{c}\mathbf{J} = \frac{1}{c}\frac{\partial \mathbf{D}}{\partial t} \tag{7.8}$$

The Fourier transform pair is defined in this book as

$$F(\mathbf{r},t) = \int_{-\infty}^{\infty} \frac{d\omega}{2\pi} \int_{-\infty}^{\infty} \frac{d^3k}{(2\pi)^3} F(\mathbf{k},\omega) e^{i\mathbf{k}\cdot\mathbf{r} - i\omega t} \tag{7.9}$$

$$F(\mathbf{k},\omega) = \int_{-\infty}^{\infty} dt \int_{-\infty}^{\infty} d^3r F(\mathbf{r},t) e^{-i\mathbf{k}\cdot\mathbf{r} + i\omega t} \tag{7.10}$$

A function and its Fourier transform will be written with the same symbol but with different argument; note that $F(t)$ and $F(\omega)$ have different dimensions. Fourier transforming (7.8) gives the relation

$$E_i(\mathbf{k},\omega) + \frac{4\pi i}{\omega} J_i(\mathbf{k},\omega) = D_i(\mathbf{k},\omega), \quad (i = 1,2,3) \tag{7.11}$$

The conductivity tensor is defined by the relation of proportionality between $\mathbf{J}$ and $\mathbf{E}$ in a matrix multiplication form:

$$J_i = \sigma_{ij} E_j \tag{7.12}$$

Then (7.11) can be written as

$$\left(\delta_{ij} + \frac{4\pi i}{\omega}\sigma_{ij}\right) E_j = \varepsilon_{ij}\ E_j = D_i \tag{7.13}$$

The quantity in the parenthesis is the dielectric tensor:

$$\varepsilon_{ij} = \delta_{ij} + \frac{4\pi i}{\omega}\sigma_{ij} \tag{7.14}$$

The susceptibility tensor is simply

$$\chi_{ij} = \frac{4\pi i}{\omega}\sigma_{ij} \tag{7.15}$$

Electric polarization (**P**) is defined by

$$\mathbf{J} = \frac{\partial \mathbf{P}}{\partial t} \tag{7.16}$$

So we have from (7.11)

$$\mathbf{D} = \mathbf{E} + 4\pi\mathbf{P} \tag{7.17}$$

Then (7.5) and (7.16) give

$$\rho = -\nabla \cdot \mathbf{P} \tag{7.18}$$

Equation (7.16) is characteristic of plasma: in plasma we assume that there is no magnetization current, which means that we take the magnetic permeability to be 1. The relationship between **D** and **E** in (7.11) is causal: it defines **D** as the effect while **E** is regarded to be the cause. The electric displacement **D** represents the response of the medium when it is acted by the electric field **E**.

Fourier transforming (7.1) and (7.2) gives

$$\mathbf{k} \times \mathbf{E} = \frac{\omega}{c}\mathbf{B} \tag{7.19}$$

$$\mathbf{k} \times \mathbf{B} = -\frac{\omega}{c}\mathbf{E} - \frac{4\pi i}{c}\mathbf{J} \tag{7.20}$$

Eliminating **B** in (7.19) and (7.20) yields

$$\mathbf{k} \times (\mathbf{k} \times \mathbf{E}) + \frac{\omega^2}{c^2}\mathbf{E} = -\frac{4\pi i\omega}{c^2}\mathbf{J} \tag{7.21}$$

Equation (7.21) can be put into the form

$$\left[n^2\left(\frac{k_i k_j}{k^2} - \delta_{ij}\right) + \varepsilon_{ij}\right] E_j = 0 \tag{7.22}$$

where n is the refractive index, $n = ck/\omega$.

The dispersion equation for determining the spectrum of eigen oscillations is the solvability condition of the homogeneous simultaneous equations in (7.22). *Viz.*, the determinant of the matrix =0:

$$\Delta(\mathbf{k},\omega) = \left| n^2\left(\frac{k_i k_j}{k^2} - \delta_{ij}\right) + \varepsilon_{ij} \right| = 0 \tag{7.23}$$

For given $\mathbf{k}$, this equation has a number of complex roots, which we denote by $\omega_n(\mathbf{k}) - i\gamma_n(\mathbf{k})$, $(n = 1, 2, \ldots)$. The real part $\omega_n(\mathbf{k})$ represents the frequency of the nth oscillation, and the imaginary part $\gamma_n(\mathbf{k})$ the accompanying damping or growth rate of the nth oscillation.

For an isotropic plasma when there are no external fields the dielectric tensor is a linear combination of two basic tensors $k_i k_j/k^2$ and $(k_i k_j/k^2 - \delta_{ij})$:

$$\varepsilon_{ij} = \frac{k_i k_j}{k^2}\varepsilon_L(\mathbf{k},\omega) + \left(\delta_{ij} - \frac{k_i k_j}{k^2}\right)\varepsilon_T(\mathbf{k},\omega) \tag{7.24}$$

where ε_L and ε_T are the *longitudinal dielectric function* and the *transverse dielectric function*, respectively. We prove (7.24) later by using the Vlasov dielectric tensor. The two basic tensors are the projection tensors as can be seen in the following. When $k_i k_j/k^2$ operates on a vector $\mathbf{A}$, we obtain

$$\frac{k_i k_j}{k^2}A_i = \frac{\mathbf{A}\cdot\mathbf{k}k_j}{k^2} = \mathbf{A}\cdot\hat{k}\hat{k} = \mathbf{A}_{\|}$$

Similarly,

$$\left(\delta_{ij} - \frac{k_i k_j}{k^2}\right)A_i = \mathbf{A} - \mathbf{A}_{\|} = \mathbf{A}_{\perp}$$

where the symbols $\|$ and $\perp$ are referred to the direction of $\mathbf{k}$.

Putting (7.24) into (7.22), we have for the wave equation

$$\left\{(\varepsilon_T - n^2)\left(\delta_{ij} - \frac{k_i k_j}{k^2}\right) + \varepsilon_L\frac{k_i k_j}{k^2}\right\}E_j = 0 \tag{7.25}$$

or equivalently we have

$$(\epsilon_T - n^2)\mathbf{E}_{\perp} + \varepsilon_L\mathbf{E}_{\|} = 0 \tag{7.26}$$

Since the two vectors $\mathbf{E}_{\perp}$ and $\mathbf{E}_{\|}$ are orthogonal, we have two separate dispersion relations

$$\varepsilon_L(\mathbf{k},\omega) = 0 \tag{7.27}$$

$$\varepsilon_T(\mathbf{k},\omega) - n^2 = 0 \tag{7.28}$$

Equation (7.27) is the dispersion relation for the longitudinal wave. Equation (7.28) is the dispersion relation for the transverse wave. When the speed of light $c \to \infty$ in (7.26), $\mathbf{E}_\perp = 0$, and we are left with $\varepsilon_L \mathbf{E}_\parallel = 0$, and the wave satisfies the longitudinal dispersion relation in (7.27). Clearly, in this case $\nabla \times \mathbf{E} = 0$. For the transverse wave, the electric vector is perpendicular to $\underline{k}$, thus $\nabla \cdot \mathbf{E} = 0$.

From (7.24) we can derive the useful relations,

$$\varepsilon_L = \frac{k_i k_j}{k^2}\varepsilon_{ij}, \quad \varepsilon_T = \frac{1}{2}(\varepsilon_{ii} - \varepsilon_L) \tag{7.29}$$

where ε_{ii} is the trace of the matrix. Also by direct expansion of the determinant, (7.23) gives

$$\Delta = \varepsilon_L(\varepsilon_T - n^2)^2 = 0 \tag{7.30}$$

Derivation of the above two equations are left as exercise.

Equation (7.13) is rewritten here:

$$D_i(\mathbf{k}, \omega) = \varepsilon_{ij}(\mathbf{k}, \omega) E_j(\mathbf{k}, \omega) \tag{7.31}$$

Using (7.24) in the above equation gives

$$\mathbf{D}(\mathbf{k}, \omega) = \varepsilon_T \mathbf{E}_\perp + \varepsilon_L \mathbf{E}_\parallel \tag{7.32}$$

Therefore $\varepsilon_{T,L}$ determines, respectively, the contribution to the electric displacement $\mathbf{D}$ along the direction perpendicular and parallel to $\mathbf{k}$. If the electric field is perpendicular to $\mathbf{k}$, $\mathbf{D} = \varepsilon_T \mathbf{E}$. If the electric field is parallel to $\mathbf{k}$, $\mathbf{D} = \varepsilon_L \mathbf{E}$.

• **Boundary conditions.** When the plasma is bounded, the Eqs. (7.1)–(7.8) must be supplemented with appropriate boundary conditions to be satisfied at the interface. The latter follow from the infinitesimal integration across the interface:

$$\lim_{\epsilon \to 0} \int_{-\epsilon}^{\epsilon} (\cdots) dn$$

with n the normal coordinate. In the following $[\cdots]$ means the jump.

Equation (7.1) yields

$$\hat{\mathbf{n}} \times [\mathbf{E}] = 0 \tag{7.33}$$

Equation (7.3) gives

$$\hat{\mathbf{n}} \cdot [\mathbf{E}] = 4\pi \lim_{\epsilon \to 0} \int_{-\epsilon}^{\epsilon} \rho \, dn = 4\pi \rho^* \tag{7.34}$$

where ρ^* is the surface charge.

Equation (7.4) gives

$$\hat{\mathbf{n}} \cdot [\mathbf{B}] = 0 \tag{7.35}$$

Equation (7.8) yields

$$\hat{\mathbf{n}} \times [\mathbf{B}] = \frac{4\pi}{c} \lim_{\epsilon \to 0} \int_{-\epsilon}^{\epsilon} \mathbf{J}\, dn = \frac{4\pi}{c} \mathbf{J}^* \tag{7.36}$$

where $\mathbf{J}^*$ is the surface current. Surface charge or surface current may or may not be existent, depending upon the problems.

Equation (7.8) gives

$$\nabla \cdot \mathbf{D} = 0 \tag{7.37}$$

Hence

$$\hat{\mathbf{n}} \cdot [\mathbf{D}] = 0 \tag{7.38}$$

7.2 Causality

Causality in electrodynamics means that the response (effect) must always follow the cause; the cause cannot be precedent to the effect. For a Fourier component of a wave with phasor

$$e^{ikx - i\omega t} \tag{7.39}$$

the principle of causality can be operationally incorporated by assuming that the frequency ω has an infinitesimally small *positive* imaginary part. Then the disturbance at infinitely remote past ($t = -\infty$) is nil in accordance with the causal notion that the time $t = -\infty$ is necessarily prior to the time of occurrence of any stimulus (cause) which gives rise to the onset of disturbance (effect). The small positive imaginary part (ν) of the frequency is called *adiabatic switching* on of the perturbation at $t = -\infty$. The small positive imaginary frequency ν arises mathematically if one solves the plasma equations as an initial value problem via Laplace transform, as Landau demonstrated [see next chapter], but it can be simply invoked in accordance with the principle of causality. Adiabatic switching supplies the first definition of causality that distinguishes the remote past from the remote future.

We mention that the phasor in the expression (7.39) is compatible with the Fourier transform convention in (7.9) and (7.10). One who uses different sign in front of the time-phase in the Fourier integral, the frequency ω should be assumed to have a negative small imaginary frequency to be compatible with the notion of causality.

Example. Solve the following ordinary differential equation by Fourier transform.

$$(1) \quad \frac{df}{dt} + a\ f(t) + b = 0$$

where a and b are real constants. Let us Fourier transform the above equation by integrating $\int_{-\infty}^{\infty} dt(\cdots)e^{i\omega t}$. The first term gives

$$(2) \quad \int_{-\infty}^{\infty} \frac{df}{dt} e^{i\omega t} dt = \left[f(t) e^{i\omega t} \right]_{-\infty}^{\infty} - i\omega f(\omega)$$

The integrated term must vanish, so we need appropriate logic to discard it. At $t = -\infty$, we can assume $f(-\infty) = 0$, physically, because the universe was non-existent at the remote past. At the upper limit $t = \infty$, the positive imaginary part of ω must be large enough to suppress the growth of $f(t)$. If $f(t) \sim e^{\alpha t}$, we should have $\mathrm{Im}\,\omega > \alpha$. So we have to work on a limited region of ω plane in the inverse transform. Equation (1) is transformed to

$$(3) \quad (\omega + ia) f(\omega) = -2\pi i\ b\delta(\omega)$$

where we used the *delta* function formula

$$(4) \quad \delta(\omega) = \frac{1}{2\pi} \int_{-\infty}^{\infty} e^{i\omega t} dt$$

The algebraic equation (3) requires cautiousness in writing as

$$(5) \quad f(\omega) = P \frac{-2\pi i\ b\delta(\omega)}{\omega + ia} + c\ \delta(\omega + ia) \quad c : \text{const.}$$

where P denotes the principal value at the singularity $\omega = -ia$. Note $(\omega + ia)\delta(\omega + ia) = 0$ identically. The omission of the term $\delta(\omega + ia)$ will result in serious defect in the solution. The inversion integral is

$$(6) \quad f(t) = \int_{-\infty}^{\infty} \frac{d\omega}{2\pi} e^{-i\omega t} \left[P \frac{-2\pi i b\ \delta(\omega)}{\omega + ia} + c\ \delta(\omega + ia) \right]$$

The value of the first integral is simply the contribution at $\omega = 0$ due to the function $\delta(\omega)$, giving $(-b/a)$. In the second integral the integration is along the real ω-axis, and we don't know how to deal with it as it is. The second integral should be done by contour integration with the contour enclosing the lower ω plane by infinite semi-circle for $t > 0$ plus the real ω line integral. But the latter contour should be indented so that the closed contour exclude the δ function singularity at $\omega = -ia$ (to work in the analytic region).

This can be done by chopping out the point $\omega = -ia$ by infinitesimal whole circle and work in the multiply-connected region. Thus we obtain

$$f(t) = -\frac{b}{a} + \text{const.} \times e^{-at} \quad (t > 0)$$

7.3 Causality and Susceptibility

In this section, we consider the dielectric permittivity or the susceptibility in general terms to see how the principle of causality is incorporated in electrodynamics.

We have for the electric displacement vector $\mathbf{D}(t)$ and the plasma current $\mathbf{J}(t)$ the following equation:

$$\frac{1}{c}\frac{\partial \mathbf{D}}{\partial t} = \frac{1}{c}\frac{\partial \mathbf{E}}{\partial t} + \frac{4\pi}{c}\mathbf{J} \tag{7.40}$$

Here, we consider a particular Fourier component $\mathbf{k}$ but it will not be explicitly written. $\mathbf{k}$ can be reinstated at appropriate stage if desirable. We integrate (7.40) with the initial conditions $\mathbf{D}(t) = 0$ and $\mathbf{E}(t) = 0$ at $t = -\infty$ to get

$$\mathbf{D}(t) = \mathbf{E}(t) + 4\pi \int_{-\infty}^{t} \mathbf{J}(t')dt' = \mathbf{E}(t) + 4\pi \int_{0}^{\infty} \mathbf{J}(t-\tau)d\tau, \tag{7.41}$$

where we made a change of variable, $t' = t - \tau$. In (7.41), $\mathbf{J}$ is related with $\mathbf{E}$ through the material relation. The displacement vector $\mathbf{D}(t)$ then represents the response of the medium, or the effect caused by the input $\mathbf{E}(t)$. Introducing the Fourier integral for $\mathbf{J}(t)$,

$$\mathbf{J}(t) = \frac{1}{2\pi}\int_{-\infty}^{\infty} \mathbf{J}(\omega)e^{-i\omega t}d\omega,$$

the last term of (7.41) can be written as

$$4\pi \int_{-\infty}^{\infty}\int_{0}^{\infty} d\tau e^{i\omega\tau}\frac{1}{2\pi}\mathbf{J}(\omega)e^{-i\omega t}d\omega.$$

The $\int d\tau$-integral gives i/ω (ω has a small positive imaginary part). Writing $J_i = \sigma_{ij}E_j$, and $\chi_{ij} = \frac{4\pi i}{\omega}\sigma_{ij}$ with σ_{ij} being the conductivity tensor, the above expression becomes

$$\frac{1}{2\pi}\int_{-\infty}^{\infty} d\omega\ \chi_{ij}(\omega)E_j(\omega)e^{-i\omega t}$$

which is the inverse Fourier transform of $\chi_{ij}(\omega)E_j(\omega)$, and can be expressed as a convolution integral. Therefore, we have

$$D_i(t) = E_i(t) + \int_{-\infty}^{\infty} \chi_{ij}(t')E_j(t-t')dt' \tag{7.42}$$

Causality demands that the kernel $\chi(t)$ in (7.42) be zero for $t < 0$. This is so because **D** at present time t should be irrelevant to the values of **E** at future times. This agrees with the basic idea of causality in natural phenomena and is vindicated by experience. Thus, we *impose* the condition,

$$\chi_{ij}(t) = 0, \quad \text{for } t < 0 \tag{7.43}$$

Then, (7.42) is rewritten as

$$\begin{aligned} D_i(t) &= E_i(t) + \int_{0}^{\infty} \chi_{ij}(t')E_j(t-t')dt'. \\ &= E_i(t) + \int_{-\infty}^{t} \chi_{ij}(t-t')E_j(t')dt' \end{aligned} \tag{7.44}$$

Fourier transforming (7.44) gives

$$D_i(\omega) = E_i(\omega) + \int_{-\infty}^{\infty} dt\; e^{i\omega t} \int_{0}^{\infty} \chi_{ij}(t')E_j(t-t')dt'.$$

Via a change of variable, $t - t' = t'', dt \to dt''$, the above equation takes the form

$$D_i(\omega) = \left(\delta_{ij} + \int_{0}^{\infty} \chi_{ij}(t')e^{i\omega t'}dt'\right) E_j(\omega) \tag{7.45}$$

This equation can be written in terms of the dielectric tensor ε_{ij} as

$$\begin{aligned} D_i(\omega) = \varepsilon_{ij}(\omega)E_j(\omega); \quad \varepsilon_{ij}(\omega) &= \delta_{ij} + \chi_{ij}(\omega), \\ \chi_{ij}(\omega) &= \int_{0}^{\infty} \chi_{ij}(t)e^{i\omega t}dt \end{aligned} \tag{7.46}$$

Clearly $\chi(\omega)$ in the above equation is analytic in the upper half ω-plane because the integrand is analytic therein and the integral is finite at the upper limit $t \to \infty$. We emphasize that the causal requirement expressed by (7.43) is an important constraint on the susceptibility function $\chi(t)$, which is defined by the inverse transform of (7.46). The constraint can be viewed as the foremost cause for the collisionless damping of plasma waves, as is discussed in the sequel.

7.4 Kramers–Kronig Relations

The Kramers–Kronig (K–K) relations relate the real (imaginary) part of the susceptibility $\chi(\omega)$ to the imaginary (real) part of $\chi(\omega)$. If one knows one of the two, the other can be obtained from the known function by Hilbert transform. Usually the K–K relations are derived by the mathematical theory of the analytic continuation from the analytic region of $\chi(\omega)$, which is the upper-half of the complex ω-plane, to the real ω-axis. In the following, we derive the K–K relations by two different methods. The first method directly uses the causal requirement expressed by (7.43), while the second method utilizes the analyticity of $\chi(\omega)$ in the upper half ω-plane.

The algebraic constraint for $\chi(t)$ in (7.43) can be written as

$$\chi(t) = H(t)\ \chi(t) \tag{7.47}$$

where $H(t)$ is Heaviside step function:

$$H(t) = \begin{cases} 1 & t > 0 \\ 0 & t < 0 \end{cases} \tag{7.48}$$

Equation (7.47) is akin to an eigenvalue equation, with an algebraic operator $H(t)$. Fourier transforming (7.47) gives

$$\chi(\omega) = \int_{-\infty}^{\infty} dt\ e^{i\omega t} H(t)\chi(t) \tag{7.49}$$

which can be written as a convolution integral

$$\chi(\omega) = \frac{1}{2\pi} \int_{-\infty}^{\infty} d\omega' H(\omega - \omega')\chi(\omega') \tag{7.50}$$

The Fourier transform of the step function $H(t)$ is (see Appendix)

$$H(\omega) = P\frac{i}{\omega} + \pi\ \delta(\omega) \tag{7.51}$$

where P denotes the *principal value* and $\delta(\omega)$ is Dirac delta function. Using (7.51) in (7.50) gives

$$\chi(\omega) = \frac{i}{\pi} P \int_{-\infty}^{\infty} d\omega' \frac{\chi(\omega')}{\omega - \omega'} \tag{7.52}$$

which is the implicit form of the K–K relations. It reads a complex function χ with a real variable ω. We put $\chi(\omega) = \mathrm{Re}\,\chi(\omega) + i\ \mathrm{Im}\,\chi(\omega)$ and separate the real and the imaginary parts to obtain the K–K relations in the following form:

$$\operatorname{Re}\chi(\omega) = \frac{1}{\pi}P\int_{-\infty}^{\infty}\frac{\operatorname{Im}\chi(\omega')}{\omega'-\omega}d\omega', \tag{7.53}$$

$$\operatorname{Im}\chi(\omega) = -\frac{1}{\pi}P\int_{-\infty}^{\infty}\frac{\operatorname{Re}\chi(\omega')}{\omega'-\omega}d\omega'. \tag{7.54}$$

The K–K relations relate the real (imaginary) part of the susceptibility function $\chi(\omega)$ to the imaginary (real) part of $\chi(\omega)$. Equations (7.53) and (7.54) are Hilbert transform pair. If one knows any one of the two, the other can be obtained by Hilbert transform as is expressed by the equations.

The second method to derive the K–K relations begins with putting Fourier inversion in (7.46):

$$\begin{aligned}
\chi(\omega) &= \int_0^{\infty} dt\ e^{i\omega t}\int_{-\infty}^{\infty}\frac{d\omega'}{2\pi}\chi(\omega')e^{-i\omega' t}\\
&= \int_{-\infty}^{\infty}\frac{d\omega'}{2\pi}\chi(\omega')\int_0^{\infty} dt\ e^{i(\omega-\omega')t}\\
&= \frac{1}{2\pi i}\int_{-\infty}^{\infty} d\omega'\frac{\chi(\omega')}{\omega-\omega'}[e^{i(\omega-\omega')t}]_0^{\infty}\\
&= \frac{1}{2\pi i}\int_{-\infty}^{\infty}\frac{\chi(\omega')}{\omega'-\omega}d\omega' \quad (\operatorname{Im}\omega > 0) \qquad (7.55)\\
&= \frac{1}{2\pi i}\oint_C\frac{\chi(\omega')}{\omega'-\omega}d\omega' \quad (\operatorname{Im}\omega > 0). \qquad (7.56)
\end{aligned}$$

In (7.56), the Cauchy integral formula, the closed contour C consists of the real ω' axis and the infinite semi-circle enclosing the *upper*-half ω' plane. Note that the integral along the infinite semi-circle vanishes. Equation (7.56) should be compared with the relation $\chi(\omega) = \int_{-\infty}^{\infty}\chi(\omega')\delta(\omega'-\omega)d\omega'$, which appears in connection with the full-range Fourier integral. The Cauchy formula holds if and only if $\chi(\omega)$ is analytic inside the closed contour. It is emphasized that Eqs. (7.55) and (7.56) are the legitimate deductions made from the causal requirement $\chi(t<0) = 0$.

Next, we analytically continue $\chi(\omega)$ in (7.55) onto the real ω axis ($\operatorname{Im}\omega \equiv \omega_i = 0$). We let the variable ω become real by descending the point ω downward so that it lands on the real axis. This limiting process is accomplished by the following steps. We put $\omega = \omega + i\nu$ ($\nu > 0$).

Then we have

$$\frac{1}{\omega' - \omega} = \frac{1}{\omega' - \omega - i\nu} = \frac{\omega' - \omega + i\nu}{(\omega' - \omega)^2 + \nu^2} = P\frac{1}{\omega' - \omega} + i\pi\ \delta(\omega' - \omega)$$

The last expression is the result of taking the limit $\nu \to 0$. Using the above equation in (7.55) gives

$$\chi(\omega) = \frac{1}{\pi i}P\int_{-\infty}^{\infty}\frac{\chi(\omega')}{\omega' - \omega}d\omega' \quad (\omega_i = 0) \tag{7.57}$$

which is the analytic continuation of the function in (7.55) onto the axis $\omega_i = 0$, and is the implicit form of the K–K relations in (7.52).

• **Hilbert transform.** The pair of equations, (7.53) and (7.54), is Hilbert transform of each other. Any one of them can be obtained from the other by *Hilbert transform*. The *Hilbert inversion* is facilitated by the identity

$$P\int_{-\infty}^{\infty}\frac{d\omega}{(\omega - \omega_1)(\omega - \omega_2)} = \pi^2 \quad \delta(\omega_1 - \omega_2) \tag{7.58}$$

which is the closure property of the Hilbert transform. This identity can be derived from (7.53) and (7.54) by eliminating either $\mathrm{Re}\,\chi(\omega)$ or $\mathrm{Im}\,\chi(\omega)$ between them.

In order to invert (7.53), let us multiply both sides by $P\frac{1}{\omega - \omega''}$ and integrate $\int_{-\infty}^{\infty} d\omega(\cdots)$:

$$P\int_{-\infty}^{\infty}\frac{\mathrm{Re}\,\chi(\omega)}{\omega - \omega''}d\omega = \frac{1}{\pi}P\int_{-\infty}^{\infty}\frac{d\omega}{\omega - \omega''}\int_{-\infty}^{\infty}\frac{\mathrm{Im}\,\chi(\omega')}{\omega' - \omega}d\omega'.$$

If we use the formula in (7.58), the righthand side of the above equation becomes $-\pi\,\mathrm{Im}\,\chi(\omega'')$, and we obtain

$$\mathrm{Im}\,\chi(\omega) = -\frac{1}{\pi}P\int_{-\infty}^{\infty}\frac{\mathrm{Re}\,\chi(\omega')}{\omega' - \omega}d\omega' \tag{7.59}$$

7.5 Susceptibility of Vlasov Plasma

In this section, we consider the susceptibility of electron plasma governed by the Vlasov and Poisson equations introduced in Section 1.3. We apply the dispersion theory developed in the earlier part of this chapter. We show that the well-known *Landau damping* is a consequence of the causality; the algebraic constraint $\chi(t < 0) = 0$.

We begin with the following one-dimensional linearized Vlasov equation and Poisson equation:

$$\frac{\partial}{\partial t}g(x,v,t)+v\frac{\partial g}{\partial x}-\frac{e}{m}E(x,t)\frac{dg_0(v)}{dv}=0 \tag{7.60}$$

$$\frac{\partial}{\partial x}E(x,t)=-4\pi e\textstyle\int_{-\infty}^{\infty}dv\ g(x,v,t) \tag{7.61}$$

where $g(x,v,t)$ is the perturbed electron distribution function, $g_0(v)$ is the zero order equilibrium electron distribution function, and the ions are assumed to form the uniform neutralizing background. We Fourier transform the above equations to obtain

$$g(k,v,\omega)=\frac{e}{m}\frac{dg_0}{dv}\frac{i}{\omega-kv}E(k,\omega) \tag{7.62}$$

$$ikE(k,\omega)=-4\pi e\int_{-\infty}^{\infty}dv\ g(k,v,\omega) \tag{7.63}$$

The electron charge density is by definition

$$\rho_c=-e\int_{-\infty}^{\infty}dvg(k,v,\omega)=-ikP_e \tag{7.64}$$

The *electric polarization* P_e takes the form

$$P_e=\frac{e^2}{km}\int_{-\infty}^{\infty}dv\frac{dg_0}{dv}\frac{1}{\omega-kv}E(k,\omega) \tag{7.65}$$

Then, the relation $E+4\pi P_e=\varepsilon E=(1+\chi)E$ gives

$$\chi(k,\omega)=\frac{\omega_{\rm pe}^2}{k}\int_{-\infty}^{\infty}dv\frac{dg_0}{dv}\frac{1}{\omega-kv}\qquad(\omega_i>0) \tag{7.66}$$

where $\omega_{\rm pe}$ is the electron plasma frequency, and the imaginary part of ω, ω_i is >0.

When the complex frequency ω becomes real, we have a singular integral in (7.66). According to Landau, the singularity can be taken care of by deforming the v-integral path to the *Landau contour* (See Chapter 8). Here we evaluate the integral in conformity with the causal constraint $\chi(t<0)=0$. We use the well-known identity

$$\frac{1}{\omega-kv}=P\frac{1}{\omega-kv}+\lambda\delta(\omega-kv) \tag{7.67}$$

where λ is an undetermined constant. Equation (7.67) is correct because of the two identities

$$(x-a)P\frac{1}{x-a} = 1, (x-a)\delta(x-a) = 0 \tag{7.68}$$

Using (7.67) in (7.66) gives

$$\chi(k,\omega) = \frac{\omega_{\text{pe}}^2}{k}\int_{-\infty}^{\infty} dv\frac{dg_0}{dv}\left[P\frac{1}{\omega - kv} + \lambda\ \delta(\omega - kv)\right] \quad (\omega_i = 0) \tag{7.69}$$

Let us invert the above equation to obtain $\chi(k,t)$:

$$\chi(k,t) = \frac{\omega_{\text{pe}}^2}{k}\int_{-\infty}^{\infty} dv\frac{dg_0}{dv}\int_{-\infty}^{\infty}\frac{d\omega}{2\pi}e^{-i\omega t}\left[P\frac{1}{\omega - kv} + \lambda\delta(\omega - kv)\right] \tag{7.70}$$

The principal part of the integral in (7.70) can be evaluated by using (A.7), and one finds that $\lambda = -i\pi$ satisfies the causality constraint $\chi(t<0) = 0$. Therefore (7.69) and (7.70) take the definite values

$$\chi(k,\omega) = \frac{\omega_{\text{pe}}^2}{k}\int_{-\infty}^{\infty} dv\frac{dg_0}{dv}\left[P\frac{1}{\omega - kv} - i\pi\ \delta(\omega - kv)\right] \quad (\omega_i = 0) \tag{7.71}$$

$$\chi(k,t) = \begin{cases} -i\dfrac{\omega_{\text{pe}}^2}{k}\displaystyle\int_{-\infty}^{\infty} dv\frac{dg_0}{dv}e^{-ikvt} & \text{if } t > 0 \\ 0 & \text{if } t < 0 \end{cases} \tag{7.72}$$

As a cross check, we can ascertain that (7.71) and (7.72) are the Fourier transform pair. We see that (7.71) agrees with the susceptibility evaluated along the Landau contour (see Chapter 8).

Separating the real and imaginary parts of $\chi(k,\omega)$ in (7.71), we have

$$\text{Re}\,\chi(\omega) = \frac{\omega_{\text{pe}}^2}{k}\int_{-\infty}^{\infty} dv\frac{dg_0}{dv}P\frac{1}{\omega - kv} \tag{7.73}$$

$$\text{Im}\,\chi(\omega) = -\pi\frac{\omega_{\text{pe}}^2}{k}\int_{-\infty}^{\infty} dv\frac{dg_0}{dv}\delta(\omega - kv) \tag{7.74}$$

Equations (7.73) and (7.74) satisfy the Kramers–Kronig relations. We see that (7.53) is trivially satisfied due to the δ function in (7.74). Equation (7.54) is also trivially satisfied by (7.73) owing to the Hilbert transform property, (7.58).

Generally, we write a Hilbert transform pair in the form

$$f(\omega) = \frac{1}{\pi} P \int_{-\infty}^{\infty} \frac{g(\omega')}{\omega' - \omega} d\omega' \tag{7.75}$$

$$g(\omega) = -\frac{1}{\pi} P \int_{-\infty}^{\infty} \frac{f(\omega')}{\omega' - \omega} d\omega' \tag{7.76}$$

Setting $g(\omega) = \delta(\omega)$ in (7.75) yields $f(\omega) = -P\frac{1}{\pi\omega}$. They satisfy (7.76) per (7.58). Therefore, $\delta(\omega)$ and $-P\frac{1}{\pi\omega}$ are Hilbert transform pair.

Equation (7.74) can be further evaluated, giving

$$\operatorname{Im} \chi(k,\omega) = -\pi \frac{\omega_{\text{pe}}^2}{k^2} \frac{dg_0}{dv}\bigg|_{v=\omega/k} \tag{7.77}$$

which is the famous Landau damping rate of the electrostatic plasma wave. It is left as an exercise to derive the same result by using (7.73) in the K–K relation in (7.54).

7.6 Dielectric Permittivity of a Maxwellian Plasma

In this section we examine the causal connection between **E** and **D** for a Maxwellian plasma

$$g_0(v) = \sqrt{\frac{m}{2\pi T}} \exp\left(-\frac{mv^2}{2T}\right) \tag{7.78}$$

Using (7.78) in (7.72) yields

$$\chi(k,t) = \begin{cases} \omega_{\text{pe}}^2 t \, e^{-\frac{T}{2m}k^2t^2} & \text{if } t > 0 \\ 0 & \text{if } t < 0 \end{cases} \tag{7.79}$$

We have the relation between **E** and **D**, Equation (7.44),

$$\mathbf{D}(t) = \mathbf{E}(t) + \int_0^{\infty} \chi(t')\mathbf{E}(t-t')dt' = \mathbf{E}(t) + \int_{-\infty}^{t} \chi(t-t')\mathbf{E}(t')dt' \tag{7.80}$$

Using (7.79) in (7.80), we write

$$D(k,t) = E(k,t) + \omega_{\text{pe}}^2 \int_{-\infty}^{t} (t-t')e^{-\frac{T}{2m}k^2(t-t')^2} E(k,t')dt' \tag{7.81}$$

In the inverse problem, this is a Voltera integral equation for $\mathbf{E}$. Now we consider its Fourier transform.

$$\begin{aligned}D(\omega) &= E(\omega) + \omega_{\text{pe}}^2 \int_{-\infty}^{\infty} e^{i\omega t} dt \int_{-\infty}^{t} (t-t')e^{-\frac{T}{2m}k^2(t-t')^2} E(t')dt' \\ &= E(\omega) + \omega_{\text{pe}}^2 \int_{-\infty}^{\infty} e^{i\omega t} dt \int_{0}^{\infty} \tau e^{-\frac{T}{2m}k^2\tau^2} E(t-\tau)d\tau\end{aligned}$$

Using the inverse relation $E(t-\tau) = \int_{-\infty}^{\infty} E(\omega')\exp[-i\omega'(t-\tau)]d\omega'/2\pi$ in the above expression yields

$$D(\omega) = E(\omega) + \omega_{\text{pe}}^2 E(\omega) \int_0^{\infty} \tau\, e^{i\omega\tau}\, e^{-\frac{T}{2m}k^2\tau^2} d\tau \tag{7.82}$$

The integral in (7.82) can be carried out in the complex τ-plane. We have a term obtained by integrating by parts plus the following integral to be done:

$$\begin{aligned}\int_0^{\infty} e^{i\omega\tau} e^{-\frac{T}{2m}k^2\tau^2} d\tau &= e^{-\frac{\omega^2}{4\alpha}} \int_{0-i\omega/2\alpha}^{\infty-i\omega/2\alpha} e^{-\alpha z^2} dz (\alpha = Tk^2/2m) \\ &= e^{-\frac{\omega^2}{4\alpha}} \left[\int_{-\omega/2\beta}^{0} i\, dy\, e^{\alpha y^2} + \int_0^{\infty} e^{-\alpha x^2} dx \right].\end{aligned}$$

Thus (7.82) becomes after collecting the terms

$$D(\omega,k) = E(\omega,k) \left[1 + \frac{1}{k^2\lambda_D^2} \left(1 + \beta e^{-\beta^2} \left[-2F(\beta) + i\sqrt{\pi} \right] \right) \right], \tag{7.83}$$

where $\lambda_D = \sqrt{T/4\pi N e^2}$, Debye length, $\beta = \omega/kv_{th}$, $mv_{th}^2/2 = T$, and the error function $F(x)$ is defined by $F(x) = \int_0^x e^{t^2} dt$. The quantity in the large bracket in (7.83) is the plasma dielectric permittivity:

$$\varepsilon_L(\omega,k) = 1 + \frac{1}{k^2\lambda_D^2} \left[1 + \beta e^{-\beta^2} \left(-2F(\beta) + i\sqrt{\pi} \right) \right] \tag{7.84}$$

Equation (7.84) agrees with the well-known result expressed in terms of the plasma dispersion function. See for example, G. Bekefi [*Radiation Processes in Plasmas*, Wiley (1966)]. We emphasize that (7.84) has been derived here by using only the causal constraint $\chi(t<0) = 0$ applied to the Vlasov equation. Equation (7.84) is identical with the familiar formula

$$\varepsilon_L(k,\omega) = 1 + \frac{1}{k^2\lambda_D^2}(1 + \beta Z(\beta)) \tag{7.85}$$

where $Z(\beta)$, the plasma dispersion function (see Section 7.9), is

$$Z(\beta) = \frac{1}{\sqrt{\pi}} \int_{-\infty}^{\infty} \frac{e^{-q^2}}{q-\beta} dq = e^{-\beta^2}\left(i\pi - 2\int_0^{\beta} e^{w^2} dw\right) \tag{7.86}$$

See Section 7.9. The direct inversion of (7.79) to derive (7.85) is left as an exercise. The equality in (7.86) will be proven later.

7.7 Asymptotic Behavior of Perturbation

Let us consider the following problem (quoted from Landau and Lifshitz *Electrodynamics of continuous media*, Pergammon (1960)):

> At a given instant $t = 0$ an electromagnetic perturbation occurs in some region of space. The perturbation is not maintained by external agencies, and is therefore damped in time. Find the damping decrement.

We restrict the problem to an electrostatic perturbation. Then by solving the dispersion relation $\varepsilon_L(\mathbf{k}, \omega) = 0$, we can find the roots $\omega_j(\mathbf{k})$ $(j = 1, 2, \ldots)$. The imaginary parts of the ω_js will be less than 0, and we conjecture: $\varepsilon_L(\mathbf{k}, \omega)$ has no zeros in the upper-half ω-plane. The asymptotic behavior of the perturbation will be determined by the mode having the least value of the imaginary frequency.

The posed problem can be approached as an initial value problem as is done in the reference. We have, for an electrostatic wave,

$$-\frac{1}{c}\frac{\partial \mathbf{B}}{\partial t} = \nabla \times \mathbf{E} = 0 \tag{7.87}$$

$$\frac{1}{c}\frac{\partial \mathbf{D}}{\partial t} = \nabla \times \mathbf{B} \tag{7.88}$$

Therefore, we have $\frac{\partial^2 D_\parallel}{\partial t^2} = 0$, where $\parallel$ refers to the direction of $\mathbf{k}$. Hence $D_\parallel = 0$. Thus, when no external currents are applied, we have $D_\parallel = 0$ and it is the initial condition for the differential equation in (7.40). In this case of electrostatic wave, we have

$$\varepsilon_L(k, \omega) E(k, \omega) = 0 \tag{7.89}$$

This homogeneous equation is solved by $E(\omega, k) = C\delta(\varepsilon_L(\omega, k))(C : \text{const.})$, or

$$E(\omega, k) = C \sum_j |\alpha_j| \delta(\omega - \omega_j) \tag{7.90}$$

$$\text{with } \alpha_j(\omega_j) = \left[\left.\frac{\partial \varepsilon_L}{\partial \omega}\right|_{\omega_j}\right]^{-1} \tag{7.91}$$

where ω_j is the zeros of ε_L. We can invert (7.90) with ease because of the delta functions to obtain $E(t,k)$. $\alpha_j s$ have no significant influence on the asymptotic behavior of the wave.

• **Finding zeros of $\varepsilon_L(\omega)$.** The method resorts to a perturbational scheme. We expect a complex ω satisfies simultaneously $\varepsilon_r(\omega_r + i\omega_i) = 0$ and $\varepsilon_i(\omega_r + i\omega_i) = 0$, where the subscript r and i denotes respectively the real and imaginary parts. We assume $\omega_r \gg \omega_i$ and Taylor expand the real and imaginary parts:

$$\varepsilon_r(\omega_r) + \nu i \frac{\partial \varepsilon_r}{\partial \omega_r}\omega_i = 0$$

$$\varepsilon_i(\omega_r) + \nu i \frac{\partial \varepsilon_i}{\partial \omega_r}\omega_i = 0$$

where ν is the small parameter: $\omega_i/\omega_r = O(\nu)$. We further assume $\varepsilon_i/\varepsilon_r = O(\nu)$. Combining the above two equations, we write

$$\varepsilon_r(\omega_r) + \nu\varepsilon_i(\omega_r) + \nu i\omega_i \frac{\partial \varepsilon_r}{\partial \omega_r} + \nu^2 i\omega_i \frac{\partial \varepsilon_i}{\partial \omega_r} = 0$$

In the zero order we have $\varepsilon_r(\omega_r) = 0$ which gives the real frequency ω_r. In the first order we have the imaginary part of the frequency as given by

$$\omega_i = -i\frac{\varepsilon_i(\omega_r)}{\partial \varepsilon_r/\partial \omega_r} \tag{7.92}$$

• **Landau damping.** As an application of (4.92), we consider the dispersion relation of electrostatic wave in a Vlasov–Poisson plasma as obtained in (7.73) and (7.74). They can be put into the following form

$$\varepsilon_r(k,\omega_r) = 1 - \frac{\omega_{\text{pe}}^2}{k^2} P\int_{-\infty}^{\infty} \frac{g_0'(v)dv}{v - \omega_r/k} = 0 \tag{7.93}$$

$$\varepsilon_i(k,\omega_r) = -\pi\frac{\omega_{\text{pe}}^2}{k^2}\left[g_0'\right]_{\frac{\omega_r}{k}} \tag{7.94}$$

where $g_0'(v) = dg_0/dv$. Then (7.92) gives

$$\omega_i = -\pi\; k[g_0']_{\frac{\omega_r}{k}} \div P\int \frac{g_0'(v)dv}{(v - \omega_r/k)^2} \tag{7.95}$$

Differentiating (7.93) with respect to k gives [we consider ω_r is a function of k]

$$P\int \frac{g_0'(v)dv}{(v-\omega_r/k)^2} = \left(\frac{\partial \omega_r}{\partial k} - \frac{\omega_r}{k}\right)^{-1} 2P\int \frac{g_0'(v)dv}{v - \omega_r/k} = \left(\frac{\partial \omega_r}{\partial k} - \frac{\omega_r}{k}\right)^{-1} \frac{2k^2}{\omega_{\text{pe}}^2}$$

Hence

$$\omega_i = \frac{\pi}{2}\frac{\omega_{\rm pe}^2}{k^2}\left(\omega_r - k\frac{\partial \omega_r}{\partial k}\right)[g_0']_{\frac{\omega_r}{k}} \simeq \frac{\pi}{2}\frac{\omega_{\rm pe}^2}{k^2}\omega_r[g_0']_{\frac{\omega_r}{k}} \tag{7.96}$$

where we used $\omega_r^2 = \omega_{\rm pe}^2(1+3k^2\lambda_D^2)$, which can be obtained from (7.93) or fluid equation. Note that $\omega_r - k\ \partial\omega_r/\partial k = \omega_{\rm pe}^2/\omega_r > 0$. According to (7.96), for an instability to occur ($\omega_i > 0$), the slope of the distribution function, must be positive ($g_0'(\omega_r/k) > 0$). In this case, the phase velocity of the wave is larger than the group velocity. Equation (7.96) is the rate of damping, known as Landau damping, of which physical origin is not due to inter-particle collisions but due to wave-particle interaction. In the process of Landau damping the entropy is conserved. It is often called *collisionless damping*. For a Maxwellian distribution $g_0(v)$, (7.96) gives

$$\omega_i = -\sqrt{\frac{\pi}{8}}\frac{\omega_{\rm pe}}{k^3\lambda_D^3}\exp\left[\frac{-1}{2k^2\lambda_D^2} - \frac{3}{2}\right] \tag{7.97}$$

Reminder: $\lambda_D = \sqrt{T_e/4\pi Ne^2}$: Debye length, $\omega_{\rm pe} = \sqrt{4\pi Ne^2/m_e}$, electron plasma frequency.

7.8 Susceptibility of a Fluid Plasma

We assume that ions are immobile and form only the neutralizing background. We begin with the following linearized equations for electrons.

$$\frac{\partial n}{\partial t} + n_0\nabla\cdot\mathbf{v} = 0 \tag{7.98}$$

$$\frac{\partial \mathbf{v}}{\partial t} = -\frac{e}{m}\mathbf{E} - \frac{T}{mn_0}\nabla n \tag{7.99}$$

$$\nabla\cdot\mathbf{E} = -4\pi e n \tag{7.100}$$

where $n(n_0)$ is perturbation (equilibrium) number density. After Fourier transforming, the above equations become

$$\left(\omega^2 - \frac{T}{m}k^2\right)n(k,\omega) = -\frac{i\ en_0}{m}\mathbf{k}\cdot\mathbf{E} \tag{7.101}$$

$$i\mathbf{k}\cdot\mathbf{E} = -4\pi e n \tag{7.102}$$

We have $-en = -i\mathbf{k}\cdot\mathbf{P}_e$, $4\pi\mathbf{P}_e = \chi\mathbf{E}$ ($\mathbf{P}_e$ is the polarization). Then (7.101) gives $\chi = 4\pi en/i\mathbf{k}\cdot\mathbf{E}$, which takes the form

$$\chi(k,\omega) = -\frac{\omega_{\rm pe}^2}{\omega^2 - \frac{T}{m}k^2} \tag{7.103}$$

The susceptibility can be separated into the real and imaginary parts as follows:

$$\operatorname{Re}\chi(k,\omega) = \frac{\omega_{\rm pe}^2}{2kV}\left(P\frac{1}{\omega+kV} - P\frac{1}{\omega-kV}\right) \tag{7.104}$$

$$\operatorname{Im}\chi(k,\omega) = -\frac{\pi\omega_{\rm pe}^2}{2kV}(\delta(\omega+kV) - \delta(\omega-kV)) \tag{7.105}$$

where $V = \sqrt{T/m}$ is the thermal velocity. In a fluid plasma, the susceptibility $\chi(k,\omega)$ is separated in operatorial form; nevertheless they satisfy the K–K relations as can be easily proven with the aid of Eq. (7.58). We can recover (7.103) from (7.104) and (7.105) as follows. From (7.104) and (7.105) we have

$$(\omega - kV)(\omega + kV)\operatorname{Re}\chi = -\omega_{\rm pe}^2, \quad (\omega - kV)(\omega + kV)i\operatorname{Im}\chi = 0.$$

Adding the these two equations yields (7.103).

The foregoing exercise indicates that when the operators $P\frac{1}{\omega-kV}$ and $\delta(\omega - kV)$ have an object to operate on, the collisionless damping can be manifested, as in a Vlasov plasma. The fluid equation results are instrumental to establish the general rule for the Fourier transforms in the complex ω plane where $\chi(\omega)$ is analytic in the upper plane $\operatorname{Im}\omega > 0$, or $\chi(t) = \chi(t)$ for $t > 0$ or $=0$ for $t < 0$. We can replace the singular integral by *Plemelj's formula*

$$\frac{1}{x - x_0 + i\nu} = P\frac{1}{x - x_0} - i\pi\delta(x - x_0) \tag{7.106}$$

with ν being a positive infinitesimal.

The susceptibility in (7.103) gives the dielectric permeability

$$\varepsilon_L(\omega,k) = 1 - \frac{\omega_{\rm pe}^2}{\omega^2 - \frac{T}{m}k^2} \tag{7.107}$$

When no external currents are introduced in the plasma, $D = 0$ and we have $\varepsilon_L(\omega,k)E(\omega,k) = 0$, of which solution can be written as

$$E(\omega,k) = C\delta[\varepsilon_L(\omega,k)] = C\sum_i |\alpha_i|\delta(\omega - \omega_i) \quad C: \text{const} \tag{7.108}$$

where $\alpha_i(\omega_i) = \left(\frac{\partial\varepsilon_L}{\partial\omega}\big|_{\omega_i}\right)^{-1}$ with ω_i being the zeros of $\varepsilon_L(\omega,k)$. We can easily find $\alpha_i = \frac{\pm\omega_{\rm pe}^2}{2\omega_0}$, and Eq. (7.108) takes the form

$$E(\omega,k) = C\frac{\omega_{\rm pe}^2}{2\omega_0}[\delta(\omega - \omega_0) + \delta(\omega + \omega_0)], \omega_0 = \sqrt{k^2V^2 + \omega_{\rm pe}^2} \tag{7.109}$$

Inverting (7.109) yields

$$E(k,t) = \frac{C}{2\pi}\frac{\omega_{\rm pe}^2}{\omega_0}\cos\omega_0 t \tag{7.110}$$

There is no damping.

7.9 Plasma Dispersion Function $Z(\zeta)$

The important function introduced in (7.86) will be considered here to prove the identity, which we encounter in the study of Vlasov plasma of Maxwellian velocity distribution.

$$Z(\zeta) = \frac{1}{\sqrt{\pi}}\int_{-\infty}^{\infty}\frac{e^{-x^2}}{x-\zeta}dx \tag{7.111}$$

where the complex number ζ has positive imaginary part. Consider the principal value of $Z(\zeta)$, Z_p when the imaginary part of $\zeta \to 0$:

$$Z_p = P\frac{1}{\sqrt{\pi}}\int_{-\infty}^{\infty}\frac{e^{-x^2}}{x-\zeta}dx$$

Differentiating gives

$$\begin{aligned}\frac{dZ_p}{d\zeta} &= \frac{1}{\sqrt{\pi}}\int_{-\infty}^{\infty}\frac{e^{-x^2}}{(x-\zeta)^2}dx = \frac{1}{\sqrt{\pi}}\int_{-\infty}^{\infty}\frac{d}{dx}\left(\frac{-1}{x-\zeta}\right)e^{-x^2}dx\\ &= \frac{-2}{\sqrt{\pi}}\int_{-\infty}^{\infty}\frac{xe^{-x^2}}{x-\zeta}dx\\ &= \frac{-2}{\sqrt{\pi}}\int_{-\infty}^{\infty}\left(1+\frac{\zeta}{x-\zeta}\right)e^{-x^2}dx\\ &= -2(1+\zeta Z_p)\end{aligned} \tag{7.112}$$

Equation (7.112) is a first order differential equation with the boundary condition $Z_p(0) = 0$ (an odd function integral) which is solved by

$$Z_p(\zeta) = -2e^{-\zeta^2}\int_0^{\zeta} e^{x^2}dx \tag{7.113}$$

So the plasma dispersion function can be written as

$$Z(\zeta) = \frac{1}{\sqrt{\pi}}\int_{-\infty}^{\infty}\frac{e^{-x^2}}{x-\zeta}dx = e^{-\zeta^2}\left(-2\int_0^{\zeta} e^{x^2}dx + i\sqrt{\pi}\right) \tag{7.114}$$

Thus (7.86) is proven. The error function integral in (7.114) can be asymptotically expanded [Arfken and Weber, *Mathematical methods for physicists*, 5th edn. Harcourt/Academic Press (2001)] for the cases of $\zeta \gg 1$ and $\zeta \ll 1$:

$$Z(\zeta) = -\frac{1}{\zeta} - \frac{1}{2\zeta^3} - \frac{3}{4\zeta^5} - \frac{15}{8}\frac{1}{\zeta^7} - \cdots + i\sqrt{\pi}e^{-\zeta^2} \quad \zeta \gg 1 \tag{7.115}$$

$$Z(\zeta) = -2\zeta\left(1 - \frac{2}{3}\zeta^2 + \frac{4}{15}\zeta^4 - \cdots\right) + i\sqrt{\pi}e^{-\zeta^2} \quad \zeta \ll 1 \tag{7.116}$$

If we use (7.115) in (7.85), we note that

$$\varepsilon_L(\mathbf{k}, \omega) = 1 - \frac{\omega_p^2}{\omega^2} \quad (\omega \to \infty) \tag{7.117}$$

See also (7.103). From (7.117) we immediately see that $(\varepsilon_L(\mathbf{k}, \omega) - 1)$ is an analytic function in the upper half ω plane that goes to zero as $\omega \to \infty$. Taking the reciprocal of (7.117), we observe that the function $\frac{1}{\varepsilon_L(\mathbf{k},\omega)} - 1$ is also an analytic function in the upper half ω plane that goes to zero as $\omega \to \infty$. Therefore, the function $\frac{1}{\varepsilon_L(\mathbf{k},\omega)} - 1$ satisfies the Cauchy integral formula at a point $\omega_i > 0$, and upon analytic continuation to the real ω axis we eventually reach the Kramer–Kronig relations for the function $\frac{1}{\varepsilon_L(\mathbf{k},\omega)} - 1$.

$$\mathrm{Re}\,\frac{1}{\varepsilon_L(\omega)} - 1 = \frac{1}{\pi}P\int_{-\infty}^{\infty}\frac{d\omega'}{\omega' - \omega}\mathrm{Im}\frac{1}{\varepsilon_L(\omega')} \tag{7.118}$$

$$\mathrm{Im}\,\frac{1}{\varepsilon_L(\omega)} = \frac{1}{\pi}P\int_{-\infty}^{\infty}\frac{d\omega'}{\omega' - \omega}\left(1 - \mathrm{Re}\frac{1}{\varepsilon_L(\omega')}\right) \tag{7.119}$$

7.10 Initial Value Problem

This section should be read as continuation of the problem posed at the beginning of Section 7.7. We consider a homogeneous infinite medium in which at $t = 0$ an electromagnetic field is produced by external sources. At subsequent times $(t > 0)$ they are turned off. Let us consider the time variation of the excited field in the medium. This problem can be approached as the initial value problem of the two Maxwell equations involving time derivative:

$$\nabla \times \mathbf{E} = -\frac{1}{c}\frac{\partial \mathbf{B}}{\partial t} \tag{7.120}$$

$$\nabla \times \mathbf{B} = \frac{1}{c}\frac{\partial \mathbf{D}}{\partial t} \tag{7.121}$$

To solve this initial value problem Laplace transform is appropriate since the initial condition at $t = 0$ is given as $\mathbf{E}(0, \mathbf{r})$ and $\mathbf{B}(0, \mathbf{r})$. The equivalent method to the Laplace transform is the half-range Fourier transform

$$\mathbf{E}(\omega, \mathbf{r}) = \int_0^\infty dt\, e^{i\omega t}\mathbf{E}(t, \mathbf{r}) \tag{7.122}$$

whose inversion is

$$\mathbf{E}(t, \mathbf{r}) = \int_{-\infty+i\alpha}^{\infty+i\alpha} \frac{d\omega}{2\pi} e^{-i\omega t}\mathbf{E}(\omega, \mathbf{r}) \tag{7.123}$$

where $i\alpha$ is such a constant that in the region $\text{Im}\,\omega > \alpha$, $\mathbf{E}(\omega, \mathbf{r})$ is analytic. For the spatial coordinate, we simply put $\nabla = i\mathbf{k}$, since the space is unbounded. To solve this problem, it is not sufficient to know the initial values of $\mathbf{E}(0, \mathbf{r})$ and $\mathbf{B}(0, \mathbf{r})$. Equation (7.122) indicates that one must know the initial value of $\mathbf{D}(0, \mathbf{r})$ also. This means that the knowledge of the history of the field $\mathbf{E}(t, \mathbf{r})$ is necessary. From (7.44) we can write

$$D_i(0, \mathbf{k}) = \int_{-\infty}^0 dt' \varepsilon_{ij}(0 - t', \mathbf{k})E_j(t', \mathbf{k}) \tag{7.124}$$

Recall $\varepsilon_{ij} = \delta_{ij} + \chi_{ij}$. The quantity $\mathbf{D}(0, \mathbf{r})$ is determined by the field $\mathbf{E}(t, \mathbf{r})$ at all the moments in the past $(t < 0)$, as we discussed earlier. The physical reason of this unsimultaneity has to do with the time dispersion or, with the inertia of the charge carriers.

Let us transform the Maxwell equations by half-range Fourier integrals.

$$\frac{\omega}{c}\mathbf{D}(\omega, \mathbf{k}) + \mathbf{k} \times \mathbf{B}(\omega, \mathbf{k}) = \frac{i}{c}\mathbf{D}(0, \mathbf{k}), \quad \mathbf{k} \cdot \mathbf{B}(\omega, \mathbf{k}) = 0 \tag{7.125}$$

$$\frac{\omega}{c}\mathbf{B}(\omega, \mathbf{k}) - \mathbf{k} \times \mathbf{E}(\omega, \mathbf{k}) = \frac{i}{c}\mathbf{B}(0, \mathbf{k}), \quad \mathbf{k} \cdot \mathbf{D}(\omega, \mathbf{k}) = 0 \tag{7.126}$$

where

$$\mathbf{B}(0, \mathbf{k}) = \int d^3r e^{-i\mathbf{k}\cdot\mathbf{r}}\mathbf{B}(t = 0, \mathbf{r}) \tag{7.127}$$

$$D_i(\omega, \mathbf{k}) = \varepsilon_{ij}(\omega, \mathbf{k})E_j(\omega, \mathbf{k}) \tag{7.128}$$

Then (7.126) and (7.127) can be arranged as inhomogeneous simultaneous algebraic equation for the electric field components:

$$\left[k^2\delta_{ij} - k_ik_j - \frac{\omega^2}{c^2}\varepsilon_{ij}(\omega, \mathbf{k})\right]E_j(\omega, \mathbf{k}) = \frac{i\omega}{c^2}D_i(0, \mathbf{k}) - \frac{i}{c}(\mathbf{k} \times \mathbf{B}(0, \mathbf{k}))_i \tag{7.129}$$

This equation can be solved by inverting the matrix on the left-hand side in the form

$$E_i(\omega, \mathbf{k}) = \frac{1}{|\Lambda_{ij}|}C_{ij}R_j \tag{7.130}$$

where

$$\Lambda_{ij} = k^2\delta_{ij} - k_i k_j - \frac{\omega^2}{c^2}\varepsilon_{ij}(\omega, \mathbf{k}) \tag{7.131}$$

is the matrix, and $|\Lambda_{ij}|$ is the determinant, and C_{ij} is the cofactor of Λ_{ij}, and R_j is the column vector corresponding to the right hand side of (7.130).

The Fourier representation of the electric field $E_i(\omega, \mathbf{k})$ in (7.131) can be inverted to $E_i(t, \mathbf{k})$ according to (7.124).

$$E_i(t, \mathbf{k}) = \int_{-\infty+i\alpha}^{\infty+i\alpha} \frac{d\omega}{2\pi} e^{-i\omega t} \frac{1}{|\Lambda_{ij}|} C_{ij} R_j \tag{7.132}$$

The inversion integral is computed by contour integration in the complex ω plane. The contour consists of the horizontal line $\omega = i\alpha$ with α greater than (above) the imaginary parts of all the singularities of the integrand, and the great half circle enclosing the lower half ω plane for $t > 0$. The singularities of the integrand are the zeros of $|\Lambda_{ij}| = 0$, which is the dispersion relation of the excited wave. By the residue theorem, we obtain

$$\mathbf{E}(t, \mathbf{k}) \sim \sum_j A_j \exp(-i\omega_j(\mathbf{k})t) \tag{7.133}$$

where $\omega_j(\mathbf{k})$ $(j = i, 2, \ldots)$ are the roots of the dispersion relation. If $\operatorname{Im}\omega_j < 0$, the excited wave damps and is stable. If $\operatorname{Im}\omega_j > 0$, the excited wave grows and is unstable—unstable plasma. If $\operatorname{Im}\omega_j = 0$, this case corresponds to persisting natural oscillation.

7.11 Ion Acoustic Wave

A plasma consists of light electrons and heavy ions with great disparity of mass ratio $m_e/m_i = 1840$. By assuming $m_i \to \infty$ (ions are immobile), we have high frequency electron wave called *Langmuir wave* whose frequency is $\omega_{\text{pe}} = \sqrt{4\pi N e^2/m_e}$, the plasma frequency. In the low frequency spectrum of plasma motion, ion inertia plays its role, and gives rise to a low frequency wave called *ion acoustic wave* whose phase velocity is

$$\frac{\omega}{k} = \sqrt{\frac{\gamma_e T_e}{m_i}} \tag{7.134}$$

where γ_e is the ratio of specific heats. The dispersion relation indicates that the wave involves the electron pressure and ion inertia. A crude picture for the ion acoustic wave is viewing the wave as the ionic space charge wave (electrostatic) in the background of Boltzmann-distributed hot electrons. The Langmuir wave is electronic space charge wave in the background of stationary neutralizing ions. In the following, we derive (7.135) by using two fluid equations and Poisson equation. We have

$$\frac{\partial n_\alpha}{\partial t} + \nabla \cdot (n_\alpha \mathbf{v}_\alpha) = 0 \quad (\alpha = e, i) \tag{7.135}$$

$$m_\alpha n_\alpha \left(\frac{\partial \mathbf{v}_\alpha}{\partial t} + \mathbf{v}_\alpha \cdot \nabla \mathbf{v}_\alpha \right) = -\nabla p_\alpha + e_\alpha n_\alpha \mathbf{E} \tag{7.136}$$

$$\nabla \cdot \mathbf{E} = 4\pi \sum_\alpha e_\alpha n_\alpha \tag{7.137}$$

$$p_\alpha = \gamma_\alpha T_\alpha n_\alpha \tag{7.138}$$

The above equations are linearized with respect to the perturbed quantities $n'_\alpha(\mathbf{r}, t) = n_\alpha - N$ and $\mathbf{v}'_\alpha = \mathbf{v}_\alpha$. Also Fourier transform by putting $\partial/\partial t = -i\omega, \nabla = i\mathbf{k}$. Then we obtain

$$-\omega n'_\alpha + N\mathbf{k} \cdot \mathbf{v}'_\alpha = 0 \tag{7.139}$$

$$N\omega \mathbf{v}'_\alpha = \frac{\gamma_\alpha T_\alpha}{m_\alpha} n'_\alpha \mathbf{k} + i \frac{e_\alpha}{m_\alpha} N\mathbf{E} \tag{7.140}$$

$$i\mathbf{k} \cdot \mathbf{E} = 4\pi e(n'_i - n'_e) \tag{7.141}$$

Eliminating $\mathbf{k} \cdot \mathbf{v}'_\alpha$ yields

$$n'_i \left(\omega^2 - k^2 \frac{\gamma_i T_i}{m_i} - \omega_{pi}^2 \right) + \omega_{pi}^2 n'_e = 0 \tag{7.142}$$

$$n'_e \left(\omega^2 - k^2 \frac{\gamma_e T_e}{m_e} - \omega_{\text{pe}}^2 \right) + \omega_{\text{pe}}^2 n'_i = 0 \tag{7.143}$$

where $\omega_{p\alpha} = \sqrt{4\pi N e^2 / m_\alpha}$, the plasma frequency. Then the dispersion relation is obtained in the form

$$(\omega^2 - \omega_e^2)(\omega^2 - \omega_i^2) = \omega_{\text{pe}}^2 \omega_{pi}^2 \tag{7.144}$$

where $\omega_\alpha^2 = \omega_{p\alpha}^2 + k^2 \gamma_\alpha T_\alpha / m_\alpha$.

Two approximate solutions of the above quadratic equation can be obtained, keeping only the first order terms in $\omega_i^2/\omega_e^2 (\simeq m_e/m_i)$:

$$\omega^2 = \omega_e^2 = \omega_{\text{pe}}^2 + k^2 \frac{\gamma_e T_e}{m_e} \tag{7.145}$$

which is the high frequency *electron plasma wave*. The other mode is the low frequency wave

$$\omega^2 = \omega_i^2 - \omega_{\text{pe}}^2 \omega_{pi}^2 / \omega_e^2 = \omega_{pi}^2 \left[1 - \frac{\omega_{\text{pe}}^2}{\omega_{\text{pe}}^2 + k^2 \gamma_e T_e / m_e} + \frac{k^2 \gamma_i T_i}{m_i \omega_{pi}^2} \right] \tag{7.146}$$

The above expression for the low frequency *ion acoustic wave* becomes

$$\omega^2 = k^2 \frac{\gamma_i T_i + \gamma_e T_e}{m_i} \tag{7.147}$$

where we neglected $k^2 \lambda_e^2 \ll 1$ (λ_e is the electron Debye length; $\lambda_e^2 = \gamma_e T_e / 4\pi N e^2$). This analysis shows that the natural oscillations of plasma in electrostatic approximation are the two waves, Langmuir wave (electron plasma wave) and ion acoustic wave. The ion acoustic wave is severely Landau-damped unless $T_e \gg T_i$, thus the propagating mode of ion acoustic wave has the dispersion relation

$$\omega^2 = k^2 \frac{\gamma_e T_e}{m_i} \tag{7.148}$$

• **Approximation of massless electron:** $m_e \to 0$. Putting $m_e = 0$ in (7.137) gives

$$\nabla p_e = e n_e \nabla \phi \tag{7.149}$$

where ϕ is the electric potential, $\mathbf{E} = -\nabla \phi$. The pressure force is balanced by the electric force. Using (7.139) in this equation gives

$$\frac{\gamma_e T_e}{e} \frac{\nabla n_e}{n_e} = \nabla \phi$$

which is integrated in the form

$$n_e = N \exp\left(\frac{e\phi}{\gamma_e T_e}\right) \tag{7.150}$$

In hot plasma, the electrons have great mobility to establish instantaneously the Boltzmann distribution in response to the electrostatic force. Now let us write the ion equations

$$\frac{\partial \mathbf{v}_i}{\partial t} = -\frac{e}{m_i} \nabla \phi \tag{7.151}$$

$$\frac{\partial n_i}{\partial t} + N \nabla \cdot \mathbf{v}_i = 0 \tag{7.152}$$

$$-\nabla^2 \phi = 4\pi e (n_i - n_e') \tag{7.153}$$

In (7.152)–(7.154), all the quantities are perturbations. From (7.151) we obtain the electron density perturbation $n_e' = n_e - N \simeq N(1 + \frac{e\phi}{\gamma_e T_e}) - N = \frac{Ne\phi}{\gamma_e T_e}$. Then Fourier analysis of the above equations yield the dispersion relation

$$\omega^2 = \frac{\omega_{pi}^2 k^2 \lambda_e^2}{1 + k^2\lambda_e^2} \tag{7.154}$$

where $\lambda_e^2 = \gamma_e T_e / 4\pi N e^2$. Neglecting $k^2\lambda_e^2 \ll 1$ gives

$$\frac{\omega^2}{k^2} = \frac{\gamma_e T_e}{m_i} \tag{7.155}$$

7.12 Plasma as a Collection of Beams

One way of describing a plasma is to view the plasma as a group of beams; beams individualized by each initial velocity $\mathbf{v}_0$. This picture is convenient when we consider the wave-particle interaction in a plasma by using fluid equations. Modeling a plasma as a group of beams with varying beam velocities was earlier adopted by Bohm and Gross.

Reference: Bohm, D. and Gross, E. P. *Theory of plasma oscillations A. Origin of medium-like behavior*, Phys. Rev. **75**, 1851 (1949); Bohm, D. and Gross, E. P. *Theory of plasma oscillations B. Excitation and damping of oscillations*, Phys. Rev. **75**, 1864 (1949).

Let us consider cold ion beams in the background of Boltzmann-distributed hot electrons. The ion equations should contain the beam velocity $\hat{\mathbf{x}}v_0$ as the zero order quantity. The one-dimensional linearized ion equations read

$$\frac{\partial n}{\partial t} + N(v_0)\frac{\partial v}{\partial x} + v_0\frac{\partial n}{\partial x} = 0 \tag{7.156}$$

$$\frac{\partial v}{\partial t} + v_0\frac{\partial v}{\partial x} = -\frac{e}{m_i}\frac{\partial \phi}{\partial x} \tag{7.157}$$

$$-\frac{\partial^2 \phi}{\partial x^2} = 4\pi e\left(n - N(v_0)\frac{e\phi}{\gamma_e T_e}\right) \tag{7.158}$$

where $N(v_0)$ is the number density of v_0-beams. Assuming a wave $\sim e^{ikx-i\omega t}$, we have

$$(\omega - kv_0)n = kN(v_0)v \tag{7.159}$$

$$(\omega - kv_0)v = \frac{e}{m_i}k\phi \tag{7.160}$$

$$k^2\phi\left(1 + \frac{4\pi e^2 N(v_0)}{k^2\gamma_e T_e}\right) = 4\pi e n \tag{7.161}$$

Equations (7.160) and (7.161) give the ion number density

$$n_i = n = \frac{k^2 e\phi N(v_0)}{m_i(\omega - kv_0)^2} \tag{7.162}$$

We have the electron number density

$$n_e = \frac{e\phi N(v_0)}{\gamma_e T_e} \tag{7.163}$$

Hence, the susceptibility $\chi(v_0) = \frac{4\pi e(n_e - n_i)}{k^2\phi}$ is obtained in the form

$$\chi(v_0) = \frac{4\pi e^2 N(v_0)}{k^2\gamma_e T_e} - \frac{4\pi e^2}{m_i}\frac{N(v_0)}{(\omega - kv_0)^2} \tag{7.164}$$

Since susceptibility is additive, the plasma susceptibility is obtained by integrating over the distribution of initial velocities by putting $N(v_0) = f(v_0)dv_0$ in the form

$$\chi(k,\omega) = \frac{1}{k^2\lambda_D^2} - \omega_{pi}^2 \int_{-\infty}^{\infty} \frac{f(v_0)dv_0}{(\omega - kv_0)^2} \tag{7.165}$$

where the Debye length is defined by

$$\lambda_D^{-2} = \frac{4\pi e^2}{\gamma_e T_e}\int_{-\infty}^{\infty} f(v_0)dv_0$$

Integrating by parts in (7.166), the susceptibility is obtained as

$$\chi(k,\omega) = \frac{1}{k^2\lambda_D^2} + \frac{\omega_{pi}^2}{k}\int_{-\infty}^{\infty} dv_0 \frac{\frac{df}{dv_0}}{\omega - kv_0} \tag{7.166}$$

The dielectric permeability of the plasma and the dispersion relation are written as

$$\varepsilon(k,\omega) = 1 + \chi(k,\omega) = 1 + \frac{1}{k^2\lambda_D^2} - \omega_{pi}^2\int_{-\infty}^{\infty}\frac{f(v_0)dv_0}{(\omega - kv_0)^2} = 0 \tag{7.167}$$

If we assume $f(v_0)$ is a Maxwellian and $\frac{\omega}{k} \gg \bar{v}_0 = \sqrt{\int v_0^2 f(v_0)dv_0}$, the dispersion relation becomes

$$1 + \frac{1}{k^2\lambda_D^2} - \frac{\omega_{pi}^2}{\omega^2} = 0 \tag{7.168}$$

which reduces to the form (7.155). For continuation, work out the following Exercises.

Exercise. Separate the real and imaginary parts of $\chi(k,\omega)$ in (7.167). Expand $\frac{1}{\omega - kv_0}$ when $\frac{\omega}{k} \gg \bar{v}_0$ and obtain the damping rate.

Exercise. Obtain the dispersion relation by using (7.162) and (7.163).

7.13 Nyquist Criterion

We have a dispersion relation for plasma waves of the form

$$\varepsilon(\omega, k) = 0 \tag{7.169}$$

We are interested in knowing whether or not the solutions of (7.170) have unstable roots ($\omega_i > 0$) without actually solving the algebraic equation $\varepsilon(\omega) = 0$ (The wave number k will be suppressed). Nyquist theorem provides means for this purpose to determine whether a dispersion relation has unstable roots in the region of interest in the ω-plane. Consider the integral in the complex ω-plane,

$$\frac{1}{2\pi i}\oint_C \frac{\varepsilon'(\omega)}{\varepsilon(\omega)} d\omega \quad \left(\varepsilon' = \frac{d\varepsilon}{d\omega}\right) \tag{7.170}$$

where C is a closed contour encircling the region of interest in the ω-plane in the positive (counterclockwise) direction. To ascertain the presence of unstable roots of (7.170), let C consist of the real ω-axis and the infinite semi-circle in the upper-half plane. We shall prove that integral (7.171) gives the number of roots of $\varepsilon(\omega) = 0$ within C. We assume that $\varepsilon(\omega)$ has no poles inside C. When $\varepsilon(\omega)$ has a zero at $\omega = \omega_1$, the integrand ε'/ε has a simple pole at $\omega = \omega_1$. By residue theorem, the number expressed by (7.171) is equal to

$$\lim_{\omega\to\omega_1} (\omega - \omega_1)\frac{\varepsilon'(\omega)}{\varepsilon(\omega)} = \varepsilon'(\omega_1) \lim_{\omega\to\omega_1} \frac{(\omega - \omega_1)}{\varepsilon(\omega)} = \frac{\varepsilon'(\omega_1)}{\varepsilon'(\omega_1)} = 1$$

which is the residue of the integral, and we used L'Hospital's rule in the algebra. Suppose $\varepsilon(\omega)$ has N zeros to be written in the form

$$\varepsilon(\omega) = (\omega - \omega_1)(\omega - \omega_2)\dots(\omega - \omega_N)g(\omega)$$

Then we have

$$\frac{\varepsilon'(\omega)}{\varepsilon(\omega)} = \frac{1}{\omega - \omega_1} + \frac{1}{\omega - \omega_2} + \cdots + \frac{1}{\omega - \omega_N} + \frac{g'(\omega)}{g(\omega)}$$

$$\frac{1}{2\pi i}\oint_C \frac{\varepsilon'(\omega)}{\varepsilon(\omega)} d\omega = N \tag{7.171}$$

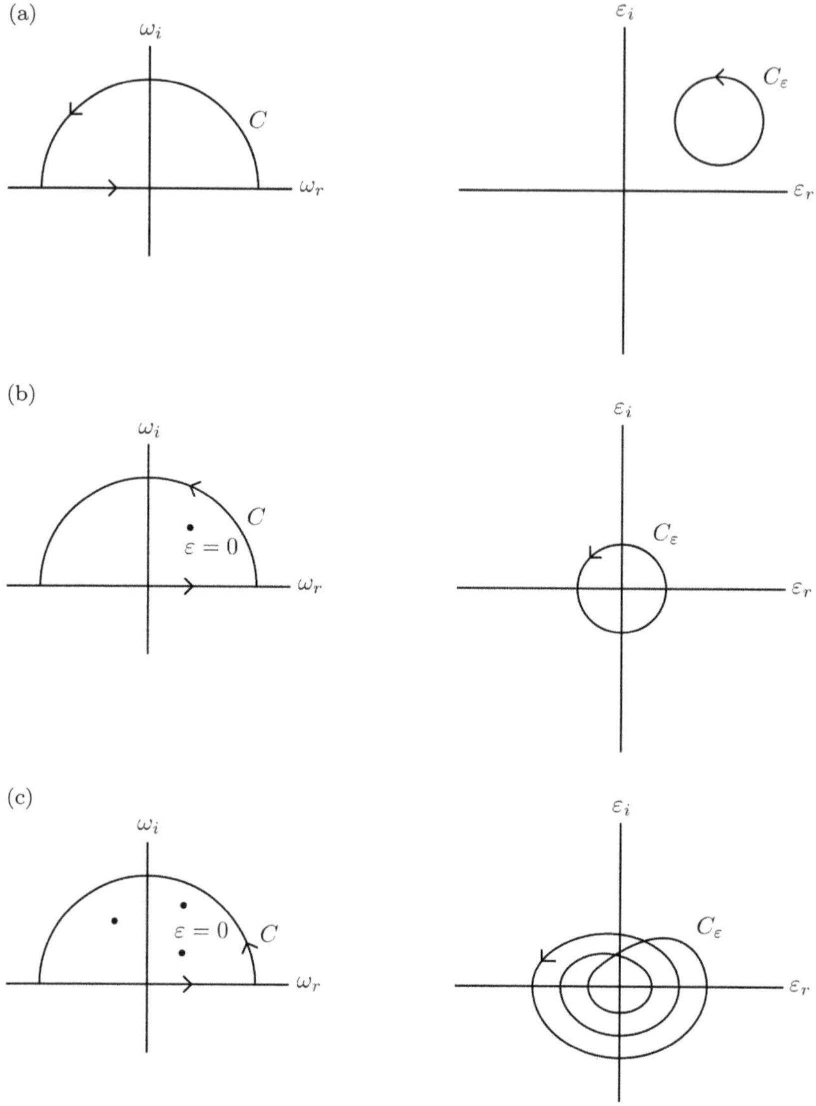

Fig. 7.1. Nyquist diagram.

When $\varepsilon(\omega)$ has multiple zeros, the multiplicity should be added on the RHS of the above equation. To show this, we write

$$\varepsilon(\omega) = (\omega - \omega_0)^m g(\omega) \quad \frac{\varepsilon'(\omega)}{\varepsilon(\omega)} = \frac{m}{\omega - \omega_0}$$

$$\frac{1}{2\pi i} \oint_C \frac{\varepsilon'(\omega)}{\varepsilon(\omega)} d\omega = m$$

We have established the identity (7.172) (*Cauchy's integral*) with the understanding that N contains the multiplicity of the roots. Next, we introduce the polar coordinates in the ω-plane,

$$\omega = re^{i\theta}, \varepsilon(\omega) = R(r,\theta)e^{i\Theta(r,\theta)}, \oint_C \frac{\varepsilon'(\omega)}{\varepsilon(\omega)}d\omega = \oint_C d[\ln\varepsilon(\omega)]$$

Here $\ln\varepsilon(\omega) = \ln R + i\Theta$. Upon introducing the polar coordinates, the contour C in the ω-plane is mapped into the contour C_ε in the (R,θ) plane (i.e. ε-plane) by the mapping functions $R(r,\theta)$ and $\Theta(r,\theta)$. Thus we have

$$\oint_C d[\ln\varepsilon(\omega)] = \oint_{C_\varepsilon} d(\ln R + i\Theta) = i\oint_{C_\varepsilon} d\Theta \qquad (7.172)$$

Note that upon completing revolutions around the closed contour, only the argument Θ changes while the modulus is intact. If the contour C_ε encircles the origin N times, the integral in (7.173) is $2\pi iN$. Now we arrive at the conclusion of the Nyquist theorem by looking at (7.172) and (7.173). If the contour C_ε in the ε plane which corresponds to the contour C in the ω plane encircles the origin of the ε plane N times, the number of the roots of $\varepsilon(\omega) = 0$ within C is N, in view of (7.172). Figures 7.1(a), 7.1(b), and 7.1(c) illustrate the Nyquist criterion as to the number of the roots of $\varepsilon(\omega) = 0$, in which (a), no unstable roots; (b), 1 unstable root; (c), 2 unstable roots.

Problems

7.1 Derive Eqs. (7.29) and (7.30).

7.2 Carry out the Fourier transform of Eq. (7.80) to obtain

$$D(\omega) = E(\omega) + \frac{1}{2\pi i}\int_{-\infty}^{\infty} d\omega' \frac{\chi(\omega')E(\omega')}{\omega' - \omega}$$

The above integral can be converted to a closed contour integral, and Cauchy integral formula can be applied, giving $D(\omega) = (1+\chi(\omega))E(\omega)$.

This exercise clearly shows that $\chi(\omega) = \int_0^\infty \chi(t)e^{i\omega t}dt$ is analytic in the region $\omega_i > 0$.

The half-sided Fourier integral along the positive t-axis in $\int_0^\infty \chi(t)e^{i\omega t}dt$ is analytic in the upper half ω- plane. An easy argument for the analyticity of $\chi(\omega)$ and $\varepsilon(\omega)$ in the region $\omega_i > 0$ can be presented by considering the Fourier integral

$$\chi(t) = \int_{-\infty}^{\infty} \chi(\omega)e^{-i\omega t}d\omega$$

In the above integral, we close the contour by encircling the upper half ω-plane when $t < 0$. Then, for this integral to be zero, $\chi(\omega)$ should be analytic in the upper half ω-plane. The reverse proposition is also true. To summarize the state of the art, the analyticity of $\varepsilon(\omega)$ in the region of $\omega_i > 0$ is a direct consequence of the causality expressed by the algebraic constraint $\chi(t < 0) = 0$.

7.3 Separating the real and imaginary parts of the dielectric permittivity

$$\varepsilon_L(\omega, k) = 1 + \frac{1}{k^2\lambda_D^2}[1 + \beta\, e^{-\beta^2}(-2F(\beta) + i\sqrt{\pi})]$$

we have

$$\mathrm{Re}\,\chi = \frac{1}{k^2\lambda_D^2}\left[1 - 2\beta\, e^{-\beta^2}\int_0^\beta e^{t^2}\, dt\right]$$

$$\mathrm{Im}\,\chi = \frac{\sqrt{\pi}}{k^2\lambda_D^2}\beta\, e^{-\beta^2}$$

Show that $\mathrm{Re}\,\chi$ and $\mathrm{Im}\,\chi$ satisfy Kramers–Kronig relation. [Reference: H. J. Lee, J. Korean Phys. Soc. **78** (2018) p. 65].

7.4 Fourier transform the implicit Kramers–Kronig relation,

$$\chi(\omega) = \frac{i}{\pi}P\int_{-\infty}^{\infty} d\omega' \frac{\chi(\omega')}{\omega - \omega'}$$

and deduce that $\chi(t) = 0$ for $t < 0$.

7.5 Solve the following equation for $F(x)$.

$$(1) \quad G(x) = \frac{1}{\pi}P\int_{-\infty}^{\infty}\frac{F(x')}{x - x'}dx'$$

Solution. Fourier transforming the above equation gives

$$G(k) = \int_{-\infty}^{\infty} G(x)e^{-ikx} = \int_{-\infty}^{\infty} e^{-ikx}\frac{1}{\pi}P\int_{-\infty}^{\infty}\frac{F(x')}{x - x'}dx'$$

Make a change of variable, $x - x' = x'', dx \to dx''$, then we have

$$G(k) = F(k)\frac{1}{\pi}P\int_{-\infty}^{\infty} dx\frac{e^{-ikx}}{x} = -iF(k)H(k)$$

Solving for $F(k)$, we have, per $1/H(k) = H(k)$ for the step function,

$$F(k) = iG(k)H(k)$$

Let us invert this equation.

$$F(x) = \frac{1}{2\pi}\int_{-\infty}^{\infty} iG(k)H(k)e^{ikx}dk$$
$$= \frac{1}{2\pi}\int_{-\infty}^{\infty} dke^{ikx}i\int_{-\infty}^{\infty} e^{-ikx''}G(x'')dx''\frac{i}{\pi}P\int_{-\infty}^{\infty} dx'\frac{e^{-ikx'}}{x'}$$

$\int dk$-integral gives δ-function. The result is

$$(2) \quad F(x) = -\frac{1}{\pi}P\int_{-\infty}^{\infty}\frac{G(x'')}{x-x''}dx''$$

Equations (1) and (2) are Hilbert transform pair. Substituting (2) into (1), we have

$$G(x) = -\frac{1}{\pi^2}P\int_{-\infty}^{\infty}dx'\int_{-\infty}^{\infty}\frac{1}{x-x'}\frac{1}{x'-x''}G(x'')dx''$$

Therefore we conclude

$$-\frac{1}{\pi^2}P\int_{-\infty}^{\infty}\frac{1}{x-x'}\frac{1}{x'-x''}dx' = \delta(x-x'')$$

This is the closure property of Hilbert transform.

7.6 Prove

$$P\int_{-\infty}^{\infty}\frac{e^{-q^2}}{q-r}dq = -2\sqrt{\pi}e^{-r^2}\int_0^r e^{w^2}dw.$$

Hint: The principal value integral in Appendix A may be used.

7.7 Equation (7.46) for the susceptibility tensor is written for a single component

$$\chi(\omega) = \int_0^{\infty}\chi(t)e^{i\omega t}dt$$

Kramers–Kronig relations are concerned with the complex function $\chi(\omega)$ for real frequency ω. It is immediately observed that

$$\chi(-\omega) = \chi^*(\omega)* : \text{complex conjugate}$$

Viz., denoting $\chi(\omega) = \chi_r(\omega) + i\chi_i(\omega)$, $\chi_r(\omega)$ is even and $\chi_i(\omega)$ is odd. Show that

$$\chi_r(\omega) = \frac{2}{\pi}P\int_0^{\infty}\frac{\omega'\chi_i(\omega')}{\omega'^2-\omega^2}d\omega'$$
$$\chi_i(\omega) = -\frac{2}{\pi}P\int_0^{\infty}\frac{\omega\chi_r(\omega')}{\omega'^2-\omega^2}d\omega'$$

7.8 Carry out Fourier inverse transform of $\chi(k,t)$ in (7.79) to derive (7.84).

7.9 Show that:

In the absence of external currents the response of a plasma against a small transverse perturbation is described by

$$\frac{\partial^2 \mathbf{D}_\perp}{\partial t^2} = c^2 k^2 \mathbf{E}_\perp$$

where we assumed the wave $\sim e^{i\mathbf{k}\cdot\mathbf{r}-i\omega t}$. Solving this equation gives

$$\mathbf{E}_\perp(\mathbf{k},t) = \mathbf{C}\int_{-\infty}^{\infty} \frac{d\omega}{2\pi} e^{-i\omega t}\delta[\omega^2\varepsilon_T(\omega) - c^2k^2]$$

where $\mathbf{C}$ is a constant vector, ε_T is the transverse dielectric permittivity. The integral can be done by contour integration with a big half circle enclosing the lower half ω plane, Clearly, the zero of $(\omega^2\varepsilon_T(\omega) - c^2k^2)$ closest to the horizontal axis determines the asymptotic behavior of the perturbation, *viz.*, the longest persisting wave.

7.10 Solve the initial value problem of Section 7.10 in regard to an electrostatic field excitation.

7.11 Show that for g_0 Maxwellian

$$\int dv \frac{\omega g_0}{kv - \omega} = \zeta Z(\zeta); \zeta = \frac{\omega/k}{(2T/m)^{1/2}}$$

7.12 $Z(\zeta)$ satisfies the same differential equation as that satisfied by $Z_p(\zeta)$ but with different boundary condition. What is it?

7.13 Show that $Z_p(\zeta) = -Z_p(-\zeta)$; odd function.

7.14 Show that the real and imaginary parts of $Z(\zeta)$ are connected by the Kramers–Kronig relations.

References

Akhiezer, A., I. Akhiezer, R. Polovin, A. Sitenko, and K. Stepanov, *Collective Oscillations in a Plasma*, Pergamon, Oxford (1967).

Bekefi, G. *Radiation Processes in Plasmas*, John Wiley, New York (1966).

Bohm, D. and E. P. Gross, *Theory of plasma oscillations. A. Origin of medium-like behavior*, Phys. Rev. **75** (1949) p. 1851.

Bohm, D. and E. P. Gross, *Theory of plasma oscillations. B. Excitation and damping of oscillations*, Phys. Rev. **75** (1949) p. 1864.

Krall, N. A. and A. W. Trivelpiece, *Principles of Plasma Physics*, San Francisco Press (1986).

Landau, L. D. and E. M. Lifshitz, *Electrodynamics of Continuous Media*, Pergamon (1960) p. 288.

Lee, H. J. and M. Y. Song, *Causality and collisionless damping in plasma,* J. Modern Phys. **4** (2013), pp. 555–558.

Lee, H. J. *Causality in plasma electrodynamics*, J. Korean Phys. Soc. **73** No. 1 (2018) pp. 65–85.

Van Kampen, N. G. and B. U. Felderhof, *Theoretical Methods in Plasma Physics*, John Wiley (1967), Chapter XV.

Chapter 8

Theory of Vlasov Equation (I)

In this chapter various mathematical methods to analyze the linearized Vlasov equation are introduced. The Vlasov equation is a *hyperbolic partial differential equation* and the *characteristic method* is a very powerful tool to obtain the general solution of Vlasov equation. We spend some pages to get familiar with the mathematics of this useful method. *Separation of variable method* is also applicable for Vlasov equation, and we shall see that the eigen functions are the *Van Kampen mode.* The Vlasov equation will be approached as an initial-value problem to follow Landau's early analysis and review Landau damping.

8.1 Vlasov Equation Revisited

A plasma can be described by Vlasov equation which governs the time rate of change of the single particle distribution function $f_\alpha(\mathbf{r}, \mathbf{v}, t)$:

$$\frac{\partial}{\partial t} f_\alpha(\mathbf{r}, \mathbf{v}, t) + \mathbf{v} \cdot \frac{\partial f_\alpha}{\partial \mathbf{r}} + \frac{e_\alpha}{m_\alpha}\left(\mathbf{E}(\mathbf{r}, \mathbf{t}) + \frac{1}{c}\mathbf{v} \times \mathbf{B}(\mathbf{r}, t)\right) \cdot \frac{\partial f_\alpha}{\partial \mathbf{v}} = 0, \quad (8.1)$$

where the subscript α takes $e(i)$ for electron (ion) distribution function. The Vlasov equation above entails an approximation which is valid only when the number of particles in the Debye sphere is much greater than unity:

$$N\lambda_{D\alpha}^3 \gg 1 \quad \left(\lambda_{D\alpha} = \sqrt{\frac{T_\alpha}{4\pi N e^2}}, \text{ Debye length; } N \text{ number density}\right) \quad (8.2)$$

This criterion can be derived from the condition that Debye cross section (λ_D^2) is much greater than Coulomb cross section $[(e^2/T)^2]$, or $\lambda_D \gg e^2/T$. Equation (8.1) is in fact a collisionless Boltzmann equation; we neglected the collision term. This is acceptable for sufficiently hot plasmas. If the particle collisions need to be included, we can use the Landau collision term (Chapter 2). A collisional plasma relaxes to a Maxwell-Boltzmann distribution in a characteristic time (*relaxation time*) $\tau = N\lambda_{D\alpha}^3/\omega_{p\alpha}$ where $\omega_{p\alpha}$ is the plasma frequency. Use of (8.1) is therefore valid when it is applied to a dynamical change which takes place in a time short as compared to τ.

The Vlasov equation describes the behavior how the aggregate of the phase points move about in the six-dimensional phase space $(\mathbf{r}, \mathbf{v})$: In terms of the hydrodynamics terminology, the Vlasov *continuum* occupying the $(\mathbf{r}, \mathbf{v})$ phase space moves like an *incompressible* fluid, i.e.

$$\frac{Df(\mathbf{r}, \mathbf{v}, t)}{Dt} = 0 \tag{8.3}$$

$$\frac{D}{Dt} = \frac{\partial}{\partial t} + \frac{d\mathbf{r}}{dt} \cdot \frac{\partial}{\partial \mathbf{r}} + \frac{d\mathbf{v}}{dt} \cdot \frac{\partial}{\partial \mathbf{v}} \tag{8.4}$$

This should be compared with the continuity equation, $\frac{d\rho}{dt} + \rho\nabla \cdot \mathbf{v} = 0$, for *compressible* fluid. The Vlasov description reflects the continuum aspect of plasma. Here we recall the fluid-analogous derivation in Section 1.3. We simply put in (8.4)

$$\frac{d\mathbf{r}}{dt} = \mathbf{v} \tag{8.5}$$

$$\frac{d\mathbf{v}}{dt} = \frac{e}{m}\left(\mathbf{E} + \frac{1}{c}\mathbf{v} \times \mathbf{B}\right) \tag{8.6}$$

then we obtain

$$\frac{\partial f}{\partial t} + \mathbf{v} \cdot \frac{\partial f}{\partial \mathbf{r}} + \frac{e}{m}\left(\mathbf{E} + \frac{1}{c}\mathbf{v} \times \mathbf{B}\right) \cdot \frac{\partial f}{\partial \mathbf{v}} = 0 \tag{8.7}$$

This derivation reflects the fluid-like behavior of the Vlasov plasma.

8.2 Maxwell Equations

In (8.1), the electric and magnetic fields and the distribution functions are connected with Maxwell equations:

$$\nabla \times \mathbf{E} = -\frac{1}{c}\frac{\partial \mathbf{B}}{\partial t} \tag{8.8}$$

$$\nabla \times \mathbf{B} = \frac{1}{c}\frac{\partial \mathbf{E}}{\partial t} + \frac{4\pi}{c}\mathbf{J} \tag{8.9}$$

where the current $\mathbf{J}$ is generated by plasma particles:

$$\mathbf{J}(\mathbf{r}, t) = \sum_{\alpha} e_\alpha \int d^3v\mathbf{v}f_\alpha(\mathbf{r}, \mathbf{v}, t) \tag{8.10}$$

Also we have the Poisson equation

$$\nabla \cdot \mathbf{E}(\mathbf{r}, t) = 4\pi \sum_{\alpha} e_\alpha \int d^3v f_\alpha(\mathbf{r}, \mathbf{v}, t) \equiv 4\pi\rho_c \tag{8.11}$$

where ρ_c is the charge density. Taking divergence of (8.8) gives $\frac{\partial}{\partial t}\nabla \cdot \mathbf{B} = 0$ which means $\nabla \cdot \mathbf{B} = 0$ always. Taking divergence of (8.9) yields

$$4\pi\nabla \cdot \mathbf{J} + \frac{\partial}{\partial t}\nabla \cdot \mathbf{E} = 0 \tag{8.12}$$

Using (8.11) in (8.12) gives

$$\nabla \cdot \mathbf{J} + \frac{\partial \rho_c}{\partial t} = 0 \tag{8.13}$$

This is the equation of charge conservation. Equations (8.1) and (8.8)–(8.13) show how f_α and the fields $\mathbf{E}$ and $\mathbf{B}$ are coupled. Here $\mathbf{E}$ and $\mathbf{B}$ are the *self-consistent* fields, since they are generated by the plasma itself.

Equation (8.1) is a nonlinear equation because of the last term which is a product of two unknown functions. The first step to investigate the Maxwell–Vlasov plasma governed by the foregoing equations is to study the linearized Vlasov equation, which we write in the following form

$$\frac{\partial}{\partial t} f_\alpha(\mathbf{r}, \mathbf{v}, t) + \mathbf{v} \cdot \frac{\partial f_\alpha}{\partial \mathbf{r}} + \frac{e_\alpha}{m_\alpha}\left(\mathbf{E}(\mathbf{r}, t) + \frac{\mathbf{v}}{c} \times \mathbf{B}(\mathbf{r}, t)\right) \cdot \frac{\partial f_{\alpha 0}}{\partial \mathbf{v}} = 0, \tag{8.14}$$

where $f_{\alpha 0}$ is the zero order equilibrium distribution function which is usually assumed to be a Maxwellian. f_α is now the perturbation. Note that the last term vanishes ($\frac{\mathbf{v}}{c} \times \mathbf{B}(\mathbf{r}, t) \cdot \frac{\partial f_{\alpha 0}}{\partial \mathbf{v}} = 0$) if we assume $f_{\alpha 0}(v)$ to be isotropic, since $\frac{\partial f_{\alpha 0}}{\partial \mathbf{v}} \sim \mathbf{v}$ in this case.

Assuming $\sim e^{i\mathbf{k}\cdot\mathbf{r} - i\omega t}$ solutions for all the unknown functions in the above, we have

$$f_\alpha(\mathbf{k}, \mathbf{v}, \omega) = \frac{-i\frac{e_\alpha}{m_\alpha}\mathbf{E}(\mathbf{k}, \omega) \cdot \frac{\partial f_{\alpha 0}}{\partial \mathbf{v}}}{\omega - \mathbf{k} \cdot \mathbf{v}} \tag{8.15}$$

$$\mathbf{k} \times \mathbf{E} = \frac{\omega}{c}\mathbf{B} \tag{8.16}$$

$$\mathbf{k} \times \mathbf{B} = -\frac{\omega}{c}\mathbf{E} - \frac{4\pi i}{c}\mathbf{J} \tag{8.17}$$

The above equations are the Fourier representation of the Vlasov equation and Maxwell equations, written down almost automatically by assuming the plane wave exponential phasor $\sim e^{i\mathbf{k}\cdot\mathbf{r} - i\omega t}$. However, for the Fourier transform to be valid, we need ω to have a small positive imaginary part, and the fraction $1/(\omega - \mathbf{k}\cdot\mathbf{v})$ in (8.15) should follow the Plemelj rule in (7.106) when ω becomes real. Eliminating $\mathbf{B}$ in (8.16) and (8.17) gives

$$\mathbf{k}\times(\mathbf{k}\times\mathbf{E}) + \frac{\omega^2}{c^2}\mathbf{E} = -\frac{4\pi i\omega}{c^2}\mathbf{J} \tag{8.18}$$

Equations (8.10) and (8.15) give

$$J_i = -iE_j\sum_\alpha \frac{e_\alpha^2}{m_\alpha}\int_{-\infty}^{\infty} d^3v \frac{v_i\frac{\partial f_{\alpha 0}}{\partial v_j}}{\omega - \mathbf{k}\cdot\mathbf{v}} \tag{8.19}$$

where the subscript i and j are cartesian indexes and the summation convention is used.

The material quantities, conductivity and the dielectric permittivity, take the following forms for Vlasov plasma. For conductivity

$$\sigma_{ij} = -i\sum_\alpha \frac{e_\alpha^2}{m_\alpha}\int d^3v \frac{v_i\frac{\partial f_{\alpha 0}}{\partial v_j}}{\omega - \mathbf{k}\cdot\mathbf{v}} \tag{8.20}$$

We have the expression for the dielectric tensor:

$$\varepsilon_{ij} = \delta_{ij} + \sum_\alpha \frac{\omega_{p\alpha}^2}{\omega}\int d^3v \frac{v_i\frac{\partial f_{\alpha 0}}{\partial v_j}}{\omega - \mathbf{k}\cdot\mathbf{v}} \tag{8.21}$$

Equation (8.18) is written as

$$\left[n^2\left(\frac{k_ik_j}{k^2} - \delta_{ij}\right) + \varepsilon_{ij}\right]E_j = 0 \quad \left(n^2 = \frac{c^2k^2}{\omega^2},\ \text{refractive index}\right) \tag{8.22}$$

We have the following important relation for ε_{ij} which is valid if $f_{\alpha 0}$ is isotropic:

$$\varepsilon_{ij} = \delta_{ij}\varepsilon_T + (\varepsilon_L - \varepsilon_T)\frac{k_ik_j}{k^2} \tag{8.23}$$

The *longitudinal dielectric permittivity* is

$$\varepsilon_L = 1 + \sum_\alpha \frac{\omega_{p\alpha}^2}{k^2}\int \frac{\mathbf{k}\cdot\frac{\partial f_{\alpha 0}}{\partial \mathbf{v}}}{\omega - \mathbf{k}\cdot\mathbf{v}} d^3v \tag{8.24}$$

The *transverse dielectric permittivity* is

$$\varepsilon_T = 1 - \sum_\alpha \frac{\omega_{p\alpha}^2}{\omega}\int \frac{f_{\alpha 0}}{\omega - \mathbf{k}\cdot\mathbf{v}} d^3v \tag{8.25}$$

Proof of (8.23) goes as follows. The integral in (8.21) can be written as

$$\int \frac{v_i \partial f_0/\partial v_j}{\omega - \mathbf{k}\cdot\mathbf{v}} d^3v = -\int f_0 \frac{\partial}{\partial v_j}\left(\frac{v_i}{\omega - \mathbf{k}\cdot\mathbf{v}}\right) d^3v$$
$$= -\delta_{ij}\int \frac{f_0 d^3v}{\omega - \mathbf{k}\cdot\mathbf{v}} - \int \frac{f_0 v_i k_j d^3v}{(\omega - \mathbf{k}\cdot\mathbf{v})^2} \tag{8.26}$$

We use the following identity

$$\frac{1}{(\omega - \mathbf{k}\cdot\mathbf{v})^2} = \frac{k_l}{k^2}\frac{\partial}{\partial v_l}\frac{1}{\omega - \mathbf{k}\cdot\mathbf{v}} \tag{8.27}$$

So the last term of (8.26)

$$-\frac{k_j k_l}{k^2}\int f_0 v_i \frac{\partial}{\partial v_l}\frac{1}{\omega - \mathbf{k}\cdot\mathbf{v}} d^3v = \frac{k_j k_l}{k^2}\int \frac{f_0\delta_{il} + v_i \partial f_0/\partial v_l}{\omega - \mathbf{k}\cdot\mathbf{v}} d^3v$$
$$= \frac{k_i k_j}{k^2}\int \frac{f_0 d^3v}{\omega - \mathbf{k}\cdot\mathbf{v}} + \frac{k_j k_l}{k^2}\int \frac{v_i \partial f_0/\partial v_l}{\omega - \mathbf{k}\cdot\mathbf{v}} d^3v \tag{8.28}$$

Here we use $\partial f_0/\partial v_l = v_l f_0'/v$ ($f_0' = \frac{df_0}{dv}$). So the last term in (8.28) becomes

$$\frac{k_j}{k^2}\int \frac{v_i}{v}\frac{\mathbf{k}\cdot\mathbf{v}}{\omega - \mathbf{k}\cdot\mathbf{v}} f_0' d^3v = \frac{k_j}{k^2}\int \frac{v_i}{v} f_0'\left(-1 + \frac{\omega}{\omega - \mathbf{k}\cdot\mathbf{v}}\right) d^3v = \frac{k_j\omega}{k^2}\int \frac{\partial f_0/\partial v_i}{\omega - \mathbf{k}\cdot\mathbf{v}} d^3v,$$

because the -1 term vanishes upon integration (odd function integral). Integrating by parts the last term gives

$$-\frac{k_i k_j \omega}{k^2}\int \frac{f_0 d^3v}{(\omega - \mathbf{k}\cdot\mathbf{v})^2} = \frac{\omega k_i k_j}{k^4}\int \frac{k_l \partial f_0/\partial v_l}{\omega - \mathbf{k}\cdot\mathbf{v}} d^3v \tag{8.29}$$

where we used (8.27) again and integrated by parts. Collecting the above results establishes (8.23).

Dielectric tensor in anisotropic plasma. Even if no static magnetic field is present, a plasma can be anisotropic when the zeroth order distribution function has a temperature anisotropy or when the plasma is drifting. In this case the wave magnetic field term in (8.1) does not vanish. Eliminating the wave magnetic field by $\mathbf{B} = \frac{c}{\omega}\mathbf{k} \times \mathbf{E}$, current $\mathbf{J}$ is obtained as

$$J_i = -iE_j \sum_\alpha \frac{e_\alpha^2}{m_\alpha \omega} \int \frac{d^3v}{\omega - \mathbf{k}\cdot\mathbf{v}} \left[(\omega - \mathbf{k}\cdot\mathbf{v}) v_i \frac{\partial f_{\alpha 0}}{\partial v_j} + v_i v_j k_l \frac{\partial f_{\alpha 0}}{\partial v_l} \right]$$

The coefficient of E_j is σ_{ij}. Then the dielectric tensor takes the form,

$$\varepsilon_{ij} = \delta_{ij} + \sum_\alpha \frac{\omega_{p\alpha}^2}{\omega^2} \int d^3v \frac{v_i [\delta_{jl}(\omega - \mathbf{k}\cdot\mathbf{v}) + v_j k_l] \frac{\partial f_{\alpha 0}}{\partial v_l}}{\omega - \mathbf{k}\cdot\mathbf{v}} \tag{8.30}$$

Exercise. Show that the above dielectric tensor ε_{ij} reduces to (8.21) if $f_{\alpha 0}$ is isotropic.

Hint: use $\frac{\partial f_{\alpha 0}}{\partial v_l} = \frac{v_l}{v} \frac{df_{\alpha 0}(v)}{dv}$ and discard the odd function integrals.

Exercise. Derive (8.24) and (8.25) from (8.23).

In summary, using (8.23) in (8.22) yields the simultaneous homogeneous algebraic equations for the Fourier components of the electric field in the form of (7.25). Thus the normal modes of small oscillations of the Vlasov–Maxwell plasma are obtained by setting the determinant of the matrix in (8.22) to zero, with ε_L and ε_T evaluated by (8.24) and (8.25). Now we must resolve the ambiguity in the *resonant denominator* $1/(\omega - \mathbf{k}\cdot\mathbf{v})$. This can be done by requiring the causal constraint on the susceptibility as expounded in Chapter 7. For further elaboration, let us change the three-dimensional integral in (8.24) to one-dimensional integral. Let us take $\mathbf{k}$ the $\hat{\mathbf{x}}$ direction. In isotropic medium, the $\hat{\mathbf{x}}$ direction is not special. Let

$$g_0(v_x) = \int_{-\infty}^{\infty} \int_{-\infty}^{\infty} dv_y dv_z f_0(v_x, v_y, v_z)$$

Then we have (8.24) in the form

$$\varepsilon_L(k, \omega) = 1 + \sum_\alpha \frac{\omega_{p\alpha}^2}{k} \int_{-\infty}^{\infty} \frac{\frac{dg_{\alpha 0}}{dv_x}}{\omega - kv_x} dv_x \tag{8.31}$$

$$= 1 - \sum_\alpha \frac{\omega_{p\alpha}^2}{k^2} \int_{-\infty}^{\infty} \frac{\frac{dg_{\alpha 0}}{dv_x}}{v_x - \omega/k} dv_x \tag{8.32}$$

When ω takes a real value, the velocity integral in (8.32) is singular. Landau evaluated the singular velocity integral by contour integration in the *complex* v_x *plane* along the *Landau contour*—See Fig. 8.1. In (8.32), ω in $\varepsilon_L(k, \omega)$ has an *inherent* positive imaginary part, and $\varepsilon_L(k, \omega)$ is regular in the region $\operatorname{Im}\omega > 0$. Landau contour gives to the singular integral the analytically continued value. What we have done in Chapter 7 was Fourier inverting

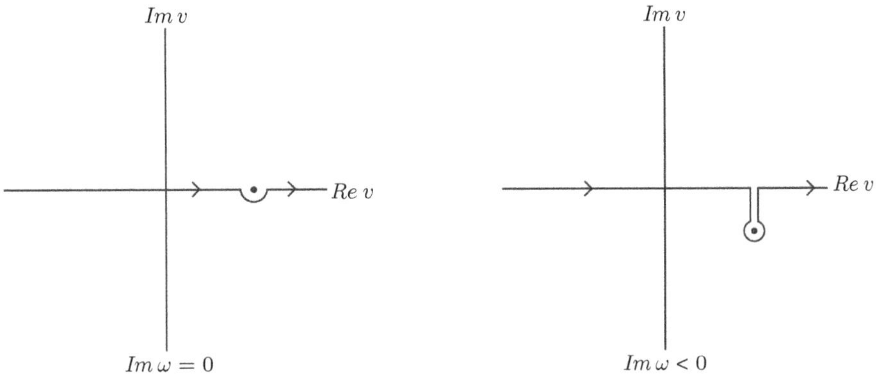

Fig. 8.1. Landau contour.

$\varepsilon_L(k,\omega)$ to $\varepsilon_L(k,t)$, and having applied the causal constraint. The two treatments in regard to the resonant denominator $1/(\omega - \mathbf{k}\cdot\mathbf{v})$ entirely agree.

When we wish to change the three-dimensional integral in (8.24) to one-dimensional integral with a specially designated $\mathbf{k}$ direction, we put

$$g(u) = \int d^3v f(\mathbf{v})\delta\left(u - \frac{\mathbf{k}\cdot\mathbf{v}}{k}\right) \tag{8.33}$$

then

$$\varepsilon_L(\omega,k) = 1 + \frac{\omega_p^2}{k^2}\int_{-\infty}^{\infty} du \frac{k\frac{\partial g}{\partial u}}{\omega - ku} \tag{8.34}$$

8.3 Initial-Value Problem: Landau's Treatment

Laplace transform is suitable for dealing with initial-value problems. Let a certain state represented by $f(t)$ is agitated by a stimulus activated at $t = 0$:

$$f(t) = 0 \quad \text{for } t \leq 0 \tag{8.35}$$

We are interested in knowing $f(t)$ for $t > 0$. Landau used Laplace transform to solve the initial-value problem of Vlasov–Poisson plasma and clarified the issue of the resonant denominator. In this book we use the modified Laplace transform, which is, roughly speaking, half-range Fourier transform. The mathematical connection of the half-range Fourier transform with the conventional Laplace transform is referred to mathematical physics book.

We define Laplace transform of $f(t)$:

$$F(\omega) = \int_0^\infty e^{i\omega t} f(t)dt \tag{8.36}$$

which can be remembered as 'half-range Fourier integral'. Then its inversion is

$$f(t) = \frac{1}{2\pi}\int_{-\infty+ic}^{\infty+ic} e^{-i\omega t}F(\omega)\,d\omega \tag{8.37}$$

In the integration path above, c must be large enough so that it runs above all the singularities of $F(\omega)$, and the contour is completed by the infinite semi-circle enclosing the lower-half ω-plane.

If we attempt to find $f(t)$ for $t < 0$ by (8.37), the contour should consists of the straight line path plus the great circle enclosing the upper ω plane. Since the enclosed region contains no singularity, the integral should be zero—in accord with (8.35).

Next, the *closure property* is considered. Putting (8.37) in (8.36) gives

$$F(\omega) = \frac{1}{2\pi}\int_{-\infty+ic}^{\infty+ic} d\omega' \,\frac{iF(\omega')}{\omega-\omega'}$$

provided $\mathrm{Im}\,(\omega-\omega') > 0$. This equation must be an identity—Cauchy integral formula. Thus $F(\omega')$ must be analytic above $\omega' = ic$.

8.3.1 *Initial value problem of Vlasov–Poisson equations*

We begin with linearized one-dimensional electron Vlasov equation (ions are immobile)

$$\frac{\partial}{\partial t}g(k,v,t) + ikvg(k,v,t) - \frac{e}{m}E(k,t)\,\frac{dg_0(v)}{dv} = 0 \tag{8.38}$$

and Poisson equation

$$ikE(k,t) = -4\pi e\int_{-\infty}^{\infty} g(k,v,t) \tag{8.39}$$

where we Fourier transformed $g(r,v,t)$ with respect to the spatial coordinate x. Transform (8.38) by performing $\int_0^\infty(\cdots)e^{i\omega t}dt$ to obtain

$$-i\omega g(k,v,\omega) - g(k,v,0) + ikv\;g(k,v,\omega) - \frac{e}{m}E(k,\omega)\frac{\partial g_0}{\partial v} = 0 \tag{8.40}$$

Hence

$$g(k,v,\omega) = \frac{g(k,v,0) + \frac{e}{m}E(k,\omega)\frac{\partial g_0}{\partial v}}{-i(\omega - kv)} \tag{8.41}$$

Transforming (8.39) gives gives

$$ikE(k,\omega) = -4\pi e \int_{-\infty}^{\infty} g(k,v,\omega)dv \tag{8.42}$$

Eliminating $g(k,v,\omega)$ in the above two equations gives

$$E(k,\omega) = \left(\frac{4\pi e n_0}{k^2}\int_{-\infty}^{\infty} dv\, \frac{g(k,v,0)}{v-\omega/k}\right) \Big/ \left(1 - \frac{\omega_{\text{pe}}^2}{k^2}\int_{-\infty}^{\infty} \frac{\frac{\partial g_0}{\partial v}}{v-\omega/k}dv\right) \tag{8.43}$$

whose inversion is

$$E(k,t) = \int_{-\infty+ic}^{\infty+ic} d\omega\, e^{-i\omega t}\, \frac{\frac{2en_0}{k^2}\int_{-\infty}^{\infty} dv\, \frac{g(k,v,0)}{v-\omega/k}}{1 - \frac{\omega_{\text{pe}}^2}{k^2}\int_{-\infty}^{\infty} \frac{\frac{\partial g_0}{\partial v}}{v-\omega/k}dv} \tag{8.44}$$

where the integration path runs from $-\infty$ to ∞ above all the singularities of the integrand in the complex ω-plane.

One immediate concern is the integral

$$\int_{-\infty}^{\infty} dv\, \frac{(\cdots)}{v-\frac{\omega}{k}} \quad \text{at } v = \frac{\omega}{k} \tag{8.45}$$

which is not well-defined. Also the zeros of the denominator are singularities in the inversion integral:

$$\varepsilon(k,\omega) \equiv 1 - \frac{\omega_{\text{pe}}^2}{k^2}\int_{-\infty}^{\infty} dv\, \frac{\frac{dg_0}{dv}}{v-\omega/k} = 1 + \chi(\omega) = 0 \tag{8.46}$$

To compute integral (8.44) we make a closed contour by enclosing the lower half ω plane for $t > 0$. Then the integral (8.44) is computed by the residue theorem as

$$E(k,t) = (2\pi i)(2en_0)\sum_j (\text{Residue})_j \text{ of } \left[\frac{e^{-i\omega t}}{k^2\varepsilon(\omega,k)}\int_{-\infty}^{\infty} \frac{g(k,v,0)}{v-\omega/k}dv\right] \tag{8.47}$$

Here the subscript j denotes the zeros of $\varepsilon(\omega,k) = 0$ scattered on the ω-plane. We note that (8.46) is the dispersion relation of the plasma waves that we have obtained from the normal mode analysis of Maxwell–Vlasov plasma.

First we take care of the singular integral in (8.45). This can be done by integrating along the Landau contour in Fig. 8.1.

When ω is real, we have

$$\int_{-\infty}^{\infty} \frac{dv}{v - \frac{\omega}{k}} = P \int_{-\infty}^{\infty} \frac{dv}{v - \frac{\omega}{k}} + \int_{ihc} \frac{dv}{v - \frac{\omega}{k}} \tag{8.48}$$

where P denotes the principal value and $\int_{ihc}$ denotes the integral along the infinitesimal half circle, which is equal to $i\pi\delta(v - \frac{\omega}{k})$. Hence we write symbolically

$$\frac{1}{v - \frac{\omega}{k}} = P\frac{1}{v - \frac{\omega}{k}} + i\pi\delta\left(v - \frac{\omega}{k}\right) \quad (\mathrm{Im}\,\omega = 0) \tag{8.49}$$

Multiplying both sides by $-1/k$, we have

$$\frac{1}{\omega - kv} = P\frac{1}{\omega - kv} - i\pi\delta(\omega - kv) \tag{8.50}$$

Using these relations in (8.46) gives

$$\chi(k,\omega) = \frac{\omega_{\mathrm{pe}}^2}{k} \int_{-\infty}^{\infty} dv \, \frac{dg_0}{dv} \left[P\frac{1}{\omega - kv} - i\pi\delta(\omega - kv)\right] \quad (\mathrm{Im}\,\omega = 0). \tag{8.51}$$

which agrees with (7.71).

Next, we consider the integral when $\mathrm{Im}\,\omega < 0$:

$$\int_{-\infty}^{\infty} dv \, \frac{(\cdots)}{v - \frac{\omega}{k}} \quad (\mathrm{Im}\,\omega < 0) \tag{8.52}$$

This integral is the sum of the principal value plus the infinitesimal whole circle contribution (the vertical line integrals cancel; see Fig. 8.1). Hence we symbolically write

$$\frac{1}{v - \frac{\omega}{k}} = P\frac{1}{v - \frac{\omega}{k}} + 2i\pi\delta\left(v - \frac{\omega}{k}\right) \quad (\mathrm{Im}\,\omega < 0) \tag{8.53}$$

Using (8.53) in (8.46) gives

$$\chi(k,\omega) = \frac{\omega_{\mathrm{pe}}^2}{k} \int_{-\infty}^{\infty} dv \, \frac{dg_0}{dv} \left[P\frac{1}{\omega - kv} - i\,2\pi\,\delta(\omega - kv)\right] \quad (\mathrm{Im}\,\omega < 0). \tag{8.54}$$

Equations (8.49) and (8.53) are the analytic continuations of the expression $\frac{1}{v-\omega/k}$, which is originally defined in the region $\mathrm{Im}\,\omega > 0$, to the regions $\mathrm{Im}\,\omega = 0$ and $\mathrm{Im}\,\omega < 0$.

Summary. The *electrostatic dielectric function* $\varepsilon(k,\omega)$, given by (8.46), is defined by three different expressions in respective regions.

$$\varepsilon(k,\omega) = 1 - \frac{\omega_{\rm pe}^2}{k^2}\int_{-\infty}^{\infty}\frac{\frac{\partial g_0}{\partial v}}{v-\omega/k}dv \quad (\omega_i > 0) \tag{8.55}$$

$$\varepsilon(k,\omega) = 1 - \frac{\omega_{\rm pe}^2}{k^2}\,P\int_{-\infty}^{\infty}\frac{\frac{\partial g_0}{\partial v}}{v-\omega/k}\,dv - i\pi\,\frac{\omega_{\rm pe}^2}{k^2}\left[\frac{\partial g_0}{\partial v}\right]_{v=\omega/k} \quad (\omega_i = 0) \tag{8.56}$$

$$\varepsilon(k,\omega) = 1 - \frac{\omega_{\rm pe}^2}{k^2}P\int_{-\infty}^{\infty}\frac{\frac{\partial g_0}{\partial v}}{v-\omega/k}\,dv - 2i\pi\,\frac{\omega_{\rm pe}^2}{k^2}\left[\frac{\partial g_0}{\partial v}\right]_{v=\omega/k} \quad (\omega_i < 0) \tag{8.57}$$

• Why Landau contour? To clarify the ambiguity in the singular integral (8.45), we pay attention to the factor $e^{-i\omega t}$ in the inversion integral (8.44). The frequency ω has an inherent small imaginary part (ν): $\omega = \omega_r + i\nu$. The singularity at $v = \omega/k$ implies $\nu \to 0$ from the above in the complex ω plane. Hence the integral in (8.45) can be evaluated by taking the limiting process

$$\frac{1}{\omega} = \lim_{\nu\to 0}\frac{1}{\omega_r + i\nu} = P\frac{1}{\omega_r} - i\pi\delta(\omega_r) \tag{8.58}$$

which can be proven algebraically as done in Chapter 7. Landau contour is novel since the algebraic process of analytic continuation is accomplished by simply indenting the integration path.

Equation (8.44) is rewritten as

$$E(k,t) = \frac{2en_0}{k^2}\int_{-\infty+ic}^{\infty+ic}\frac{d\omega\, e^{-i\omega t}}{\varepsilon(k,\omega)}\int_{-\infty}^{\infty}dv\ g(k,v,0)\left[P\,\frac{1}{\omega-kv} - i\pi\delta(\omega-kv)\right] \tag{8.59}$$

In the above Laplace inversion, the zeros of $\varepsilon(\omega,k) = 0$ correspond to the normal modes of $E(k,t)$ and Landau damps. The pole at the δ function term $\omega = kv$ gives an additional term which goes to zero asymptotically:

$$\sim \int_{-\infty}^{\infty} dvg(k,v,0)\frac{e^{-ikvt}}{\varepsilon(k,kv)} \to 0 \quad \text{as } t\to\infty \tag{8.60}$$

As to the distribution function, use of (8.43) in (8.41) to eliminate $E(k,\omega)$ gives

$$g(k,v,\omega) = \frac{i}{\omega - kv}\left[g(k,v,0) + \frac{\omega_{\text{pe}}^2}{k^2}\frac{\partial g_0}{\partial v}\frac{1}{\varepsilon(\omega,k)}\int_{-\infty}^{\infty}\frac{g(k,v,0)}{v-\omega/k}\,dv\right]$$

whose inversion is

$$g(k,v,t) = \frac{i}{2\pi}\int_{-\infty+ic}^{\infty+ic}\frac{d\omega\, e^{-i\omega t}}{\omega - kv}\left[g(k,v,0) + \frac{\omega_{\text{pe}}^2}{k^2}\frac{\partial g_0}{\partial v}\frac{1}{\varepsilon(\omega,k)}\int_{-\infty}^{\infty}\frac{g(k,v,0)}{v-\omega/k}dv\right] \tag{8.61}$$

The pole at $\omega = kv$ gives the term $\sim e^{-ikvt}$ due to the δ function, which is called *free streaming or ballistic* term. This term persists endlessly and becomes more and more oscillatory as time increases. But $\int dv$ will give almost complete cancellations as in (8.60), yielding no appreciable effect on macroscopic variables such as density perturbation. This term exists because the collisionless Vlasov particles carry the memory of their initial perturbations with them for all subsequent time, but their memory will be wiped out if particle collisions enter the dynamics (mathematically putting, $\omega = kv$ is no longer singular). The imaginary part of $\varepsilon(k,\omega)$ in (8.56) is due to the *resonant particles* whose velocities are $v = \omega/k$ and which absorb the wave energy.

8.4 Evaluation of Maxwellian Dispersion Relations

In this section we compute the dispersion relations of the waves, electrostatic and electromagnetic, in the plasma of Maxwellian distribution whose dielectric tensor is as given by (8.21). The electrostatic dispersion relation is $\varepsilon_L = 0$ (Eq. (7.27)), or per (7.29),

$$\begin{aligned}\varepsilon_L(\mathbf{k},\omega) = \varepsilon_{ij}\,\frac{k_i k_j}{k^2} &= 1 + \sum_\alpha \frac{\omega_{p\alpha}^2}{k^2\omega}\int d^3v\frac{\mathbf{k}\cdot\mathbf{v}\;k_j\frac{\partial f_{0\alpha}}{\partial v_j}}{\omega - \mathbf{k}\cdot\mathbf{v}}\\ &= 1 + \sum_\alpha \frac{\omega_{p\alpha}^2}{k^2}\int d^3v\frac{k_j\frac{\partial f_{0\alpha}}{\partial v_j}}{\omega - \mathbf{k}\cdot\mathbf{v}} = 0\end{aligned} \tag{8.62}$$

The electromagnetic dispersion relation is (7.28):

$$\varepsilon_T - n^2 = 0 \tag{8.63}$$

where we recall (7.29):

$$\varepsilon_T = \frac{1}{2}(\varepsilon_{ii} - \varepsilon_L) \tag{8.64}$$

We calculate

$$\varepsilon_{ii} = 3 + \sum_{\alpha} \frac{\omega_{p\alpha}^2}{\omega} \int d^3v \frac{v_j \frac{\partial f_{0\alpha}}{\partial v_j}}{\omega - \mathbf{k}\cdot\mathbf{v}} \tag{8.65}$$

Hence

$$\varepsilon_T = 1 + \frac{1}{2}\sum_{\alpha} \frac{\omega_{p\alpha}^2}{k^2\omega} \int d^3v \frac{k^2 v_j \frac{\partial f_{0\alpha}}{\partial v_j} - (\mathbf{k}\cdot\mathbf{v})k_j \frac{\partial f_{\alpha 0}}{\partial v_j}}{\omega - \mathbf{k}\cdot\mathbf{v}} \tag{8.66}$$

After integration by parts, we get

$$\varepsilon_T = 1 - \sum_{\alpha} \frac{\omega_{p\alpha}^2}{\omega} \int d^3v \frac{f_{\alpha 0}}{\omega - \mathbf{k}\cdot\mathbf{v}} \tag{8.67}$$

as we already have obtained in (8.24) and (8.25).

Without loss of generality, we take $\mathbf{k} = \hat{\mathbf{x}}k$ and integrate (8.62) $\int dv_y \int dv_z(\cdots)$ to obtain

$$\varepsilon_L = 1 + \sum_{\alpha} \frac{\omega_{p\alpha}^2}{k} \int_{-\infty}^{\infty} dv \frac{\frac{\partial g_{0\alpha}}{\partial v}}{\omega - kv} \tag{8.68}$$

where

$$g_{0\alpha} = \sqrt{\frac{m_\alpha}{2\pi T_\alpha}} \exp\left(-\frac{m_\alpha v^2}{2T_\alpha}\right) \tag{8.69}$$

Using $dg_{\alpha 0}/dv = -\frac{m_\alpha}{T_\alpha} v g_{\alpha 0}$, we can write

$$\varepsilon_L = 1 + \sum_{\alpha} \frac{\omega_{p\alpha}^2}{k^2} \frac{m_\alpha}{T_\alpha} \left(1 + \int \frac{\omega\, g_{\alpha 0}}{kv - \omega} dv\right) \tag{8.70}$$

A useful relation is

$$\int dv \frac{\omega g_0}{kv - \omega} = \zeta Z(\zeta); \quad \zeta = \frac{\omega/k}{(2T/m)^{1/2}} \tag{8.71}$$

where $Z(\zeta)$ is the plasma dispersion function (7.114). Hence

$$\varepsilon_L = 1 + 2\sum_{\alpha} \frac{\omega_{p\alpha}^2}{\omega^2} \zeta_\alpha^2 (1 + \zeta_\alpha Z(\zeta_\alpha)) = 1 + \sum_{\alpha} \frac{1}{k^2 \lambda_{D\alpha}^2} (1 + \zeta_\alpha Z(\zeta_\alpha)) \tag{8.72}$$

with

$$\zeta_\alpha = \frac{\omega/k}{(2T_\alpha/m_\alpha)^{1/2}}, \quad \lambda_{D\alpha}^2 = \frac{T_\alpha}{4\pi N e^2} \tag{8.73}$$

Analogously we obtain

$$\varepsilon_T = 1 + \sum_\alpha \frac{\omega_{p\alpha}^2}{\omega^2} \zeta_\alpha Z(\zeta_\alpha) \tag{8.74}$$

• **Cold plasma wave.** We first consider a *cold plasma*. If we take the limit $T_\alpha \to 0$ in (8.69), the Maxwellian becomes

$$g_{\alpha 0}(v) = \delta(v) \tag{8.75}$$

Remember that

$$\delta(x) = \lim_{n\to\infty} \frac{n}{\sqrt{\pi}} \exp(-n^2 x^2) \tag{8.76}$$

Neglect ion contribution in (8.68), which is small by the mass ratio m_e/m_i, and integrate by parts to obtain

$$\varepsilon_L = 1 - \frac{\omega_{\text{pe}}^2}{\omega^2} = 0 \quad \text{(electrostatic, cold)} \tag{8.77}$$

In a cold plasma, it is easy to obtain by using (8.67) $\varepsilon_T = 1 - \Sigma\, \frac{\omega_{p\alpha}^2}{\omega^2} = \varepsilon_L$, so the electromagnetic dispersion relation becomes

$$\omega^2 = \omega_{\text{pe}}^2 + c^2k^2 \quad \text{(electromagnetic, cold)} \tag{8.78}$$

• **Langmuir wave.** The wave velocity of a Langmuir wave (electron plasma wave) is much greater than the electron thermal velocity, and so $\zeta_e \gg 1$. The ions are stationary in this wave, and the ion contribution to the dielectric function is negligible. Then we obtain

$$\varepsilon_L = 1 + \frac{1}{k^2\lambda_{De}^2}(1 + \zeta_e Z(\zeta_e)) \tag{8.79}$$

We use (7.115) for the asymptotic series for $Z(\zeta_e)$. Or we can use (7.84).

The following is another way of analysis of the electron plasma wave. We separate (8.68) into real and imaginary parts:

$$\text{Re}\, \varepsilon_L \equiv \varepsilon_r = 1 + \frac{\omega_{\text{pe}}^2}{k} P \int_{-\infty}^{\infty} dv\, \frac{dg_0}{dv} \frac{1}{\omega - kv} \tag{8.80}$$

$$\text{Im}\, \varepsilon_L \equiv \varepsilon_i = -i\pi\, \frac{\omega_{\text{pe}}^2}{k} \int_{-\infty}^{\infty} dv\, \frac{dg_0}{dv} \delta(\omega - kv) = -i\pi\, \frac{\omega_{\text{pe}}^2}{k^2} \left[\frac{\partial g_0}{\partial v}\right]_{v=\omega/k} \tag{8.81}$$

Landau damping of electron plasma wave. The wave velocity of electron plasma wave is much greater than the thermal velocity of plasma:

$\omega/k \gg v_{th}$. In this case, the wave velocity can be thought as situated somewhere on the Maxwellian tail, and the integrand in (8.80) can be expanded in the powers of vk/ω. Insignificant error is introduced by taking only a few terms in the expansion. Then the real part of the dielectric constant in (8.80) becomes

$$\begin{aligned}\varepsilon_r(k,\omega) &= 1 + \frac{\omega_{\rm pe}^2}{k\omega}\int dv\,\frac{\partial g_0}{\partial v}\left(1 + \frac{vk}{\omega} + \frac{v^2k^2}{\omega^2} + \frac{v^3k^3}{\omega^3} + \cdots\right)\\ &= 1 - \frac{\omega_{\rm pe}^2}{\omega^2}\int dv g_0\left(1 + 3v^2\frac{k^2}{\omega^2} + \cdots\right)\end{aligned} \tag{8.82}$$

where we integrated by parts. The average of kinetic energy $\frac{mv^2}{2}$ in one-dimension is $\frac{T}{2}$, and so

$$\int_{-\infty}^{\infty} g_0(v)\, v^2 dv = \frac{T}{m} \equiv v_{th}^2. \tag{8.83}$$

Also introducing the Debye length,

$$\lambda_{De}^2 = \frac{T_\alpha}{4\pi n_0 e^2} = \frac{v_{th}^2}{\omega_{\rm pe}^2} \tag{8.84}$$

we can write

$$\varepsilon_r(\omega,k) = 1 - \frac{\omega_{\rm pe}^2}{\omega^2} - 3\frac{\omega_{\rm pe}^4}{\omega^4}k^2\lambda_{De}^2 \tag{8.85}$$

$\varepsilon_r(\omega,k) = 0$ is solved by

$$\omega_r^2 - \omega_{\rm pe}^2 - 3\frac{\omega_{\rm pe}^4}{\omega_r^2}k^2\lambda_{De}^2 = 0 \tag{8.86}$$

If $k^2\lambda_{De}^2 \ll 1$, $\omega_r^2 \simeq \omega_{\rm pe}^2$ is the approximate solution of (8.86). Using this, we obtain

$$\omega_r^2 = \omega_{\rm pe}^2(1 + 3k^2\lambda_{De}^2) \quad \text{or} \quad \omega_r = \omega_{\rm pe}\left(1 + \frac{3}{2}k^2\lambda_{De}^2\right) \tag{8.87}$$

Equations (8.81) and (8.87) give the imaginary frequency, the Landau damping rate of electron plasma wave:

$$\omega_i = -\sqrt{\frac{\pi}{8}}\,\frac{\omega_{\rm pe}}{k^3\lambda_{De}^3}\exp\left[\frac{-1}{2k^2\lambda_{De}^2} - \frac{3}{2}\right] \tag{8.88}$$

where we used (7.92). The imaginary frequency ω_i vanishes for $k \to 0$ and increases rapidly as k increases. The damping is comparable to $\omega_{\rm pe}$ for $k\lambda_{De} > 0.3$ so that such waves are not observed.

• **Electromagnetic Wave.** To obtain the electromagnetic dispersion relation, integrate (8.25) with respect to the velocity components perpendicular to **k**:

$$\varepsilon_T = 1 + \Sigma \frac{\omega_{p\alpha}^2}{k\omega}\left[P\int \frac{f_{0\alpha}}{v-\omega/k}dv - i\pi\left[f_{0\alpha}\right]_{v=\omega/k}\right] \tag{8.89}$$

where f_0 is one-dimensional distribution function. Clearly the ion contribution is negligible. The Landau damping term corresponds to a velocity far down on the Maxwellian tail at $v = \omega/k \simeq c$ where f_0-value is negligibly small. Expanding in the powers of vk/ω gives the dispersion relation

$$\frac{c^2k^2}{\omega^2} = 1 - \frac{\omega_{\rm pe}^2}{\omega^2}\int_{-\infty}^{\infty} f_0(1 + kv/\omega + k^2v^2/\omega^2 + \cdots)dv \tag{8.90}$$

Since $v_{th}^2 \ll c^2$, the third term can be neglected as compared to the first, giving the same result as (8.78), which is the electromagnetic wave (light wave) in a plasma.

Summary. The Landau damping of the electromagnetic wave in a plasma is nil because the phase velocity of the wave is greater than c, and there are no resonant particles.

Temperature Correction. The electromagnetic wave dispersion relation $\varepsilon_T = n^2$ reads

$$n^2 = \frac{c^2k^2}{\omega^2} = 1 + \frac{\omega_{\rm pe}^2}{\omega^2}\zeta_e Z(\zeta_e) \tag{8.91}$$

Expanding $Z(\zeta_e)$, neglecting the Landau damping, gives the temperature correction for the refractive index:

$$\omega^2 = \omega_{\rm pe}^2 + c^2k^2 + \frac{T_e}{m_e}\frac{\omega_{\rm pe}^2}{\omega^2}k^2 \tag{8.92}$$

$$n^2 = \frac{1 - \frac{\omega_{\rm pe}^2}{\omega^2}}{1 + \frac{T_e}{mc^2}\frac{\omega_{\rm pe}^2}{\omega^2}} \tag{8.93}$$

8.4.1 *Ion acoustic wave*

Confer Sections 7.11 and 7.12. Here we work out the ion acoustic wave by using (8.72). Ion acoustic wave which is a low frequency electrostatic wave is peculiar and characteristic of plasma in that the restoring force is provided by electron pressure while the inertia by ion. Both electron and ion equations should be used, and the wave velocity is such that $v_{thi} \ll \omega/k \ll v_{the}$, or $\zeta_e \ll 1$ and $\zeta_i \gg 1$. Using (7.115) for ions and (7.116) for electrons give

$$\varepsilon_L = 1 + \frac{1}{k^2\lambda_{De}^2}\left(1 - \frac{T_e}{m_i}\,\frac{k^2}{\omega^2} + i\sqrt{\pi}\,\zeta_e\, e^{-\zeta_e^2} + i\sqrt{\pi}\,\zeta_i\,\frac{T_e}{T_i}\,e^{-\zeta_i^2}\right)$$

Separating the real and imaginary parts, we obtain

$$\varepsilon_r = 1 + \frac{1}{k^2\lambda_{De}^2}\left(1 - \frac{T_e}{m_i}\,\frac{k^2}{\omega^2}\right) \tag{8.94}$$

$$\varepsilon_i = \frac{\sqrt{\pi}}{k^2\lambda_{De}^2}\left(\zeta_e\, e^{-\zeta_e^2} + \zeta_i \frac{T_e}{T_i}\, e^{-\zeta_i^2}\right)$$

$$= \sum_{\alpha=e,i} \frac{\sqrt{\pi}}{k^2\lambda_\alpha^2}\,\frac{\omega}{k}\sqrt{\frac{m_\alpha}{2T_\alpha}}\exp\left[\frac{-m_\alpha\omega^2}{2T_\alpha k^2}\right] \tag{8.95}$$

The real part gives the dispersion relation,

$$\omega_r^2 = \frac{T_e}{m_i}\,\frac{k^2}{1 + k^2\lambda_{De}^2} \tag{8.96}$$

Using (7.92), we obtain the imaginary frequency, the Landau damping rate

$$\omega_i = -\sqrt{\frac{\pi}{8}}\frac{\omega_r}{(1 + k^2\lambda_{De}^2)^{3/2}}\left[\sqrt{\frac{m_e}{m_i}} + \left(\frac{T_e}{T_i}\right)^{3/2}\exp\left(\frac{-T_e/T_i}{2[1 + k^2\lambda_{De}^2]}\right)\right] \tag{8.97}$$

In the long wave region where the charge neutrality holds, we can set $k^2\lambda_{De}^2 \ll 1$ in (8.96) and (8.97). For future reference, we write explicitly

$$\omega_r^2 = \frac{T_e}{m_i}k^2, \quad \omega_i = -\sqrt{\frac{\pi}{8}}\omega_r\left(\frac{T_e}{T_i}\right)^{3/2}\exp\left(\frac{-T_e}{2T_i}\right) \tag{8.98}$$

In (8.97) the mass ratio term $\sqrt{\frac{m_e}{m_i}}$ comes from the electron Landau damping. The ion Landau damping dominates over this, so we say that ion acoustic wave Landau damps on ions. If $T_e \simeq T_i$, the exponential term (ion term) also becomes small. Only the ion acoustic wave in a plasma of $T_e \gg T_i$ can propagate without suffering from strong Landau damping. As T_i approaches T_e, ion acoustic wave severely Landau damps. This is anticipated: in the latter case, the wave velocity ω/k lies well in the *body* of the ion distribution; if $T_e \gg T_i$, the wave velocity would be on the *tail* portion.

Exercise. If you retain one more term in the expansion of $Z(\zeta_i)$, ε_L has an extra term, $-3\frac{T_i}{m_i}k^2\omega_{pi}^2/\omega^4$. How does this term change the real and imaginary frequency?

Answer.

$$\omega_r^2 = \frac{T_e}{m_i}\frac{k^2}{1+k^2\lambda_{De}^2} + 3k^2\frac{T_i}{m_i}$$

8.4.2 *Ion-acoustic instability as a two-stream instability*

The two-stream instability can arise when there is a relative motion between two species of plasma; for example, an electron beam moving through the ions. We shall consider the ions to be cold and stationary: $f_{0i} = \delta(v)$. On the other hand, the electron beam can be represented by a drifting Maxwellian,

$$f_{0e} = \left[\frac{m_e}{2\pi T_e}\right]^{\frac{1}{2}} \exp\left[-\frac{m_e(v-v_0)^2}{2T_e}\right] \tag{8.99}$$

We use these distribution functions in the dispersion relation,

$$\varepsilon_L = 1 + \sum_\alpha \frac{\omega_{p\alpha}^2}{k^2}\int \frac{k\frac{\partial f_{\alpha 0}}{\partial v}}{\omega - kv}dv = 0 \tag{8.100}$$

The ion term gives after integrating by parts, $-\omega_{pi}^2/\omega^2$. Consider the electron integral,

$$\int_{-\infty}^{\infty}\frac{\partial f_{0e}/\partial v}{v-\omega/k}dv = P\int_{-\infty}^{\infty}\frac{\partial f_{0e}/\partial v}{v-\omega/k}dv + i\pi[f_{0e}']_{v=\omega/k} \tag{8.101}$$

The principal part becomes

$$-\frac{m_e}{T_e}\int\frac{(v-v_0)f_{e0}}{v-\omega/k}dv = -\frac{m_e}{T_e}\int\frac{uf_{e0}(u)}{u+v_0-\omega/k}du \tag{8.102}$$

where $u = v - v_0$ and $f_{e0}(u)$ is now the stationary Maxwellian. The drift velocity v_0 and the phase velocity ω/k in the denominator will be neglected because we consider they are much smaller than the thermal velocity. Then the integral (8.102) simply gives the value $-m_e/T_e$. Then the dispersion relation (8.100) reads

$$1 - \frac{\omega_{pi}^2}{\omega^2} + \frac{m_e}{T_e}\frac{\omega_{pe}^2}{k^2} - i\pi\frac{\omega_{pe}^2}{k^2}[f_{e0}']_{u=\omega/k} = 0 \tag{8.103}$$

If we neglect the imaginary part, the dispersion relation gives the steady ion acoustic wave. The slope of the distribution function in the imaginary part

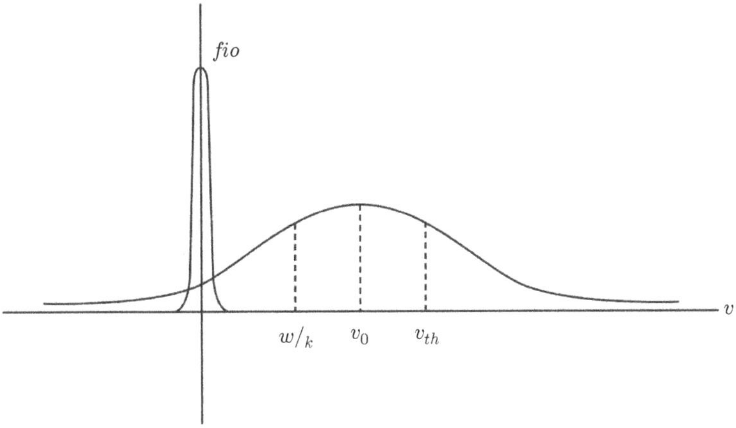

Fig. 8.2. Two stream instability of cold ions and moving Maxwellian electrons.

of (8.103) can be different in sign from the slope of stationary distribution function ($v_0 = 0$). In the stationary plasma, the slope of the Maxwellian at the wave velocity is negative (damping); in (8.103) in which f_{e0} is a moving Maxwellian, the slope can be positive, giving rise to instability if the drift velocity is greater than the ion acoustic wave speed:

$$v_0 > \sqrt{T_e/m_i} = \omega_r/k \quad \text{(for instability)} \tag{8.104}$$

The growth rate ω_i can be calculated with the aid of (7.92). We have $\partial\varepsilon_r/\partial\omega_r = 2\omega_{pi}^2/\omega_r^3$, $\varepsilon_i = -\pi(\omega_{\rm pe}^2/k^2)(\frac{\omega_r}{k}-v_0)\sqrt{\frac{m_e}{2\pi T_2}}(-)T_e/m_e\exp(\cdots)$. Neglecting ω_r/k as compared to v_0 and putting the exponential factor 1, we obtain (see Fig. 8.2).

$$\omega_i = \sqrt{\frac{\pi}{8}}\sqrt{\frac{m_e}{T_e}}\,\omega_r v_0 \sim \omega_r\sqrt{\frac{\pi}{8}}\sqrt{\frac{m_e}{m_i}} \tag{8.105}$$

8.5 Wave-Particle Interaction: Physics of Landau Damping

In the foregoing development in Chapter 7, we discussed Landau damping in view of the causality. Mathematically, Landau damping resulted in by evaluating the singular dielectric permittivity along the Landau contour. Although the causal requirement can be considered to be the foremost origin of Landau damping, we need to have the wave–particle interaction operative in order for the collisionless damping to actually occur. The wave–particle interaction results in energy exchange between the wave and the *resonant particles* whose velocities are close to the wave velocity. The orbits of such particles are periodic in the *wave frame* because they are subject to back and forth motion between two turning points. They are confined in the potential well and the particles are said to be *trapped* in the wave.

Resonant Particles. Consider the equation of motion of an electron moving in electric field $E(x,t) = E_0 \sin(k_0 x - \omega_0 t)$,

$$\frac{dv}{dt} = -\frac{e}{m} E_0 \ \sin(k_0 x - \omega_0 t) \tag{8.106}$$

This is *Lagrangian* description of the particle trajectory; it implicitly means that x in the above equation is the particle position at time t: $x(t)$. In this sense v is a function of t only: $v(t)$. In *Eulerian* description, the same equation is written as

$$\frac{\partial v}{\partial t} + v\frac{\partial v}{\partial x} = -\frac{e}{m} E_0 \ \sin(k_0 x - \omega_0 t) \tag{8.107}$$

where x and t are independent, and $v = v(x,t)$. Equation (8.107) is nonlinear, and the solution is not easily obtained. We consider the linearized equation

$$\frac{\partial v}{\partial t} + v_0 \ \frac{\partial v}{\partial x} = -\frac{e}{m} E_0 \ \sin(k_0 x - \omega_0 t) \tag{8.108}$$

where v_0 is the initial velocity; $v_0 = v(x,0)$.

We shall solve (8.108) as an initial value problem. First let us Fourier transform it with respect to the spatial coordinate x to write

$$\frac{\partial v(k,t)}{\partial t} + ikv_0 v(k,t) = \frac{ie\pi}{m} E_0 [e^{-i\omega_0 t}\delta(k-k_0) - e^{i\omega_0 t}\delta(k+k_0)] \tag{8.109}$$

This is a first order linear differential equation for t; its solution can be written in the form

$$v(k,t) = v(k,0)e^{-ikv_0 t} - \frac{ie\pi}{m} E_0 \left[\frac{\delta(k+k_0)}{i(\omega_0 + kv_0)}\left(e^{i\omega_0 t} - e^{-ikv_0 t}\right) + \frac{\delta(k-k_0)}{i(\omega_0 - kv_0)}\left(e^{-i\omega_0 t} - e^{-ikv_0 t}\right)\right] \tag{8.110}$$

Fourier inversion of (8.110) gives

$$v(x,t) = v(x - v_0 t, 0) - \frac{e}{m} \frac{E_0}{\omega_0 - k_0 v_0} [\cos(k_0 x - \omega_0 t) - \cos k_0 (x - v_0 t)] \tag{8.111}$$

This is the particular solution of (8.108) and shows the resonant behavior. $\omega_0 = k_0 v_0$ appears to be a singular point but it is only apparent. If $\omega_0 \to k_0 v_0$, $v(x,t) = v(x - v_0 t, 0) - \frac{e}{m} E_0 \ t \sin(k_0(x - v_0 t))$, i.e. the particle velocity increases as $\sim t$. The secular behavior $\sim t$ is manifested in the particular solution by the *resonant particles.* The resonant particles whose velocities are close to the wave velocity ω_0/k_0 exchange energy effectively with the wave and are responsible for Landau damping.

The homogeneous solution of (8.108) is $v_h(x,t) = F(x - v_0 t)$ with $F(t)$ being an arbitrary function. The complete solution of (8.108) is written as

$$v(x,t) = F(x - v_0 t) - \frac{eE_0}{m} P \frac{\cos(k_0 x - \omega_0 t)}{\omega_0 - k_0 v_0} \tag{8.112}$$

where P denotes the principal value. By choosing

$$F(x - v_0 t) = v(x - v_0 t, 0) + \frac{eE_0}{m} P \frac{\cos(k_0(x - v_0 t))}{\omega_0 - k_0 v_0} \tag{8.113}$$

Equation (8.112) becomes (8.111). The principal part solution in (8.112) applies to *nonresonant particles.*

The Eulerian solution (8.111) can be obtained from the Lagrangian equation (8.106) by using the *unperturbed orbit,*

$$x = v_0 t + x_0 \tag{8.114}$$

where x_0 and v_0 are the initial position and velocity, respectively. Then (8.106) becomes

$$\frac{dv}{dt} = -\frac{e}{m} E_0 \sin(k_0 x_0 + k_0 v_0 t - \omega_0 t) \tag{8.115}$$

which is integrated to yield

$$v(t) = v_0 - \frac{e}{m} E_0 \left[\frac{\cos(k_0 x_0 + k_0 v_0 t - \omega_0 t) - \cos(k_0 x_0)}{\omega_0 - k_0 v_0} \right] \tag{8.116}$$

If we go back to the *Eulerian coordinate* x, by putting $x_0 = x - v_0 t$, the above equation becomes

$$v(x,t) = v_0 - \frac{e}{m} E_0 \left[\frac{\cos(k_0 x - \omega_0 t) - \cos(k_0(x - v_0 t))}{\omega_0 - k_0 v_0} \right] \tag{8.117}$$

which is (8.111). The initial position x_0 serves as the *Lagrangian coordinate.* This recovery is anticipated since (8.106) and (8.107) are two alternative descriptions of the same motion. Linearization in the Eulerian equation corresponds to use of unperturbed orbit in the Lagrangian equation.

Next, we derive the Landau damping rate by considering the rate of change of the kinetic energy owned by the resonant particles. At early time of interaction between the resonant particles and the wave, we can assume, in (8.117), that v_0 is much greater than the E_0-term (perturbation). Thus we have for the kinetic energy,

$$\frac{m}{2}v^2 = -eE_0 v_0 \frac{\cos(k_0 x - \omega_0 t) - \cos(k_0 x - k_0 v_0 t)}{\omega_0 - k_0 v_0}$$

where we omitted $mv_0^2/2$ which is only an additive constant. Expanding the zero order distribution function in the vicinity of $v_0 = \omega_0/k_0$,

$$f_0(v) = f_0(v_0) + (v - v_0)\left[\frac{\partial f_0}{\partial v}\right]_{v=v_0} + \cdots$$

Hence

$$\int \frac{m}{2}v^2 f_0(v)dv = \frac{e^2 E_0^2}{m}\int dv_0\, v_0 \left[\frac{\partial f_0}{\partial v}\right]_{v_0} \frac{(\cos(k_0 x - \omega_0 t) - \cos(k_0 x - k_0 v_0 t))^2}{(\omega_0 - k_0 v_0)^2} \tag{8.118}$$

Note that $\int \frac{m}{2}v^2 f_0(v_0)dv_0$ vanishes upon spatial average. Also note that the integrand is not singular at $\omega_0 = k_0 v_0$. The dominant contribution to the integral comes from the interval around $v_0 = \omega_0/k_0$, and thus $\left[\frac{\partial f_0}{\partial v}\right]_{v_0}$ can be taken out of the integral, and the limits of the integration can be extended to $\pm\infty$. Using $\cos(kx - \omega t) - \cos(kx - kv_0 t) = 2\sin[(\omega - kv_0)t/2]\sin[kx - (\omega + kv_0)t/2]$ and taking spatial average, we obtain

$$\int \frac{m}{2}v^2 f_0(v)dv = \frac{e^2 E_0^2}{m}\left[\frac{\partial f_0}{\partial v}\right]_{\frac{\omega_0}{k_0}} \int_{-\infty}^{\infty} dv_0\; v_0 \frac{\sin^2\left[\frac{(k_0 v_0 - \omega_0)t}{2}\right]}{(\omega - kv_0)^2} \equiv K \tag{8.119}$$

Using $\int_{-\infty}^{\infty} dx\, \frac{\sin^2 x}{x^2} = \pi$, the kinetic energy of the resonant particles becomes

$$K = -\frac{\pi t}{2}\frac{e^2 E_0^2}{m}\frac{\omega_0}{k_0^2}\left[\frac{\partial f_0}{\partial v}\right]_{\frac{\omega_0}{k_0}} \tag{8.120}$$

Note that K increases linearly with t. It is readily seen that the rate of the kinetic energy increase dK/dt is equal to the Landau damping rate of the wave within a numerical factor (when the slope of f_0 is negative).

• **Nonlinear Landau damping.** A repetitive damped oscillation of wave in a collisionless plasma can occur due to repetitive linear Landau dampings. This phenomena was predicted by O'Neil (1965) in terms of the elliptic function solution of the nonlinear orbit [Reference: O'Neil, T. *Collisionless damping of nonlinear plasma oscillations*, Phys. Fluids **8** (1965), p. 2255]. Without resorting to the elliptic function, we deal with the equation of motion for an electron in a sinusoidal electric field in the *wave frame* as follows:

$$\frac{d^2x}{dt^2} = -\frac{e}{m}E(k,\omega)\sin(kx(t)) = \frac{dv}{dt} \tag{8.121}$$

or

$$\frac{d^2v}{dt^2} = -akv\cos(kx(t)) \quad \left(a = \frac{eE}{m}\right) \tag{8.122}$$

Multiplying by dv/dt and using $\cos kx = \pm\sqrt{1 - \frac{1}{a^2}(\frac{dv}{dt})^2}$, (8.122) becomes

$$\frac{d}{dt}\left(\frac{dv}{dt}\right)^2 = \pm ka\sqrt{1 - \frac{1}{a^2}\left(\frac{dv}{dt}\right)^2}\,\frac{dv^2}{dt} \tag{8.123}$$

Equation (8.123) can be integrated to yield

$$\sqrt{1 - \frac{1}{a^2}\left(\frac{dv}{dt}\right)^2} = \pm\frac{k}{2a}v^2 + c_1 \tag{8.124}$$

where c_1 is an integration constant. This equation can be arranged in the form after squaring

$$\left(\frac{dv}{dt}\right)^2 + V(v) = a^2 \tag{8.125}$$

where

$$V(v) = \left(\frac{kv^2}{2} + c_1 a\right)^2 \tag{8.126}$$

is the potential when we view (8.125) as the energy integral of a particle motion with correspondence $v \to x$, the particle position. By inspecting the shape of the potential $V(v)$, we can figure out qualitatively the behavior of $v(t)$. $V(v)$ is W-shaped, symmetric with respect to the axis $v = 0$. $v(t)$ draws anharmonic motion between two turning points or even a solitary pulse behavior, depending upon the value of 'energy' a^2. While the particles execute a round trip between two turning points, the Landau wave-particle resonance condition can be met, and the electric field Landau damps. In the next cycle of round trip, the resonance condition is met over a somewhat lowered energy level, and the wave Landau damps again. The repetitive damping due to the *trapped particles* bring about damped oscillations of the electric field, i.e. *nonlinear Landau damping.* O'Neil (1965) performed detailed nonlinear analysis of the Vlasov–Poisson plasma impressed by the sinusoidal electric field in the form of (8.121) in terms of the elliptic function, and derived the nonlinear Landau damping rate.

8.6 Small Amplitude Progressive Wave

Let us use the wave forms of progressive wave, $g(x,t,v) = \widehat{g}(v) f(\xi)$, $E(x,t) = E(\xi)$ with $\xi = x - v_0 t$, in the Vlasov and Poisson equations

$$\frac{\partial}{\partial t} g(x,t,v) + v\frac{\partial g}{\partial x} - \frac{e}{m} E(x,t)\ \frac{dg_0}{dv} = 0 \tag{8.127}$$

$$\frac{\partial E}{\partial x} = -4\pi e \int_{-\infty}^{\infty} g(x,t,v) dv \tag{8.128}$$

to write

$$(v - v_0)\widehat{g}(v)\ \frac{\partial f}{\partial \xi} - \frac{e}{m} \frac{dg_0}{dv} E(\xi) = 0 \tag{8.129}$$

$$\frac{\partial E}{\partial \xi} = -4\pi e f(\xi) \tag{8.130}$$

where we put by normalization

$$\int_{-\infty}^{\infty} \widehat{g}(v) dv = 1 \tag{8.131}$$

Equations (8.129) and (8.130) give

$$\frac{\partial^2 f}{\partial \xi^2} + k^2 f = 0 \tag{8.132}$$

$$\frac{\partial^2 E}{\partial \xi^2} + k^2 E = 0 \tag{8.133}$$

where

$$k^2 = \frac{\omega_p^2 g_0'}{(v - v_0)\widehat{g}(v)} \quad \left(g_0' = \frac{dg_0}{dv}\right) \tag{8.134}$$

or

$$\widehat{g}(v) = \frac{\omega_p^2}{k^2}\ \frac{g_0'}{v - v_0} \tag{8.135}$$

In summary, the progressive waves propagate with the phasor $\sim e^{\pm ik(x - v_0 t)}$, where the 'eigenvalue' k and the 'eigenfunction' $\widehat{g}(v)$ are related by (8.135).

Hence

$$g(x,t,v) = \frac{\omega_p^2}{k^2}\frac{g_0'(v)}{v-v_0}\,e^{ik(x-v_0t)} \tag{8.136}$$

Due to the normalization, k is not arbitrary:

$$\frac{\omega_p^2}{k^2}\int_{-\infty}^{\infty}\frac{\frac{dg_0}{dv}}{v-v_0}dv = 1 \tag{8.137}$$

Hence, k is a function of the beam velocity v_0. If we rename $v_0 \to \frac{\omega}{k}$, we write

$$g(x,t,v) = \frac{\omega_p^2}{k^2}\frac{g_0'(v)}{v-\frac{\omega}{k}}\,e^{ikx-i\omega t} \tag{8.138}$$

Then the normalization requires

$$\frac{\omega_p^2}{k^2}\int_{-\infty}^{\infty}\frac{g_0'(v)}{v-\frac{\omega}{k}}dv = 1 \tag{8.139}$$

which is the relation to be satisfied by k and ω. Recall the susceptibility

$$\chi(k,\omega) = \frac{\omega_p^2}{k}\int_{-\infty}^{\infty}\frac{g_0'(v)dv}{\omega-kv} \tag{8.140}$$

Then (8.139) is the dispersion relation $1+\chi(k,\omega) = \varepsilon(k,\omega) = 0$.

• **Van Kampen's eigenfunction.** Equation (8.135) can be written as

$$\widehat{g}(v) = P\,\frac{\eta(v)}{v-v_0} + \lambda\,\eta(v)\delta(v-v_0) \tag{8.141}$$

where

$$\eta(v) = \frac{\omega_p^2}{k^2}g_0'(v) \tag{8.142}$$

and λ is determined by normalization of $\widehat{g}(v)$, giving

$$\widehat{g}(v,v_0) = -P\,\frac{\eta(v)}{v_0-v} + \delta(v-v_0)\left[1+P\int_{-\infty}^{\infty}\frac{\eta(v')}{v_0-v'}dv'\right] \tag{8.143}$$

The eigenvalue v_0 is also hidden in η through k.

An arbitrary function of v, say $g(v)$, can be expanded in terms of the Van Kampen's eigenfunction $\widehat{g}(v)$ as

$$g(v) = \int_{-\infty}^{\infty} dv_0\, c(v_0)\, \widehat{g}(v, v_0) \tag{8.144}$$

or

$$g(v) = \int_{-\infty}^{\infty} dv_0\, c(v_0) \left[P\, \frac{\eta(v)}{v - v_0} + \delta(v - v_0) \left(1 + P \int_{-\infty}^{\infty} \frac{\eta(v')}{v_0 - v'} dv' \right) \right] \tag{8.145}$$

The integral involving the δ-function can be carried out to write

$$g(v) = \eta(v) P \int_{-\infty}^{\infty} dv_0\, \frac{c(v_0)}{v - v_0} + c(v) \left(1 + P \int_{-\infty}^{\infty} \frac{\eta(v')dv'}{v - v'} \right) \tag{8.146}$$

Van Kampen solved this integral equation to find the expansion coefficient $c(v_0)$ as follows.

We decompose all the functions in the above equation into the corresponding half-range Fourier integrals (denoted by superscripts $\pm$) as defined by

$$c^+(v) = \int_0^{\infty} d\tau c(\tau) e^{iv\tau}, \quad c^-(v) = \int_{-\infty}^{0} d\tau c(\tau) e^{iv\tau} \tag{8.147}$$

We also can deduce the following relations from Cauchy's integral formula (see Appendix A):

$$c^+(v) = \frac{1}{2\pi i} \left[P \int_{-\infty}^{\infty} \frac{c(\nu)d\nu}{\nu - v} + i\pi c(v) \right] \tag{8.148}$$

$$c^-(v) = \frac{-1}{2\pi i} \left[P \int_{-\infty}^{\infty} \frac{c(\nu)d\nu}{\nu - v} - i\pi c(v) \right] \tag{8.149}$$

Then clearly we have

$$c^+(v) + c^-(v) = c(v) \tag{8.150}$$

$$c^+(v) - c^-(v) = \frac{1}{\pi i} P \int_{-\infty}^{\infty} \frac{c(\nu)d\nu}{\nu - v} \tag{8.151}$$

Likewise, we have similar expressions for other quantities in the integral equation (8.146):

$$\eta^+(v) + \eta^-(v) = \eta(v) \tag{8.152}$$

$$\eta^+(v) - \eta^-(v) = \frac{1}{\pi i} P \int_{-\infty}^{\infty} \frac{\eta(\nu)d\nu}{\nu - v} \tag{8.153}$$

$$g^+(v) + g^-(v) = g(v) \tag{8.154}$$

$$g^+(v) - g^-(v) = \frac{1}{\pi i} P \int_{-\infty}^{\infty} \frac{g(\nu)d\nu}{\nu - v} \tag{8.155}$$

Next, we use the foregoing decomposition relations to write Eq. (8.146) in the form

$$g^+(v) + g^-(v) = (\eta^+(v) + \eta^-(v))\; P \int_{-\infty}^{\infty} d\nu \frac{c(\nu)}{v - \nu} + (c^+(v) + c^-(v)) \left[1 + P \int_{-\infty}^{\infty} \frac{\eta(\nu)}{v - \nu} d\nu \right]$$
$$= (\eta^+(v) + \eta^-(v))(-i\pi)(c^+(v) - c^-(v)) + (c^+(v) + c^-(v))[1 - i\pi(\eta^+(v) - \eta^-(v))]$$

After some cancellations, we obtain

$$g^+(v) + g^-(v) = c^+(v) + c^-(v) - 2i\pi(\eta^+(v)c^+(v) - \eta^-(v)c^-(v))$$
$$= c^+(v)(1 - 2i\pi\eta^+(v)) + c^-(v)(1 + 2i\pi\eta^-(v))$$

Separating the $\pm$ quantities, which are mutually independent, yield the following relations:

$$g^+(v) = c^+(v)(1 - 2\pi i\eta^+(v)) \tag{8.156}$$

$$g^-(v) = c^-(v)(1 + 2\pi i\eta^-(v)) \tag{8.157}$$

Equations (8.156) and (8.157) connect $c(v)$ with the function $g(v)$:

$$c(v) = c^+(v) + c^-(v) = \frac{g^+(v)}{1 - 2\pi i\eta^+(v)} + \frac{g^-(v)}{1 + 2\pi i\eta^-(v)} \tag{8.158}$$

This is the desired solution of Eq. (8.146).

Next, let us take a look at the η^+-equation

$$\eta^+(v) = \frac{1}{2\pi i}\left[P \int_{-\infty}^{\infty} \frac{\eta(\nu)d\nu}{\nu - v} + i\pi\eta(v)\right]$$

Changing $v \to \omega/k$, dummy $\nu \to v$, we have

$$\eta^+\left(\frac{\omega}{k}\right) = \frac{1}{2\pi i}\frac{\omega_p^2}{k^2}\left[P \int_{-\infty}^{\infty} \frac{g_0'(v)dv}{v - \frac{\omega}{k}} + i\pi g_0'\left(\frac{\omega}{k}\right)\right] = -\frac{\chi(k,\omega)}{2\pi i} \tag{8.159}$$

where we recalled (8.51). So η^+ corresponds in essence to the dielectric susceptibility χ.

• **General solution of progressive wave.** The initial profile of the wave $g(x, v, 0)$ changes in time according to the translation $g(x - v_0 t, v)$. Thus, for a beam of velocity v_0, we can write for the Vlasov distribution function

$$g(x, v, t) = e^{ik(x-v_0 t)}\, \widehat{g}(v, v_0) \tag{8.160}$$

where we recalled (8.136) and $\widehat{g}$ is given by (8.143). The general solution is obtained by integrating over various beams the above single beam distribution function

$$g(x, v, t) = e^{ikx} \int_{-\infty}^{\infty} dv_0\, e^{-ikv_0 t}\, c(v_0)\, \widehat{g}(v, v_0) \tag{8.161}$$

Putting $t = 0$ in the above gives

$$g(x, v, 0) = e^{ikx} \int_{-\infty}^{\infty} dv_0\, c(v_0)\, \widehat{g}(v, v_0) \tag{8.162}$$

Given $g(x, v, 0)$, $c(v_0)$ is obtained by (8.158). Thus, we write the general solution in the form

$$\begin{aligned} g(x, v, t) = &\int_{-\infty}^{\infty} dv_0\, e^{ik(x-v_0 t)} \left(\frac{g^+(v_0, 0)}{1 - 2\pi i \eta^+(v_0)} + \frac{g^-(v_0, 0)}{1 - 2\pi i \eta^-(v_0)} \right) \\ &\times \left[P\, \frac{\eta(v)}{v - v_0} + \delta(v - v_0) \left(1 - P \int_{-\infty}^{\infty} \frac{\eta(v')dv'}{v' - v_0} \right) \right] \end{aligned} \tag{8.163}$$

For $t > 0$, the contour integration path of the above integral encloses the lower half complex v_0-plane. The g^- term vanishes since that term has no singularities in the lower half plane. Hence

$$\begin{aligned} g(x, t, v) = &\, e^{ikx} \eta(v) P \int_{-\infty}^{\infty} dv_0\, e^{-ikv_0 t}\, \frac{c^+(v_0)}{v - v_0} - e^{ikx - ikvt} c^+(v) \\ &\times P \int_{-\infty}^{\infty} \frac{\eta(v')dv'}{v' - v} + e^{ikx - ikvt} c^+(v) \end{aligned} \tag{8.164}$$

where $c^+(v_0) = \frac{g^+(v_0, 0)}{1 - 2\pi i \eta^+(v_0)}$. The last term $(e^{ikx - ikvt} c^+(v))$ in the RHS of (8.164), which is called the *free streaming term*, oscillates more and more rapidly as $t \to \infty$, and it vanishes asymptotically upon velocity integration. Its origin traces back to the term $v\partial f / \partial x$ in the Vlasov equation.

• **Electric field.** Now what is the electric field at t in terms of $g(v,x,0)$? Since the phasor of the wave is $e^{ik(x-v_0t)}$, (8.128) gives

$$\begin{aligned} E(x,t) &= \frac{4\pi ie}{k}\,e^{ikx}\int_{-\infty}^{\infty}dv\int_{-\infty}^{\infty}dv_0\,e^{-ikv_0t}\,c(v_0)\,\widehat{g}(v,v_0) \\ &= \frac{4\pi ie}{k}\,e^{ikx}\int_{-\infty}^{\infty}dv\int_{-\infty}^{\infty}dv_0\,e^{-ikv_0t}\,c(v_0) \\ &\quad\times\left[P\,\frac{\eta(v)}{v-v_0}+\delta(v_0-v)\left(1-P\int\frac{\eta(v')dv'}{v'-v_0}\right)\right] \end{aligned} \tag{8.165}$$

where we used (8.143) and (8.161).

Integrals involving the principal value cancel upon $\int dv$-integration, and we are left with:

$$E(x,t)=\frac{4\pi ie}{k}\,e^{ikx}\int_{-\infty}^{\infty}dv\int_{-\infty}^{\infty}dv_0\,e^{-ikv_0t}\delta(v_0-v)\,c(v_0) \tag{8.166}$$

Using (8.158) in the above equation gives

$$E(x,t)=\frac{4\pi ie}{k}\,e^{ikx}\int_{-\infty}^{\infty}dv_0\,e^{-ikv_0t}\left[\frac{g^+(v_0,0)}{1-2\pi i\eta^+(v_0)}+\frac{g^-(v_0,0)}{1+2\pi i\eta^-(v_0)}\right] \tag{8.167}$$

For $t>0$, the second term has no singularities in the lower half plane, and is integrated to yield zero. Thus we finally obtain

$$\begin{aligned} E(x,t) &= \frac{4\pi ie}{k}\,e^{ikx}\int_{-\infty}^{\infty}dv_0\,e^{-ikv_0t}\,\frac{g^+(v_0,0)}{\varepsilon(k,v_0)} \\ &= \frac{4\pi ie}{k}\,e^{ikx}\int_{-\infty}^{\infty}dv_0\,e^{-ikv_0t}\,\frac{P\int\frac{g(v,0)dv}{v-v_0}+i\pi g(v_0,0)}{2\pi i\;\varepsilon(k,v_0)} \end{aligned}$$

where we used (8.159). Changing $v_0\to\omega/k$ in the above gives

$$E(x,t)=\frac{2e}{k^2}\,e^{ikx}\int d\omega\,e^{-i\omega t}\,\frac{P\int\frac{g(v,0)dv}{v-\omega/k}+i\pi\;g(\frac{\omega}{k},0)}{\varepsilon(k,\omega)} \tag{8.168}$$

Note that $g^+(v_0,0)$ is given analogously to (8.148).

Equation (8.168) agrees with Landau's solution. This completes the demonstration that the two methods for analysis of the linear Vlasov equation, Van Kampen's integral equation method and Landau's Laplace transform method, yield entirely equivalent results.

8.7 Relativistic Vlasov Equation

In this section, the subject of relativistic hot plasmas will be dealt with by Vlasov equation. The relativistic Vlasov equation has two versions: p-version and v-version. The former one is entirely written in terms of the relativistic momentum $\mathbf{p}$, whereas the latter employs the more familiar variable, velocity $\mathbf{v}$. The crucial factor is the relativistic Boltzmann distribution, which incurs considerable complexity when the velocity moments are evaluated. The v-version is more convenient when one compares the relativistic waves with the nonrelativistic waves, since the normalization of the zero order Boltzmann distribution is more familiar in the v-version. However the p-version has a more formal similarity with the nonrelativistic Vlasov theory.

8.7.1 *Basic equations*

We begin with the relativistic equation of motion for electrons

$$\frac{d\mathbf{p}}{dt} = -e\left(\mathbf{E} + \frac{1}{c}\mathbf{v} \times \mathbf{B}\right) \tag{8.169}$$

where $\mathbf{p}$, the relativistic momentum, is

$$\mathbf{p} = m\,\gamma\,\mathbf{v} \tag{8.170}$$

with m being the rest mass and $\mathbf{v}$ being the particle velocity, and

$$\gamma = \left(1 - \frac{v^2}{c^2}\right)^{-1/2} \tag{8.171}$$

The relativistic Vlasov equation reads

$$\frac{\partial}{\partial t}f(\mathbf{r},\mathbf{p},t) + \mathbf{v}\cdot\frac{\partial f}{\partial \mathbf{r}} - e\left(\mathbf{E}(\mathbf{r},t) + \frac{1}{c}\mathbf{v}\times\mathbf{B}\right)\cdot\frac{\partial f}{\partial \mathbf{p}} = 0 \tag{8.172}$$

The Poisson equation reads

$$\nabla\cdot\mathbf{E}(\mathbf{r},t) = -4\pi e\int f(\mathbf{r},\mathbf{p},t)d^3p \tag{8.173}$$

This is the equation giving the charge density. The current is calculated by

$$\begin{aligned}\mathbf{J}(\mathbf{r},t) &= -eN\int \mathbf{v}f(\mathbf{r},\mathbf{p},t)d^3p\\ &= -\frac{eN}{m}\int \frac{\mathbf{p}}{\gamma}f(\mathbf{r},\mathbf{p},t)d^3p\end{aligned} \tag{8.174}$$

In the above p-equations, $\mathbf{v}$ and γ should be expressed as functions of $\mathbf{p}$. One can calculate from (8.170) and (8.171):

$$\gamma(p) = \sqrt{1 + \frac{p^2}{m^2c^2}} \tag{8.175}$$

$$\mathbf{v}(p) = \frac{c\,\mathbf{p}}{\sqrt{c^2m^2 + p^2}} \tag{8.176}$$

All the above equations plus the Maxwell equations constitute the basic relativistic plasma equations.

• **v-version equations.** In (8.172), we change $f(\mathbf{r}, \mathbf{p}, t) \to f(\mathbf{r}, \mathbf{p}(\mathbf{v}), t)$. Therefore we need the transform $\frac{\partial f}{\partial p_i} = \frac{\partial f}{\partial v_j}\frac{\partial v_j}{\partial p_i}$ which should be substituted into (8.172). We have

$$\frac{\partial v_j}{\partial p_i} = \frac{1}{m\gamma(v)}\left(\delta_{ij} - \frac{v_i v_j}{c^2}\right) \tag{8.177}$$

To prove this, differentiate (8.176) and use (8.170). The derivation is left as an exercise. Then (8.172) is written in terms of $\mathbf{v}$ as

$$\frac{\partial f}{\partial t} + \mathbf{v}\cdot\frac{\partial f}{\partial \mathbf{r}} - \frac{e}{m\gamma}\left(\mathbf{E}(\mathbf{r}, t) + \frac{1}{c}\mathbf{v}\times\mathbf{B}\right)\cdot\left(\frac{\partial f}{\partial \mathbf{v}} - \frac{\mathbf{v}}{c^2}\mathbf{v}\cdot\frac{\partial f}{\partial \mathbf{v}}\right) = 0 \tag{8.178}$$

This is the v-version of the relativistic Vlasov equation.

• **Relativistic Boltzmann distribution.** We have

$$f_0 = A\exp\frac{-\varepsilon}{T} \tag{8.179}$$

where ε is the relativistic energy,

$$\begin{aligned}\varepsilon &= \gamma mc^2 \\ &= \sqrt{m^2c^4 + p^2c^2}\end{aligned} \tag{8.180}$$

where we used (8.175).

• **Normalization.** Let N be the number density. Then we normalize $f_0(p)$ so that $\int d^3p f_0$=N. Viz.,

$$\frac{N}{A} = 4\pi\int_0^\infty p^2\, e^{-\mu\gamma}dp \tag{8.181}$$

$$\text{where}\quad \mu = \frac{mc^2}{T} \tag{8.182}$$

Using (8.175) in the above and putting $p = mc\,\bar{p}$ give

$$\frac{N}{A} = 4\pi m^3 c^3 \int_0^\infty \bar{p}^2\, e^{-\mu\sqrt{1+\bar{p}^2}}\, d\bar{p} \tag{8.183}$$

Transforming $\bar{p}^2 = \xi$ gives

$$\frac{N}{A} = 4\pi m^3 c^3 \int_0^\infty \frac{1}{2}\sqrt{\xi}\, e^{-\mu\sqrt{1+\xi}}\, d\xi$$

$$= 4\pi m^3 c^3 \frac{K_2(\mu)}{\mu} \tag{8.184}$$

where

$$K_2(\mu) = \frac{\mu}{2}\int_0^\infty d\xi\, \sqrt{\xi}\, e^{-\mu\sqrt{1+\xi}} \tag{8.185}$$

is the *modified Bessel function of second kind* of order 2. Thus we have the *relativistic Maxwellian distribution*

$$f_0(p) = \frac{N \exp -\frac{\varepsilon}{T}}{4\pi T m^2 c\; K_2(\frac{mc^2}{T})} \tag{8.186}$$

Compare this with the nonrelativistic Maxwellian distribution

$$f_0(v) = N\left(\frac{m}{2\pi T}\right)^{3/2} e^{-\frac{mv^2}{2T}} \tag{8.187}$$

• **Modified Bessel function of second kind.**

$$K_\nu(\mu) = \frac{(\frac{\mu}{2})^\nu \Gamma(\frac{1}{2})}{\Gamma(\nu + \frac{1}{2})}\int_1^\infty e^{-\mu s}(s^2 - 1)^{\nu - \frac{1}{2}}\, ds \tag{8.188}$$

where

$$\text{Re}\left(\nu + \frac{1}{2}\right) > 0, \quad |\arg\mu| < \frac{\pi}{2} \tag{8.189}$$

It is left as an exercise to show that $K_2(\mu)$ in (8.188) agrees with (8.185). *Hint*: put in (8.185) $\sqrt{1+\xi} = s$ and $s\sqrt{s^2-1} = \frac{1}{3}\frac{d}{ds}(s^2-1)^{3/2}$ and integrate by parts.

The asymptotic series of $K_\nu(z)$ for $z \gg 1$ reads

$$K_\nu(z) = \sqrt{\frac{\pi}{2z}}\, e^{-z}\left[1 + \frac{4\nu^2 - 1}{8z} + \cdots\right] \tag{8.190}$$

The above formulas for Bessel function can be found in the book by G. Arfken *Mathematical methods for physicists* Academic Press; M. Abramowitz and I. Stegun *Handbook of mathematical functions* Dover.

• **Classical limit.** In the classical region, $v^2/c^2 \ll 1$ and $mc^2 \gg k_B T$, or $\mu \gg 1$. Then we approximate $\gamma \simeq 1 + v^2/2c^2$ and use (8.190) in (8.186). We then obtain

$$f_0(p) = \frac{1}{(2\pi T m)^{3/2}}\, e^{-\frac{mv^2}{2T}} \tag{8.191}$$

In view of the equality $f_0(p)d^3p = f_0(v)d^3v$ and $d^3p = m^3 d^3v$, we can write

$$f_0(v) = \left(\frac{m}{2\pi T}\right)^{3/2} e^{-\frac{mv^2}{2T}} \tag{8.192}$$

• **Jacobian.** In (8.173) or (8.174), we wish to change the volume integral $\int d^3p F(p) \to \int d^3v F(p(v))$ as the variable p is transformed to v. Then we use the rule by introducing Jacobian determinant J

$$d^3p = J d^3v, \quad J = \left|\frac{\partial p_i}{\partial v_j}\right| \tag{8.193}$$

The Jacobian can be found from the relation $p_i = m\gamma v_i$

$$\begin{aligned} \frac{\partial p_i}{\partial v_j} &= m\left(\gamma \frac{\partial v_i}{\partial v_j} + v_i \frac{\partial \gamma}{\partial v_j}\right) \\ &= m\gamma\left(\delta_{ij} + \frac{\gamma^2}{c^2} v_i v_j\right) \end{aligned} \tag{8.194}$$

where we used

$$\frac{\partial \gamma}{\partial v_j} = \frac{\gamma^3}{c^2} v_j \tag{8.195}$$

From the array in (8.194), we obtain the Jacobian by direct expansion

$$J = m^3 \gamma^5 \tag{8.196}$$

The reciprocal Jacobian $\bar{J} = |\frac{\partial v_i}{\partial p_j}| = J^{-1}$.

8.7.2 *Relativistic Langmuir wave: Kinetic theory*

The linearized relativistic Vlasov equation reads

$$\frac{\partial f(\mathbf{r}, \mathbf{p}, t)}{\partial t} + \mathbf{v} \cdot \frac{\partial f}{\partial \mathbf{r}} - e\, \mathbf{E} \cdot \frac{\partial f_0(p)}{\partial \mathbf{p}} = 0 \tag{8.197}$$

where

$$f_0(p) = A\, e^{-\mu\gamma}; \quad A = \frac{N}{4\pi T m^2 c\, K_2(\mu)} \tag{8.198}$$

We have the Poisson equation

$$\nabla \cdot \mathbf{E} = -4\pi e \int f d^3p \tag{8.199}$$

In Fourier variables, we have

$$f = ie\mathbf{E} \cdot \frac{\frac{\partial f_0}{\partial \mathbf{p}}}{\omega - \mathbf{k} \cdot \mathbf{v}} \tag{8.200}$$

Putting this equation into the Poisson equation, we obtain the dispersion relation in the form

$$\varepsilon_L = 1 + \frac{4\pi e^2}{k^2} \int \frac{\mathbf{k} \cdot \frac{\partial f_0(p)}{\partial \mathbf{p}}}{\omega - \mathbf{k} \cdot \mathbf{v}} d^3p = 0 \tag{8.201}$$

where ε_L is the dielectric conductivity. Compare with the nonrelativistic expression:

$$\varepsilon_L = 1 + \frac{4\pi e^2}{mk^2} \int \frac{\mathbf{k} \cdot \frac{\partial f_0(v)}{\partial \mathbf{v}}}{\omega - \mathbf{k} \cdot \mathbf{v}} d^3v = 0 \tag{8.202}$$

So far, the relativistic dielectric permittivity looks simple and analogous to the nonrelativistic ε_L. However, the actual evaluation of it is very complex. Equation (8.175) gives

$$\frac{\partial \gamma}{\partial p_i} = \frac{p_i}{m^2c^2\gamma} \tag{8.203}$$

Differentiating (8.198), we have

$$\frac{\partial f_0}{\partial p_i} = -\frac{\mu f_0}{m^2c^2}\frac{p_i}{\gamma} = -\frac{A}{mT}\frac{p_i}{\gamma}\, e^{-\mu\gamma} \tag{8.204}$$

Hence

$$\begin{aligned} \varepsilon_L &= 1 - \frac{4\pi A e^2}{Tk^2} \int \frac{\gamma \mathbf{k} \cdot \mathbf{p}}{m\gamma\omega - \mathbf{k} \cdot \mathbf{p}}\, e^{-\mu\gamma} d^3p \\ &= 1 - \frac{4\pi A e^2}{Tk^2} \int \gamma\, e^{-\mu\gamma} \left(\frac{m\gamma\omega}{m\gamma\omega - \mathbf{k} \cdot \mathbf{p}} - 1 \right) d^3p \end{aligned} \tag{8.205}$$

Let us introduce the spherical coordinates in the $\mathbf{p}$ space and take $\mathbf{k} = \hat{\mathbf{z}}k$ to write the integral in the form

$$\begin{aligned} \text{Integral} &= \int_0^\infty dp\, p^2 \int_{-\pi}^{\pi} d\theta \sin\theta \int_0^{2\pi} d\varphi\, \gamma\, e^{-\mu\gamma} \left(\frac{m\gamma\omega}{m\gamma\omega - kp\cos\theta} - 1 \right) \\ &= 2\pi \int_0^\infty dp \gamma p^2\, e^{-\mu\gamma} \int_{-1}^{1} dx \left(\frac{1}{1 - \frac{kpx}{m\gamma\omega}} - 1 \right) \end{aligned} \tag{8.206}$$

Assuming $kp/m\gamma\omega \ll 1$ (electron thermal speed $\ll \omega/k$), the fraction is expanded in the series and we obtain

$$\varepsilon_L = 1 - \frac{4\pi Ae^2}{Tk^2} 4\pi \int_0^\infty dp\, \gamma\, p^2\, e^{-\mu\gamma} \left(\frac{1}{3} \frac{k^2p^2}{m^2\gamma^2\omega^2} + \frac{1}{5} \frac{k^4p^4}{m^4\gamma^4\omega^4} + \cdots \right) \tag{8.207}$$

In a weakly relativistic plasma, we approximate γ in (8.175) as $\gamma = 1 + p^2/2m^2c^2$, so

$$e^{-\mu\gamma} = e^{-\mu} e^{-\frac{p^2}{2mT}} \tag{8.208}$$

$$\frac{1}{\gamma} = 1 - \frac{p^2}{2m^2c^2}, \quad \frac{1}{\gamma^3} = 1 - \frac{3}{2} \frac{p^2}{m^2c^2} \tag{8.209}$$

Then the $\int dp$-integrals in (8.207) are Gaussian moment integrals, and can be carried out straightforwardly with the aid of the formula

$$\int_{-\infty}^{\infty} x^{2n}\, e^{-ax^2} dx = \frac{1 \cdot 3 \cdots (2n-1)}{2^n} \sqrt{\frac{\pi}{a^{2n+1}}} \tag{8.210}$$

where $n = 1, 2, 3, \ldots$. Using (8.198) for A and (8.190), we obtain after some algebra

$$\varepsilon_L = 1 - \frac{\omega_p^2}{\omega^2} \left(1 - \frac{5}{2} \frac{T}{mc^2} + \frac{3k^2}{\omega^2} \frac{T}{m} \right) \tag{8.211}$$

This result is valid for a weakly relativistic plasma ($\frac{v}{c} \ll 1$ or equivalently $p \ll mc$) in the classical limit $\mu \gg 1$ (we used the approximation in (8.190)). Identical result for a weakly relativistic plasma wave is obtained in the text by P. C. Clemmow and J. P. Dougherty, *Electrodynamics of Particles and Plasmas*, Addison-Wesley, 1969.

The dispersion relation of relativistic electromagnetic wave can be derived analogously. This problem is left as an exercise.

If one uses the relativistic distribution function

$$f_0(p) = A \exp\left(-\mu\sqrt{1 + \frac{p^2}{m^2c^2}} \right) \tag{8.212}$$

the complicated dispersion relation is expressed in terms of the Bessel function $K_n(\mu)$ series. See M. Brambilla, *Kinetic Theory of Plasma waves*, Clarendon Press (1998).

Problems

8.1 Laplace-invert $\chi(\omega)$ in (8.46).

$$\chi(t) = \int_{-\infty+ic}^{\infty+ic} \frac{d\omega}{2\pi} e^{-i\omega t}\chi(\omega)$$
$$= -\frac{\omega_{\rm pe}^2}{k^2}\int_{-\infty+ic}^{\infty+ic} \frac{d\omega}{2\pi} e^{-i\omega t}\int_{-\infty}^{\infty} dv\, \frac{g_0'(v)}{v-\omega/k}$$

Do the $\int d\omega$ integral first. By contour integration enclosing the lower half ω plane for $t > 0$, obtain

$$\chi(t) = -i\frac{\omega_{\rm pe}^2}{k}\int_{-\infty}^{\infty} e^{-ikvt} g_0'(v)dv$$

which agrees with (7.72). Use $\delta(v - \frac{\omega}{k}) = k\delta(\omega - kv)$.

8.2 Use (8.21) and (7.29) to show for isotropic Maxwellian that

$$\varepsilon_L = 1 - \sum_\alpha \frac{1}{\omega k^2\lambda_\alpha^2}\int d^3v \frac{(\mathbf{k}\cdot\mathbf{v})^2}{\omega - \mathbf{k}\cdot\mathbf{v}} f_{0\alpha}$$

$$\varepsilon_T = 1 - \frac{1}{2}\sum_\alpha \frac{1}{\omega k^2\lambda_\alpha^2}\int d^3v \frac{(\mathbf{k}\times\mathbf{v})^2}{\omega - \mathbf{k}\cdot\mathbf{v}} f_{0\alpha}$$

8.3 Let us begin with one-dimensional Vlasov–Poisson equations:

$$(1)\quad \frac{\partial}{\partial t}g(x,v,t) + v\frac{\partial g}{\partial x} - \frac{e}{m}E(x,t)\frac{dg_0(v)}{dv} = 0$$

$$(2)\quad \frac{\partial E(x,t)}{\partial x} = -4\pi e\int_{-\infty}^{\infty} dv\; g(x,v,t))$$

Check the following steps to deal with the above simultaneous equations. Note that (1) is a first order linear differential equation in the variable t. Let us Fourier-transform (1) and (2) with respect to the spatial coordinate x by performing the integral $\int(\cdots)e^{-ikx}dx$.

$$(3)\quad \frac{\partial}{\partial t}g(k,v,t) + ikv\; g(k,v,t) - \frac{e}{m}E(k,t)\frac{dg_0(v)}{dv} = 0$$

$$(4)\quad ikE(k,t) = -4\pi e\int_{-\infty}^{\infty} dv\; g(k,v,t)$$

Equation (3) is solved in the following form

$$(5)\quad g(k,v,t) = \frac{e}{m}\frac{dg_0(v)}{dv}\int_{t_0}^{t} E(k,t')e^{ikv(t'-t)}dt' + g(k,v,t_0)e^{ikv(t_0-t)}$$

where the last term is called the *free-streaming* term. Let us put $t_0 = -\infty$ and assume

$$g(k, v, t = -\infty) = 0$$

The above condition means that there are no perturbations in the infinitely remote past which is often referred to as the *causality* and amounts to assuming, in our definition of Fourier transform, that ω has a small positive imaginary part. Then we have

$$(6) \quad g(k, v, t) = \frac{e}{m}\frac{dg_0(v)}{dv}\int_{-\infty}^{t} E(k, t')e^{ikv(t'-t)}dt'$$

Fourier transform (6) to obtain

$$g(k, v, \omega) = \frac{e}{m}\frac{dg_0(v)}{dv}\frac{i}{(\omega - kv)}E(k, \omega)$$

Next, (4) is Fourier transformed in the form

$$ikE(k, \omega) = -4\pi e\int_{-\infty}^{\infty} g(k, v, \omega)dv$$

Elimination of $E(k, \omega)$ in the above two equations yield the dispersion relation

$$(7) \quad 1 + \frac{\omega_p^2}{k}\int_{-\infty}^{\infty} dv\, \frac{dg_0}{dv}\frac{1}{\omega - kv} = 0$$

If the initial condition is given as $g(k, v, t = 0) = 0$, we have, instead of (5),

$$(8) \quad g(k, v, t) = \frac{e}{m}\frac{dg_0(v)}{dv}\int_{0}^{t} E(k, t')e^{ikv(t'-t)}dt'$$

Laplace transform (4) and (8) and obtain

$$ikE_k(\omega) = -\omega_p^2\int_{-\infty}^{\infty} dv\, \frac{dg_0(v)}{dv}\int_0^{\infty} dt\, e^{i(\omega - kv)t}\int_0^t E_k(t')e^{ikvt'}dt'$$

Put in the above

$$e^{i(\omega - kv)t} = \frac{1}{i(\omega - kv)}\frac{d}{dt}\, e^{i(\omega - kv)t}$$

and integrate by parts. The integrated term vanishes at both limits of $t = \infty$ and $t = 0$. Note that the $\int dv$-integral containing the factor e^{-ikvt} vanishes as $t \to \infty$, because the phase oscillates rapidly. The result is again the dispersion relation (7).

8.4 Explain why the slope of distribution function $f(v)$ at the resonant velocity must be negative for the Landau damping to occur.

8.5 Check the steps in the following development. Electron distribution function $f(x,t,v)$ in a sinusoidal wave train

$$(1)\quad E(x,t) = E_0 \sin(k_0 x - \omega_0 t)$$

is governed by

$$(2)\quad \frac{\partial f}{\partial t} + v\frac{\partial f}{\partial x} - \frac{e}{m}E_0 \sin(k_0 x - \omega_0 t)\frac{\partial f_0(v)}{\partial v} = 0$$

Fourier transform of $f(x,t,v)$ gives

$$(3)\quad f(k,\omega,v) = 2\pi^2 \frac{eE_0}{m}\frac{\partial f_0}{\partial v}\frac{\delta(k-k_0)\delta(\omega-\omega_0) - \delta(k+k_0)\delta(\omega+\omega_0)}{\omega - kv}$$

Inverting this,

$$(4)\quad f_{\mathrm{nr}}(x,t,v) = \frac{eE_0}{m}\frac{\partial f_0}{\partial v}P\frac{\cos(k_0 x - \omega_0 t)}{\omega_0 - k_0 v}$$

where P denotes the principal value and the subscript 'nr' 'nonresonant'. Suppose that the wave is turned on at $t = 0$ and initially $f(t=0) = 0$. Using this initial condition, solve (2) to obtain

$$(5)\quad f_r(x,t,v) = \frac{eE_0}{m}\frac{\partial f_{0r}}{\partial v}\frac{\cos(k_0 x - \omega_0 t) - \cos(k_0 x - k_0 v t)}{\omega_0 - k_0 v}$$

Note that (5) satisfies the prescribed initial condition. Also note that f_r in (5) shows the $\sim t$ behavior as $\omega_0 \to k_0 v$. The resonant particle distribution f_r should be obtained as an initial value problem.

To see the electron energy transfer to the wave from particles or vice versa, we consider the electrostatic Maxwell equation

$$(6)\quad \frac{\partial E}{\partial t} = 4\pi e \int dv v\ (f_r + f_{\mathrm{nr}})$$

Multiplying by $E(x,t)$,

$$(7)\quad \frac{\partial E}{\partial t} = 4\pi e E \int dv v\ (f_r + f_{\mathrm{nr}})$$

which expresses the rate of wave energy transfer to all the electrons. Now we substitute (4) and (5) for f_{nr} and f_r in the right hand side and take the spatial average (denoted by $<>$) by integrating over a wavelength $2\pi/k_0$. Noting $< \sin(k_0 x - \omega_0 t)\cos(k_0 x - \omega_0 t) >= 0$, the f_{nr}-term in the above vanishes upon spatial average. This means that the non-resonant electrons do not change the average electrical energy. Using

$$< \cos(k_0 x - k_0 vt)\sin(k_0 x - \omega_0 t) >= \frac{1}{2}\sin(k_0 vt - \omega_0 t)$$

Equation (5.148) is written

$$(8) \quad \frac{\partial}{\partial t} < E^2 >= E_0(t)\frac{\partial E_0}{\partial t} = \frac{1}{2}\omega_{\rm pe}^2 E_0^2 \int_{-\infty}^{\infty} dv v f'_{0r}(v)\frac{\sin(\omega_0 t - k_0 vt)}{\omega_0 - k_0 v}$$

where we consider E_0 is a slowly varying function of t as the result of energy exchange with the electrons. Note that (8) is not singular at $\omega_0 = k_0 v$. The integration interval in (8) is the narrow band centered about $v = \omega_0/k_0$, and therefore $f'_{0r}(v)$ can be taken out of the integral with $v = \omega_0/k_0$. Furthermore the value of the integral mainly comes from the region $v \simeq \omega_0/k_0$, and we can change the integration limits from $-\infty$ to ∞ without introducing much error.

$$(9) \quad \frac{1}{E_0}\frac{\partial E_0}{\partial t} = \frac{\omega_{\rm pe}^2}{2}[f'_{0r}(v)]_{v=\omega_0/k_0} \int_{-\infty}^{\infty} dv\ v\frac{\sin(\omega_0 t - k_0 vt)}{\omega_0 - k_0 v}$$

where the integral $\int_{-\infty}^{\infty} dv(\cdots)$ is equal to

$$-\frac{1}{k_0}\int_{-\infty}^{\infty} dy\left(y + \frac{\omega_0}{k_0}\right)\frac{\sin(yt)}{y} = -\frac{\omega_0}{k_0^2}\pi \left(\text{per} \int_{-\infty}^{\infty}\frac{\sin(x)}{x}dx = \pi\right)$$

$$(10) \quad \frac{1}{E_0}\frac{\partial E_0}{\partial t} = -\frac{\pi}{2}\omega_{\rm pe}^2\frac{\omega_0}{k_0^2}[f'_{0r}(v)]_{v=\omega_0/k_0}$$

which shows that the electric wave damps according to the rate given by the right hand side. Equation (11) agrees with the Landau damping rate (8.88).

Reference: D. C. Montgomery and D. A. Tidman, *Plasma Kinetic Theory*, McGraw-Hill, New York (1964); O'Neil, T. *Collisionless damping of nonlinear plasma oscillations*, Phys. Fluids **3** (1965), 2225.

8.6 Derive (8.34) from (8.24).

8.7 **Review Landau contour.** When ω becomes real in (8.46), the v-integral can be carried out along the Landau contour (Fig. 8.1), which analytically continue the function $\chi(\omega)$ as originally defined in the region $\omega_i > 0$ to the region $\omega_i = 0$. Let us write

$$\chi(\omega, k) = -\frac{\omega_{\rm pe}^2}{k^2}\int_{-\infty}^{\infty} dv\, \frac{dg_0}{dv}\frac{1}{v - \frac{\omega}{k}}$$

where ω is now real, so $v = \omega/k$ is somewhere on the integral path. The v-integral is to be carried out on the complex v plane, along the

Landau contour, which consists of the principal part plus the integral along the infinitesimal semi-circle of concave down shape (c). Let

$$I_c = \int_c dv \, \frac{dg_0}{dv} \frac{1}{v - \frac{\omega}{k}}$$

Let us put $v = \frac{\omega}{k} + r\, e^{i\theta}$ on c. And take the limit $r \to 0$. Then

$$I_c = \left[\frac{dg_0}{dv}\right]_{v=\frac{\omega}{k}} \int_{-\pi}^{0} i\, d\theta = i\,\pi \left[\frac{dg_0}{dv}\right]_{v=\frac{\omega}{k}}$$

Hence

$$\chi(\omega, k) = -\frac{\omega_{\rm pe}^2}{k^2} \int_{-\infty}^{\infty} dv\, \frac{dg_0}{dv} \left[P \frac{1}{v - \frac{\omega}{k}} + i\,\pi\,\delta\left(v - \frac{\omega}{k}\right)\right] \quad (\omega_i = 0)$$

Note that $\delta(\omega - kv) = \frac{\delta(v - \frac{\omega}{k})}{k}$. Next, we analytically continue $\chi(\omega, k)$ into the region $\omega_i < 0$. The job can be carried out by integration along the second Landau contour in the Fig. 8.1. The integrals along the two anti-parallel vertical lines cancel. Then, the integral is the sum of the principal part and the whole small circle contribution, which is $\int_{-\pi}^{\pi} i d\theta = 2\pi i$. Hence

$$\chi(\omega, k) = -\frac{\omega_{\rm pe}^2}{k^2} \int_{-\infty}^{\infty} dv\, \frac{dg_0}{dv} \left[P \frac{1}{v - \frac{\omega_r}{k}} + i\,\pi\,\delta\left(v - \frac{\omega}{k}\right)\right] \quad (\omega_i < 0)$$

where ω_r is the real part of ω.

References

Akhiezer, A., I. Akhiezer, R. Polovin, A. Sitenko, and K. Stepanov, *Collective Oscillations in a Plasma*, Pergamon, Oxford (1967).

Alexandrov, A. F., L. S. Bogdankevich, and A. A. Rukhadze, *Principles of Plasma Electrodynamics*, Springer-Verlag, New York (1984).

Brambilla, M. *Kinetic Theory of Plasma Waves*, Clarendon Press (1998).

Landau, L. *On the vibrations of the electronic plasma*, J. Physics (USSR), **10**, (1946) p. 25.

Lee, H. J. *Causality in plasma electrodynamics*, J. Korean Phys. Soc. **73** No. 1 (2018) pp. 65–85.

Montgomeery, D. C. and D. A. Tidman, *Plasma Kinetic Theory*. McGraw-Hill, New York, (1964).

Montgomery, D. C. *Theory of the Unmagnetized Plasma*, Gordon and Beach, New York, (1971).

O'Neil, T. *Collisionless damping of nonlinear plasma oscillations*, Phys. Fluids **8** (1965) p. 2255.

Van Kampen, N. G. and B. U. Felderhof, *Theoretical Methods in Plasma Physics*, John Wiley (1967), Chapter XV.

Chapter 9

Theory of Vlasov Equation (II)

In this chapter, we deal with plasma waves in magnetized plasmas. One of the methods to solve the Vlasov equation for a magnetized plasma uses a special technique called *characteristic method.* The Vlasov equation is a *hyperbolic partial differential equation* and the characteristic method is a very powerful tool to obtain the general solution of hyperbolic partial differential equations. We spend some pages to get familiar with the mathematics of this useful method. The dielectric conductivity tensor in a magnetized plasma is very complex, but it can be nicely formulated. Kinetic theory of the magnetohydrodynamics will be dealt with another method which may be called the method of harmonic expansion of the distribution function.

9.1 Characteristic Method

The characteristic method is closely related with Lagrangian orbit of particle dynamics and is known by the name *integration along the unperturbed orbit.* The characteristic equations of the Vlasov and the Eulerian equations are identical; thus the solutions of the two equations considerably overlap.

9.1.1 *Mathematics of characteristic method*

The one-dimensional form of Vlasov equation for electrons reads

$$\frac{\partial}{\partial t}g(x,v,t)+v\frac{\partial g}{\partial x}-\frac{e}{m}E(x,t)\frac{\partial g}{\partial v}=0 \tag{9.1}$$

This equation is mathematically a quasi-linear partial differential equation of first order. Let us consider a three-dimensional space consisting of three coordinate axes, (t,x,v). Then g is a function of these three coordinates, and we define the gradient operator in this space:

$$\nabla=\hat{\mathbf{t}}\frac{\partial}{\partial t}+\hat{\mathbf{x}}\frac{\partial}{\partial x}+\hat{\mathbf{v}}\frac{\partial}{\partial v}$$

Also we define the vector

$$\mathbf{V} = \hat{\mathbf{t}}\ 1 + \hat{\mathbf{x}}v - \hat{\mathbf{v}}\frac{e}{m}E(x,t)$$

Then (9.1) is

$$\mathbf{V} \cdot \nabla g = 0 \tag{9.2}$$

Since ∇g is a vector normal to the surface $g = c = \text{const.}$, (9.2) states that the vector $\mathbf{V}$ lies in the tangent plane of $g = c$. That is, at any point, the vector $\mathbf{V}$ is tangent to a curve in the integral surface $g = c$ which passes through that point. Equation (9.1) may be considered as specifying the direction of the vector $\mathbf{V}$ at a point in (t, x, v)-space. Imagine that $\mathbf{V}$ is the velocity of a particle in (t, x, v) space. If a particle moves from a given initial point in the direction of $\mathbf{V}$ and continues its motion along the direction given by $\mathbf{V}$, a space curve is drawn which we call the *characteristic curve* of the partial differential equation (9.1). Any surface which is made up from such characteristic curves contains the vector $\mathbf{V}$ on the tangent plane of that surface, and that surface (consisting of the characteristic curves) is an integral surface of (9.1). [The ordinary differential equation, $dy/dx = f(x, y)$ defines the tangential line at the point (x, y) which is tangent to the integral curve $y(x)$. Integral curve can be obtained by connecting such infinitesimal tangential segments.]

If $d\mathbf{r}$ is the difference vector between two neighboring points on a characteristic curve, $d\mathbf{r}$ is parallel to $\mathbf{V}$, or $d\mathbf{r} \times \mathbf{V} = 0$. This condition gives the equation for the characteristic curve which can be written in the form

$$\frac{dt}{1} = \frac{dx}{v} = \frac{dv}{-\frac{e}{m}E(x,t)} \tag{9.3}$$

These relations are equivalent to two ordinary differential equations. Let the solutions of two independent equations of (9.3) be denoted by

$$u_1(x,t,v) = c_1 \quad \text{and} \quad u_2(x,t,v) = c_2$$

with c_1 and c_2 independent constants. Then the intersection of these two surfaces for a given value of c_1 and c_2 is a characteristic curve. The intersections of the two families of the surfaces $u_1 = c_1$ and $u_2 = c_2$ generate a host of characteristic curves which form the integral surface of (9.1). Thus the general solution of (9.1) can be represented by the equation of the form

$$F(c_1, c_2) = 0 \quad \text{or} \quad F\big(u_1(x,t,v), \quad u_2(x,t,v)\big) = 0 \tag{9.4}$$

Next we shall consider the linearized Vlasov equation,

$$\frac{\partial}{\partial t} g(x, v, t) + v \frac{\partial g}{\partial x} - \frac{e}{m} E(x, t) \frac{d g_0(v)}{dv} = 0 \tag{9.5}$$

We assume the solution of this equation in the form

$$u(t, x, g(t, x)) = c \tag{9.6}$$

in the sense that (9.6) determines g as a function of x and t which satisfies (9.5). It should be noted that v plays the role of a parameter only, not a variable in (9.5). Differentiating both sides of (9.6), we have

$$\frac{\partial u}{\partial t} + \frac{\partial u}{\partial g} \frac{\partial g}{\partial t} = 0, \qquad \frac{\partial u}{\partial x} + \frac{\partial u}{\partial g} \frac{\partial g}{\partial x} = 0$$

Solving for $\partial g / \partial t$ and $\partial g / \partial x$ in the above equations and substituting the results into (9.6) give

$$\frac{\partial u}{\partial t} + v \frac{\partial u}{\partial x} + \frac{e}{m} E(x, t) \frac{d g_0(v)}{dv} \frac{\partial u}{\partial g} = 0 \tag{9.7}$$

which is of the same form as (9.1), and the relations for the characteristic curve is

$$\frac{dt}{1} = \frac{dx}{v} = \frac{dg}{\frac{e}{m} E(x, t) \frac{d g_0(v)}{dv}} \tag{9.8}$$

In this case the perturbed distribution function g can be found by integrating the relation

$$dg = \frac{e}{m} E(x, t) \frac{d g_0(v)}{dv} \, dt \tag{9.9}$$

where x and t are related by the equation,

$$dx = v \, dt \tag{9.10}$$

where v should be regarded as a parameter (constant).

The solution of (9.5) is obtained in terms of the constants of integration that are derived from integrating the characteristic equations (9.9) and (9.10) as will be illustrated in the following. Integrating (9.10) gives

$$x = vt + c_1 \quad (c_1 = \text{const.}) \tag{9.11}$$

Integrating (9.9), we write

$$g(t) = \frac{e}{m} \frac{d g_0(v)}{dv} \int_{-\infty}^{t} E(x(t'), t') dt' \tag{9.12}$$

where we put $g(t = -\infty) = 0$. Now for further progress, we give specific functional form for the function $E(x,t)$. We assume $E(x,t) = E(k,\omega)$ $e^{i(kx-\omega t)}$. Then (9.12) is written as

$$\begin{aligned} g(t) &= \frac{e}{m}\frac{dg_0}{dv}\, E(k,\omega)\int_{-\infty}^{t} \exp\left(ikx(t') - i\omega t'\right) dt' \\ &= \frac{e}{m}\frac{dg_0}{dv}\, E(k,\omega)e^{ikc_1}\int_{-\infty}^{t} e^{ikvt' - i\omega t'}\, dt' \\ &= \frac{e}{m}\frac{dg_0}{dv}\, E(k,\omega)e^{ik(x-vt)}\int_{-\infty}^{t} e^{ikvt' - i\omega t'}\, dt' \end{aligned}$$

where we used (9.11). Carrying out the integral gives

$$g(t,x,v) = \frac{e}{m}\frac{dg_0}{dv}\frac{E(k,\omega)}{i(kv-\omega)} e^{ikx - i\omega t} \tag{9.13}$$

In (9.13), the coefficient of $e^{ikx-i\omega t}$ is the Fourier coefficient $g(k,\omega,v)$:

$$g(k,\omega,v) = \frac{e}{m}\frac{dg_0}{dv}\frac{E(k,\omega)}{i(kv-\omega)} \tag{9.14}$$

which agrees with Fourier transform solution of (9.5). This result is due to the specification $E(x,t) = E(k,\omega)e^{i(kx-\omega t)}$.

In (9.12), we specified the integration constant as 0 at $t = -\infty$. Instead, we can set up more general solution by introducing the second integration constant from (9.9):

$$g(t) = \frac{e}{m}\frac{dg_0}{dv}\int^{t} E\left(x(t'),t'\right) dt' + c_2 \tag{9.15}$$

Then the general solution of (9.5) can be written as $G(c_1,c_2) = 0$, or

$$g(x,v,t) = \frac{e}{m}\frac{dg_0}{dv}\int^{t} E\left(x - vt + vt', t'\right) dt' + F(x-vt) \tag{9.16}$$

where we eliminated c_1 by (9.11). We note that $F(x-vt)$ is the homogeneous solution of (9.5).

Example. Solve the following equation

$$\frac{\partial}{\partial t} f(x,v,t) + v\frac{\partial f}{\partial x} = \frac{e}{m}E_0 \sin(k_0x - \omega_0 t)\frac{df_0(v)}{dv} \tag{9.17}$$

with the initial condition $f(x,v,t=0) = f(v,0)\cos(k_0 x)$ by characteristic method.

The characteristic equation of (9.17) reads

$$\frac{dt}{1} = \frac{dx}{v} = \frac{df}{\frac{e}{m}E_0\frac{df_0}{dv}\,\sin(k_0x - \omega_0 t)} \tag{9.18}$$

We have

$$x - vt = c_1 \tag{9.19}$$

$$df = \frac{e}{m}E_0\frac{df_0}{dv}\sin(k_0x - \omega_0 t)dt \tag{9.20}$$

Integrating gives

$$\begin{aligned} f(t' = t) - f(t' = 0) &= \frac{eE_0}{m}\frac{df_0}{dv}\int_0^t \sin\left(k_0x'(t') - \omega_0 t'\right)dt' \\ &= \frac{eE_0}{m}\frac{df_0}{dv}\int_0^t \sin\left(k_0(c_1 + vt') - \omega_0 t'\right)dt' \end{aligned}$$

Carrying out the dt' integral, we obtain

$$f(x,t,v) - f(t' = 0) = \frac{eE_0}{m}\frac{df_0}{dv}\frac{1}{\omega_0 - k_0 v}[\cos(k_0x - \omega_0 t) - \cos k_0(x - vt)] \tag{9.21}$$

where we used (9.19) and

$$f(t' = 0) = f(v,0)\cos[k_0x'(0)] = f(v,0)\cos[k_0c_1] = f(v,0)\cos[k_0(x - vt)] \tag{9.22}$$

The above two equations supply the solution. Readers are advised to solve this problem by Fourier transform and ascertain that the results agree.

• Let us consider the particular case $E = 0$ in the linear equation (9.5) and the corresponding characteristic equation (9.8). Then the characteristic equation

$$\frac{dt}{1} = \frac{dx}{v} = \frac{dg}{0} \tag{9.23}$$

solves the equation

$$\frac{\partial}{\partial t}g(x,v,t) + v\,\frac{\partial g}{\partial x} = 0 \tag{9.24}$$

which generates the unperturbed orbits $x - vt = \text{const} = c$. Equations (9.23) and (9.24) simply state that $g = \text{const} = b$ along the characteristic line $x - vt = c$. Then the general solution is $F(b,c) = 0$ with F an arbitrary function. Therefore $g = f(c) = f(x - vt)$ is the general solution, with f an arbitrary function. This trivial example is generalized in the following.

Remark on the general solution of Vlasov equation (9.1). The characteristic relations associated with the homogeneous equation (9.1) can be augmented:

$$\frac{dt}{1} = \frac{dx}{v} = \frac{dv}{-\frac{e}{m}E(x,t)} = \frac{dg}{0} \tag{9.25}$$

which immediately states that $g(x, v, t) = \text{const} = c$ along the intersections of any two integral surfaces of (9.25): c is an arbitrary function of any two independent constants derived from (9.25). For example, (9.25) gives three independent constants of integral:

$$c_1(x,t) = x - vt, \quad c_2(v,t) = v + \frac{e}{m}\int^t E(c_1 + vt', t')dt',$$

$$c_3(x,v) = \frac{v^2}{2} - \frac{e}{m}\phi(x)$$

where $-\partial\phi(x)/\partial x = E(x)$. Then (9.1) is solved by $H(c, c_1, c_2) = 0$ with H an arbitrary function, which can be written as

$$g(x,v,t) = F_1(c_1(x,t), c_2(v,t)) \tag{9.26}$$

Likewise (9.1) is solved by

$$g(x,v,t) = F_2(c_1(x,t), c_3(x,v)) \quad \text{or} \quad F_3(c_2(v,t), c_3(x,v)) \tag{9.27}$$

with arbitrary $F's$.

Exercise. Show (9.26) and (9.27) satisfy (9.1) by differentiation.

Example. Find solution of the equation

$$\frac{\partial\rho(x,t)}{\partial t} + a(\rho)\frac{\partial\rho}{\partial x} = 0 \tag{9.28}$$

The characteristic equation is

$$\frac{dt}{1} = \frac{dx}{a(\rho)} = \frac{d\rho}{0} \tag{9.29}$$

We have $\rho(x,t) = c_1$ and $x = \int^t a[\rho(x',t')]dt' + c_2$ with c_1 and c_2 constant. Since ρ is constant, $c_2 = x - a(\rho)t$. General solution of (9.28) is a function of c_1 and c_2:

$$\rho = f[x - t\, a(\rho(x,t))]$$

Exercise. Show by direct differentiation that the above equation satisfies (9.28).

One who is not entirely happy with the augmented characteristic relation (9.25) can take the following direct view of the characteristics. The left hand side of (9.28) (as well as (9.1)) is a directional derivative, and thus (9.28) (as well as (9.1)) states that ρ is constant along the direction parametrically represented by the curve $x = x(s)$ and $t = t(s)$ [this curve is the characteristic curve as would be turned out to be]. We have

$$\frac{\partial \rho(x,t)}{\partial t} + a(\rho)\frac{\partial \rho}{\partial x} = \frac{d}{ds}\rho(x(s), t(s)) = 0$$

$$\frac{d\rho}{ds} = \frac{\partial \rho}{\partial t}\frac{dt}{ds} + \frac{\partial \rho}{\partial x}\frac{dx}{ds}$$

$$\frac{dt}{ds} = 1, \quad \frac{dx}{ds} = a(\rho), \quad \frac{d\rho}{ds} = 0$$

which yield respectively

$$t = s \quad (t(s=0) \equiv 0), \quad x = \int^{s'=t} a\big(\rho(s')\big)ds' + x_0,$$

Because ρ is constant, the second equation gives $x = a(\rho)t + x_0$. Therefore, we obtain

$$\rho = \text{const} \equiv \rho(x_0) = \rho\big(x - t\, a(\rho)\big) \tag{9.30}$$

Clearly (9.30) is equivalent to (9.29).

Example. Solve the initial value problem of the equation

$$\frac{\partial v}{\partial t} + v\,\frac{\partial v}{\partial x} = -\gamma v \tag{9.31}$$

with $v(x,0) = f(x)$ and $\gamma > 0$.

First, let us transform $v = w\, e^{-\gamma t}$. Then we have

$$\frac{\partial w}{\partial t} + v\,\frac{\partial w}{\partial x} = 0 \tag{9.32}$$

whose characteristic equation is

$$\frac{dt}{1} = \frac{dx}{v} = \frac{dw}{0} \tag{9.33}$$

which states that w is constant along the orbit $v = dx/dt$. Equation (9.31) is simply

$$\frac{dv}{dt} = -\gamma v \tag{9.34}$$

where $\frac{d}{dt}$ is the Lagrangian derivative. Hence we can write the characteristic equation of (9.31) in the form

$$\frac{dt}{1} = \frac{dx}{v} = \frac{dv}{-\gamma v} \tag{9.35}$$

Equation (9.34) is integrated to

$$v = e^{-\gamma t}\ f(x_0) \tag{9.36}$$

Then the orbit equation is

$$\frac{dx}{dt} = e^{-\gamma t} f(x_0) \tag{9.37}$$

and we require $x = x_0$ at $t = 0$. The solution is

$$x = x_0 + \frac{1 - e^{-\gamma t}}{\gamma} f(x_0) \tag{9.38}$$

The nonlinearity vdv/dx gives the typical steepening of the wave profile, but the wave is damped due to the damping term $-\gamma v$.

9.1.2 *Vlasov equation vs. fluid equation*

Let us consider the Eulerian equation of motion

$$\frac{\partial v(x,t)}{\partial t} + v\frac{\partial v}{\partial x} = -\frac{e}{m}E(x,t) \tag{9.39}$$

We assume that solution v is obtained in implicit form: $g(x, t, v(x, t)) = c$ is an integral of (9.39) in the sense that the function g determines v as a function of x and t. Then by partial differentiation

$$\frac{\partial g}{\partial t} + \frac{\partial g}{\partial v}\frac{\partial v}{\partial t} = 0, \qquad \frac{\partial g}{\partial x} + \frac{\partial g}{\partial v}\frac{\partial v}{\partial x} = 0$$

Solving for $\frac{\partial v}{\partial x}$ and $\frac{\partial v}{\partial t}$, and substituting the results into (9.39) give

$$\frac{\partial g}{\partial t} + v\frac{\partial g}{\partial x} - \frac{e}{m}E(x,t)\frac{\partial g}{\partial v} = 0 \tag{9.40}$$

which is (9.1), Vlasov equation. We note that (9.39) and (9.40) have the same characteristic relation. Equation (9.39) determines the particle trajectory in *Eulerian coordinates*. This example shows that the characteristics of Vlasov equation are the particle trajectories. Conversely, if $g\big(x, v(x,t), t\big) = c$ solves (9.40), we can show that $v(x,t)$ satisfies (9.39). Therefore, Euler equation and Vlasov equation imply each other.

9.1.3 *Oscillating frame*

If a plasma is irradiated by an oscillating electric field, the zero order equilibrium distribution function is determined by

$$\frac{\partial F_\alpha(t, \mathbf{v})}{\partial t} + \frac{e_\alpha}{m_\alpha} \sin(\omega_0 t) \mathbf{E_0} \cdot \frac{\partial F_\alpha}{\partial \mathbf{v}} = 0 \tag{9.41}$$

For simplicity, we take $\mathbf{E_0} = \hat{\mathbf{x}} E_0$. Then we have

$$\frac{\partial G_\alpha(t, v)}{\partial t} + \frac{e_\alpha}{m_\alpha} \sin(\omega_0 t) E_0 \frac{\partial G_\alpha}{\partial v} = 0$$

where G_α is the one-dimensional distribution function. The characteristic relation is

$$\frac{dt}{1} = \frac{dv}{E_0 \sin(\omega_0 t) e_\alpha / m_\alpha} = \frac{dG_\alpha}{0} \tag{9.42}$$

which gives two constants of integral $G_\alpha = c_1$ and $v + \frac{e_\alpha}{m_\alpha \omega_0} E_0 \cos(\omega_0 t) = c_2$. Therefore $G_\alpha(v, t) = G_\alpha(v + \frac{e_\alpha}{m_\alpha \omega_0} E_0 \cos(\omega_0 t))$ is the general solution of (9.42). By defining a new velocity, $u_\alpha = v + \frac{e_\alpha}{m_\alpha \omega_0} E_0 \cos(\omega_0 t)$, one goes into the *oscillating frame* (Dawson and Oberman, 1962; Sanmartin, 1970). If $\mathbf{E}_0$ has all the three components, the characteristic relation reads

$$\frac{dt}{1} = \frac{dv_x}{E_{0x} \sin(\omega_0 t) e_\alpha / m_\alpha} = \frac{dv_y}{E_{0y} \sin(\omega_0 t) e_\alpha / m_\alpha} = \frac{dv_z}{E_{0z} \sin(\omega_0 t) e_\alpha / m_\alpha} = \frac{dF_\alpha}{0}$$

which gives four constants of integral: $F_\alpha = c$ and

$$v_i + \frac{e_\alpha}{m_\alpha \omega_0} E_{0i} \cos(\omega_0 t) = c_i \qquad (i = x, y, z)$$

Thus c expressed in terms of c_i is a solution of (9.41):

$$F_\alpha(\mathbf{v}, t) = F_\alpha(c_x, c_y, c_z) = F_\alpha \left(\mathbf{v} + \frac{e_\alpha}{m_\alpha \omega_0} \mathbf{E}_0 \cos\ \omega_0 t \right) \tag{9.43}$$

9.2 Initial Value Problem of Euler–Poisson Plasma

The following electron equations are the object of our nonlinear analysis in this section:

$$\frac{\partial v}{\partial t} + v \frac{\partial v}{\partial x} = -\frac{e}{m} E(x, t), \tag{9.44}$$

$$\frac{\partial n}{\partial t} + \frac{\partial}{\partial x}(nv) = 0 \tag{9.45}$$

$$\frac{\partial E}{\partial x} = 4\pi e(n_0 - n) \tag{9.46}$$

In the above Poisson equation, n_0 is the uniform background ion number density. After elimination of n between the continuity equation and the Poisson equations, we have

$$\frac{\partial}{\partial x}\left[\frac{\partial E}{\partial t}+v\frac{\partial E}{\partial x}-4\pi e n_0 v\right]=0, \tag{9.47}$$

which can immediately be integrated in the form

$$\frac{\partial E}{\partial t}+v\frac{\partial E}{\partial x}=4\pi e n_0 v+f_1(t), \tag{9.48}$$

where $f_1(t)$ is an arbitrary function of t. E can be easily eliminated between (9.44) and (9.48):

$$\left(\frac{\partial}{\partial t}+v\,\frac{\partial}{\partial x}\right)\left(\frac{\partial v}{\partial t}+v\,\frac{\partial v}{\partial x}\right)+\omega_p^2\,v=f(t), \tag{9.49}$$

where $f(t)=-ef_1(t)/m$ and ω_p is the electron plasma frequency. Equation (9.49) is exact; no approximation is involved in the expression. As a special case, we shall consider the case of $f(t)=0$. By inspection, we see that

$$\frac{\partial v}{\partial t}+v\,\frac{\partial v}{\partial x}=\pm i\omega_p v \tag{9.50}$$

solves (9.49) when $f(t)=0$. The characteristic equation of (9.50) reads

$$\frac{dt}{1}=\frac{dx}{v}=\frac{dv}{i\omega_p v}, \tag{9.51}$$

which gives

$$v=c_1e^{i\omega_p t},\quad x=\int v dt+c_2=\frac{c_1}{i\omega_p}e^{i\omega_p t}+c_2. \tag{9.52}$$

Here and in the following, the letter c denotes a constant of motion. The way to read the solution in (9.52) is that v is oscillatory, in the fashion of the first equation, along the orbit (characteristic curve), as given by the second equation. In (9.52), we simply take the real parts. The second characteristic equation, $dv/dx=i\omega_p$, is recovered from the first equation of (9.52) upon using the relation $dv/dx=(dv/dt)(dt/dx)=v^{-1}dv/dt$. The general solution to (9.50) can be constructed as an arbitrary function of the constants c_1 and c_2. We have $c_1=v\ exp(-i\omega_p t),\quad c_2=x-v/i\omega_p$. Assuming $c_1=f(c_2)$, we write

$$v(x,t)=e^{i\omega_p t}\ f\left(x-\frac{v}{i\omega_p}\right). \tag{9.53}$$

Checking that (9.53) satisfies (9.50) by direct substitution is instructive. Inspecting (9.53) at $t = 0$, we see that the function f, whatever the argument is, is the initial value of $v(x, t)$ at $t = 0$, which we write $v(x_0)$, and so it can be written as

$$v(x, t) = v(x_0)e^{i\omega_p t}. \tag{9.54}$$

The second equation of (9.52) gives, at $t = 0$, $x_0 = v(x_0)/i\omega_p + c_2$, which determines c_2 in terms of the initial values. Using this result in (9.52) yields

$$x(t) = x_0 + \frac{e^{i\omega_p t} - 1}{i\ \omega_p} v(x_0). \tag{9.55}$$

In (9.54), the x-dependence is implicit; it is given by (9.55). With the aid of the initial data $v(x_0)$, the above two equations determine $v(x, t)$ in the Eulerian sense, i.e. determine $v(x, t)$ as a field variable.

As another application of (9.53), let us put therein

$$f\left(x - \frac{v}{i\omega_p}\right) = A\ \exp\left[-ik\left(x - \frac{v}{i\omega_p}\right)\right],$$

where A, k are constants. Then, we have, letting $\alpha = -ikx + i\omega_p t$,

$$Ae^{\alpha} = v\ e^{-\frac{kv}{\omega_p}} = v\ (1 - kv/\omega_p + k^2v^2/2\omega_p^2 + \cdots)$$

The above series can be inverted in the form

$$v = Ae^{\alpha} + \frac{k}{\omega_p}A^2\ e^{2\alpha} + \frac{k^2}{2\omega_p^2}A^3 e^{3\alpha} + \cdots. \tag{9.56}$$

This indicates the well-known nonlinear harmonic generation, which is responsible for the wave steepening. Recall that the wave steepening is due to the nonlinear term $v\partial v/\partial x$ in the Euler equation.

9.3 Characteristics Calculus-Initial Value Problem

In the last section, we invoked the theory of hyperbolic partial differential equations according to which general solutions can be composed by combinations of the integration constants that are derived from the characteristic equations. In this section, we resort to an alternative analysis that solves the characteristic equations with simple calculus by employing the characteristic coordinate 's'. This method is tantamount to the *Lagrangian* solution of the given Eulerian equations. Use of the characteristic equations and ensuing simple algebraic development might dispel the mystique that loom in the

Lagrangian dynamics. The characteristic equations of (9.44) and (9.48) with $f_1(t) = 0$ read

$$\frac{dt}{1} = \frac{dx}{v} = \frac{dv}{-eE/m} = ds, \tag{9.57}$$

$$\frac{dt}{1} = \frac{dx}{v} = \frac{dE}{4\pi e n_0 v} = ds. \tag{9.58}$$

These equations give the time rates of change dv and dE along the *common* characteristics that we designated with the symbol s. The characteristic curve on the (x, t) plane will be drawn with the aid of the parameter s having the relations $x = x(s)$ and $t = t(s)$. The Eulerian increment of $v(x, t)$ is equal to the increment along the curve s, the Lagrangian increment. In fact, the parameter s is nothing but the Lagrangian time, which is not different from the Eulerian time ($ds = dt$). However, if preferable, we stick to the notation s to engrave the nature of an ordinary differentiation along the curve s. Let us consider dv/ds along the curve s:

$$\frac{d}{ds} v\Big(t(s), x(s)\Big) = \frac{\partial v}{\partial t}\frac{dt}{ds} + \frac{\partial v}{\partial x}\frac{dx}{ds} = \frac{\partial v}{\partial t} + v\frac{\partial v}{\partial x},$$

which is the Eulerian time rate of change of v.

Equations (9.57) and (9.58) take the forms

$$\frac{dv}{ds} = -\frac{e}{m}E(s), \quad \frac{dE}{ds} = 4\pi e n_0 v. \tag{9.59}$$

On combining these two, we have

$$\frac{d^2 v}{ds^2} + \omega_p^2 v = 0, \quad \frac{d^2 E}{ds^2} + \omega_p^2 E = 0. \tag{9.60}$$

These are solved to give

$$E(s) = A\ \cos\omega_p s + B\ \sin\omega_p s, \tag{9.61}$$

$$v(s) = \frac{e}{m\omega_p}[-A\ \sin\omega_p s + B\ \cos\omega_p s], \tag{9.62}$$

where A and B are constants that are independent of s, but can be functions of the initial value of x, i.e. $x(s = 0) \equiv x_0$. Integrating $dx = v ds$ gives

$$x(s) = \frac{e}{m\omega_p^2}\big[A(x_0)\ \cos\omega_p s + B(x_0)\ \sin\omega_p s\big] + C. \tag{9.63}$$

The constant C can be expressed in terms of x_0, yielding

$$x(s) = \frac{e}{m\omega_p^2}\big[A(x_0)(-1 + \cos\ \omega_p s) + B(x_0)\ \sin\omega_p s\big] + x_0. \tag{9.64}$$

A and B can be expressed in terms of the initial values $v(x_0)$ and $E(x_0)$: $A(x_0) = E(x_0)$, $B(x_0) = m\omega_p v(x_0)$. We write

$$E(s) = E(x_0)\ \cos\omega_p s + \frac{m\omega_p}{e} v(x_0)\ \sin\omega_p s, \tag{9.65}$$

$$v(s) = -\frac{e}{m\omega_p} E(x_0)\ \sin\omega_p s + v(x_0)\ \cos\omega_p s, \tag{9.66}$$

$$x(s) = \frac{e}{m\omega_p^2} E(x_0)(-1 + \cos\omega_p s) + \frac{v(x_0)}{\omega_p}\ \sin\omega_p s + x_0. \tag{9.67}$$

Equation (9.65) expresses $E(x,t)$ in which the x-dependence obeys Eq. (9.64).

In order to obtain the Eulerian representation of $E(x,t)$ in (9.67), we must invert the initial distribution $E(x_0)$ to a function of x by means of the relation in (9.67). This is a way to pose a Cauchy's initial-value problem. For simplicity, the velocity distribution is assumed to be zero. Let us assume the initial electric field to be

$$E(x_0) = a\ \sin kx_0 \quad (a = \text{const.}). \tag{9.68}$$

Then (9.67) can be written as

$$kx = kx_0 - \alpha(t)\ \sin kx_0, \quad \alpha(t) = \frac{eak}{m\omega_p^2}(1 - \cos\omega_p t). \tag{9.69}$$

Our task is to invert (9.69) to put $(\sin kx_0)$ into a function of x. We follow Infeld [E. Infeld and G. Rowlands, *Nonlinear waves, solitons and chaos*, Cambridge University Press, Cambridge (1990)]. This can be done by expanding in a Fourier sine series

$$\sin kx_0 = \sum_{n=1}^{\infty} b_n(t)\ \sin nkx, \tag{9.70}$$

where the expansion coefficients b_n is calculated according to

$$b_n = \frac{k}{\pi} \int_0^{2\pi/k} \sin kx_0\ \sin nkx\ dx. \tag{9.71}$$

Using (9.69) in (9.71) gives, per $kx_0 = y$,

$$b_n = \frac{1}{\pi} \int_0^{2\pi} \sin y\ \sin(ny - \alpha n\ \sin y)\ (1 - \alpha\ \cos y)\ dy. \tag{9.72}$$

This becomes, after integration by parts,

$$\begin{aligned} b_n &= \frac{1}{n\pi}\int_0^{2\pi} \cos y \cos(ny - \alpha n \sin y)dy \\ &= \frac{1}{2n\pi}\int_0^{2\pi} dy\Big(\cos[(n+1)y - \alpha n \sin y] + \cos[(n-1)y - \alpha n \sin y]\Big). \end{aligned} \tag{9.73}$$

With the aid of the Bessel function integral formula

$$\frac{1}{2\pi}\int_0^{2\pi} \cos(ny - \beta\ \sin y)dy = J_n(\beta), \tag{9.74}$$

we obtain

$$b_n = \frac{1}{n}\ [J_{n+1}(n\alpha) + J_{n-1}(n\alpha)] = \frac{2}{n\alpha} J_n(n\alpha). \tag{9.75}$$

Therefore, we finally obtain the Eulerian expression for E(x,t):

$$E(x,t) = 2a\ \cos\omega_p t \sum_{n=1}^{\infty} \frac{J_n(n\alpha(t))}{n\alpha(t)}\ \sin nkx. \tag{9.76}$$

In summary, Eq. (9.76) solves the simultaneous equations Eqs. (9.44) and (9.48) (with $f_1(t) = 0$) under the initial-boundary condition $v(x_0, t = 0) = 0$ and $E(x_0, t = 0) = a\ \sin kx_0$. Note that the specification of the initial distribution of the electric field (Eq. (9.68)) determines the subsequent particle distribution by the laws of mechanics.

9.4 Kinetic Theory of Waves in Magnetized Plasma

Cosmic or fusion plasma requires consideration of the effect of a magnetic field imbedded in the plasma, hence the name magnetized plasma. Plasma normal modes in a magnetized plasma have more diversity as compared to unmagnetized plasma. Vlasov equation for a magnetized plasma can be solved after exercising more labor as compared with the case without a static magnetic field therein. In isotropic plasmas, $\frac{\partial f_0(v)}{\partial \mathbf{v}} \sim \mathbf{v}$, so the magnetic force term becomes zero. However, a static magnetic field $\mathbf{B}_0$ introduces anisotropy in the plasma, and consequently, the wave solutions of Vlasov equation manifest a great complexity in the dispersion relation. Algebraically, Bessel functions, which may be considered as extended sinusoidal functions, make appearance in the analysis of magnetized plasmas, both in Vlasov equation and fluid equation. This is because we need to expand the function $\exp(i\ \sin kx)$ in a Bessel series. The *Bernstein waves* are interesting because they have no direct counterparts in the fluid plasmas and make an important entry in the list of normal modes of magnetized plasma. *Cyclotron damping (or growing)*, which is the namesake of the Landau damping (growing) in magnetized plasmas, is an important mechanism for energy transfer between the plasma particles and the electric field in magnetized plasmas. *Whistler instability* is important for explaining the very low frequency (VLF) emissions in the magnetosphere.

9.4.1 *Electrostatic wave in a magnetized plasma*

To obtain solutions of Vlasov equation for a magnetized plasma requires complicated algebra which is not trivial. We first consider an electrostatic wave propagating in a magnetized plasma with $\mathbf{B_0} = \hat{\mathbf{z}}B_0$.

$$\frac{\partial}{\partial t}f_\alpha(\mathbf{r},\mathbf{v},t) + \mathbf{v}\cdot\frac{\partial f_\alpha}{\partial \mathbf{r}} + \frac{e_\alpha}{m_\alpha}\left[\mathbf{E}(\mathbf{r},t) + \frac{\mathbf{v}}{c}\times\mathbf{B}_0\right]\cdot\frac{\partial f_\alpha}{\partial \mathbf{v}} = 0 \quad (\alpha = e,\ i) \tag{9.77}$$

We linearize this equation by putting $f = f_0 + f_1$:

Zero order:

$$\mathbf{v}\cdot\frac{\partial f_{\alpha 0}}{\partial \mathbf{r}} + \frac{e_\alpha}{m_\alpha}\frac{\mathbf{v}\times\mathbf{B_0}}{c}\cdot\frac{\partial f_{\alpha 0}}{\partial \mathbf{v}} = 0 \tag{9.78}$$

First order:

$$\frac{\partial}{\partial t}f_{\alpha 1}(\mathbf{r},\mathbf{v},t) + \mathbf{v}\cdot\frac{\partial f_{\alpha 1}}{\partial \mathbf{r}} + \frac{e_\alpha}{m_\alpha}\frac{\mathbf{v}\times\mathbf{B_0}}{c}\cdot\frac{\partial f_{\alpha 1}}{\partial \mathbf{v}} = -\frac{e_\alpha}{m_\alpha}\mathbf{E}(\mathbf{r},t)\cdot\frac{\partial f_{\alpha 0}}{\partial \mathbf{v}} \tag{9.79}$$

The static magnetic field introduces anisotropy for the plasma since the motion of charged particles along the magnetic field is different from the motion across the magnetic field. A two-temperature Maxwellian distribution which is anisotropic in the three dimensional space but is isotropic in the plane perpendicular to $\mathbf{B_0}$ is often used for such a magnetized plasma:

$$f_{\alpha 0}(v_x,v_y,v_z) = \left[\frac{m_\alpha}{2\pi T_{\alpha\|}}\right]^{\frac{1}{2}} exp\left[-\frac{m_\alpha v_z^2}{2T_{\alpha\|}}\right]\frac{m_\alpha}{2\pi T_{\alpha\perp}}exp\left[-\frac{m_\alpha(v_x^2+v_y^2)}{2T_{\alpha\perp}}\right] \tag{9.80}$$

It is readily seen that (9.80) satisfies (9.78).

9.4.2 *Solution by characteristic method*

We integrate (9.79) by means of the characteristic method. The characteristic method applied to Vlasov equation is known as *integration along the unperturbed orbit.* Here 'unperturbed orbit' means the orbit obtained from the zero order solution. The characteristic equations of (9.79) reads

$$\frac{dt}{1} = \frac{d\mathbf{r}}{\mathbf{v}} = \frac{d\mathbf{v}}{\omega_{c\alpha}\mathbf{v} \times \hat{\mathbf{z}}} = \frac{df_{\alpha 1}}{-\frac{e_\alpha}{m_\alpha}\mathbf{E} \cdot \frac{\partial f_{\alpha 0}}{\partial \mathbf{v}}} \tag{9.81}$$

where $\omega_{c\alpha} = \frac{e_\alpha B_0}{m_\alpha c}$, cyclotron frequency.

Assuming the form

$$\mathbf{E}(\mathbf{r}, t) = -i\mathbf{k}\ \phi(\mathbf{k}, \omega)e^{i\mathbf{k}\cdot\mathbf{r} - i\omega t} \tag{9.82}$$

the last relation of (9.81) gives

$$f_\alpha(t, \mathbf{r}, \mathbf{v}) = \frac{ie_\alpha}{m_\alpha}\ \phi \int_{-\infty}^{t} dt'\ \mathbf{k} \cdot \frac{\partial f_{\alpha 0}(v')}{\partial \mathbf{v}'} e^{i\mathbf{k}\cdot\mathbf{r}' - i\omega t'} \tag{9.83}$$

where the primed variables should be obtained by solving the other characteristic equations

$$\frac{d\mathbf{r}'(t')}{dt'} = \mathbf{v}', \quad \frac{d\mathbf{v}'(t')}{dt'} = \omega_{c\alpha}\mathbf{v}' \times \hat{\mathbf{z}} \tag{9.84}$$

Let the integration constant of the above equations be such that at t'=t

$$\mathbf{r}' = \mathbf{r} \tag{9.85}$$

$$\mathbf{v}' = \mathbf{v} \tag{9.86}$$

Remember that the unprimed quantities are the integration constants. To determine $\mathbf{r}'$ and $\mathbf{v}'$, (9.84) is written in components:

$$\frac{dv'_x}{dt'} = \omega_{c\alpha}v'_y, \quad \frac{dv'_y}{dt'} = -\omega_{c\alpha}v'_x, \quad \frac{dv'_z}{dt'} = 0 \tag{9.87}$$

which are solved by, according to the initial conditions in (9.85) and (9.86),

$$v'_x(t') = v_\perp \cos[\omega_{c\alpha}(t - t') + \varphi] \tag{9.88}$$

$$v'_y(t') = v_\perp \sin[\omega_{c\alpha}(t - t') + \varphi] \tag{9.89}$$

$$v'_z(t') = v_z \equiv v_\parallel : \quad \text{constant} \tag{9.90}$$

where $v_\perp$ and φ are two integration constants that are related with the integration constants v_x and v_y: $v_x = v_\perp \cos\varphi,\ v_y = v_\perp \sin\varphi$. Integrating the above three equations gives

$$x'(t') = x - \frac{v_\perp}{\omega_{c\alpha}}[\sin(\omega_{c\alpha}(t - t') + \varphi) - \sin\varphi] \tag{9.91}$$

$$y'(t') = y + \frac{v_\perp}{\omega_{c\alpha}}[\cos(\omega_{c\alpha}(t - t') + \varphi) - \cos\varphi] \tag{9.92}$$

$$z'(t') = z - v_z(t - t') \tag{9.93}$$

Here again the unprimed quantities are the integration constants. Without loss of generality, we put $\mathbf{k} = \hat{\mathbf{x}}k_\perp + \hat{\mathbf{z}}k_\parallel$, where the symbols $\perp$ and $\parallel$ are referred to the direction of the magnetic field. Substituting the above solutions into (9.83) and noting that $f_{\alpha 0}$ consists of the constants of motion, v_z and $v_\perp$, and introducing a new variable $\tau = t - t'$, we write

$$f_{\alpha 1}(\mathbf{r}, t, \mathbf{v}) = \frac{ie_\alpha}{m_\alpha}\phi(\mathbf{k}, \omega)e^{i\mathbf{k}\cdot\mathbf{r} - i\omega t}\int_0^\infty d\tau\ e^{i(\omega - k_\parallel v_\parallel)\tau}\left[k_\parallel \frac{\partial f_{\alpha 0}}{\partial v_\parallel} + k_\perp \frac{\partial f_{\alpha 0}}{\partial v_\perp}\cos(\omega_{c\alpha}\tau + \varphi)\right]$$
$$\times\ exp\left[-i\frac{k_\perp v_\perp}{\omega_{c\alpha}}[\sin(\omega_{c\alpha}\tau + \varphi) - \sin\varphi]\right] \tag{9.94}$$

In obtaining (9.94), we used the following relations for the term in (9.83)

$$\frac{\partial f_{\alpha 0}}{\partial v'_x} = \frac{\partial f_{\alpha 0}(v_\perp, v_\parallel)}{\partial v_\perp}\frac{\partial v_\perp}{\partial v'_x} = \frac{\partial f_{\alpha 0}(v_\perp, v_\parallel)}{\partial v_\perp}\ \frac{v'_x}{v_\perp}$$

In (9.94), the cos-term integral ($\equiv I$) reads

$$I = \frac{1}{2}\int_0^\infty d\tau e^{i(\omega - k_\parallel v_\parallel)\tau}\left[e^{i(\omega_c\tau + \varphi)} + e^{-i(\omega_c\tau + \varphi)}\right]e^{ia\ \sin\varphi}\ e^{-ia\sin(\omega_c\tau + \varphi)}$$

where

$$a = \frac{k_\perp v_\perp}{\omega_{c\alpha}} \tag{9.95}$$

Using the Bessel function identity,

$$e^{ia\ \sin\varphi} = \sum_{n=-\infty}^{\infty} J_n(a)e^{in\varphi}$$

we write

$$I = \frac{1}{2}\int_0^\infty d\tau \sum_l \sum_n J_l(a)J_n(a)e^{i(l-n\pm 1)\varphi}e^{i\tau(\omega - k_\parallel v_\parallel - n\omega_c \pm \omega_c)}$$

where $\pm$-terms should be summed over. Thus, we obtain

$$I = \frac{1}{2}\int_0^\infty d\tau \sum_l J_l \sum_{n'} (J_{n'+1} + J_{n'-1})\ e^{i(l-n')\varphi}\ e^{i\tau(\omega - k_\parallel v_\parallel - n'\omega_c)}$$

Using the Bessel function identity

$$J_{n+1}(a) + J_{n-1}(a) = \frac{2n}{a}J_n(a)$$

we have

$$I = \sum_l \sum_n J_l J_n \frac{n}{a} e^{i(l-n)\varphi} \int_0^\infty d\tau e^{i\tau(\omega - k_\| v_\| - n\omega_c)}$$

At later stage, we integrate over the velocity. Recall that φ is the azimuthal angle in the velocity space and only the $l = n$ terms survive in the integral I. Thus we put $l = n$ in the above to write

$$I = \sum_{n=-\infty}^{\infty} \frac{i}{\omega - k_\| v_\| - n\omega_c} \frac{n\omega_c}{k_\perp v_\perp} J_n^2(a)$$

In (9.94), $k_\|$-term can be computed similarly. We obtain the Fourier amplitude, the coefficient of the phasor $e^{i\mathbf{k}\cdot\mathbf{r} - i\omega t}$ in (9.94):

$$f_{\alpha 1}(\mathbf{k}, \omega, \mathbf{v}) = -\frac{e_\alpha}{m_\alpha} \phi(\mathbf{k}, \omega) \sum_{n=-\infty}^{\infty} \left[k_\| \frac{\partial f_{\alpha 0}}{\partial v_\|} + \frac{n\omega_{c\alpha}}{v_\perp} \frac{\partial f_{\alpha 0}}{\partial v_\perp} \right] \frac{[J_n(\frac{k_\perp v_\perp}{\omega_{c\alpha}})]^2}{\omega - k_\| v_\| - n\omega_{c\alpha}} \tag{9.96}$$

When $k_\perp = 0$, only the n=0 term is nonzero with $J_0(0) = 1$, giving

$$f_{\alpha 1}(\mathbf{k}, \omega, \mathbf{v}) = -\frac{e_\alpha}{m_\alpha} \phi(\mathbf{k}, \omega) \frac{\mathbf{k} \cdot \frac{\partial f_{\alpha 0}}{\partial \mathbf{v}}}{\omega - \mathbf{k} \cdot \mathbf{v}}$$

In this parallel propagation, (9.96) is identical with the electrostatic perturbation when $B_0 = 0$; the particles are not aware of the presence of the static magnetic field.

Equation (9.96) has the resonant structure at the Doppler-shifted wave frequency equal to the harmonics of the cyclotron frequency. The fraction can be written due to the causal principle, as

$$\frac{1}{\omega - k_\| v_\| - n\omega_{c\alpha}} = P \frac{1}{\omega - k_\| v_\| - n\omega_{c\alpha}} - i\pi\delta(\omega - n\omega_{c\alpha} - k_\| v_\|)$$

The delta function term above gives the *cyclotron damping* of the wave.

Dispersion relation is obtained by putting (9.96) into the Poisson equation (8.11):

$$\varepsilon_L = 1 + \sum_\alpha \frac{\omega_{p\alpha}^2}{k^2} \sum_{n=-\infty}^{\infty} \int d^3v \left[k_\| \frac{\partial f_{\alpha 0}}{\partial v_\|} + \frac{n\omega_{c\alpha}}{v_\perp} \frac{\partial f_{\alpha 0}}{\partial v_\perp} \right] \frac{[J_n(\frac{k_\perp v_\perp}{\omega_{c\alpha}})]^2}{\omega - k_\| v_\| - n\omega_{c\alpha}} = 0 \tag{9.97}$$

where $d^3v = 2\pi v_\perp dv_\perp dv_\parallel$. Here we designated the quantity in (9.97) as ε_L, the longitudinal dielectric constant defined by (7.29). This usage of the term can be justified when we have all the tensor elements of ε_{ij} of the magnetized plasma now under consideration. The entire tensor elements of ε_{ij} will be derived later.

To further calculate (9.97), we use (9.80) for $f_{\alpha 0}$ in which $v_x^2 + v_y^2 = v_\perp^2$. With the aid of another Bessel function identity,

$$\int_0^\infty e^{-a^2x^2} J_n^2(bx)\, x\, dx = \frac{1}{2a^2} e^{-b^2/2a^2} I_n\left(\frac{b^2}{2a^2}\right)$$

where I_n is the modified Bessel function of the nth order, the $\int dv_\perp$-integral can be carried out, giving

$$\varepsilon_L = 1 + \sum_\alpha \frac{\omega_{p\alpha}^2}{k^2} \sum_{n=-\infty}^{\infty} e^{-\mu_\alpha} I_n(\mu_\alpha) \int_{-\infty}^{\infty} dv_\parallel \frac{k_\parallel \dfrac{\partial f_{0\parallel}^\alpha}{\partial v_\parallel} - n\,\omega_{c\alpha}\, f_{0\parallel}^\alpha \, \frac{m_\alpha}{T_{\alpha\perp}}}{\omega - k_\parallel v_\parallel - n\omega_{c\alpha}}$$

$$with \quad \mu_\alpha = \frac{k_\perp^2 T_{\alpha\perp}}{\omega_{c\alpha}^2 m_\alpha} = k_\perp^2 r_{L\alpha}^2, \quad r_{L\alpha} = \sqrt{\frac{T_{\alpha\perp}}{m_\alpha \omega_{c\alpha}^2}} : \ \text{Larmor radius} \tag{9.98}$$

The integral in ε_L given by (9.98) consists of two terms. The first integral, $k_\parallel$-term, is obtained, in terms of the plasma dispersion function,

$$2\sum_\alpha \frac{\omega_{p\alpha}^2}{k^2 v_{\alpha\parallel}^2} \sum_n e^{-\mu_\alpha} I_n(\mu_\alpha) \left[1 + \frac{\omega - n\omega_{c\alpha}}{k_\parallel v_{\alpha\parallel}} Z\left(\frac{\omega - n\omega_{c\alpha}}{k_\parallel v_{\alpha\parallel}}\right)\right]$$

The second integral, $\omega_{c\alpha}$-term, is obtained

$$2\sum_\alpha \frac{\omega_{p\alpha}^2}{k^2 v_{\alpha\perp}^2} \sum_n e^{-\mu_\alpha} I_n(\mu_\alpha) \frac{n\omega_{c\alpha}}{k_\parallel v_{\alpha\parallel}} Z\left(\frac{\omega - n\omega_{c\alpha}}{k_\parallel v_{\alpha\parallel}}\right)$$

where $v_\alpha^2 = 2T_\alpha/m_\alpha$. Adding the above two equations give

$$\varepsilon_L = 1 + \sum_\alpha \frac{1}{k^2 \lambda_{D\alpha}^2} \sum_{n=-\infty}^{\infty} e^{-\mu_\alpha} I_n(\mu_\alpha)$$
$$\times \left[1 + \left(\omega - n\omega_{c\alpha} + \frac{T_{\alpha\parallel}}{T_{\alpha\perp}} n\omega_{c\alpha}\right) \frac{\sqrt{m_\alpha}}{k_\parallel \sqrt{2T_{\alpha\parallel}}} Z\left(\frac{\omega - n\omega_{c\alpha}}{k_\parallel v_{\alpha\parallel}}\right)\right] \tag{9.99}$$

where $\lambda_{D\alpha}^2 = T_{\alpha\parallel}/\omega_{p\alpha}^2$, Debye length squared.

Summary. For **parallel propagation** ($k_\perp = 0$), μ_α=0 and $I_n(0) = 0$ for all n except $n = 0$. Since $I_0(0) = 1$, (9.99) reduces to

$$\varepsilon_L = 1 + \sum_\alpha \frac{\omega_{p\alpha}}{k^2} \int_{-\infty}^{\infty} dv_\| \frac{k_\| \frac{\partial f_{0\|}^\alpha}{\partial v_\|}}{\omega - k_\| v_\|}$$

$$= 1 + \sum_\alpha \frac{1}{k^2 \lambda_{D\alpha}^2} \left[1 + \frac{\omega}{k_\| \sqrt{2T_\alpha/m_\alpha}} Z\left(\frac{\omega}{k_\| \sqrt{2T_\alpha/m_\alpha}} \right) \right] \tag{9.100}$$

which is (8.72).

For **perpendicular propagation** ($k_\| = 0$), (9.97) reduces to

$$\varepsilon_L = 1 - 2\sum_\alpha \omega_{p\alpha}^2 \frac{e^{-\mu_\alpha}}{\mu_\alpha} \sum_{n=1}^{\infty} \frac{I_n(\mu_\alpha) n^2}{\omega^2 - n^2 \omega_{c\alpha}^2} \tag{9.101}$$

where we used the Bessel function relation $I_n(x) = I_{-n}(x)$. Putting ε_L equal to zero in (9.101), we obtain the dispersion relation of *Bernstein waves* (cyclotron harmonic waves):

$$1 - 2\sum_\alpha \omega_{p\alpha}^2 \frac{e^{-\mu_\alpha}}{\mu_\alpha} \sum_{n=1}^{\infty} \frac{I_n(\mu_\alpha) n^2}{\omega^2 - n^2 \omega_{c\alpha}^2} = 0 \tag{9.102}$$

To help cross reference of (9.102), we refer to the following references: D. G. Swanson, *Plasma Waves*, Academic Press, New York (1989), Eq. (4.177); G. Bekefi, *Radiation Processes in Plasmas*, Wiley, New York (1966), Eq. (7.32); and P. C. Clemmow and J. P. Dougherty, *Electrodynamics of Particles and Plasmas*, Addison-Wesley, London (1969), Eq. (9.36). The electron Bernstein waves are obtained by putting $\omega_{pi} = 0$ in (9.102). The ω *vs* $k_\perp$ curves (suitably non-dimensionalized; ω/ω_{ce} versus $\sqrt{\mu}$) are available in Fig. 7.12, Bekefi (1966). By neglecting the higher order terms of the mass ratio m_e/m_i in (9.102) we can obtain the ion Bernstein waves. The dispersion curves of the latter are available in Figure 4.5 in Swanson (1989).

If we use the one-dimensional Maxwellian for $f_{0\|}$ with $T_\| = T_\perp$, (9.99) becomes

$$\varepsilon_L = 1 + \sum_\alpha \frac{1}{k^2 \lambda_{D\alpha}^2} \sum_{n=\infty}^{\infty} e^{-\mu_\alpha} I_n(\mu_\alpha)$$

$$\times \left[1 + \frac{\omega}{\sqrt{2\pi T_\alpha/m_\alpha}} \int_{-\infty}^{\infty} \frac{e^{-m_\alpha v_\|^2/2T_\alpha}}{k_\| v_\| - \omega + n\omega_{c\alpha}} dv_\| \right] \tag{9.103}$$

where the integral path goes along the Landau contour. When the Doppler-shifted frequency $(\omega - k_\parallel v_\parallel)$ is equal to the harmonics of the cyclotron frequency, there occurs a damping. However, unlike to the Landau damping, this *cyclotron damping* does not require the negative slope of $f_{0\alpha}$. Only the aforementioned resonance condition suffices to giving rise to the damping.

ε_L in (9.98) contains a new physics. The perpendicular component of the wave vector gives rise to the *finite Larmor radius effect* which is represented by the terms $I_n(\mu_\alpha)$. This effect owes to the presence of the harmonics of the cyclotron frequency $(n\omega_{c\alpha})$ inside the velocity integral. This effect gives rise to the new family of waves known as Bernstein waves which have no corresponding waves in cold plasmas.

• **Asymptotic behavior of dispersion relation.** We use the following asymptotic expansions for $I_n(x)$ to investigate the behavior of the dispersion relation (9.101) for large and small values of $k_\perp r_L$:

$$I_n(x) = \left(\frac{x}{2}\right)^n \sum_{m=0}^{\infty} \frac{(x/2)^{2m}}{m!(m+n)!} \qquad (x \ll 1)$$

$$I_n(x) \sim \frac{e^x}{\sqrt{2\pi x}} \qquad (x \to \infty)$$

Assume the ions are immobile $(\omega_{pi} = 0)$. When $k_\perp r_L \ll 1$, we have from (9.101)

$$1 - \frac{\omega_{\mathrm{pe}}^2}{\omega^2 - \omega_{ce}^2} - \sum_{n=2}^{\infty} \frac{1}{n!} \frac{n^2\omega_{\mathrm{pe}}^2}{\omega^2 - n^2\omega_{ce}^2} (k_\perp^2 r_L^2/2)^{n-1} = 0$$

In the limit $k_\perp \to 0$, the solutions of the above equation are

$$\omega = \sqrt{\omega_{\mathrm{pe}}^2 + \omega_{ce}^2}, \qquad \omega = n\,\omega_{ce} \tag{9.104}$$

When $k_\perp r_L \gg 1$, we have

$$1 - \sqrt{\frac{2}{\pi}}\,\mu_e^{-3/2} \sum_{n=1}^{\infty} \frac{n^2\omega_{\mathrm{pe}}^2}{\omega^2 - n^2\omega_{ce}^2} = 0$$

In the limit of $k_\perp r_L \to \infty$, the solutions of the above dispersion relation are

$$\omega = n\,\omega_{ce} \tag{9.105}$$

• **More about the Bernstein wave dispersion relation.** Let us consider the electron wave in (9.99) by putting $\omega_{pi} = 0$. Also $T_\perp = T_\parallel$. $\varepsilon_L = 0$ gives the dispersion relation

$$1+\lambda_D^2(k_\perp^2+k_\parallel^2)+\sqrt{\frac{m}{2T}}\frac{\omega}{k_\parallel}\sum_{n=-\infty}^{\infty}I_n(\mu)Z\left(\frac{\omega-n\omega_c}{k_\parallel\sqrt{\frac{2T}{m}}}\right)=0 \tag{9.106}$$

where we used the identity [see Arfken and Weber, 2001]

$$\sum_{n=-\infty}^{\infty}I_n(x)=e^x$$

To obtain the perpendicular dispersion relation, we take the limit $k_\parallel \to 0$. Taking the first term in the asymptotic expansion of the Z-function, $Z \simeq -k_\parallel\sqrt{\frac{2T}{m}}/(\omega-n\omega_c)$. Then (9.106) becomes

$$1+\lambda_D^2k_\perp^2-e^{-\mu}I_0(\mu)-\omega e^{-\mu}\sum_{n=1}^{\infty}I_n(\mu)\left(\frac{1}{\omega-n\omega_c}+\frac{1}{\omega+n\omega_c}\right)=0 \tag{9.107}$$

where we used $I_n(x)=I_{-n}(x)$. Equation (9.107) is the original Bernstein's equation for the Bernstein wave dispersion relation. [See I. B. Bernstein, *Waves in a plasma in a magnetic field*, Phys. Rev. **109** (1958), p. 10; also T. H. Stix, *Waves in Plasmas*, AIP, New York (1992) Eq. (11–87), p. 295.] The function $\lambda_D^2k_\perp^2$ is plotted versus ω/ω_c for various values of μ in the Bernstein's paper (1958).

9.4.3 *Cyclotron resonance*

Consider the following equation of motion which describes the motion of a positively charged particle acted on by an electric wave in a magnetized plasma.

$$\frac{d\mathbf{v}}{dt}=\hat{\mathbf{x}}\,\frac{e}{m}E_0\cos(k_0x-\omega_0t)+\omega_c\mathbf{v}\times\,\hat{\mathbf{z}} \tag{9.108}$$

This is a Lagrangian description, and x and $\mathbf{v}$ in the RHS are functions of t only. In the zeroth order, the particle orbit is the gyration around $\mathbf{B}_0$: $\frac{d\mathbf{v}}{dt}=\omega_c\mathbf{v}\times\hat{z}$. This is solved by $v_x=v_\perp\cos(\omega_ct+\theta), x=x_0+(v_\perp/\omega_c)\sin(\omega_ct+\theta)$ with $v_\perp$ and θ constants of integration. Using this orbit, the first order velocity is obtained by the integration,

$$v_1(t)-v_1(0)=\frac{e}{m}E_0\int_0^t\cos[k_0x(t')-\omega_0t']dt' \tag{9.109}$$

where $x(t') = x_0 + (v_\perp/\omega_c)\sin(\omega_c t' + \theta)$. Here we choose $x_0 = x$ and $\theta = -\omega_c t$ so that $x(t') = x$ when $t' = t$. This orbit is obtained by putting $\varphi = 0$ in (9.98) and (9.91). The cos-integral in (9.109) can be obtained by taking the real part of the integral,

$$\int_0^t e^{ik_0x - i\omega_0 t' + ia\sin[\omega_c(t'-t)]} = \sum_n J_n(a) e^{ik_0 x - in\omega_c t} \int_0^t e^{-i\omega_0 t' + in\omega_c t'} dt'$$

where $a = k_0 v_\perp/\omega_c$. Carrying out the integral and taking the real part yield

$$v_1(t) - v_1(0) = \frac{e}{m} E_0 \sum_n J_n(a) \frac{\sin(k_0 x - n\omega_c t) - \sin(k_0 x - \omega_0 t)}{\omega_0 - n\omega_c} \tag{9.110}$$

If $\omega_0 = n\omega_c$, $v_1(t)$ is not singular but increases as $\sim t\cos(k_0 x - \omega_0 t)$ which is the familiar resonance behavior. (Physically this is unacceptable since the velocity increases indefinitely. This is due to mathematically too a crude perturbation scheme.) In this *cyclotron resonance*, the wave energy is transferred to the kinetic energy of particles, thus increasing kinetic temperature of plasma. This method of heating is known as *radio frequency heating* of plasma by cyclotron resonance.

Remark. The same problem can be described by Eulerian equation,

$$\frac{\partial \mathbf{v}}{\partial t} + \mathbf{v} \cdot \frac{\partial \mathbf{v}}{\partial \mathbf{r}} = \hat{x}\frac{e}{m} E_0 \cos(k_o x - \omega_0 t) + \omega_c \mathbf{v} \times \hat{z} \tag{9.111}$$

In the zeroth order, we have

$$\frac{\partial \mathbf{v}^0}{\partial t} = \omega_c \mathbf{v}^0 \times \hat{z}$$

In the first order,

$$\frac{\partial \mathbf{v}^1}{\partial t} + \mathbf{v}^0 \cdot \frac{\partial \mathbf{v}^1}{\partial \mathbf{r}} = \hat{x}\frac{e}{m} E_0 \cos(k_o x - \omega_0 t)$$

whose characteristic relations are

$$\frac{dt}{1} = \frac{d\mathbf{r}}{\mathbf{v}^0} = \frac{d\mathbf{v}^1}{\frac{e}{m}\hat{\mathbf{x}} E_0 \cos(k_0 x - \omega_0 t)}$$

which is the same as (9.108).

• **Cold plasma limit.** We calculate the dielectric permittivity in the cold plasma limit from (9.97). We put the distribution function in the form

$$f_{\alpha 0}(\mathbf{v}) = f_{\alpha 0\|}(v_\|) f_{\alpha 0\perp}(v_\perp) \tag{9.112}$$

with the cold plasma distribution functions

$$f_{\alpha 0\|}(v_\|) = \delta(v_\|), \quad f_{\alpha 0}(v_\perp) = \frac{1}{2\pi v_\perp}\delta(v_\perp) \tag{9.113}$$

In the limit of $a \to 0$, we use the asymptotic relation

$$\left(\frac{nJ_n(a)}{a}\right)^2 \to \frac{1}{4}(\delta_{n,1} + \delta_{n,-1}) \quad as \quad a \to 0 \tag{9.114}$$

Then, (9.97) becomes

$$\varepsilon_L = 1 + \sum_\alpha \frac{\omega_{p\alpha}^2}{k^2} \sum_n \int d^3v \left[k_\| \frac{\partial f_{\alpha 0}(\mathbf{v})}{\partial v_\|} + \frac{n\omega_{c\alpha}}{v_\perp}\frac{\partial f_{\alpha 0}(\mathbf{v})}{\partial v_\perp}\right] \frac{(\delta_{n,1} + \delta_{n,-1})}{\omega - k_\| v_\| - n\omega_{c\alpha}} \frac{a^2}{4n^2} \tag{9.115}$$

where $d^3v = 2\pi v_\perp dv_\perp dv_\|$. The first integral vanishes due to the delta function. Then, we have

$$\varepsilon_L = 1 + \sum_\alpha \frac{\omega_{p\alpha}^2}{k^2} \frac{k_\perp^2}{4\omega_{c\alpha}} \sum_n \int d^3v \ v_\perp f_{\alpha 0\|} \frac{\partial f_{\alpha 0\perp}}{\partial v_\perp} \frac{1}{n} \frac{(\delta_{n,1} + \delta_{n,-1})}{\omega - k_\| v_\| - n\omega_{c\alpha}}$$

After some algebra and integrating by parts, one obtains

$$\varepsilon_L = 1 - \sum_\alpha \frac{k_\perp^2}{k^2} \frac{\omega_{p\alpha}^2}{\omega^2 - \omega_{c\alpha}^2} \tag{9.116}$$

For parallel propagation ($k_\perp = 0$), $J_n(0) = \delta_{n0}$. Then (9.97) becomes

$$\begin{aligned}
\varepsilon_L &= 1 + \sum_\alpha \frac{\omega_{p\alpha}^2}{k_\|} \int d^3v f_{\alpha 0\perp} \frac{\partial f_{\alpha 0\|}}{\partial v_\|} \frac{1}{\omega - k_\| v_\|} \\
&= 1 + \sum_\alpha \frac{\omega_{p\alpha}^2}{k_\|} \int_{-\infty}^{\infty} dv_\| \frac{\partial f_{\alpha 0\|}}{\partial v_\|} \frac{1}{\omega - k_\| v_\|}
\end{aligned}$$

After integrating by parts, one obtains

$$\varepsilon_L = 1 - \sum_\alpha \frac{\omega_{p\alpha}^2}{\omega^2} \tag{9.117}$$

• **Zero Larmor radius limit.** In (9.97). $k_\parallel$-term is calculated by using $J_n(0) = \delta_{n0}$, yielding

$$\sum_{n=-\infty}^{\infty} \int d^3v \, k_\parallel \frac{\partial f_{\alpha 0}(\mathbf{v})}{\partial v_\parallel} \frac{[J_n(\frac{k_\perp v_\perp}{\omega_{c\alpha}})]^2}{\omega - k_\parallel v_\parallel - n\omega_{c\alpha}} = -k_\parallel^2 \int_{-\infty}^{\infty} dv_\parallel \frac{f_{\alpha 0\parallel}(v_\parallel)}{(\omega - k_\parallel v_\parallel)^2}$$

where we integrated by parts. For the second integral in (9.97), we use (9.114), obtaining after some algebra

$$\sum_{n=-\infty}^{\infty} \int d^3v \, \frac{n\omega_{c\alpha}}{v_\perp} \frac{\partial f_{\alpha 0}}{\partial v_\perp} \frac{[J_n(\frac{k_\perp v_\perp}{\omega_{c\alpha}})]^2}{\omega - k_\parallel v_\parallel - n\omega_{c\alpha}} = -k_\perp^2 \int_{-\infty}^{\infty} dv_\parallel \frac{f_{\alpha 0\parallel}(v_\parallel)}{(\omega - k_\parallel v_\parallel)^2 - \omega_{c\alpha}^2}$$

Adding the above two equations, we obtain

$$\varepsilon_L = 1 - \sum_\alpha \frac{\omega_{p\alpha}^2}{k^2} \int_{-\infty}^{\infty} dv_\parallel f_{\alpha 0\parallel}(v_\parallel) \left[\frac{k_\parallel^2}{(\omega - k_\parallel v_\parallel)^2} + \frac{k_\perp^2}{(\omega - k_\parallel v_\parallel)^2 - \omega_{c\alpha}^2} \right] \tag{9.118}$$

• **Cyclotron emission derived from fluid equation.** In a magnetized plasma, the wave vector $\mathbf{k}$ must be assumed wholly three-dimensional: $\mathbf{k} = \hat{\mathbf{x}}k_x + \hat{\mathbf{y}}k_y + \hat{\mathbf{z}}k_z \equiv \hat{\mathbf{k}}_\perp k_\perp + \hat{\mathbf{z}}k_z$ with $\hat{\mathbf{z}} = \mathbf{B}_0/B_0$ where $\mathbf{B}_0$ designate the imbedded static magnetic field. The electrodynamic material relation takes the matrix multiplication form with the dielectric tensor $\varepsilon_{ij}(\omega, \mathbf{k})$

$$D_i(\omega, \mathbf{k}) = \varepsilon_{ij}(\omega, \mathbf{k}) E_j(\omega, \mathbf{k}) \tag{9.119}$$

$$\varepsilon_{ij}(\omega, \mathbf{k}) = \int_0^{\infty} dt \int_{-\infty}^{\infty} d^3r \; \varepsilon_{ij}(t, \mathbf{r}) \; e^{i\omega t - i\mathbf{k}\cdot\mathbf{r}} \tag{9.120}$$

where

$$\varepsilon_{ij}(t, \mathbf{r}) = \delta_{ij}\delta(\mathbf{r})\delta(t) + \chi_{ij}(t, \mathbf{r}) \tag{9.121}$$

$$\varepsilon_{ij}(\omega, \mathbf{k}) = \delta_{ij} + \chi_{ij}(\omega, \mathbf{k}) \tag{9.122}$$

The causality imposes the constraint $\chi_{ij}(t, \mathbf{r}) = 0$ for $t < 0$. Hence,

$$D_i(\mathbf{r}, t) = E_i(\mathbf{r}, t) + \int_{-\infty}^{t} \chi_{ij}(\mathbf{r}, t - t') \; E_j(\mathbf{r}, t')dt' \tag{9.123}$$

Cyclotron emission is the inverse Landau process in a magnetized plasma which is usually described in terms of the kinetic equation. Here, we show that the fluid equations can serve the purpose by adopting the view that a plasma is a collection of non-interacting beams which are distributed over the initial beam velocities. To calculate ε_{ij}, we use the following beam fluid equations. The advantage of using the fluid equations is algebraic convenience in as much as we don't have to solve the 'magnetized Vlasov equation' by introducing the Bessel functions.

$$\frac{\partial n}{\partial t} + n_0(v_0)\nabla \cdot \mathbf{v} + \mathbf{v}_0 \cdot \nabla n = 0 \tag{9.124}$$

$$\frac{\partial \mathbf{v}}{\partial t} + \mathbf{v}_0 \cdot \nabla \mathbf{v} = -\frac{e}{m}\left(\mathbf{E} + \frac{1}{c}\mathbf{v} \times \mathbf{B}_0\right) \tag{9.125}$$

When we write (9.125), we assumed that the zero order magnetic force is nil per $\mathbf{v}_0 \| \mathbf{B}_0$. The above two equations should be connected with the Poisson equation to form a complete set:

$$\nabla \cdot \mathbf{E} = -4\pi e \sum_{v_0} n(\mathbf{v}_0) = -4\pi e \int d^3 v_0 n(\mathbf{v}_0) g(\mathbf{v}_0) \tag{9.126}$$

where we introduced the beam distribution $g(\mathbf{v}_0)$. Equations (9.124)–(9.126) are reduced to $\varepsilon_L(\mathbf{k},\omega)\phi(\mathbf{k},\omega) = 0$ where $\mathbf{E} = -\nabla\phi$ and ε_L is the longitudinal dielectric permittivity

$$\varepsilon_L(\mathbf{k},\omega) = 1 - \frac{4\pi e^2}{m}\int_{-\infty}^{\infty} d^3 v_0 g(\mathbf{v}_0)\,\frac{1}{(\omega - \mathbf{k}\cdot\mathbf{v}_0)^2 - \omega_c^2}\left[1 - \frac{k_z^2}{k^2}\,\frac{\omega_c^2}{(\omega - \mathbf{k}\cdot\mathbf{v}_0)^2}\right] \tag{9.127}$$

Equation (9.127) can be recast into the form

$$\varepsilon_L(\mathbf{k},\omega) = 1 - \frac{4\pi e^2}{mk^2}\int_{-\infty}^{\infty} d^3 v_0 g(\mathbf{v}_0)\left[\frac{k_\perp^2}{(\omega - \mathbf{k}\cdot\mathbf{v}_0)^2 - \omega_c^2} + \frac{k_z^2}{(\omega - \mathbf{k}\cdot\mathbf{v}_0)^2}\right] \tag{9.128}$$

which agrees with (9.118). Since we have assumed $\mathbf{v}_0 \| \mathbf{B}_0$, the integral in (9.128) reduces to one-dimensional integral. Further reduction of the integral yields for a Maxwellian distribution $g(\mathbf{v}_0) = g(v_{0z})g_\perp(v_{0\perp})$

$$\begin{aligned}\varepsilon_L(\mathbf{k},\omega) = 1 + \frac{1}{k^2\lambda_D^2} - \int_{-\infty}^{\infty} dv_{0z} g(v_{0z})\Bigg[\frac{\omega}{k^2\lambda_D^2}\,\frac{1}{\omega - k_z v_{0z}} \\ + \frac{\omega_{pe}^2 k_\perp^2}{2\omega_c k^2}\left(\frac{1}{\omega - k_z v_{0z} - \omega_c} - \frac{1}{\omega - k_z v_{0z} + \omega_c}\right)\Bigg]\end{aligned} \tag{9.129}$$

where we note $\int d^2v_{0\perp}g_\perp(v_{0\perp}) = 1$. The fractional integrand in the above expression consists of the real part (principal value) and the imaginary part (δ-function part), from which the cyclotron emission rate can be calculated.

9.5 Dielectric Tensor for Electromagnetic Wave in Magnetized Plasma: Circular Polarization

In this section, the general form of the dielectric tensor ε_{ij} for electromagnetic wave will be derived by solving the Vlasov equation. Before doing this, we derive ε_{ij} from cold fluid equation. The fluid equation result is shown to be derived by taking the cold fluid limit of the kinetic equation version of ε_{ij}.

9.5.1 *Cold plasma dielectric tensor for electromagnetic wave*

The linearized equation of motion for cold electrons (ions are immobile) in static magnetic field $\mathbf{B_0} = \hat{\mathbf{z}}B_0$ reads

$$-i\,\omega\,\mathbf{v} = -\frac{e}{m}\mathbf{E} - \omega_c\mathbf{v}\times\hat{\mathbf{z}} \tag{9.130}$$

We have the current density

$$\mathbf{J} = -eN\mathbf{v} \tag{9.131}$$

where N is the equilibrium number density. In the above, the variables are the Fourier amplitudes with phasor $\exp(i\mathbf{k}\cdot\mathbf{r} - i\omega t)$. We also have the Maxwell equations in (8.8)–(8.11).

Solving (9.130) for $\mathbf{v}$ and using the result in (9.131), one obtains

$$J_x = \frac{ie^2N\omega}{m(\omega^2-\omega_c^2)}\left(E_x - i\frac{\omega_c}{\omega}E_y\right) \tag{9.132}$$

$$J_y = \frac{ie^2N\omega}{m(\omega^2-\omega_c^2)}\left(E_y + i\frac{\omega_c}{\omega}E_x\right) \tag{9.133}$$

$$J_z = \frac{ie^2N}{m\omega}E_z \tag{9.134}$$

Or in matrix form, we have

$$\begin{pmatrix} J_x \\ J_y \\ J_z \end{pmatrix} = \begin{pmatrix} \frac{ie^2N\omega}{m(\omega^2-\omega_c^2)} & \frac{e^2N\omega_c}{m(\omega^2-\omega_c^2)} & 0 \\ \frac{-e^2N\omega_c}{m(\omega^2-\omega_c^2)} & \frac{ie^2N\omega}{m(\omega^2-\omega_c^2)} & 0 \\ 0 & 0 & \frac{ie^2N}{m\omega} \end{pmatrix} \begin{pmatrix} E_x \\ E_y \\ E_z \end{pmatrix}$$

The matrix above is the conductivity tensor (σ_{ij}). The relation $\varepsilon_{ij} = \delta_{ij} + \frac{4\pi i}{\omega}\sigma_{ij}$ gives the dielectric tensor in the form

$$\varepsilon_{ij} = \begin{pmatrix} 1 - \frac{\omega_{\rm pe}^2}{\omega^2 - \omega_c^2} & i\frac{\omega_c}{\omega}\frac{\omega_{\rm pe}^2}{\omega^2-\omega_c^2} & 0 \\ -i\frac{\omega_c}{\omega}\frac{\omega_{\rm pe}^2}{\omega^2-\omega_c^2} & 1 - \frac{\omega_{\rm pe}^2}{\omega^2-\omega_c^2} & 0 \\ 0 & 0 & 1 - \frac{\omega_{\rm pe}^2}{\omega^2} \end{pmatrix}$$

where $\omega_c = eB_0/mc > 0$.

• **Dispersion relation.** Recall Eqs. (7.19)–(7.22). Putting the determinant in (7.22) to zero gives the dispersion relation of the electromagnetic wave propagating in the cold magnetized plasma. Equivalently we have

$$\mathbf{k} \times (\mathbf{k} \times \mathbf{E}) + \frac{\omega^2}{c^2}\varepsilon_{ij} E_j = 0 \tag{9.135}$$

This equation is obtained from (7.21) by using $J_i = \sigma_{ij}E_j$ and $\sigma_{ij} = \frac{\omega}{4\pi i}(\varepsilon_{ij} - \delta_{ij})$.

9.5.2 *Kinetic theory*

In this section, the general form of the dielectric tensor for electromagnetic wave will be derived by solving the Vlasov equation

$$\begin{aligned} &\frac{\partial}{\partial t} f_{\alpha 1}(\mathbf{r}, \mathbf{v}, t) + \mathbf{v} \cdot \frac{\partial f_{\alpha 1}}{\partial \mathbf{r}} + \frac{e_\alpha}{m_\alpha}\frac{\mathbf{v} \times \mathbf{B_0}}{c} \cdot \frac{\partial f_{\alpha 1}}{\partial \mathbf{v}} \\ &\quad = -\frac{e_\alpha}{m_\alpha}\left(\mathbf{E}(\mathbf{r}, t) + \frac{1}{c}\mathbf{v} \times \mathbf{B}(\mathbf{r}, t)\right) \cdot \frac{\partial f_{\alpha 0}}{\partial \mathbf{v}} \end{aligned} \tag{9.136}$$

where the zero order distribution $f_{\alpha 0}(\mathbf{v})$ is given by (9.80), the two-temperature Maxwellian. The above equation includes the magnetic field perturbation while (9.77) does not. The algebra is quite complicated, but we derive the well-documented result by employing the mathematical method used in the previous section.

• **Whistler wave.** As a special polarization (the direction of $\mathbf{E}$ with respect to $\mathbf{k}$) of the electromagnetic wave, we take up the *whistler wave*. The whistler mode is a typical electromagnetic wave propagating in magnetized plasmas, and is important in the ionosphere and magnetosphere research. Whistler mode is a *circularly polarized* electromagnetic wave that propagates along the direction of the static magnetic field ($\mathbf{B}_0 = \hat{\mathbf{z}}B_0$). The electric and magnetic fields of a whistler mode can be represented by

$$\mathbf{E}(z,t) = E_1[\hat{\mathbf{x}}\ \sin(kz-\omega t) + \hat{\mathbf{y}}\ \cos(kz-\omega t)] \quad (E_1 \text{ is real}) \tag{9.137}$$

$$\mathbf{B}(z,t) = B_1[\hat{\mathbf{x}}\ \cos(kz-\omega t) - \hat{\mathbf{y}}\ \sin(kz-\omega t)] \tag{9.138}$$

We have $B_1 = -\frac{ck}{\omega}E_1$ per Maxwell equation $\nabla\times\mathbf{E} = -\frac{1}{c}\frac{\partial\mathbf{B}}{\partial t}$. Note $\mathbf{E}\cdot\mathbf{B} = 0$. Alternatively we can write, putting $\omega t \to \omega t - \pi/2$

$$\mathbf{E}(z,t) = E_1[\hat{\mathbf{x}}\ \cos(kz-\omega t) - \hat{\mathbf{y}}\sin(kz-\omega t)] \tag{9.139}$$

$$\mathbf{B}(z,t) = -B_1[\hat{\mathbf{x}}\ \sin(kz-\omega t) + \hat{\mathbf{y}}\ \cos(kz-\omega t)] \tag{9.140}$$

The tip of the electric vector draws a circle on the (x,y) plane in the counter clockwise direction with the wave vector $\mathbf{k} = \hat{\mathbf{z}}k \parallel \mathbf{B}_0$ (Wave of this polarization is called *R-wave*). The electric vector in (9.139) can be obtained from

$$\mathbf{E}(z,t) = \mathrm{Re}(\hat{\mathbf{x}} + i\hat{\mathbf{y}})E_1\ e^{ikz-i\omega t} \tag{9.141}$$

where we take the real part. Before starting kinetic analysis, we work out the whistler wave dispersion relation from the fluid equation. The equation of motion for electrons (ions are assumed immobile) reads

$$\frac{d\mathbf{v}}{dt} = -\frac{e}{m}\mathbf{E} - \frac{e}{mc}\mathbf{v}\times\mathbf{B}_0 \tag{9.142}$$

In the dipole approximation, since ($E(z,t)$ does not change much over the wavelengths or wave periods, the equation of motion is solved by assuming $E \sim e^{-i\omega t}$. In components,

$$-i\omega v_x = -\frac{e}{m}E_x - \omega_{ce}v_y \tag{9.143}$$

$$-i\omega v_y = -\frac{e}{m}E_y + \omega_{ce}v_x \tag{9.144}$$

Solving the above equations,

$$v_x = \frac{-ieE_1/m}{\omega-\omega_{ce}}, \quad v_y = \frac{eE_1/m}{\omega-\omega_{ce}} \tag{9.145}$$

Thus the current is obtained as

$$\mathbf{J} = -eN\mathbf{v} = (\hat{\mathbf{x}} + i\hat{\mathbf{y}})\frac{iE_1}{\omega-\omega_{ce}}\frac{Ne^2}{m} \tag{9.146}$$

On the other hand, (7.21) with ($\mathbf{k} \cdot \mathbf{E} = 0$) gives

$$(1 - n^2)\ \mathbf{E} = -\frac{4\pi i}{\omega}\mathbf{J} \quad (n^2 = c^2k^2/\omega^2) \tag{9.147}$$

Substituting (9.141) and (9.146) into (9.147) yields the whistler dispersion relation,

$$n^2 = 1 - \frac{\omega_{\text{pe}}^2}{\omega(\omega - \omega_{ce})} \tag{9.148}$$

Exercise. When the electric vector is polarized as

$$\mathbf{E} = \text{Re}(\hat{\mathbf{x}} - i\hat{\mathbf{y}})E_1\ e^{ikz - i\omega t} \tag{9.149}$$

determine the sense of rotation and repeat the above algebra to derive the dispersion relation of the *L-wave*,

$$n^2 = 1 - \frac{\omega_{\text{pe}}^2}{\omega(\omega + \omega_{ce})} \tag{9.150}$$

Kinetic analysis. Integrating (9.136) gives

$$f_\alpha(t, \mathbf{r}, \mathbf{v}) = -\frac{e_\alpha}{m_\alpha}\int_{-\infty}^{t} dt' \left(\mathbf{E}' + \frac{1}{c}\mathbf{v}' \times \mathbf{B}'\right) \cdot \frac{\partial f_{\alpha 0}(\mathbf{v}')}{\partial \mathbf{v}'} \tag{9.151}$$

where the primed quantities are functions of t'. The wave magnetic field corresponding to the electric field (9.141) is

$$\mathbf{B}(z, t) = \text{Re}\ B_1(i\hat{\mathbf{x}} - \hat{\mathbf{y}})e^{ikz - i\omega t} \quad \left(B_1 = -\frac{ck}{\omega}E_1\right) \tag{9.152}$$

Equations (9.141) and (9.152) give

$$\mathbf{E}' + \frac{1}{c}\mathbf{v}' \times \mathbf{B}' = \left[\hat{\mathbf{x}}\left(1 - \frac{kv_z'}{\omega}\right) + i\hat{\mathbf{y}}\left(1 - \frac{kv_z'}{\omega}\right) + \hat{\mathbf{z}}\ \frac{k}{\omega}\ (v_x' + iv_y')\right] E_1 e^{ikz' - i\omega t'} \tag{9.153}$$

Where we use the orbits in (9.88)–(9.90):

$$v_x'(t') = v_\perp\ \cos[\omega_{c\alpha}(t - t') + \varphi] \tag{9.154}$$

$$v_y'(t') = v_\perp\ \sin[\omega_{c\alpha}(t - t') + \varphi] \tag{9.155}$$

$$v_z'(t') = v_z \equiv v_\parallel :\ \text{constant} \tag{9.156}$$

$$z'(t') = z - v_z(t - t') \tag{9.157}$$

Using (9.80) and (9.153) in (9.151) gives

$$f_\alpha(t,\mathbf{r},\mathbf{v}) = -\frac{e_\alpha}{m_\alpha}E_1\left[\frac{k}{\omega}\frac{\partial f_{\alpha 0}}{\partial v_\parallel} + \left(1-\frac{kv_\parallel}{\omega}\right)\frac{1}{v_\perp}\frac{\partial f_{\alpha 0}}{\partial v_\perp}\right]\int_{-\infty}^{t} dt'\ (v_x' + iv_y')e^{ikz'-i\omega t'}$$

Substituting the orbits into the time integral, we obtain

$$\int_{-\infty}^{t} dt'(\cdots) = \frac{iv_\perp e^{i\varphi}}{\omega+\omega_{c\alpha}-kv_\parallel}e^{ikz-i\omega t}$$

Note that we have no Bessel function expansions in this parallel propagation. Thus the current is obtained as

$$\begin{aligned}\mathbf{J} = {} & -iE_1 e^{ikz-i\omega t}\Sigma_\alpha \frac{e_\alpha^2}{m_\alpha}\int v_\perp dv_\perp dv_\parallel d\varphi\ (\hat{\mathbf{x}}v_x + \hat{\mathbf{y}}v_y + \hat{\mathbf{z}}v_\parallel)\frac{v_\perp e^{i\varphi}}{\omega+\omega_{c\alpha}-kv_\parallel} \\ & \times\left[\frac{k}{\omega}\frac{\partial f_{\alpha 0}}{\partial v_\parallel} + \left(1-\frac{kv_\parallel}{\omega}\right)\frac{1}{v_\perp}\frac{\partial f_{\alpha 0}}{\partial v_\perp}\right] \end{aligned} \tag{9.158}$$

Note that the z-component of the current vanishes because of the $\int d\varphi$-integral. Putting $v_x = v_\perp \cos\varphi$ and $v_y = v_\perp \sin\varphi$, the $\int d\varphi$-integral can be carried out:

$$\begin{aligned}\mathbf{J} = {} & -i\pi E_1\ (\hat{\mathbf{x}} + i\hat{\mathbf{y}})e^{ikz-i\omega t}\Sigma_\alpha \frac{e_\alpha^2}{m_\alpha}\int \frac{v_\perp^3 dv_\perp dv_\parallel}{\omega+\omega_{c\alpha}-kv_\parallel} \\ & \times\left[\frac{k}{\omega}\frac{\partial f_{\alpha 0}}{\partial v_\parallel} + \left(1-\frac{kv_\parallel}{\omega}\right)\frac{1}{v_\perp}\frac{\partial f_{\alpha 0}}{\partial v_\perp}\right] \end{aligned} \tag{9.159}$$

Substituting (9.159) into (9.147) yields the whistler dispersion relation,

$$\frac{c^2k^2}{\omega^2} = 1 + \pi\sum_\alpha \frac{\omega_{p\alpha}^2}{\omega}\int \frac{v_\perp^3 dv_\perp dv_\parallel}{\omega \pm \omega_{c\alpha} - kv_\parallel}\left[\frac{k}{\omega}\frac{\partial f_{\alpha 0}}{\partial v_\parallel} + \left(1-\frac{kv_\parallel}{\omega}\right)\frac{1}{v_\perp}\frac{\partial f_{\alpha 0}}{\partial v_\perp}\right] \tag{9.160}$$

where $\pm$ signs correspond respectively to $R(+)$ and $L(-)$ wave.

Cold plasma limit. The equilibrium distribution function $f_{\alpha 0}$ is given by (9.80). Consider the zero temperature limit. The first term in the integral ($\sim \partial f_0/\partial v_\parallel$) vanishes. The $kv_\parallel/\omega$-term also vanishes. Integrating by parts with respect to $v_\perp$ and using $\int_0^\infty v_\perp dv_\perp f_{\alpha 0}(v_\perp) = 1/2\pi$, the cold plasma limit of (9.160) is obtained to be

$$1 - \frac{c^2k^2}{\omega^2} - \Sigma_\alpha \frac{\omega_{p\alpha}^2}{\omega(\omega \pm \omega_{c\alpha})} = 0 \tag{9.161}$$

which agrees with (9.150). Let us consider the *R*-wave:

$$1 - \frac{c^2k^2}{\omega^2} - \frac{\omega_{pi}^2}{\omega(\omega + \omega_{ci})} - \frac{\omega_{\mathrm{pe}}^2}{\omega(\omega - |\omega_{ce}|)} = 0$$

Recall that $\omega_{c\alpha}$ contains sign. In the low frequency domain, $\omega \ll \omega_{ci}$, binomial expansion of the above gives

$$1 - \frac{k^2c^2}{\omega^2} - \frac{1}{\omega}\left[\frac{\omega_{pi}^2}{\omega_{ci}} - \frac{\omega_{\mathrm{pe}}^2}{|\omega_{ce}|}\right] + \frac{\omega_{pi}^2}{\omega_{ci}^2} + \frac{\omega_{\mathrm{pe}}^2}{\omega_{ce}^2} = 0$$

Note that the square bracket term is zero. Then the above equation becomes

$$1 - \frac{k^2c^2}{\omega^2} + \frac{c^2}{v_A^2} = 0 \qquad \left(v_A = \sqrt{\frac{B_0^2}{4\pi N m_i}},\ \text{Alfven velocity}\right) \tag{9.162}$$

which is the dispersion relation of the *Alfven wave.*

Example. Solve (9.142) without using the dipole approximation and modify dispersion relation (9.148).

Take the x-component of (9.142),

$$\frac{dv_x}{dt} = -\frac{e}{m}E_x - \omega_{ce}v_y$$

Integrating this,

$$v_x(t) = \int_{-\infty}^{t}\left[-\frac{e}{m}E_x(t') - \omega_{ce}v_y(t')\right]dt' = \int_{-\infty}^{t}\left[-\frac{e}{m}\tilde{E}_x - \omega_{ce}\tilde{v}_y\right]e^{ikz(t')-i\omega t'}dt'$$

where the tilded quantities are Fourier coefficients. Since v_z is a constant of motion, we have $z(t') = v_\parallel t' + z - v_\parallel t$. Using this the integral can be done:

$$v_x(z,t) = i\, e^{ikz - i\omega t}\, \frac{\frac{e}{m}\tilde{E}_x + \omega_{ce}\tilde{v}_y}{kv_\parallel - \omega} \tag{9.163}$$

Likewise we obtain

$$v_y(z,t) = i\ e^{ikz-i\omega t}\ \frac{\frac{e}{m}\tilde{E}_y - \omega_{ce}\tilde{v}_x}{kv_\parallel - \omega} \tag{9.164}$$

Without the exponential factors, the above equations give the Fourier coefficients, $\tilde{v}_{x,y}$. Solving for these, we obtain

$$\tilde{v}_x = -\frac{e}{m}\ \frac{i\omega'\tilde{E}_x - \omega_{ce}\tilde{E}_y}{\omega'^2 - \omega_{ce}^2} \tag{9.165}$$

$$\tilde{v}_y = +\frac{e}{m}\ \frac{-i\omega'\tilde{E}_y + \omega_{ce}\tilde{E}_x}{\omega'^2 - \omega_{ce}^2} \tag{9.166}$$

where $\omega' = \omega - kv_\parallel$ is the Doppler-shifted frequency. Using these the current **J** can be obtained. Then (9.147) can be written in components,

$$(1-n^2)\tilde{E}_x = \frac{-i\omega_{\rm pe}^2}{\omega(\omega'^2 - \omega_{ce}^2)}(i\omega'\tilde{E}_x + \omega_{ce}\tilde{E}_y) \tag{9.167}$$

$$(1-n^2)\tilde{E}_y = \frac{-i\omega_{\rm pe}^2}{\omega(\omega'^2 - \omega_{ce}^2)}(i\omega'\tilde{E}_y - \omega_{ce}\tilde{E}_x) \tag{9.168}$$

These are two simultaneous equations for the quantities $\tilde{E}_{x,\ y}$. Nontrivial solutions are obtained when the determinant of the coefficients are zero, giving

$$\left[1 - n^2 - \frac{\omega'\omega_{\rm pe}^2}{\omega(\omega'^2 - \omega_{ce}^2)}\right]^2 = \left[\frac{\omega_{ce}\omega_{\rm pe}^2}{\omega(\omega'^2 - \omega_{ce}^2)}\right]^2 \tag{9.169}$$

Equation (9.169) shows that the electromagnetic wave propagating along the direction of the static magnetic field has always two modes which corresponds to the $\pm$ signs when the square root is taken in the above. We obtain

$$1 - n^2 = \frac{\omega_{\rm pe}^2}{\omega(\omega - kv_\parallel \pm \omega_{ce})} \tag{9.170}$$

which are R (+sign) and L (−sign) waves. Note that the Doppler-shifted frequency enter in the dynamics.

If (9.142) is cast into the form of the corresponding Eulerian equation, we should solve

$$\frac{\partial \mathbf{v}}{\partial t} + v_\parallel \frac{\partial \mathbf{v}}{\partial z} = -\frac{e}{m}\mathbf{E} - \frac{e}{mc}\mathbf{v}\times\mathbf{B}_0 \tag{9.171}$$

This equation yields the identical results, in easier way. Clearly the Doppler-shifted frequency emerges from the left hand side of (9.171).

Exercise. In the kinetic dispersion relation (9.160), put

$$f_{\alpha 0} = f_{\alpha 0}(v_\parallel)\frac{1}{v_\perp}\delta(v_\perp) \tag{9.172}$$

where $f_{\alpha 0}(v_\parallel)$ is the one-dimensional Maxwellian. Obtain the corresponding dispersion relation.

9.6 Dielectric Tensor of Magnetized Plasma: General Polarization

In this section, the dielectric tensor ε_{ij} of electromagnetic plasma waves propagating in general direction with general polarization ($\mathbf{k}$ not necessarily parallel to $\mathbf{B}_0$ as in whistler) is obtained by solving (9.136). The starting equation is (9.136) or (9.151) with the zero order distribution function (9.80). In terms of Fourier coefficients, we have $\mathbf{B} = \frac{c}{\omega}\mathbf{k} \times \mathbf{E}$. Using this in (9.151) gives

$$f_\alpha(t, \mathbf{r}, \mathbf{v}) = -\frac{e_\alpha}{m_\alpha}\int_{-\infty}^{t} dt'\left[E_j\frac{\partial f_{\alpha 0}(\mathbf{v}')}{\partial v_j'}\left(1 - \frac{k_l v_l'}{\omega}\right) + \frac{k_l}{\omega}\frac{\partial f_{\alpha 0}(\mathbf{v}')}{\partial v_l'}v_j' E_j\right] e^{i\mathbf{k}\cdot\mathbf{r}' - i\omega t'} \tag{9.173}$$

where the primed quantities are functions of t', and E_j is the Fourier coefficient. Therefore the current becomes

$$J_i(\mathbf{r}, t) = -E_j\sum_\alpha \frac{e_\alpha^2}{m_\alpha}\int d^3v v_i\int_{-\infty}^{t} dt'\left[\frac{\partial f_{\alpha 0}(\mathbf{v}')}{\partial v_j'}\left(1 - \frac{k_l v_l'}{\omega}\right) + \frac{k_l}{\omega}\frac{\partial f_{\alpha 0}(\mathbf{v}')}{\partial v_l'}v_j'\right] e^{i\mathbf{k}\cdot\mathbf{r}' - i\omega t'} \tag{9.174}$$

As we experienced in (9.83), t and $\mathbf{r}$ are the constants of motion of the orbit equation (9.81) or (9.84); $\mathbf{r}' = \mathbf{r}$ and $\mathbf{v}' = \mathbf{v}$ at $t' = t$. We have already obtained the orbit in (9.88)–(9.93). Without loss of generality, we set $\mathbf{k} = \hat{\mathbf{x}}k_\perp + \hat{\mathbf{z}}k_\parallel$. Then the array in the square bracket above has components,

$$[\cdots]_x = \frac{U}{\omega v_\perp}v_x', \quad [\cdots]_y = \frac{U}{\omega v_\perp}v_y', \quad [\cdots]_z = \frac{\partial f_{\alpha 0}}{\partial v_\parallel} + \Gamma\, v_x'$$

where

$$U(v_\perp, v_\parallel) = \left(\omega - k_\parallel v_\parallel\right) \frac{\partial f_{\alpha 0}}{\partial v_\perp} + k_\parallel v_\perp \frac{\partial f_{\alpha 0}}{\partial v_\parallel}$$

$$\Gamma = \frac{k_\perp v_\parallel}{\omega} \left(\frac{1}{v_\perp} \frac{\partial f_{\alpha 0}}{\partial v_\perp} - \frac{1}{v_\parallel} \frac{\partial f_{\alpha 0}}{\partial v_\parallel} \right) \tag{9.175}$$

Upon making a change of variable, $t - t' = \tau$, we have

$$J_i = -E_j \; \Sigma_\alpha \frac{e_\alpha^2}{m_\alpha} \int d^3v \; v_i \; e^{-i\omega t} \int_0^\infty d\tau \; \left[\cdots\right]_j e^{ik_\perp x' + ik_\parallel z' + i\omega\tau} \tag{9.176}$$

The orbit to be used here is

$$v_x'(\tau) = v_\perp \; \cos(\omega_{c\alpha}\tau + \varphi), \quad v_y'(\tau) = v_\perp \; \sin(\omega_{c\alpha}\tau + \varphi), \quad v_z'(\tau) = v_\parallel : \text{ const.}$$

$$x'(\tau) = x - \frac{v_\perp}{\omega_{c\alpha}} [\sin(\omega_{c\alpha}\tau + \varphi) - \sin\varphi], \quad z'(\tau) = z - v_\parallel \tau$$

To evaluate (9.176), we need three kinds of integral:

$$(I_1, \; I_{vx}, \; I_{vy}) \equiv \int_0^\infty (1, \; v_x', \; v_y') e^{ik_\perp x' + ik_\parallel z' + i\omega\tau} \; d\tau$$

These are calculated by means of the type of algebra used in (9.94).

$$I_1 = ie^{ik_\perp x + ik_\parallel z} \; \Sigma_n \Sigma_l J_n(a_\alpha) J_l(a_\alpha) \; \frac{e^{i(l-n)\varphi}}{\omega - k_\parallel v_\parallel - n\omega_{c\alpha}} \tag{9.177}$$

$$I_{vx} = iv_\perp e^{ik_\perp x + ik_\parallel z} \; \Sigma_n \Sigma_l \frac{n}{a_\alpha} J_n(a_\alpha) J_l(a_\alpha) \; \frac{e^{i(l-n)\varphi}}{\omega - k_\parallel v_\parallel - n\omega_{c\alpha}} \tag{9.178}$$

$$I_{vy} = v_\perp e^{ik_\perp x + ik_\parallel z} \; \Sigma_n \Sigma_l J_n'(a_\alpha) J_l(a_\alpha) \; \frac{e^{i(l-n)\varphi}}{\omega - k_\parallel v_\parallel - n\omega_{c\alpha}} \tag{9.179}$$

where

$$a_\alpha = \frac{k_\perp v_\perp}{\omega_{c\alpha}}, \quad J_n'(a_\alpha) = \frac{dJ_n(a_\alpha)}{da_\alpha}$$

We also used

$$J_{n+1}(a) + J_{n-1}(a) = \frac{2n}{a} J_n(a), \quad J_{n+1}(a) - J_{n-1}(a) = 2J_n'(a) \tag{9.180}$$

Now we are ready to derive the conductivity tensor σ_{ij} from (9.176). The coefficient of $E_j \; e^{ik_\perp x + ik_\parallel z - i\omega t}$ is the tensor element of σ_{ij}, corresponding to the term $v_i[\cdots]_j$. We have

$$\sigma_{xx} = -i\ \Sigma \frac{e_\alpha^2}{\omega\ m_\alpha} \int d^3 v v_x U \Sigma_n \Sigma_l \frac{n}{a_\alpha} J_n J_l \frac{e^{i(n-l)\varphi}}{\omega - k_\parallel v_\parallel - n\omega_{c\alpha}}$$

Putting $v_x = v_\perp(e^{i\varphi} + e^{-i\varphi})/2$, the $\int d\varphi$ integral yields two terms $\delta_{l,\ n-1}$ and $\delta_{l,\ n+1}$. Combining these terms by (9.180) gives

$$\sigma_{xx} = -i\Sigma_\alpha \frac{e_\alpha^2}{\omega\ m_\alpha}\ \Sigma_n \int d^3 v\ v_\perp \frac{U[\frac{n}{a_\alpha} J_n]^2}{\omega - k_\parallel v_\parallel - n\omega_{c\alpha}} \tag{9.181}$$

Likewise we obtain

$$\sigma_{xy} = -\Sigma_\alpha \frac{e_\alpha^2}{\omega\ m_\alpha}\ \Sigma_n \int d^3 v\ v_\perp \frac{U\ \frac{n}{a_\alpha} J_n\ J_n'}{\omega - k_\parallel v_\parallel - n\omega_{c\alpha}} \tag{9.182}$$

$$\sigma_{yx} = -\sigma_{xy} \tag{9.183}$$

We have, using (9.175),

$$\sigma_{xz} = -\Sigma_\alpha \frac{e_\alpha^2}{m_\alpha} \Sigma_n \Sigma_l \int d^3 v v_x\ i\ J_n J_l \left[\frac{\partial f_{\alpha 0}}{\partial v_\parallel} + \Gamma v_\perp \frac{n}{a_\alpha}\right] \frac{e^{i(l-n)\varphi}}{\omega - k_\parallel v_\parallel - n\omega_{c\alpha}}$$

$$= -i\ \Sigma_\alpha \frac{e_\alpha^2}{m_\alpha \omega} \Sigma_n \int d^3 v v_\perp\ \frac{\frac{n}{a_\alpha}\ W\ J_n^2}{\omega - k_\parallel v_\parallel - n\omega_{c\alpha}} \tag{9.184}$$

where

$$W(v_\perp, v_\parallel) = (\omega - n\omega_{c\alpha}) \frac{\partial f_{\alpha 0}}{\partial v_\parallel} + n\omega_{c\alpha} \frac{v_\parallel}{v_\perp} \frac{\partial f_{\alpha 0}}{\partial v_\perp} \tag{9.185}$$

$$\sigma_{zx} = -i\ \Sigma_\alpha\ \frac{e_\alpha^2}{\omega m_\alpha}\ \Sigma_n \int d^3 v\ v_\parallel\ \frac{\frac{n}{a_\alpha}\ U\ J_n^2}{\omega - k_\parallel v_\parallel - n\omega_{c\alpha}} \tag{9.186}$$

$$\sigma_{yy} = -\Sigma \frac{e_\alpha^2}{\omega m_\alpha} \int d^3 v v_y U \Sigma_n \Sigma_l J_n' J_l \frac{e^{i(n-l)\varphi}}{\omega - k_\parallel v_\parallel - n\omega_{c\alpha}}$$

$$= -i\ \Sigma \frac{e_\alpha^2}{\omega m_\alpha}\ \Sigma_n \int d^3 v\ v_\perp \frac{U(J_n')^2}{\omega - k_\parallel v_\parallel - n\omega_{c\alpha}} \tag{9.187}$$

$$\sigma_{zz} = -\Sigma_\alpha \frac{e_\alpha^2}{m_\alpha} \Sigma_n \Sigma_l \int d^3 v v_\parallel\ i\ J_n J_l \left[\frac{\partial f_{\alpha 0}}{\partial v_\parallel} + \Gamma v_\perp \frac{n}{a_\alpha}\right] \frac{e^{i(l-n)\varphi}}{\omega - k_\parallel v_\parallel - n\omega_{c\alpha}}$$

$$= -i\ \Sigma_\alpha\ \frac{e_\alpha^2}{m_\alpha \omega}\ \Sigma_n \int d^3 v\ v_\parallel\ \frac{W\ J_n^2}{\omega - k_\parallel v_\parallel - n\omega_{c\alpha}} \tag{9.188}$$

We have, using (9.175),

$$\sigma_{yz} = -\Sigma_\alpha \frac{e_\alpha^2}{m_\alpha} \Sigma_n \Sigma_l \int d^3v \; v_y \; i \; J_n J_l \left[\frac{\partial f_{\alpha 0}}{\partial v_\parallel} + \Gamma v_\perp \frac{n}{a_\alpha} \right] \frac{e^{i(l-n)\varphi}}{\omega - k_\parallel v_\parallel - n\omega_{c\alpha}}$$

$$= \Sigma_\alpha \frac{e_\alpha^2}{m_\alpha \omega} \Sigma_n \int d^3v \; v_\perp \frac{W \; J_n J_n'}{\omega - k_\parallel v_\parallel - n\omega_{c\alpha}} \tag{9.189}$$

$$\sigma_{zy} = - \Sigma_\alpha \frac{e_\alpha^2}{\omega m_\alpha} \Sigma_n \int d^3v \; v_\parallel \frac{U \; J_n J_n'}{\omega - k_\parallel v_\parallel - n\omega_{c\alpha}} \tag{9.190}$$

Equations (9.181)–(9.190) give all the nine elements of the conductivity tensor. The above results are collected to give the following expression for the dielectric tensor

$$\varepsilon_{ij} = \delta_{ij} + \sum_\alpha \frac{\omega_{p\alpha}^2}{\omega^2} \sum_{n=-\infty}^{\infty} \int d^3v \frac{S_{ij}}{\omega - k_\parallel v_\parallel - n\omega_{c\alpha}} \tag{9.191}$$

where the tensor elements S_{ij} are given by

$$\mathbf{S_{ij}} = \begin{pmatrix} v_\perp U \left(\frac{nJ_n}{a_\alpha}\right)^2 & -iv_\perp U \; \frac{n}{a_\alpha} J_n J_n' & v_\perp W \; \frac{n}{a_\alpha} J_n^2 \\ iv_\perp U \frac{n}{a_\alpha} J_n J_n' & v_\perp U J_n'^2 & iv_\perp W J_n J_n' \\ v_\parallel U \; \frac{n}{a_\alpha} J_n^2 & -iv_\parallel U J_n J_n' & v_\parallel W J_n^2 \end{pmatrix}$$

Exercise. Show that $v_\perp W$ in S_{13} and S_{23} can be replaced by $v_\parallel U$, so that the tensor S_{ij} satisfies Onsager's relation.

Exercise. Equation (9.191) for ε_{ij} can also be put into the following form:

$$\varepsilon_{ij} = \left(1 - \sum_\alpha \frac{\omega_{p\alpha}^2}{\omega^2}\right) \delta_{ij} + \sum_\alpha \frac{\omega_{p\alpha}^2}{\omega^2} \sum_{n=-\infty}^{\infty} \int d^3v \frac{k_\parallel \frac{\partial f_{\alpha 0}}{\partial v_\parallel} + \frac{n\omega_{c\alpha}}{v_\perp} \frac{\partial f_{\alpha 0}}{\partial v_\perp}}{\omega - k_\parallel v_\parallel - n\omega_{c\alpha}} S_{ij}' \tag{9.192}$$

where S_{ij}' are given by

$$\mathbf{S_{ij}'} = \begin{pmatrix} \left(\frac{n\omega_{c\alpha}}{k_\perp} J_n\right)^2 & iv_\perp \frac{n\omega_{c\alpha}}{k_\perp} J_n J_n' & \frac{n\omega_{c\alpha}}{k_\perp} v_\parallel J_n^2 \\ -iv_\perp \frac{n\omega_{c\alpha}}{k_\perp} J_n J_n' & (v_\perp J_n')^2 & -iv_\perp v_\parallel J_n J_n' \\ \frac{n\omega_{c\alpha}}{k_\perp} v_\parallel J_n^2 & iv_\parallel v_\perp J_n J_n' & (v_\parallel J_n)^2 \end{pmatrix}$$

Hint: The following would be useful.

$$\int_0^\infty e^{-a^2x^2} J_n^2(bx)\ x\ dx = \frac{1}{2a^2}e^{-\lambda}\ I_n(\lambda) \quad (\lambda = b^2/2a^2),$$

$$\sum_{n=-\infty}^{\infty} n^2\ I_n(\lambda) = \lambda\ e^\lambda$$

$$\sum_{n=\infty}^{\infty} n^2 \int d^3v v_\perp \frac{J_n^2(a)}{a^2}\frac{\partial f_0}{\partial v_\perp} = -1$$

The dispersion relation is obtained by substituting (9.192) into Maxwell's equation:

$$\left[n^2\left(\frac{k_ik_j}{k^2} - \delta_{ij}\right) + \varepsilon_{ij}\right] E_j = 0 \tag{9.193}$$

Specializing without loss of generality as $\mathbf{k} = (k\ \sin\theta, 0, k\ \cos\theta)$, the determinant of the coefficients of E_j can be written as, after a long algebra,

$$An^4 + Bn^2 + C = 0$$

$$A = k_ik_j\varepsilon_{ij}/k^2, \quad C = det(\varepsilon_{ij})$$

$$B = \varepsilon_{xz}^2 - \varepsilon_{xx}\varepsilon_{zz} + 2\sin\theta\cos\theta\ (\varepsilon_{xy}\varepsilon_{yz} - \varepsilon_{yy}\varepsilon_{zx})$$
$$- \cos^2\theta(\varepsilon_{yy}\varepsilon_{zz} - \varepsilon_{yz}^2) - \sin^2\theta\ (\varepsilon_{xy}^2 + \varepsilon_{xx}\varepsilon_{yy})$$

It can be shown that $|E_\parallel| \gg |E_\perp|$ provided $A = 0$, where the symbols $\parallel$ and $\perp$ refer to the direction of the static magnetic field. In this sense of approximation, the electrostatic dispersion relation in a magnetized plasma reads $A = 0$. It is to be noted that the purely longitudinal wave in the strict sense ($|E_\perp| = 0$) exists only when $\theta = 0$.

Exercise. Using the tensor elements of ε_{ij}, show that Eq. (9.97) is indeed the electrostatic dielectric constant because it is equal to $\varepsilon_L = k_ik_j\varepsilon_{ij}/k^2$.

$$\begin{aligned}
\varepsilon_L &= \frac{1}{k^2}(k_x^2\varepsilon_{xx} + 2k_xk_z\varepsilon_{xz} + k_z^2\varepsilon_{zz})\\
&= 1 + \sum_\alpha \frac{\omega_{p\alpha}^2}{k^2\omega^2}\sum_n \int \frac{d^3v}{\omega - k_\parallel v_\parallel - n\omega_{c\alpha}}\\
&\quad \times \left(\frac{n^2\omega_{c\alpha}^2}{v_\perp}J_n^2\ U + \frac{2k_\parallel v_\parallel n\omega_{c\alpha}}{v_\perp}J_n^2\ U + k_\parallel^2 v_\parallel J_n^2\ W\right)
\end{aligned}$$

which reduces to (9.97).

9.6.1 *Alternative integration*

We have the Vlasov equation (9.136) which we write here again

$$\frac{\partial}{\partial t}f_\alpha(\mathbf{r},\mathbf{v},t)+\mathbf{v}\cdot\frac{\partial f_\alpha}{\partial \mathbf{r}}+\frac{e_\alpha}{m_\alpha}\frac{\mathbf{v}\times\mathbf{B_0}}{c}\cdot\frac{\partial f_\alpha}{\partial \mathbf{v}}=-\frac{e_\alpha}{m_\alpha}\left(\mathbf{E}(\mathbf{r},t)+\frac{1}{c}\mathbf{v}\times\mathbf{B}(\mathbf{r},t)\right)\cdot\frac{\partial f_{\alpha 0}}{\partial \mathbf{v}} \tag{9.194}$$

In the foregoing development, this equation was solved by characteristic method. Alternatively, we can deal with it in the cylindrical velocity space as a first order differential equation with respect to the angular coordinate φ. We Fourier-transform for the variables $\mathbf{r}$ and t. Putting $f_\alpha \sim e^{i\mathbf{k}\cdot\mathbf{r}-i\omega t}$, we have

$$(-i\omega+i\mathbf{k}\cdot\mathbf{v})f(\mathbf{k},\mathbf{v},\omega)+\omega_c\mathbf{v}\times\hat{\mathbf{z}}\cdot\frac{\partial f}{\partial \mathbf{v}}=-\frac{e}{m}\left(\mathbf{E}(\mathbf{k},\omega)+\frac{1}{c}\mathbf{v}\times\mathbf{B}(\mathbf{k},\omega)\right)\cdot\frac{\partial f_0}{\partial \mathbf{v}} \tag{9.195}$$

where we omitted the subscript α denoting the species, which will be reinstated at the final result. Introducing the cylindrical coordinates in the velocity space

$$\mathbf{v}=\hat{\mathbf{v}}_\perp v_\perp+\hat{\mathbf{z}}v_z$$

where

$$\hat{\mathbf{v}}_\perp=\hat{\mathbf{x}}\ \cos\varphi+\hat{\mathbf{y}}\ \sin\varphi$$
$$\hat{\varphi}=-\hat{\mathbf{x}}\ \sin\varphi+\hat{\mathbf{y}}\ \cos\varphi$$

and the $\hat{\mathbf{z}}$-direction is the direction of the static magnetic field $\mathbf{B}_0$. Then we have

$$\frac{\partial f}{\partial \mathbf{v}}=\hat{\mathbf{v}}_\perp\frac{\partial f}{\partial v_\perp}+\frac{\hat{\varphi}}{v_\perp}\frac{\partial f}{\partial\varphi}+\hat{\mathbf{z}}\frac{\partial f}{\partial v_z}$$

Hence, (9.195) takes the form

$$\frac{\partial f}{\partial\varphi}+\frac{i(\omega-\mathbf{k}\cdot\mathbf{v})}{\omega_c}f=\frac{e}{m\omega_c}\left(\mathbf{E}(\mathbf{k},\omega)+\frac{1}{c}\mathbf{v}\times\mathbf{B}(\mathbf{k},\omega)\right)\cdot\frac{\partial f_0}{\partial \mathbf{v}} \tag{9.196}$$

This equation is an ordinary first order differential equation in the angular coordinate φ. Without loss of generality, we set $\mathbf{k}=\hat{\mathbf{x}}k_x+\hat{\mathbf{z}}k_z$. Multiplying the integration factor, the above equation can be written as

$$\frac{d}{d\varphi}\left(f_\alpha\ e^{-i\Psi_\alpha(\varphi)}\right)=\frac{e_\alpha}{m_\alpha\omega_{c\alpha}}e^{-i\Psi_\alpha(\varphi)}\ Q_\alpha(\varphi) \tag{9.197}$$

$$\Psi_\alpha(\varphi) = \frac{-(\omega - k_z v_z)\varphi + k_x v_\perp \ \sin\varphi}{\omega_{c\alpha}}$$

$$Q_\alpha(\varphi) = \left(\mathbf{E}(\mathbf{k},\omega) + \frac{1}{c}\mathbf{v} \times \mathbf{B}(\mathbf{k},\omega)\right) \cdot \frac{\partial f_{\alpha 0}}{\partial \mathbf{v}}$$

Integrating (9.197) from $\varphi = \varphi_0$ to φ gives

$$\left[f_\alpha \ e^{-i\Psi_\alpha(\varphi)}\right]_{\varphi_0}^{\varphi} = \frac{e_\alpha}{m_\alpha \omega_{c\alpha}} \int_{\varphi_0}^{\varphi} e^{-i\Psi_\alpha(\varphi')} \ Q_\alpha(\varphi')d\varphi' \tag{9.198}$$

The exponential term $\exp(-i\Psi_\alpha(\varphi))$ vanishes at $\varphi = +\infty$ ($\varphi = -\infty$) when $\omega_{c\alpha} > 0$ ($\omega_{c\alpha} < 0$) if we assume ω has a small positive imaginary part. Thus we obtain

$$f_\alpha(\varphi) = \frac{e_\alpha}{m_\alpha \omega_{c\alpha}} e^{i\Psi_\alpha(\varphi)} \int_{\pm\infty}^{\varphi} e^{-i\Psi_\alpha(\varphi')} Q_\alpha(\varphi')d\varphi' \tag{9.199}$$

where $+\infty(-\infty)$ corresponds to ion (electron) equation.

Using (9.199), we can determine the dielectric permittivity tensor. The algebra along this line is referred to the references:

A. Akhiezer, I. Akhiezer, R. Polovin, A. Sitenko, K. Stepanov, *Collective Oscillations in a plasma*, Pergamon, Oxford (1967).

D. C. Montgomery and D. A. Tidman, *Plasma kinetic theory*, McGraw-Hill (1964).

Exercise. Carry out the integral in (9.199) to show that the result agrees with (9.96).

9.6.2 *Cold plasma limit of the kinetic dielectric tensor*

Since the Larmor radius $v_\perp/\omega_{c\alpha} \to 0$, in the cold plasma, we first should take the limit $a_\alpha \to 0$ in the matrix S_{ij} in (9.191). Use can be made of: $J_n(a) \to \delta_{n,0}$, $\frac{n}{a}J_n(a) = \frac{1}{2}[J_{n+1}(a) + J_{n-1}(a)] \to \frac{1}{2}(\delta_{n,1} + \delta_{n,-1})$, $J_n'(a) = \frac{1}{2}[J_{n-1}(a) - J_{n+1}(a)] \to \frac{1}{2}(\delta_{n,1} - \delta_{n,-1})$, $[\frac{nJ_n(a)}{a}]^2 \to \frac{1}{4}(\delta_{n,1} + \delta_{n,-1})$, $\frac{n}{a}J_n(a)J_n'(a) \to \frac{1}{4}(\delta_{n,1} - \delta_{n,-1})$, $\frac{n}{a}J_n(a)J_n(a) \to 0$, $J_n(a)J_n'(a) \to 0$.

Thus we have in (9.191)

$$\mathbf{S_{ij}} = \begin{pmatrix} v_\perp U \ \frac{1}{4}(\delta_{n,1} + \delta_{n,-1}) & iv_\perp U \ \frac{1}{4}(\delta_{n,1} - \delta_{n,-1}) & 0 \\ iv_\perp U \ \frac{1}{4}(\delta_{n,-1} - \delta_{n,1}) & v_\perp U \ \frac{1}{4}(\delta_{n,-1} + \delta_{n,1}) & 0 \\ 0 & 0 & v_\parallel W \delta_{n,0} \end{pmatrix}$$

$$U(v_\perp, v_\parallel) = \left(\omega - k_\parallel v_\parallel\right) \frac{\partial f_{\alpha 0}}{\partial v_\perp} + k_\parallel v_\perp \frac{\partial f_{\alpha 0}}{\partial v_\parallel}$$

$$W(v_\perp, v_\parallel) = (\omega - n\omega_{c\alpha})\frac{\partial f_{\alpha 0}}{\partial v_\parallel} + n\omega_{c\alpha}\frac{v_\parallel}{v_\perp}\frac{\partial f_{\alpha 0}}{\partial v_\perp}$$

This form of S_{ij} is also valid for the parallel propagation, $\mathbf{k} = k\mathbf{B}_0/B_0$. The operations of taking the limit $a \to 0$ and $\int d^3v$ with the delta function distribution is commutative. Next the tensor elements ε_{ij} are evaluated by using the delta function distribution in cylindrical coordinates,

$$f_0(\mathbf{v}) = \delta(\mathbf{v}) = \frac{1}{v_\perp}\delta(v_\perp)\delta(\varphi)\delta(v_\parallel)$$

Using the above equations in (9.191) gives, for example,

$$\varepsilon_{11} = 1 + \frac{1}{4}\Sigma\,\frac{\omega_{p\alpha}^2}{\omega^2}\int d^3v \left[(\omega - k_\parallel v_\parallel)\frac{\partial f_{0\alpha}}{\partial v_\perp} + k_\parallel v_\perp \frac{\partial f_{0\alpha}}{\partial v_\parallel}\right]$$
$$\times\left(\frac{1}{\omega - k_\parallel v_\parallel - \omega_{c\alpha}} + \frac{1}{\omega - k_\parallel v_\parallel + \omega_{c\alpha}}\right)$$

where $d^3v = v_\perp dv_\perp dv_\parallel d\varphi$. The second integral vanishes. Integrating by parts in the first integral, one obtains $\varepsilon_{11} = 1 - \Sigma\frac{\omega_{p\alpha}^2}{\omega^2 - \omega_{c\alpha}^2}$. The rest of the matrix elements are likewise evaluated, giving

$$\varepsilon_{\mathbf{ij}} = \begin{pmatrix} 1 - \Sigma\frac{\omega_{p\alpha}^2}{\omega^2-\omega_{c\alpha}^2} & -i\,\Sigma\,\frac{\omega_{c\alpha}}{\omega}\,\frac{\omega_{p\alpha}^2}{\omega^2-\omega_{c\alpha}^2} & 0 \\ i\,\Sigma\,\frac{\omega_{c\alpha}}{\omega}\,\frac{\omega_{p\alpha}^2}{\omega^2-\omega_{c\alpha}^2} & 1 - \Sigma\frac{\omega_{p\alpha}^2}{\omega^2-\omega_{c\alpha}^2} & 0 \\ 0 & 0 & 1 - \,\Sigma\,\frac{\omega_{p\alpha}^2}{\omega^2} \end{pmatrix}$$

which agrees with the earlier result in Section 9.5.1.

9.7 Kinetic Theory of Magnetohydrodynamic Waves

Reference: S. I. Braginskii and A. P. Kazantsev, *Magnetohydrodynamic waves in a rarefied plasma*, (work done in 1957; the Journal of this article is unknown to the author of this book.)

In this section, we derive MHD waves from Vlasov equation. MHD waves look somewhat detached from the kinetic waves derived from the Vlasov–Maxwell equations in a magnetized plasma. In fact, MHD waves are subset of the general kinetic waves whose frequency range is restricted to the low frequency $\omega \ll \omega_{ci}$, the ion cyclotron frequency. To appreciate this structure of plasma physics, it may be worthy to derive the MHD waves from the Vlasov equation. This can be done by calculating the low frequency limit of the dielectric tensor obtained in Section 9.6; but the calculation involves power series expansion of the the plasma dispersion function and the asymptotic expansion of Bessel functions, thus working with the double series renders the algebra quite cumbersome. Here we present Braginskki and Kazantsev's derivation which solves the Vlasov equation in a magnetized plasma by expanding the distribution function in the harmonic series of the azimuthal angle.

We begin with the linearized Vlasov equation governing a magnetized plasma

$$\frac{\partial}{\partial t} f_\alpha(\mathbf{r}, \mathbf{v}, t) + \mathbf{v} \cdot \frac{\partial f_\alpha}{\partial \mathbf{r}} + \frac{e_\alpha}{m_\alpha} \frac{\mathbf{v} \times \mathbf{B_0}}{c} \cdot \frac{\partial f_\alpha}{\partial \mathbf{v}} = -\frac{e_\alpha}{m_\alpha} \mathbf{E}(\mathbf{r}, t) \cdot \frac{\partial F_\alpha}{\partial \mathbf{v}} \tag{9.200}$$

In fully electromagnetic analysis, the wave magnetic field term $(\mathbf{v} \times \mathbf{B})$ should be retained on the right hand side, but for MHD wave that term is neglected assuming that $v/c \ll 1$; MHD is a slow process and v can be assumed to be $\ll c$. Fourier transforming with respect to the spatial coordinates, we write

$$-i\omega f_\alpha(\mathbf{k}, \omega, \mathbf{v}) + i\mathbf{k} \cdot \mathbf{v} f_\alpha + \omega_{c\alpha} \mathbf{v} \times \hat{\mathbf{z}} \cdot \frac{\partial f_\alpha}{\partial \mathbf{v}} = -\frac{e_\alpha}{m_\alpha} \mathbf{E}(\mathbf{k}, \omega) \cdot \frac{\partial F_\alpha}{\partial \mathbf{v}} \tag{9.201}$$

where we assumed $\mathbf{B}_0 = \hat{\mathbf{z}} B_0$ and F_α is zero order equilibrium distribution function

$$F_\alpha(\mathbf{v}) = F_{\alpha\|}(v_z) F_{\alpha\perp}(v_\perp) \tag{9.202}$$

with

$$F_{\alpha\|}(v_z) = \sqrt{\frac{m_\alpha}{2\pi T_\alpha}} e^{-\frac{m_\alpha v_z^2}{2T_\alpha}}, \quad F_{\alpha\perp}(v_\perp) = \frac{m_\alpha}{2\pi T_\alpha} e^{-\frac{m_\alpha v_\perp^2}{2T_\alpha}} \tag{9.203}$$

where $v_\perp^2 = v_x^2 + v_y^2$. We also assume that the y coordinate is ignorable, thus $\mathbf{k} = \hat{\mathbf{x}} k_x + \hat{\mathbf{z}} k_z$. Introducing the cylindrical coordinate in the velocity space, (9.201) becomes

$$(-i\omega + ik_x v_\perp \cos\varphi + ik_z v_z) f_\alpha - \omega_{c\alpha} \frac{\partial f_\alpha}{\partial \varphi} = \frac{e_\alpha}{T_\alpha} F_\alpha (v_\perp E_x \cos\varphi + v_\perp E_y \sin\varphi + E_z v_z) \tag{9.204}$$

Let us expand f_α in the series of harmonic functions:

$$f_\alpha = \sum_{m=-\infty}^{\infty} f_m(v_\perp, v_z) e^{im\varphi}$$

Substituting this series into (9.204) and applying the orthogonality of the harmonic function $e^{im\varphi}$ give

$$e^{i0} : -i(\omega - k_z v_z) f_0 + \frac{i}{2} k_x v_\perp (f_1 + f_{-1}) = \frac{e_\alpha}{T_\alpha} F_\alpha E_z v_z \tag{9.205}$$

$$e^{i\varphi} : \ (-i\omega + ik_z v_z - i\omega_c) f_1 + \frac{i}{2} k_x v_\perp (f_0 + f_2) = \frac{e_\alpha}{T_\alpha} \frac{v_\perp}{2} F_\alpha (E_x - iE_y) \tag{9.206}$$

$$e^{-i\varphi} : \ (-i\omega + ik_z v_z + i\omega_c) f_{-1} + \frac{i}{2} k_x v_\perp (f_0 + f_{-2}) = \frac{e_\alpha}{T_\alpha} \frac{v_\perp}{2} F_\alpha (E_x + iE_y) \tag{9.207}$$

and so on. We truncate the infinite coupled equations by setting $f_{\pm 2} = f_{\pm 3} = \cdots = 0$. By the truncation of the higher harmonics terms, we exclude the *finite Larmor radius effect*. Putting $f_{\pm 2} = 0$ in (9.205)–(9.207), we have three equations for three unknowns f_0 and $f_{\pm 1}$. Let

$$f_x = \frac{1}{2}(f_1 + f_{-1}), \quad f_y = \frac{i}{2}(f_1 - f_{-1}) \tag{9.208}$$

Then (9.205)–(9.207) give

$$\omega' f_x - i\omega_c \, f_y - \frac{k_x v_\perp}{2} f_0 = \frac{ie_\alpha}{2T_\alpha} v_\perp F_\alpha E_x \tag{9.209}$$

$$i\omega_c f_x + \omega' f_y = \frac{ie_\alpha}{2T_\alpha} v_\perp F_\alpha E_y \tag{9.210}$$

$$\omega' f_0 + k_x v_\perp f_x = \frac{ie_\alpha}{T_\alpha} v_z F_\alpha E_z \tag{9.211}$$

where $\omega' = \omega - k_z v_z$ is the Doppler-shifted frequency.

The above equations are solved for f_0, f_x, f_y, giving

$$f_0 = \frac{e_\alpha F_\alpha}{T_\alpha} \left[-i \frac{k_x v_\perp^2}{2\omega_{c\alpha}^2} E_x + \frac{k_x v_\perp^2}{2\omega' \omega_{c\alpha}} E_y + i \frac{v_z}{\omega'} E_z \right] \tag{9.212}$$

$$f_x = \frac{e_\alpha F_\alpha v_\perp}{2T_\alpha \omega_{c\alpha}} \left[-i \frac{\omega'}{\omega_{c\alpha}} E_x + E_y - i \frac{k_x v_z}{\omega_{c\alpha}} E_z \right] \tag{9.213}$$

$$f_y = \frac{-e_\alpha F_\alpha v_\perp}{2T_\alpha \omega_{c\alpha}} \left[E_x + i \frac{\omega'}{\omega_{c\alpha}} E_y + \frac{k_x v_z}{\omega'} E_z \right] \tag{9.214}$$

where we neglected $k_x^2 v_\perp^2$, ω'^2 as compared with $\omega_{c\alpha}^2$. This approximation is consistent with neglecting the finite Larmor radius effect.

Next, we calculate the current

$$\mathbf{J} = \sum_\alpha e_\alpha \int_0^{2\pi} d\varphi \int_{-\infty}^{\infty} dv_z \int_0^{\infty} v_\perp dv_\perp \mathbf{v} f_\alpha \tag{9.215}$$

where

$$\begin{aligned} f_\alpha &= f_0 + f_1 e^{i\varphi} + f_{-1} e^{-i\varphi} \\ &= f_0 + 2 f_x \cos\varphi + 2 f_y \sin\varphi \end{aligned}$$

Thus we can write the relation with the conductivity σ_{ij} in the form

$$J_x = 2\pi \sum_\alpha e_\alpha \int dv_z \int dv_\perp v_\perp^2 f_x = \sigma_{xx} E_x + \sigma_{xy} E_y + \sigma_{xz} E_z \tag{9.216}$$

$$J_y = 2\pi \sum_\alpha e_\alpha \int dv_z \int dv_\perp v_\perp^2 f_y = \sigma_{yx} E_x + \sigma_{yy} E_y + \sigma_{yz} E_z \tag{9.217}$$

$$J_z = 2\pi \sum_\alpha e_\alpha \int dv_z v_z \int dv_\perp v_\perp f_0 = \sigma_{zx} E_x + \sigma_{zy} E_y + \sigma_{zz} E_z \tag{9.218}$$

In the following, the conductivity tensor components are calculated by using (9.212)–(9.214).

$$\sigma_{xx} = -i\pi \sum_\alpha \frac{e_\alpha^2}{T_\alpha \omega_{c\alpha}^2} \int_{-\infty}^{\infty} dv_z \omega' \int_0^{\infty} dv_\perp v_\perp^3 F_\alpha \tag{9.219}$$

$$\sigma_{xy} = \pi \sum_\alpha \frac{e_\alpha^2}{T_\alpha \omega_{c\alpha}} \int_{-\infty}^{\infty} dv_z \int_0^{\infty} dv_\perp v_\perp^3 F_\alpha \tag{9.220}$$

$$\sigma_{xz} = 0 \tag{9.221}$$

$$\sigma_{yx} = -\sigma_{xy} \tag{9.222}$$

$$\sigma_{yy} = \sigma_{xx} \tag{9.223}$$

$$\sigma_{yz} = -\pi\, k_x \sum_\alpha \frac{e_\alpha^2}{T_\alpha \omega_{c\alpha}} \int_{-\infty}^{\infty} dv_z \frac{v_z}{\omega'} \int_0^{\infty} dv_\perp v_\perp^3 F_\alpha \tag{9.224}$$

$$\sigma_{zx} = 0 \tag{9.225}$$

$$\sigma_{zy} = -\sigma_{yz} \tag{9.226}$$

$$\sigma_{zz} = 2\pi i \sum_\alpha \frac{e_\alpha^2}{T_\alpha} \int_{-\infty}^{\infty} dv_z \frac{v_z^2}{\omega'} \int_0^{\infty} dv_\perp v_\perp F_\alpha \tag{9.227}$$

For the velocity integrals, the following formulas are useful:

$$I(n) \equiv \int_0^\infty e^{-\frac{m}{2T}x^2} x^n dx$$

$$I(0) = \frac{1}{2}\sqrt{\frac{2\pi T}{m}}, \quad I(1) = \frac{T}{m}, \quad I(2) = \frac{T}{2m}\sqrt{\frac{2\pi T}{m}}, \quad I(3) = 2\left(\frac{T}{m}\right)^2$$

$$F_0(v_z) = \sqrt{\frac{m}{2\pi T}} e^{-\frac{mv_z^2}{2T}}$$

$$A(n) \equiv \int_{-\infty}^\infty dv_z \, \frac{v_z^n F_0(v_z)}{\omega - k_z v_z}, \quad \zeta = \sqrt{\frac{m}{2T}} \frac{\omega}{k_z}$$

$$A(0) = -\frac{1}{k_z}\sqrt{\frac{m}{2T}} Z(\zeta), \quad A(1) = -\frac{1}{k_z}(1 + \zeta Z(\zeta)), \quad A(2) = -\frac{\omega}{k_z^2}(1 + \zeta Z(\zeta))$$

where

$$Z(\zeta) = \frac{1}{\sqrt{\pi}} \int_{-\infty}^\infty \frac{e^{-x^2}}{x - \zeta} \, dx$$

is the plasma dispersion function. Using the above formulas, one obtains

$$\sigma_{xx} = -i\omega \sum_\alpha \frac{e_\alpha^2}{m_\alpha \omega_{c\alpha}^2} \tag{9.228}$$

$$\sigma_{xy} = \sum_\alpha \frac{e_\alpha^2}{m_\alpha \omega_{c\alpha}} = 0 \tag{9.229}$$

$$\sigma_{yz} = \frac{k_x}{k_z} \sum_\alpha \frac{e_\alpha^2}{m_\alpha \omega_{c\alpha}} \zeta_\alpha Z(\zeta_\alpha) \tag{9.230}$$

$$\sigma_{zz} = -i \, \frac{\omega}{k_z^2} \sum_\alpha \frac{e_\alpha^2}{T_\alpha}(1 + \zeta_\alpha Z(\zeta_\alpha)) \tag{9.231}$$

Then the dielectric tensor

$$\varepsilon_{ij} \equiv \frac{4\pi i}{\omega} \sigma_{ij} \tag{9.232}$$

are obtained as

$$\varepsilon_{xx} = \sum_\alpha \frac{\omega_{p\alpha}^2}{\omega_{c\alpha}^2} \simeq \frac{\omega_{pi}^2}{\omega_{ci}^2} = \varepsilon_{yy} \tag{9.233}$$

$$\varepsilon_{xy} = \varepsilon_{yx} = \varepsilon_{xz} = \varepsilon_{zx} = 0 \tag{9.234}$$

$$\varepsilon_{yz} = -\varepsilon_{zy} = i\frac{k_x}{k_z}\sum_\alpha \frac{\omega_{p\alpha}^2}{\omega\omega_{c\alpha}}\zeta_\alpha Z(\zeta_\alpha) \tag{9.235}$$

$$\varepsilon_{zz} = \frac{1}{k_z^2}\sum_\alpha \frac{4\pi N e_\alpha^2}{T_\alpha}(1+\zeta_\alpha Z(\zeta_\alpha)) \tag{9.236}$$

We further evaluate the dielectric tensor components in (9.235) and (9.236) in the frequency range

$$\sqrt{\frac{T_i}{m_i}} \ll \frac{\omega}{k} \ll \sqrt{\frac{T_e}{m_e}} \tag{9.237}$$

where $\zeta_i \gg 1$ and $\zeta_e \ll 1$. The corresponding asymptotic series for the plasma dispersion function $Z(\zeta)$ are as follows:

$$Z(\zeta_i) \simeq -\frac{1}{\zeta_i} - \frac{1}{2}\frac{1}{\zeta_i^3} \ (\zeta_i \gg 1)\,, \quad Z(\zeta_e) \simeq -2\zeta_e\left(1-\frac{2}{3}\zeta_e^2\right) \ \ (\zeta_e \ll 1)$$

Then, we obtain

$$\varepsilon_{yz} \simeq -i\frac{k_x}{k_z}\frac{\omega_{pi}^2}{\omega\omega_{ci}} \tag{9.238}$$

$$\varepsilon_{zz} \simeq -\frac{\omega_{pi}^2}{\omega^2} + \frac{1}{k_z^2\lambda_e^2} \tag{9.239}$$

where $\lambda_e = \sqrt{T_e/4\pi N e^2}$ is the electron Debye length.

In summary, we obtain the magnetohydrodynamic dielectric tensor in the form

$$\varepsilon_{ij} = \begin{pmatrix} \epsilon_{xx} & 0 & 0 \\ 0 & \epsilon_{yy} & \epsilon_{yz} \\ 0 & -\epsilon_{yz} & \epsilon_{zz} \end{pmatrix} \tag{9.240}$$

where the tensor elements are given by (9.233), (9.238), and (9.239).

• **Cold plasma.** We use the cold plasma distribution function in the above conductivity tensor.

$$F_\alpha(\mathbf{v}) = \delta(v_z)\,\frac{\delta(v_\perp)}{2\pi v_\perp} \tag{9.241}$$

We can show from (41) that $\varepsilon_{yz} = 0$. In (42), we expand $Z(\zeta) = -\frac{1}{\zeta} - \frac{1}{2}\frac{1}{\zeta^3}$ for both electrons and ions. Then we obtain $\varepsilon_{zz} = -\omega_{\rm pe}^2/\omega^2$. Hence, for a cold plasma

$$\varepsilon_{ij} = \begin{pmatrix} \frac{\omega_{pi}^2}{\omega_{ci}^2} & 0 & 0 \\ 0 & \frac{\omega_{pi}^2}{\omega_{ci}^2} & 0 \\ 0 & 0 & -\frac{\omega_{pe}^2}{\omega^2} \end{pmatrix} \tag{9.242}$$

Equation (9.242) should be compared with the general expression for the cold plasma dielectric tensor.

$$\varepsilon_{ij} = \begin{pmatrix} 1 - \Sigma\frac{\omega_{p\alpha}^2}{\omega^2-\omega_{c\alpha}^2} & i\Sigma\frac{\omega_{c\alpha}}{\omega}\frac{\omega_{p\alpha}^2}{\omega^2-\omega_{c\alpha}^2} & 0 \\ -i\Sigma\frac{\omega_{c\alpha}}{\omega}\frac{\omega_{p\alpha}^2}{\omega^2-\omega_{c\alpha}^2} & 1 - \Sigma\frac{\omega_{p\alpha}^2}{\omega^2-\omega_{c\alpha}^2} & 0 \\ 0 & 0 & 1 - \Sigma\frac{\omega_{p\alpha}^2}{\omega^2} \end{pmatrix} \tag{9.243}$$

Using the approximation $\omega^2 \ll \omega_{c\alpha}^2$ and $m_e \ll m_i$ in (9.243) gives (9.242). The unity in the diagonal elements is due to the high frequency term $\partial \mathbf{E}/\partial t$ that we omitted in the Maxwell equation.

• **Dispersion relation.** Dispersion relation of the magnetosonic wave is obtained from the relation

$$k^2\mathbf{E} - \mathbf{k}\,\mathbf{k}\cdot\mathbf{E} = \frac{4\pi i\omega}{c^2}\mathbf{J} = \frac{\omega^2}{c^2}\varepsilon_{ij}E_j \tag{9.244}$$

which gives in components

$$\left(k_z^2 - \frac{\omega^2}{c^2}\varepsilon_{xx}\right)E_x - k_x k_z E_z = 0$$

$$\left(k^2 - \frac{\omega^2}{c^2}\varepsilon_{yy}\right)E_y - \frac{\omega^2}{c^2}\varepsilon_{yz}E_z = 0$$

$$-k_x k_z E_x - \frac{\omega^2}{c^2}\varepsilon_{zy}E_y + \left(k_x^2 - \frac{\omega^2}{c^2}\varepsilon_{zz}\right)E_z = 0$$

The solvability condition for the above simultaneous equations is that the determinant of the coefficients be zero, viz.,

$$\Delta = \left(k_z^2 - \frac{\omega^2}{c^2}\varepsilon_{xx}\right)\left[\left(k^2 - \frac{\omega^2}{c^2}\varepsilon_{yy}\right)\left(k_x^2 - \frac{\omega^2}{c^2}\varepsilon_{zz}\right) + \frac{\omega^4}{c^4}\varepsilon_{yz}^2\right] - k_x^2 k_z^2\left(k^2 - \frac{\omega^2}{c^2}\varepsilon_{yy}\right) = 0 \tag{9.245}$$

Note $\varepsilon_{yy} = \varepsilon_{xx}$. In (9.245), we neglect

$$\frac{c^2 k_x^2}{\omega^2} \ll \varepsilon_{zz}$$

by assuming that the wave propagates predominantly along the direction of the static magnetic field; keep only k_x^2 term. Furthermore we shall assume $\varepsilon_{zz} \gg \varepsilon_{xx}$, which is valid in the low frequency $\omega \ll \omega_{ci}$ and for $k_z \lambda_e \ll 1$. Then the determinant equation becomes

$$\left(k_z^2 - \frac{\omega^2}{c^2}\varepsilon_{xx}\right)\left[\left(\varepsilon_{xx} - \frac{c^2 k^2}{\omega^2}\right)\varepsilon_{zz} + \varepsilon_{yz}^2\right] = 0 \tag{9.246}$$

Using the relations

$$\frac{\omega_{pi}^2}{\omega_{ci}^2} = \frac{c^2}{v_A^2}, \quad \lambda_e^2 \omega_{pi}^2 = v_s^2$$

where v_s is the sound speed $\sqrt{T_e/m_i}$, the dispersion relation becomes

$$(k_z^2 v_A^2 - \omega^2)\left[(\omega^2 - k^2 v_A^2)(\omega^2 - v_s^2 k_z^2) - k_x^2 \omega^2 v_s^2\right] = 0 \tag{9.247}$$

The first factor gives the Alfven wave

$$\omega^2 = v_A^2 k_z^2 \tag{9.248}$$

The second bracket gives the equation of the dispersion relation of the magnetosonic wave in the form

$$(\omega^2 - k^2 v_A^2)(\omega^2 - k_z^2 v_s^2) - \omega^2 k_x^2 v_s^2 = 0 \tag{9.249}$$

which agrees with (3.100) in Chapter 3.

Here we used sinusoidal harmonic functions instead of Bessel functions to solve the Vlasov equation in the region of low frequency that is much smaller than the ion cyclotron frequency. It is interesting that a few number of terms in the series of harmonic functions can describe the low frequency MHD waves kinetically. The approximation is quite satisfactory. The mathematical basis underlying this analysis is that the Bessel functions are a kind of extended sinusoidal functions.

Problems

9.1 We have in (9.151)

$$f_\alpha(t, \mathbf{r}, \mathbf{v}) = -\frac{e_\alpha}{m_\alpha}\int_{-\infty}^{t} dt' \; \mathbf{E}'(\mathbf{r}', t') \cdot \frac{\partial f_{\alpha 0}(\mathbf{v}')}{\partial \mathbf{v}'}$$

Fourier transform this equation to obtain

$$f_\alpha(\omega, \mathbf{k}, \mathbf{v}) = -\frac{e_\alpha}{m_\alpha}\, \mathbf{E}((\mathbf{k}, \omega) \cdot \int_0^\infty d\tau \exp\Big(-i\mathbf{k}\cdot\mathbf{q}(\tau) + i\omega\tau\Big)\frac{\partial f_{\alpha 0}(\mathbf{v}')}{\partial \mathbf{v}'}$$

where

$$\mathbf{q} = \mathbf{r} - \mathbf{r}', \quad \tau = t - t'$$

and $\mathbf{r}$ and τ are the integration constants of the orbit $\mathbf{r}'(t')$ of the characteristic equation.

9.2 Prove that in an extremely strong magnetic field the dielectric tensor reduces to

$$\varepsilon_{\mathbf{ij}} = \begin{pmatrix} 1 & 0 & 0 \\ 0 & 1 & 0 \\ 0 & 0 & 1 - \frac{\omega_p^2}{\omega^2} \end{pmatrix}$$

9.3 In the parallel propagation ($\mathbf{k} \parallel \mathbf{B}_0$), work out the wave equation (9.135) to obtain the simultaneous equations

$$AE_x + iBE_y = 0$$
$$-iBE_x + AE_y = 0$$

where

$$A = 1 - \frac{c^2k^2}{\omega^2} - \frac{\omega_p^2}{\omega^2 - \omega_c^2}, \quad B = \frac{\omega_c}{\omega}\frac{\omega_p^2}{\omega^2 - \omega_c^2}$$

Obtain the R-wave ($A = B$ case) dispersion relation in the form (9.148).

Obtain the L-wave ($A = -B$ case) dispersion relation in the form (9.150).

Obtain the wave numbers

$$k_\pm = \frac{\omega}{c}\sqrt{1 - \frac{\omega_p^2}{\omega(\omega \mp \omega_c)}}$$

where the upper (lower) sign corresponds to R(L) wave.
Explain the terms *right hand circularly polarized wave* and *left hand circularly polarized wave*.

9.4 **Faraday rotation.** A linearly polarized electromagnetic wave propagating along the direction of the magnetic field can be decomposed into a pair of circularly polarized waves, R- and L-waves, as can be seen by adding or subtracting Eqs. (9.141) and (9.149). The circularly polarized waves are characterized by different wave vectors $k_\pm$ in Problem 9.3. The linearly polarized wave can be written as

$$\mathbf{E}(z,t) = [\hat{\mathbf{x}}(E_+\, e^{ik_+z} + E_-\, e^{ik_-z}) + i\hat{\mathbf{y}}(E_+\, e^{ik_+z} - E_-\, e^{ik_-z})]\, e^{-i\omega t}$$

where the subscripts $\pm$ correspond to R- and L-waves. Show that for $E_+ = E_-$

$$\frac{E_y}{E_x} = \tan\, \frac{k_- - k_+}{2}\, z$$

Since $k_+ \neq k_-$, the angle of rotation $\Psi = \frac{k_- - k_+}{2}\, z$ changes as the wave propagates.

9.5 Derive Eq. (9.233).

9.6 Work out the dispersion relation in (9.135) with ε_{ij} in the cold magnetized plasma (at the end of Section 9.6.2). Show that the dispersion relation reduces to

$$k^2c^2 = \omega^2(\varepsilon_{xx} \pm i\varepsilon_{yx})$$

for the transverse mode. The two signs correspond to two independent modes with different polarization. The left-handed mode with $E_y/E_x = -i$ is called the *ordinary wave*. The right-handed mode with $E_y/E_x = i$ is called the *extraordinary wave*.

References

Akhiezer, A., I. Akhiezer, R. Polovin, A. Sitenko, and K. Stepanov, *Collective Oscillations in a Plasma*, Pergamon, Oxford (1967).

Bekefi, G. *Radiation Processes in Plasmas*, Wiley, New York (1966).

Bernstein, I. B. *Waves in a plasma in a magnetic field*, Phys. Rev. **109**, p. 10 (1958).

Braginskii, S. I. and A. P. Kazantsev, *Magnetohydrodynamic waves in a rarefied plasma*, (1957). (The Journal name is unknown to the author.)

Clemmow, P. C. and J. P. Dougherty, *Electrodynamics of Particles and Plasmas*, Addison-Wesley, London, (1969).

Dawson, J. and C. Oberman, *High-frequency conductivity and the emission and absorption coefficients of a fully ionized plasmas*, Phys. Fluids **5** p. 517 (1962).

Harris, E. G. *Plasma instabilities*, Physics of Hot plasmas Scottish Universities' Summer School 1968 Ed. B. J. Rye and J. C. Taylor Plenum (1968).

Hildebrand, F. B. *Advance Calculus for Applications*, Prentice-Hall, Englewood Cliffs, N.J. (1976).

Infeld, E. and G. Rowlands, *Nonlinear Waves, Solitons and Chaos*, (Cambridge University Press, Cambridge (1990).

Lee, H. J. *Nonlinear analysis of the cold fluid-Poisson plasma by using the characteristic method*, J. Korean Phys. Soc. **69** No. 7 pp. 1191–1211 (2016).
Montgomery, D. C. and D. A. Tidman, *Plasma Kinetic Theory*, McGraw-Hill (1964).
Sanmartin, J. R. *Electrostatic plasma instabilities excited by a high frequency electric field*, Phys. Fluids **13** pp. 1533–1542 (1970).
Stix, T. H. *Waves in Plasmas*, AIP, New York (1992).
Swanson, D. G. *Plasma Waves*, Academic Press, New York (1989).

Chapter 10

Theory of Dispersive Waves

A dispersion relation specifies a certain functional relationship between ω and $\mathbf{k}$, which are, respectively, the Fourier variables conjugate to t and $\mathbf{r}$ in ordinary space. The underlying physical assumption in the wave analysis general is that a disturbance in a medium can be represented by Fourier integral in the form

$$f(x,t) = \int_{-\infty}^{\infty} F(k)\exp(ikx - i\omega(k)t)dk \tag{10.1}$$

This is a mathematical representation of a wave packet. The asymptotic evaluation of the integral in (10.1) reveals the kinematic aspect of the wave packet. The important characteristics of the wave packet in (10.1) is that the oscillation of the wave has two time scales; propagating with a slowly varying amplitude and rapidly changing phasor. The group velocity of the wave emerges in the course of evaluation of the integral. Plasmas are dispersive media whose dielectric permittivity $\varepsilon(\mathbf{k},\omega)$ depends on the wave frequency and wave vector. The electromagnetic processes in plasmas as dispersive media are important subject in plasma physics. Without knowing the specific form of $\omega(k)$, we can extract significant general information from this integral. Also a dispersion relation itself contains information concerning the propagation of the waves and the properties of the medium supporting the waves.

We first sort out a few salient general features which can be extracted from each case of the relation $\omega = \omega(k)$.

(i) ω is real for real k.
This implies that the energy of the wave is conserved and the propagation is not inhibited.

(ii) ω is complex or purely imaginary for real k.
The energy of the wave is not conserved. Usually this means that dissipative mechanism is at work in the medium and the wave becomes

damped as it propagates. But if there is some mechanism of free energy liberation from the medium, nonreal ω implies instability.

(iii) k is complex for real ω.
This is a conservative system in which a number of modes propagate with possible exchange among them (mode conversion). A mode with a complex wavenumber would thus lose its energy to (or absorb energy from) other modes at appropriate rate to conserve the total energy. This situation occurs, for example, for certain modes of electromagnetic waves in plasmas.

(iv) $k = 0$ for real ω.
This case occurs when k changes from real values to imaginary with the corresponding ω varying continuously. The particular frequency yielding zero k is called *cutoff frequency.* The wave evanesces when k becomes purely imaginary. What actually happens is that the wave is mostly reflected at cutoff where the phase velocity of the wave is infinite.

(v) k is infinite for real ω.
This is where resonances take place. Since the phase velocity is nearly zero, the wave can spend long enough time at one locality to be absorbed by the medium.

10.1 Asymptotic Evaluation of Fourier Integral and Group Velocity

Actual signals propagating in a plasma are not monochromatic but have finite spread of frequencies or wavelengths. Such disturbances or wavepackets can be synthesized as Fourier integral, and we find, in the asymptotic evaluation of the integral for large time, that the group velocity emerges to be an important physical variable. Let us consider a wavepacket superposed about a well-defined central wavenumber k_0 and the corresponding frequency $\omega(k_0)$ with a finite spread of wavenumber Δk:

$$f(x,t) = \frac{1}{\Delta k}\int_{k_0-\Delta k}^{k_0+\Delta k} F(k)e^{i[kx-\omega(k)t]}dk \tag{10.2}$$

The factor $\frac{1}{\Delta k}$ is introduced so that f will remain nonzero as $\Delta k \to 0$. Expanding $\omega(k)$ in the vicinity of k_0 as a power series,

$$\omega(k) = \omega(k_0) + (k-k_0)\left(\frac{\partial \omega}{\partial k}\right)_{k_0}$$

Equation (10.2) becomes

$$f(x,t) = \frac{F(k_0)}{\Delta k} e^{i[k_0 x - \omega(k_0)t]} \int_{k_0-\Delta k}^{k_0+\Delta k} e^{i(k-k_0)x - i(k-k_0)\left(\frac{\partial \omega}{\partial k}\right)_{k_0} t}\, dk$$

$$= 2F(k_0)\frac{\sin[\Delta k(x-\omega_0' t)]}{\Delta k(x-\omega_0' t)} e^{i(k_0 x - \omega_0 t)} \tag{10.3}$$

where $\omega_0 = \omega(k_0), \omega_0' = \left(\frac{\partial \omega}{\partial k}\right)_{k_0}$. This is a wave train whose wavenumber and frequency are respectively the central wavenumber k_0 and the corresponding frequency ω_0 of the constituent Fourier wave trains, and having its amplitude modulated by a slowly varying envelope ($\sin[\cdots]term$) to form a system of wavepackets (see Fig. 10.1). The packets travel with the *group velocity* $\left(\frac{\partial \omega}{\partial k}\right)_{k_0}$ while the wave train contained inside the envelope propagate with the phase velocity ω_0/k_0. The amplitude of the modulation is a varying function. The maximum packet is the one centered about the position $x = \omega_0' t$ where the amplitude is $2F(k_0)$.

Reference: S. A. Thau *Linear dispersive waves* in Nonlinear waves, Ed. S. Leibovich and A. R. Seebass, Cornell University Press (1970).

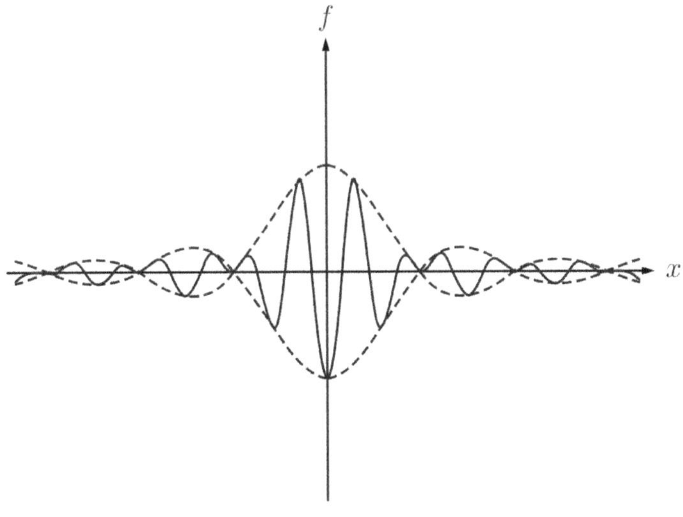

Fig. 10.1. Wave packet.

Going back to (10.1) to carry out the integral asymptotically, we write the phase as $t\,\varphi(k)$, where

$$\varphi(k) = i\left[\frac{kx}{t} - \omega(k)\right] \tag{10.4}$$

Then (10.1) takes the form

$$f(x,t) = \int_{-\infty}^{\infty} F(k)e^{t\varphi(k)}dk \tag{10.5}$$

We evaluate the integral in (10.5) when t is large. This asymptotic evaluation will reveal the hidden characteristics of the Fourier-synthesized wave, or wave packet in general.

Method of steepest descent

Reference: Arfken, G. B. and H. J. Weber, *Mathematical methods for physicists*, Elsevier (2005).

Lighthill, E. J. *Waves in Fluids*, Cambridge University Press, Cambridge (1978).

Integral (10.5) is amenable to asymptotic evaluation by the method of steepest descent (or *saddle point method*). Since the integrand is analytic in the complex k-plane, the integration path which is the k-axis from $-\infty$ to ∞ can be deformed to a path which we can choose as we like with the end points fixed. This is legitimate according to Cauchy's theorem. We assume here that the ends of the path do not contribute significantly to the integral. When t is large, the exponential function in the integrand of (10.5) has a rapidly oscillating phase which makes almost exact cancellations in the integral. However, this is not true when there are points of stationary phase where $\varphi'(k) \equiv d\varphi/dk = 0$. Let us write the phase function which we assume to be analytic as

$$\varphi(z) = u(x,y) + iv(x,y) \quad z \equiv k = k_r + ik_i, \quad (x = k_r,\ y = k_i)$$

where we designated the complex k-plane as the more friendly z-plane. Most of the contribution to f in (10.5) comes from the section centered about the point where u is largest (local maximum). [The local maximum of u is also at the same time the point of stationary phase.] The idea of the method of steepest descent is to deform the integration path so that the region of large u is compressed into as short a path as possible. At the stationary phase point $z = z_0$, $\varphi'(z_0) = 0$, which implies $\frac{\partial u}{\partial x} = \frac{\partial v}{\partial x} = 0$ and $\frac{\partial u}{\partial y} = \frac{\partial v}{\partial y} = 0$ at $z_0 = (x_0, y_0)$. This is the necessary condition that $u(x,y)$ has a local maximum or a local minimum at z_0. Since $\varphi(z)$ is analytic, we have from Cauchy-Riemann condition,

$$\frac{\partial^2 u}{\partial x^2} + \frac{\partial^2 u}{\partial y^2} = 0 \tag{10.6}$$

Thus $\frac{\partial^2 u}{\partial x^2} > 0$ implies $\frac{\partial^2 u}{\partial y^2} < 0$. In words, $u(x, y)$ is maximum at $z = z_0$ along the x-coordinate, but minimum along the y-coordinate. And vice versa. Therefore the stationary phase point z_0 is a saddle point when we view it in the z-plane. This is the consequence of the analyticity of $\varphi(z)$. By Taylor expansion in the vicinity of $z = z_0$, we have

$$\varphi(z) = \varphi(z_0) + \frac{1}{2}\varphi''(z_0)(z - z_0)^2$$

We write

$$\varphi''(z_0) = \varrho e^{i\vartheta}, \; z - z_0 = \gamma e^{i\phi}, \; \varphi(z) = \varphi(z_0) + \frac{1}{2}\varrho\,\gamma^2 e^{i\vartheta + 2i\phi} \tag{10.7}$$

which gives

$$u(x, y) = u(x_0, y_0) + \frac{1}{2}\varrho\gamma^2 \;\cos(\vartheta + 2\phi) \tag{10.8}$$

$$v(x, y) = v(x_0, y_0) + \frac{1}{2}\varrho\gamma^2 \;\sin(\vartheta + 2\phi) \tag{10.9}$$

Note that ρ and ϑ are now fixed coordinates, and γ and ϕ are variables. Furthermore, we impose the additional requirement that the imaginary part v is constant while traversing the segment of the deformed contour centered about the saddle point. If v is not constant, the rapidly oscillating exponential factor will make the integral vanishingly small, but it is not actually nil. This inaccuracy will be more and more augmented as the time t becomes large. Thus we put $\sin(\vartheta + 2\phi) = 0$, which gives

$$\vartheta + 2\phi = 0, \;\; \pm\,\pi$$

The case where $\vartheta + 2\phi = 0$ should be rejected since $u(x_0, y_0)$ in this case does not correspond to the maximum of u. Therefore ϕ should be one of the two

$$\phi = -\frac{\vartheta}{2} \pm \frac{\pi}{2} \tag{10.10}$$

So our deformed path centered about the saddle point is described by

$$u(x, y) = u(x_0, y_0) - \frac{1}{2}\varrho\gamma^2 \tag{10.11}$$

$$v = v(x_0, y_0) = \text{const.} \tag{10.12}$$

When u and v are harmonic, satisfying (10.6), the curves $u(x, y) =$ const. and the curves $v(x, y) =$ const. are orthogonal. Since ∇u is perpendicular to $u =$ const., ∇u and the curves $v =$ const. are parallel. Thus the integration path $v =$ const. is the path of steepest descent. Now (10.7) becomes (reinstating k in place of z)

$$\varphi(k) = \varphi(k_0) - \frac{1}{2}\varrho\gamma^2$$

$$dk = dz = e^{i\phi} d\gamma$$

where the angle ϕ is constant as one of the two in (10.10). So we have

$$f(x,t) = F(k_0) \int e^{t\varphi(k)} dk = F(k_0) e^{i\phi} \int \exp(t\varphi(k_0) - \frac{\varrho}{2}\gamma^2 t) d\gamma \tag{10.13}$$

In (10.13), the limits of the integration can be taken from $-\infty$ to $+\infty$ because the integral has non-zero contribution only from the segment of the steepest descent. Then the $d\gamma$-integral gives $\sqrt{2\pi/t\varrho}$. Recalling $\varrho = |\varphi''(k_0)|$, (10.13) becomes

$$f(x,t) = F(k_0)\sqrt{\frac{2\pi}{t}}\, e^{i\phi} |\varphi''(k_0)|^{-1/2} e^{t\varphi(k_0)} \tag{10.14}$$

The above formula is quite general but the angle ϕ must be determined case by case according to the problem under consideration.

Determination of the angle ϕ. Going back to (10.4) and (10.5), it is more convenient to define

$$\Psi(k) = \frac{kx}{t} - \omega(k) : \; f(x,t) = \int_{-\infty}^{\infty} F(k) e^{it\Psi(k)} dk \tag{10.15}$$

Suppose $\Psi'(k) > 0$ everywhere (this may occur for large x/t). We slightly raise the path by an amount $i\varepsilon (\varepsilon > 0)$ above the real k-axis. Then along the raised path, Ψ has the value

$$\Psi(k + i\varepsilon) = \Psi(k) + i\varepsilon\Psi'(k) \; (k, \text{ real})$$

The positive imaginary part of the phase function Ψ, which we write $\delta = \varepsilon\Psi'(k)$, makes integral (10.15) exponentially small for large t.

When $\Psi'(k) < 0$ everywhere (this is the case where $x = 0$ and $d\omega/dk > 0$), integral (10.15) is again vanishingly small ($\sim e^{-t\delta}$), as can be seen by lowering the path by an amount $\varepsilon = \delta/\Psi'(k)$. Note that $\Psi'(k)$ changes sign upon crossing the stationary phase point $k = k_0$ where $\Psi'(k_0) = 0$. Therefore the raised (lowered) path on one side of k_0 should be

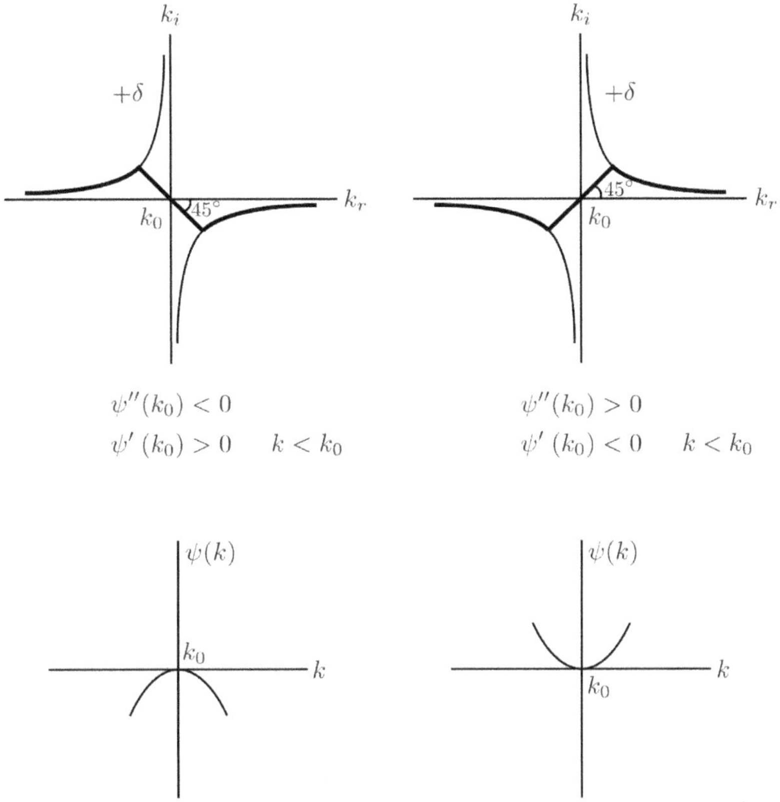

Fig. 10.2. Path of steepest descent.

connected to the lowered (raised) path on the other side of k_0 (see Fig. 10.2). Now we have to consider the second derivative term in the expansion of $\Psi(k)$ near k_0:

$$\Psi(k) = \Psi(k_0) + \frac{1}{2}\Psi''(k_0)(k - k_0)^2 \ (k = k_r + ik_i)$$

which gives

$$\mathrm{Im}[\Psi(k)] = k_i(k_r - k_0)\Psi''(k_0)$$

The contours of $\mathrm{Im}[\Psi(k)] = \delta$ are the hyperbolas (see Fig. 10.2),

$$k_i = \frac{\delta}{\Psi''(k_0)(k_r - k_0)} \tag{10.16}$$

It is immediately seen that the shortest straight line which connects the two branches of the hyperbola (two contours above and below) and passes through $k = k_0$ is either $\phi = \pi/4$ or $-\pi/4$.

Now $f(x,t)$ in (10.14) is determined without ambiguity, and can be written as a single equation for both cases of (a) and (b) in Fig. 10.2:

$$f(x,t) = F(k_0)\left[\frac{2\pi}{t|\omega''(k_0)|}\right]^{1/2} \exp\left[i(k_0x - \omega_0 t) - \frac{i\pi}{4}\frac{\omega''(k_0)}{|\omega''(k_0)|}\right] \tag{10.17}$$

where $\omega_0 = \omega(k_0)$ and k_0 is the wavenumber for which

$$\omega'(k_0) = \frac{x}{t} \tag{10.18}$$

The preceding method of evaluating integral (10.15) fails if $\Psi''(k_0) = 0$ for the stationary phase point k_0. Readers who are interested in the asymptotic evaluation of the integral when $\Psi''(k_0) = 0$ are referred to the book by Lighthill (1978) *Waves in Fluids.* If there are more than one stationary phase point satisfying (10.18), the asymptotic value of the integral should be a sum of the terms each of which is similar to the right hand side of (10.17).

Let us note that the stationary phase point k_0, which is defined by (10.18), is the same for two sets of (x,t) if $\frac{x_1}{t_1} = \frac{x_2}{t_2} = c$ (const.). In fact, along the straight line defined by $\frac{x}{t} = c$, the stationary phase point k_0 is the same. The straight line is a line of characteristics along which the k_0-value is constant. Therefore the k_0-value propagates along the line with the velocity

$$\frac{x_2 - x_1}{t_2 - t_1} = \frac{x_2}{t_2} = \frac{x_1}{t_1} = \omega'(k_0) \tag{10.19}$$

So the stationary phase point k_0 for which $\omega'(k_0) = x/t$ propagates with the group velocity $\omega'(k_0)$ in (10.19). The integral $f(x,t)$ in (10.5) has a non-zero value only in the vicinity of k_0, so it is quite obvious that a physical quantity associated with a wavepacket, such as energy, will propagate with the group velocity.

We write the asymptotic result (10.17) in the form,

$$f(x,t) = A(t)e^{i\theta(x,t)} \tag{10.20}$$

$$\theta(x,t) = k_0x - \omega(k_0)t = t\,(k_0\omega'(k_0) - \omega(k_0)) \tag{10.21}$$

$$A(t) = F(k_0)\left[\tfrac{2\pi}{t|\omega''(k_0)|}\right] \exp\left(\tfrac{i\pi}{4}\tfrac{\omega''(k_0)}{|\omega''(k_0)|}\right) \tag{10.22}$$

Since the value of k_0 can be determined from (10.18) at any point (x,t), it is a function of (x,t). Consequently, $\omega(k_0)$ is also a function of (x,t) through the dispersion relation. Therefore the amplitude A is also a function of (x, t). Equation (10.20) indicates that $f(x, t)$ is represented by a 'monochromatic' wave whose amplitude, wavenumber, and frequency are all slowly varying functions of (x, t). This result is interesting since the Fourier superposition ends up with a monochromatic plane wave expression (albeit with varying parameters) after going through all the constructive and destructive interferences. The last paragraph is quoted from Lighthill (1978). In conclusion, we can state that the group velocity of a wavepacket constructed by Fourier synthesis (0.1) is the velocity of the constructive interference maximum of the wave. When $F(k)$ in (10.1) is a slowly varying function, the maximum at each instant of time occurs at the point of stationary phase which moves with the group velocity $\omega'(k_0)$. This result also applies to the electromagetic processes of the plasma waves.

Differentiating (10.21) with respect to x and t gives, respectively,

$$\frac{\partial \theta}{\partial x} = k + (x - \omega'(k)t)\,\frac{\partial k}{\partial x} \tag{10.23}$$

$$\frac{\partial \theta}{\partial t} = -\omega(k) + (x - \omega'(k)t)\,\frac{\partial k}{\partial t} \tag{10.24}$$

where we omitted the subscript '0'. Since k is the stationary phase point, the above equations reduce to

$$\frac{\partial \theta}{\partial x} = k(x, t) \tag{10.25}$$

$$\frac{\partial \theta}{\partial t} = -\omega(x, t) \tag{10.26}$$

Therefore the slowly varying phase function $\theta(x, t)$ is the eikonal whose spatial and temporal derivatives give, respectively, the local wavenumber and the local frequency in a nonuniform wave train $\sim e^{ik(x,t)x - i\omega(x,t)t}$. The above two equations yield

$$\frac{\partial k}{\partial t} + \frac{\partial \omega}{\partial x} = 0 \quad \text{or} \quad \frac{\partial k}{\partial t} + \omega'(k)\frac{\partial k}{\partial x} = 0 \tag{10.27}$$

Using the dispersion relation, $k = k(\omega)$, we also have

$$\frac{\partial \omega}{\partial t} + \omega'(k)\,\frac{\partial \omega}{\partial x} = 0 \tag{10.28}$$

Equations (10.27) and (10.28) have the solutions in the form

$$k(x, t) = k(x - \omega'(k)t) \tag{10.29}$$

$$\omega(x, t) = \omega(x - \omega'(k)t) \tag{10.30}$$

The above equations indicate that the wavenumber k and the frequency ω remain constant along the *line of characteristics* defined by

$$\frac{dx}{dt} = \omega'(k)$$

10.2 Kinematics of Wave, Group Velocity in Three-Dimension

In the preceding discussion, we showed that the group velocity of a wavepacket that is formed of Fourier integral is the velocity of the moving stationary phase point in the (x, t)-plane. In fact, the concept of the group velocity can be established independently of Fourier integral, even valid in inhomogeneous media and for nonlinear waves, for both of which the Fourier superposition is not appropriate. In this Section, we derive the group velocity independently of the Fourier integral representation of a wavepacket, extending also to three-dimensional propagation. First, we assume that a wave quantity $f(x, t)$ is written as

$$f(\mathbf{x}, t) = A(\mathbf{x}, t)e^{i\theta(\mathbf{x},t)} \tag{10.31}$$

Compare this expression with (10.20). For a wave propagating in an inhomogeneous medium whose property is slowly changing (the fractional change of the property in one local wavelength is much smaller than unity), the amplitude A can be assumed to be a slowly varying function of $\mathbf{x}$ and t. Over a small region of space and short interval of time, the phase *(eikonal)* $\theta(\mathbf{x}, t)$ may be expanded in Taylor series in the vicinity of $\mathbf{x} = 0$ and $t = 0$.

$$\theta(\mathbf{x}, t) = \theta_0 + \mathbf{x} \cdot \nabla\theta + t\,\frac{\partial\theta}{\partial t} \tag{10.32}$$

Over a small region of space and short duration of time, the wave (10.31) can be regarded as a plane wave. Therefore, we can define the wave vector and the frequency at each point as

$$\mathbf{k} = \nabla\theta,\ \omega = -\frac{\partial\theta}{\partial t} \tag{10.33}$$

Compare (10.33) with (10.25) and (10.26). Combining the two equations in (10.33) gives

$$\frac{\partial\mathbf{k}}{\partial t} = -\nabla\omega \tag{10.34}$$

In a homogeneous medium, the frequency of a wave depends on $\mathbf{k}$ only: $\omega = \omega(\mathbf{k})$. Therefore (10.34) is a differential equation for $\mathbf{k}$:

$$\frac{\partial\mathbf{k}}{\partial t} + \frac{\partial\omega}{\partial\mathbf{k}} \cdot \nabla\mathbf{k} = 0$$

By solving the differential equation, we obtain the temporal change of $\mathbf{k}(\mathbf{x}, t)$. In one-dimensional case, (10.34) takes a simple form

$$\frac{\partial k}{\partial t} + \frac{\partial \omega}{\partial k}\frac{\partial k}{\partial x} = 0$$

which is (10.27) and indicates that the wavenumber k is merely transported with the group velocity. In an inhomogeneous medium, the dispersion relation takes the form:

$$\omega = \omega(\mathbf{k}(\mathbf{x}, t), \mathbf{x}, t) \tag{10.35}$$

Substitution of (10.35) into (10.34) yields

$$\frac{\partial k_i}{\partial t} + \frac{\partial \omega}{\partial k_j}\frac{\partial k_j}{\partial x_i} + \frac{\partial \omega}{\partial x_i} = 0$$

Upon using $\frac{\partial k_i}{\partial x_j} = \frac{\partial k_j}{\partial x_i}$, which is $\nabla \times \mathbf{k} = 0$, the above equation is written as

$$\frac{\partial \mathbf{k}}{\partial t} + \mathbf{v}_g \cdot \nabla \mathbf{k} = -\nabla \omega \tag{10.36}$$

where $\mathbf{v}_g = \partial\omega/\partial\mathbf{k}$ is the three-dimensional group velocity. Equation (10.36) is also a differential equation for $\mathbf{k}$ because ω is a function of $(\mathbf{x}, t)$ per (10.35). The above equation is written in the characteristic form as

$$\frac{d\mathbf{k}}{dt} = -\nabla \omega \tag{10.37}$$

where the total derivative $d/dt = \partial/\partial t + \mathbf{v}_g \cdot \nabla$ is the directional derivative taken along the characteristic line

$$\frac{d\mathbf{x}}{dt} = \mathbf{v}_g = \frac{\partial \omega}{\partial \mathbf{k}} \tag{10.38}$$

When the medium is homogeneous, $\partial\omega/\partial\mathbf{x} = 0$, and $\partial\omega/\partial\mathbf{k}$ is a constant vector. Then the characteristics are straight lines in $(\mathbf{x}, t)$ space, and each value of $\mathbf{k}$ propagates with the corresponding constant group velocity $\mathbf{v}_g$. In an inhomogeneous medium, the characteristics are not straight lines, and also the value of $\mathbf{k}$ varies as they propagate along the characteristics.

The total derivative of frequency ω taken along the characteristics is

$$\frac{d\omega}{dt} = \frac{\partial \omega}{\partial t} + \frac{\partial \omega}{\partial \mathbf{k}} \cdot \frac{d\mathbf{k}}{dt} + \frac{\partial \omega}{\partial \mathbf{x}} \cdot \frac{d\mathbf{x}}{dt}$$

where the last two terms add to zero owing to (10.37) and (10.38), yielding

$$\frac{d\omega}{dt} = \frac{\partial \omega}{\partial t} \tag{10.39}$$

Thus the frequency is constant along each characteristics in a steady state ($\partial\omega/\partial t = 0$). It is interesting that (10.37) and (10.38) are formally identical with Hamilton's equations in mechanics

$$\dot{\mathbf{p}} = -\frac{\partial H}{\partial \mathbf{x}}\ \dot{\mathbf{x}} = \frac{\partial H}{\partial \mathbf{p}}$$

with the correspondence, momentum $\mathbf{p} \to$ wave vector $\mathbf{k}$ and Hamiltonian (energy)$\to$ frequency ω. The second equation of (10.33) gives, upon introducing the dispersion relation,

$$\frac{\partial \theta}{\partial t} + \omega\left(\frac{\partial \theta}{\partial \mathbf{x}}, \mathbf{x}, t\right) = 0 \tag{10.40}$$

This is the Hamilton–Jacobi equation in which the eikonal $\theta(\mathbf{x}, t)$ plays the role of the action of the particle. In a sound or electromagnetic wave whose dispersion relation is of the form $\omega = ck$, (10.40) takes the form

$$(\nabla\theta)^2 - \frac{1}{c^2}\left(\frac{\partial \theta}{\partial t}\right)^2 = 0 \tag{10.41}$$

This is the basic equation of geometrical acoustics or optics. The characteristics defined by (10.38) are often called '*rays*' in geometrical acoustics.

10.3 Wave Energy

Let us consider how the wave energy propagates. Here we use the Fourier integral result obtained in (10.20)–(10.22). The wave energy contained between two planes $x = x_1$ and $x = x_2$ is proportional to

$$Q(t) = \int_{x_1}^{x_2} |A|^2 dx = 2\pi \int_{x_1}^{x_2} \frac{F(k_0)F^*(k_0)}{t|\omega''(k_0)|} dx$$

In this integral, k_0 is given by the relation, $\omega'(k_0) = x/t$. Here the two planes $x = x_1$ and $x = x_2$ move with the group velocities $\omega'(k_{01})$ and $\omega'(k_{02})$, respectively. It is advantageous to change the integration variable from x to k_0, using the relation,

$$dx = t\ \omega''(k_0) dk_0$$

Then we have

$$Q(t) = \pm 2\pi \int_{k_{01}}^{k_{02}} F(k_0)F^*(k_0) dk_0$$

where $\pm$ signs correspond to $\omega'' > 0$ and $\omega'' < 0$, respectively, and k_{01} and k_{02} are defined by

$$x_1 = t\ \omega'(k_{01}), \quad x_2 = t\ \omega'(k_{02})$$

Since k_{01} and k_{02} are fixed wavenumbers, $Q(t)$ is a constant. Therefore we have

$$\frac{dQ}{dt} = \frac{d}{dt}\int_{x_1(t)}^{x_2(t)} |A|^2 dx = 0$$

which gives

$$\int_{x_1}^{x_2} \frac{\partial |A|^2}{\partial t} dx + |A|^2_{x_2}\omega'(k_{02}) - |A|^2_{x_1}\omega'(k_{01}) = 0$$

We have thus proved that the total amount of energy between any pair of planes moving with respective group velocities remains constant. The above integral relation can be converted to a differential equation by taking the limit $x_2 - x_1 \to 0$:

$$\frac{\partial |A|^2}{\partial t} + \frac{\partial}{\partial x}(\omega'(k_0)|A|^2) = 0 \tag{10.42}$$

This equation is a conservation equation relating the energy density $|A|^2$ with the energy flux $\omega'(k_0)|A|^2$. This equation shows that the group velocity $\omega'(k_0)$ is the propagation velocity of the energy. The three-dimensional extension of the energy equation takes the form

$$\frac{\partial |A|^2}{\partial t} + \nabla \cdot \left(\mathbf{v}_g |A|^2\right) = 0 \tag{10.43}$$

The three-dimensional energy equation (10.43) (as well as (10.42)) can be derived without resorting to Fourier integral but directly from the quasi-monochromatic wave expression (10.31). [For example, Reference: B. B. Kadomtsev, *Cooperative effects in plasmas*, Reviews of plasma physics **22** Kluwer Academic/Consultants Bureau, New York (2001), Eq. (5.113), where the electromagnetic field energy can be expressed in such a form.]

10.4 Nonlinear Waves; Distortion of Wave Profile

We consider a model equation

$$\frac{\partial \rho}{\partial t} \pm c_0 \frac{\partial \rho}{\partial x} = 0 \tag{10.44}$$

where c_0 is a constant. More general wave equation

$$\frac{\partial^2 \rho}{\partial t^2} - c_0^2 \frac{\partial^2 \rho}{\partial x^2} = 0$$

can be factored into

$$\left(\frac{\partial \rho}{\partial t} + c_0 \frac{\partial \rho}{\partial x}\right)\left(\frac{\partial \rho}{\partial t} - c_0 \frac{\partial \rho}{\partial x}\right) = 0$$

which reduces to the model equation (10.44). The solution to (10.44) is $\rho(x,t) = f(x \mp c_0 t)$, where $f(x)$ is an arbitrary function which is required only to satisfy the prescribed initial conditions. The wave motion governed by (10.44) is parallel translation of initial shape $f(x)$ along the x-direction with the constant velocity $\pm c_0$. Now we consider a little more complicated equation

$$\frac{\partial \rho}{\partial t} + c(\rho)\,\frac{\partial \rho}{\partial x} = 0 \tag{10.45}$$

where the velocity c is not constant but a function of ρ. The characteristic equation associated with (10.45) is

$$\frac{dt}{1} = \frac{dx}{c(\rho)} = \frac{d\rho}{0}$$

which states that $\rho(x,t) \;=\;$ const. along the characteristic line:

$$\rho(x,t) = \text{const.:}\ \frac{dx}{dt} = c(\rho) \tag{10.46}$$

This statement can be put into a mathematical form

$$\rho(x,t) = f[x - c(\rho(x,t))t] \tag{10.47}$$

By direct substitution one can show that (10.47) satisfies (10.45). The proof goes as follows.

$$\frac{\partial \rho}{\partial t} = f' \cdot \left[-c - t\,\frac{dc}{d\rho}\frac{\partial \rho}{\partial t}\right], \quad \frac{\partial \rho}{\partial x} = f' \cdot \left[1 - t\,\frac{dc}{d\rho}\frac{\partial \rho}{\partial x}\right]$$

where f' is the derivative with respect to the argument. Therefore we have

$$\frac{\partial \rho}{\partial t} + c\frac{\partial \rho}{\partial x} = \left(\frac{\partial \rho}{\partial t} + c\frac{\partial \rho}{\partial x}\right)\left(-\frac{dc}{d\rho}\right)\ t\ f'$$

$$\text{or}\quad \left(\frac{\partial \rho}{\partial t} + c\frac{\partial \rho}{\partial x}\right)(1 + \frac{dc}{d\rho}\ tf') = 0$$

The second factor is not zero for an arbitrary t, and we arrive at the equality given by (10.45). It is observed from (10.47) that ρ is constant when the argument of f is constant ($\equiv \xi$):

$$\rho = f(\xi) = \text{const.};\ x = \xi + c(f(\xi))t \tag{10.48}$$

Note that $f(\xi)$ is just $\rho(x,0)$, the initial distribution of density ρ. We can summarize (10.48) by saying that the initial value of ρ corresponding to the initial position ξ is conserved along the straight line on (x,t) plane given by the second equation in (10.48). Equation (10.48) that paraphrases (10.47) constitute an equivalent representation of the implicit relation (10.47). The straight lines on the (x, t) plane in (10.48) are the characteristics of the *hyperbolic* partial differential equation (10.45). According to (10.48), the initial shape $\rho(x, 0) = f(x)$ is translated along the positive x-direction with velocity $c(\rho)$. In contrast to (10.44), the translation velocity c is not constant but depends on ρ. Therefore the initial shape of ρ will be distorted. This distortion of the wave shape is due to the nonlinearity of (10.45). When $dc/d\rho > 0$, ρ of higher value propagates faster than ρ of smaller value. Eventually, ρ of a higher value catches up the "runners" of smaller ρ ahead. The wave profile ultimately develops a vertical slope, where shock or discontinuity is formed.

References: G. B. Whitham, *Linear and Nonlinear Waves*, John Wiley, New York (1974) Chapter 2.

R. K. Dodd, J. C. Eilbeck, J. D. Gibbon and H. C. Morris, *Solitons and Nonlinear wave equations*, Academic Press, New York (1982).

10.5 Inclusion of Dispersion; Steepening vs. Dispersion

Reference: B. B. Kadomtsev, *Cooperative effects in plasmas*, Reviews of Plasma Physics Vol 22 Ed. V. D. Shafranov, Consultant Bureau, New York (2001).

If a dispersion term is added in (10.45), that term can counter-balance the wave steepening to form a stationary wave profile; a solitary pulse (or soliton) in the form of a progressive wave. The dispersion term is characterized by $\sim\partial^3/\partial x^3$ [see (10.58) below]. As a specific example, we consider ion acoustic wave in a plasma. The ion acoustic wave is associated with the ions which provide the inertia of the medium supporting the wave, on the background of Boltzmann-distributed electrons (massless electrons are assumed). The electron pressure provides the restoring force which is necessary for the oscillation. The ion and electron equations, for one-dimensional motion, are

$$\frac{dv}{dt} = \frac{\partial v}{\partial t} + v\,\frac{\partial v}{\partial x} = \frac{e}{m_i}E \tag{10.49}$$

$$\frac{\partial n_i}{\partial t} + \frac{\partial}{\partial x}(n_i v) = 0 \tag{10.50}$$

$$T_e\,\frac{\partial n_e}{\partial x} = -eEn_e \tag{10.51}$$

$$\frac{\partial E}{\partial x} = 4\pi e(n_i - n_e) \tag{10.52}$$

In (10.51), we neglected the electron inertial force because the electrons can be considered to have infinite mobility during the much slower ion oscillations. Also the electrons have enough time to attain the equilibrium temperature (T_e) during the slow ion motion. The above equations are linearized with respect to the equilibrium state, $v = 0, n_e = n_i = n_0 = \text{const.}$ Denoting the perturbed quantities by prime, we have from (10.51) and (10.52), assuming the phasor in the form $\sim e^{ikx-i\omega t}$,

$$ikT_e n_e' = -eEn_0$$

$$ikE = 4\pi e(n_i' - n_e')$$

The above two equations yield, after eliminating E,

$$n_i' = (1 + k^2\lambda_e^2)n_e' \quad \lambda_e = \sqrt{\frac{T_e}{4\pi n_0 e^2}} \tag{10.53}$$

Linearizing (10.50) gives

$$-i\omega n_i' + n_0 \frac{\partial v}{\partial x} = 0 \tag{10.54}$$

Eliminating n_e' and n_i' in the above equations gives

$$E = -\frac{T_e}{e} \frac{1}{1 + k^2\lambda_e^2} \frac{1}{v_{ph}} \frac{\partial v}{\partial x} \tag{10.55}$$

where $v_{ph} = \omega/k$ is the phase velocity. Now (10.49) becomes upon using (10.55)

$$\frac{dv}{dt} + \frac{c_0^2}{v_{ph}(1 + k^2\lambda_e^2)} \frac{\partial v}{\partial x} = 0 \tag{10.56}$$

where $c_0 = \sqrt{T_e/m_i}$ is the sound velocity. This is the equation of nonlinear ion acoustic wave which was derived by calculating the driving force (the electric field term) linearly. In the linear approximation, $dv/dt = \partial v/\partial t = -i\omega v$, and (10.56) yields the linear dispersion relation.

$$\omega^2 = \frac{c_0^2 k^2}{1 + k^2\lambda_e^2}$$

When $k^2\lambda_e^2 \ll 1$, the ion acoustic wave has no dispersion: $\omega/k = c_0 = \text{const.}$ We take into account weak dispersion effects. For ion acoustic waves propagating in the positive x-direction, the phase velocity is equal to $v_{ph} = c_0(1 + k^2\lambda_e^2)^{-1/2}$. Substituting this value into (10.53), we obtain

$$\frac{\partial v}{\partial t} + v\frac{\partial v}{\partial x} + c_0\left(1 - \frac{1}{2}k^2\lambda_e^2\right)\frac{\partial v}{\partial x} = 0 \tag{10.57}$$

Here we introduce a new frame of reference moving with velocity c_0 : $v = u - c_0$. Also we reinstate $\partial/\partial x$ by putting $k^2 = -\partial^2/\partial x^2$. Then (10.57) becomes

$$\frac{\partial u}{\partial t} + u\,\frac{\partial u}{\partial x} + \frac{1}{2}\,c_0\lambda_e^2\frac{\partial^3 u}{\partial x^3} = 0 \tag{10.58}$$

Korteweg-de Vries equation. Equation (10.58) is the famous Korteweg-de Vries equation which is dealt with by many authors in the literature. The title of the original paper written by the two authors is interesting: D. J. Korteweg and G. de Vries, *On the change of form of long waves advancing in a rectangular canal, and on a new type of long stationary waves*, Phil. Mag., **39** (1895) pp. 422–443. Now we investigate what effect is brought about by the dispersion term. We seek the solution of (10.58) in the form of progressive wave,

$$u(x,t) = u(\xi);\ \xi = x - ct$$

Then we have

$$-c\frac{du}{d\xi} + u\frac{du}{d\xi} + \mu\frac{d^3u}{d\xi^3} = 0\ \left(\mu = \frac{1}{2}c_0\lambda_e^2\right)$$

Integrating this equation gives

$$\mu\,\frac{d^2u}{d\xi^2} + \frac{u^2}{2} - cu = \ \text{const.} \equiv 0$$

$$\text{or}\ \mu\,\frac{d^2u}{d\xi^2} = cu - \frac{u^2}{2} \equiv -\frac{d}{du}\phi(u) \tag{10.59}$$

$$\text{where}\ \phi(u) = -\frac{c}{2}u^2 + \frac{u^3}{6} + c'\ (c' = \text{const.}) \tag{10.60}$$

Equation (10.59) should be compared with Newton's equation for a particle of mass μ subject to a force of potential $\phi(x)$: $\mu\,\frac{d^2x}{dt^2} = -\frac{d\phi}{dx}$ with the correspondence $u \to x$ and $\xi \to t$. Equation (10.59) allows for an energy integral:

$$\frac{\mu}{2}\left(\frac{du}{d\xi}\right)^2 + \phi(u) = \text{const.} \equiv E \tag{10.61}$$

Letting the constant E ('energy') absorb the constant c' in (10.60), we write the 'potential' in the form

$$\phi(u) = -\frac{c}{2}u^2 + \frac{u^3}{6} \tag{10.62}$$

The potential $\phi(u)$ is plotted in Fig. 10.3.

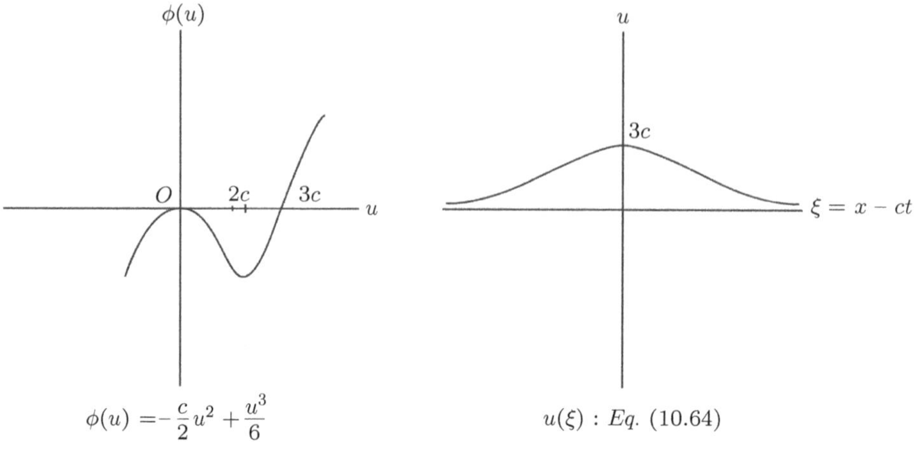

Fig. 10.3. 'Potential' and the orbit.

The function ϕ has a maximum at $u = 0$, a minimum at $u = 2c$, and $\phi = 0$ at $u = 0$ and $3c$. Depending upon the values of E, different kinds of solution emerge. When the energy E takes the value somewhere between 0 and $\phi_{\min}$ (corresponding to $u = 2c$), the particle executes asymmetric oscillation in (u, ξ)-plane. This is because the right slope of the potential well is steeper than the left slope. A more interesting solution emerges when $E = 0$. In this case the particle starts to move near the point $u = 0$ (where $\phi = 0$), rolls down to the bottom of the well, and climbs up the hill until to reach the point $u = 3c$ which is the turning point. The turned-around particle takes the backward course to reach the point $u = 0$, taking infinite time. Thus the particle orbit has only one crest ($u = 3c$) in this case. When $E = 0$, (10.61) gives

$$\sqrt{\frac{c}{\mu}}\xi = \int \frac{du}{u\sqrt{1 - \frac{u}{3c}}} \equiv I \tag{10.63}$$

To calculate the integral I, we put

$$\sqrt{1 - \frac{u}{3c}} = \psi, \quad u = 3c(1 - \psi^2)$$

Then we have

$$I = -2\int \frac{d\psi}{1 - \psi^2} = -\int d\psi \left(\frac{1}{1 - \psi} + \frac{1}{1 + \psi}\right)$$

Note that $1 - u/3c > 0$ and also $u/3c > 0$. Therefore $\psi < 1$. Thus we obtain

$$I = \ln \frac{1-\psi}{1+\psi},$$

$$\psi = \frac{1 - e^{\sqrt{\frac{c}{\mu}}\xi}}{1 + e^{\sqrt{\frac{c}{\mu}}\xi}}, \qquad u = \frac{3c}{\cos h^2 \left(\frac{1}{2}\sqrt{\frac{c}{\mu}}\ \xi\right)} \tag{10.64}$$

A plot u vs. ξ is shown in Fig. 10.3.

It is suggestive to write (10.64) in the form

$$u = \frac{u_0}{\cos h^2 \frac{\xi}{\Delta}}$$

with $u_0 = 3c$: soliton amplitude (height), and $\Delta = 2\sqrt{\frac{\mu}{c}}$: soliton thickness. Since c denotes the soliton speed, a soliton which is moving faster has taller height. Also we have $u_0\Delta^2 = 12\mu = 6c_0\lambda_e^2 =$ const; the product of the square of the thickness and the amplitude is constant. If we have a *positive dispersion*, the LHS of (10.59) should have the negative sign (change $\mu \to -\mu$). In this case, we still have the soliton solution in the form of (10.64) provided the sign of c is changed. Thus, the shape of the soliton is upsided-down, and moves toward the negative direction.

A useful reference for Korteweg-de Vries (KdV) soliton is: Kadomtsev (2001).

Boussinesq equation. Equation (10.57) can be converted to a differential equation by putting $k = -i\frac{\partial}{\partial x}$ and $\omega = i\frac{\partial}{\partial t}$ without making the weak dispersion approximation, $k^2\lambda_e^2 \ll 1$. In this case, we obtain

$$\left(\frac{\partial^2}{\partial t^2} - \lambda_e^2 \frac{\partial^2}{\partial x^2}\frac{\partial^2}{\partial t^2} - c_0^2 \frac{\partial^2}{\partial x^2}\right) v = \left(1 - \lambda_e^2 \frac{\partial^2}{\partial x^2}\right) \frac{\partial^2}{\partial t \partial x}\frac{v^2}{2} \tag{10.65}$$

The operator acting on v in the left hand side of (10.65) is the Boussinesq wave operator. This equation describes the waves propagating both $\pm x$ direction while the KdV equation (10.58) is relevant to the wave propagating only $+x$ direction.

Remark. Equation (10.51) can be integrated upon introducing the electric potential $\mathbf{E} = -\nabla\phi$:

$$n_e = n_0\ e^{\frac{e\phi}{T_e}} \quad \text{or} \quad \phi = \frac{T_e}{e}\ ln\frac{n_e}{n_0} \tag{10.66}$$

Then the Poisson equation reads

$$\frac{\partial^2 \phi}{\partial x^2} = 4\pi e \left(n_0\, e^{\frac{e\phi}{T_e}} - n_i\right) \tag{10.67}$$

Now we have the ion continuity equation

$$\frac{\partial n_i}{\partial t} + \frac{\partial}{\partial x}(n_i v) = 0 \tag{10.68}$$

and the ion equation of motion

$$\frac{\partial v}{\partial t} + v\frac{\partial v}{\partial x} = -\frac{e}{m_i}\frac{\partial \phi}{\partial x} \tag{10.69}$$

The above three equations are fully nonlinear and include the dispersion effect (i.e. effect of charge separation) for the ion acoustic wave. Equations (10.67)–(10.69) have been analyzed by various perturbation methods. In Appendix C, it is shown that the method of *strained coordinates* derives the KdV equation from (10.67)–(10.69).

Note. If we assume the charge neutrality, putting $n_i \simeq n_e$ and using (10.43) and (10.67) give

$$\frac{\partial v}{\partial t} + v\frac{\partial v}{\partial x} = -\frac{T_e}{m_i}\frac{1}{n_i}\frac{\partial n_i}{\partial x} \tag{10.70}$$

(10.68) and (10.70) are the equations of ideal gas dynamics for an adiabatic law of $\gamma = 1$ (γ is the ratio of specific heats). The wave governed by (10.68) and (10.70) steepens its own wave profile and is developed into a shock wave. This can be rigorously proved by using *Riemann invariants.*

Reference: H. J. Lee, *Nonlinear analysis of the cold fluid-Poisson plasma by using the characteristic method*, J. Korean Phys. Soc. **69** (2016) pp. 1191–1211.

Sagdeev potential. We look for an exact nonlinear solution of the ion acoustic wave equations (10.67)–(10.69) under given boundary conditions. The solution is sought in the form of a progressive wave. Write $\phi(x,t) = \phi(\xi)$ with $\xi = x - ut$. Then, $\partial/\partial x = \partial/\partial\xi$ and $\partial/\partial t = -u\,\partial/\partial\xi$, and (10.68) gives

$$-u\,\frac{\partial n_i}{\partial \xi} + \frac{\partial}{\partial \xi}(n_i v) = 0 \longrightarrow n_i(v-u) = \text{ const.} \equiv c_1$$

The boundary conditions are: $n_i = n_0; v = 0$ at $\xi \to \infty$, which gives

$$n_i(v-u) = -n_0 u = c_1 \tag{10.71}$$

From (10.69) we obtain

$$-u\,\frac{\partial v}{\partial \xi}+v\frac{\partial v}{\partial \xi}=-\frac{e}{m_i}\frac{\partial \phi}{\partial \xi}\longrightarrow (v-u)^2+2\,\frac{e\phi}{m_i}=u^2 \tag{10.72}$$

where we used the boundary condition, $\phi = v = 0$ at $\xi = \infty$. Eliminating v from (10.71) and (10.72), we obtain

$$n_i=\frac{n_0 u}{\sqrt{u^2-\frac{2e\phi}{m_i}}} \tag{10.73}$$

Substituting (10.73) into (10.67), we can write

$$\frac{d^2\phi}{d\xi^2}=4\pi e n_0\left[\exp\left(\frac{e\phi}{T_e}\right)-\frac{u}{\sqrt{u^2-\frac{2e\phi}{m_i}}}\right]\equiv-\frac{dV(\phi)}{d\phi} \tag{10.74}$$

Equation (10.74) is again Newton's equation of a particle moving in a potential $V(\phi)$, which is called the *Sagdeev potential* and takes the form

$$V(\phi)=4\pi e n_0\left[-\frac{T_e}{e}e^{\frac{e\phi}{T_e}}-\frac{m_i u^2}{e}\sqrt{1-\frac{2e\phi}{m_i u^2}}\right]+C$$

where C is an integration constant. Let us determine C so that $V(\phi = 0) = 0$. Thus we obtain

$$V(\phi)=-4\pi n_0\left[T_e\,e^{\frac{e\phi}{T_e}}+m_i u^2\sqrt{1-\frac{2e\phi}{m_i u^2}}-T_e-m_i u^2\right] \tag{10.75}$$

Note that $V(\phi)$ in (10.75) satisfies $dV/d\phi = 0$ at $\phi = 0$. So it is possible to have the shape of the potential *well* as shown in Fig. 10.4. [A shape of potential *hill* is also possible.] A particle situated at the origin $V = \phi = 0$ which is the point of unstable equilibrium draws a 'trajectory' of a solitary

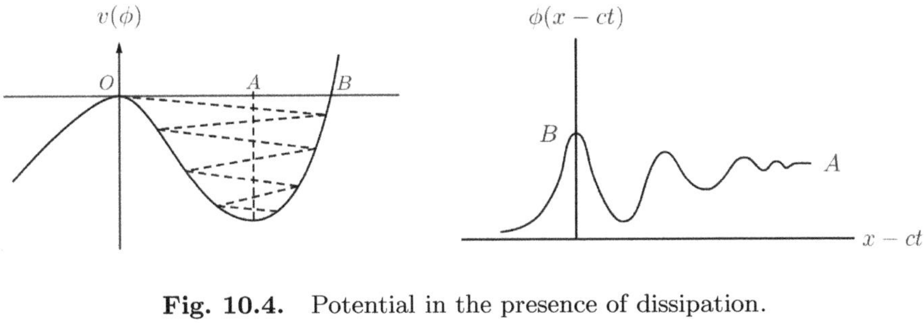

Fig. 10.4. Potential in the presence of dissipation.

pulse. When there are dissipations, the particle loses energy while it bounces back and forth in the potential well. After each cycle, the amplitude of oscillation decreases, and these ever diminishing oscillations repeat several times until the particle settles down at at the bottom of the well. The plot ϕ vs. ξ is shown in Fig. 10.5. It represents the quasi-shock structure which is called the *ion acoustic shock*.

When the potential has two maxima as shown in Fig. 10.5, the corresponding 'motion' is described by a hyperbolic tangent function. The shape of the curve is a *shock-like transition*.

Nonlinear Schrödinger equation. The nonlinear Schrödinger equation describes amplitude modulation of an otherwise monochromatic wave.

The amplitude of a wave can be modulated by interactions among different Fourier components. Parametric interaction among waves in plasma often results in amplitude-modulated wave of normal mode. Here we derive heuristically the nonlinear Schrödinger equation by *derivative expansion* technique applied to a *nonlinear dispersion relation*. In a weakly nonlinear system, the dispersion relation can be amplitude-dependent form

$$\omega = \omega(\mathbf{k}, |A|^2) \tag{10.76}$$

Such an amplitude-dependent dispersion relation emerges in nonlinear optics as well as in plasmas when the dielectric permeability of the medium depends on the electric field of the wave propagated in it. Suppose that the wave function $\psi(\mathbf{r}, t)$ takes the form

$$\psi(\mathbf{r}, t) = A(\mathbf{r}, t)e^{i\mathbf{k}_0\cdot\mathbf{r} - i\omega_o(\mathbf{k}_0)t} \tag{10.77}$$

where ω_0 corresponds to the eigen frequency of the plane harmonic wave with wave vector $\mathbf{k}_0$. We assume that the wave propagates along the x-direction:

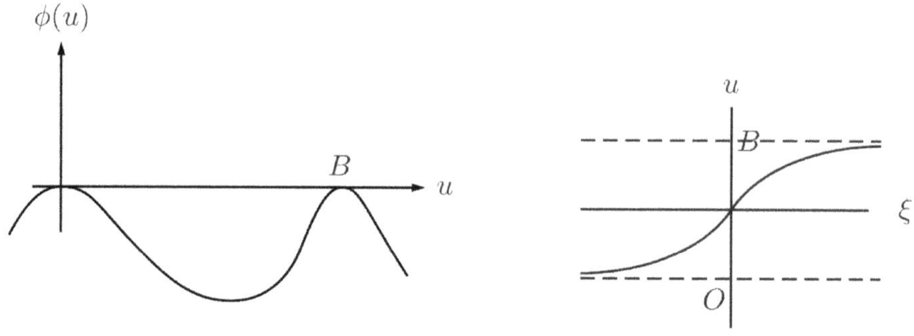

Fig. 10.5. Shock-like transition.

$\mathbf{k}_0 = \hat{\mathbf{x}} k_0$. In a nonlinear medium, the wave amplitude $A(\mathbf{r}, t)$ depends not only on x but also on the perpendicular coordinates. So we rewrite (10.77) as

$$\psi(\mathbf{r}, t) = A(x, \mathbf{r}_\perp, t) e^{ik_0 x - i\omega_o(\mathbf{k}_0)t} \tag{10.78}$$

We regard ω and $\mathbf{k}$ in the nonlinear dispersion relation (10.76) as corresponding to the operators: $\omega \to i\frac{\partial}{\partial t}$, and $\mathbf{k} \to -i\nabla$. Differentiating (10.78) gives

$$i\frac{\partial \psi}{\partial t} = \left[\omega_0 A + i\left(\frac{\partial}{\partial t}\right)_\epsilon A\right] e^{-i\omega_0 t + ik_0 x}$$

$$-i\nabla\psi = (\hat{\mathbf{x}} k_0 A - i\nabla_\epsilon A) e^{-i\omega_0 t + ik_0 x}$$

where $\left(\frac{\partial}{\partial t}\right)_\epsilon$ and ∇_ϵ denote the slow changes of the amplitude $A(\mathbf{r}, t)$. Explicitly, we have the correspondence:

$$\omega = \omega_0 + i\left(\frac{\partial}{\partial t}\right)_\epsilon, \quad \mathbf{k} = \hat{\mathbf{x}} k_0 - i\nabla_\epsilon$$

And the slow changes can be found from the nonlinear dispersion relation (10.76) after Taylor expanding $\omega(\mathbf{k}, |A|^2)$ around the wave number k_0 and the frequency ω_0. We assume that the wave propagates in an isotropic medium, then the dispersion relation reads

$$\omega = \omega(k, |A|^2) \tag{10.79}$$

where we assume that the wave vector $\mathbf{k}$ is predominantly along the $\hat{\mathbf{x}}$-direction:

$$k = \sqrt{k_x^2 + k_\perp^2} = k_x\sqrt{1 + \frac{k_\perp^2}{k_x^2}} \simeq k_x\left(1 + \frac{k_\perp^2}{2k_x^2}\right) = k_x + \frac{1}{2}\frac{k_\perp^2}{k_x}$$

$$k \simeq k_0 + \left(k_x - k_0 + \frac{1}{2}\frac{k_\perp^2}{k_0}\right) \tag{10.80}$$

Note that the quantity in the parenthesis is much smaller than k_0. Expanding the right-hand side of (10.79) in a Taylor series, we have

$$\omega = \omega_0 + \left(\frac{\partial \omega}{\partial k}\right)_{k_0}(k - k_0) + \frac{1}{2}\left(\frac{\partial^2 \omega}{\partial k^2}\right)_{k_0}(k - k_0)^2 + \left(\frac{\partial \omega}{\partial |A|^2}\right)_{k_0}|A|^2 + \cdots \tag{10.81}$$

Using (10.80) in this equation gives

$$\omega = \omega_0 + \left(\frac{\partial \omega}{\partial k}\right)_{k_0}\left(k_x - k_0 + \frac{1}{2}\frac{k_\perp^2}{k_0}\right) + \frac{1}{2}\left(\frac{\partial^2 \omega}{\partial k^2}\right)_{k_0}(k_x - k_0)^2 + \left(\frac{\partial \omega}{\partial |A|^2}\right)_{k_0}|A|^2 + \cdots \tag{10.82}$$

where $\omega_0 = \omega(k_0, |A|^2 = 0)$. Now the operator equation

$$\omega(k, |A|^2)\ \psi(\mathbf{r}, t) = 0$$

is converted into the following differential equation, by invoking the correspondence in (10.79), viz., $\omega - \omega_0 = i(\frac{\partial}{\partial t})_\epsilon$, $k_x - k_0 = -i\left(\frac{\partial}{\partial x}\right)_\epsilon$, and $k_\perp^2 = -\nabla_\perp^2$. We obtain the equation for the slow change of the amplitude $A(\mathbf{r}, t)$:

$$i\left[\frac{\partial}{\partial t} + \left(\frac{\partial \omega}{\partial k}\right)_{k_0} \frac{\partial}{\partial x}\right] A + \frac{1}{2}\left(\frac{\partial^2 \omega}{\partial k^2}\right)_{k_0} \frac{\partial^2 A}{\partial x^2} + \frac{1}{2}\left(\frac{\partial \omega}{\partial k}\right)_{k_0} \frac{\nabla_\perp^2 A}{k_0} - \left(\frac{\partial \omega}{\partial |A|^2}\right)_{k_0} |A|^2 A = 0 \tag{10.83}$$

Equation (10.83) is called *nonlinear parabolic equation* [Reference: V. I. Karpman, *Non-linear waves in dispersive media*, Pergamon (1975)]. If we neglect the variation of A in the perpendicular direction ($\nabla_\perp^2 A = 0$), (10.83) is the nonlinear Schrödinger equation.

10.6 Interaction of Langmuir Wave and Ion Acoustic Wave; Zakharov Equation

In the last section, we discussed the low frequency ion acoustic wave in the background of thermalized (Boltzmann-distributed) electrons to see that the nonlinear ion acoustic wave is described by the KdV equation. In this section, we consider the mutual interaction between the ion acoustic wave (long wave) and the Langmuir wave (short wave) which are the two branches belonging to one dispersion relation of the electrostatic plasma wave. The Zakharov equation describes the nonlinear coupling of the two waves. We shall consider the linear waves first which can be derived from the following linearized two fluid equations.

$$\frac{\partial n_\alpha}{\partial t} + n_0 \frac{\partial v_\alpha}{\partial x} = 0\ (\alpha = e,\ i) \tag{10.84}$$

$$\frac{\partial v_\alpha}{\partial t} = -\frac{\gamma_\alpha T_\alpha}{m_\alpha n_0} \frac{\partial n_\alpha}{\partial x} + \frac{e_\alpha}{m_\alpha} E \tag{10.85}$$

$$\frac{\partial E}{\partial x} = 4\pi e(n_i - n_e) \tag{10.86}$$

We have the electrostatic Maxwell equation,

$$\frac{\partial E}{\partial t} + 4\pi n_0 \sum_\alpha e_\alpha v_\alpha = 0 \tag{10.87}$$

Assuming that the wave quantities, n_i, n_e, v_i, v_e, E are $\sim e^{ikx-i\omega t}$, we obtain the dispersion relation

$$1 - \frac{\omega_{\text{pi}}^2}{\omega^2 - k^2\frac{\gamma_i T_i}{m_i}} - \frac{\omega_{\text{pe}}^2}{\omega^2 - k^2\frac{\gamma_e T_e}{m_e}} = 0 \tag{10.88}$$

The algebraic equation (10.88) can be approximately solved for ω by expanding it in the powers of m_e/m_i to obtain the following two modes (See Section 7.11)

$$\omega^2 = \omega_{\text{pe}}^2 + \frac{\gamma_e T_e}{m_e}k^2 \tag{10.89}$$

$$\omega^2 = \frac{k^2}{m_i}\left[\gamma_i T_i + \frac{\gamma_e T_e}{1+\gamma_e k^2\lambda_e^2}\right] \tag{10.90}$$

where λ_e is the electron Debye length. Equation (10.89) represents the Langmuir wave and (10.90) the ion acoustic wave. In (10.90), T_i term can be neglected since the condition $T_e \gg T_i$ is necessary in order for the ion acoustic wave to be sustained as a normal mode. Otherwise it is heavily Landau-damped. In one-dimensional motion, the value of the specific heat ratio is: $\gamma_i = 1$ and $\gamma_e = 3$. [For adiabatic process, $\gamma = 3$. In a fast wave, the electrons are not fast enough to transport heat, so the process is regarded as adiabatic. For isothermal process, $\gamma = 1$. In a slow wave, the particles have ample time to equalize temperature, and the process is regarded as isothermal.]

The two waves are coupled if nonlinear terms are introduced. The important nonlinearities are: the high frequency nonlinear electron current in the Langmuir wave equation and the low frequency ponderomotive force acting on the low frequency ion acoustic perturbations. The ponderomotive force acting on the ion wave is the radiation pressure that is given rise to by the beat of two high frequency electric fields associated with the Langmuir waves. The underlying physics owes to the fact that the Langmuir wave frequency $\omega(k)$ is much greater than the ion acoustic wave frequency, $\sqrt{T_e/m_i}\, k'$, where k' is the wavenumber of the ion wave:

$$\omega^2(k) = \omega_{\text{pe}}^2 + 3k^2\,\frac{T_e}{m_e} \gg k'^2\,\frac{T_e}{m_i} \quad (k \gg k')$$

In this case, the electron Langmuir mode is treated as propagating in a slowly varying, weakly inhomogeneous background of ions. The ion background is not uniform because of the reaction exerted by the electron plasma waves propagated in it. The ions see effects averaged over many periods of the rapid electron oscillation. The coupling is similar to the interaction in fluid which is known as the *long wave short wave interaction* [Benny, D. J., *A general theory for interactions between long and short waves*, Studies in Appl. Math., **56**, pp. 81–94]. In order to formulate the interaction of the two waves, we separate the high frequency equations and the low frequency equations. First, in the Langmuir plus the ion acoustic wave fields, we write for the various fluid quantities,

$$v_e = v_e^h + v_e^l, \quad n_e = n_0 + n_e^h + n_e^l, \quad n_i = n_0 + n_i^l, \quad v_i = v_i^l, \quad E = E^h + E^l$$

where the superscript 'h' denotes the high frequency quantities associated with the Langmuir wave and 'l' denotes the low frequency ion acoustic wave fields. The 'h' and 'l' quantities oscillate with respective time scales of great difference. Note that the electron quantities have both the low and high frequency components while the ion quantities have only the low frequency components.

High frequency electron equation of motion:

$$\frac{\partial v_e^h}{\partial t} = -\frac{e}{m_e}E^h - \frac{3T_e}{n_0 m_e}\frac{\partial n_e^h}{\partial x} \tag{10.91}$$

High frequency electron continuity equation:

$$\frac{\partial n_e^h}{\partial t} + n_0\frac{\partial v_e^h}{\partial x} = 0 \tag{10.92}$$

Low frequency electron equation of motion:

$$-\frac{e}{m_e}E^l - \frac{T_e}{n_0 m_e}\frac{\partial\, n_e^l}{\partial x} = \left[v_e^h \frac{\partial v_e^h}{\partial x}\right]_l \tag{10.93}$$

Low frequency electron continuity equation:

$$\frac{\partial n_e^l}{\partial t} + n_0\frac{\partial v_e^l}{\partial x} = 0 \tag{10.94}$$

Low frequency ion equation of motion:

$$\frac{\partial v_i^l}{\partial t} = \frac{e}{m_i}E^l \tag{10.95}$$

Low frequency ion continuity equation:

$$\frac{\partial n_i^l}{\partial t} + n_0 \frac{\partial v_i^l}{\partial x} = 0 \tag{10.96}$$

We neglected the electron inertia in (10.93) and the last term of it is derived from the low frequency beat of two high frequency waves, i.e. the electron ponderomotive force. The ion ponderomotive force which might be added to the ion equation is negligible by the mass ratio. In addition to the above fluid equations, we have Maxwell's equations:

$$\frac{\partial E^h}{\partial t} = 4\pi e v_e^h (n_0 + n_e^l) \tag{10.97}$$

$$\frac{\partial E^l}{\partial t} = -4\pi e n_0 (v_i^l - v_e^l) \tag{10.98}$$

In (10.97), the last term $v_e^h n_e^l$ represents the nonlinear beat current producing a high frequency component. Differentiating (10.91) and using (10.92) and (10.97) yield

$$\left(\frac{\partial^2}{\partial t^2} + \omega_{\mathrm{pe}}^2 - \frac{3T_e}{m_e}\frac{\partial^2}{\partial x^2}\right) v_e^h = -\omega_{\mathrm{pe}}^2 v_e^h \frac{n_e^l}{n_0} \tag{10.99}$$

Differentiating (10.93) and using (10.95) for E^l give

$$\frac{m_i}{m_e}\frac{\partial}{\partial t}\frac{\partial v_i^l}{\partial x} - \frac{T_e}{n_0 m_e}\frac{\partial^2 n_e^l}{\partial x^2} = \frac{\partial}{\partial x}\left[v_e^h \frac{\partial v_e^h}{\partial x}\right]_l$$

For the first term of the above equation, we use (10.96) and put $n_i^l \simeq n_e^l$ (in the low frequency region the charge neutrality holds) to obtain

$$\left(\frac{\partial^2}{\partial t^2} - \frac{T_e}{m_i}\frac{\partial^2}{\partial x^2}\right)\frac{n_e^l}{n_0} = \frac{m_e}{m_i}\frac{\partial}{\partial x}\left[v_e^h \frac{\partial v_e^h}{\partial x}\right]_l \tag{10.100}$$

Equations (10.99) and (10.100) are the result of reduction of (10.91)–(10.98). Note that we used the charge neutrality relation and did not use (10.94) and (10.98). Combining (10.98) with (10.94) and (10.96) gives the Poisson equation

$$\frac{\partial E^l}{\partial x} = 4\pi e (n_i^l - n_e^l) \tag{10.101}$$

The Zakharov equations can be derived from (10.99) and (10.100) as follows. We write the high frequency electron fluid velocity as

$$v_e^h(x,t) = v_0(x,t)e^{-i\omega_{\text{pe}}t} + c.c. \tag{10.102}$$

The slowly varying amplitude $v_0(x,t)$ owes its variation to the nonlinear coupling which makes the plasma slightly 'inhomogeneous'. If this expression is substituted into (10.99), it becomes

$$i\omega_{\text{pe}}\frac{\partial v_0}{\partial t} + \frac{3}{2}\frac{T_e}{m_e}\frac{\partial^2 v_0}{\partial x^2} = \frac{1}{2}\omega_{\text{pe}}^2 v_0\,\frac{n_e^l}{n_0} \tag{10.103}$$

where we neglected the second temporal derivative of v_0, and now (10.103) is a low frequency equation for the slowly varying envelope $v_0(x,t)$ of the rapidly oscillating plasma wave $v_e^h(x,t)$. The low frequency ponderomotive force in (10.100) drives a density modulation in the ion time scale. Usually, the low frequency component of the force is obtained by taking average over many periods of the Langmuir wave,

$$\left\langle v_e^h\frac{\partial v_e^h}{\partial x}\right\rangle = \frac{1}{2}\frac{\partial}{\partial x}\left\langle (v_e^h)^2\right\rangle = \frac{\partial}{\partial x}|v_0(x,t)|^2$$

where $v_0^2e^{-2i\omega_{\text{pe}}t}$, etc. average to zero. Then (10.100) becomes

$$\left(\frac{\partial^2}{\partial t^2} - \frac{T_e}{m_i}\frac{\partial^2}{\partial x^2}\right)\frac{n_e^l}{n_0} = \frac{m_e}{m_i}\frac{\partial^2}{\partial x^2}|v_0(x,t)|^2 \tag{10.104}$$

This is the ion acoustic wave equation driven by the ponderomotive force. Equations (10.103) and (10.104) are the Zakharov equations.

Note. If we use the expression, $v_e^h(x,t) = \frac{1}{2}[v_0(x,t)e^{-i\omega_{\text{pe}}t} + v_0^*(x,t)\ e^{i\omega_{\text{pe}}t}]$ (this amounts to using $\cos(\omega_{\text{pe}}t)$ for the rapidly oscillating function), instead of $v_e^h(x,t) = v_0(x,t)e^{-i\omega_{\text{pe}}t} + v_0^*(x,t)e^{i\omega_{\text{pe}}t}$ (without the factor 1/2, see (10.102)), we have $<(v_e^h)^2> = v_0^2/2$. So the right hand side of (10.104) should be multiplied by a factor $\frac{1}{4}$.

In order to write (10.104) in terms of the electric field, we use (10.91) with the pressure term neglected:

$$\frac{\partial v_e^h}{\partial t} = -\frac{e}{m_e}E^h$$

Using $v_e^h = v_0(x,t)e^{-i\omega_{\rm pe}t}$ in the above equation gives

$$-\frac{e}{m_e}E^h = -i\omega_{\rm pe}v_0(x,t)e^{-i\omega_{\rm pe}t}$$

where we neglected $\frac{\partial v_0}{\partial t}$. So the slowly varying part of E^h corresponding to $v_0(x,t)$ is

$$v_0(x,t) = \frac{-ie}{m_e\omega_{\rm pe}}\tilde{E}(x,t) \quad (E^h(x,t) = \tilde{E}(x,t)e^{-i\omega_{\rm pe}t})$$

which gives

$$\left(\frac{\partial^2}{\partial t^2} - \frac{T_e}{m_i}\frac{\partial^2}{\partial x^2}\right)\frac{n_e^l}{n_0} = \frac{m_e}{m_i}\cdot\frac{e^2}{m_e^2\omega_{\rm pe}^2}\frac{\partial^2}{\partial x^2}|\tilde{E}(x,t)|^2 \tag{A}$$

The right hand side of (A) should be multiplied by factor $\frac{1}{16}$ if factor $\frac{1}{2}$ is used in v_0.

If we neglect the second temporal derivative in (10.104), as has been done in (10.103), we have

$$i\frac{\partial v_0}{\partial t} + \frac{3}{2\omega_{\rm pe}}\frac{T_e}{m_e}\frac{\partial^2 v_0}{\partial x^2} + \frac{\omega_{\rm pe}}{2}\frac{m_e}{T_e}|v_0|^2 v_0 = 0 \tag{10.105}$$

This is a nonlinear Schrödinger equation.

Equation (10.105) allows for a soliton solution, which is an *envelope soliton* in this case. Introducing the dimensionless variables,

$$v_0' = \sqrt{\frac{m_e}{T_e}}\, v_0,\ t' = \frac{\omega_{\rm pe}}{2}t,\ x' = \frac{x}{\sqrt{3}\,\lambda_e} \tag{10.106}$$

Equation (10.105) takes the form

$$i\,\frac{\partial v_0'}{\partial t'} + \frac{\partial^2 v_0'}{\partial x'^2} + |v_0'|\,^2 v_0' = 0 \tag{10.107}$$

Exercise. Show that

$$v_0(x,t) = \sqrt{2}e^{i[\frac{1}{2}bx - (\frac{1}{4}b^2 - a^2)t]}\sec h[a(x-bt)]$$

satisfies (10.107) and the boundary condition $v_0 \to 0$ as $|x| \to \infty$ where a and b are arbitrary constants. Note that the $\sec h$ function part is an envelope to the oscillatory part. This is a progressive wave solution.

Exercise. Show that

$$v_0(x,t) = \sqrt{2A}\,e^{iAt}\,\sec h(\sqrt{A}\,x)$$

satisfies (10.107) and the boundary condition $v_0 \to 0$ as $|x| \to \infty$ where A is an arbitrary constant. This solution is called a *Langmuir soliton*. This is a standing wave solution.

10.7 Modulational Instability

Let us take a look at Eq. (10.100). On the right hand side of it, beating two high frequency waves can produce the low frequency component that resonate with the low frequency ion wave on the left hand side, thus driving an instability—the envelope of the low frequency wave can grow, hence the name.

Consider two side bands about a central wavenumber k in the Langmuir wavepacket with $k_+ = k + \Delta k$, $k_- = k - \Delta k$ ($\Delta k \ll k$), and $\omega_+ = \omega(k_+)$, $\omega_- = \omega(k_-)$:

$$\phi(k_+, \omega_+) = A_+ e^{ik_+x - i\omega_+t} + A_+^* e^{-ik_+x + i\omega_+t}$$

$$\phi(k_-, \omega_-) = A_- e^{ik_-x - i\omega_-t} + A_-^* e^{-ik_-x + i\omega_-t}$$

If these side band components are mixed in a nonlinear term, the sum and difference frequencies and wavenumbers are generated:

$$\phi(k_+, \omega_+)\phi(k_-, \omega_-) = A_+A_- e^{i(k_+ + k_-)x - i(\omega_+ + \omega_-)t} + A_+A_-^* e^{i(k_+ - k_-)x - i(\omega_+ - \omega_-)t} + \text{c.c.}$$

where c.c. is the complex conjugate. We pay attention to the second term whose wavenumber and frequency are

$$k_+ - k_- = 2\Delta k$$

$$\omega_+ - \omega_- = \omega(k + \Delta k) - \omega(k - \Delta k) \simeq \frac{\partial \omega}{\partial k} 2\Delta k$$

Under the resonance conditions,

$$2\Delta k = k_l$$

$$2\Delta k \frac{\partial \omega}{\partial k} = \omega_l$$

where k_l and ω_l are, respectively, the wavenumber and frequency of the low frequency ion wave. It is seen that the resonance condition takes the form

$$\frac{\omega_l}{k_l} = \frac{\partial \omega}{\partial k} \tag{10.108}$$

at some k. In words, the resonance condition states that the phase velocity of the low frequency ion wave should be equal to the group velocity of the high frequency Langmuir wave at some wavenumber. The feasibility of this resonance condition is graphically shown in Fig. 10.6. Describing this instability, the terminology, *long wave short wave resonance*, appears to be coined by Benny who investigated the coupling of the gravity and capillary modes of water waves.

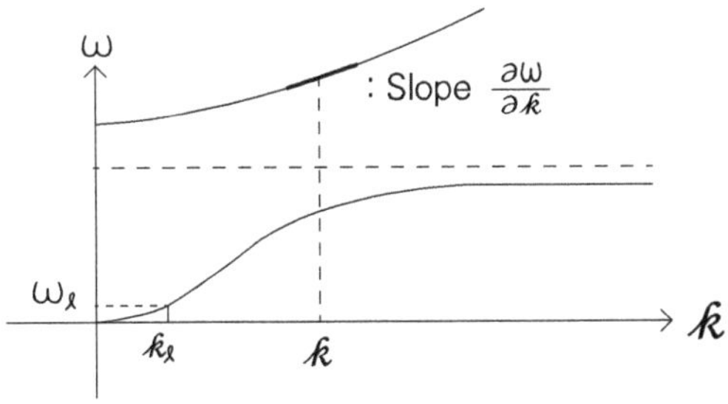

Fig. 10.6. Long wave and short wave resonance.

• **Modulational instability.** Instability of the slowly changing envelope. In the case that the amplitude $\psi(x,t)$ asymptotically approaches ψ_0 as $t \to \infty$, the corresponding nonlinear Schroedinger equation takes the form

$$i\left(\frac{\partial \psi}{\partial t}+\frac{\partial \omega}{\partial k}\frac{\partial \psi}{\partial x}\right)+\frac{1}{2}\frac{\partial^2 \omega}{\partial k^2}\frac{\partial^2 \psi}{\partial x^2}-\frac{\partial \omega}{\partial |\psi|^2}\left(|\psi|^2-|\psi_0|^2\right)\psi=0 \tag{10.109}$$

which corresponds to the expansion of the nonlinear dispersion relation $\omega(k,|\psi|^2)$ in the vicinity of $k=k_0$ and $|\psi|^2=|\psi_0|^2$. All the derivatives of ω in (10.109) are understood being evaluated at $k=k_0$, the center wavenumber. The modulational instability is investigated by performing the linear stability analysis of (10.109). Let us write the complex amplitude of the amplitude function ψ in the form

$$\psi(x,t)=\rho(x,t)\;e^{i\varphi(x,t)}$$

where ρ and φ are real. Upon substitution of the above expression into (10.109), we obtain two equations for $\rho(x,t)$ and $\varphi(x,t)$.

$$\left(\frac{\partial}{\partial t}+\frac{\partial \omega}{\partial k}\frac{\partial}{\partial x}\right)\varphi-\frac{1}{2\rho}\frac{\partial^2 \omega}{\partial k^2}\frac{\partial^2 \rho}{\partial x^2}+\frac{1}{2}\frac{\partial^2 \omega}{\partial k^2}\left(\frac{\partial \varphi}{\partial x}\right)^2+q(\rho^2-|\psi_0|^2)=0 \tag{10.110}$$

where $q=\frac{\partial \omega}{\partial |\psi|^2}$ is the nonlinear frequency shift.

$$\left(\frac{\partial}{\partial t}+\frac{\partial \omega}{\partial k}\frac{\partial}{\partial x}\right)\rho+\frac{\partial^2 \omega}{\partial k^2}\left(\frac{\partial \rho}{\partial x}\frac{\partial \varphi}{\partial x}+\frac{\rho}{2}\frac{\partial^2 \varphi}{\partial x^2}\right)=0 \tag{10.111}$$

The slow variation of the amplitude in space and time can be described in terms of small perturbations in ρ and φ:

$$\rho=\rho_0+\rho',\quad \varphi=\varphi_0+\varphi'$$

where primed quantities are perturbations. Linearizing (10.110) and (10.111) gives

$$\rho_0\left(\frac{\partial}{\partial t}+\frac{\partial\omega}{\partial k}\frac{\partial}{\partial x}\right)\varphi'-\frac{1}{2}\frac{\partial^2\omega}{\partial k^2}\frac{\partial^2\rho'}{\partial x^2}+2q\rho_0^2\,\rho'=0 \tag{10.112}$$

$$\left(\frac{\partial}{\partial t}+\frac{\partial\omega}{\partial k}\frac{\partial}{\partial x}\right)\rho'+\frac{\rho_0}{2}\frac{\partial^2\omega}{\partial k^2}\frac{\partial^2\varphi'}{\partial x^2}=0 \tag{10.113}$$

The above equations are solved by the form $\exp(i\kappa x - i\nu t)$ to obtain the dispersion relation

$$\nu=\kappa\frac{\partial\omega}{\partial k}\pm\sqrt{\kappa^2\frac{\partial^2\omega}{\partial k^2}\left(q\rho_0^2+\frac{\kappa^2}{4}\frac{\partial^2\omega}{\partial k^2}\right)} \tag{10.114}$$

Growing solution is obtained only when

$$q\,\frac{\partial^2\omega}{\partial k^2}=\frac{\partial\omega}{\partial|\psi|^2}\frac{\partial^2\omega}{\partial k^2}<0 \tag{10.115}$$

Viz., only when the group dispersion $\frac{\partial^2\omega}{\partial k^2}$ and the nonlinear frequency shift $\frac{\partial\omega}{\partial|\psi|^2}$ have opposite signs. This condition is known as the *Lighthill criterion.*

10.8 Electromagnetic Processes in Dispersive Plasma Waves

References: L. D. Landau and E. M. Lifshitz, *Electrodynamics of continuous media*, Pergamon Press, Oxford (1966) Section 61; G. Bekefi, *Radiation processes in plasmas*, Wiley, New York (1966); B. B. Kadomtsev, *Plasma turbulence*, Academic Press, London (1965).

Plasmas are dispersive media whose dielectric permittivity $\varepsilon(\mathbf{k},\omega)$ depends on the wave frequency and the wave vector. Dispersion of a wave in general accompanies a dissipation of energy, which means that a dispersive medium absorbs energy from the wave. The imaginary part of $\varepsilon(\mathbf{k},\omega)$ determines the absorption of the wave energy by the medium. Here we recall that the real and the imaginary parts of $\varepsilon(\mathbf{k},\omega)$ are connected by the Kramers–Kronig relations. Thus, the latter relations may be useful in calculating the energy dissipation. We have the *poynting theorem*

$$-\nabla\cdot\frac{c}{4\pi}(\mathbf{E}\times\mathbf{B})=\frac{1}{4\pi}\left[\mathbf{E}\cdot\frac{\partial\mathbf{D}}{\partial t}+\mathbf{B}\cdot\frac{\partial\mathbf{B}}{\partial t}\right]$$

By averaging with respect to time, the RHS yield the energy production rate in the medium, i.e. heat generated per unit time and volume. For a monochromatic wave of single frequency, this time average vanishes, and the entropy is conserved. Here we quote Landau and Lifshitz (1966): "the strict periodicity results in no steady accumulation of electromagnetic energy". To have correspondence with real irreversible process, we must consider a multi-chromatic wave or a wavepacket. We consider a simple form of wave with a slowly changing amplitude in the form (10.122). Mathematically, we have a two-time scale problem in calculating the electromagnetic field energy in plasma.

10.8.1 *Field energy of a dispersive wave*

In this section, we consider how an external forces (currents) changes the electromagnetic energy of a plasma. To calculate the work done by the external current, we write

$$\nabla \times \mathbf{B} = \frac{1}{c}\frac{\partial \mathbf{D}}{\partial t} + \frac{4\pi}{c}\mathbf{J}_{\text{ext}} \tag{10.116}$$

$$\nabla \times \mathbf{E} = -\frac{1}{c}\frac{\partial \mathbf{B}}{\partial t} \tag{10.117}$$

The scalar product of (10.116) with $\mathbf{E}$ gives the work performed by the external current. Also forming the divergence of the Poynting vector $\frac{c}{4\pi}\mathbf{E}\times\mathbf{B}$, we obtain

$$\frac{\partial W}{\partial t} \equiv \frac{1}{4\pi}\left[\mathbf{E}\cdot\frac{\partial \mathbf{D}}{\partial t} + \mathbf{B}\cdot\frac{\partial \mathbf{B}}{\partial t}\right] = -\nabla\cdot\frac{c}{4\pi}(\mathbf{E}\times\mathbf{B}) - \mathbf{E}\cdot\mathbf{J}_{\text{ext}} \tag{10.118}$$

The above equation defines the *field energy* W. When $\mathbf{J}_{\text{ext}} = 0$, $\int W d^3r =$ const, that is, the total field energy W contained in the whole space does not change with time. Note that $\mathbf{E}\cdot\frac{\partial \mathbf{D}}{\partial t}$ term is the sum of the electric wave energy plus the particle kinetic energy that enters through the polarization current. This quantity has a meaning only when it is averaged over the period of the wave.

For simplicity we shall assume that the response vector $\mathbf{D}$ is directed along the direction of the input field $\mathbf{E}$. So we consider

$$D(t) = E(t) + \int_0^\infty \chi(\tau)E(t-\tau)d\tau \tag{10.119}$$

Fourier-transforming gives

$$D(\omega) = \left(1 + \int_0^\infty dt\chi(t)e^{i\omega t}\right)E(\omega) = \varepsilon(\omega)E(\omega) \tag{10.120}$$

From the knowledge of the dielectric permeability $\varepsilon(\omega)$, we can determine $D(t)$ by inverting the above equation. Differentiating $D(t) = \int D(\omega)e^{-i\omega t}d\omega/2\pi$ gives

$$\frac{\partial D}{\partial t} = \frac{-i}{2\pi}\int_{-\infty}^{\infty} \omega\ \varepsilon(\omega)E(\omega)e^{-i\omega t}d\omega \tag{10.121}$$

Here we assume that $E(t)$ in (10.119) is quasi-monochromatic, i.e. it oscillates with a certain frequency ω_0 with a slowly changing amplitude:

$$E(t) = E_0(\nu t)e^{-i\omega_0 t} \tag{10.122}$$

where νt indicates the slow time with $\nu \ll \omega_0$. It should be emphasized that $E_0(\nu t)$ varies much more slowly than the phase $e^{-i\omega_0 t}$ $(\frac{\partial E_0}{\partial t}\frac{1}{E_0} \ll \omega_0)$. The slow modulation of the amplitude occurs in reality due to interaction between waves or due to change of the medium property.

If E_0 is a strict constant, we have $E(\omega) = 2\pi E_0\delta(\omega - \omega_0)$. Thus (10.120) gives

$D(\omega) = 2\pi E_0\delta(\omega - \omega_0)\varepsilon(\omega_0)$. In this case, the response function $D(\omega)$ is contributed by $\varepsilon(\omega_0)$, solely at the frequency $\omega = \omega_0$, sharp. However, for a slowly modulated electric field, $D(\omega)$ is expected to be contributed by the values of $\varepsilon(\omega)$ over some narrow range of ω centered about ω_0. Let us put the slowly modulated amplitude as

$$E_0(\nu t) = \overline{E}_0 e^{i\nu t}$$

where $\overline{E}_0$ is constant and $\nu \ll \omega_0$. Then we have

$$E(t) = \overline{E}_0\ e^{i(\nu-\omega_0)t}$$

$$\text{Hence}\quad E(\omega) = \overline{E}_0 2\pi\delta(\omega - \omega_0 + \nu)$$

Therefore (10.121) gives

$$\frac{\partial D}{\partial t} = i\overline{E}_0\ (-\omega_0 + \nu)\ \varepsilon(\omega_0 - \nu)\ e^{-i(\omega_0-\nu)t} \tag{10.123}$$

Expand $\varepsilon(\omega_0 - \nu) = \varepsilon(\omega_0) - \nu\ \frac{\partial\varepsilon}{\partial\omega_0} + \cdots$, then (10.124) can be arranged into the form

$$\frac{\partial D}{\partial t} = -i\omega_0\varepsilon(\omega_0)E_0(t)e^{-i\omega_0 t} + e^{-i\omega_0 t}\ \frac{\partial}{\partial\omega_0}(\omega_0\varepsilon(\omega_0))\ \frac{\partial}{\partial t}E_0(t) \tag{10.124}$$

where we put $i\nu\overline{E}_0 e^{i\nu t} = \partial E_0(t)/\partial t$, and neglected ν^2-term.

Equation (10.124) can be interpreted operationally as follows. In the equation, $D(\omega) = \varepsilon(\omega)E(\omega)$, ω contains Fourier frequency ω_0 and a slow time variation $\partial/\partial t_1$, i.e.

$$\omega = \omega_0 + i\frac{\partial}{\partial t_1} \quad \text{with } \omega_0 \gg \frac{\partial}{\partial t_1} \tag{10.125}$$

where the slow time variable t_1 is the time variable of the amplitude $E_0(t)$. Then we have

$$\varepsilon(\omega) = \varepsilon\left(\omega_0 + i\frac{\partial}{\partial t_1}\right) = \varepsilon(\omega_0) + \frac{\partial\varepsilon}{\partial\omega_0}i\frac{\partial}{\partial t_1}$$

$$\begin{aligned}\frac{\partial D}{\partial t} &= -i\omega D(\omega) = \left(-i\omega_0 + \frac{\partial}{\partial t_1}\right)\left(\varepsilon(\omega_0) + i\frac{\partial\varepsilon}{\partial\omega_0}\frac{\partial}{\partial t_0}\right)E_0(t_1)e^{-i\omega_0 t}\\ &= e^{-i\omega_0 t}\left(-i\omega_0\varepsilon(\omega_0) + \omega_0\frac{\partial\varepsilon}{\partial\omega_0}\frac{\partial}{\partial t_1} + \varepsilon(\omega_0)\frac{\partial}{\partial t_1}\right)E_0(t_1)\end{aligned} \tag{10.126}$$

which agrees with (10.124). In (10.126), we neglected $\partial^2/\partial t_1^2$.

For a quasi-monochromatic plane wave, we now calculate the time average (denoted by $< \cdots >$) of $\mathbf{E}\cdot\frac{\partial\mathbf{D}}{\partial t}$ in (10.118).

$$\begin{aligned}\mathbf{E}\cdot\frac{\partial\mathbf{D}}{\partial t} &\to \frac{1}{4}(\mathbf{E}+\mathbf{E}^*)\cdot\left(\frac{\partial\mathbf{D}}{\partial t} + \frac{\partial\mathbf{D}^*}{\partial t}\right)\\ &= \frac{1}{4}\left(\mathbf{E}\cdot\frac{\partial\mathbf{D}}{\partial t} + \mathbf{E}^*\cdot\frac{\partial\mathbf{D}^*}{\partial t} + \mathbf{E}^*\cdot\frac{\partial\mathbf{D}}{\partial t} + \mathbf{E}\cdot\frac{\partial\mathbf{D}^*}{\partial t}\right)\end{aligned}$$

The first two terms in the above equation vanish upon time-averaging, since they vary as $e^{\pm 2i\omega_0 t}$. Using (10.122) and (10.124) yields for the electric energy (W_E)

$$\begin{aligned}\frac{dW_E}{dt} &= \frac{1}{4\pi}\left\langle\mathbf{E}\cdot\frac{\partial\mathbf{D}}{\partial t}\right\rangle = \frac{1}{16\pi}\left[\frac{\partial}{\partial\omega}(\omega\varepsilon(\omega))\right]_{\omega=\omega_0}\left(\mathbf{E}_0\cdot\frac{\partial\mathbf{E}_0^*}{\partial t} + \mathbf{E}_0^*\cdot\frac{\partial\mathbf{E}_0}{\partial t}\right)\\ &= \frac{1}{16\pi}\left[\frac{\partial}{\partial\omega}(\omega\varepsilon(\omega))\right]_{\omega=\omega_0}\frac{\partial}{\partial t}|E_0|^2 = \frac{1}{16\pi}\left[\frac{\partial}{\partial\omega}(\omega\varepsilon(\omega))\right]_{\omega=\omega_0}\frac{\partial}{\partial t}|E|^2\end{aligned} \tag{10.127}$$

In obtaining (10.127), we assumed $\varepsilon(\omega)$ is real (*transparent region*). It is seen that only the terms involving $\frac{\partial}{\partial\omega}(\omega\varepsilon)$ survive the time-averaging. $E_0(t)$ terms cancel. From the above equation, it is natural to define the electric energy as

$$W_E = \frac{1}{16\pi}\left[\frac{\partial}{\partial\omega}(\omega\varepsilon(\omega))\right]_{\omega=\omega_0}|E|^2 \tag{10.128}$$

The magnetic field energy is simply

$$W_B = \frac{1}{16\pi}|B|^2 \tag{10.129}$$

so that the field energy W in (10.118) is $W = W_E + W_B$. In W_E, the particle kinetic energy is included through the polarization current.

In general, the displacement vector is connected to the electric field by the dielectric tensor: $D_i(\omega) = \varepsilon_{ij}(\omega)E_j(\omega)$ (neglecting the spatial dispersion). In this case, we have

$$\begin{aligned}\frac{\partial D_i}{\partial t} &= -i\omega D_i(\omega) = \left(-i\omega_0 + \frac{\partial}{\partial t_0}\right)\left[\varepsilon_{ij}(\omega_0) + i\,\frac{\partial\varepsilon_{ij}}{\partial\omega_0}\frac{\partial}{\partial t_0}\right]E_{j0}(t_0)e^{-i\omega_0 t}\\ &= -i\omega_0\varepsilon_{ij}(\omega_0)E_{j0}(t_0)e^{-i\omega_0 t} + \left[\frac{\partial}{\partial\omega}(\omega\varepsilon_{ij}(\omega))\right]_{\omega=\omega_0}\frac{\partial E_{j0}}{\partial t_0}e^{-i\omega_0 t}\end{aligned} \tag{10.130}$$

Thus we have

$$\begin{aligned}\left\langle \mathbf{E}\cdot\frac{\partial \mathbf{D}}{\partial t}\right\rangle &= \frac{1}{4}\,i\,\omega_0\varepsilon^*_{ij}(\omega_0)E_{i0}E^*_{j0} - \frac{1}{4}\,i\,\omega_0\varepsilon_{ij}(\omega_0)E^*_{i0}E_{j0}\\ &+\frac{1}{4}\left[\frac{\partial}{\partial\omega}(\omega\varepsilon_{ij}(\omega))\right]_{\omega=\omega_0}E^*_{i0}\frac{\partial E_{j0}}{\partial t_0} + \frac{1}{4}\left[\frac{\partial}{\partial\omega}(\omega\varepsilon^*_{ij}(\omega))\right]_{\omega=\omega_0}E_{i0}\frac{\partial E^*_{j0}}{\partial t_0}\end{aligned} \tag{10.131}$$

• **Hermitian and Anti-Hermitian Tensor.** For further development of (10.131), the dielectric tensor, which is not real in general, is separated into non-dissipative and dissipative part; they are respectively termed Hermitian (H) and anti-Hermitian (A) part. Anti-Hermitian part is the relevant generalization of the imaginary part in tensors or matrices. They are defined as

$$\varepsilon^H_{ij} = \frac{1}{2}(\varepsilon_{ij} + \varepsilon^*_{ji}), \quad \varepsilon^A_{ij} = \frac{1}{2i}(\varepsilon_{ij} - \varepsilon^*_{ji}) \tag{10.132}$$

Note that $\varepsilon^H_{ij}, \varepsilon^A_{ij}$, themselves are complex quantities in general. From the above definition, we can derive

$$\varepsilon_{ij} = \varepsilon^H_{ij} + i\,\varepsilon^A_{ij}, \ \varepsilon^*_{ji} = \varepsilon^H_{ij} - i\,\varepsilon^A_{ij} \tag{10.133}$$

The matrix ε_{ji}^*, the *adjoint matrix* of ε_{ij}, is obtained by taking the complex conjugate after transposing ε_{ij}. If $\varepsilon_{ji}^* = \varepsilon_{ij}$, ε_{ij} is self-adjoint or the medium is *loss-free* or *non-dissipative*. As can be seen from (10.133), the dielectric tensor is loss free if $\varepsilon_{ij}^H \gg \varepsilon_{ij}^A$. From (10.132) one can show that

$$\varepsilon_{ij}^H = \varepsilon_{ji}^{H*}, \ \varepsilon_{ij}^A = \varepsilon_{ji}^{A*}$$

That is, ε_{ij}^H, ε_{ij}^A themselves are self-adjoint but ε_{ij} is not self-adjoint in general. In **transparent region** of $(\omega, \mathbf{k})$, $\varepsilon_{ij}^H \gg \varepsilon_{ij}^A$ (the medium is loss-free), the waves are only weakly damped. Dissipation of waves arises from the anti-Hermitian part ε_{ij}^A or the imaginary frequency ω_i. In the transparent region, the imaginary part of the dispersion relation determinant (Eq. (7.30)) is much less than the real part (Im $\Delta \ll$ Re Δ). Therefore, the dispersion relation is approximately written as

$$\text{Re}\ \Delta(\omega_r, \mathbf{k}) = 0 \quad (\text{in transparent region})$$

The damping rate can be obtained from the formula (7.92) in the form

$$\text{damping rate} = -\frac{\text{Im}\ \Delta(\omega_r)}{\partial \text{Re}\ \Delta(\omega_r)/\partial \omega_r}$$

Going back to (10.131), the first two terms add to

$$\frac{i}{4}\omega_0 E_{j0} E_{i0}^* (\varepsilon_{ji}^* - \varepsilon_{ij}) = \frac{i}{4}\omega_0 E_{j0} E_{i0}^* (-2i)\varepsilon_{ij}^A = \frac{\omega_0}{2} E_{i0} E_{jo}^* \varepsilon_{ji}^A \quad (per\ i \longleftrightarrow j)$$

In adding up the last two terms in (10.131), we assume $\varepsilon_{ij}^H \gg \varepsilon_{ij}^A$. Also interchanging the dummy indexes, $i \longleftrightarrow j$, when necessary, one obtains for the sum of the last two terms in (10.131)

$$\frac{1}{4}\frac{\partial}{\partial t}\left[E_{j0}^* \frac{\partial}{\partial \omega}\left(\omega \varepsilon_{ji}^H\right) E_{i0}\right]$$

Then averaging (10.118) leads to the energy equation

$$\frac{dW}{dt} + \left\langle \nabla \cdot \frac{c}{4\pi}\mathbf{E} \times \mathbf{B} \right\rangle = -\frac{\omega_0}{8\pi} E_i E_j^* \varepsilon_{ji}^A \tag{10.134}$$

where the wave energy W is defined by

$$W = \frac{1}{16\pi}\left(|B|^2 + E_j^* \left[\frac{\partial}{\partial \omega}\left(\omega \varepsilon_{ji}^H(\omega)\right)\right]_{\omega=\omega_0} E_i\right) \tag{10.135}$$

Note that W consists of three components: the magnetic field energy $|B|^2$, the electric field energy and coherent(oscillating) particle kinetic energy which are both contained in the second term of W. Using $\varepsilon_{ij} = \delta_{ij} + \frac{4\pi i}{\omega}\sigma_{ij}$, we have $E_j^* \frac{\partial}{\partial\omega}(\omega\varepsilon_{ji}^H)E_i = E_j E_j^* + 4\pi i E_i E_j^* \sigma_{ij}^H$. The first term is the electric field energy while the second term is seen to be the particle mechanical energy. In (10.134), note that plasma wave energy W may either decrease or increase depending upon the anti-Hermitian part ε_{ji}^A. For a *negative energy wave*, W in (10.135) can be negative if the particles have some source of non-equilibrium kinetic energy in the unperturbed state (as in a beam).

Remark. For a stationary cold plasma, the dielectric constant is $\varepsilon(\omega) = 1 - \omega_p^2/\omega^2$ for plasma oscillation, and $\frac{\partial}{\partial\omega}(\omega\varepsilon(\omega)) \equiv U = 1 + \frac{\omega_p^2}{\omega^2} > 0$ (positive energy wave). If the plasma is moving with a drifting velocity v_0, the corresponding wave has

$$\varepsilon(\omega) = 1 - \frac{\omega_p^2}{(\omega - kv_0)^2}$$

Show that $U < 0$ (negative energy wave) for $kv_0 > \omega_p$.

Exercise. In non-absorbing isotropic medium, use (7.24) in (10.135) for ε_{ji}^H to show that

$$W = \frac{1}{16\pi}\frac{1}{k^2}\left[|\mathbf{k}\cdot\mathbf{E}|^2 \frac{\partial}{\partial\omega}\omega\varepsilon_L(\omega,k) + |\mathbf{k}\times\mathbf{E}|^2 \frac{\partial}{\partial\omega}\omega\left(\varepsilon_T(\omega,k) - \frac{c^2k^2}{\omega^2}\right)\right] \tag{10.136}$$

Hint: use $|\mathbf{k}\times\mathbf{E}|^2 = k^2\ |\mathbf{E}|^2 - (\mathbf{k}\cdot\mathbf{E})(\mathbf{k}\cdot\mathbf{E}^*)$

10.8.2 *Spatial dispersion: k-dependence of* D

Electric displacement vector $\mathbf{D}(\mathbf{r},t)$ is defined by [see Eq. (7.41)]

$$\mathbf{D}(\mathbf{r},t) = \mathbf{E}(\mathbf{r},t) + 4\pi\int_{-\infty}^{t} \mathbf{J}(\mathbf{r},t')dt' \tag{10.137}$$

This form of equation suggests the name *electric displacement.* When the medium is homogeneous in time and space, the conductivity tensor σ_{ij} depends on $(t-t')$ and $(\mathbf{r}-\mathbf{r}')$, and consequently, the current is obtained by

$$J_i(\mathbf{r},t) = \int_{-\infty}^{t} dt' \int_{-\infty}^{\infty} d^3r' \sigma_{ij}(t-t',\mathbf{r}-\mathbf{r}')E_j(t',\mathbf{r}') \tag{10.138}$$

This relation is called *material relation.* The integral is a convolution integral. One can show that the Fourier transform of (10.138) becomes

$$J_i(\mathbf{k},\omega) = \sigma_{ij}(\mathbf{k},\omega)E_j(\mathbf{k},\omega) \tag{10.139}$$

Let us consider the following expression that is to be used in (10.137):

$$J_i(\mathbf{r},t') = \left(\frac{1}{2\pi}\right)^4 \int e^{i\mathbf{k}\cdot\mathbf{r}-i\omega t'} J_i(\mathbf{k},\omega) d^3k d\omega$$

It is left as an exercise to show

$$4\pi \int_{-\infty}^{t} dt' J_i(\mathbf{r},t') = \int_{-\infty}^{\infty} d^3r' \int_{-\infty}^{\infty} dt' \chi_{ij}(t-t',\mathbf{r}-\mathbf{r}') E_j(t',\mathbf{r}') \tag{10.140}$$

where $\chi_{ij}(\mathbf{k},\omega) = \frac{4\pi i}{\omega}\sigma_{ij}(\mathbf{k},\omega)$ and (10.139) would be useful. Therefore, (10.137) can be written as

$$D_i(\mathbf{r},t) = \int_{-\infty}^{\infty} d^3r' \int_{-\infty}^{\infty} dt' \varepsilon_{ij}(t-t',\mathbf{r}-\mathbf{r}') E_j(t',\mathbf{r}') \tag{10.141}$$

$$\varepsilon_{ij}(t-t',\mathbf{r}-\mathbf{r}') = \delta_{ij}\delta(\mathbf{r}-\mathbf{r}')\delta(t-t') + \chi_{ij}(t-t',\mathbf{r}-\mathbf{r}') \tag{10.142}$$

The above equation gives the Fourier transform of $\varepsilon_{ij}(\mathbf{r},t)$ in the form

$$\varepsilon_{ij}(\mathbf{k},\omega) = \delta_{ij} + \chi_{ij}(\mathbf{k},\omega) \tag{10.143}$$

because the Fourier transform of the delta function is unity. (10.141) is a convolution integral. It's Fourier transform is the product of each Fourier transform:

$$D_i(\mathbf{k},\omega) = \varepsilon_{ij}(\mathbf{k},\omega) E_j(\mathbf{k},\omega) \tag{10.144}$$

Exercise. Prove the following relations:

$$\begin{gathered}\varepsilon^H_{ij}(\omega,\mathbf{k}) = \varepsilon^H_{ji}(-\omega,-\mathbf{k}), \quad \varepsilon^A_{ij}(\omega,\mathbf{k}) = -\varepsilon^A_{ji}(-\omega,-\mathbf{k}), \\ \varepsilon_{ij}(-\omega,-\mathbf{k}) = \varepsilon^*_{ij}(\omega,\mathbf{k})\end{gathered} \tag{10.145}$$

10.8.3 *Field energy of electromagnetic wave with slow temporal and spatial modulation*

In this section, we generalize the earlier result for the field energy of wavepacket of one-dimensional electric field with a slow temporal modulation. The key quantity in the development is $\partial\mathbf{D}(\mathbf{r},t)/\partial t$. We consider an electric field:

$$\mathbf{E}(\mathbf{r},t) = \mathbf{E}_0(\mathbf{r},t) e^{i\mathbf{k}_0\cdot\mathbf{r}-i\omega_0 t} \tag{10.146}$$

where the amplitude $\mathbf{E}_0$ is a slow function of $\mathbf{r}$ and t. The analysis is facilitated by putting

$$\mathbf{E}_0(\mathbf{r}, t) = \mathbf{A}e^{-i\mathbf{q}\cdot\mathbf{r}+i\nu t} \tag{10.147}$$

with $q \ll k_0$ and $\nu \ll \omega_0$ and $\mathbf{A}$ a constant vector. Then we have

$$\frac{\partial E_{0i}}{\partial t} = i\nu E_{0i}, \quad \frac{\partial E_{0i}}{\partial r_j} = -iq_j E_{0i} \tag{10.148}$$

Fourier transforming (10.146) gives

$$\mathbf{E}(\mathbf{k}, \omega) = \mathbf{A}\ (2\pi)^4\ \delta(\omega - \omega_0 + \nu)\ \delta(\mathbf{k} - \mathbf{k}_0 + \mathbf{q}) \tag{10.149}$$

We work with the expression

$$\frac{\partial D_i(\mathbf{r}, t)}{\partial t} = -\frac{i}{(2\pi)^4} \int_{-\infty}^{\infty} d\omega \int_{-\infty}^{\infty} d^3k\ e^{i\mathbf{k}\cdot\mathbf{r}-i\omega t}\omega\ \varepsilon_{ij}(\omega, \mathbf{k}) E_j(\omega, \mathbf{k}) \tag{10.150}$$

Using (10.149) in (10.150) gives

$$\begin{aligned}\frac{\partial D_i(\mathbf{r}, t)}{\partial t} &= -i\ E_{0j}\ e^{i\mathbf{k}_0\cdot\mathbf{r}-i\omega_0 t}\ (\omega_0 - \nu)\ \varepsilon_{ij}(\omega_0 - \nu,\ \mathbf{k}_0 - \mathbf{q}) \\ &= -i\ E_{0j}\ e^{i\mathbf{k}_0\cdot\mathbf{r}-i\omega_0 t} \left[\omega_0\varepsilon_{ij}(0) - \nu\frac{\partial}{\partial\omega_0}(\omega_0\varepsilon_{ij}(0)) - q_l\omega_0\frac{\partial\varepsilon_{ij}(0)}{\partial k_{0l}}\right]\end{aligned} \tag{10.151}$$

where we expanded ε_{ij}, and the argument $(0) \equiv (\mathbf{k}_0,\ \omega_0)$. Using the relations (10.148) in the above gives

$$\begin{aligned}\frac{\partial D_i(\mathbf{r}, t)}{\partial t} = \Bigg[&-i\omega_0\varepsilon_{ij}(\mathbf{k}_0, \omega_0)E_{0j} + \frac{\partial}{\partial\omega_0}(\omega_0\varepsilon_{ij}(\mathbf{k}_0, \omega_0))\frac{\partial E_{0j}}{\partial t} \\ &- \omega_0\frac{\partial\varepsilon_{ij}(\mathbf{k}_0, \omega_0)}{\partial k_{0l}}\ \frac{\partial E_{0j}}{\partial r_l}\Bigg]\ e^{i\mathbf{k}_0\cdot\mathbf{r}-i\omega_0 t}\end{aligned} \tag{10.152}$$

Compare the above equation with (10.130). The last term in (10.152) is a new term derived from the spatial variation. With the inclusion of the spatial modulation, averaging (10.118) by the analogous way that led to (10.134) amends the energy equation in the form

$$\frac{dW}{dt} + \left\langle\nabla\cdot\frac{c}{4\pi}\mathbf{E}\times\mathbf{B}\right\rangle = -\frac{\omega_0}{8\pi}E_iE_j^*\varepsilon_{ji}^A - \frac{\omega_0}{8\pi}\ \frac{\partial E_i^*}{\partial r_l}\ \frac{\partial\varepsilon_{ij}^H}{\partial k_l}\ E_j \tag{10.153}$$

$$W = \frac{1}{16\pi}\left(|B|^2 + E_j^*\left[\frac{\partial}{\partial\omega}(\omega\varepsilon_{ji}^H(\omega))\right]_{\omega=\omega_0} E_i\right) \tag{10.154}$$

The new term represents the mechanical energy flux due to particles flowing coherently with the wave.

10.8.4 *Compact form of wave energy equation*

We rederive (10.153) and (10.154) by the technique of *derivative expansion*, and cast them into a compact form. In Fourier representation, Maxwell equations are

$$\mathbf{k} \times \mathbf{E} = \frac{\omega}{c}\,\mathbf{B} \tag{10.155}$$

$$\mathbf{k} \times \mathbf{B} = -\frac{\omega}{c}\mathbf{D} = -\frac{\omega}{c}\varepsilon_{ij}E_j \tag{10.156}$$

Here we consider the field components $\mathbf{k}$ and ω are complex quantities: $\mathbf{k} = \mathbf{k}_r + i\mathbf{k}_i$, $\omega = \omega_r + i\omega_i$. Performing (10.155)$\cdot$ $\mathbf{B}^* +$ (10.156)$^* \cdot \mathbf{E}$ ($*$ denotes the complex conjugate) yields

$$\mathbf{E}\cdot\mathbf{k}^* \times \mathbf{B}^* + \mathbf{B}^* \cdot \mathbf{k} \times \mathbf{E} = \frac{\omega}{c}B^2 - \frac{\omega^*}{c}E_i\varepsilon_{ij}^*E_j^* \tag{10.157}$$

Taking the complex conjugate of the above equation gives

$$\mathbf{E}^* \cdot \mathbf{k} \times \mathbf{B} + \mathbf{B} \cdot \mathbf{k}^* \times \mathbf{E}^* = \frac{\omega^*}{c}B^2 - \frac{\omega}{c}E_i^*\varepsilon_{ij}E_j \tag{10.158}$$

Subtracting (10.158) from (10.157) yields

$$2i\ \mathbf{k}_i \cdot (\mathbf{E} \times \mathbf{B}^* + \mathbf{E}^* \times \mathbf{B}) = 2i\omega_i B^2 + \frac{1}{c}E_i^*(\omega\varepsilon_{ij} - \omega^*\varepsilon_{ji}^*)E_j \tag{10.159}$$

In the last term we interchanged the dummy indexes. We assume $\omega_r \gg \omega_i$, $k_r \gg k_i$, and Taylor expand

$$\omega\varepsilon_{ij}(\omega, \mathbf{k}) = \omega_r\varepsilon_{ij}(\omega_r, \mathbf{k}_r) + i\ \mathbf{k}_i \cdot \frac{\partial}{\partial \mathbf{k}_r}(\omega_r\varepsilon_{ij}) + i\omega_i\frac{\partial}{\partial\omega_r}(\omega_r\varepsilon_{ij}) \tag{10.160}$$

where we understand that ω_r is independent of $\mathbf{k}_r$. The subscripts r and i at $\mathbf{k}$ and ω are not cartesian indexes; they are not subject to summation convention. Taking complex conjugate of the above equation and interchanging the indexes $i \leftrightarrow j$ yields

$$\omega^*\varepsilon_{ji}^*(\omega, \mathbf{k}) = \omega_r\varepsilon_{ji}^*(\omega_r, \mathbf{k}_r) - i\ \mathbf{k}_i \cdot \frac{\partial}{\partial \mathbf{k}_r}(\omega_r\varepsilon_{ji}^*) - i\omega_i\frac{\partial}{\partial\omega_r}(\omega_r\varepsilon_{ji}^*) \tag{10.161}$$

Subtraction gives, with using (10.132),

$$\omega\varepsilon_{ij} - \omega^*\varepsilon_{ji}^* = 2i\omega_r\varepsilon_{ij}^A + 2i\mathbf{k}_i \cdot \frac{\partial}{\partial \mathbf{k}_r}(\omega_r\varepsilon_{ij}^H) + 2i\omega_i\frac{\partial}{\partial\omega_r}\left(\omega_r\varepsilon_{ij}^H\right) \tag{10.162}$$

Using (10.162) in (10.159) yields

$$\mathbf{k}_i \cdot \left[\mathbf{E} \times \mathbf{B}^* + \mathbf{E}^* \times \mathbf{B} - \frac{1}{c} E_i^* \frac{\partial}{\partial \mathbf{k}_r} (\omega_r \varepsilon_{ij}^H) E_j\right]$$

$$= \omega_i \left[B^2 + \frac{1}{c} E_i^* \frac{\partial}{\partial \omega_r} (\omega_r \varepsilon_{ij}^H) E_j\right] + \frac{1}{c} E_i^* \omega_r \varepsilon_{ij}^A E_j \tag{10.163}$$

Physical meaning of each term of the energy equation (10.163) goes as follows. The sum of the first two vector product terms on the LHS is the time-averaged energy flux (poynting vector). The second term represents the non-electromagnetic energy flux due to particles flowing coherently with the wave. In a cold plasma, ε_{ij} is independent of $\mathbf{k}$, and this term vanishes. In a purely electrostatic wave, the electromagnetic energy flux (the sum of the first two terms) vanishes, and all the energy flow is contained in the second term. On the RHS of (10.163), the first term is just the time average of the magnetic field energy. The second term in the bracket is the wave energy which is the sum of electric field energy and the oscillating particle kinetic energy. The last term in (10.163) represents the dissipation of the electric field energy.

Equation (10.163) can be converted to the differential form of energy equation which applies to a single wave of the form $\mathbf{E}(\mathbf{r}, t)e^{i\mathbf{k}\cdot\mathbf{r} - i\omega t}$ where the modulus is slowly varying as compared to the phase in $\mathbf{r}$ and t. We have the correspondence

$$\mathbf{k}_i \to -\frac{\partial}{\partial \mathbf{r}}, \quad \mathbf{k}_r \to \mathbf{k}, \quad \omega_i \to \frac{\partial}{\partial t}, \quad \omega_r \to \omega$$

Then (10.163) takes the form

$$\frac{\partial W}{\partial t} + \omega E_i^* \varepsilon_{ij}^A E_j = c\,\frac{\partial}{\partial \mathbf{r}} \cdot (\mathbf{E} \times \mathbf{B}^* + \mathbf{E}^* \times \mathbf{B}) - \omega \frac{\partial}{\partial \mathbf{r}} \cdot E_i^* \frac{\partial \varepsilon_{ij}^H}{\partial \mathbf{k}} E_j \tag{10.164}$$

$$\begin{aligned} W &= |\mathbf{B}|^2 + \frac{\partial}{\partial \omega}(\omega \varepsilon_{ij}^H)\; E_i^* E_j \\ &= \frac{c^2}{\omega^2} k^2 |\mathbf{E}|^2 - \frac{c^2}{\omega^2}\,(\mathbf{k} \cdot \mathbf{E})\,(\mathbf{k} \cdot \mathbf{E}^*) + \frac{\partial}{\partial \omega}(\omega \varepsilon_{ij}^H)\; E_i^* E_j \end{aligned}$$

Here we define:

$$G = \omega \varepsilon_{ij}^H E_i^* E_j - \frac{c^2}{\omega} k^2 |\mathbf{E}|^2 + \frac{c^2}{\omega} (\mathbf{k} \cdot \mathbf{E})(\mathbf{k} \cdot \mathbf{E}^*) \tag{10.165}$$

Then we have

$$W = \frac{\partial G}{\partial \omega} \tag{10.166}$$

Using $\mathbf{B} = \frac{c}{\omega}\mathbf{k} \times \mathbf{E}$ in the RHS of (10.164), we obtain

$$\text{RHS of (10.164)} = \frac{\partial}{\partial \mathbf{r}} \cdot \frac{\partial G}{\partial \mathbf{k}}$$

$$= -\frac{\partial}{\partial \mathbf{r}} \cdot \left[\frac{c^2}{\omega}[-2|\mathbf{E}|^2\mathbf{k} + (\mathbf{k}\cdot\mathbf{E})\mathbf{E}^* + (\mathbf{k}\cdot\mathbf{E}^*)\mathbf{E}] + E_i^* E_j \frac{\partial\left(\omega \varepsilon_{ij}^H\right)}{\partial \mathbf{k}} \right]$$

Therefore (10.164) can be written as

$$\frac{\partial W}{\partial t} + \frac{\partial}{\partial \mathbf{r}} \cdot \mathbf{S} = -\omega \varepsilon_{ij}^A E_i^* E_j; \quad \mathbf{S} = \frac{\partial G}{\partial \mathbf{k}}$$

or

$$\frac{\partial}{\partial t}\frac{\partial G}{\partial \omega} + \frac{\partial}{\partial \mathbf{r}} \cdot \frac{\partial G}{\partial \mathbf{k}} = -\omega \varepsilon_{ij}^A E_i^* E_j \tag{10.167}$$

[This form of energy equation would be obtained alternatively by calculating the poynting vector term in (10.153).]

The wave equation in Fourier space in isotropic plasma can be written as

$$\left[\frac{c^2 k^2}{\omega^2} \left(\frac{k_i k_j}{k^2} - \delta_{ij} \right) + \varepsilon_{ij} \right] E_j \equiv \Lambda_{ij}\ E_j = 0$$

In the transparent region ($\text{Re}\ \varepsilon_{ij} \gg \text{Im}\ \varepsilon_{ij}$ or $\varepsilon_{ij}^H \gg \varepsilon_{ij}^A$), the quantity G in (10.165) is found to be

$$G = \Lambda_{ij} E_i^* E_j = \omega \varepsilon_{ij} E_i^* E_j - \frac{c^2}{\omega} k^2 |\mathbf{E}|^2 + \frac{c^2}{\omega}(\mathbf{k}\cdot\mathbf{E})(\mathbf{k}\cdot\mathbf{E}^*) \tag{10.168}$$

Let us introduce the polarization vector $\hat{\mathbf{a}}$:

$$\mathbf{E}(\mathbf{k},\omega) = \hat{\mathbf{a}}|\mathbf{E}(\mathbf{k},\omega)| = \hat{\mathbf{a}}E(\mathbf{k},\omega) \tag{10.169}$$

It is the unit vector directed along the direction of the electric field. In general, it is a complex quantity normalized so that $\hat{\mathbf{a}}\cdot\hat{\mathbf{a}}^* = 1$.

Using (10.169) in (10.168) gives

$$G = \left[\omega \varepsilon_{ij} a_i^* a_j - \frac{c^2}{\omega} k^2 + \frac{c^2}{\omega} |\mathbf{k}\cdot\hat{\mathbf{a}}|^2 \right] |E|^2 \equiv G_1 |E|^2$$

Now the RHS of (10.167) is $|E|^2$ times a small imaginary part of G_1 (in the transparent region). Thus, (10.167) can be written as

$$\frac{\partial I_k}{\partial t} + \frac{\partial \omega(\mathbf{k})}{\partial \mathbf{k}} \cdot \frac{\partial I_k}{\partial \mathbf{r}} = 2\gamma_k I_k \tag{10.170}$$

$$I_k = |E_k|^2, \ \frac{\partial \omega(\mathbf{k})}{\partial \mathbf{k}} = \frac{\frac{\partial G_1}{\partial \mathbf{k}}}{\frac{\partial G_1}{\partial \omega}}, \ 2\gamma_k = -\frac{\text{Im}\ G_1}{\frac{\partial G_1}{\partial \omega}} \tag{10.171}$$

where I_k is the spectral distribution of the electric field energy.

Reference: B. B. Kadomtsev, *Cooperative effects in plasmas*, Reviews of plasma physics **22**, Kluwer Academic/Consultants Bureau, New York (2001).

10.8.5 *Ponderomotive force*

A charged particle in a high frequency inhomogeneous electric field experiences, in addition to a high frequency oscillatory force, a zero frequency (d.c.) *ponderomotive force* (radiation pressure) which can be represented by $-\nabla\Phi$ (Φ is the ponderomotive force potential). It is a nonlinear force that survives the time average in a sinusoidal force field. Use of the ponderomotive force makes it easier to analyze parametric instabilities and wave-particle interactions in general. The ponderomotive force causes a slow variation of the density of the medium in which the high frequency electromagnetic wave is propagated (electric striction).

We consider an electron in external high frequency electric field with inhomogeneous amplitude $\mathbf{E}(\mathbf{r},t) = \overline{\mathbf{E}}(\mathbf{r})\sin\,\omega_0 t$. The magnetic field accompanying this electric field is $\mathbf{B}(\mathbf{r},t) = \overline{\mathbf{B}}(\mathbf{r})\cos\,\omega_0 t$, with $\overline{\mathbf{B}}(\mathbf{r}) = \frac{c}{\omega_0}\nabla\times\overline{\mathbf{E}}(\mathbf{r})$. In Eulerian description, the equation of motion reads

$$\frac{\partial}{\partial t}\mathbf{v}(\mathbf{r},t) + \mathbf{v}\cdot\nabla\mathbf{v} = \frac{e}{m}\left[\mathbf{E}(\mathbf{r},t) + \frac{1}{c}\mathbf{v}\times\mathbf{B}(\mathbf{r},t)\right] \tag{10.172}$$

In the zero order linear approximation with respect to the external field $\overline{\mathbf{E}}(\mathbf{r})\sin\,\omega_0 t$, we have

$$\frac{\partial}{\partial t}\mathbf{v}_0(\mathbf{r},t) = \frac{e}{m}\overline{\mathbf{E}}(\mathbf{r})\sin\,\omega_0 t,$$

$$\mathbf{v}_0(\mathbf{r},t) = \frac{-e}{m\omega_0}\overline{\mathbf{E}}(\mathbf{r})\cos\,\omega_0 t$$

Note that the magnetic field term is of higher order ($\frac{v}{c} \ll 1$). In the order of $|\mathbf{E}(\mathbf{r})|^2$, we have

$$\frac{\partial}{\partial t}\mathbf{v}_1(\mathbf{r},t) = -\mathbf{v}_0\cdot\nabla\mathbf{v}_0 + \frac{e}{mc}\mathbf{v}_0\times\overline{\mathbf{B}}(\mathbf{r})\cos\,\omega_0 t$$

Here the nonlinear terms on RHS produce the zero frequency component. The nonlinear terms have rapidly changing phases, and only the average taken over a period has physical meaning. The ponderomotive force is the force surviving the averaging:

$$\begin{aligned}\mathbf{F}_p &= \frac{\omega_0}{2\pi}\int_0^{\frac{2\pi}{\omega_0}} dt\; m\frac{\partial}{\partial t}\mathbf{v}_1(\mathbf{r},t) = \frac{-e^2}{2m\omega_0^2}(\overline{\mathbf{E}}\cdot\nabla\overline{\mathbf{E}} + \overline{\mathbf{E}}\times\nabla\times\overline{\mathbf{E}}) \\ &= \frac{-e^2}{4m\omega_0^2}\nabla\overline{E}^2(\mathbf{r})\end{aligned} \tag{10.173}$$

The steady d.c. ponderomotive force repels electrons and ions (regardless of the charge sign) from the region of a strong field to the region of weaker field.

It would be instructive to solve the same problem of finding the average force due to the inhomogeneous high frequency electric field by means of Lagrangian equation:

$$\ddot{\mathbf{r}} = \frac{e}{m}\left[\mathbf{E}(\mathbf{r})\sin\ \omega_0 t + \frac{1}{c}\ \dot{\mathbf{r}} \times \mathbf{B}(\mathbf{r})\cos\omega_0 t\right] \left(\cdot\ = \frac{d}{dt}\right) \tag{10.174}$$

where $\mathbf{r}$ is the particle position. The oscillation starts at some position $\mathbf{r}_0$. This *oscillation center* slowly drifts (like a guiding center) due to the inhomogeneity of $\mathbf{E}(\mathbf{r})$. So we can write the particle position as

$$\mathbf{r}(t) = \mathbf{r}_0(t) + \mathbf{r}_1(t)$$

where $\mathbf{r}_1$ is the oscillating displacement vector of frequency ω_0 and the oscillation center $\mathbf{r}_0(t)$ is a slow function of t, much slower than ω_0. Assuming that the electric field does not carry the particle too far from the oscillation center in a period, we can expand the electric and magnetic fields about the oscillation center,

$$\mathbf{E}(\mathbf{r}) = \mathbf{E}(\mathbf{r}_0) + \mathbf{r}_1 \cdot \nabla_0 \mathbf{E}(\mathbf{r}_0) + \cdots$$
$$\mathbf{B}(\mathbf{r}) = \mathbf{B}(\mathbf{r}_0) + \mathbf{r}_1 \cdot \nabla_0 \mathbf{B}(\mathbf{r}_0) + \cdots$$

Then we have

$$\begin{aligned}\ddot{\mathbf{r}_0} + \ddot{\mathbf{r}_1} = \frac{e}{m}\Big[&\mathbf{E}(\mathbf{r}_0)\sin\omega_0 t + \mathbf{r}_1 \cdot \nabla_0 \mathbf{E}(\mathbf{r}_0)\sin\omega_0 t + \frac{1}{c}\dot{\mathbf{r}_0} \times \mathbf{B}(\mathbf{r}_0)\cos\omega_0 t \\ &+ \frac{1}{c}\dot{\mathbf{r}_1} \times \mathbf{B}(\mathbf{r}_0)\cos\omega_0 t + \frac{1}{c}\dot{\mathbf{r}_0} \times (\mathbf{r}_1 \cdot \nabla_0)\mathbf{B}(\mathbf{r}_0)\cos\omega_0 t + \frac{1}{c}\dot{\mathbf{r}_1} \times (\mathbf{r}_1 \cdot \nabla_0)\mathbf{B}(\mathbf{r}_0)\cos\omega_0 t\Big]\end{aligned} \tag{10.175}$$

In the zero order, the two large terms are balanced:

$$\ddot{\mathbf{r}_1} = \frac{e}{m}\mathbf{E}(\mathbf{r}_0)\sin\ \omega_0 t, \quad \mathbf{r}_1 = \frac{-e}{m\omega_0^2}\mathbf{E}(\mathbf{r}_0)\sin\omega_0 t$$

Clearly the last two terms in (10.175) are of higher order and neglected. Taking the time average of the remaining terms, we obtain for the time average of $(m\ddot{\mathbf{r}_0})$ the same result as (6.170). This ponderomotive force causes *electric striction* which gives rise to variation of the density of the medium when an intense high frequency electromagnetic wave propagates in it.

Equation (10.173) is the force acting on an electron, and the force density ($\mathbf{f}_p$) in high frequency inhomogeneous electric field may be defined in terms of electron number density (n),

$$\mathbf{f}_p = n\mathbf{F}_p = -\frac{ne^2}{4m\omega^2}\nabla E^2 = -\frac{\omega_p^2}{\omega^2}\nabla\frac{E^2}{16\pi}$$

where $E^2/16\pi$ is recognized as the mean electric pressure.

The preceding derivation of ponderomotive force indicates that the origin of the ponderomotive force is the low frequency remnant resulted in by beating of two high frequency quantities in the terms of $\mathbf{v}\cdot\mathbf{v}$ and $\mathbf{v}\times\mathbf{B}$. The low frequency contribution is obtained from the average taken over the fast oscillations.

Note in the above that there are two distinctive time scales: the fast change $\omega_0\mathbf{v}_1$ and the slow change $\partial v_1/\partial t$. We show in the following that the ponderomotive force is a dynamical consequence of two distinctive time scales involved in the dynamics. Let us consider the electron equation of motion

$$\frac{\partial\mathbf{v}}{\partial t} + \nabla\frac{v^2}{2} - \mathbf{v}\times(\nabla\times\mathbf{v}) = \frac{e}{m}\left(\mathbf{E} + \frac{1}{c}\mathbf{v}\times\mathbf{B}\right) \tag{10.176}$$

where we used the identity $\mathbf{v}\cdot\nabla\mathbf{v} = \nabla\frac{v^2}{2} - \mathbf{v}\times(\nabla\times\mathbf{v})$. Suppose that the plasma field consists of two oscillations of very different time scales: $\mathbf{v} = \mathbf{v}_0 + \mathbf{v}_1$ where the subscripts 0 and 1 respectively mean the fast and slowly varying quantities. We also have Maxwell's equation

$$\frac{\partial\mathbf{B}}{\partial t} = -c\nabla\times\mathbf{E} \tag{10.177}$$

For the time derivative of the fast quantity $\mathbf{v}_0$, or $\frac{\partial\mathbf{v}_0}{\partial t}$, we can write

$$\frac{\partial\mathbf{v}_0}{\partial t} = \frac{e}{m}\mathbf{E}_0 \tag{10.178}$$

because all the high frequency beat terms v_0^2, v_0B_0, generate only the second harmonics of high frequency or zero frequency, and do not contribute to $\frac{\partial\mathbf{v}_0}{\partial t}$ (in other words they are non-resonant). The above two equations give

$$\nabla\times\mathbf{v}_0 = -\frac{e}{mc}\mathbf{B}_0 \tag{10.179}$$

The change of the slow field, $\partial\mathbf{v}_1/\partial t$, can be obtained by averaging (10.176) over many periods of fast oscillation; or by inspection, retaining the nonlinear beat terms of two fast oscillations. Since we have $-\mathbf{v}_0\times(\nabla\times\mathbf{v}_0) = \frac{e}{mc}\mathbf{v}_0\times\mathbf{B}_0$, we are left with

$$\frac{\partial\mathbf{v}_1}{\partial t} + \frac{1}{2}\nabla v_0^2 = \frac{e}{m}\mathbf{E}_1 \tag{10.180}$$

When the external electric field is a high frequency wave $\sim$ sin $\omega_0 t$ as in (10.172), put $\mathbf{E}_1 = 0$ and recover (10.173) upon averaging. Note that terms like $v_0 v_1$, $v_1 v_1$ are non-resonant and contribute very little to $\frac{\partial \mathbf{v}_1}{\partial t}$. Equation (10.180) is valid both for electromagnetic and electrostatic fields [if electrostatic, $\mathbf{B} = 0$ and $\nabla \times \mathbf{v} = 0$]. The ion ponderomotive force is smaller than the electron ponderomotive force by the factor of mass ratio m_e/m_i.

Problems

10.1 Method of Stationary Phase. Put Eqs. (10.4) and (10.5) in the following form by expanding the phase $\varphi(k)$ near the stationary phase point k_0:

$$f(x,t) = F(k_0)\ e^{t\varphi(k_0)} \int_{-\infty}^{\infty} \exp\left(-i\frac{t}{2}\omega''(k_0)(k-k_0)^2\right) dk$$

Assuming k and k_0 are real, carry out the integral to obtain

$$f(x,t) = F(k_0)\ e^{t\varphi(k_0)}\ e^{\mp\frac{i\pi}{4}} \sqrt{\frac{\pm 2\pi}{t\ \omega''(k_0)}}$$

where the upper sign corresponds to the case of $\omega''(k_0) > 0$; the lower sign corresponds to the case of $\omega''(k_0) < 0$. The result agrees with (10.17). The method of stationary phase is applicable when the integration variable is real. Use

$$\int_{-\infty}^{\infty} e^{\pm i\alpha u^2} du = \sqrt{\frac{\pi}{\alpha}}\ e^{\pm\frac{i\pi}{4}}$$

$$\int_{-\infty}^{\infty} \cos\ t^2 dt = \int_{-\infty}^{\infty} \sin\ t^2 dt = \sqrt{\frac{\pi}{2}}$$

which can be proven by contour integration.

10.2 The original KdV equation reads in non-dimensional form

$$\frac{\partial u}{\partial t} + \frac{\partial u}{\partial x} + 12u\ \frac{\partial u}{\partial x} + \frac{\partial^3 u}{\partial x^3} = 0$$

Assume the solution in the form $u(x,t) = u(\xi)$, where $\xi = kx - \omega t$ and $u(\xi) \to 0$ as $\xi \to \infty$. Obtain the solution

$$u(x,t) = \frac{a^2}{4}\ \sec h^2\ \frac{a}{2}\left(x - (1+a^2)t\right)$$

where $a = \sqrt{\frac{\omega}{k} - 1}$. The wave velocity is $(1+a^2)$ which is amplitude-dependant, and therefore a taller wave moves faster than a smaller one. *Hint*: make correspondence with the solution in the text.

10.3 Prove the following statement. In terms of the dielectric constant, $\varepsilon = 1 - \frac{\omega_p^2}{\omega^2}$, the force density of ponderomotive force is the average of the divergence of the Maxwell stress tensor:

$$\begin{aligned} f_k &= \frac{\varepsilon - 1}{16\pi} \frac{\partial}{\partial x_k} <E^2> \\ &= \frac{1}{8\pi} \frac{\partial}{\partial x_i} \left[\varepsilon <E_i E_k> + <B_i B_k> - \frac{\delta_{ik}}{2} (<E^2> + <B^2>) \right] \end{aligned}$$

Hint: use $\nabla \times \mathbf{E} = -\frac{1}{c} \frac{\partial \mathbf{B}}{\partial t}$, $\nabla \times \mathbf{B} = \frac{\partial}{\partial t} (\varepsilon \mathbf{E})$, $\nabla \cdot \mathbf{B} = 0$, $\nabla \cdot (\varepsilon \mathbf{E}) = 0$.

References

Arfken, G. B. and H. J. Weber, *Mathematical Methods for Physicists*, Elsevier (2005).

Bekefi, G. *Radiation Processes in Plasmas*, Wiley, New York (1966).

Benny, D. J. *A general theory for interactions between long and short waves*, Studies in Appl. Math., **56** pp. 81–94.

Dodd, R. K., J. C. Eilbeck, J. D. Gibbon, and H. C. Morris, *Solitons and Nonlinear Wave Equations*. Academic Press, New York (1982).

Kadomtsev, B. B. *Cooperative Effects in Plasmas*, Reviews of Plasma Physics, Vol 22 Ed. V. D. Shafranov, Consultant Bureau, New York (2001).

Kadomtsev, B. B. *Plasma Turbulence*, Academic Press, London (1965).

Karpman, V. I. *Nonlinear Waves in Dispersive Media*, Pergamon (1975).

Korteweg, D. J. and G. de Vries, *On the change of form of long waves advancing in a rectangular canal, and on a new type of long stationary waves*, Phil. Mag., **39** (1895), pp. 422–443.

Landau, L. D. and E. M. Lifshitz, *Electrodynamics of Continuous Media*, Pergamon Press, Oxford (1966) § 61.

Lee, H. J. *Nonlinear analysis of the cold fluid-Poisson plasma by using the characteristic method*, J. Korean Phys. Soc. **69** No. 7 (2016) pp. 1191–1211.

Lighthill, E. J. *Waves in Fluids* Cambridge University Press, Cambridge, (1978).

Thau, S. A. *Linear Dispersive Waves*, in Nonlinear waves, Ed. S. Leibovich and A. R. Seebass, Cornell University Press (1970).

Whitham, G. B. *Linear and Nonlinear Waves*, Wiley, New York (1974).

Chapter 11

Discrete Charge Effects

Viewing a plasma as a fluid-like medium is the lowest order kinetic description of the actual plasma. The Vlasov equation describe the plasma as a fluid in $(\mathbf{r}, \mathbf{v})$ phase space. The characteristic equations of the Vlasov equation and Euler equation of fluid are the same. In Vlasov fluid picture, the plasma electric field is the ensemble-averaged field that averages out the fluctuating electric fields associated with individual particles. However, in a real plasma, the individual particles excite field fluctuations and emit radiations. These effects are not accounted for by Vlasov theory. In the next higher order, the individual discrete particle effect should be included to describe a plasma more realistically in the events such as radiation and scattering. The simplest way to include the discrete particle effect is to treat the plasma particles as *dressed particles*. In this chapter, we show that the calculation of electric field fluctuation is facilitated by the dressed particle approach. The result agrees with the *fluctuation-dissipation theorem*. It is shown that the dressed particle method can be successfully employed to derive the electromagnetic fluctuation as well as the fluctuation in a magnetized plasma.

The fluctuations of the electric field and other plasma properties such as number density and current are all connected with the imaginary part of the plasma dielectric function. This connection is manifestation of the fluctuation–dissipation theorem.

The Cerenkov radiation is emitted by moving charges of constant velocities when the Landau wave-particle resonance conditions are met. It is the inverse process of Landau damping. It is a collective effect, and the emission of the radiation requires the whole plasma medium. The Cerenkov emission from a plasma can be fully accounted for by treating the plasma as aggregate of dressed particles. In earlier chapter, we derived the Landau damping from the Kramers–Kronig relations and also directly from the causal requirement. It is shown that the Cerenkov emission power is the fluctuation of the electric field generated by dressed particles, and the fundamental reason of the Cerenkov emission can be traced to the causality. According to the fluctuation–dissipation theorem, the nonzeroness of fluctuation is due to the nonzeroness of the imaginary part of the linear dielectric function, which is thus fundamentally attributed to the causality.

11.1 Self-Consistent Potential in Vlasov Equation

It would be useful to review the Vlasov theory which neglects entirely the individual particle effects before we attempt to introduce the new factor associated with the discreteness of the medium.

The electric field $\mathbf{E}$ in the Poisson equation

$$\nabla \cdot \mathbf{E}(\mathbf{r}, t) = -\nabla^2 \phi = 4\pi \sum_\alpha e_\alpha \int f_\alpha(\mathbf{r}, \mathbf{v}, t) d^3 v \tag{11.1}$$

is the self-consistent field, which is generated by the plasma particles and at the same time influence particle distribution by the Vlasov equation

$$\frac{\partial f_\alpha}{\partial t} + \mathbf{v} \cdot \frac{\partial f_\alpha}{\partial \mathbf{r}} + \frac{e_\alpha}{m_\alpha} \mathbf{E} \cdot \frac{\partial f_\alpha}{\partial \mathbf{v}} = 0 \tag{11.2}$$

To see this more clearly, we calculate the potential $\phi(\mathbf{r}, t)$ made at $\mathbf{r}$ by all the plasma particles of the system which consists of electrons and ions. The potential at $\mathbf{r}$ made by the particles located at $\mathbf{r}'$ is

$$\sum_\alpha \frac{e_\alpha}{|\mathbf{r} - \mathbf{r}'|} \int d^3 v' f_\alpha(\mathbf{r}', \mathbf{v}', t)$$

The potential at $\mathbf{r}$ due to all the particles in the system is

$$\phi(\mathbf{r}, t) = \sum_\alpha \int d^3 r' \frac{e_\alpha}{|\mathbf{r} - \mathbf{r}'|} \int d^3 v' f_\alpha(\mathbf{r}', \mathbf{v}', t) \tag{11.3}$$

$$\mathbf{E}(\mathbf{r}, t) = -\nabla \phi = \sum_\alpha e_\alpha \int d^3 r' \frac{\mathbf{r} - \mathbf{r}'}{|\mathbf{r} - \mathbf{r}'|^3} \int d^3 v' f_\alpha(\mathbf{r}', \mathbf{v}', t) \tag{11.4}$$

where we used the relation

$$\nabla_r \frac{1}{|\mathbf{r} - \mathbf{r}'|} = -\frac{\mathbf{r} - \mathbf{r}'}{|\mathbf{r} - \mathbf{r}'|^3}$$

Taking divergence of (11.4) gives (11.1):

$$\begin{aligned} \nabla \cdot \mathbf{E}(\mathbf{r}, t) &= \sum_\alpha e_\alpha \int d^3 r' \nabla_r \cdot \frac{\mathbf{r} - \mathbf{r}'}{|\mathbf{r} - \mathbf{r}'|^3} \int d^3 v' f_\alpha(\mathbf{r}', \mathbf{v}', t) \\ &= 4\pi \sum_\alpha e_\alpha \int d^3 v' f_\alpha(\mathbf{r}, \mathbf{v}', t) \end{aligned} \tag{11.5}$$

where we used

$$\nabla_r \cdot \frac{\mathbf{r}-\mathbf{r}'}{|\mathbf{r}-\mathbf{r}'|^3} = -\nabla_r^2 \frac{1}{|\mathbf{r}-\mathbf{r}'|} = 4\pi\delta(\mathbf{r}-\mathbf{r}') \tag{11.6}$$

Now (11.2) takes the form

$$\frac{\partial f_\alpha}{\partial t} + \mathbf{v}\cdot\frac{\partial f_\alpha}{\partial \mathbf{r}} + \frac{e_\alpha}{m_\alpha}\frac{\partial f_\alpha}{\partial \mathbf{v}}\cdot\sum_\alpha e_\alpha \int d^3r' \frac{\mathbf{r}-\mathbf{r}'}{|\mathbf{r}-\mathbf{r}'|^3}\int d^3v' f_\alpha(\mathbf{r}',\mathbf{v}',t) = 0 \tag{11.7}$$

Equation (11.7) has the virtue of having gotten rid of the electric field.

Example. Solve (11.1) for $\mathbf{E}$ in reverse order. To solve Laplace equation with a source term

$$\nabla^2\phi(\mathbf{r},t) = -4\pi\sum e_\alpha \int d^3v f_\alpha(\mathbf{r},\mathbf{v},t)$$

recall Green function of Laplace equation is,

$$G(\mathbf{r},\mathbf{r}') = \frac{-1}{4\pi|\mathbf{r}-\mathbf{r}'|}$$

which satisfies

$$\nabla^2 G(\mathbf{r},\mathbf{r}') = \delta(\mathbf{r}-\mathbf{r}')$$

Therefore we calculate:

$$\phi(\mathbf{r},t) = \int d^3r' G(\mathbf{r},\mathbf{r}') \times \text{ source term } (\mathbf{r}') = \int \frac{d^3r'}{|\mathbf{r}-\mathbf{r}'|}\sum e_\alpha \int d^3v f_\alpha(\mathbf{r}',\mathbf{v},t)$$

which is (11.3).

• **Convolution.** We now calculate the Fourier transform of (11.4) in which the $\int d^3r'$-integral is a *convolution.* Convolution integral of two functions is

$$(f*g)_\mathbf{r} = \int_{-\infty}^{\infty} g(\mathbf{r}')f(\mathbf{r}-\mathbf{r}')d^3r' = \int_{-\infty}^{\infty} f(\mathbf{r}')g(\mathbf{r}-\mathbf{r}')d^3r'$$

It can be easily proved that

$$\int e^{-i\mathbf{k}\cdot\mathbf{r}}(f*g)_\mathbf{r}\, d^3r = \tilde{f}(\mathbf{k})\,\tilde{g}(\mathbf{k})$$

In words, the Fourier transform of a convolution integral is the product of each Fourier transform. The inverse of $\tilde{f}(\mathbf{k})\ \tilde{g}(\mathbf{k}) =$

$$\left(\frac{1}{2\pi}\right)^3 \int e^{i\mathbf{k}\cdot\mathbf{r}} \tilde{f}(\mathbf{k})\ \tilde{g}(\mathbf{k}) d^3k = (f * g)_{\mathbf{r}}$$

In words, the Fourier transform (or the inverse transform) of the product of two functions is the convolution. So much for the mathematics of convolution.

Next, Fourier transforming (11.4) gives

$$\mathbf{E}(\mathbf{k}, t) = \sum_\alpha e_\alpha \left[\int e^{-i\mathbf{k}\cdot\mathbf{r}} \frac{\mathbf{r}}{r^3} d^3r\right] \int d^3v' f_\alpha(\mathbf{k}, \mathbf{v}', t)$$

The Fourier integral [...] can be computed by introducing the spherical coordinates

$$\mathbf{r} = \hat{\mathbf{r}} r = r(\hat{\mathbf{x}}\ \sin\theta\ \cos\varphi + \hat{\mathbf{y}}\ \sin\theta\ \sin\varphi + \hat{\mathbf{z}} \cos\theta)$$

Also we take $\mathbf{k} = \hat{\mathbf{z}} k$ without loss of generality. Then after $\int d\varphi$ integral we have

$$\int e^{-i\mathbf{k}\cdot\mathbf{r}} \frac{\mathbf{r}}{r^3} d^3r = \hat{\mathbf{z}}\ 2\pi \int_0^\infty dr \int_0^\pi d\theta\ \sin\theta\ \cos\theta\ e^{-ikr\cos\theta}$$

It is left as an exercise to carry out this integral to obtain

$$\int e^{-i\mathbf{k}\cdot\mathbf{r}} \frac{\mathbf{r}}{r^3} d^3r = \hat{\mathbf{z}} \frac{-4\pi i}{k} = -\frac{4\pi i}{k^2}\ \mathbf{k}$$

Hence

$$\mathbf{E}(\mathbf{k}, t) = -\frac{4\pi i}{k^2}\ \mathbf{k} \sum_\alpha e_\alpha \int d^3v' f_\alpha(\mathbf{k}, \mathbf{v}', t) \tag{11.8}$$

This result can be obtained directly by Fourier transforming (11.1).

11.2 Potential Due to a Point Charge Placed in a Plasma

This is a well-known problem. The answer is Debye potential. Here we consider this problem from two different points of view.

First, we begin with an external charge q located at $\mathbf{r}_0$, which is represented by charge density

$$\rho_0(\mathbf{r}) = q\delta(\mathbf{r} - \mathbf{r}_0) \tag{11.9}$$

The potential at $\mathbf{r}$ made by the particles located at $\mathbf{r}'$ is

$$\sum_\alpha \frac{e_\alpha}{|\mathbf{r}-\mathbf{r}'|}\int d^3v f_\alpha(\mathbf{r}',\mathbf{v},t) + \frac{q\,\delta(\mathbf{r}'-\mathbf{r}_0)}{|\mathbf{r}-\mathbf{r}'|}$$

Integrating this over the whole space $\int d^3r'$ gives the potential at $\mathbf{r}$ made by the plasma and the single charge:

$$\phi(\mathbf{r},t) = \int d^3r' \sum_\alpha \frac{e_\alpha}{|\mathbf{r}-\mathbf{r}'|}\int d^3v f_\alpha(\mathbf{r}',\mathbf{v},t) + \frac{q}{|\mathbf{r}-\mathbf{r}_0|} \tag{11.10}$$

Hence

$$\nabla\cdot\mathbf{E} = -\nabla^2\phi = 4\pi\sum_\alpha e_\alpha \int d^3v f_\alpha(\mathbf{r},\mathbf{v},t) + 4\pi q\delta(\mathbf{r}-\mathbf{r}_0) \tag{11.11}$$

where we used (11.6). The first term on the right side is the charge density ρ of the plasma. Equation (11.11) is rewritten as

$$\nabla\cdot\mathbf{E} = 4\pi(\rho(\mathbf{r},t) + q\delta(\mathbf{r}-\mathbf{r}_0)) \tag{11.12}$$

We also have

$$\mathbf{E} + 4\pi\mathbf{P} = \mathbf{D} = \varepsilon_L\mathbf{E} \tag{11.13}$$

and

$$\rho = -\nabla\cdot\mathbf{P} \tag{11.14}$$

where $\mathbf{P}$ is the polarization. The above three equations give

$$\nabla\cdot\mathbf{E} = \frac{4\pi q}{\varepsilon_L}\delta(\mathbf{r}-\mathbf{r}_0) \tag{11.15}$$

Fourier transform of this equation gives

$$\phi(\mathbf{k}) = \frac{4\pi q}{k^2\varepsilon_L}e^{-i\mathbf{k}\cdot\mathbf{r}_0} \tag{11.16}$$

Alternatively, in view of the definition of $\mathbf{D}$, we have

$$\nabla\cdot\mathbf{D} = 4\pi q\;\delta(\mathbf{r}-\mathbf{r}_0) \tag{11.17}$$

Fourier transforming gives

$$ik_iD_i = 4\pi q\; e^{-i\mathbf{k}\cdot\mathbf{r}_0}$$

Since we have $D_i = \varepsilon_{ij} E_j, \quad E_j = -ik_j \phi$, it follows that

$$ik_i D_i = k_i k_j \varepsilon_{ij} \phi = k^2 \; \varepsilon_L \; \phi = 4\pi q \; e^{-i\mathbf{k}\cdot\mathbf{r}_0}$$

which agrees with (11.16).

This is a static problem. Using (8.70) with $\omega = 0$ (with $F_{\alpha 0}$ Maxwellian) gives

$$\varepsilon_L = 1 + \sum_\alpha \frac{1}{k^2 \lambda_{D\alpha}^2} \equiv 1 + \frac{1}{k^2 \lambda_D^2} \tag{11.18}$$

$$\phi(\mathbf{k}) = \frac{4\pi q \lambda_D^2}{1 + k^2 \lambda_D^2} e^{-i\mathbf{k}\cdot\mathbf{r}_0} \tag{11.19}$$

Fourier inversion gives

$$\phi(\mathbf{r}) = \left(\frac{1}{2\pi}\right)^3 4\pi q \int_{-\infty}^{\infty} d^3k \frac{e^{i\mathbf{k}\cdot(\mathbf{r}-\mathbf{r}_0)}}{k^2 + \lambda_D^{-2}}$$

Taking $\mathbf{k}$ along z-direction, using spherical coordinate, and putting $\mathbf{r}_0 = 0$,

$$\phi(\mathbf{r}) = \left(\frac{q}{2\pi^2}\right) \int_0^\infty k^2 dk \int \sin\theta \; d\theta \frac{e^{ikr\cos\theta}}{k^2 + \lambda_D^{-2}} \int d\varphi = \text{etc.} = \frac{2q}{\pi r} \int_0^\infty dk \frac{k \sin kr}{k^2 + \lambda_D^{-2}}$$

Picking up the residue at the pole $k = i/\lambda_D$, we finally obtain

$$\phi(\mathbf{r}) = q \; \frac{e^{-r/\lambda_D}}{r} \tag{11.20}$$

The result is due to the shielding of the point charge field by plasma particles. Comparison with the field in vacuum ($\sim 1/r^2$) indicates that the point charge field in plasma is *less* far reaching (*Debye shielding or screening*).

Secondly, more direct way of treating this problem goes as follows. In a neutral plasma, an external charge q is introduced; after a sufficient time a thermal equilibrium state is reached in which the electrons and ions are Boltzmann distributed. Then we have the Poisson equation,

$$\nabla^2 \phi = -4\pi e n_0 (e^{-e\phi/T_i} - e^{e\phi/T_e}) - 4\pi q \delta(\mathbf{r}) \tag{11.21}$$

Expanding in terms of $e\phi/T$,

$$\nabla^2 \phi = \frac{\phi}{\lambda_D^2} - 4\pi q \delta(\mathbf{r}) \tag{11.22}$$

Fourier transforming gives

$$\phi(k) = \frac{4\pi q}{k^2 + \lambda_D^{-2}} \tag{11.23}$$

which is identical with (11.19).

Example. Potential due to an oscillating charge in a plasma.

When an oscillating charge of high frequency is introduced in a Maxwellian plasma, what is the response? Assume the external charge density in the form

$$\rho_0(\mathbf{r}, t) = q\delta(\mathbf{r} - \mathbf{r}_0)\cos(\omega_0 t) \tag{11.24}$$

Hence

$$\nabla \cdot \mathbf{D}(\mathbf{r}, t) = 4\pi q\delta(\mathbf{r} - \mathbf{r}_0)\cos(\omega_0 t) \tag{11.25}$$

Fourier transforming gives

$$ik_i D_i(\mathbf{k}, \omega) = 4\pi q e^{-ik\cdot\mathbf{r}_0}\pi\ (\delta(\omega - \omega_0) + \delta(\omega + \omega_0)) \tag{11.26}$$

Using the relation $D_i = \varepsilon_{ij}E_j = \varepsilon_{ij}(-ik_j)\phi(\mathbf{k}, \omega)$ in the above gives

$$\phi(\mathbf{k}, \omega) = (2\pi)^2\ q\ \frac{e^{-ik\cdot\mathbf{r}_0}[\delta(\omega - \omega_0) + \delta(\omega + \omega_0)]}{k^2\varepsilon_L(k, \omega)} \tag{11.27}$$

Fourier inversion in t gives

$$\phi(\mathbf{k}, t) = 4\pi\ q\ \cos(\omega_0 t)\ \frac{e^{-ik\cdot\mathbf{r}_0}}{k^2\varepsilon_L(k, \omega_0)} \tag{11.28}$$

where we used $\varepsilon_L(\mathbf{k}, \omega_0) = \varepsilon_L(\mathbf{k}, -\omega_0)$. We assume ω_0 is a high frequency; $\omega_0/k \gg v_{th\alpha}$. Then the ionic response is negligible, and we have $\varepsilon_L(k, \omega_0) = 1 - \omega_0^2/\omega_{\text{pe}}^2$. Fourier inverting (11.28) gives, using spherical coordinates in k-space,

$$\phi(\mathbf{r}, t) = \frac{q}{|\mathbf{r} - \mathbf{r}_0|}\frac{\cos(\omega_0 t)}{\left(1 - \frac{\omega_0^2}{\omega_{\text{pe}}^2}\right)} \tag{11.29}$$

where the integral $\int_{-\infty}^{\infty} dx\ \sin x/x = \pi$ was used. Two points are worthy of note: first, in a high frequency field electrons do not form Debye screening; secondly, if the impressing frequency ω_0 is $\sim \omega_{\text{pe}}$, the charge resonantly invokes a large amplitude response. The low frequency response is the Debye shielding.

Example (Electric field due to a charge introduced in a magnetized plasma with $\hat{\mathbf{z}}B_0$).
The charge density due to the single charge can be represented by $\rho_s(\mathbf{r},t) = e\ \delta\ (\mathbf{r} - \mathbf{r}_s(t))$ where $\mathbf{r}_s(t)$ is the orbit of the charge. We have

$$\nabla \cdot \mathbf{D}(\mathbf{r},t) = 4\ \pi e\ \delta\ (\mathbf{r} - \mathbf{r}_s(t))$$

Fourier transforming,

$$\begin{aligned} ik_i D_i(\mathbf{k},\omega) &= 4\pi e \int e^{-i\mathbf{k}\cdot\ \mathbf{r}+i\omega t}\delta\ (\mathbf{r} - \mathbf{r}_s(t))d^3r\ dt = ik_i\varepsilon_{ij}E_j \\ &= k_i k_j\ \varepsilon_{ij}\ \phi(\mathbf{k},\omega) = k^2\varepsilon_L\ \phi(\mathbf{k},\omega) \\ \phi(\mathbf{k},\omega) &= \frac{4\pi e}{k^2\varepsilon_L} \int e^{-i\mathbf{k}\cdot\ \mathbf{r}_s(t)+i\omega t}dt \end{aligned} \tag{11.30}$$

To find the orbit, we solve

$$\frac{d\mathbf{r}}{dt} = \omega_c\ \mathbf{r}\times\hat{\mathbf{z}}$$

to get

$$(x(t),y(t),z(t)) = (r_L\ \sin(\omega_c t),\ r_L\ \cos(\omega_c t),\ v_{\parallel}t) \quad (r_L,\ \ v_{\parallel},\ \ \text{const.})$$

In the integral in (11.30), we put $\mathbf{k} = \hat{x}k_\perp + \hat{z}k_\parallel$ without loss of generality. Then

$$\begin{aligned} \int e^{-i\mathbf{k}\cdot\ \mathbf{r}_s(t)+i\omega t}dt &= \sum_{n=-\infty}^{\infty} J_n(k_\perp r_L)\int_{-\infty}^{\infty} e^{i\omega t - ik_\parallel v_\parallel t - in\omega_c t}dt \\ &= 2\pi \sum_{n=-\infty}^{\infty} J_n(k_\perp r_L)\ \delta(\omega - k_\parallel v_\parallel - n\omega_c) \end{aligned}$$

$$\mathbf{E}(\mathbf{k},\omega) = -ie8\pi^2 \frac{\mathbf{k}}{k^2\varepsilon_L(\mathbf{k},\omega)} \sum_{n=-\infty}^{\infty} J_n(k_\perp r_L)\ \delta(\omega - k_\parallel v_\parallel - n\omega_c) \tag{11.31}$$

• **A moving external charge introduced into plasma.** For simplicity, we assume that the external charge moves with a constant velocity $\mathbf{v}_0$. It would be instructive to solve this problem in two ways: i) with the aid of the displacement vector $\mathbf{D}$ and ii) from the Poisson equation. If the initial position of the charge (q) is $\mathbf{r}_0$, the charge density associated with the moving charge is represented by

$$\rho_{\text{ext}}(\mathbf{r}, t) = q\ \delta(\mathbf{r} - \mathbf{r}_0 - \mathbf{v}_0 t) \tag{11.32}$$

Fourier transforming (11.32) gives

$$\begin{aligned}\rho_{\text{ext}}(\mathbf{k}, \omega) &= q \int e^{-i\mathbf{k}\cdot\mathbf{r}+i\omega t}\delta(\mathbf{r} - \mathbf{r}_0 - \mathbf{v}_0 t) d^3r\ dt \\ &= 2\pi q e^{-i\mathbf{k}\cdot\mathbf{r}_0}\delta(\omega - \mathbf{k}\cdot\mathbf{v}_0)\end{aligned} \tag{11.33}$$

Now, we use the equation

$$\nabla \cdot \mathbf{D} = 4\pi\rho_{\text{ext}} = 4\pi q\ \delta(\mathbf{r} - \mathbf{r}_0 - \mathbf{v}_0 t) \tag{11.34}$$

Fourier transforming the above equation gives

$$i\ \mathbf{k}\cdot\mathbf{D}(\mathbf{k}, \omega) = 4\pi q\ e^{-i\mathbf{k}\cdot\mathbf{r}_0} 2\pi\delta(\omega - \mathbf{k}\cdot\mathbf{v}_0) \tag{11.35}$$

$$D_i = \varepsilon_{ij}E_j,\ E_j = -ik_j\phi(\mathbf{k}, \omega),\quad k_ik_j\varepsilon_{ij} = k^2\varepsilon_L$$

Therefore we obtain

$$\phi(\mathbf{k}, \omega) = \frac{8\pi^2 q\ e^{-i\mathbf{k}\cdot\mathbf{r}_0}\delta(\omega - \mathbf{k}\cdot\mathbf{v}_0)}{k^2\varepsilon_L(\omega, \mathbf{k})} \tag{11.36}$$

It is to be noted that, in this derivation, the dielectric function $\varepsilon_L(\mathbf{k}, \omega)$ can take the form obtained either from kinetic equation or fluid equation.

Alternatively, let us use the Poisson equation,

$$\nabla \cdot \mathbf{E}(\mathbf{r}, t) = 4\pi \sum_\alpha e_\alpha \int f_\alpha(\mathbf{r}, \mathbf{v}, t)\ d^3v + 4\pi\ \rho_{\text{ext}}(\mathbf{r}, t) \tag{11.37}$$

We assume that the influence of ρ_{ext} is weak enough that the response of the plasma is described by the linear equation,

$$\begin{aligned}(-i\omega + i\mathbf{k}\cdot\mathbf{v})f_\alpha(\mathbf{k}, \mathbf{v}, \omega) + \tfrac{e_\alpha}{m_\alpha}\mathbf{E}(\mathbf{k}, \omega)\cdot\tfrac{\partial f_{\alpha 0}}{\partial \mathbf{v}} &= 0 \\ \text{or}\quad f_\alpha(\mathbf{k}, \omega) = -\phi(\mathbf{k}, \omega)\ \tfrac{e_\alpha}{m_\alpha}\ \frac{\mathbf{k}\cdot\frac{\partial f_{0\alpha}}{\partial \mathbf{v}}}{\omega - \mathbf{k}\cdot\mathbf{v}}&\end{aligned} \tag{11.38}$$

Fourier transforming the Poisson equation (11.37) gives

$$\phi(\mathbf{k}, \omega) = \frac{8\pi^2 q\ e^{-i\mathbf{k}\cdot\mathbf{r}_0}\delta(\omega - \mathbf{k}\cdot\mathbf{v}_0)}{k^2\varepsilon_L(\omega, \mathbf{k})} \tag{11.39}$$

where ε_L is given by (8.24):

$$\varepsilon_L(\mathbf{k}, \omega) = 1 + \sum_\alpha \frac{\omega_{p\alpha}^2}{k^2} \int d^3v\ \frac{\mathbf{k}\cdot\frac{\partial f_{\alpha 0}}{\partial \mathbf{v}}}{\omega - \mathbf{k}\cdot\mathbf{v}} \tag{11.40}$$

To properly deal with the singularity in ε_L, a convenient way is to insert a small positive imaginary part in ω, in accordance with the causal requirement or by the analytic continuation. Then, Fourier inversion of the above equation gives

$$\phi(\mathbf{r},t) = \frac{q}{2\pi^2} \int d^3k \frac{e^{i\mathbf{k}\cdot(\mathbf{r}-\mathbf{r}_0-\mathbf{v}_0 t)}}{k^2 \varepsilon_L(\mathbf{k}\cdot\mathbf{v}_0 + i\epsilon,\ \mathbf{k})} \tag{11.41}$$

If $\mathbf{v}_0 = 0$, we recover the static solution.

Dressed particle. In the preceding examples, we calculated the potentials made by a single charged particle (it is called 'test particle') introduced in the plasma. The potential is the sum of two contributions: (i) the potential generated by the whole plasma surrounding the test charge and (ii) the potential made by the test charge itself as if it were in vacuum–potential of the 'bare charge'. The former (i) is the self-consistent potential generated by the polarization cloud surrounding the test charge. The sum of (i) and (ii) is called the potential of *dressed particle*, which is quite a descriptive name coined by Rosenbluth and Rostoker. The Debye potential is the dressed particle potential in the static situation. Introduction of a test charge in a plasma mathematically implies that the Vlasov–Poisson system of differential equations becomes an inhomogeneous one, and we have no dispersion relation. Let us take a look at (11.36). If we put $q = 0$ therein, we must have $\varepsilon_L(\mathbf{k},\omega) = 0$, which is the dispersion relation of the homogeneous differential equation system when the potential is the self-consistent one. The important point of development is that we regard a plasma as an aggregate of dressed particles. This view is analogous to the view that a Vlasov fluid is a collection of beams individualized by the initial velocities. Now every particle in a plasma is a dressed particle individualized by the initial velocities. Treating a plasma as a collection of mutually non-interacting dressed particles facilitates calculation of the problems such as scattering and radiation. In fact, the Vlasov equation is the lowest order (in the small parameter $\epsilon = 1/\lambda_D^3 n$) equation for the Liouville distribution function, and the crudest one. In the next higher order approximation, the dressed particle description is thought to be useful.

11.3 Initial Value Problem of a Test Charge

Here we consider the dynamic screening of plasma against a moving test charge, a dressed particle. In the last section, we obtained the potential due to a moving test charge by considering the displacement vector $\mathbf{D}$—(11.41).

Here, we approach the same problem as an initial value problem via Fourier–Laplace transform. Assume that an external charge q is suddenly introduced into the plasma at position $\mathbf{r}_0$ with velocity $\mathbf{v}_0$. The subsequent position is given by $\mathbf{r} = \mathbf{r}_0 + \mathbf{v}_0 t$, and the charge density associated with the test particle is represented by $q\delta(\mathbf{r} - \mathbf{r}_0 - \mathbf{v}_0 t)$. The potential made by plasma particles and the test charge is

$$\nabla \cdot \mathbf{E} = -\nabla^2 \phi = 4\pi \sum_\alpha e_\alpha \int d^3 v f_\alpha(\mathbf{r}, \mathbf{v}, t) + 4\pi q \delta(\mathbf{r} - \mathbf{r}_0 - \mathbf{v}_0 t) \tag{11.42}$$

In the above f_α is the perturbed distribution function (assuming that the perturbation excited by the introduction of the test charge is small) which satisfies the linearized equation

$$\frac{\partial f_\alpha}{\partial t} + \mathbf{v} \cdot \frac{\partial f_\alpha}{\partial \mathbf{r}} + \frac{e_\alpha}{m_\alpha} \mathbf{E} \cdot \frac{\partial f_{\alpha 0}}{\partial \mathbf{v}} = 0 \tag{11.43}$$

with the initial condition

$$f_\alpha(\mathbf{r}, \mathbf{v}, 0) = 0 \tag{11.44}$$

Equation (11.44) means that the test charge is suddenly introduced into the plasma at $t = 0$. The above equations are Fourier–Laplace transformed by performing $\int_{-\infty}^{\infty} d^3 r e^{-i\mathbf{k}\cdot\mathbf{r}} \int_0^\infty e^{i\omega t}(\cdots)$ with the initial condition (11.44). We obtain

$$k^2 \phi(\mathbf{k}, \omega) = 4\pi \sum_\alpha e_\alpha \int d^3 v f_\alpha(\mathbf{k}, \mathbf{v}, \omega) + 4\pi i q \frac{e^{-i\mathbf{k}\cdot\mathbf{r}_0}}{\omega - \mathbf{k}\cdot\mathbf{v}_0} \tag{11.45}$$

$$f_\alpha(\mathbf{k}, \mathbf{v}, \omega) = -\frac{e_\alpha}{m_\alpha} \phi(\mathbf{k}, \omega) \frac{\mathbf{k} \cdot \frac{\partial f_{\alpha 0}}{\partial \mathbf{v}}}{\omega - \mathbf{k}\cdot\mathbf{v}} \tag{11.46}$$

Using (11.46) in (11.45) yields

$$k^2 \phi(\mathbf{k}, \omega) = 4\pi i q \frac{e^{-i\mathbf{k}\cdot\mathbf{r}_0}}{\varepsilon(\mathbf{k}, \omega)(\omega - \mathbf{k}\cdot\mathbf{v}_0)} \tag{11.47}$$

$$\text{where } \varepsilon(\mathbf{k}, \omega) = 1 + \Sigma \frac{\omega_{p\alpha}^2}{k^2} \int d^3 v \frac{\mathbf{k} \cdot \frac{\partial f_{\alpha 0}}{\partial \mathbf{v}}}{\omega - \mathbf{k}\cdot\mathbf{v}} \tag{11.48}$$

is the plasma dielectric constant. Equation (11.47) should be compared with (11.36). We see that $\frac{1}{\omega - \mathbf{k}\cdot\mathbf{v}_0}$ in the Laplace formulation in (11.47) corresponds to $\delta(\omega - \mathbf{k}\cdot\mathbf{v}_0)$ in the Fourier formulation. After the inversion in t, the two agrees. The correspondence between the two is readily recognized by invoking the Plemelj formula in (11.47) with neglect of the singularity in ε. Laplace-inversion of (11.47) is written as

$$\phi(\mathbf{k},t) = \int_L \frac{d\omega}{2\pi} e^{-i\omega t} \frac{4\pi iq\ e^{-i\mathbf{k}\cdot\mathbf{r}_0}}{k^2\varepsilon(\mathbf{k},\omega)(\omega - \mathbf{k}\cdot\mathbf{v}_0)} \tag{11.49}$$

When $q = 0$, in order for (11.49) to be valid we must have $\varepsilon(\mathbf{k},\omega) = 0$ which gives the normal modes of the perturbed plasma. Here in the inversion integral of (11.49), note that we have two kinds of singularities: one kind corresponding to $\omega's$ which are the roots of $\varepsilon(\mathbf{k},\omega) = 0$, and the other corresponding to $\omega = \mathbf{k}\cdot\mathbf{v}_0$. We take advantage of the fact that the normal modes are Landau-damped [the associated singularities are in the lower-half ω-plane]. Thus the singularity on the real ω-axis, $\omega = \mathbf{k}\cdot\mathbf{v}_0$, which is uppermost, determines the long time behavior of $\phi(\mathbf{k},t)$. So we asymptotically write

$$\phi(\mathbf{k},t) = 4\pi q\ \frac{e^{-i\mathbf{k}\cdot(\mathbf{r}_0+\mathbf{v}_0 t)}}{k^2\varepsilon(\mathbf{k},\omega)|_{\omega=\mathbf{k}\cdot\mathbf{v}_0}} \tag{11.50}$$

This equation, the main result, gives the potential screened by the moving point charge. Equation (11.50) gives the electric field

$$\mathbf{E}(\mathbf{k},t) = -i\mathbf{k}\phi(\mathbf{k},t) = -4\pi iq\ \frac{\mathbf{k}\ e^{-i\mathbf{k}\cdot(\mathbf{r}_0+\mathbf{v}_0 t)}}{k^2\varepsilon(\mathbf{k},\omega)|_{\omega=\mathbf{k}\cdot\mathbf{v}_0}} \tag{11.51}$$

Inverting (11.51) yields

$$\mathbf{E}(\mathbf{r},t) = \int \frac{d^3k}{8\pi^3} e^{i\mathbf{k}\cdot\mathbf{r}}\mathbf{E}(\mathbf{k},t) = -\frac{iq}{2\pi^2}\int d^3k \frac{\mathbf{k}\ e^{i\mathbf{k}\cdot\mathbf{r}}e^{-i\mathbf{k}\cdot(\mathbf{r}_0+\mathbf{v}_0 t)}}{k^2\varepsilon(\mathbf{k},\omega)|_{\omega=\mathbf{k}\cdot\mathbf{v}_0}} \tag{11.52}$$

Equation (11.52) gives the electric field at point $\mathbf{r}$ due to the plasma *and* the test particle whose orbit is $\mathbf{r}_0 + \mathbf{v}_0 t$. If we add up the fields generated at the position $\mathbf{r}$ due to all the *test particles* which are distributed according to a distribution $f_0(\mathbf{r}_0,\mathbf{v}_0)$, we have the ensemble-averaged electric field

$$<\mathbf{E}(\mathbf{r},t)> = \int d^3v_0 \int d^3r_0\ \mathbf{E}(\mathbf{r},t;\mathbf{r}_0) f_0(\mathbf{r}_0,\mathbf{v}_0) \tag{11.53}$$

In a uniform and isotropic plasma, there is no favored direction and $f(\mathbf{r}_0,\mathbf{v}_0)$ takes the form of isotropic $f_0(\mathbf{v}_0)$, so (11.53) is expected to vanish. This can be proved by substituting (11.52) into (11.53) and using a Maxwellian distribution for $f_0(\mathbf{v}_0)$:

$$<\mathbf{E}(\mathbf{r},t)> = \int d^3v_0 f_0(v_0) \int d^3r_0 \left(-\frac{iq}{2\pi^2}\right) \int d^3k \frac{\mathbf{k}\, e^{i\mathbf{k}\cdot\mathbf{r}} e^{-i\mathbf{k}\cdot(\mathbf{r}_0+\mathbf{v}_0 t)}}{k^2\varepsilon(\mathbf{k},\omega)|_{\omega=\mathbf{k}\cdot\mathbf{v}_0}} \tag{11.54}$$

In the above, integral $\int d^3r_0$ yields $\delta(\mathbf{k})$, and $\varepsilon(\mathbf{k},\omega)$ can be evaluated by its static form, $\varepsilon(\mathbf{k},\omega=0) = 1 + 1/k^2\lambda_D^2$. Thus it is immediately seen that (11.54) vanishes upon $\int d^3k$ integration.

11.4 Fluctuation of Electric Field Energy

Next, we consider the average of E^2, the fluctuating electric field energy of the plasma. It can be calculated in the same fashion as done in (11.53) by taking the ensemble average:

$$\begin{aligned} W &= \frac{1}{8\pi} <\mathbf{E}(\mathbf{r},t)\cdot\mathbf{E}(\mathbf{r},t)> = \frac{1}{8\pi} <\mathbf{E}(\mathbf{r},t)\cdot\mathbf{E}^*(\mathbf{r},t)> \\ &= \frac{1}{8\pi}\int d^3v_0 f_0(\mathbf{v}_0)\int d^3r_0 \left(\frac{q}{2\pi^2}\right)^2 \int d^3k\, \frac{\mathbf{k}e^{i\mathbf{k}\cdot\mathbf{r}}e^{-i\mathbf{k}\cdot\mathbf{r}_0}e^{-i\mathbf{k}\cdot\mathbf{v}_0 t}}{k^2\varepsilon(\mathbf{k},\omega)|_{\omega=\mathbf{k}\cdot\mathbf{v}_0}} \\ &\quad\cdot \int d^3k' \frac{\mathbf{k}'\, e^{-i\mathbf{k}'\cdot\mathbf{r}}e^{i\mathbf{k}'\cdot\mathbf{r}_0}e^{i\mathbf{k}'\cdot\mathbf{v}_0 t}}{k'^2\varepsilon^*(\mathbf{k}',\omega)|_{\omega=\mathbf{k}'\cdot\mathbf{v}_0}} \end{aligned} \tag{11.55}$$

where we used $\mathbf{E}^* = \mathbf{E}$ because $\mathbf{E}(\mathbf{r},t)$ is a real quantity. The integral $\int d^3r_0$ yields $(2\pi)^3\delta(\mathbf{k}-\mathbf{k}')$, and (11.55) becomes after $\int d^3k'$ integration

$$\begin{aligned} W &= \frac{q^2}{4\pi^2}\int d^3v_0 f_0(\mathbf{v}_0)\int \frac{d^3k}{k^2|\varepsilon_L(\mathbf{k},\omega)|^2_{\omega=\mathbf{k}\cdot\mathbf{v}_0}} \\ &= \frac{q^2}{4\pi^2}\int d^3v_0 f_0(\mathbf{v}_0)\int \frac{d^3k}{k^2}\int d\omega \frac{\delta(\omega-\mathbf{k}\cdot\mathbf{v}_0)}{|\varepsilon_L(\mathbf{k},\omega)|^2} \\ &= \frac{q^2}{4\pi^2}\int du g_0(u)\int \frac{d^3k}{k^2}\int d\omega \frac{\delta(\omega-ku)}{|\varepsilon_L(k,\omega)|^2} \\ &= \frac{Nq^2}{4\pi^2}\int_{-\infty}^{\infty} d\omega \int d^3k\, \frac{g_0\left(\frac{\omega}{k}\right)}{k^3|\varepsilon_L(k,\omega)|^2}, \end{aligned} \tag{11.56}$$

where we have used

$$\int d^3v f_0(\mathbf{v})\delta(\omega-\mathbf{k}\cdot\mathbf{v}) = \int du g_0(u)\delta(\omega-ku), \tag{11.57}$$

with $g_0(u)$ being the corresponding one-dimensional distribution function.

On the other hand, let us recall (8.21) which we write in the form

$$\varepsilon_{ij} = \delta_{ij} + \frac{\omega_{\mathrm{pe}}^2}{\omega}\int d^3v v_i \frac{\partial f_0}{\partial v_j}\left[P\frac{1}{\omega - \mathbf{k}\cdot\mathbf{v}} - i\pi\delta(\omega - \mathbf{k}\cdot\mathbf{v})\right] \tag{11.58}$$

Using the Maxwellian relation

$$\frac{\partial f_0}{\partial v_j} = -\frac{m}{T}v_j f_0 \tag{11.59}$$

in the above gives

$$\mathrm{Im}(\varepsilon_{ij}) = \pi\frac{4\pi N q^2}{T\omega}\int d^3v v_i v_j \delta(\omega - \mathbf{k}\cdot\mathbf{v}) f_0(\mathbf{v}) \tag{11.60}$$

Hence, using the relation $\varepsilon_L = k_i k_j \varepsilon_{ij}/k^2$ in the above gives

$$\begin{aligned}\mathrm{Im}(\varepsilon_L(\omega,\mathbf{k})) &= \pi\frac{4\pi N q^2}{T\omega}\int d^3v \frac{(\mathbf{k}\cdot\mathbf{v})^2}{k^2}\delta(\omega - \mathbf{k}\cdot\mathbf{v}) f_0(\mathbf{v})\\ &= \pi\frac{4\pi N q^2}{T\omega}\frac{\omega^2}{k^2}\int d^3v \delta(\omega - \mathbf{k}\cdot\mathbf{v}) f_0(\mathbf{v})\\ &= \pi\frac{\omega_p^2}{k^2}\frac{\omega m}{T}\int_{-\infty}^{\infty} du g_0(u)\delta(\omega - ku)\\ &= \pi\frac{\omega_p^2}{k^2}\frac{m}{T}\frac{\omega}{k}\, g_0\left(\frac{\omega}{k}\right)\end{aligned} \tag{11.61}$$

Using (11.61) in (11.56) gives

$$W = \frac{1}{8\pi}<E^2(\mathbf{r},t)> = T\int\frac{d^3k}{(2\pi)^3}\int_{-\infty}^{\infty}\frac{d\omega}{2\pi}\frac{\mathrm{Im}\,\varepsilon_L}{\omega|\varepsilon_L(k,\omega)|^2}. \tag{11.62}$$

Equation (11.55) can be put into the form

$$<E^2(\mathbf{r},t)> = \int\frac{d^3k}{(2\pi)^3}\int_{-\infty}^{\infty}\frac{d\omega}{2\pi}<|E(\mathbf{k},\omega)|^2>, \tag{11.63}$$

with

$$<|E(\mathbf{k},\omega)|^2> = \frac{8\pi\, T\,\mathrm{Im}\,\varepsilon}{\omega|\varepsilon_L(k,\omega)|^2} = \frac{8\pi T}{\omega}\,\mathrm{Im}\left(\frac{1}{\varepsilon_L(\mathbf{k},\omega)}\right). \tag{11.64}$$

Equations (11.63) and (11.64) express the *fluctuation-dissipation theorem* for the electric field fluctuation. Equation (11.64) represents the *spectral density* of the electric field fluctuation. This equation shows that the fluctuations in the plasma are proportional to its temperature and to the imaginary part of the dielectric permittivity which is the cause of the electromagnetic field dissipation in the plasma.

The following is Rostoker's derivation of the fluctuation-dissipation theorem. The *impedance* (Z) connects the external current to the electric field by the relation

$$\mathbf{J}_{\text{ext}} = \frac{\mathbf{E}}{Z}. \tag{11.65}$$

Equation (11.34) gives

$$\mathbf{E}(\mathbf{k}, \omega) = \frac{-4\pi i\, \mathbf{k}}{k^2 \varepsilon_L(\mathbf{k}, \omega)}\, \rho_{\text{ext}}. \tag{11.66}$$

The charge conservation equation for the external current takes the form

$$\rho_{\text{ext}} = -\frac{1}{\omega}\, \mathbf{k} \cdot \mathbf{J}_{\text{ext}}. \tag{11.67}$$

Using (11.65) and (11.67) in (11.66) gives

$$\mathbf{E}(\mathbf{k}, \omega) = \frac{4\pi i}{\omega}\, \frac{\mathbf{k}}{k^2 \varepsilon_L}\, \frac{\mathbf{k} \cdot \mathbf{E}}{Z}. \tag{11.68}$$

Therefore, we have

$$Z = \frac{4\pi i}{\omega\, \varepsilon_L}. \tag{11.69}$$

This equation expresses the impedance in terms of the familiar dielectric function. In terms of the impedance Z, (11.64) takes the form

$$<|E(\mathbf{k}, \omega)|^2> = 2T \text{ Real part } Z = \frac{8\pi T}{\omega}\, \mathrm{Im}\left(\frac{1}{\varepsilon_L}\right). \tag{11.70}$$

The first equality above is the form of the fluctuation-dissipation theorem derived by Rostoker by using the dressed particle approach.

Summary. The fluctuations of the plasma electric field can be calculated by regarding the plasma as an aggregate of non-interacting dressed particles. The distribution function f_α and the electromagnetic field quantities are all can be determined in terms of ρ_{ext} as given in (11.32) in which $\mathbf{r}_0$ and $\mathbf{v}_0$ are the initial position and velocity. We introduce the initial distribution function $f_0(\mathbf{r}_0, \mathbf{v}_0)$ of the plasma to take the ensemble average of physical quantities. In fact, f_0 is the distribution of the dressed particles. Using (11.36) in

(11.38) gives

$$f_\alpha(\mathbf{k},\omega) = -4\pi\,\frac{e_\alpha}{m_\alpha}\,\frac{1}{k^2\varepsilon_L(\mathbf{k},\omega)}\,\frac{\mathbf{k}\cdot\frac{df_{\alpha 0}}{d\mathbf{v}}}{\omega-\mathbf{k}\cdot\mathbf{v}}\rho_{\text{ext}}(\mathbf{k},\omega) \tag{11.71}$$

where $\rho_{\text{ext}}(\mathbf{k},\omega)$ is given by (11.33). For another derivation of this equation via Klimontovich equation refer to: A. G. Sitenko, *Fluctuations and nonlinear wave interactions in plasmas*, Pergamon, Oxford (1982), note Sitenko's $\rho^0_{k,\omega} = \rho_{\text{ext}}(\mathbf{k},\omega)$ here. See also N. A. Krall and A. W. Trivelpiece, *Principles of plasma physics*, McGraw-Hill (1973), Eq. (11.1.11).

- **Density fluctuation.** In terms of the electric field, (11.34) is written as

$$ik_i\varepsilon_{ij}E_j(\mathbf{k},\omega) = 4\pi\rho_{\text{ext}}(\mathbf{k},\omega) \tag{11.72}$$

Using $E_j(\mathbf{k},\omega) = -ik_j\phi(\mathbf{k},\omega)$ in the above yields

$$\phi(\mathbf{k},\omega) = \frac{4\pi\rho_{\text{ext}}(\mathbf{k},\omega)}{k^2\varepsilon_L(\mathbf{k},\omega)} \tag{11.73}$$

where we used $k_ik_j\varepsilon_{ij} = \varepsilon_L$, the dielectric function. Equation (11.73) immediately gives

$$\mathbf{E}(\mathbf{k},\omega) = -4\pi i\mathbf{k}\,\frac{\rho_{\text{ext}}(\mathbf{k},\omega)}{k^2\varepsilon_L(\mathbf{k},\omega)} \tag{11.74}$$

Then the Poisson equation $\nabla\cdot\mathbf{E} = 4\pi\rho$ gives

$$\rho(\mathbf{k},\omega) = \frac{\rho_{\text{ext}}(\mathbf{k},\omega)}{\varepsilon_L(\mathbf{k},\omega)} \tag{11.75}$$

Using (11.33) in the above equation, we can write for the *density fluctuation*

$$\begin{aligned}
<\rho(\mathbf{k},\omega)\rho^*(\mathbf{k}',\omega')> &= \frac{4\pi^2q^2}{\varepsilon_L(\mathbf{k},\omega)\varepsilon_L^*(\mathbf{k}',\omega')}\int d^3r_0\int d^3v_0 f_0(v_0)e^{-i(\mathbf{k}-\mathbf{k}')\cdot\mathbf{r}_0} \\
&\quad\times\delta(\omega-\mathbf{k}\cdot\mathbf{v}_0)\delta(\omega'-\mathbf{k}'\cdot\mathbf{v}_0) \\
&= \frac{(2\pi)^3 4\pi^2q^2}{\varepsilon_L(\mathbf{k},\omega)\varepsilon_L^*(\mathbf{k}',\omega')}\delta(\mathbf{k}-\mathbf{k}')\int d^3v_0 f_0(v_0)\delta(\omega-\mathbf{k}\cdot\mathbf{v}_0)\delta(\omega'-\mathbf{k}'\cdot\mathbf{v}_0) \\
&= \frac{(2\pi)^3 4\pi^2q^2}{\varepsilon_L(\mathbf{k},\omega)\varepsilon_L^*(\mathbf{k}',\omega')}\delta(\mathbf{k}-\mathbf{k}')\delta(\omega-\omega')\int d^3v_0 f_0(v_0)\delta(\omega-\mathbf{k}\cdot\mathbf{v}_0)
\end{aligned}$$

Now we put $\mathbf{k} = \mathbf{k}'$ and $\omega = \omega'$ in the above equation. In doing so, note that $2\pi\delta(\omega - \omega') \to 1$ and $(2\pi)^3\delta(\mathbf{k} - \mathbf{k}') \to 1$. Then we obtain

$$<|\rho(\mathbf{k}, \omega)|^2> = \frac{2\pi q^2}{|\varepsilon_L(\mathbf{k}, \omega)|^2} \int du g_0(u)\delta(\omega - ku) = \frac{Tk^2}{2\pi\omega} \frac{\text{Im } \varepsilon_L(\mathbf{k}, \omega)}{|\varepsilon_L(\mathbf{k}, \omega)|^2} \tag{11.76}$$

This result can be obtained more rapidly, putting the relation, $\rho = i\mathbf{k} \cdot \mathbf{E}/4\pi$ in (11.64).

Using the relation $i\omega\mathbf{E} = 4\pi\mathbf{J}$ in (11.64) gives the spectral density of *current fluctuation*,

$$<|\mathbf{J}(\mathbf{k}, \omega)|^2> = \frac{T\omega}{2\pi} \text{Im}\left(\frac{1}{\varepsilon_L(\mathbf{k}, \omega)}\right) \tag{11.77}$$

Now it is clear that we can write

$$<|\mathbf{E}(\mathbf{k}, \omega)|^2> = \frac{16\pi^2}{k^2} \frac{<|\rho_{\text{ext}}(\mathbf{k}, \omega)|^2>}{|\varepsilon_L|^2} \tag{11.78}$$

$$<|\rho(\mathbf{k}, \omega)|^2> = \frac{<|\rho_{\text{ext}}(\mathbf{k}, \omega)|^2>}{|\varepsilon_L|^2} \tag{11.79}$$

$$<|\rho_{\text{ext}}(\mathbf{k}, \omega)|^2> = 2\pi q^2 \int du g_0(u)\delta(\omega - ku) = \frac{Tk^2}{2\pi\omega} \text{Im } \varepsilon_L(\mathbf{k}, \omega) \tag{11.80}$$

The preceding development indicates that the ensemble average in calculating various fluctuations is performed over the initial (or external) distribution of the *non-interacting* dressed particles.

In the above development, it is useful to know

$$\iiint f_0(\mathbf{v}_0)\delta(\omega - \mathbf{k} \cdot \mathbf{v}_0)d^3v_0 = \int g_0(u)\delta(\omega - ku)du, \ k = \sqrt{k_x^2 + k_y^2 + k_z^2} \tag{11.81}$$

11.5 Cerenkov Emission

The Cerenkov radiation is emitted by charges moving with constant velocities; this contrasts with the usual emission that is emitted by accelerated charges. The Cerenkov radiation of waves is a collective effect, and the emission process does not occur in a vacuum. It requires a medium and is emitted when the Landau wave-particle resonance conditions are met. It is the inverse process of Landau damping. The plasma electrons loses a fraction of their energy, and the wave is strengthened by satisfying the Landau resonance condition. The Cerenkov emission of a plasma medium can be fully accounted for by treating the plasma particles as dressed particles, as developed in the previous section. The electric field of a Cerenkov wave is given by (11.52), and the electric field energy of Cerenkov waves in a plasma can be expressed by (11.63) and (11.64). The source of thermal noise in the plasma is known to be spontaneous emission, which is Cerenkov emission due to the thermal motions of plasma particles. In the following, we calculate the rate of change of the average kinetic energy of the plasma particles due to the Cerenkov emission. The following development owes to Nishikawa and Wakatani. See References: A. F. Alexandrov, L. S. Bogdankevich, and A. A. Rukhadze, *Principles of plasma electrodynamics*, Springer-Verlag, Heidelberg (1984); K. Nishikawa and M. Wakatani, *Plasma Physics*, Springer-Verlag, Heidelberg (1990).

Let us calculate

$$\int d^3v_0 f(\mathbf{v}_0)\frac{d}{dt}\frac{1}{2}mv^2 = -e\int d^3v_0 f(\mathbf{v}_0)\mathbf{E}\cdot\mathbf{v}_0. \tag{11.82}$$

Here we consider only the electron emission and assume that the ions are immobile. Using (11.52) in (11.82) gives

$$-e\mathbf{E}\cdot\mathbf{v}_0 = -\frac{ie^2}{2\pi^2}\int d^3k\frac{\mathbf{k}\cdot\mathbf{v}_0\ e^{i\mathbf{k}\cdot\mathbf{r}}e^{-i\mathbf{k}\cdot\mathbf{r}_0}e^{-i\mathbf{k}\cdot\mathbf{v}_0 t}}{k^2\varepsilon_L(\mathbf{k},\omega)|_{\omega=\mathbf{k}\cdot\mathbf{v}_0}}. \tag{11.83}$$

The integrand in the above equation without the exponential phasor is the Fourier amplitude in **k**-space of the quantity $(-e\mathbf{E}\cdot\mathbf{v}_0)$, which is the time rate of the kinetic energy increase. Thus, the ensemble average (or time average) of this quantity is the sum of all the Fourier amplitudes over **k**-space. Thus, we consider the quantity

$$<-e\mathbf{E}\cdot\mathbf{v}_0> = \frac{e^2}{2\pi^2}\int d^3v_0 f(\mathbf{v}_0)\int d^3k\frac{-i\ \mathbf{k}\cdot\mathbf{v}_0}{k^2\varepsilon_L(\mathbf{k},\omega)|_{\omega=\mathbf{k}\cdot\mathbf{v}_0}}, \tag{11.84}$$

whose real part is given by

$$\begin{aligned}<-e\mathbf{E}\cdot\mathbf{v}_0> &= \frac{e^2}{2\pi^2}\int d^3v_0 f(\mathbf{v}_0)\int d^3k\frac{\mathbf{k}\cdot\mathbf{v}_0}{k^2}\mathrm{Im}\left(\frac{1}{\varepsilon_L(\mathbf{k},\omega)|_{\omega=\mathbf{k}\cdot\mathbf{v}_0}}\right)\\ &= \frac{e^2}{2\pi^2}\int d^3v_0 f(\mathbf{v}_0)\int\frac{d^3k}{k^2}\int_{-\infty}^{\infty}d\omega\ \omega\ \delta(\omega-\mathbf{k}\cdot\mathbf{v}_0)\ \mathrm{Im}\left(\frac{1}{\varepsilon_L(\mathbf{k},\omega)}\right)\\ &= \frac{e^2}{2\pi^2}\int du g_0(u)\int\frac{d^3k}{k^2}\int_{-\infty}^{\infty}d\omega\ \omega\ \delta(\omega-ku)\ \mathrm{Im}\left(\frac{1}{\varepsilon_L(k,\omega)}\right).\end{aligned} \tag{11.85}$$

On the other hand, we have

$$\text{Im}\ \frac{1}{\varepsilon_L(k,\omega)} = \frac{-\varepsilon_i(k,\omega)}{|\varepsilon_L(k,\omega)|^2} = \frac{-\varepsilon_i(\mathbf{k},\omega)}{\varepsilon_r^2(\mathbf{k},\omega)+\varepsilon_i^2(\mathbf{k},\omega)}, \tag{11.86}$$

where

$$\varepsilon_r = \text{Re}\ \varepsilon_L,\ \ \varepsilon_i = \text{Im}\ \varepsilon_L.$$

Let us expand $\varepsilon_r(\mathbf{k},\omega)$ in the vicinity of $\omega = \omega_r$, the root of $\varepsilon_r(\mathbf{k},\omega) = 0$:

$$\varepsilon_r(\mathbf{k},\omega) \doteq \frac{\partial\varepsilon_r(\mathbf{k},\omega_r)}{\partial\omega_r}(\omega-\omega_r).$$

The Cerenkov emission condition is $\omega_r = \mathbf{k}\cdot\mathbf{v}_0$, which is also the Landau resonance condition. Also introducing the Landau damping rate $\gamma = \varepsilon_i/\frac{\partial\varepsilon_r}{\partial\omega_r}$, Equation (11.86) becomes, for $\gamma \ll \omega_r$,

$$\text{Im}\ \frac{1}{\varepsilon_L(k,\omega)} = -\left(\frac{\partial\varepsilon_r(\mathbf{k},\omega_r)}{\partial\omega_r}\right)^{-1}\frac{\gamma}{(\omega-\omega_r)^2+\gamma^2} = -\left(\frac{\partial\varepsilon_r(\mathbf{k},\omega_r)}{\partial\omega_r}\right)^{-1}\pi\delta(\omega-\omega_r). \tag{11.87}$$

Therefore, (11.85) is calculated to yield

$$<-e\mathbf{E}\cdot\mathbf{v}_0> = \frac{e^2}{2\pi^2}\int\frac{d^3k}{k^2}\int d\omega\ \frac{\omega}{k}\ g_0\left(\frac{\omega}{k}\right)\ \pi\ \frac{\delta(\omega-\omega_r)}{\partial\varepsilon_r/\partial\omega_r},$$

which reduces to

$$<-e\mathbf{E}\cdot\mathbf{v}_0> = T\int\frac{d^3k}{(2\pi)^3}\gamma(k,\omega_r), \tag{11.88}$$

where we have used (11.61). Equation (11.88) is also a form of the fluctuation-dissipation theorem.

- **A useful formula.** Let us write $\varepsilon = \varepsilon' + i\varepsilon''$ $(\varepsilon'' \ll \varepsilon')$ in (11.87) to write

$$\frac{1}{\varepsilon(\omega,k)} = \lim_{\varepsilon''\to 0}\frac{1}{\varepsilon'+i\varepsilon''} = P\ \frac{1}{\varepsilon'(k,\omega_r)} - i\pi\ \frac{\delta(\omega-\omega_r)}{\partial\varepsilon'/\partial\omega_r} \tag{11.89}$$

where $\varepsilon(\omega,k) = 0$ at $\omega = \omega_r$. This formula is often seen in plasma physics literature. This formula is useful for theoretical development of resonant wave-wave interaction and for formulation of the wave action kinetic equation. See M. Camac, A. R., Kantrowitz, M. M., Litvak, R. M., Patrick, H. E., and Petschek, *Shock waves in collision-free plasmas*, Nuc. Fusion Supplement, **Pt. 2**, (1962), pp. 423–444.

11.6 Electromagnetic Fluctuation

The current associated with a dressed particle can be represented as

$$\mathbf{J}_{\text{ext}}(\mathbf{r},t) = q\mathbf{v}_0\delta(\mathbf{r}-\mathbf{r}_0-\mathbf{v}_0 t) \tag{11.90}$$

The full electromagnetic Maxwell equations should be used with the above *external* current as the source term. Then, the electromagnetic fields determined in terms of $\mathbf{J}_{\text{ext}}$ will be ensemble-averaged over the distribution of dressed particles which is the equilibrium plasma distribution itself. We have

$$\nabla \times \mathbf{E} = -\frac{1}{c}\frac{\partial \mathbf{B}}{\partial t} \tag{11.91}$$

$$\nabla \times \mathbf{B} = \frac{1}{c}\frac{\partial \mathbf{E}}{\partial t} + \frac{4\pi}{c}\sum_\alpha e_\alpha \int \mathbf{v} f_\alpha d^3v + \frac{4\pi}{c}\mathbf{J}_{\text{ext}} \tag{11.92}$$

The plasma distribution function f_α obeys the Vlasov equation

$$\frac{\partial}{\partial t} f_\alpha(\mathbf{r},\mathbf{v},t) + \mathbf{v}\cdot\frac{\partial f_\alpha}{\partial \mathbf{r}} + \frac{e_\alpha}{m_\alpha}\left[\mathbf{E}(\mathbf{r},t) + \frac{\mathbf{v}}{c}\times\mathbf{B}(\mathbf{r},t)\right]\cdot\frac{\partial f_{\alpha 0}}{\partial \mathbf{v}} = 0. \tag{11.93}$$

After Fourier transforming the above equations and eliminating $\mathbf{B}$ we obtain

$$\mathbf{k}\times(\mathbf{k}\times\mathbf{E}) + \frac{\omega^2}{c^2}\mathbf{E} = -\frac{4\pi i\omega}{c^2}(\mathbf{J}+\mathbf{J}_{\text{ext}}) \tag{11.94}$$

where

$$\mathbf{J} = -i\ E_j \sum_\alpha \frac{e_\alpha^2}{m_\alpha}\int d^3v\ \frac{\mathbf{v}\ \frac{\partial f_{\alpha 0}}{\partial v_j}}{\omega - \mathbf{k}\cdot\mathbf{v}} \tag{11.95}$$

where index j is summed over.

$$J_{\text{ext}}(\mathbf{k},\omega) = 2\pi q\mathbf{v}_0 e^{-i\mathbf{k}\cdot\mathbf{r}_0}\delta(\omega - \mathbf{k}\cdot\mathbf{v}_0) \tag{11.96}$$

Equation (11.94) can be put into the form,

$$\left[n^2\left(\frac{k_i k_j}{k^2} - \delta_{ij}\right) + \varepsilon_{ij}\right]E_j = -\frac{4\pi i}{\omega}J_i^{\text{ext}} \tag{11.97}$$

where $n^2 = \frac{c^2k^2}{\omega^2}$ is the refractive index, and ε_{ij} is the dielectric tensor: here we recall Eqs. (8.22)–(8.25).

$$\varepsilon_{ij} = \delta_{ij} + \sum \frac{\omega_{p\alpha}^2}{\omega}\int d^3v\ \frac{v_i\ \frac{\partial f_{\alpha 0}}{\partial v_j}}{\omega - \mathbf{k}\cdot\mathbf{v}} \tag{11.98}$$

$$\varepsilon_{ij} = \delta_{ij}\varepsilon_T + (\varepsilon_L - \varepsilon_T)\frac{k_i k_j}{k^2} \tag{11.99}$$

$$\varepsilon_L = 1 + \Sigma\frac{\omega_{p\alpha}^2}{k^2}\int \frac{\mathbf{k}\cdot\frac{\partial f_{\alpha 0}}{\partial \mathbf{v}}}{\omega - \mathbf{k}\cdot\mathbf{v}} d^3v \tag{11.100}$$

$$\varepsilon_T = 1 - \Sigma\frac{\omega_{p\alpha}^2}{\omega}\int \frac{f_{\alpha 0}}{\omega - \mathbf{k}\cdot\mathbf{v}} d^3v \tag{11.101}$$

Equation (11.97) can be inverted in the form

$$E_i(\mathbf{k},\omega) = -\frac{4\pi i}{\omega}\, G_{ij}(\mathbf{k},\omega)\; J_j^{\text{ext}}(\mathbf{k},\omega) \tag{11.102}$$

with

$$G_{ij} = \frac{k_i k_j}{k^2}\,\frac{1}{\varepsilon_L} + \left(\delta_{ij} - \frac{k_i k_j}{k^2}\right)\frac{1}{\varepsilon_H},\ (\varepsilon_H = \varepsilon_T - n^2) \tag{11.103}$$

We form the correlation spectral density

$$<E_i(\mathbf{k},\omega)E_j^*(\mathbf{k}',\omega')> = \frac{16\pi^2}{\omega\omega'}\int d^3r_0 \int d^3v_0 f_0(v_0) \\ \times G_{il}(\mathbf{k},\omega)J_l^{\text{ext}}(\mathbf{k},\omega)G_{jm}^*(\mathbf{k}',\omega')J_m^{\text{ext}*}(\mathbf{k}',\omega') \tag{11.104}$$

Using (11.96) and carrying out the $\int d^3r_0$-integral, the RHS of the above equation becomes

$$\frac{16\pi^2}{\omega^2}(2\pi)^5 q^2 \delta(\mathbf{k}-\mathbf{k}')\delta(\omega-\omega')\frac{T\omega}{m\pi\omega_p^2}G_{il}(\mathbf{k},\omega)G_{jm}^*(\mathbf{k}',\omega')\text{Im}\ \varepsilon_{lm} \tag{11.105}$$

where we used

$$\int d^3v f_0(v) v_l v_m \delta(\omega - \mathbf{k}\cdot\mathbf{v}) = \frac{T\omega n_0}{m\pi\omega_p^2}\ \text{Im}\ \varepsilon_{lm} \tag{11.106}$$

which comes from the definition of ε_{ij}. Equation (11.104) becomes upon putting $\mathbf{k} = \mathbf{k}'$ and $\omega = \omega'$

$$<E_i(\mathbf{k},\omega)E_j^*(\mathbf{k},\omega)> = \frac{8\pi T}{\omega}\left[\frac{k_i k_l}{k^2}\,\frac{1}{\varepsilon_L} + \left(\delta_{il} - \frac{k_i k_l}{k^2}\right)\frac{1}{\varepsilon_H}\right] \\ \times \left[\frac{k_j k_m}{k^2}\,\frac{1}{\varepsilon_L^*} + \left(\delta_{jm} - \frac{k_j k_m}{k^2}\right)\frac{1}{\varepsilon_H^*}\right]\left[\delta_{lm}\text{Im}\ \varepsilon_T + \frac{k_l k_m}{k^2}(\text{Im}\ \varepsilon_L - \text{Im}\ \varepsilon_T)\right] \tag{11.107}$$

where we used (11.99) for Im ε_{lm}. It is straightforward to carry out the above multiplication. At the expense of some algebra, we obtain the final result

$$<E_i(\mathbf{k},\omega)E_j^*(\mathbf{k},\omega)> = \frac{8\pi T}{\omega}\left[\frac{k_ik_j}{k^2}\frac{\text{Im }\varepsilon_L(\mathbf{k},\omega)}{|\varepsilon_L(\mathbf{k},\omega)|^2} + \left(\delta_{ij} - \frac{k_ik_j}{k^2}\right)\frac{\text{Im }\varepsilon_T(\mathbf{k},\omega)}{|\varepsilon_H(\mathbf{k},\omega)|^2}\right] \tag{11.108}$$

[Cf. Eq. (10.3.2), A. F. Alexandrov, L. S. Bogdankevich, A. A. Rukhadze, *Principles of plasma electrodynamics*, Springer, Berlin (1984)].

- **Electron density fluctuations in Maxwellian plasma.** In Section 11.4, we obtained the spectral density of the electron density fluctuation $<|\rho(\mathbf{k},\omega)|^2>$:

$$<|\rho(\mathbf{k},\omega)|^2> = \frac{Tk^2}{2\pi\omega}\frac{\text{Im }\varepsilon_L(\mathbf{k},\omega)}{|\varepsilon_L(\mathbf{k},\omega)|^2} \tag{11.109}$$

This quantity determines the scattering cross section of the incoherent scattering of electromagnetic waves by plasma. Let us further calculate (11.102) for a Maxwellian plasma. Let us recall Eqs. (8.72)–(8.74) to write

$$\varepsilon_L(\mathbf{k},\omega) = 1 + \frac{1}{k^2\lambda_e^2}[1 - \varphi(\zeta) + i\sqrt{\pi}\zeta e^{-\zeta^2} - \varphi(\mu\zeta) + i\sqrt{\pi}\ \mu\ \zeta e^{-\mu^2\zeta^2}] \tag{11.110}$$

$$\varepsilon_T(\mathbf{k},\omega) = 1 - \frac{\omega_{\text{pe}}^2}{\omega^2}[\ \varphi(\zeta) - i\sqrt{\pi}\zeta e^{-\zeta^2}]7 \tag{11.111}$$

where

$$\varphi(\zeta) = 2\zeta e^{-\zeta^2}\int_0^\zeta dx\ e^{x^2} = -\zeta Z(\zeta) + i\sqrt{\pi}\zeta e^{-\zeta^2} \tag{11.112}$$

with $Z(\zeta)$ the plasma dispersion function. In (11.110), $\mu = m_e/m_i$, and $\mu\zeta$ terms are the ion contributions which should be included in the low frequency region. Using (11.110) and (11.112) in (11.109) gives

$$<|\rho(\mathbf{k},\omega)|^2> = \sqrt{2\pi}n_0\frac{\lambda_e^3k^3}{\omega_{\text{pe}}}\frac{e^{-\zeta^2} + \mu e^{-\mu^2\zeta^2}}{[2 + \lambda_e^2k^2 - \varphi(\zeta) - \varphi(\mu\zeta)]^2 + \pi\zeta^2[e^{-\zeta^2} + \mu e^{-\mu^2\zeta^2}]^2} \tag{11.113}$$

The spectral density in (11.113) is shown in Fig. 11.1: after reference [A. Sitenko and V. Malnev, *Plasma physics theory*, Chapman and Hall, London (1995) p. 270]; our λ_e and ζ correspond, respectively, a and z in the reference.

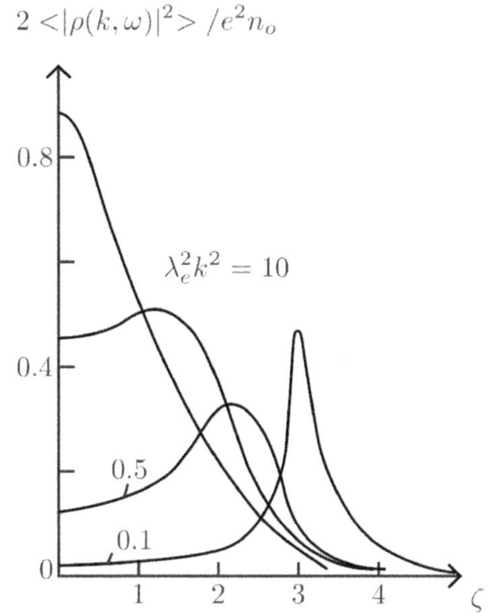

Fig. 11.1. The spectral distribution of the charge density (after A. Sitenko and V. Malnev, *Plasma Physics Theory*, Chapman and Hall, London, 1995, p. 270).

• **Current fluctuations.** Let us derive the spectral density of the electric current fluctuation $<J_i(\mathbf{k},\omega)J_j^*(\mathbf{k},\omega)>$. First, we can verify the following equation by using (11.95) and (11.98):

$$J_i = \frac{-i\omega}{4\pi}(\varepsilon_{ij} - \delta_{ij})E_j = \frac{-i\omega}{4\pi}\left[(\varepsilon_T - 1)\delta_{ij} + (\varepsilon_L - \varepsilon_T)\frac{k_i k_j}{k^2}\right]E_j \quad (11.114)$$

where we used (11.99). Using (11.102) in (11.114) gives

$$J_i = \left[(1 - \varepsilon_T)\delta_{ij} + (\varepsilon_T - \varepsilon_L)\frac{k_i k_j}{k^2}\right] G_{jl}J_l^{\text{ext}} \equiv H_{jl}J_l^{\text{ext}}$$

where

$$H_{ij} = \frac{1-\varepsilon_L}{\varepsilon_L}\frac{k_i k_j}{k^2} + \left(\delta_{ij} - \frac{k_i k_j}{k^2}\right)\frac{1-\varepsilon_T}{\varepsilon_H}, \quad (\varepsilon_H = \varepsilon_T - n^2)$$

Analogously with (11.104) we calculate

$$<J_i(\mathbf{k},\omega)J_j^*(\mathbf{k}',\omega')> = \int d^3r_0 \int d^3v_0 f_0(v_0) H_{il}(\mathbf{k},\omega)J_l^{\text{ext}}(\mathbf{k},\omega)H_{jm}^*(\mathbf{k}',\omega')J_m^{ext*}(\mathbf{k}',\omega') \quad (11.115)$$

Carrying out further algebra analogous with (11.105) and (11.106), we obtain the intermediate result

$$<J_i(\mathbf{k},\omega)J_j^*(\mathbf{k},\omega)> = \frac{\omega T}{2\pi}\left[\frac{1-\varepsilon_L}{\varepsilon_L}\frac{k_ik_j}{k^2} + \left(\delta_{ij} - \frac{k_ik_j}{k^2}\right)\frac{1-\varepsilon_T}{\varepsilon_H}\right]$$
$$\times \left[\frac{1-\varepsilon_L^*}{\varepsilon_L^*}\frac{k_ik_j}{k^2} + \left(\delta_{ij} - \frac{k_ik_j}{k^2}\right)\frac{1-\varepsilon_T^*}{\varepsilon_H^*}\right]$$
$$\times \left[\delta_{lm}\mathrm{Im}\ \varepsilon_T + \frac{k_lk_m}{k^2}(\mathrm{Im}\ \varepsilon_L - \mathrm{Im}\ \varepsilon_T)\right] \quad (11.116)$$

After straightforward algebra with (11.116) we obtain

$$<J_i(\mathbf{k},\omega)J_j^*(\mathbf{k},\omega)> = \frac{\omega T}{2\pi}\left[\frac{k_ik_j}{k^2}\mathrm{Im}\ \varepsilon_L\frac{|1-\varepsilon_L|^2}{|\varepsilon_L(\mathbf{k},\omega)|^2} + \left(\delta_{ij} - \frac{k_ik_j}{k^2}\right)\mathrm{Im}\ \varepsilon_T\frac{|1-\varepsilon_T|^2}{|\varepsilon_H(\mathbf{k},\omega)|^2}\right] \quad (11.117)$$

11.7 Fluctuation of Electrostatic Field in a Magnetized Plasma

The particle orbit of species α in a magnetized plasma with the static magnetic field $\mathbf{B}_0 = \hat{\mathbf{z}}B_0$ is described by

$$\mathbf{r}(t) = \mathbf{r}_0 + \mathbf{R}_\alpha(t,\mathbf{v}_0) \quad (11.118)$$

$$R_x = \frac{1}{\omega_{c\alpha}}[v_{0x}\sin\ \omega_{c\alpha}t - v_{0y}(\cos\ \omega_{c\alpha}t - 1)] \quad (11.119)$$

$$R_y = \frac{1}{\omega_{c\alpha}}[v_{0x}(\cos\ \omega_{c\alpha}t - 1) + v_{0y}\ \sin\ \omega_{c\alpha}t] \quad (11.120)$$

$$R_z = v_{0z}t \quad (11.121)$$

where the subscript 0 quantities are the initial values at $t = 0$. The charge density associated with a dressed particle can be written as

$$\rho_{\mathrm{ext}}(\mathbf{r},t) = q\delta(\mathbf{r} - \mathbf{r}(t)) = q\delta(\mathbf{r} - \mathbf{r}_0 - \mathbf{R}_\alpha(t,\mathbf{v}_0)) \quad (11.122)$$

Its Fourier transform is

$$\rho_{\mathrm{ext}}(\mathbf{k},\omega) = q\ e^{-i\mathbf{k}\cdot\mathbf{r}_0}\int_{-\infty}^{\infty} dt\ \exp[i\omega t - i\mathbf{k}\cdot\mathbf{R}_\alpha(t,\mathbf{v}_0)] \quad (11.123)$$

We have the Poisson equation for the single dressed particle

$$\nabla\cdot\mathbf{E}(\mathbf{r},t) = 4\pi\sum_\alpha\int f_\alpha(\mathbf{r},\mathbf{v},t)d^3v + 4\pi\rho_{\mathrm{ext}}(\mathbf{r},t) \quad (\alpha = e,\ i) \quad (11.124)$$

where f_α is governed by the linearized Vlasov equation

$$\frac{\partial}{\partial t} f_\alpha(\mathbf{r},\mathbf{v},t) + \mathbf{v}\cdot\frac{\partial f_\alpha}{\partial \mathbf{r}} + \frac{e_\alpha}{m_\alpha}\frac{\mathbf{v}\times\mathbf{B_0}}{c}\cdot\frac{\partial f_\alpha}{\partial \mathbf{v}} = -\frac{e_\alpha}{m_\alpha}\mathbf{E}(\mathbf{r},t)\cdot\frac{\partial f_{\alpha 0}}{\partial \mathbf{v}} \tag{11.125}$$

Equation (11.125) can be solved by characteristic method with the result expressed by Bessel function series. Consulting Chapter 9, we obtain the following solution:

$$f_\alpha(\mathbf{k},\omega,\mathbf{v}) = -\frac{e_\alpha}{m_\alpha}\phi(\mathbf{k},\omega)\sum_{n=-\infty}^{\infty}\left[k_\parallel\frac{\partial f_{\alpha 0}}{\partial v_\parallel} + \frac{n\omega_{c\alpha}}{v_\perp}\frac{\partial f_{\alpha 0}}{\partial v_\perp}\right]\frac{[J_n\left(\frac{k_\perp v_\perp}{\omega_{c\alpha}}\right)]^2}{\omega - k_\parallel v_\parallel - n\omega_{c\alpha}} \tag{11.126}$$

Putting the above equation into the Fourier-transformed version of (11.124) yields for the electric potential contributed by a single dressed particle

$$\phi(\mathbf{k},\omega) = \frac{4\pi}{k^2}\frac{\rho_{\text{ext}}(\mathbf{k},\omega)}{\varepsilon_L(\mathbf{k},\omega)} \tag{11.127}$$

where the dielectric function ε_L is

$$\varepsilon_L = 1 + \sum_\alpha \frac{\omega_{p\alpha}^2}{k^2}\sum_{n=-\infty}^{\infty}\int d^3v\left[k_\parallel\frac{\partial f_{\alpha 0}}{\partial v_\parallel} + \frac{n\omega_{c\alpha}}{v_\perp}\frac{\partial f_{\alpha 0}}{\partial v_\perp}\right]\frac{\left[J_n\left(\frac{k_\perp v_\perp}{\omega_{c\alpha}}\right)\right]^2}{\omega - k_\parallel v_\parallel - n\omega_{c\alpha}} \tag{11.128}$$

where $d^3v = 2\pi v_\perp dv_\perp dv_\parallel$.

Next, we calculate the correlation spectral density $<\phi(\mathbf{k},\omega)\ \phi^*(\mathbf{k}',\omega')>$ where the ensemble average is taken over the entire plasma particles, both species of $\alpha = e,\ i$. Using (11.123) and (11.127), we can write after $\int d^3r_0$-integral

$$\begin{aligned}
&<\phi(\mathbf{k},\omega)\phi^*(\mathbf{k}',\omega')> \\
&= \frac{16\pi^2}{k^4}(2\pi)^3\delta(\mathbf{k}-\mathbf{k}')\frac{1}{\varepsilon_L(\mathbf{k},\omega)}\frac{1}{\varepsilon_L^*(\mathbf{k},\omega')}\sum_\alpha\int d^3v_0 f_{0\alpha}(\mathbf{v}_0) \\
&\quad\times\int_{-\infty}^{\infty} dt\ \exp\ [i\omega t - i\mathbf{k}\cdot\mathbf{R}_\alpha(\mathbf{v}_0,t)] \\
&\quad\times\int_{-\infty}^{\infty} dt'\ \exp\ [-i\omega' t' + i\mathbf{k}\cdot\mathbf{R}_\alpha(\mathbf{v}_0,t')]
\end{aligned} \tag{11.129}$$

To carry out the time-integral in (11.129), we calculate

$$\mathbf{k}\cdot\mathbf{R}_\alpha(t,\mathbf{v}_0)=k_z v_{0z}t+\frac{1}{\omega_{c\alpha}}[(k_x v_{0x}+k_y v_{0y})\sin\ \omega_{c\alpha}t\\ +(\cos\ \omega_{c\alpha}t-1)(k_y v_{0x}-k_x v_{0y})]$$

Introduce cylindrical coordinates:

$$v_{0x}=v_\perp\cos\ \varphi_1,\quad v_{0y}=v_\perp\sin\ \varphi_1,\quad k_x=k_\perp\ \cos\ \varphi_2,\quad k_y=k_\perp\ \sin\varphi_2 \tag{11.130}$$

Then, we have

$$\mathbf{k}\cdot\mathbf{R}(t,\mathbf{v}_0)=k_z v_{0z}t+a_\alpha[\sin\ (\omega_{c\alpha}t+\varphi)-\sin\ \varphi] \tag{11.131}$$

$$\text{where}\ \ \varphi=\varphi_2-\varphi_1,\ a_\alpha=\frac{k_\perp v_\perp}{\omega_{c\alpha}} \tag{11.132}$$

Then the product of two time-integral in (11.129) can be written as

$$\int_{-\infty}^{\infty}dte^{i(\omega-k_z v_{0z})t}\sum_{n=-\infty}^{\infty}J_n(a_\alpha)e^{in(\omega_{c\alpha}t+\varphi)}\int_{-\infty}^{\infty}dt'e^{-i(\omega'-k_z v_{0z})t'}\sum_{m=-\infty}^{\infty}J_m(a_\alpha)e^{-im(\omega_{c\alpha}t'+\varphi)}$$

Note that only the $n=m$ terms survive in the double series because of the $\int d\varphi$-integral. Thus, the above expression becomes

$$\sum_{n=-\infty}^{\infty}J_n^2(a_\alpha)(2\pi)^2\ \delta(\omega-k_z v_{0z}+n\omega_{c\alpha})\delta(\omega'-k_z v_{0z}+n\omega_{c\alpha})$$

Therefore, (11.129) takes the form

$$<\phi(\mathbf{k},\omega)\phi^*(\mathbf{k}',\omega')>=\frac{16\pi^2}{k^4}(2\pi)^3\delta(\mathbf{k}-\mathbf{k}')\delta(\omega-\omega')\frac{1}{|\varepsilon_L(\mathbf{k},\omega)|^2}\\ \times\sum_\alpha q_\alpha^2\int d^3v_0 f_{0\alpha}(v_{\perp 0},v_{\parallel 0})\sum_{n=-\infty}^{\infty}J_n^2(a_\alpha)(2\pi)^2\delta(\omega-k_z v_{0z}+n\omega_{c\alpha}) \tag{11.133}$$

Putting $\omega=\omega'$ and $\mathbf{k}=\mathbf{k}'$ in the above equation, we obtain

$$<|\phi(\mathbf{k},\omega)|^2>=\frac{16\pi^2 2\pi}{k^4}\frac{1}{|\varepsilon_L(\mathbf{k},\omega)|^2}\\ \times\sum_\alpha q_\alpha^2\int d^3v_0 f_{0\alpha}(v_{\perp 0},v_{\parallel 0})\sum_{n=-\infty}^{\infty}J_n^2(a_\alpha)\ \delta(\omega-k_z v_{0\parallel}+n\omega_{c\alpha}) \tag{11.134}$$

This is the spectral density of the electric field fluctuation in a magnetized plasma.

11.8 Fluctuation Derived from Quasilinear Theory

In this section the electric fluctuation will be derived from the quasilinear equations of Liouville equation. For this subject, useful references are:

D. C. Montgomery, *Theory of unmagnetized plasma*, Gordon and Beach, New York (1971).

E. M. Lifshitz and L. P. Pitaevskii, *Physical kinetics*, Landau and Lifshitz Course of theoretical physics, Volume 10, Butterworth-Heinemann (1981).

Consider a stable plasma that consists of N_α ($\alpha = e,\ i$) particles and has average field $\mathbf{E} = 0$ (the plasma is homogeneous). The quasilinear theory of Liouville equation derived the following equations in Chapter 1, (1.98)–(1.100):

$$F_\alpha(\mathbf{k}, \mathbf{v}, t) = \sum_{i=1}^{N_\alpha} \delta(\mathbf{v} - \mathbf{v}_{i0}) e^{-i\mathbf{k}\cdot(\mathbf{r}_{i0}+\mathbf{v}t)}$$

$$- \frac{1}{k^2} \sum_\alpha \omega_{p\alpha}^2 \sum_{i=1}^{N_\alpha} \frac{\mathbf{k} \cdot \frac{\partial F_\alpha(\mathbf{k}=0,\mathbf{v},\epsilon t)}{\partial \mathbf{v}}}{\mathbf{k}\cdot\mathbf{v} - \mathbf{k}\cdot\mathbf{v}_{i0}} \left(\frac{e^{-i\mathbf{k}\cdot(\mathbf{r}_{i0}+\mathbf{v}t)}}{\varepsilon(\mathbf{k}, \mathbf{k}\cdot\mathbf{v})} - \frac{e^{-i\mathbf{k}\cdot(\mathbf{r}_{i0}+\mathbf{v}_{i0}t)}}{\varepsilon(\mathbf{k}, \mathbf{k}\cdot\mathbf{v}_{i0})} \right). \tag{11.135}$$

$$\mathbf{E}(\mathbf{k}, t) = \frac{4\pi i \mathbf{k}}{k^2} \sum_\alpha n_{0\alpha} e_\alpha \sum_{i=1}^{N_\alpha} \frac{e^{-i\mathbf{k}\cdot(\mathbf{r}_{i0}+\mathbf{v}_{i0}t)}}{\varepsilon(\mathbf{k},\ \mathbf{k}\cdot\mathbf{v}_{i0})}. \tag{11.136}$$

$$\frac{\partial}{\partial t} F_\alpha(\mathbf{k}=0, \mathbf{v}, \epsilon t) = -\frac{\partial}{\partial \mathbf{v}} \cdot \sum_{\mathbf{k}} \frac{e_\alpha}{m_\alpha} <\mathbf{E}(-\mathbf{k}, t) F_\alpha(\mathbf{k}, \mathbf{v}, t)>, \tag{11.137}$$

where the ensemble average should be taken as is indicated by the angular bracket. Equations (11.135)–(11.137) are the quasilinear equations of the Liouville equation which will be used in the subsequent development to calculate the electric field fluctuation.

We consider fluctuation of the electric field in a spatially uniform plasma whose average electric field vanishes: $<\mathbf{E}> = 0$. Also $\mathbf{E}(t)$ is assumed to be a stationary random process, so that fluctuation $<\mathbf{E}(\mathbf{r}, t)\mathbf{E}(\mathbf{r}', t')>$ depends only upon $\tau = t - t'$ and $\mathbf{R} = \mathbf{r} - \mathbf{r}'$. Therefore, we have

$$<\mathbf{E}(\mathbf{r}, t)\ \mathbf{E}(\mathbf{r}', t')> = <\mathbf{E}(\mathbf{r}, t)\ \mathbf{E}(\mathbf{r} - \mathbf{R}, t - \tau)>,$$

which is independent of $\mathbf{r}$ and t, but is dependent on $\mathbf{R}$ and τ only. Because $<\mathbf{E}> = 0$, $\mathbf{E} = <\mathbf{E}> + \delta\mathbf{E} = \delta\mathbf{E}$, and thus we have

$$<\mathbf{E}(\mathbf{r},t)\ \mathbf{E}(\mathbf{r}-\mathbf{R},t-\tau)> = <\delta\mathbf{E}(\mathbf{r},t)\ \delta\mathbf{E}(\mathbf{r}-\mathbf{R},t-\tau)>. \quad (11.138)$$

Let us evaluate the above quantity in the limit of $t \to \infty$ by using (11.136). The ensemble average in the above expression is taken over the distribution

$$F^{(N)}(\mathbf{x}_{10}, \mathbf{x}_{20}, \ldots, \mathbf{x}_{N0}) = F^{(1)}(\mathbf{v}_{10})F^{(1)}(\mathbf{v}_{20})\ldots F^{(1)}(\mathbf{v}_{N0}),$$

where $F^{(N)}$ is the N-particle Liouville distribution function, and we have assumed that the ensemble is spatially uniform and is free of correlation. Then (11.136) gives

$$<\delta\mathbf{E}(\mathbf{k},t)\ \delta\mathbf{E}(\mathbf{k}',t')> = \left\langle i\frac{4\pi e\mathbf{k}}{k^2}\sum_{i=1}^{N}\frac{e^{-i\mathbf{k}\cdot(\mathbf{r}_{i0}+\mathbf{v}_{i0}t)}}{\varepsilon(\mathbf{k},\ \mathbf{k}\cdot\mathbf{v}_{i0})}\ i\ \frac{4\pi e\mathbf{k}'}{k'^2}\sum_{j=1}^{N}\frac{e^{-i\mathbf{k}'\cdot(\mathbf{r}_{j0}+\mathbf{v}_{j0}t')}}{\varepsilon(\mathbf{k}',\ \mathbf{k}'\cdot\mathbf{v}_{j0})}\right\rangle.$$

The operation $<\cdots>$ includes a spatial integration from $-\infty$ to $+\infty$ (with density $n_0 = N/V$ constant while both N and V go to ∞). Thus, we have the relations

$$<\mathbf{k}e^{-i\mathbf{k}\cdot\mathbf{r}_{i0}}> \sim \mathbf{k}\ \delta(\mathbf{k}) = 0, \quad <\mathbf{k}e^{-i\mathbf{k}\cdot\mathbf{r}_{i0}}e^{-i\mathbf{k}'\cdot\mathbf{r}_{j0}}> = 0 \quad \text{when } i \neq j$$

and

$$<\mathbf{k}e^{-i\mathbf{k}\cdot\mathbf{r}_{i0}}e^{-i\mathbf{k}'\cdot\mathbf{r}_{j0}}> \sim \mathbf{k}\ \delta(\mathbf{k}+\mathbf{k}') \quad \text{when } i = j.$$

Therefore, only products of terms $i = j$ survive the averaging operation to give

$$\begin{aligned}
&<\delta\mathbf{E}(\mathbf{k},t)\ \delta\mathbf{E}(\mathbf{k}',t')> \\
&= -\frac{4\pi e\mathbf{k}}{k^2}\frac{4\pi e\mathbf{k}'}{k'^2}\delta(\mathbf{k}+\mathbf{k}')\sum_{i=1}^{N}\frac{e^{-i\mathbf{k}\cdot\mathbf{v}_{i0}t}}{\varepsilon(\mathbf{k},\ \mathbf{k}\cdot\mathbf{v}_{i0})}\frac{e^{-i\mathbf{k}'\cdot\mathbf{v}_{i0}t'}}{\varepsilon(\mathbf{k}',\ \mathbf{k}'\cdot\mathbf{v}_{i0})} \\
&\quad \times \int F^{(1)}(\mathbf{v}_{10})F^{(1)}(\mathbf{v}_{20})\ldots F^{(1)}(\mathbf{v}_{N0})d^3v_{10}d^3v_{20}\ldots d^3v_{N0}.
\end{aligned}$$

The summation yields N identical terms. Note that $\mathbf{v}_{i0}$ is a dummy integration variable, which we can put as $\mathbf{v}$. With a normalization constant (volume)$^{-1}$, which we have omitted so far, the integration yields the factor n_0, the density, and we can write

$$<\delta\mathbf{E}(\mathbf{k},t)\ \delta\mathbf{E}(\mathbf{k}',t-\tau)> = -\delta(\mathbf{k}+\mathbf{k}')\left(\frac{4\pi e}{k^2}\right)^2 n_0\mathbf{k}\mathbf{k}'\int d^3vF(\mathbf{v})\frac{e^{-i\mathbf{k}\cdot\mathbf{v}\tau}}{|\varepsilon(\mathbf{k},\ \mathbf{k}\cdot\mathbf{v})|^2} \tag{11.139}$$

The integral in (11.139) can be written as

$$\int du\int d^3v\frac{e^{-iku\tau}}{|\varepsilon(\mathbf{k},\ ku)|^2}F(\mathbf{v})\delta\left(u-\frac{\mathbf{k}\cdot\mathbf{v}}{k}\right).$$

Then, (11.139) reads

$$<\delta\mathbf{E}(\mathbf{k},t)\ \delta\mathbf{E}(\mathbf{k}',t')> = \delta(\mathbf{k}+\mathbf{k}')\left(\frac{4\pi e}{k^2}\right)^2 n_0\mathbf{k}\mathbf{k}'\int_{-\infty}^{\infty} dug_0(u)\ \frac{e^{-iku\tau}}{|\varepsilon(\mathbf{k},ku)|^2},$$

where

$$g_0(u) = \int F(\mathbf{v})\delta\left(u-\frac{\mathbf{k}\cdot\mathbf{v}}{k}\right)d^3v$$

Making a change of variable $u\rightarrow\omega/k$ in the above yields

$$<\delta\mathbf{E}(\mathbf{k},t)\ \delta\mathbf{E}(\mathbf{k}',t-\tau)> = -\delta(\mathbf{k}+\mathbf{k}')\left(\frac{4\pi e}{k^2}\right)^2 n_0\frac{\mathbf{k}\mathbf{k}'}{k}\int_{-\infty}^{\infty} d\omega\ g_0\left(\frac{\omega}{k}\right)\ \frac{e^{-i\omega\tau}}{|\varepsilon(\mathbf{k},\omega)|^2} \tag{11.140}$$

Up to here, the result in (11.140) owes to the quasilinear theory of the Liouville equation. Random process analysis by using this equation can derive the fluctuation-dissipation theorem.

If the ensemble is homogeneous both in space and time, the *correlation tensor* is defined by

$$C_{ij}(\mathbf{R},\tau) = <\delta E_i(\mathbf{r},t)\delta E_j(\mathbf{r}-\mathbf{R},t-\tau)>. \tag{11.141}$$

It is reminded that the bracket $<\cdots>$ indicates that the spatial and time average have been taken. The Fourier transform of $C_{ij}(\mathbf{R},\tau)$ is the *spectral density* $C_{ij}(\mathbf{k},\omega)$:

$$C_{ij}(\mathbf{k},\omega) = \int d^3R\int d\tau C_{ij}(\mathbf{R},\tau)e^{-i\mathbf{k}\cdot\mathbf{R}+i\omega\tau}. \tag{11.142}$$

Note that (11.140) represents $C_{ij}(\mathbf{k},\tau)$; thus, we can write

$$C_{ij}(\mathbf{R},\tau) = \int\frac{d^3k}{(2\pi)^3}e^{i\mathbf{k}\cdot\mathbf{R}}\left(\frac{4\pi e}{k^2}\right)^2\ n_0\ \frac{k_ik_j}{k}\int_{-\infty}^{\infty} d\omega\ g_0\left(\frac{\omega}{k}\right)\ \frac{e^{-i\omega\tau}}{|\varepsilon(\mathbf{k},\omega)|^2}. \tag{11.143}$$

Comparing the above two equations enables one to identify

$$C_{ij}(\mathbf{k},\omega) = n_0 \left(\frac{4\pi e}{k^2}\right)^2 \frac{k_i k_j}{k}\, 2\pi\, \frac{g_0\left(\frac{\omega}{k}\right)}{|\varepsilon(\mathbf{k},\omega)|^2}. \tag{11.144}$$

This form of the spectral density tensor can be obtained by solving two-time Liouville distributions with the aid of the Laplace transform applied to the Vlasov operator [see Eq. (8.37) by Montgomery and Tidman, *Plasma kinetic theory*, McGraw-Hill, (1964)].

From the correlation tensor defined in (11.143), we can calculate the average energy of the particle executing random motion as a consequence of being acted by a randomly fluctuating electric field. From the trace C_{ii},

$$C_{ii}(\mathbf{R},\tau) = \langle E_i(\mathbf{r},t)E_i(\mathbf{r}-\mathbf{R},t-\tau)\rangle, \tag{11.145}$$

we obtain the electric fluctuation, which is the trace of the correlation tensor at $\mathbf{R}$ and $t = 0$:

$$C_{ii}(\mathbf{R},\tau)\Big|_{R=\tau=0} = C_{ii}(0,0) = \langle|\mathbf{E}(\mathbf{r},t)|^2\rangle. \tag{11.146}$$

Integrating (11.159) over $d\omega$ and d^3k yields

$$\int\!\!\int C_{ij}(\mathbf{k},\omega)\frac{d^3k}{(2\pi)^3}\frac{d\omega}{2\pi} = \int\!\!\int d\tau d^3R\; C_{ij}(\mathbf{R},\tau)\delta(\tau)\delta(\mathbf{R}) = C_{ij}(0,0), \tag{11.147}$$

and contraction gives

$$C_{ii}(0,0) = \int\!\!\int C_{ii}(\mathbf{k},\omega)\frac{d^3k}{(2\pi)^3}\frac{d\omega}{2\pi}, \tag{11.148}$$

On the other hand, (11.146) gives

$$C_{ii}(0,0) = \left\langle \int \frac{d^3k}{(2\pi)^3}\frac{d\omega}{2\pi}\int \frac{d^3k'}{(2\pi)^3}\frac{d\omega'}{2\pi}\mathbf{E}(\mathbf{k},\omega)\cdot\mathbf{E}^*(\mathbf{k}',\omega')\exp\left[i(\mathbf{k}-\mathbf{k}')\cdot\mathbf{r} - i(\omega-\omega')t\right]\right\rangle \tag{11.149}$$

where we used

$$\mathbf{E}(\mathbf{r},t) = \int \frac{d^3k}{(2\pi)^3}\,\frac{d\omega}{2\pi}\;\mathbf{E}(\mathbf{k},\omega)\; e^{i\mathbf{k}\cdot\mathbf{r}-i\omega t} \tag{11.150}$$

Therefore, we have the equality,

$$\langle\mathbf{E}(\mathbf{k},\omega)\cdot\mathbf{E}^*(\mathbf{k}',\omega')\rangle = C_{ii}(\mathbf{k},\omega)(2\pi)^4\delta(\omega-\omega')\delta(\mathbf{k}-\mathbf{k}'). \tag{11.151}$$

In particular, when $\omega = \omega'$ and $\mathbf{k} = \mathbf{k}'$, we have

$$<|\mathbf{E}(\mathbf{k},\omega)|^2> = C_{ii}(\mathbf{k},\omega) = n_0 2\pi k \left(\frac{4\pi e}{k^2}\right)^2 \frac{g_0\left(\frac{\omega}{k}\right)}{|\varepsilon(\mathbf{k},\omega)|^2} = 8\pi\ \frac{T}{\omega}\ \frac{\text{Im}\ \varepsilon(\mathbf{k},\omega)}{|\varepsilon(\mathbf{k},\omega)|^2} \tag{11.152}$$

where we have used (11.144).

Summary. The preceding development that ended up with (11.152) for the spectral density, starting from the correlation tensor (11.140), is repetitive to readers who are familiar with the *Wiener–Khintchine theorem* in the random variable theory. For a one-dimensional random variable $E(k,t)$, the correlation tensor as given by (11.140) reduces to

$$\begin{aligned} C(k,\tau) &= <E(k,t)E(k,t-\tau)> \\ &= <\delta E(k,t)\ \delta E(k,t-\tau)> \\ &= \left(\frac{4\pi e}{k^2}\right)^2\ n_0 k \int_{-\infty}^{\infty} d\omega\ g_0\left(\frac{\omega}{k}\right) \frac{e^{-i\omega\tau}}{|\varepsilon(\mathbf{k},\omega)|^2} \end{aligned} \tag{11.153}$$

The Wiener–Khintchine theorem states that the correlation function $C(\tau)$ and the spectral density $<|\mathbf{E}(\mathbf{k},\omega)|^2>$ are a Fourier transform pair:

$$C(\tau) = \int_{-\infty}^{\infty} <|\mathbf{E}(\mathbf{k},\omega)|^2> e^{-i\omega t}\ \frac{d\omega}{2\pi} \tag{11.154}$$

$$C(\omega) = <|\mathbf{E}(\mathbf{k},\omega)|^2> = \int_{-\infty}^{\infty} C(\tau) e^{i\omega t}\ d\tau \tag{11.155}$$

Substituting (11.153) into (11.155) and carrying out the integral, one immediately obtains (11.169). For the proof of the Wiener–Khintchine theorem, refer to Hee J. Lee *Fundamentals of Theoretical Plasma Physics*, World Scientific, 2019.

Problems

11.1 Show that the Fourier transform of $\frac{\mathbf{r}}{r^3}$ is $-4\pi i \frac{\mathbf{k}}{k^2}$. Work in the spherical coordinates.

11.2 Prove Eq. (11.102).

11.3 Prove the Wiener–Khintchine theorem.

11.4 Write (11.22) in spherical coordinates and solve the differential equation to derive the Debye potential.

11.5 Prove by direct calculation

$$I \equiv \int dv_x \int dv_z g_0(v_x) g_0(v_z) \delta(\omega - k_x v_x - k_z v_z) = \int du g_0(u) \delta(\omega - ku)$$

where g_0 is one-dimensional Maxwellian and $k = \sqrt{k_x^2 + k_z^2}$.
Hint: $\delta(\omega - k_x v_x - k_z v_z) = \delta(v_x - \frac{\omega - k_z v_z}{k_x})/|k_x|$. And carry out the integral

$$I = \frac{m}{2\pi T} \exp\left(\frac{m}{2T}\frac{\omega^2}{k^2}\right) \int_{-\infty}^{\infty} dv_z \exp\left[-\frac{m}{2T}\frac{k^2}{k_x^2}\left(v_z - \frac{\omega k_z}{k^2}\right)^2\right]$$

References

Alexandrov, A. F., L. S. Bogdankevich, and A. A. Rukhadze, *Principles of Plasma Electrodynamics*, Springer-Verlag, Heidelberg, (1984).

Camac, M., A. R. Kantrowitz, M. M. Litvak, R. M. Patrick, and H. E. Petschek, *Shock waves in collision-free plasmas*, Nuc. Fusion Supplement, **Pt. 2**, (1962), pp. 423–444.

Clemmow, P. C. and J. P. Daugherty, *Electrodynamics of Particles and Plasmas*, Addison-Wesley London (1969).

Krall, N. A. and A. W. Trivelpiece, *Principles of Plasma Physics*, McGraw-Hill (1973).

Landau and Lifshitz, Course of theoretical physics Volume 10, Butterworth-Heinemann (1981).

Lee, H. J. *Fundamentals of Theoretical Plasma Physics*, World Scientific (2019).

Lee, H. J. *Fluctuation of the electric field in a plasma*, J. Korean Phys. Soc. **66** No. 8 (2015), pp. 1167–1185.

Lifshitz, E. M. and L. P. Pitaevskii, *Physical Kinetics*, Montgomery, D. C. and D. A, Tidman *Plasma Kinetic Theory*, McGraw-Hill (1964).

Montgomery, D. C. *Theory of Unmagnetized Plasma*, Gordon and Beach, New York (1971).

Nicholson, D. R. *Introduction to Plasma Theory*, Wiley (1983).

Nishikawa, K. and M. Wakatani, *Plasma Physics*, Springer-Verlag, Heidelberg (1990).

Rostoker, N. and M. N. Rosenbluth, Phys. Fluids **3** (1960) p. 1.

Rostoker, N. Nucl. Fusion **1**, (1960), p. 101.

Sitenko, A. G. *Fluctuations and non-Linear Wave Interactions in Plasmas*, Pergamon, Oxford (1982).

Sitenko, A. G. and V. Malnev, *Plasma Physics Theory*, Chapman and Hall, London (1995).

Chapter 12

Surface Waves

Surface waves propagate along the interface of two different media; the amplitude of a surface wave is rapidly decreasing in the perpendicular direction to the interface. Therefore the associated wave energy propagates along the interface, narrowly confined about the interfacial layer. Laboratory and natural plasmas are bounded in some form or another. The normal modes arising in bounded plasmas by satisfying appropriate boundary conditions may be called *surface waves*. The localized character of the surface wave energy is relevant to heating of the solar corona. Surface waves have application in the low temperature surface physics.

If we deal with a bounded plasma, the governing equations should be solved, respectively, for the plasma region and the non-plasma region, and the solutions thus obtained should be matched through appropriate matching conditions. Applying correct boundary conditions at the interface is important to identify legitimate surface wave modes. The boundary conditions are two kinds; electromagnetic and kinematic. Additional dynamic boundary condition must be considered if we use warm fluid equation since the pressure becomes another dynamical variable. If Vlasov equation is used, the kinematic boundary condition will be taken as the *specular reflection condition*, which means $f(v_x, v_y, v_z, x, y, z = 0) = f(v_x, v_y, -v_z, x, y, z = 0)$ where $z = 0$ is the interface. The electromagnetic field components also should satisfy certain jump conditions across the interface. As a rule, we can obtain meaningful jump conditions from the governing equations by integrating the latter equations across the interface over the infinitesimally thin layer. This method is applicable when the interfacial layer is sharp. Assuming a sharp *fixed* boundary is an idealization of a real interfacial boundary because the surface is undulating simultaneously with the wave motion. Assuming a fixed interface amounts to linearizing the nonlinear kinematic boundary condition.

If we use a *warm* fluid equation, the wave equation is of a higher order differential equation as compared with the *cold* fluid wave equation. Thus the number of boundary conditions (kinematic plus dynamic) for a warm plasma increases by one as compared with a cold plasma. This mathematical distinction can be conveniently translated into physical terms which states that *surface charges* are formed on a cold plasma boundary while such entity is nonexistent on a warm boundary. The extra necessary boundary condition for a warm plasma is the fluid version of the specular reflection condition.

It is convenient to write the Maxwell equations by decomposing the electromagnetic fields into two directions: perpendicular and parallel components with respect to the interface. The Maxwell equations then decouple as TM (*transverse magnetic*) and TE (*transverse electric*) modes in isotropic plasma.

12.1 Surface Waves in Fluid Plasma

First, let us investigate surface waves in fluid plasmas in which electrons participate in wave motion while ions are stationary. We assume the following step function type density profile:

$$N(x) = \begin{cases} N_0 \text{ if } x > 0 & \text{plasma} \\ 0 \text{ if } x < 0 & \text{vacuum} \end{cases} \tag{12.1}$$

The governing fluid equations for electrostatic mode are:

$$-i\omega \mathbf{v}(x) = \frac{e}{m}\,\nabla\phi(x) - \frac{\beta^2}{N(x)}\nabla n \tag{12.2}$$

$$-i\omega n + N(x)\nabla\cdot\mathbf{v} + v_x\,\frac{dN}{dx} = 0 \tag{12.3}$$

$$\nabla^2\phi = 4\pi e n \tag{12.4}$$

where β is the electron thermal velocity, and the variables are Fourier amplitudes: for example, the electrostatic potential ($\mathbf{E} = -\nabla\phi$),

$$\phi(x,z,t) = \phi(x)\; e^{ik_z z - i\omega t} \tag{12.5}$$

In components, we have

$$v_x = \frac{i}{\omega}\left(\frac{e}{m}\,\frac{\partial\phi}{\partial x} - \frac{\beta^2}{N}\,\frac{\partial n}{\partial x}\right) \tag{12.6}$$

$$v_z = \frac{k_z}{\omega}\left(-\frac{e}{m}\,\phi + \frac{\beta^2}{N}\,n\right) \tag{12.7}$$

• **Cold plasma.** First, let us consider a cold plasma ($\beta = 0$), in which case we have

$$v_x = \frac{i}{\omega}\frac{e}{m}\frac{\partial \phi}{\partial x} \tag{12.8}$$

$$v_z = -\frac{k_z}{\omega}\frac{e}{m}\,\phi \tag{12.9}$$

Equation (12.4) gives after some algebra

$$\frac{\partial^2 \phi}{\partial x^2} - k_z^2\left(1 - \frac{\omega_p^2}{\omega^2}\right)\phi - \frac{4\pi e^2}{m\omega^2}\frac{\partial}{\partial x}\left(N\,\frac{\partial \phi}{\partial x}\right) = 0 \tag{12.10}$$

where ω_p is the plasma frequency. We shall examine the above three equations to extract boundary conditions that must be satisfied at $x = 0$.

Boundary conditions can be mathematically worked out from the governing equations themselves if the density gradient across the plasma and the other side is very steep. We assume "a sharp interface" which implies a theoretically infinite density gradient across the boundary. In this case, the connection formula matching the quantities of both sides can be obtained by "infinitesimal integration" across the interface, which is the operation performed on a certain relevant equation in the manner $\int_{-\epsilon}^{\epsilon}(\cdots)dx$, where ϵ is a positive infinitesimal. The well-known Stokes theorem can be derived this way. If the quantity $(\cdots)$ is a perfect differential, this operation yields a non-vanishing surface term that contributes to the connection formula. Usually the surface term is surface charge or surface current, In this way, the well-known electromagnetic and dynamic boundary conditions on the boundary can be derived. The difference between 0^- and 0^+ is the jump of the quantity across the boundary. Equation (12.8) gives

$$[\phi] = 0 \quad \text{or} \quad \phi|_{0^+} = \phi|_{0^-} \tag{12.11}$$

where $[\cdots]$ denotes the jump of the quantity across the interface. Since $E_z = -ik_z\phi$, (12.11) is equivalent to

$$[E_z] = 0 \tag{12.12}$$

Equation (12.9) gives $0 = 0$. Likewise, (12.10) yields

$$\left.\frac{\partial \phi}{\partial x}\right|_{0^+} - \left.\frac{\partial \phi}{\partial x}\right|_{0^-} = \left.\frac{\omega_p^2}{\omega^2}\frac{\partial \phi}{\partial x}\right|_{0^+}$$

$$\text{or } [\varepsilon E_x] = 0 \tag{12.13}$$

where $\varepsilon = 1 - \frac{\omega_p^2}{\omega^2}$ is the dielectric constant. Equations (12.12) and (12.13) are the two cold plasma boundary conditions that must be satisfied at the interface.

Wave equation (12.10) becomes

$$\varepsilon \nabla^2 \phi = 0 \quad (x > 0) \tag{12.14}$$

$$\nabla^2 \phi = 0 \quad (x < 0) \tag{12.15}$$

Solving the Laplace equation, $\partial^2\phi/\partial x^2 - k_z^2\phi = 0$, in each region gives

$$\phi_2 = A\, e^{-k_z x} \quad (x > 0), \quad \phi_1 = B\, e^{k_z x}\ (x < 0)$$

where A and B are constants and the subscripts 1 and 2 denote, respectively, the vacuum and plasma sides. Applying the two boundary conditions, (12.11) and (12.13), yields the dispersion relation of the surface wave in a cold electron plasma,

$$\omega = \omega_p/\sqrt{2}$$

which is known as *Ritchie's formula.*

- **Warm plasma.** Equation (12.12) is valid even in a warm plasma, since $\frac{1}{N(x)}\frac{dn}{dx}$ is not a perfect differential:

$$[E_z] = 0$$

Let us write (12.7) in the form

$$N(x)\ v_z = \frac{k_z}{\omega}\left(-\frac{e}{m}N\phi + \frac{\beta^2}{4\pi e}\nabla^2\phi\right)$$

Infinitesimal integration of this equation gives $\left[\frac{\partial\phi}{\partial x}\right] = 0$, or,

$$[E_x] = 0 \tag{12.16}$$

This result contrasts with the cold plasma boundary condition in (12.13). In cold plasma, $[E_x]$ is not zero. The presence of the dielectric constant ε in the bracket is due to the surface charge that can be formed in cold plasmas. Equations (12.4)–(12.7) are combined to yield the equation

$$(\omega^2 - \omega_p^2)\nabla^2\phi + \beta^2\nabla^2\nabla^2\phi = \frac{\partial\omega_p^2}{\partial x}\frac{\partial\phi}{\partial x} = \frac{\partial}{\partial x}\left(\omega_p^2\frac{\partial\phi}{\partial x}\right) - \omega_p^2\frac{\partial^2\phi}{\partial x^2} \tag{12.17}$$

Using $\nabla^2 = \frac{\partial^2}{\partial x^2} - k_z^2$ in the first term gives

$$\omega^2 \nabla^2 \phi + k_z^2 \omega_p^2 \phi + \beta^2 \nabla^2 \nabla^2 \phi = \frac{\partial}{\partial x} \left(\omega_p^2 \frac{\partial \phi}{\partial x} \right) \tag{12.18}$$

Performing the infinitesimal integration over this equation gives

$$(\omega^2 - \beta^2 k_z^2) \left[\frac{\partial \phi}{\partial x} \right]_{0-}^{0+} + \beta^2 \left[\frac{\partial}{\partial x} \nabla^2 \phi \right]_{0-}^{0+} = \omega_p^2 \left. \frac{\partial \phi}{\partial x} \right|_{0+} \tag{12.19}$$

Using (12.16) in the above, we have

$$\beta^2 \left[\frac{\partial}{\partial x} \nabla^2 \phi \right]_{0-}^{0+} - \omega_p^2 \left. \frac{\partial \phi}{\partial x} \right|_{0+} = 0 \tag{12.20}$$

Let us explicitly denote the plasma and the vacuum solutions by ϕ_2 and ϕ_1, respectively. Then the above equation takes the form

$$\beta^2 \frac{\partial}{\partial x} \nabla^2 \phi_2 - \omega_p^2 \frac{\partial \phi_2}{\partial x} = 0 \quad \text{at } x = 0 \tag{12.21}$$

because $\nabla^2 \phi_1 = 0$. Equation (12.21) is the third boundary condition for a warm plasma electrostatic wave. We see that (12.6) takes the form, with the aid of (12.5),

$$v_x = \frac{ie}{m\omega} \left(\frac{\partial \phi_2}{\partial x} - \frac{\beta^2}{\omega_p^2} \frac{\partial}{\partial x} \nabla^2 \phi_2 \right) \tag{12.22}$$

Thus (12.21) can be expressed as

$$v_x|_0 = 0 \tag{12.23}$$

which is known as the specular reflection condition in a fluid plasma.

Now the wave equation in each region becomes from (12.16)

$$\nabla^2 \phi_1 = 0 \quad x < 0 \tag{12.24}$$

$$(\beta^2 \nabla^2 + \omega^2 - \omega_p^2) \nabla^2 \phi_2 = 0 \quad x > 0 \tag{12.25}$$

The vacuum and plasma solutions are, respectively,

$$\phi_1 = A\, e^{k_z x} \tag{12.26}$$

$$\phi_2 = C\, e^{-\gamma x} + D\, e^{-k_z x} \tag{12.27}$$

with A, C, and D being constants and

$$\gamma = \left(k_z^2 + \frac{\omega_p^2 - \omega^2}{\beta^2} \right)^{\frac{1}{2}} \tag{12.28}$$

Applying the above three boundary conditions, (12.12), (12.16), and (12.23), gives the surface wave dispersion relation in a warm plasma

$$\frac{\gamma}{k_z} = \frac{\omega_p^2}{2\omega^2 - \omega_p^2} \tag{12.29}$$

The cold plasma limit of (12.29) gives $\omega^2 = \omega_p^2/2$ per $\gamma \to \infty$. Finally, a comment is in order with regard to the condition in (12.16). The continuity of E_x in a warm plasma is equivalent to the continuity of D_x, the normal component of the displacement vector. By definition, $D_x = E_x + \frac{4\pi i}{\omega} J_x$ where J_x is the current. Since $v_x = 0$ at $x = 0$ (Eq. (12.23)), the continuity condition in (12.16) is equivalent to $[D_x] = 0$ in a warm plasma.

- **Surface charges.** When the number density $N(x)$ integrated over infinitesimally thin layer across the interface does not vanish, we have surface charges on the thin layer about $x = 0$. In a cold plasma, such surface charges are formed on the interface. In a warm plasma, surface charges are non-existent. To see this, let us integrate the Poisson equation (12.4) over the interface $x = 0$ to obtain the electron surface charge density σ,

$$\sigma = -e \int_{-\epsilon}^{\epsilon} n dx = -\frac{1}{4\pi}\left[\frac{\partial \phi}{\partial x}\right] = \frac{1}{4\pi}[E_x] \tag{12.30}$$

where the square bracket signifies the jump. Hence, in a warm plasma, $\sigma = 0$ in view of (12.16). For a cold plasma, using (12.13) in (12.30) yields

$$\sigma = \frac{1}{4\pi}\frac{\omega_p^2}{\omega^2} E_x(0^+) \tag{12.31}$$

Equation (12.30) states that the cold plasma electric field across the interface is discontinuous by the amount $4\pi\times$ *surface charges.*

The mathematical difference between cold and warm plasmas is represented by the respective wave equations. Equation (10), the cold plasma wave equation, is of second order differential equation with the highest order term $\partial^2\phi/\partial x^2$, and requires two boundary conditions. The warm plasma wave equation in (12.18), is of fourth order containing $\partial^4\phi/\partial x^4$. After satisfying the boundary condition at $x = \pm\infty$, i.e. $\phi(\pm\infty) = 0$, we need three more conditions as is seen in (12.26) and (12.27).

- **Fourier transform method for the half-space problem.** In the preceding development for the semi-infinite plasma, we solved the surface wave differential equation by directly integrating the fourth order differential equation. In this article, we shall solve the equation by Fourier transform. This approach is facilitated by extending the plasma region ($x > 0$) to the entire region $-\infty < x < \infty$, so that Fourier transform method is amenable. In extending the region, the plasma equations should assume the same form in the extended region ($x < 0$) as in the original region $x > 0$. After putting $\partial/\partial t = -i\omega$, $\partial/\partial z = ik_z$, the warm fluid equations read

$$-i\omega v_x(\omega, k_z, x) = -\frac{e}{m}E_x(\omega, k_z, x) - \frac{\beta^2}{N_0}\frac{\partial n(\omega, k_z, x)}{\partial x} \tag{12.32}$$

$$-i\omega v_z(\omega, k_z, x) = -\frac{e}{m}E_z(\omega, k_z, x) - \frac{\beta^2}{N_0}ik_z n(\omega, k_z, x) \tag{12.33}$$

$$-i\omega n + N_0\left(\frac{\partial v_x}{\partial x} + ik_z v_z\right) = 0 \tag{12.34}$$

$$\frac{\partial E_x}{\partial x} + ik_z E_z = -4\pi e n \tag{12.35}$$

The rule of extension is as follows: we reflect $x \to -x$ and require $v_x(x)$ and $E_x(x)$ be oddly continued and the rest of the variables, $v_z(x)$ and $E_z(x)$ and $n(x)$, be evenly continued.

$$v_x(-x) = -v_x(x), \quad E_x(-x) = -E_x(x) \tag{12.36}$$

$$v_z(-x) = v_z(x), \quad E_z(-x) = E_z(x), \quad n(-x) = n(x) \tag{12.37}$$

We can easily check by inspection that (12.32)–(12.35) are *invariant* under the reflections (12.36)–(12.37). Then, (12.32)–(12.35) are amenable to Fourier transform in the range $-\infty < x < \infty$. In Fourier transforming, care should be exercised for the terms $\frac{\partial E_x}{\partial x}$ and $\frac{\partial v_x}{\partial x}$.

The oddly continued functions $v_x(x)$ and $E_x(x)$ have jumps at $x = 0$:

$$v_x(0^-) - v_x(0^+) = A, \quad E_x(0^-) - E_x(0^+) = B \tag{12.38}$$

where A and B are undetermined constants, and $0^\pm = 0 \pm \epsilon$ (ϵ: infinitesimal). The evenly continued functions are continuous at $x = 0$. By performing $\times \int_{-\infty}^{\infty} dx\, e^{-ik_x x}(\cdots)$ in (12.32)–(12.35), the Fourier transformed equations read

$$-i\omega v_x(k_x) = -\frac{e}{m}E_x(k_x) - \frac{\beta^2}{N_0}ik_x n(k_x) \tag{12.39}$$

$$-i\omega v_z(k_x) = -\frac{e}{m}E_z(k_x) - \frac{\beta^2}{N_0}ik_z n(k_x) \tag{12.40}$$

$$-i\omega n(k_x) + N_0(ik_x v_x(k_x) + ik_z v_z(k_x) + A) = 0 \tag{12.41}$$

$$ik_x E_x(k_x) + ik_z E_z(k_x) + B = -4\pi e n(k_x) \tag{12.42}$$

Also we have, for an electrostatic wave,

$$k_z E_x(k_x) = k_x E_z(k_x) \tag{12.43}$$

The vacuum field also contains unknown integration constants, which should be connected with the above unknown constants A and B through the boundary conditions at $x = 0$. Then the solvability condition for the undetermined constants yield the surface wave dispersion relation. The plasma variables in (12.39)–(12.43) can be obtained in terms of A and B:

$$v_x(k_x) = \frac{1}{\Delta}\frac{ie}{m\omega}\left[\frac{(k^2\beta^2+\omega_p^2)}{\omega^2}A' + B\right] \tag{12.44}$$

$$E_x(k_x) = -\frac{1}{\Delta}\left[\frac{\omega_p^2}{\omega^2}A' + B\left(1-\frac{k^2\beta^2}{\omega^2}\right)\right] \tag{12.45}$$

$$E_z(k_x) = -\frac{1}{\Delta}\frac{k_z}{k_x}\left[\frac{\omega_p^2}{\omega^2}A' + B\left(1-\frac{k^2\beta^2}{\omega^2}\right)\right] \tag{12.46}$$

$$\Delta = \frac{ik^2}{k_x}\left[1-\frac{\omega_p^2+k^2\beta^2}{\omega^2}\right] \tag{12.47}$$

where we put $A = \frac{ie}{m\omega}A'$, and $k^2 = k_x^2 + k_z^2$. The above functions should be Fourier inverted to obtain x-dependence and apply the boundary conditions at $x = 0$.

Let us first invert $v_x(k_x)$:

$$v_x(k_x) = -\frac{e\omega}{m\beta^2}\frac{k_x}{(k_x^2+k_z^2)(k_x^2+\gamma^2)}\left[\frac{\beta^2(k_x^2+\gamma^2)}{\omega^2}A' + A' + B\right] \tag{12.48}$$

where γ is defined in (12.28). By contour integral in k_x-plane, we evaluate

$$\int_{-\infty}^{\infty}\frac{dk_x}{2\pi}\,e^{ik_x x}\,\frac{k_x}{k_x^2+k_z^2} = \frac{i}{2}\,e^{-k_z x} \tag{12.49}$$

$$\int_{-\infty}^{\infty}\frac{dk_x}{2\pi}\,e^{ik_x x}\,\frac{k_x}{(k_x^2+k_z^2)(k_x^2+\gamma^2)} = \frac{i}{2}\,\frac{1}{\gamma^2-k_z^2}(e^{-k_z x}-e^{-\gamma x}) \tag{12.50}$$

Hence

$$v_x(x) = -\frac{ie\omega}{2m\beta^2}\left[A'\,\frac{\beta^2}{\omega^2}\,e^{-k_z x} + (A'+B)\,\frac{1}{\gamma^2-k_z^2}(e^{-k_z x}-e^{-\gamma x})\right] \tag{12.51}$$

By taking the limit $x \to 0$ in (12.51), the boundary condition in (12.23) is satisfied by $A' = 0$, which will be used in (12.45) and (12.46). Analogous contour integrations yield

$$E_x(x) = -\frac{B}{2}\,\frac{1}{\beta^2(\gamma^2 - k_z^2)}(-\omega^2\, e^{-k_z x} + \omega_p^2\, e^{-\gamma x}) \tag{12.52}$$

$$E_z(x) = \frac{iB}{2}\,\frac{1}{\beta^2(\gamma^2 - k_z^2)}(-\omega^2\, e^{-k_z x} + \omega_p^2\,\frac{k_z}{\gamma}\, e^{-\gamma x}) \tag{12.53}$$

Hence

$$E_x(x)|_{0^+} = -\frac{B}{2} \tag{12.54}$$

$$E_z(x)|_{0^+} = \frac{iB}{2\beta^2}\,\frac{1}{(\gamma^2 - k_z^2)}\left(-\omega^2 + \frac{k_z\omega_p^2}{\gamma}\right) \tag{12.55}$$

• **Vacuum solution and dispersion relation.** The vacuum solution, (12.26), gives

$$E_x(x) = -\frac{\partial \phi_1}{\partial x} = -k_z A\, e^{k_z x} \tag{12.56}$$

$$E_z(x) = -\frac{\partial \phi_1}{\partial z} = -ik_z A\, e^{k_z x} \tag{12.57}$$

Then the continuity of E_x and E_z at $x = 0$ gives

$$k_z A = \frac{B}{2} \tag{12.58}$$

$$-k_z A = \frac{B}{2}\,\frac{1}{\beta^2(\gamma^2 - k_z^2)}\left(-\omega^2 + \frac{k_z\omega_p^2}{\gamma}\right) \tag{12.59}$$

The above two equations yield the surface wave dispersion relation in (12.29).

The Fourier transform method is rather complicated as compared with directly solving the wave differential equation, in the simple geometry of semi-infinite plasma. However, Fourier transform is more useful in kinetic theory of surface wave and for more complex geometry such as plasma slab or duct.

• **Fictitious surface charges in a warm plasma.** In the preceding development, the jump of $E_x(x)$ at $x = 0$ is given rise to by the oddly continuation of the perpendicular to the interface component of the electric field, $E_x(x)$. The jump is equivalent to the effect of fictitious surface charges placed on the plane $x = 0$. Thus the Poisson equation is modified to include the fictitious surface charges. This is the economical way to account for the boundary effect in the semi-infinite plasma. Now the Poisson equation reads

$$\nabla \cdot \mathbf{E} = -4\pi en + B\ \delta(x) \tag{12.60}$$

where B is the unknown strength of the surface charges. Equation (12.60) replaces (12.35). Equations (12.32), (12.33), (12.34), and (12.60) are the governing equations for a semi-infinite plasma. These equations are Fourier-transformed as if they are valid in the entire region $-\infty < x < \infty$ to yield the final result, the dispersion relation in (12.29). From the foregoing algebra, we readily see that the specular reflection condition in (12.23) is automatically satisfied.

12.2 Kinetic Theory of Surface Waves

We consider a *semi-bounded plasma* which occupies the region $(x > 0)$ bounded by a vacuum $(x < 0)$. We use the linearized Vlasov equation for electrons and ions.

$$\frac{\partial}{\partial t} f_\alpha(\mathbf{r}, \mathbf{v}, t) + \mathbf{v} \cdot \frac{\partial f_\alpha}{\partial \mathbf{r}} + \frac{e_\alpha}{m_\alpha}\left(\mathbf{E} + \frac{1}{c}\mathbf{v} \times \mathbf{B}\right) \cdot \frac{\partial f_{\alpha 0}}{\partial \mathbf{v}} = 0 \tag{12.61}$$

The plasma current

$$\mathbf{J} = \sum_\alpha e_\alpha \int \mathbf{v} f_\alpha d^3v \tag{12.62}$$

gives rise to electric and magnetic fields

$$\nabla \times \mathbf{E} = -\frac{1}{c}\frac{\partial \mathbf{B}}{\partial t} \tag{12.63}$$

$$\nabla \times \mathbf{B} = \frac{1}{c}\frac{\partial \mathbf{E}}{\partial t} + \frac{4\pi}{c}\mathbf{J} \tag{12.64}$$

Solving (12.61) in a finite plasma in general requires a kinematic boundary condition on f at the boundary (at the interface $x = 0$ here). We assume the *specular reflection condition* in which the particles undergo a mirror reflection such that

$$f(v_x, v_y, v_z, t, x = 0, y, z) = f(-v_x, v_y, v_z, t, x = 0, y, z) \tag{12.65}$$

Equation (12.65) gives

$$\int_{-\infty}^{\infty} v_x f(v_x) dv_x = 0 \quad \text{at } x = 0$$

Hence, the normal component of the *fluid velocity* at the interface is zero; recall (12.23).

The distribution function f, which satisfies the following reflectional symmetry

$$f(v_x, v_y, v_z, t, x, y, z) = f(-v_x, v_y, v_z, t, -x, y, z) \tag{12.66}$$

satisfies (12.65). To examine this reflectional symmetry, we fully write (12.61), exhibiting all the terms in components:

$$\frac{\partial f}{\partial t} + v_x \frac{\partial f}{\partial x} + v_y \frac{\partial f}{\partial y} + v_z \frac{\partial f}{\partial z} - \frac{e}{m}\left[E_x \frac{\partial f_0}{\partial v_x} + E_y \frac{\partial f_0}{\partial v_y} + E_z \frac{\partial f_0}{\partial v_z}\right]$$

$$-\frac{e}{mc}\left[(v_y B_z - B_y v_z)\frac{\partial f_0}{\partial v_x} + (v_z B_x - v_x B_z)\frac{\partial f_0}{\partial v_y} + (v_x B_y - v_y B_x)\frac{\partial f_0}{\partial v_z}\right] = 0 \tag{12.67}$$

Now we see that the following prescription makes (12.67) invariant under the reflection ($x \to -x,\ v_x \to -v_x$):

$$E_x(-x) = -E_x(x),\ E_y(-x) = E_y(x),\quad E_z(-x) = E_z(x) \tag{12.68}$$

$$B_x(-x) = B_x(x),\ B_y(-x) = -B_y(x),\quad B_z(-x) = -B_z(x) \tag{12.69}$$

The parallel components of $\mathbf{E}$ (with respect to the interface) are continued evenly; the perpendicular component E_x oddly; the parallel components of $\mathbf{B}$ are continued oddly; the perpendicular component B_x evenly. We can now solve (12.61) by Fourier transform which is valid in the entire region $-\infty < x < \infty$, with the extended electric and magnetic fields. In the mean time, the artificial continuation prescribed in (12.68) and (12.69) should be carefully incorporated in the transformations of the Maxwell equations (12.63) and (12.64). Instead of dealing with the recipe of the field extensions, we can introduce equivalently the fictitious surface currents that stem from the surface charges placed on $x = 0$, on the warm plasma interface. See Equation (12.60). Accordingly, (12.64) is replaced by

$$\nabla \times \mathbf{B} = \frac{1}{c}\frac{\partial \mathbf{E}}{\partial t} + \frac{4\pi}{c}\mathbf{J} + \hat{\mathbf{z}}A(z,t)\delta(x) \tag{12.70}$$

where A is unknown magnitude of the surface currents. The fictitious surface charges and currents should be related by the charge conservation equation in the form

$$\nabla \cdot \mathbf{J}_s + \frac{\partial \rho_s}{\partial t} = 0$$

$$\text{or} \quad \frac{\partial J_{sx}}{\partial x} + ik_z J_{sz} = -i\omega A\delta(x)$$

Since $J_{sx} = 0$ at $x = 0$, we have $J_{sz} \sim \delta(x)$. Hence we write (12.70).

We suppress the y-coordinate which has the translational invariance and is ignored while we assume the wave propagates along the z-direction ($\sim e^{ik_z z - i\omega t}$). Also put $E_y = 0$ which decouples from the TM mode set (E_x, E_z, B_y). In general, surface waves in a bounded plasma are separated into two modes, transverse magnetic (TM) and transverse electric (TE) modes. TM mode consists of the field components (B_y, E_x, E_z). TE mode consists of the field components (E_y, B_x, B_z). Maxwell equations are decoupled in terms of the two groups of field components. In plasma-vacuum surface waves, only the TM mode wave can propagate. Now we shall investigate the TM mode surface wave by putting (E_y, B_x, B_z) $= 0$.

Fourier transforms of (12.63) and (12.70) can be combined in the form

$$\mathbf{k} \times (\mathbf{k} \times \mathbf{E}) + \frac{\omega^2}{c^2}\mathbf{E} + \frac{4\pi i\omega}{c^2}\mathbf{J} = \hat{\mathbf{z}} M(k_z, \omega) \tag{12.71}$$

where $M(k_z, \omega) = -\frac{i\omega}{c} A(k_z, \omega)$ is the boundary term which is absent in an infinite plasma.

Fourier transform of (12.61) gives

$$f_\alpha(\mathbf{k}, \mathbf{v}, \omega) = -i\frac{e_\alpha}{m_\alpha}\frac{\mathbf{E}(\mathbf{k}, \omega)}{\omega - \mathbf{k}\cdot\mathbf{v}} \cdot \frac{\partial f_{\alpha 0}}{\partial \mathbf{v}} \tag{12.72}$$

Now the plasma fields will be determined in terms of $\hat{\mathbf{z}} M(k_z, \omega)$ in (12.71). Introducing the conductivity tensor per $J_i = \sigma_{ij} E_j$, the dielectric tensor $\varepsilon_{ij} = \delta_{ij} + \frac{4\pi i}{\omega}\sigma_{ij}$ is obtained as

$$\varepsilon_{ij} = \delta_{ij} + \sum_\alpha \frac{\omega_{p\alpha}^2}{\omega} \int \frac{v_i}{\omega - \mathbf{k}\cdot\mathbf{v}} \frac{\partial f_{\alpha 0}}{\partial v_j}\, d^3v \tag{12.73}$$

Equation (10.71) can be written as

$$\mathbf{k} \times (\mathbf{k} \times \mathbf{E}) + \frac{\omega^2}{c^2}\varepsilon_{ij} \cdot \mathbf{E}(k_x, k_z, \omega) = \hat{\mathbf{z}} M(k_x, k_z, \omega) \tag{12.74}$$

Further manipulation of the integral in (12.73), including integration by parts, allows us to write it in the form

$$\varepsilon_{ij} = \varepsilon_L \frac{k_i k_j}{k^2} + \varepsilon_T \left(\delta_{ij} - \frac{k_i k_j}{k^2}\right)$$

where ε_L and ε_T are given in (8.62) and (8.67). Equation (12.74) is easily inverted to give

$$E_x(\mathbf{k},\omega) = -\frac{M}{\Delta}\left(\varepsilon_{xz} + \frac{c^2}{\omega^2}k_x k_z\right) \tag{12.75}$$

$$E_z(\mathbf{k},\omega) = \frac{M}{\Delta}\left(\varepsilon_{xx} - \frac{c^2}{\omega^2}k_z^2\right) \tag{12.76}$$

$$\Delta = \varepsilon_L\left(\varepsilon_T - \frac{c^2k^2}{\omega^2}\right); \quad k^2 = k_x^2 + k_z^2 \tag{12.77}$$

Using the above equations in (12.63) yields

$$B_y(\mathbf{k},\omega) = -\frac{M}{\Delta}\,\frac{c}{\omega}k_x\varepsilon_L \tag{12.78}$$

To reinstate the x-dependence, the Fourier inversion integral $\int dk_x$ must be performed on (12.75)–(12.78). This can be done by using the contour integral and evaluating the residues at the poles. To ensure, say, $B_y \to 0$ as $x \to +\infty$, the inversion integral should be carried out by encircling the upper-half k_x-plane.

The vacuum solutions satisfy $\nabla\cdot\mathbf{E} = 0$ and $\nabla^2\mathbf{E} + (\omega^2/c^2)\mathbf{E} = 0$, and we obtain

$$E_z(x, k_z, \omega) = i\;\lambda F e^{\lambda x} \qquad (F:\ \text{const.}) \tag{12.79}$$

$$\text{with}\quad \lambda = \sqrt{k_z^2 - \omega^2/c^2} \tag{12.80}$$

$$E_x = k_z F e^{\lambda x} \tag{12.81}$$

$$B_y(x, k_z, \omega) = \frac{\omega}{c}F e^{\lambda x} \tag{12.82}$$

where we omitted the phasor $\exp(ik_z z - i\omega t)$.

The plasma solutions and the vacuum solutions should be matched at the interface $x = 0$ according to the *electromagnetic boundary conditions*: the continuity of E_z and B_y. We have two undetermined constants, F and M (k_x-independent). Enforcing the two boundary conditions, we have

$$\int_{-\infty}^{\infty}\frac{dk_x}{\Delta}\left[\varepsilon_{xx} - \frac{c^2}{\omega^2}k_z^2 + i\lambda\frac{c^2}{\omega^2}\varepsilon_L k_x\right] = 0 \tag{12.83}$$

Using (12.77) and the relation $\varepsilon_{xx} - c^2k_z^2/\omega^2 = (k_x^2\varepsilon_L + k_z^2(\varepsilon_T - c^2k^2/\omega^2))/k^2$ in the above equation gives the dispersion relation in the form

$$\int_{-\infty}^{\infty}\frac{dk_x}{k^2}\left(\frac{k_x^2}{\varepsilon_T - c^2k^2/\omega^2} + \frac{k_z^2}{\varepsilon_L}\right) + i\lambda\,\frac{c^2}{\omega^2}\int_{-\infty}^{\infty}\frac{dk_x\;k_x}{\varepsilon_T - c^2k^2/\omega^2} = 0 \tag{12.84}$$

This is the general dispersion relation of electromagnetic surface wave propagating in the semi-bounded plasma.

Electromagnetic boundary conditions: the continuity of E_z and B_y. The two curl equations, (12.63) and (12.64), contain the perfect differential terms $\frac{\partial E_z}{\partial x}$ and $\frac{\partial B_y}{\partial x}$. Thus the continuity of E_z and B_y immediately follows after the infinitesimal integration. However, in a drifting plasma (beam), B_y is not continuous.

Invariance of Maxwell equations. In extending the half space ($x > 0$) to the entire space $-\infty < x < \infty$ under the simultaneous reflection ($x \to -x$, $v_x \to -v_x$), Maxwell equations must also be invariant. We have

$$J_x(x) = -e\int_{-\infty}^{\infty} v_x f(v_x, x)dv_x, \quad J_z(x) = -e\int_{-\infty}^{\infty} v_z f(v_x, x)dv_x$$

According to (12.66), the currents J_x and J_z satisfy

$$J_x(-x) = -J_x(x), \quad J_z(-x) = J_z(x) \tag{12.85}$$

Equations (12.68), (12.69), and (12.85) satisfy the invariance of Maxwell equations.

TM mode surface wave in a cold plasma. For a cold plasma, we have $\varepsilon_L = \varepsilon_T \equiv \varepsilon(\omega) = 1 - \frac{\omega_p^2}{\omega^2}$. Using this expression in (12.84) gives

$$\int_{-\infty}^{\infty} \frac{dk_x}{k^2}\left(\frac{-\omega^2 k_x^2/c^2}{k_x^2 + \alpha^2} + \frac{k_z^2}{\varepsilon(\omega)}\right) - i\lambda\int_{-\infty}^{\infty} \frac{k_x dk_x}{k_x^2 + \alpha^2} = 0 \tag{12.86}$$

where

$$\alpha = \sqrt{k_z^2 + \frac{\omega_p^2 - \omega^2}{c^2}} \tag{12.87}$$

The first integral in (12.86) yields two terms due to the simple poles at $k_x = ik_z$ and $k_x = i\alpha$. It can be seen that the contribution from the pole at ik_z cancels the second integral, giving the result

$$\alpha + \lambda\, \varepsilon(\omega) = 0 \tag{12.88}$$

The wave represented by the dispersion relation (12.88) is known as the *surface polariton*. The dispersion in (12.88) is similar to that of the two-stream instability in (18.16). Squaring and solving by quadrature give

$$\omega^2 = c^2k^2 + \frac{\omega_{\rm pe}^2}{2} - \frac{1}{2}\sqrt{\omega_{\rm pe}^2 + 4c^4k^4} \tag{12.88$'$}$$

Electrostatic dispersion relation. The electrostatic surface wave dispersion relation in a semi-infinite plasma can be derived by taking the limit $c \to \infty$ in (12.84) which is the electromagnetic surface wave dispersion relation in a semi-infinite plasma. Note λ in (12.84) becomes k_z in the $c \to \infty$ limit. Thus (12.84) yields

$$\int dk_x \frac{k_z^2}{k^2 \varepsilon_L} - ik_z \int dk_x \frac{k_x}{k^2} = 0$$

The second term can be contour-integrated, and the above equation reduces to

$$\int_{-\infty}^{\infty} dk_x \frac{k_z}{k^2 \varepsilon_L} + \pi = 0 \tag{12.89}$$

Independent derivation of the electrostatic surface wave dispersion relation in semi-infinite plasma by using electrostatic Maxwell equations is left as an exercise.

Exercise. Prove the relation

$$\varepsilon_{xx} - \frac{c^2 k_z^2}{\omega^2} = \frac{1}{k^2}\left(k_x^2 \varepsilon_L + k_z^2\left(\varepsilon_T - \frac{c^2 k^2}{\omega^2}\right)\right) \tag{12.90}$$

Show that

$$M(k_z, \omega) = 2ik_z E_x(x = 0^-, k_z, \omega) - 2\left[\frac{\partial}{\partial x} E_z(x, k_z, \omega)\right]_{x=0^-} \tag{12.91}$$

12.3 Surface Wave in a Relativistic Cold Plasma

A relativistic plasma means a moving plasma. Its perturbation analysis is therefore relates the interaction between the beam (zero order velocity of the plasma as a whole) and the plasma electromagnetic fields. We investigate TM mode surface wave in a relativistically streaming plasma in a half-space bounded by vacuum. The governing fluid equations are:

$$\frac{\partial n}{\partial t} + \nabla \cdot (n\mathbf{v}) = 0 \tag{12.92}$$

$$\frac{\partial \mathbf{p}}{\partial t} + (\mathbf{v} \cdot \nabla)\mathbf{p} = -e\left(\mathbf{E} + \frac{1}{c}\mathbf{v} \times \mathbf{B}\right) \tag{12.93}$$

where $\mathbf{p}$ is the relativistic momentum,

$$\mathbf{p} = m\gamma\mathbf{v} \tag{12.94}$$

m is the rest mass, and

$$\gamma = \frac{1}{\sqrt{1-\frac{v^2}{c^2}}} \tag{12.95}$$

One can easily calculate

$$\mathbf{v} = \frac{c\,\mathbf{p}}{\sqrt{m^2c^2+p^2}} \tag{12.96}$$

$$p^2 = m^2c^2(\gamma^2-1) \tag{12.97}$$

$$v^2 = c^2\left(1-\frac{1}{\gamma^2}\right) \tag{12.98}$$

The characteristic equation of (12.93) reads

$$\frac{dt}{1} = \frac{d\mathbf{r}}{\mathbf{v}} = \frac{d\mathbf{p}}{-e\left(\mathbf{E}+\frac{1}{c}\mathbf{v}\times\mathbf{B}\right)} \tag{12.99}$$

which gives the Lagrangian relativistic momentum equation

$$\frac{d\mathbf{p}}{dt} = -e\left(\mathbf{E}+\frac{1}{c}\mathbf{v}\times\mathbf{B}\right), \quad \mathbf{v} = \frac{d\mathbf{r}}{dt} \tag{12.100}$$

whose Eulerian version is (12.93). In some literature, the momentum equation is written as $\frac{d}{dt}(n\mathbf{p}) = -e\;n\;\left(\mathbf{E}+\frac{1}{c}\mathbf{v}\times\mathbf{B}\right)$, which is incorrect. To see (12.99), write $(\mathbf{v}\cdot\nabla)\mathbf{p} = (\mathbf{v}\cdot\frac{\partial}{\partial\mathbf{r}})\mathbf{p}$.

Also we have Maxwell equations for the TM mode components, (12.71)–(12.73)

$$ik_zE_x - \frac{\partial E_z}{\partial x} = \frac{i\omega}{c}B_y \tag{12.101}$$

$$-\frac{i\omega}{c}E_x + \frac{4\pi}{c}J_x = -ik_zB_y \tag{12.102}$$

$$-\frac{i\omega}{c}\;E_z + \frac{4\pi}{c}J_z = \frac{\partial B_y}{\partial x} \tag{12.103}$$

$$J_x = -\mathrm{env}_x \tag{12.104}$$

$$J_z = -\mathrm{env}_z \tag{12.105}$$

In components, (12.92) and (12.93) give

$$\frac{\partial n}{\partial t} + \frac{\partial}{\partial x}(nv_x) + \frac{\partial}{\partial z}(nv_z) = 0 \tag{12.106}$$

$$\frac{\partial}{\partial t}(\gamma v_x) + v_x\frac{\partial}{\partial x}(\gamma v_x) + v_z\frac{\partial}{\partial z}(\gamma v_x) = -\frac{e}{m}\left(E_x - \frac{1}{c}B_y v_z\right) \tag{12.107}$$

$$\frac{\partial}{\partial t}(\gamma v_z) + v_x\frac{\partial}{\partial x}(\gamma v_z) + v_z\frac{\partial}{\partial z}(\gamma v_z) = -\frac{e}{m}\left(E_z + \frac{1}{c}B_y v_x\right) \tag{12.108}$$

Equations (12.101)–(12.108) govern the dynamics of the cold relativistic plasma occupying the region $x > 0$, bounded by vacuum in the region $x < 0$. We can see by inspection that (12.101)–(12.108) are invariant under the reflections $x \to -x$ if we extend the plasma region into the region $x < 0$ by the recipe in (12.36), (12.37), (12.68), and (12.69). This observation is helpful if one solves the plasma equations by Fourier transform.

Now we shall linearize the plasma equations by putting

$$n = N(x) + n_1(x)\, e^{ik_z z - i\omega t}, \quad \mathbf{v} = \hat{\mathbf{z}}v_0 + \mathbf{v}_1(x)\, e^{ik_z z - i\omega t} \tag{12.109}$$

where the subscript 1 denotes perturbation. First, let us expand $\gamma(\mathbf{v})$:

$$\gamma = \left(1 - \frac{1}{c^2}(v_0\hat{\mathbf{z}} + \mathbf{v}_1)^2\right)^{-1/2} = \left(1 - \frac{v_0^2}{c^2} - 2\frac{v_0 v_{z1}}{c^2}\right)^{-1/2}$$

After binomial expansion, we have

$$\gamma_0 = \frac{1}{\sqrt{1 - \frac{v_0^2}{c^2}}}, \quad \gamma_1 = \gamma_0^3\, \frac{v_0 v_{z1}}{c^2} \tag{12.110}$$

Then the linearized equations of (12.106)–(12.108) are written in terms of the Fourier variables (ω, k_z):

$$-i\,\omega' n_1 + ik_z N v_{z1} + \frac{\partial}{\partial x}(N v_{x1}) = 0 \tag{12.111}$$

$$-i\gamma_0\,\omega' v_{x1} = -\frac{e}{m}\left(E_x - \frac{v_0}{c}B_y\right) \tag{12.112}$$

$$\omega' v_{z1} = -\frac{i}{\gamma_0^3}\,\frac{e}{m}E_z \tag{12.113}$$

where $\omega' = \omega - k_z v_0$ is the Doppler-shifted frequency. For the current,

$$J_x = -eNv_{x1} \tag{12.114}$$

$$J_z = -e(Nv_{z1} + n_1 v_0) \tag{12.115}$$

Also we have the Poisson equation

$$n_1 = -\frac{1}{4\pi e}\left(\frac{\partial E_x}{\partial x} + ik_z E_z\right) \tag{12.116}$$

Having linearized, we will drop the subscript 1 for notational simplicity.

Equations (12.112) and (12.114) give

$$J_x = \frac{ie^2N(x)}{m\gamma_0\omega'}\left(E_x - \frac{v_0}{c}B_y\right) \tag{12.117}$$

Equations (12.113) and (12.115) give

$$J_z = -env_0 + \frac{ie^2N(x)}{m\gamma_0^3\omega'}E_z \tag{12.118}$$

Equations (12.102) and (12.117) give

$$\left(k_z - \frac{v_0\omega_p^2(x)}{c^2\gamma_0\omega'}\right)B_y = \frac{\omega}{c}\left(1 - \frac{\omega_p^2(x)}{\gamma_0\omega\omega'}\right)E_x \tag{12.119}$$

Equations (12.103) and (12.118) give

$$\frac{\partial B_y}{\partial x} = \frac{v_0}{c}\frac{\partial E_x}{\partial x} - i\,\frac{\omega'}{c}\left(1 - \frac{\omega_p^2(x)}{\omega'^2\gamma_0^3}\right)E_z \tag{12.120}$$

where we used (12.116). If we use (12.111), we get

$$\frac{\partial B_y}{\partial x} = \frac{i4\pi ev_0}{c\,\omega'}\frac{\partial}{\partial x}(Nv_x) - i\,\frac{\omega}{c}\left(1 - \frac{\omega_p^2(x)}{\omega'^2\gamma_0^3}\right)E_z \tag{12.121}$$

Equations (12.120) and (12.121) are equivalent.

Boundary conditions. We can derive the boundary conditions from the governing equations by performing the infinitesimal integration across the interface $x = 0$ with $N(x)$ being specialized as a step function. We extract the perfect differential terms. Equation (12.101) gives

$$[E_z] = E_z(0^+) - E_z(0^-) = 0 \tag{12.122}$$

That is, the tangential component of the electric field is continuous. Using (120) yields the important relation

$$[B_y] = \frac{v_0}{c}[E_x] \tag{12.123}$$

Analogously, using (12.121) gives

$$[B_y] = \frac{v_0\omega_p^2}{c\gamma_0\omega'^2}\left(E_x - \frac{v_0}{c}B_y\right)\Bigg|_{0+} \tag{12.124}$$

where we used (12.112). Equations (12.123) and (12.124) are equivalent. The boundary condition (12.123) is characteristic of a beam plasma, in which surface currents are generated and using $[B_y] = 0$ gives incorrect result.

Wave equation. Eliminating B_y between (12.101) and (12.119) gives

$$E_x \left(k_z^2 - \frac{\omega^2 - \omega_p^2/\gamma_0}{c^2} \right) = -i \left(k_z - \frac{v_0 \, \omega_p^2/\gamma_0}{c^2 \omega'} \right) \frac{\partial E_z}{\partial x} \tag{12.125}$$

Eliminating B_y between (12.119) and (12.120) gives

$$\frac{\partial}{\partial x} \left[\frac{\left(1 - \frac{\omega_p^2}{\gamma_0^3 \omega'^2}\right)}{\left(k_z - \frac{v_0 \omega_p^2}{c^2 \gamma_0 \omega'}\right)} E_x \right] + i \, E_z \left(1 - \frac{\omega_p^2}{\gamma_0^3 \omega'^2} \right) = 0 \tag{12.126}$$

One can ascertain that eliminating B_y between (12.119) and (12.121) also yields (12.126). Using (12.125) in (12.126) gives

$$\frac{\partial}{\partial x} \left\{ \frac{\left(1 - \frac{\omega_p^2}{\gamma_0^3 \omega'^2}\right)}{\left(k_z^2 - \frac{\omega^2 - \omega_p^2/\gamma_0}{c^2}\right)} \frac{\partial E_z}{\partial x} \right\} = \left(1 - \frac{\omega_p^2}{\gamma_0^3 \omega'^2} \right) E_z \tag{12.127}$$

Equation (12.127) gives another form of boundary condition:

$$\left[\frac{\left(1 - \frac{\omega_p^2}{\gamma_0^3 \omega'^2}\right)}{\left(k_z^2 - \frac{\omega^2 - \omega_p^2/\gamma_0}{c^2}\right)} \frac{\partial E_z}{\partial x} \right]_{0^-}^{0^+} = 0 \tag{12.128}$$

The jump conditions (12.123), (12.124), and (12.128) are all equivalent; any one of them is acceptable. Now we shall solve (12.127) in the plasma region $x > 0$ where ω_p is constant. Then (12.127) reduces to

$$\frac{\partial^2 E_z}{\partial x^2} = \left(k_z^2 - \frac{\omega^2 - \omega_p^2/\gamma_0}{c^2} \right) E_z \tag{12.129}$$

Let

$$\Lambda = \sqrt{k_z^2 + \frac{\omega_p^2/\gamma_0 - \omega^2}{c^2}} \tag{12.130}$$

Then

$$E_z(x) = G \, e^{-\Lambda x}, \; G: \; \text{const.} \tag{12.131}$$

from which it follows that

$$E_x(x) = \frac{i}{\Lambda}\left(k_z - \frac{v_0\omega_p^2/\gamma_0}{c^2\omega'}\right) G\, e^{-\Lambda x} \tag{12.132}$$

$$B_y(x) = \frac{i\omega}{c\Lambda}\left(1 - \frac{\omega_p^2/\gamma_0}{\omega\omega'}\right) G\, e^{-\Lambda x} \tag{12.133}$$

Vacuum solution and dispersion relation. The wave equation in vacuum is obtained by putting $\omega_p = 0$ in the corresponding plasma equations. One obtains

$$B_y(x) = F\, e^{\lambda x} \quad F: \text{ const.} \tag{12.134}$$

$$\text{where} \quad \lambda = \sqrt{k_z^2 - \omega^2/c^2} \tag{12.135}$$

$$E_z(x) = \frac{icF}{\omega}\lambda\, e^{\lambda x} \tag{12.136}$$

$$E_x(x) = \frac{cF}{\omega} k_z\, e^{\lambda x} \tag{12.137}$$

The connection formula (12.122) gives

$$\frac{ic\lambda}{\omega} F = G \tag{12.138}$$

Equation (12.123) gives

$$F = \frac{i\omega}{c\Lambda}\left(1 - \frac{\omega_p^2}{\gamma_0^3\omega'^2}\right) G \tag{12.139}$$

From the above two equations, we finally obtain the dispersion relation

$$\Lambda + \lambda\left(1 - \frac{\omega_p^2}{\omega'^2\gamma_0^3}\right) = 0 \tag{12.140}$$

Using either (12.124) or (12.128) in place of (12.123) yields the same dispersion relation in (12.140).

• **Surface charge and surface current in a streaming cold plasma.** In a cold plasma, *real* surface charges are formed in the thin layer of interface (not the fictitious surface charge that is not considered here). Furthermore, surface currents are generated if the plasma is drifting. The boundary condition (12.123) implies that the tangential component of magnetic field is discontinuous by the amount of the surface current. The infinitesimal integration of Poisson equation (12.116) gives

$$E_x(x)\Big|_{0^-}^{0^+} = [E_x] = -4\pi e \int_{-\epsilon}^{\epsilon} n dx \equiv 4\pi\sigma \tag{12.141}$$

where σ is the surface charge density. Using (12.111) in the above equation and picking up the perfect differential term, we find

$$\sigma = -e \int_{-\epsilon}^{\epsilon} n dx = \frac{ieN}{\omega'} v_x(0^+) \tag{12.142}$$

Using (12.112) and (12.142) in (12.124) gives

$$[B_y] = \frac{v_0}{c} 4\pi\sigma \tag{12.143}$$

which is readily seen to be equivalent with (12.123).

We can derive the important relation (12.123) directly from the Maxwell equation and by calculating the surface current. Performing the infinitesimal integration on the Maxwell equation

$$\nabla \times \mathbf{B} = \frac{4\pi}{c}\mathbf{J} + \frac{1}{c}\frac{\partial \mathbf{E}}{\partial t} \tag{12.144}$$

we have

$$[B_y] = \frac{4\pi}{c} \int_{-\epsilon}^{\epsilon} J_z \, dx \tag{12.145}$$

In a streaming plasma, the right side can be non-vanishing, indicating the presence of surface current. Using (12.115) and (12.116) in the above equation, we obtain

$$[B_y] = \frac{v_0}{c}[E_x] \tag{12.146}$$

The x component of (12.144) reads

$$B_y = \frac{\omega}{ck_z}\left(E_x + \frac{4\pi i}{\omega} J_x\right) = \frac{\omega}{ck_z} D_x \tag{12.147}$$

Hence, the normal component of the displacement vector $\mathbf{D}$ has the jump in a streaming plasma, since $[B_y] \neq 0$.

$$[D_x] = \frac{ck_z}{\omega}[B_y] \tag{12.148}$$

The casual use of $[D_x] = 0$ in a streaming plasma leads to erroneous result.

The formulas obtained in this section are applicable for a non-relativistic plasma with a drifting velocity $v_0 \ll c$, if one puts $\gamma_0 = 1$. In particular, the boundary conditions (12.122), (12.123), (12.143), and (12.148) are valid in non-relativistic plasmas as well as in relativistic plasmas.

• **Fourier transform method.** By extending the plasma region $x > 0$ to the whole region $-\infty < x < \infty$ with the odd-even reflectional stipulation [Equations (12.36), (12.37) and (12.68), (12.69)], the plasma differential equations can be transformed to algebraic equations by Fourier transform. Then, the set of the algebraic equations involves a single undetermined constant 'A' which comes from

$$\frac{\partial}{\partial x} B_y(x) \to ik_x B_y(k_x) + A; \; A = B_y(0^-) - B_y(0^+) \tag{12.149}$$

since the oddly continued function $B_y(x)$ is discontinuous at $x = 0$. For evenly-continued function $E_z(x)$, $\partial/\partial x = ik_x$, simply, since $E_x(x)$ is continuous at $x = 0$. All the plasma Fourier variables are determined in terms of A. By the same token, the vacuum region $x < 0$ is extended to the whole region $-\infty < x < \infty$ with the same odd-even reflectional stipulation. Then, the Fourier-transformed vacuum equations are obtained from the plasma equations by replacing $A \to -A$ and putting $\omega_p = 0$. Hence, the whole of plasma-vacuum system involves a single undetermined constant, A. Consequently, we need only one connection formula, which is $[E_z] = 0$. The rest of boundary conditions are *automatically satisfied.* In fact, this is another way of deriving the boundary conditions. One can easily check this property of the Fourier transform solution from Maxwell equations by specializing to a cold plasma. The boundary conditions for TM mode surface wave in a non-relativistic cold plasma was derived earlier by this method [H. J. Lee and S. H. Cho, J. Plasma Phys. **58**, part 3, (1997), pp. 409–419].

12.4 Kinetic Theory of Relativistic Surface Waves: Semi-Infinite Plasma

We begin with the relativistic equation of motion for electrons

$$\frac{d\mathbf{p}}{dt} = -e\left(\mathbf{E} + \frac{1}{c}\mathbf{v} \times \mathbf{B}\right) \tag{12.150}$$

where $\mathbf{p}$, the relativistic momentum, is

$$\mathbf{p} = m\,\gamma\,\mathbf{v} \tag{12.151}$$

with m being the rest mass and $\mathbf{v}$ being the particle velocity, and

$$\gamma = \left(1 - \frac{v^2}{c^2}\right)^{-1/2} \tag{12.152}$$

The above equations have been already introduced in Section 12.3.

An easy way to write down the relativistic Vlasov equation is to note that the particle orbit as given by (12.150) is the characteristics of the Vlasov equation. Equation (12.150) is the characteristic equation of the following relativistic Vlasov equation:

$$\frac{\partial}{\partial t} f(\mathbf{r}, \mathbf{p}, t) + \mathbf{v} \cdot \frac{\partial f}{\partial \mathbf{r}} - e\left(\mathbf{E}(\mathbf{r}, t) + \frac{1}{c}\mathbf{v} \times \mathbf{B}\right) \cdot \frac{\partial f}{\partial \mathbf{p}} = 0 \tag{12.153}$$

This is the mostly used form of the relativistic Vlasov equation. Linearizing (12.153) gives

$$\frac{\partial}{\partial t} f_1(\mathbf{r}, \mathbf{p}, t) + \mathbf{v} \cdot \frac{\partial f_1}{\partial \mathbf{r}} - e\left(\mathbf{E}(\mathbf{r}, t) + \frac{1}{c}\mathbf{v} \times \mathbf{B}\right) \cdot \frac{\partial f_0(\mathbf{p})}{\partial \mathbf{p}} = 0 \tag{12.154}$$

where $f_0(\mathbf{p})$ is the zero order equilibrium distribution function which will be specified later. The perturbed distribution function f_1 will be simply denoted as f by omitting the subscript henceforth. We also have the Maxwell equations for electrons. Ions are assumed to be stationary and only form the neutralizing background.

$$\nabla \times \mathbf{E} = -\frac{1}{c}\frac{\partial \mathbf{B}}{\partial t} \tag{12.155}$$

$$\nabla \times \mathbf{B} = \frac{4\pi}{c}\mathbf{J} + \frac{1}{c}\frac{\partial \mathbf{E}}{\partial t} + \hat{\mathbf{z}} J_s(\mathbf{r}, t) \tag{12.156}$$

where $\hat{\mathbf{z}} J_s$ is the fictitious surface current (cf. (12.70)) and the plasma current $\mathbf{J}$ is

$$\begin{aligned}\mathbf{J}(\mathbf{r}, t) &= -eN \int \mathbf{v}\ f(\mathbf{r}, \mathbf{p}, t)\ d^3p \\ &= -\frac{eN}{m} \int d^3p\ f(\mathbf{r}, \mathbf{p}, t)\ \frac{\mathbf{p}}{\gamma}\end{aligned} \tag{12.157}$$

with N being the zero order equilibrium number density. We also have

$$\nabla \cdot \mathbf{B} = 0 \tag{12.158}$$

$$\nabla \cdot \mathbf{E} = -4\pi e \int f d^3p + S(\mathbf{r}, t) \tag{12.159}$$

where S is the fictitious surface charge. The fictitious surface current J_s corresponding to the surface charge S, placed on $x = 0$, takes the form

$$J_s = A(z,t)\ \delta(x) \tag{12.160}$$

We Fourier transform all the dependent variables in the basic equations, including f, $\mathbf{E}$, and $\mathbf{B}$, and assume the form $\sim \exp(i\mathbf{k}\cdot\mathbf{r} - i\omega t)$. Then Eq. (12.154) gives

$$f(\omega,\mathbf{k},\mathbf{p}) = i\frac{e}{m}\frac{1}{\omega - \mathbf{k}\cdot\mathbf{v}}\left(\mathbf{E} + \frac{1}{c}\mathbf{v}\times\mathbf{B}\right)\cdot\frac{\partial f_0}{\partial \mathbf{p}} \tag{12.161}$$

where $\mathbf{v}$ reads as the function of $\mathbf{p}$ as given in (12.151). Using (12.161) in the Fourier transform of (12.157) yields

$$J_l = -\frac{ie^2N}{m}\int d^3p\ p_l\left[\frac{E_i + \frac{1}{c}e_{ijk}v_jB_k}{\gamma(\omega - \mathbf{k}\cdot\mathbf{v})}\right]\frac{\partial f_0}{\partial p_i} \tag{12.162}$$

where l is the cartesian index and e_{ijk} is the cartesian tensor called Levi-Civita symbol, and repeated indexes are summed over.

Cold plasma dielectric tensor. We evaluate the current in (12.162) for a cold streaming plasma whose zero order distribution function is

$$f_0(\mathbf{p}) = \delta(p_x)\delta(p_y)\delta(p_z - p_0) \tag{12.163}$$

$$\text{where } p_0 = \gamma_0 mu, \quad \gamma_0 = \frac{1}{\sqrt{1 - \frac{u^2}{c^2}}} \tag{12.164}$$

which corresponds to the streaming velocity $\mathbf{v}_0 = \hat{\mathbf{z}}u$. Integrating by parts (12.162) gives

$$J_l = \frac{ie^2N}{m}\int d^3p f_0\left(\delta_{il} + p_l\frac{\partial}{\partial p_i}\right)\left[\frac{E_i + \frac{1}{c}e_{ijk}v_jB_k}{\gamma(\omega - \mathbf{k}\cdot\mathbf{v})}\right] \tag{12.165}$$

The following relations are useful for evaluating the above integral:

$$\gamma(p) = \sqrt{1 + \frac{p^2}{m^2c^2}}, \quad \frac{\partial\gamma}{\partial p_j} = \frac{p_j}{mc\sqrt{m^2c^2 + p^2}} \tag{12.166}$$

$$v_i = \frac{cp_i}{\sqrt{m^2c^2 + p^2}}, \quad \frac{\partial v_i}{\partial p_j} = \frac{c}{\sqrt{m^2c^2 + p^2}}\left(\delta_{ij} - \frac{p_ip_j}{m^2c^2 + p^2}\right) \tag{12.167}$$

Using (12.163) in (12.165), we obtain after a long algebra,

$$\mathbf{J}(\mathbf{k},\omega)=\frac{ie^2N}{m\gamma_0\omega'}\left[\mathbf{E}+\frac{u}{c}\hat{\mathbf{z}}\times\mathbf{B}+\hat{\mathbf{z}}\left\{-\frac{u^2}{c^2}E_z+\frac{u}{\omega'}\quad\left(\mathbf{k}\cdot\left(\mathbf{E}+\frac{u}{c}\hat{\mathbf{z}}\times\mathbf{B}\right)-\frac{u^2}{c^2}k_zE_z\right)\right\}\right] \tag{12.168}$$

where $\omega'=\omega-k_zu$ is Doppler-shifted frequency.

Using $\mathbf{B}=\frac{c}{\omega}\mathbf{k}\times\mathbf{E}$ and writing in components, we have

$$J_x=\frac{ie^2N}{m\omega\gamma_0}\left(E_x+\frac{uk_x}{\omega'}E_z\right) \tag{12.169}$$

$$J_y=\frac{ie^2N}{m\omega\gamma_0}\left(E_y+\frac{uk_y}{\omega'}E_z\right) \tag{12.170}$$

$$J_z=\frac{ie^2N}{m\omega\gamma_0}\left[\frac{u}{\omega'}(k_xE_x+k_yE_y)+\frac{E_z}{\omega'^2}\left(u^2(k_x^2+k_y^2)+\frac{\omega^2}{\gamma_0^2}\right)\right] \tag{12.171}$$

The array formed by the coefficients of the electric field components is the conductivity tensor, as defined by $J_i=\sigma_{ij}E_j$.

Combining (12.155) and (12.156) with the aid of (12.169)–(12.171) yields a vector equations for $\mathbf{E}(\mathbf{k},\omega)$ in the form

$$\mathbf{k}\times(\mathbf{k}\times\mathbf{E})+\frac{\omega^2}{c^2}\varepsilon_{ij}E_j=-\hat{\mathbf{z}}\,\frac{i\omega}{c}J_s \tag{12.172}$$

where

$$\varepsilon_{ij}=\delta_{ij}+\frac{4\pi i}{\omega}\sigma_{ij}=\left(1-\frac{\omega_p^2}{\gamma_0\omega^2}\right)\delta_{ij}-\frac{\omega_p^2}{\gamma_0\omega^2}U_{ij} \tag{12.173}$$

is the dielectric tensor, and ω_p is the plasma frequency, and

$$U_{ij}=\begin{pmatrix}0 & 0 & \frac{uk_x}{\omega'}\\ 0 & 0 & \frac{uk_y}{\omega'}\\ \frac{uk_x}{\omega'} & \frac{uk_y}{\omega'} & U_{zz}\end{pmatrix}$$

$$U_{zz}=\frac{u^2}{\omega'^2}\left(k^2-\frac{\omega^2}{c^2}\right)+\frac{2k_zu}{\omega'}$$

Putting (12.173) into (12.172) gives a 3 by 3 matrix equation for E_j:

$$\left[\left(1-n^2-\frac{\omega_p^2}{\gamma_0\omega^2}\right)\delta_{ij}+\frac{k_ik_j}{k^2}n^2-\frac{\omega_p^2}{\gamma_0\omega^2}U_{ij}\right]E_j=-\frac{ic}{\omega}\hat{\mathbf{z}}J_s \tag{12.174}$$

where $n^2 = c^2k^2/\omega^2$ is the refractive index, and J_s is the Fourier transform of the fictitious surface currents in (12.160):

$$J_s(\mathbf{k}, \omega) = \delta(k_y) A_1 \tag{12.175}$$

where A_1 is undetermined Fourier coefficient of the surface charge which should be determined by satisfying the boundary condition at the interface.

Equation (12.174) can be inverted after spending a considerable amount of time to solve for $\mathbf{E}(\mathbf{k}, \omega)$:

$$E_x = -\frac{J_s}{\Delta} \frac{c^2 k_x}{\omega^2} \left(k_z - \frac{u}{c^2} \frac{\omega_p^2}{\gamma_0 \omega'} \right) \tag{12.176}$$

$$E_y = -\frac{J_s}{\Delta} \frac{c^2 k_y}{\omega^2} \left(k_z - \frac{u}{c^2} \frac{\omega_p^2}{\gamma_0 \omega'} \right) \tag{12.177}$$

$$E_z = \frac{J_s}{\Delta} \left(1 - \frac{\omega_p^2}{\gamma_0 \omega^2} - \frac{c^2 k_z^2}{\omega^2} \right) \tag{12.178}$$

where

$$\Delta = \left(1 - \frac{\omega_p^2}{\gamma_0 \omega^2} - \frac{c^2 k^2}{\omega^2} \right) \left[1 - \frac{\omega_p^2}{\gamma_0 \omega^2} - \frac{\omega_p^2}{\gamma_0 \omega^2} \left(U_{zz} - \frac{u^2}{\omega'^2} (k_x^2 + k_y^2) \right) \right] \tag{12.179}$$

After further algebra, Δ becomes

$$\Delta = \left(1 - \frac{\omega_p^2}{\gamma_0 \omega^2} - \frac{c^2 k^2}{\omega^2} \right) \left(1 - \frac{\omega_p^2}{\gamma_0^3 \omega'^2} \right) \tag{12.180}$$

Also we obtain

$$B_x = \frac{c}{\omega} \frac{J_s}{\Delta} k_y \left(1 - \frac{\omega_p^2}{\gamma_0 \omega \omega'} \right) \tag{12.181}$$

$$B_y = -\frac{c}{\omega} \frac{J_s}{\Delta} k_x \left(1 - \frac{\omega_p^2}{\gamma_0 \omega \omega'} \right) \tag{12.182}$$

In addition, we put $B_z = 0$ since we investigate the transverse magnetic mode.

Boundary equations. In order to apply the boundary conditions, the electric and magnetic field components in the Fourier $\mathbf{k}$ space should be inverted to the fields in the ordinary $\mathbf{r}$ space performing the integral

$\int_{-\infty}^{\infty} dk_x \int_{-\infty}^{\infty} dk_y \; e^{ik_x x + ik_y y}(\cdots)$. The boundary values are obtained by putting $x = 0$ in the inverted integrals.

$$E_x(0,z) = -\frac{c^2}{\omega^2}\left(k_z - \frac{u\omega_p^2}{c^2\omega'\gamma_0}\right)\int_{-\infty}^{\infty} dk_x \int_{-\infty}^{\infty} dk_y \frac{k_x}{\Delta}\delta(k_y)A_1$$
$$= -\frac{c^2}{\omega^2}\left(k_z - \frac{u\omega_p^2}{c^2\omega'\gamma_0}\right)A_1 \int_{-\infty}^{\infty} dk_x \; \frac{k_x}{\Delta} \tag{12.183}$$

where Δ is given by (12.180) and $k^2 = k_x^2 + k_z^2$.

$$B_y(0,z) = -\frac{c}{\omega}\left(1 - \frac{\omega_p^2}{\omega\omega'\gamma_0}\right)A_1 \int_{-\infty}^{\infty} dk_x \frac{k_x}{\Delta} \tag{12.184}$$

$$E_z(0,z) = \left(1 - \frac{\omega_p^2}{\gamma_0\omega^2} - \frac{c^2k_z^2}{\omega^2}\right)A_1 \int_{-\infty}^{\infty} dk_x \int_{-\infty}^{\infty} dk_y \frac{\delta(k_y)}{\Delta}$$
$$= \left(1 - \frac{\omega_p^2}{\gamma_0\omega^2} - \frac{c^2k_z^2}{\omega^2}\right)A_1 \int_{-\infty}^{\infty} \frac{dk_x}{\Delta} \tag{12.185}$$

The integrals in the above can be evaluated by contour integration on complex k_x plane:

$$\int_{-\infty}^{\infty} dk_x \; \frac{k_x}{\Delta} = \frac{i\pi}{\beta} \tag{12.186}$$

$$\int_{-\infty}^{\infty} dk_x \; \frac{1}{\Delta} = \frac{\pi}{\Lambda\beta} \tag{12.187}$$

where

$$\Lambda = \sqrt{k_z^2 - \frac{\omega^2 - \omega_p^2/\gamma_0}{c^2}} \tag{12.188}$$

$$\beta = -\frac{c^2}{\omega^2}\left(1 - \frac{\omega_p^2}{\gamma_0^3\omega'^2}\right) \tag{12.189}$$

Vacuum solutions for $x < 0$. The vacuum solutions are provided in (12.79)–(12.82):

$$E_z(x, k_z, \omega) = i\;\lambda F e^{\lambda x} \tag{12.190}$$

$$\lambda = \sqrt{k_z^2 - \omega^2/c^2} \tag{12.191}$$

$$E_x = k_z F e^{\lambda x} \tag{12.192}$$

$$B_y(x, k_z, \omega) = \frac{\omega}{c} F e^{\lambda x} \tag{12.193}$$

Dispersion relation. We enforce the following boundary conditions to connect the plasma and the vacuum fields: $[E_z] = 0$, $[B_y] = \frac{u}{c}[E_x]$ $[E_z] = 0$ gives

$$-\frac{c^2}{\omega^2}\frac{\Lambda}{\beta}\pi A_1 = i\lambda F \tag{12.194}$$

$[B_y] = \frac{u}{c}[E_x]$ gives

$$\frac{i\pi}{\beta}A_1\left[-\frac{c}{\omega}\left(1 - \frac{\omega_p^2}{\omega\omega'\gamma_0}\right) + \frac{cu}{\omega^2}\left(k_z - \frac{u\omega_p^2}{c^2\omega'\gamma_0}\right)\right] = \frac{\omega'}{c}F \tag{12.195}$$

The above two equations give the dispersion relation in the form

$$1 - \frac{\omega_p^2}{\gamma_0^3\omega'^2} + \frac{\Lambda}{\lambda} = 0 \tag{12.196}$$

which agrees with (12.140).

12.5 Relativistic Surface Waves in a Plasma Slab

Let us consider a plasma slab, the region between two infinite planes $x = 0$ and $x = L$, interfaced by vacuums. The surface waves are assumed to propagate along the parallel direction, the $\hat{\mathbf{z}}$ direction. The $\hat{\mathbf{y}}$ direction can be ignored. We investigate the normal modes of the relativistic plasma in the slab by kinetic equation. The basic equations are already introduced in Section 12.4; we use them as they are except the recipe of the surface currents. Since we have two bounding surfaces, we assume the surface currents in the form

$$J_s = A_1 \sum_{n=0,1,2,\ldots} \delta(x \pm 2nL) + A_2 \sum_{n=1,2,\ldots} \delta(x \pm (2n-1)L) \tag{12.197}$$

where the double signs are summed over. The fictitious surface currents above take care of the jumps arising in the continuations of $\mathbf{E}$ and $\mathbf{B}$:

$$E_x(-x) = -E_x(x), \quad E_x(2L - x) = -E_x(x) \tag{12.198}$$

$$E_z(-x) = E_z(x), \quad E_z(2L - x) = E_z(x) \tag{12.199}$$

$$B_y(-x) = -B_y(x), \quad B_y(2L - x) = -B_y(x) \tag{12.200}$$

With the above extensions, the specular reflection condition on $x = 0,\ L$ planes are satisfied. The oddly continuation of E_x and B_y indicates that these field components have jumps at $x = 2nL$ and $x = (2n-1)L$ of different magnitude (see Fig. 12.1). The discontinuities should be carefully taken into account when the fields are Fourier-transformed. We can significantly alleviate the algebraic burden of Fourier transform of the piecewise periodic functions in Fig. 12.1, simply by introducing the surface currents as in (12.197) in the $\nabla \times \mathbf{B}$-equation. The point of fact is that the discontinuities that are present in the extended field components $E_x(x)$ and $B_y(x)$ at the locations $x = \pm 2nL$ and $x = \pm(2n-1)L$ are mathematically (as well as physically) tantamount to placing fictitious surface charges at the corresponding jump locations in the form

$$\sigma(x) = C_1 \sum_{n=0,1,2,\dots} \delta(x \pm 2nL) + C_2 \sum_{n=1,2,\dots} \delta(x \pm (2n-1)L)$$

The surface charges are associated with the surface currents by satisfying the charge conservation equation, hence Eq. (12.197).

Then Fourier transform of (12.197) gives

$$J_s(\mathbf{k},\omega) = \delta(k_y)\left[A_1 \sum_{n=0,1,2,\dots} e^{\pm i2nLk_x} + A_2 \sum_{n=1,2,\dots} e^{\pm i(2n-1)Lk_x}\right] \tag{12.201}$$

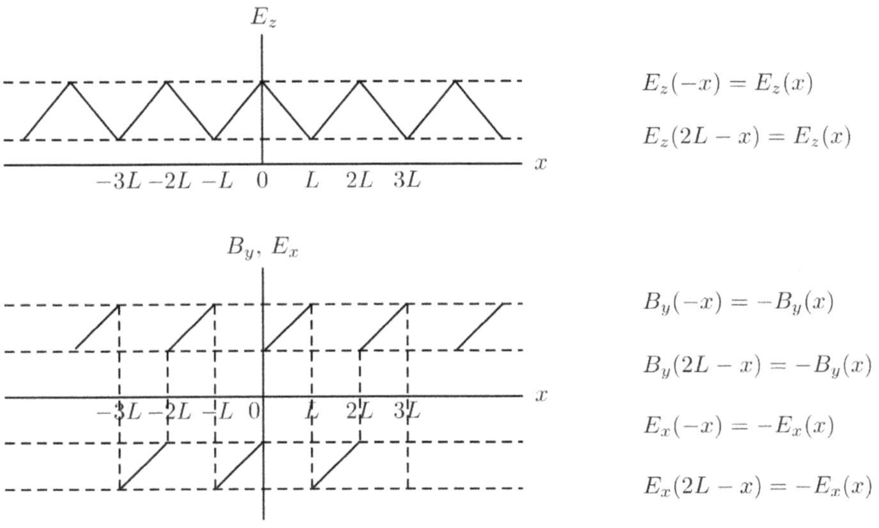

Fig. 12.1. Extension of electric and magnetic field in a plasma slab ($0 < x < L$).

where the double signs are summed over. Equations (12.176)–(12.182) are valid with J_s therein prescribed as (12.201).

Boundary equations. In order to apply the boundary conditions, the electric and magnetic field components in the Fourier $\mathbf{k}$ space should be inverted to the fields in the ordinary $\mathbf{r}$ space by performing $\int_{-\infty}^{\infty} dk_x \int_{-\infty}^{\infty} dk_y e^{ik_x x + ik_y y}(\cdots)$. The boundary values are obtained by putting $x = 0$ and $x = L$ in the inverted integrals. The integrals involve infinite series through the surface charge J_s, but the infinite series are nicely summed at the particular positions corresponding to $x = 0,\ L$. When the inversion integrals are performed, the following formulas are useful, which can be verified by a simple change of variable, as is shown in earlier report [Hee J. Lee, *Fundamentals of theoretical plasma physics*, World Scientific (2019)].

We have integrals of the type in the inversion integrals

$$Q(x) = \int_{-\infty}^{\infty} dk_x k_x \Phi(k)\, e^{ik_x x} \left[A_1 \sum_{n=0,1,2,\ldots} e^{\pm i2nLk_x} + A_2 \sum_{n=1,2,\ldots} e^{\pm i(2n-1)Lk_x}\right] \tag{12.202}$$

where $\Phi(k)$ is an even function of k_x, per $k = \sqrt{k_x^2 + k_z^2}$. Then, we have

$$Q(0) = A_1 \int_{-\infty}^{\infty} dk_x k_x \Phi(k) \tag{12.203}$$

$$Q(L) = -A_2 \int_{-\infty}^{\infty} dk_x k_x \Phi(k) \tag{12.204}$$

We also have integrals of the type

$$R(x) = \int_{-\infty}^{\infty} dk_x \Phi(k)\, e^{ik_x x} \left[A_1 \sum_{n=0,1,2,\ldots} e^{\pm i2nLk_x} + A_2 \sum_{n=1,2,\ldots} e^{\pm i(2n-1)Lk_x}\right] \tag{12.205}$$

Then, we have

$$R(0) = 2\int_{-\infty}^{\infty} dk_x \Phi(k)(A_1 S_1 + A_2 S_2) \tag{12.206}$$

$$R(L) = 2\int_{-\infty}^{\infty} dk_x \Phi(k)(A_1 S_2 + A_2 S_1) \tag{12.207}$$

where

$$S_1 = \frac{1}{2} + e^{2iLk_x} + e^{4iLk_x} + \cdots \tag{12.208}$$

$$S_2 = e^{iLk_x} + e^{3iLk_x} + \cdots \tag{12.209}$$

The above formulas are useful for evaluating the integrals.

Let us evaluate:

$$\begin{aligned} E_x(0,z) &= -\frac{c^2}{\omega^2}\left(k_z - \frac{u\omega_p^2}{c^2\omega'\gamma_0}\right)\int_{-\infty}^{\infty} dk_x \int_{-\infty}^{\infty} dk_y \frac{k_x}{\Delta}\delta(k_y) \\ &\quad \times \left(A_1 \sum_0 e^{\pm i2nLk_x} + A_2 \sum_1 e^{\pm i(2n-1)Lk_x}\right) \\ &= -\frac{c^2}{\omega^2}\left(k_z - \frac{u\omega_p^2}{c^2\omega'\gamma_0}\right) A_1 \int_{-\infty}^{\infty} dk_x \frac{k_x}{\Delta} \end{aligned} \tag{12.210}$$

where we used (12.203). In the integral, k^2 hidden in Δ is $k^2 = k_z^2 + k_x^2$. Next,

$$\begin{aligned} E_x(L,z) &= -\frac{c^2}{\omega^2}\left(k_z - \frac{u\omega_p^2}{c^2\omega'\gamma_0}\right)\int_{-\infty}^{\infty} dk_x e^{ik_xL} \int_{-\infty}^{\infty} dk_y \frac{k_x}{\Delta}\delta(k_y) \\ &\quad \times \left(A_1 \sum_0 e^{\pm i2nLk_x} + A_2 \sum_1 e^{\pm i(2n-1)Lk_x}\right) \\ &= -\frac{c^2}{\omega^2}\left(k_z - \frac{u\omega_p^2}{c^2\omega'\gamma_0}\right) (-A_2) \int_{-\infty}^{\infty} dk_x \frac{k_x}{\Delta} \end{aligned} \tag{12.211}$$

where we used (12.204). Analogously, we obtain

$$B_y(0,z) = -\frac{c}{\omega}\left(1 - \frac{\omega_p^2}{\omega\omega'\gamma_0}\right) A_1 \int_{-\infty}^{\infty} dk_x \frac{k_x}{\Delta} \tag{12.212}$$

$$B_y(L,z) = -\frac{c}{\omega}\left(1 - \frac{\omega_p^2}{\omega\omega'\gamma_0}\right) (-A_2) \int_{-\infty}^{\infty} dk_x \frac{k_x}{\Delta} \tag{12.213}$$

Next, we have

$$E_z(0,z) = \left(1 - \frac{\omega_p^2}{\gamma_0\omega^2} - \frac{c^2k_z^2}{\omega^2}\right)\int_{-\infty}^{\infty} \frac{dk_x}{\Delta}\left(A_1 \sum_0 e^{\pm i2nLk_x} + A_2 \sum_1 e^{\pm i(2n-1)Lk_x}\right) \tag{12.214}$$

which we write in the form

$$E_z(0,z) = 2\left(1 - \frac{\omega_p^2}{\gamma_0\omega^2} - \frac{c^2k_z^2}{\omega^2}\right)\int_{-\infty}^{\infty}\frac{dk_x}{\Delta}(A_1S_1(Lk_x) + A_2S_2(Lk_x)) \tag{12.215}$$

where we used (12.206), and

$$S_1(\xi) = \frac{1}{2} + e^{2i\xi} + e^{4i\xi} + \cdots, \quad S_2(\xi) = e^{i\xi} + e^{3i\xi} + \cdots \tag{12.216}$$

Analogously, we obtain

$$E_z(L,z) = 2\left(1 - \frac{\omega_p^2}{\gamma_0\omega^2} - \frac{c^2k_z^2}{\omega^2}\right)\int_{-\infty}^{\infty}\frac{dk_x}{\Delta}(A_1S_2(Lk_x) + A_2S_1(Lk_x)) \tag{12.217}$$

where we used (12.207).

Next we turn to the vacuum solutions in the region $x < 0$ and $x > L$. The vacuum equation

$$\nabla^2\mathbf{E} + \frac{\omega^2}{c^2}\mathbf{E} = 0$$

yields the following solutions:

$$E_z(x,k_z,\omega) = -i\lambda F_1\, e^{-\lambda x} \quad (x > L) \tag{12.218}$$

$$E_z(x,k_z,\omega) = i\lambda F_2\, e^{\lambda x} \quad (x < 0) \tag{12.219}$$

$$E_x(x,k_z,\omega) = k_zF_1\, e^{-\lambda x} \quad (x > L) \tag{12.220}$$

$$E_x(x,k_z,\omega) = k_zF_2\, e^{\lambda x} \quad (x < 0) \tag{12.221}$$

$$B_y(x,k_z,\omega) = \frac{\omega}{c}F_1\, e^{-\lambda x} \quad (x > L) \tag{12.222}$$

$$B_y(x,k_z,\omega) = \frac{\omega}{c}F_2\, e^{\lambda x} \quad (x < 0) \tag{12.223}$$

where $\lambda = \sqrt{k_z^2 - \omega^2/c^2}$—Eq. (12.191).

Dispersion relation. The matching conditions are $[E_z] = 0$ and $[B_y] = \frac{u}{c}[E_x]$ at $x = 0$ and $x = L$.

$$\xi(A_1I_1 + A_2I_2) = i\lambda F_2 \tag{12.224}$$

$$\xi(A_1I_2 + A_2I_1) = -i\lambda F_1\, e^{-\lambda L} \tag{12.225}$$

where

$$\xi = 1 - \frac{\omega_p^2}{\gamma_0 \omega^2} - \frac{c^2 k_z^2}{\omega^2} \tag{12.226}$$

$$I_i = 2\int_{-\infty}^{\infty} \frac{dk_x}{\Delta} S_i(Lk_x) \quad (i = 1, 2) \tag{12.227}$$

$$\frac{c}{\omega}\left(1 - \frac{\omega_p^2}{\omega\omega'\gamma_0}\right) A_1 Q + \frac{\omega}{c} F_2 = \frac{u}{c}\left\{\frac{c^2}{\omega^2}\left(k_z - \frac{u\omega_p^2}{c^2\omega'\gamma_0}\right) A_1 Q + k_z F_2\right\} \tag{12.228}$$

$$\frac{c}{\omega}\left(1 - \frac{\omega_p^2}{\omega\omega'\gamma_0}\right) A_2 Q - \frac{\omega}{c} F_1 \, e^{-\lambda L} = \frac{u}{c}\left\{\frac{c^2}{\omega^2}\left(k_z - \frac{u\omega_p^2}{c^2\omega'\gamma_0}\right) A_2 Q - k_z F_1 \, e^{-\lambda L}\right\} \tag{12.229}$$

where

$$Q = \int_{-\infty}^{\infty} dk_x \, \frac{k_x}{\Delta} \tag{12.230}$$

After some algebra (12.228) and (12.229) reduce to

$$Q\eta A_1 + \frac{\omega^2}{c^2} F_2 = 0 \tag{12.231}$$

$$Q\eta A_2 - \frac{\omega^2}{c^2} F_1 \, e^{-\lambda L} = 0 \tag{12.232}$$

where

$$\eta = 1 - \frac{\omega_p^2}{\omega'^2 \gamma_0^3} \tag{12.233}$$

Equations (12.224), (12.225), (12.231), and (12.232) yield the dispersion relation. Eliminating F_1 and F_2 gives

$$A_1\left(I_1 + \frac{i\lambda}{\xi}\,\frac{c^2}{\omega^2} Q\eta\right) + A_2 I_2 = 0$$

$$A_1 I_2 + A_2\left(I_1 + \frac{i\lambda}{\xi}\,\frac{c^2}{\omega^2} Q\eta\right) = 0$$

which gives the two mode dispersion relation in the form

$$I_1 \pm I_2 + \frac{i\lambda}{\xi}\,\frac{c^2}{\omega^2} Q\eta = 0 \tag{12.234}$$

We can easily show

$$I_1 \pm I_2 = \int_{-\infty}^{\infty} \frac{dk_x}{\Delta} \frac{1 \pm e^{iLk_x}}{1 \mp e^{iLk_x}} \tag{12.235}$$

Hence the dispersion relation in (12.234) takes the form

$$\int_{-\infty}^{\infty} \frac{dk_x}{\Delta} \left[\frac{1 \pm e^{iLk_x}}{1 \mp e^{iLk_x}} + \frac{i\lambda\eta}{\xi} \frac{c^2}{\omega^2} k_x \right] = 0 \tag{12.236}$$

The lower (upper) sign corresponds to symmetric (anti-symmetric) mode. This equation can be further evaluated by contour integration, conferring (12.186) and (12.187). The apparent singularities associated with $1 \pm e^{ik_x L}$ should be simply ignored, since the series $S_1 \pm S_2$ (12.208, 12.209) are regular in the upper k_x plane. We obtain the *slab dispersion relation* in the form

$$\Lambda \frac{1 \pm e^{-\Lambda L}}{1 \mp e^{-\Lambda L}} + \eta \, \lambda = 0 \tag{12.237}$$

where Λ is defined by (12.188), λ by (12.191), η by (12.233). If $L \to \infty$, the slab dispersion relation reduces to the semi-infinite dispersion relation, $\Lambda + \eta\lambda = 0$. See Eq. (12.196).

• **Propagation in a duct.** We consider an infinite duct formed by intersections of four planes: $x = 0$, a and $y = 0$, b, with $-\infty < z < \infty$. A duct is formed by placing two pairs of slab. The specular reflection conditions require for the distribution function $f(\mathbf{r}, \mathbf{v}, t)$ to satisfy

$$\begin{aligned}
f(0, y, z, v_x, v_y, v_z, t) &= f(0, y, z, -v_x, v_y, v_z, t) \quad \text{on } x = 0 \text{ plane} \\
f(a, y, z, v_x, v_y, v_z, t) &= f(a, y, z, -v_x, v_y, v_z, t) \quad \text{on } x = a \text{ plane} \\
f(x, 0, z, v_x, v_y, v_z, t) &= f(x, 0, z, v_x, -v_y, v_z, t) \quad \text{on } y = 0 \text{ plane} \\
f(x, b, z, v_x, v_y, v_z, t) &= f(x, b, z, v_x, -v_y, v_z, t) \quad \text{on } y = b \text{ plane.}
\end{aligned}$$

In a duct-bounded plasma, the above kinematic conditions on the four planes are satisfied by introducing extended electric field in the fashion:

$$\begin{aligned}
E_x(-x, y, z) &= -E_x(x, y, z), \quad E_x(2a - x, y, z) = -E_x(x, y, z) \\
E_y(x, -y, z) &= -E_y(x, y, z), \quad E_y(x, 2b - y, z) = -E_y(x, y, z)
\end{aligned}$$

For a duct, we place fictitious surface currents at the corresponding jump locations in the form $\hat{\mathbf{z}}J_s$ with

$$J_s(x,y,z,t) = A_1 \sum_{n=0,1,2,\ldots} \delta(x \pm 2na) + A_2 \sum_{n=1,2,\ldots} \delta(x \pm (2n-1)a) + B_1 \sum_{n=0,1,2,\ldots} \delta(y \pm 2nb) + B_2 \sum_{n=1,2,\ldots} \delta(y \pm (2n-1)b) \tag{12.238}$$

Analogous analysis with the surface current in (12.238) as performed in the case of slab, the dispersion relation for the duct surface wave is obtained in the form

$$\frac{b}{\sqrt{a^2+b^2}} \int_{-\infty}^{\infty} \frac{dk_x}{\Delta} \frac{1 \pm e^{iak_x}}{1 \mp e^{iak_x}} + \frac{a}{\sqrt{a^2+b^2}} \int_{-\infty}^{\infty} \frac{dk_y}{\Delta} \frac{1 \pm e^{ibk_y}}{1 \mp e^{ibk_y}} + i\lambda \frac{c^2}{\omega^2} \frac{\eta}{\xi} \int_{-\infty}^{\infty} k_x \frac{dk_x}{\Delta} = 0 \tag{12.239}$$

If either a or $b \to \infty$, we recover the slab dispersion relation

$$\int_{-\infty}^{\infty} \frac{dk_x}{\Delta} \frac{1 \pm e^{iak_x}}{1 \mp e^{iak_x}} + i\lambda \frac{c^2}{\omega^2} \frac{\eta}{\xi} \int_{-\infty}^{\infty} k_x \frac{dk_x}{\Delta} = 0 \tag{12.240}$$

which agrees with (12.236).

Note in (12.239): $k^2 = k_z^2 + k_x^2$ in $\int dk_x$-integral and $k^2 = k_z^2 + k_y^2$ in $\int dk_y$-integral. Thus, let us change the integration variables, both k_x and k_y, in (12.239) to κ:

$$\frac{b}{\sqrt{a^2+b^2}} \int_{-\infty}^{\infty} \frac{d\kappa}{\Delta} \frac{1 \pm e^{ia\kappa}}{1 \mp e^{ia\kappa}} + \frac{a}{\sqrt{a^2+b^2}} \int_{-\infty}^{\infty} \frac{d\kappa}{\Delta} \frac{1 \pm e^{ib\kappa}}{1 \mp e^{ib\kappa}} + i\lambda \frac{c^2}{\omega^2} \frac{\eta}{\xi} \int_{-\infty}^{\infty} \kappa \frac{d\kappa}{\Delta} = 0 \tag{12.241}$$

where $k^2 = k_z^2 + \kappa^2$.

The duct dispersion relation in (12.239) can be contour-integrated, giving for the lower sign

$$\frac{b\Lambda}{\sqrt{a^2+b^2}} \tan h\frac{a\Lambda}{2} + \frac{a\Lambda}{\sqrt{a^2+b^2}} \tan h\frac{b\Lambda}{2} + \sqrt{k_z^2 - \frac{\omega^2}{c^2}} \left(1 - \frac{\omega_p^2}{\gamma_0^3 \omega'^2}\right) = 0 \tag{12.242}$$

where $\Lambda = \sqrt{k_z^2 - \frac{\omega^2 - \omega_p^2/\gamma_0}{c^2}}$ (see (12.130)). For the upper sign, tan h-function above is replaced by cot$h$-function.

12.6 Surface Ion Acoustic Wave

The plasma consisting of hot electrons and cold ions occupies the region $x > 0$, interfaced by vacuum in the region $x < 0$. The surface wave is assumed to propagate along the z-direction with phasor $e^{ik_z z - i\omega t}$. The x-direction is inhomogeneous since it is bounded at $x = 0$. Due to the inhomogeneity, the electric potential, for example, is of the form $\phi(x, z, t) = \phi(x)e^{ik_z z - i\omega t}$. The plasma field should be connected with the vacuum field by appropriate boundary conditions. The electrons are assumed Boltzmann-distributed:

$$n_e = N\ e^{\frac{e\phi}{T_e}} \approx N\left(1 + \frac{e\phi}{T_e}\right) \tag{12.243}$$

where the second term is the perturbation. The ion equations are

$$-i\omega v_z(x) = -\frac{e}{m_i} ik_z\phi(x) \tag{12.244}$$

$$-i\omega v_x(x) = -\frac{e}{m_i}\frac{\partial\phi}{\partial x} \tag{12.245}$$

$$-i\omega n_i + N\left(ik_z v_z + \frac{\partial v_x}{\partial x}\right) = 0 \tag{12.246}$$

We have the Poisson equation

$$\nabla^2\phi = -4\pi e\left(n_i - N\ \frac{e\phi}{T_e}\right) \tag{12.247}$$

The above equations are arranged in the form

$$\frac{\partial^2\phi}{\partial x^2} - \gamma^2\phi = 0 \tag{12.248}$$

where

$$\gamma = \sqrt{k_z^2 + \frac{1}{\varepsilon\lambda_e^2}} \tag{12.249}$$

with $\varepsilon = 1 - \omega_{pi}^2/\omega^2$, $\lambda_e = \sqrt{4\pi N e^2/T_e}$, the electron Debye length. Equation (12.248) is solved in the plasma region to give

$$\phi(x) = A\ e^{-\gamma x} \quad A : \text{const.} \quad (x > 0) \tag{12.250}$$

In the vacuum ($x < 0$), we have from $\nabla^2\phi = 0$

$$\phi(x) = B\, e^{k_z x} \quad B: \text{ const.} \quad (x < 0) \tag{12.251}$$

Continuity of ϕ at $x = 0$ gives $A = B$. The normal component of the electric displacement (**D**) is continuous, which means the continuity of

$$D_x = E_x + \frac{4\pi i}{\omega} J_x = \varepsilon\, \frac{\partial\phi}{\partial x} \tag{12.252}$$

In the vacuum, $\varepsilon = 1$. We finally obtain the dispersion relation of the surface ion acoustic wave,

$$1 + \varepsilon\, \frac{\gamma}{k_z} = 0 \tag{12.253}$$

- **Kinetic dispersion relation of surface ion acoustic wave.** We obtained in (12.89) the electrostatic surface wave dispersion relation

$$k_z \int_{-\infty}^{\infty} \frac{dk_x}{k^2 \varepsilon_L} + \pi = 0 \tag{12.254}$$

For the ion acoustic wave, we put therein

$$\varepsilon_L = \varepsilon + \frac{1}{\lambda_e^2 k^2} \tag{12.255}$$

with $k^2 = k_x^2 + k_z^2$. The above two equations take the form

$$\pi + \frac{k_z}{\varepsilon} \int_{-\infty}^{\infty} \frac{dk_x}{k_x^2 + \gamma^2} = 0 \tag{12.256}$$

for a semi-infinite plasma. For a slab plasma of width L, the exponential factors are attached:

$$\pi + \frac{k_z}{\varepsilon} \int_{-\infty}^{\infty} \frac{dk_x}{k_x^2 + \gamma^2} \left(\frac{1 \pm e^{ik_x L}}{1 \mp e^{ik_x L}} \right) = 0 \quad \text{(slab)} \tag{12.257}$$

Calculating the residues at $k_x = i\gamma$, we obtain the dispersion relations

$$\varepsilon + \frac{k_z}{\gamma} = 0 \quad \text{(semi-infinite)} \tag{12.258}$$

$$\varepsilon + \frac{k_z}{\gamma}\ \tan h \left(\frac{\gamma L}{2} \right) = 0 \quad \text{(slab, symmetric)} \tag{12.259}$$

$$\varepsilon + \frac{k_z}{\gamma}\ \cot h \left(\frac{\gamma L}{2} \right) = 0 \quad \text{(slab, antisymmetric)} \tag{12.260}$$

• **Landau damping of surface ion acoustic wave.** In this section, the electrostatic dispersion relation (12.254) for the semi-bounded plasma is evaluated more exactly by taking the imaginary part of ε_L into account. The expression in (12.255) omitted the imaginary part entirely. In the low frequency range, we have [see Eqs. (8.94) and (8.95)] $\varepsilon_L = \varepsilon_r + i\varepsilon_i$ with

$$\varepsilon_r = 1 + \frac{1}{k^2\lambda_e^2}\left(1 - \frac{T_e}{m_i}\frac{k^2}{\omega^2}\right) = 1 - \frac{\omega_{pi}^2}{\omega^2} + \frac{1}{k^2\lambda_e^2} \tag{12.261}$$

$$\varepsilon_i = \sum_{\alpha=e,i} \frac{\sqrt{\pi}}{k^2\lambda_\alpha^2}\frac{\omega}{k}\sqrt{\frac{m_\alpha}{2T_\alpha}}\exp\left[\frac{-m_\alpha\omega^2}{2T_\alpha k^2}\right] \tag{12.262}$$

Our procedure in evaluating the surface wave dispersion relation and the Landau damping is a perturbational approach, assuming $\varepsilon_r \gg \varepsilon_i$: the zero order dispersion relation would be obtained by neglecting ε_i in (12.254). We put $\varepsilon_L = \varepsilon_r(1 + i\varepsilon_i/\varepsilon_r)$ where the second term is $\ll 1$. Then the dispersion relation integral in (12.254) becomes

$$1 + \frac{k_z}{\pi}\int \frac{dk_x}{k^2\varepsilon_r}\left(1 - i\frac{\varepsilon_i}{\varepsilon_r}\right) = 0 \tag{12.263}$$

Approximate eigenfrequency is obtained from the real part,

$$D_r \equiv 1 + \frac{k_z}{\pi}\int \frac{dk_x}{k^2\varepsilon_r} = 0 \tag{12.264}$$

Using (12.261) for ε_r, (12.264) becomes (12.256), and the integral is evaluated to yield (12.258). We rewrite it in the following form by changing notation $\varepsilon \to q$:

$$D_r = 1 + \frac{1}{q(\omega)}\sqrt{\frac{q\kappa}{1 + q\kappa}} = 0 \tag{12.265}$$

where $\kappa = k_z^2\lambda_e^2$. Clearly, $q(\omega)$ should be negative to have the wave supported. Also we need $1 + q\kappa < 0$:

$$q < 0, \quad q + \frac{1}{\kappa} < 0 \tag{12.266}$$

Squaring (12.265) gives $\kappa q^2 + q - \kappa = 0$ which is solved by

$$q = \frac{1}{2\kappa}(-1 \pm \sqrt{1 + 4\kappa^2}) \tag{12.267}$$

Here we consider the limiting case of $\kappa \ll 1$.

In this case, we have $q \simeq \frac{1}{2\kappa}[-1 \pm (1+2\kappa^2)]$ in which the positive sign should be rejected in view of (12.266), while the (−) sign is in accord with (12.266). Neglecting the κ^2 term, we obtain

$$\omega^2 = \frac{k_z^2 \lambda_e^2 \omega_{pi}^2}{1 + k_z^2 \lambda_e^2} \quad \left(q \simeq -\frac{1}{\kappa},\ \kappa \ll 1\right) \tag{12.268}$$

Equation (12.268) is the well-known dispersion relation. The surface ion acoustic wave dispersion relation is of the same form as the bulk ion acoustic wave.

We write for future reference

$$\frac{\gamma}{k_z} = \sqrt{\frac{\kappa^2}{1+\kappa^2}} \ll 1 \ (\kappa \ll 1) \tag{12.269}$$

Next we calculate the imaginary part of (12.263):

$$D_i \equiv -i\frac{k_z}{\pi} \int dk_x \ \frac{\varepsilon_i}{k^2 \varepsilon_r^2}$$

Using the relation $k^2 \varepsilon_r = q(k_x^2 + \gamma^2)$ and (12.262), the above expression can be put into the form

$$D_i = -i\frac{k_z}{\sqrt{\pi}} \frac{\omega}{q^2} \int_{-\infty}^{\infty} \frac{dk_x}{(k_x^2+\gamma^2)^2 \sqrt{k_z^2+k_x^2}} \sum_\alpha \frac{1}{\lambda_\alpha^2} \sqrt{\frac{m_\alpha}{2T_\alpha}} \exp\left(-\frac{m_\alpha \omega^2}{2T_\alpha (k_z^2+k_x^2)}\right) \tag{12.270}$$

We first deal with the ion term, and consider integral

$$I_i = \int_{-\infty}^{\infty} \frac{dk_x \ \exp[-\frac{\xi_i}{k_z^2+k_x^2}]}{(k_x^2+\gamma^2)^2 \sqrt{k_z^2+k_x^2}}, \quad \left(\xi_i = \frac{m_i \omega^2}{2T_i}\right) \tag{12.271}$$

We integrate (12.271) along the real k_x-axis without resorting to contour integration. The following change of variable is very convenient to carry out the asymptotic evaluation of the integral in (12.271):

$$y = \frac{k_x}{\sqrt{k_x^2+\gamma^2}}, \quad \text{or } k_x = \frac{\gamma y}{\sqrt{1-y^2}} \tag{12.272}$$

Then, integral I_i becomes

$$I_i = \frac{2}{\gamma^3} \int_0^1 dy \frac{(1-y^2)\exp\left[-\frac{\xi_i(1-y^2)}{y^2(\gamma^2-k_z^2)+k_z^2}\right]}{\sqrt{y^2(\gamma^2-k_z^2)+k_z^2}} \tag{12.273}$$

Equation (12.273) is appropriate for asymptotic evaluation for the limiting cases.

• **Case of $\kappa \ll 1$.** In this case, we have $\gamma \ll k_z$. Neglecting γ^2 as compared to k_z^2, we obtain

$$I_i = \frac{2I_0}{k_z\gamma^3}\exp\left(-\frac{\xi_i}{k_z^2}\right), \quad I_0 = \int_0^1 \sqrt{(1-y^2)}\,dy = \frac{\pi}{4}$$

Here we have

$$\frac{\xi_i}{k_z^2} = \frac{m_i}{2T_i}\lambda_e^2\omega_{pi}^2 = \frac{T_e}{2T_i}$$

Thus

$$I_i = \frac{\pi}{2k_z\gamma^3}\exp\left(-\frac{T_e}{2T_i}\right) \tag{12.274}$$

The corresponding electron integral is calculated similarly by replacing $\xi_i \to \xi_e = \frac{m_e\omega^2}{2T_e}$ in (12.273). Here $\frac{\xi_e}{k_z^2} = \frac{m_e}{2m_i} \ll 1$ and the exponential term can be put to unity. We obtain

$$I_e = \frac{\pi}{2k_z\gamma^3} \tag{12.275}$$

Then (12.270), (12.274), (12.275), and the relations $q = -1/\kappa, \omega = \omega_{pi} k_z\lambda_e, \gamma = \kappa k_z$ give

$$D_i = -i\sqrt{\frac{\pi}{8}}\frac{1}{k_z^4\lambda_e^4}\left[\left(\frac{T_e}{T_i}\right)^{\frac{3}{2}} e^{-\frac{T_e}{2T_i}} + \sqrt{\frac{m_e}{m_i}}\right] \tag{12.276}$$

The Landau damping rate is obtained from the formula

$$\omega_i = -\frac{D_i(\omega)}{\partial D_r(\omega)/\partial\omega} \tag{12.277}$$

We calculate from (12.265):

$$\frac{\partial D_r}{\partial\omega} = \frac{\partial D_r}{\partial q}\frac{\partial q}{\partial\omega}$$

$$\frac{\partial D_r}{\partial q} = \frac{1}{q} - \frac{1}{2\kappa} \simeq -\frac{1}{2\kappa}, \quad \frac{\partial q}{\partial\omega} \simeq \frac{2}{\omega_{pi}k_z^3\lambda_e^3}$$

Using the above relations in (12.277) yield the Landau damping rate for $\kappa \ll 1$

$$\omega_i = -\omega_{pi}k_z\lambda_e\sqrt{\frac{\pi}{8}}\left[\left(\frac{T_e}{T_i}\right)^{\frac{3}{2}} e^{-\frac{T_e}{2T_i}} + \sqrt{\frac{m_e}{m_i}}\right] \tag{12.278}$$

In (12.278), the mass ratio term comes from the electron term in ε_i. It is interesting to see that surface ion acoustic wave Landau damping rate in (12.278) is equal to the corresponding Landau damping rate of the bulk ion acoustic wave.

12.7 Surface Electron Plasma Wave and Landau Damping

For the high frequency wave ($\omega \gg \mathbf{k} \cdot \mathbf{v}_e$), $\varepsilon_L = \varepsilon_r + i\varepsilon_i$ has the expression [consult (8.80)–(8.82)]

$$\varepsilon_L(\omega, k) = 1 - \frac{\omega_{\rm pe}^2}{\omega^2} - 3\,\frac{\omega_{\rm pe}^4}{\omega^4} k^2 \lambda_e^2 - i\pi\,\frac{\omega_{\rm pe}^2}{k^2}\left[\frac{\partial f_0}{\partial v}\right]_{v=\frac{\omega}{k}} \tag{12.279}$$

$$\varepsilon_r = 1 - \frac{\omega_{\rm pe}^2}{\omega^2} - 3\frac{\omega_{\rm pe}^4}{\omega^4} k^2 \lambda_e^2 \tag{12.280}$$

$$\varepsilon_i = \sqrt{\frac{\pi}{2}}\left(\frac{m_e}{T_e}\right)^{\frac{3}{2}} \frac{\omega\omega_{\rm pe}^2}{k^3} \exp\left[-\frac{m_e}{2T_e}\frac{\omega^2}{k^2}\right] \tag{12.281}$$

The real part of dispersion integral (12.263) can be evaluated as follows.

$$\int \frac{dk_x}{k^2\varepsilon_r} = \int \frac{dk_x}{k^2\left(1 - \frac{\omega_{\rm pe}^2}{\omega^2} - 3\frac{\omega_{\rm pe}^4}{\omega^4}k^2\lambda_e^2\right)} = \frac{-\omega^4}{3\omega_{\rm pe}^4\lambda_e^2}\int \frac{dk_x}{(k_z^2 + k_x^2)(k_x^2 + \gamma^2)}$$

$$\gamma = \sqrt{k_z^2 - \frac{\omega^4\left(1 - \frac{\omega_{\rm pe}^2}{\omega^2}\right)}{3\omega_{\rm pe}^4\lambda_e^2}} \tag{12.282}$$

Picking up the residues at the poles $k_x = ik_z$ and $k_x = i\gamma$, the integral has the value

$$\pi\,\frac{1/k_z - 1/\gamma}{1 - \omega_{\rm pe}^2/\omega^2}$$

Thus, the real part of the dispersion relation reads

$$D_r = 1 + \frac{1}{q(\omega)}\left(1 - \frac{k_z}{\gamma}\right), \quad q(\omega) = 1 - \frac{\omega_{\rm pe}^2}{\omega^2} \tag{12.283}$$

The real frequency is obtained from $D_r = 0$. We seek the solution with the condition $q < 0$, which insures that the quantity in the square root of (12.282) is positive. Accordingly, we have $\gamma > k_z$. The relation $D_r = 0$ gives after squaring

$$-1 + (2-x)^2 - \frac{(1-x)(2-x)^2}{3k_z^2\lambda_e^2\, x^2} = 0 \qquad \left(x = \frac{\omega_{\rm pe}^2}{\omega^2}\right)$$

It is seen that $x = 2$ is an approximate solution when $k_z^2\lambda_e^2 \ll 1$. Note that $x = 1$ or $q = 0$ (in this case $\gamma = k_z$) is not a root of $D_r = 0$. Let us look for a solution of the above equation in the vicinity of $x = 2$. So put $x = 2 + \delta$ ($\delta \ll 1$), and obtain δ in the region of $k_z\lambda_e \ll 1$. One obtains

$$\omega = \frac{\omega_{\rm pe}}{\sqrt{2}}\left(1 + \frac{\sqrt{3}}{2}\, k_z\lambda_e\right) \tag{12.284}$$

This frequency is called *Ritchie mode.* Using (12.284) in (12.282) gives

$$\frac{k_z}{\gamma} = 2\sqrt{3}\, k_z\lambda_e \ll 1 \tag{12.285}$$

Next we calculate the Landau damping corresponding to the mode (12.284). The imaginary part of the dispersion integral (12.263) upon using (12.280) and (12.281) takes the form

$$D_i = \frac{-i\, I}{9\sqrt{2\pi}}\sqrt{\frac{m_e}{T_e}}\,\frac{k_z}{\lambda_e^6}\,\frac{\omega^9}{\omega_{\rm pe}^8} \tag{12.286}$$

where

$$I = \int_{-\infty}^{\infty} dk_x\, \frac{\exp\,[-\frac{\alpha}{k_z^2+k_x^2}]}{(k_z^2+k_x^2)^{\frac{5}{2}}(k_x^2+\gamma^2)^2} \qquad \left(\alpha = \frac{m_e\omega^2}{2T_e}\right) \tag{12.287}$$

It is useful to make a change of variable

$$y = \frac{k_x}{\sqrt{k_z^2+k_x^2}} \quad \text{or} \quad k_x = \frac{k_z y}{\sqrt{1-y^2}}$$

Then the integral I in (12.287) becomes

$$I = \frac{2}{k_z^4}e^{-x^2}\int_0^1 dy(1-y^2)^3\frac{e^{x^2y^2}}{[k_z^2y^2+\gamma^2(1-y^2)]^2} \qquad \left(x^2 = \frac{\alpha}{k_z^2}\right) \tag{12.288}$$

In the region where $k_z \ll \gamma$, integral I takes the simple form

$$I = \frac{2}{k_z^4\gamma^4}e^{-x^2}J, \quad J = \int_0^1 dy(1-y^2)\, e^{x^2y^2} \tag{12.289}$$

where

$$x = \frac{\sqrt{\alpha}}{k_z} = \frac{1}{2}\frac{1}{k_z\lambda_e} \gg 1 \tag{12.290}$$

The first integral in J becomes via change of variable $y = t/x$

$$\int_0^1 dy\ e^{x^2y^2} = \frac{1}{x}\int_0^x e^{t^2}dt = \frac{F(x)}{x} \tag{12.291}$$

where the integral, $F(x)$, is the error function. The second integral in J can be written as

$$\begin{aligned}\int_0^1 dyy^2\ \exp(x^2y^2) &= \frac{1}{2x^2}\int_0^1 dyy\frac{\partial}{\partial y}\ \exp(x^2y^2)\\ &= \frac{1}{2x^2}\left[e^{x^2} - \frac{F(x)}{x}\right]\end{aligned} \tag{12.292}$$

where we integrated by parts. Hence, one obtains

$$J = \frac{1}{x}F(x)\left(1 + \frac{1}{2x^2}\right) - \frac{1}{2x^2}e^{x^2}$$

The error function $F(x)$ has asymptotic expansion for $x \gg 1$:

$$F(x) = \int_0^x e^{t^2}dt \simeq \frac{e^{x^2}}{2x}\left(1 + \frac{1}{2x^2} + \frac{3}{4x^4} + \cdots\right) \quad (x \gg 1)$$

Hence, we obtain

$$J = \frac{e^{x^2}}{2x^4}\left(1 + \frac{1}{x^2}\right) \tag{12.293}$$

Using (12.293) in (12.289) yields

$$I = \frac{1}{k_z^4\gamma^4x^4}\left(1 + \frac{1}{x^2}\right) = 9 \cdot 2^8 \cdot \lambda_e^8(1 + 4k_z^2\lambda_e^2)$$

where the last term will be neglected. Then (12.286) gives, using (12.284),

$$D_i = -i\frac{8}{\sqrt{\pi}}k_z\lambda_e \tag{12.294}$$

Landau damping rate will be obtained by the formula (12.277). Equation (12.283) gives

$$\frac{\partial D_r}{\partial\omega} = -\frac{1}{q^2}\frac{\partial q}{\partial\omega}\left(1 - \frac{k_z}{\gamma}\right) + \frac{k_z}{q}\frac{1}{\gamma^2}\frac{\partial\gamma}{\partial\omega} \tag{12.295}$$

We have $\frac{\partial q}{\partial \omega} \simeq \frac{4\sqrt{2}}{\omega_{\mathrm{pe}}}$, $q \simeq -1$, $\gamma \gg k_z$. The first term of (12.295) is $\simeq -\frac{4\sqrt{2}}{\omega_{\mathrm{pe}}}$, and the second term is $\simeq \frac{12\sqrt{2}}{\omega_{\mathrm{pe}}} k_z^2 \lambda_e^2$ which will be neglected. So we finally obtain for the Landau damping rate of the surface wave

$$\omega_i = -\sqrt{\frac{2}{\pi}} k_z \lambda_e \omega_{\mathrm{pe}} \tag{12.296}$$

Problems

12.1 In the coordinate system where the interface normal is x-direction and the surface wave propagates in the z-direction, the y-coordinate is ignorable. When the electric (magnetic) field of the electromagnetic surface wave is parallel to the y-direction the polarization is called TE (TM) wave. In the book by L. D. Landau and E. M. Lifshitz, *Electrodynamics of continuous media*, Pergamon, (1960), p. 288, TE mode is E-wave and TM mode is H-wave [In plasma, the magnetic permeability $\mu = 1$, and $H = B$]. Derive the dispersion relation of TM mode surface wave in *isotropic medium.*

Solution. The Maxwell equations can be written as

$$\nabla \times \mathbf{E} = \frac{i\omega}{c} \mathbf{B}$$

$$\nabla \times \mathbf{B} = -\frac{i\omega}{c} \mathbf{D} = -\frac{i\omega}{c} \varepsilon \mathbf{E}$$

where $D_i = \varepsilon_{ij} E_j = \varepsilon \delta_{ij} E_j = \varepsilon E_i$ with ε dielectric constant, in isotropic medium. The above two equations yield

$$\nabla^2 \mathbf{E} + \frac{\omega^2}{c^2} \varepsilon \mathbf{E} - \nabla \nabla \cdot \mathbf{E} = 0 \tag{12.297}$$

$$\nabla^2 \mathbf{B} + \frac{\omega^2}{c^2} \varepsilon \mathbf{B} = 0 \tag{12.298}$$

Here we assume that the interface $x = 0$ separates two media of different dielectric constant, $\varepsilon = \varepsilon_1$ $(x > 0)$, and $\varepsilon = \varepsilon_2$ $(x < 0)$ and focus on the TM-wave. Putting $\mathbf{B} = \hat{\mathbf{y}} B_y(x) e^{ikz - i\omega t}$ in (12.298) gives

$$\left(\frac{\partial^2}{\partial x^2} - k^2 \right) B_y + \frac{\omega^2}{c^2} \varepsilon_1 B_y = 0$$

which is solved by (dropping the phasor)

$$B_y = A_1 e^{-\alpha_1 x}$$

$$E_z = \frac{ic}{\omega\varepsilon_1}\frac{\partial B_y}{\partial x} = -A_1\frac{ic\alpha_1}{\omega\varepsilon_1}e^{-\alpha_1 x} \quad (x > 0)$$

In the medium 2, $(x < 0)$, one obtains

$$B_y = A_2 e^{\alpha_2 x}$$

$$E_z = A_2\frac{ic\alpha_2}{\omega\varepsilon_2}e^{\alpha_2 x}$$

$$\alpha_{1,2} = \sqrt{k^2 - \frac{\omega^2}{c^2}\varepsilon_{1,2}} \tag{12.299}$$

The matching condition, $[B_y] = 0$ yields $A_1 = A_2$. $[E_z] = 0$ gives

$$-\frac{\alpha_1}{\varepsilon_1} = \frac{\alpha_2}{\varepsilon_2} \tag{12.300}$$

Note that a plasma of stationary ions has $\varepsilon = 1-\frac{\omega_{\mathrm{pe}}^2}{\omega^2}$. If medium 2 (or 1) is vacuum, (12.300) agrees with (12.88). In order to satisfy (12.300), ε_1 and ε_2 should have opposite signs: $\varepsilon_1\varepsilon_2 < 0$. Then (12.300) can be satisfied if

$$\varepsilon_1 < |\varepsilon_2|$$

The dispersion relation (12.300) can be written as

$$c^2k^2 = \omega^2\varepsilon_1|\varepsilon_2|(|\varepsilon_2| - \varepsilon_1) \tag{12.301}$$

12.2 Verify that TE mode surface wave does not exist in isotropic plasmas.

References

Barr, H. C. and T. J. M. Boyd, *Surface waves in hot plasmas*, J. Phys. A: Gen. Phys. **5**, (1972), pp. 1108–1118.

Buti, B. *Nonlinear Surface Phenomena in Plasmas*, Advances in space plasma physics, World Scientific Publishing Co. Pte. Ltd: Singapore, (1985), p. 167.

Gradov, O. M. and L. Stenflo, *Linear theory of a cold bounded plasma* Phys. Rep. **94** (1983), pp. 111–137.

Kaw, P. K. and J. B. McBride, *Surface waves on a plasma,* Phys. Fluids **13** (1970), pp. 1784–1793.

Lee, H. J. and S. H. Cho, *Surface waves in a streaming plasma propagating in a duct: Kinetic theory,* J. Plasma Phys. **88** (2022), pp. 1–14.

Lee, H. J. *Electrostatic surface waves in a magnetized plasma propagating in channels*, J. Plasma Phys. **90**, Online on 16 December 2024, 905900613.

Lee, H. J. and M. Lee, *Surface wave echo in a semi-bounded plasma*, J. Modern Phys. **7** (2016), pp. 1400–1412.

Lee, H. J. and Y. K. Lim, *Surface waves in a relativistic plasma stream propagating in a duct: Kinetic theory*, J. Modern Phys. **13** (2022), pp. 1065–1079.

McBride, J. B. and P. K. Kaw, *Low frequency surface waves on a warm plasma*, Phys. Lett. **33A** (1970), pp. 72–73.

Chapter 13

Surface Wave Echo

Plasma echoes have long been known theoretically as well as experimentally. The plasma echoes are due to the quadratic term in the Vlasov equation, thus a nonlinear phenomenon. The perturbed distribution function of plasma particles $f(k, \omega, v)$ in Fourier space has a singularity at $\omega = kv$, in addition to other singularities such as the zeros of the dielectric permittivity. In its inversion to (x, t) variables, this singularity modulates the distribution function with the exponential phase $e^{ik(x-vt)}$ or $e^{-i\omega(t-x/v)}$. This modulation occurs because of the *free streaming term* in the Vlasov equation, which is so called since $x = vt$ is the characteristic line of a free particle. This term makes the modulation of the distribution function more and more oscillatory as t or x increases, and consequently, $\int f dv$ will become vanishingly small due to almost complete cancellations in the integral (phase-mixing). Therefore, the free streaming term yields no appreciable effect on the macroscopic variables such as density perturbation. However, if two free streaming terms are multiplied in nonlinear mixing of the second order electric field, it is evident from the expression $e^{ik_1(x_1-vt_1)}e^{ik_2(x_2-vt_2)}$ that a constructive interference can result in at a certain time (temporal echo) or at a certain spot (spatial echo) such that $k_1x_1 + k_2x_2 = v(k_1t_1 + k_2t_2)$. In this case, the phase of the exponential of the second order distribution function vanishes, and thus the corresponding velocity integral does not phase-mix, resulting in an echo.

13.1 Echos in Infinite Plasma

For simplicity, we consider echoes in a plasma with ions stationary and forming the uniform background. We also specialize on one-dimensional electrostatic perturbations. Suppose that two pulses are externally applied at time $t = \tau_1$ and $t = \tau_2$ to perturb the plasma:

$$\phi^{\text{ext}}(x, t) = \Phi_1\, e^{ik_1x}\delta[\omega_p(t - \tau_1)] + \Phi_2\, e^{ik_2x}\delta[\omega_p(t - \tau_2)] \tag{13.1}$$

In the electrostatic approximation the Vlasov equation reads

$$\frac{\partial}{\partial t}f(x,v,t)+v\frac{\partial f}{\partial x}+\frac{e}{m}\frac{\partial\phi}{\partial x}\frac{\partial f}{\partial v}=0 \tag{13.2}$$

where ϕ is the total electric potential, the sum of the external and plasma fields: $\phi=\phi^{\text{ext}}+\phi^{pla}$. We have the Poisson equation which connects the electron distribution and the plasma electric field:

$$\frac{\partial^2}{\partial x^2}(\phi-\phi^{\text{ext}})=4\pi e\int f\ d^3v \tag{13.3}$$

We attempt to solve the simultaneous equations (13.2) and (13.3) for a given ϕ^{ext} as prescribed by (13.1). In mathematical terms, we have an inhomogeneous system, driven by the source term (13.1); the response f and ϕ should be determined by ϕ^{ext}. We solve the foregoing equations by Fourier–Laplace transform and its inversion which are defined by

$$f(k,v,\omega)=\int_{-\infty}^{\infty}dx\int_0^{\infty}dt\ e^{-ikx+i\omega t}f(x,v,t)$$

$$f(x,v,t)=\frac{1}{4\pi^2}\int_{-\infty}^{\infty}dk\int_{-\infty+ic}^{\infty+ic}d\omega\ e^{ikx-i\omega t}f(k,v,\omega)$$

where c is large enough so that it runs above all the singularities of f in the complex ω-plane, and the contour is completed by winding the lower-half ω-plane. Then (13.3) is transformed to

$$k^2\phi(k,\omega)=-4\pi e\int dvf(k,v,\omega)+k^2\phi^{\text{ext}}(k,\omega) \tag{13.4}$$

The Fourier–Laplace transform of (13.2) goes as follows. First, multiply (13.2) by e^{-ikx} and integrate with respect to x from $-\infty$ to ∞. The Fourier transform of a quadratic term is a convolution integral:

$$\begin{aligned}\int_{-\infty}^{\infty}e^{-ikx}f(x)\frac{\partial\phi}{\partial x}dx&=\frac{1}{2\pi}\int dk'f(k')\ i\ (k-k')\phi(k-k')\\&=\frac{1}{2\pi}\int dk'f(k-k')\ i\ k'\phi(k')\end{aligned} \tag{13.5}$$

Likewise, we have for Laplace transform:

$$\text{Laplace Tr. }[A(t)B(t)]=\int_L\frac{d\omega'}{2\pi}B(\omega')A(\omega-\omega')$$

where the subscript L means the Laplace contour.

In (13.2), we put f as the sum of the equilibrium distribution $f_0(v)$ and the perturbation f' (the prime on the f' will be omitted in the following). Then the Fourier–Laplace transform of (13.2) reads

$$-i(\omega - kv)f(k,v,\omega) + \frac{e}{m}\, ik\, \phi(k,\omega)\,\frac{\partial f_0}{\partial v} + \frac{i}{4\pi^2}\frac{e}{m}\int_{-\infty}^{\infty} dk'(k-k')$$

$$\times \int_{ic-\infty}^{ic+\infty} d\omega'\, \frac{\partial}{\partial v} f(k',\omega',v)\phi(k-k',\omega-\omega') = 0 \tag{13.6}$$

where we put $f(k,v,t=0) = 0$ and c is above all the singularities of $f(k',\omega',v)$ on the complex ω'-plane. The simultaneous equations (13.4) and (13.7) will be solved by successive approximations. We treat the external potential in (13.4) as a first order quantity. The first order solutions will be used to iterate for the second order quadratic term. We have from (13.4)

$$k^2\phi^{(1)}(k,\omega) = -4\pi e \int dv f^{(1)}(k,v,\omega) + k^2\phi^{\text{ext}}(k,\omega) \tag{13.7}$$

where

$$\phi^{\text{ext}}(k,\omega) = \frac{2\pi}{\omega_p}[\Phi_1\delta(k-k_1)\, e^{i\omega\tau_1} + \Phi_2\delta(k-k_2)\, e^{i\omega\tau_2}] \tag{13.8}$$

The first order equation of (13.6) is

$$-i(\omega - kv)f^{(1)}(k,v,\omega) + \frac{e}{m}\, ik\, \phi^{(1)}(k,\omega)\,\frac{df_0}{dv} = 0 \tag{13.9}$$

Hence

$$f^{(1)}(k,v,\omega) = \frac{\frac{e}{m}k\phi^{(1)}(k,\omega)f_0'(v)}{\omega - kv} \quad \left(f_0' = \frac{df_o}{dv}\right) \tag{13.10}$$

Using (13.10) in (13.7) gives

$$\phi^{(1)}(k,\omega) = \frac{\phi^{\text{ext}}(k,\omega)}{\varepsilon(k,\omega)} \tag{13.11}$$

where ε is the dielectric permittivity

$$\varepsilon(k,\omega) = 1 + \frac{\omega_{\text{pe}}^2}{k}\int dv \frac{f_0'(v)}{\omega - kv} \tag{13.12}$$

Using (13.11) in (13.10) yields

$$f^{(1)}(k,v,\omega) = \frac{e}{m}k\,\frac{\phi^{\text{ext}}(k,\omega)}{\varepsilon(k,\omega)}\frac{f_0'(v)}{\omega - kv} \tag{13.13}$$

Equations (13.11) and (13.13) are the first order solutions which are to be substituted into the second order equations.

The second order Poisson equation reads

$$k^2\phi^{(2)}(k,\omega) = -4\pi e \int dv f^{(2)}(k,v,\omega) \tag{13.14}$$

The second order equation from (13.6) is

$$-i(\omega - kv)f^{(2)}(k,v,\omega) + \frac{e}{m}\, ik\, \phi^{(2)}(k,\omega)\, \frac{df_0}{dv} + \frac{i}{4\pi^2}\frac{e}{m}\int_{-\infty}^{\infty} dk'(k-k') \times \int_{ic-\infty}^{ic+\infty} d\omega'\, \frac{\partial}{\partial v} f^{(1)}(k',\omega',v)\phi^{(1)}(k-k',\omega-\omega') = 0 \tag{13.15}$$

The above two equations yield

$$\phi^{(2)}(k,\omega) = \frac{4\pi i e}{k^2\varepsilon(k,\omega)}\int \frac{dv}{\omega - kv}\left[\frac{i}{4\pi^2}\frac{e}{m}\int_{-\infty}^{\infty} dk'(k-k') \times \int_{ic-\infty}^{ic+\infty} d\omega'\, \frac{\partial}{\partial v} f^{(1)}(k',\omega',v)\phi^{(1)}(k-k',\omega-\omega')\right] \tag{13.16}$$

The second order solution $\phi^{(2)}$ gives the responses at various k values through the combinations of k_1 and k_2 because of the δ functions in (13.8). We are interested in the response at the beat terms. From (13.8) we have

$$\phi^{\text{ext}}(k',\omega')\, \phi^{\text{ext}}(k-k',\omega-\omega') = \left(\frac{2\pi}{\omega_p}\right)^2 \Phi_1\Phi_2\delta(k-k_1-k_2)[\delta(k'-k_1)\exp(i\omega'\tau_1 + i(\omega-\omega')\tau_2) + 1 \leftrightarrow 2] \tag{13.17}$$

where $1 \leftrightarrow 2$ signifies that the subscripts 1 and 2 are interchanged in the preceding term.

Therefore we obtain from (13.16)

$$\phi^{(2)}(k,\omega) = -\frac{e}{m}\,\frac{\Phi_1\Phi_2}{k^2\varepsilon(k,\omega)}\,\delta(k-k_1-k_2)\int_{-\infty}^{\infty}\frac{dv}{\omega - kv}\int_{-\infty}^{\infty} dk'\; k'\; (k-k') \times \int_{ic-\infty}^{ic+\infty}\frac{d\omega'}{\varepsilon(k',\omega')\varepsilon(k-k',\omega-\omega')}\frac{\partial}{\partial v}\,\frac{f_0'(v)}{\omega' - k'v} \times \left[\delta(k'-k_1)\;\exp(i\omega'\tau_1 + i(\omega-\omega')\tau_2) + 1 \leftrightarrow 2\right] \tag{13.18}$$

Integrating by parts with respect to v yields

$$\phi^{(2)}(k,\omega) = \frac{e}{m}\,\frac{\Phi_1\Phi_2}{k\varepsilon(k,\omega)}\,\delta(k-k_1-k_2)\int_{-\infty}^{\infty} dv\,\frac{f_0'(v)}{(\omega-kv)^2}\int_{-\infty}^{\infty} dk'\,\frac{k'(k-k')}{\omega'-k'v}$$
$$\times\int_{ic-\infty}^{ic+\infty}\frac{d\omega'}{\varepsilon(k',\omega')\varepsilon(k-k',\omega-\omega')}$$
$$\times\left[\delta(k'-k_1)\,\exp(i\omega'\tau_1+i(\omega-\omega')\tau_2)+1\leftrightarrow 2\right] \tag{13.19}$$

Fourier inversion with respect to k gives

$$\phi^{(2)}(x,\omega) = \frac{e\Phi_1\Phi_2}{2\pi m}\,\frac{e^{ik_3x}}{k_3\varepsilon(k_3,\omega)}\int_{-\infty}^{\infty} dv\,\frac{f_0'(v)}{(\omega-k_3v)^2}$$
$$\times\int_{-\infty}^{\infty} dk'\,\frac{k'(k_3-k')}{\omega'-k'v}\int_{ic-\infty}^{ic+\infty}\frac{d\omega'}{\varepsilon(k',\omega')\varepsilon(k_3-k',\omega-\omega')}$$
$$\times\left[\delta(k'-k_1)\,\exp\Big(i\omega'\tau_1+i(\omega-\omega')\tau_2\Big)+1\leftrightarrow 2\right] \tag{13.20}$$

where

$$k_3 = k_1+k_2.$$

Carrying out the $\int dk'$ integral yields

$$\phi^{(2)}(x,\omega) = \frac{e\Phi_1\Phi_2}{2\pi m}\,\frac{k_1k_2e^{ik_3x}}{k_3\varepsilon(k_3,\omega)}\int_{-\infty}^{\infty} dv\,\frac{f_0'(v)}{(\omega-k_3v)^2}$$
$$\times\left[\int_{ic-\infty}^{ic+\infty}\frac{d\omega'}{(\omega'-k_1v)\varepsilon(k_1,\omega')\varepsilon(k_2,\omega-\omega')}\,\exp(i\omega'\tau_1+i(\omega-\omega')\tau_2)+1\leftrightarrow 2\right] \tag{13.21}$$

We need Laplace inversion by performing $\int_{ib-\infty}^{ib+\infty}\frac{d\omega}{2\pi}\,e^{-i\omega t}(\cdots)$ on the above equation:

$$\phi^{(2)}(x,t) = \frac{e\Phi_1\Phi_2k_1k_2}{4\pi^2mk_3}\,e^{ik_3x}\int_{-\infty}^{\infty} dv\,f_0'(v)\int_{ib-\infty}^{ib+\infty}\frac{d\omega\,e^{-i\omega t}}{\varepsilon(k_3,\omega)(\omega-k_3v)^2}$$
$$\times\left[\int_{ic-\infty}^{ic+\infty}\frac{d\omega'}{(\omega'-k_1v)\varepsilon(k_1,\omega')\varepsilon(k_2,\omega-\omega')}\right.$$
$$\left.\times\exp(i\omega'\tau_1+i(\omega-\omega')\tau_2)+1\leftrightarrow 2\right] \tag{13.22}$$

where b is above all the singularities of the integrand $(0<c<b)$.

If $(\tau_1 - \tau_2)$ is long compared with Landau damping time (inverse of Landau damping rate), poles at the roots of the dielectric functions contribute negligibly to the integral, and we neglect them. Picking up the poles at $\omega' = -k_1 v$ and $\omega' = k_2 v$, $\int d\omega'$-integral can be done:

$$\phi^{(2)}(x,t) = \frac{i\, e\Phi_1\Phi_2 k_1 k_2}{2\pi m k_3}\, e^{ik_3 x} \int_{-\infty}^{\infty} dv\, f_0'(v) \int_{ib-\infty}^{ib+\infty} \frac{d\omega\; e^{-i\omega t}}{\varepsilon(k_3,\omega)(\omega - k_3 v)^2}$$
$$\times \left[\frac{e^{i\omega\tau_2}\; e^{ik_1 v(\tau_1-\tau_2)}}{\varepsilon(k_1,k_1 v)\varepsilon(k_2,\omega - k_1 v)} + \frac{e^{i\omega\tau_1}\; e^{ik_2 v(\tau_2-\tau_1)}}{\varepsilon(k_2,k_2 v)\varepsilon(k_1,\omega - k_2 v)} \right] \tag{13.23}$$

In $\int d\omega$-integral, $\omega = k_3 v$ is a double pole, and the residue is calculated by the formula

$$\text{Res} = \frac{d}{d\omega}\left(\frac{e^{-i\omega t}}{\varepsilon(k_3,\omega)}[\cdots] \right)_{\omega=k_3 v} \approx -i\, t\, \left(\frac{e^{-i\omega t}}{\varepsilon(k_3,\omega)}[\cdots] \right)_{\omega=k_3 v}$$

where we retained only the $\sim t$ term since we look for the time asymptotic response. Then, we obtain

$$\phi^{(2)}(x,t) = \frac{t\, e\Phi_1\Phi_2 k_1 k_2}{2\pi m k_3}\, e^{ik_3 x} \int_{-\infty}^{\infty} dv\; f_0'(v)$$
$$\times \left[\frac{e^{-ik_3 vt} e^{ik_3 v\tau_2}\; e^{ik_1 v(\tau_1-\tau_2)}}{\varepsilon(k_1,k_1 v)\varepsilon(k_2,k_2 v)\varepsilon(k_3,k_3 v)} + \frac{e^{-ik_3 vt} e^{ik_3 v\tau_1}\; e^{ik_2 v(\tau_2-\tau_1)}}{\varepsilon(k_1,k_1 v)\varepsilon(k_2,k_2 v)\varepsilon(k_3,k_3 v)} \right] \tag{13.24}$$

The above velocity integral does not phase-mix to zero, but results in echo when the exponential phases add to zero, or

$$t = \frac{k_1\tau_1 + k_2\tau_2}{k_1 + k_2} \tag{13.25}$$

This is the echo time for the *temporal echo* that we are looking for.

Reference: R. W. Gould, T. M. O'Neil, and J. H. Malmberg, *Plasma wave echo*, Phys. Rev. Lett. **19**, (1967), pp. 219–222.

- **Spatial echo.** Suppose that two oscillating charges are externally introduced in two locations $x = L_1$ and $x = L_2$:

$$\rho_0(x,t) = \rho_1 e^{-i\omega_1 t}\delta[k_0(x - L_1)] + \rho_2 e^{-i\omega_2 t}\; \delta[k_0(x - L_2)] \tag{13.26}$$

We have the nonlinear Vlasov equation and the Poisson equation to describe the electrostatic perturbation:

$$\frac{\partial}{\partial t} f(v,x,t) + v\frac{\partial f}{\partial x} - \frac{e}{m}E(x,t)\,\frac{\partial f}{\partial v} = 0 \tag{13.27}$$

$$\frac{\partial E}{\partial x} = 4\pi\left(-e\int dvf + \rho_0(x,t)\right) \tag{13.28}$$

Let us Fourier transform the above equations with respect to t and x to write

$$\rho_0(k,\omega) = \frac{2\pi}{k_0}\left[\rho_1\delta(\omega-\omega_1)\,e^{-ikL_1} + \rho_2\delta(\omega-\omega_2)\,e^{-ikL_2}\right] \tag{13.29}$$

$$\begin{aligned}&-i(\omega-kv)f(v,k,\omega) - \frac{e}{m}\,E(k,\omega)\,\frac{\partial f_0}{\partial v}\\ &-\frac{e}{4\pi^2 m}\int_{-\infty}^{\infty}dk'\int_{-\infty}^{\infty}d\omega' E(k-k',\omega-\omega')\,\frac{\partial}{\partial v}f(k',\omega',v) = 0\end{aligned} \tag{13.30}$$

$$ikE(k,\omega) = 4\pi\left[-e\int dvf(k,\omega,v) + \rho_0(k,\omega)\right] \tag{13.31}$$

The first order equation of (13.30) is

$$-i(\omega-kv)f^{(1)}(k,v,\omega) - \frac{e}{m}\,E^{(1)}(k,\omega)\,\frac{df_0}{dv} = 0 \tag{13.32}$$

Hence

$$f^{(1)}(k,v,\omega) = \frac{i\,\frac{e}{m}E^{(1)}(k,\omega)f_0'(v)}{\omega-kv} \qquad \left(f_0' = \frac{df_o}{dv}\right) \tag{13.33}$$

Using (13.33) in (13.31) gives

$$\begin{aligned}E^{(1)}(k,\omega) &= -4\pi i\,\frac{\rho_0(k,\omega)}{k\varepsilon(k,\omega)}\\ &= \frac{-8\pi^2 i}{kk_0\varepsilon(k,\omega)}\left[\rho_1\delta(\omega-\omega_1)\,e^{-ikL_1} + \rho_2\delta(\omega-\omega_2)\,e^{-ikL_2}\right]\end{aligned} \tag{13.34}$$

where ε is the dielectric permittivity as given by (13.12). Using (13.34) in (13.33) yields

$$f^{(1)}(k,v,\omega) = \frac{4\pi e\;\rho_0(k,\omega)}{mk\;\varepsilon(k,\omega)}\,\frac{f_0'(v)}{\omega-kv} \tag{13.35}$$

Equations (13.34) and (13.35) are the first order solutions which are to be substituted into the second order equations. The second order equation can be developed analogously with the last Section, yielding

$$E^{(2)}(k,\omega) = \frac{-4\pi e^2}{mk\varepsilon(k,\omega)} \int_{-\infty}^{\infty} \frac{dv}{\omega - kv} \int_{-\infty}^{\infty} dk' \int_{-\infty}^{\infty} d\omega' E^{(1)}(k-k',\omega-\omega') \frac{\partial}{\partial v} f^{(1)}(k',v,\omega') \tag{13.36}$$

At the beat frequencies, the product of the external charges gives

$$\begin{aligned} &\rho_0(k',\omega')\rho_0(k-k',\omega-\omega') \\ &= \frac{4\pi^2}{k_0^2}\rho_1\rho_2[\delta(\omega'-\omega_1)\delta(\omega-\omega'-\omega_2)\exp(-ikL_2+ik'(L_2-L_1)) + \ 1 \leftrightarrow 2] \end{aligned} \tag{13.37}$$

Using (13.34), (13.35), and (13.37) in (13.36) gives

$$\begin{aligned} E^{(2)}(k,\omega) &= i\pi(4\pi)^4 \frac{e^3\rho_1\rho_2}{m^2k_0^2\varepsilon(k,\omega)} \int_{-\infty}^{\infty} \frac{dv f_0'(v)}{(\omega-kv)^2} \\ &\times \int_{-\infty}^{\infty} dk' \int_{-\infty}^{\infty} d\omega' \frac{\delta(\omega'-\omega_1)\delta(\omega-\omega'-\omega_2)\ e^{-ikL_2+ik'(L_2-L_1)} + 1 \leftrightarrow 2}{k'(k-k')\varepsilon(k',\omega')\varepsilon(k-k',\omega-\omega')(\omega'-k'v)} \end{aligned} \tag{13.38}$$

After $\int d\omega'$ integral, we have

$$\begin{aligned} E^{(2)}(k,\omega) &= i\pi(4\pi)^4 \frac{e^3\rho_1\rho_2}{m^2k_0^2\varepsilon(k,\omega)} \int_{-\infty}^{\infty} \frac{dv f_0'(v)}{(\omega-kv)^2} \\ &\times \int_{-\infty}^{\infty} dk' \left[\frac{\delta(\omega-\omega_1-\omega_2)\ e^{-ikL_2+ik'(L_2-L_1)}}{k'(k'-k)\varepsilon(k',\omega_1)\varepsilon(k-k',\omega_2)(-\omega_1+k'v)} + 1 \leftrightarrow 2 \right] \end{aligned} \tag{13.39}$$

In $\int dk'$-integral, poles at $k' = 0$ and $k' = k$ do not produce the echo resonance terms; poles at the roots of the dielectric functions contribute negligibly to the integral, long after the Landau damping time; poles at $k' = \omega_1/v$ and $k' = \omega_2/v$ produce the resonance terms. Therefore, picking up the residues at $k' = \omega_1/v$ and $k' = \omega_2/v$, $\int dk'$-integral can be done, giving

$$E^{(2)}(k,\omega) = i\pi(4\pi)^4 \, \frac{e^3\rho_1\rho_2\delta(\omega-\omega_1-\omega_2)}{m^2k_0^2\varepsilon(k,\omega)}$$
$$\times \int_{-\infty}^{\infty} \frac{dv f_0'(v)}{(\omega-kv)^2}\left[\frac{e^{-ikL_2+i\frac{\omega_1}{v}(L_2-L_1)}}{\omega_1\left(\frac{\omega_1}{v}-k\right)\varepsilon\left(\frac{\omega_1}{v},\omega_1\right)\varepsilon(k-\frac{\omega_1}{v},\omega_2)} + 1 \leftrightarrow 2\right] \tag{13.40}$$

$E^{(2)}(x,t)$ can be obtained by inverting the above equation. $\int d\omega$-integral can be done immediately due to the δ-function, giving

$$E^{(2)}(x,t) = (4\pi)^3 \frac{e^3\rho_1\rho_2}{m^2k_0^2}\, e^{-i\omega_3 t}\int_{-\infty}^{\infty} dv\,\frac{df_0}{dv}\int_{-\infty}^{\infty} dk\,\frac{e^{ikx}}{\varepsilon(k,\omega_3)(\omega_3-kv)^2}$$
$$\times\left[\frac{e^{-ikL_2+i\frac{\omega_1}{v}(L_2-L_1)}}{\omega_1\left(\frac{\omega_1}{v}-k\right)\varepsilon\left(\frac{\omega_1}{v},\omega_1\right)\varepsilon(k-\frac{\omega_1}{v},\omega_2)} + \frac{e^{-ikL_1+i\frac{\omega_2}{v}(L_1-L_2)}}{\omega_2\left(\frac{\omega_2}{v}-k\right)\varepsilon\left(\frac{\omega_2}{v},\omega_2\right)\varepsilon(k-\frac{\omega_2}{v},\omega_1)}\right] \tag{13.41}$$

where $\omega_3 = \omega_1+\omega_2$, and $1 \leftrightarrow 2$ is written explicitly in the last term. In the above $\int dk$-integral, the non-phase-mixing contribution comes from the double pole at $k=\omega_3/v$. The residue is calculated by

$$\text{Res} = \frac{d}{dk}\left(\frac{e^{ik(x-L_2)}}{\varepsilon(k,\omega_3)}\Big[\cdots\Big]\right)_{k=\omega_3/v}$$

Here we differentiate only the exponential factor since it gives the asymptotically dominant term. Hence

$$\text{Res} \simeq i(x-L_2)\left(\frac{e^{ik(x-L_2)}}{\varepsilon(k,\omega_3)}\Big[\cdots\Big]\right)_{k=\omega_3/v}$$

The two terms in the large bracket in (13.41) are nicely combined upon putting $k=\omega_3/v$ to yield the final result

$$E^{(2)}(x,t) = 2\pi(4\pi)^3\frac{e^3\rho_1\rho_2}{m^2k_0^2}\frac{(2x-L_2-L_1)}{\omega_1\omega_2}$$
$$\times\, e^{-i\omega_3 t}\int_{-\infty}^{\infty}\frac{dv}{v}\frac{df_0}{dv}\frac{\exp(-i\frac{\omega_1L_1}{v}-i\frac{\omega_2L_2}{v}+i\frac{\omega_3x}{v})}{\varepsilon(\omega_1,\frac{\omega_1}{v})\varepsilon(\omega_2,\frac{\omega_2}{v})\varepsilon(\omega_3,\frac{\omega_3}{v})} \tag{13.42}$$

Echo is given rise to where the exponential argument becomes zero, thus the velocity integral is not phase-mixed:

$$x_{\text{echo}} = \frac{\omega_1 L_1 + \omega_2 L_2}{\omega_3} \tag{13.43}$$

$$E^{(2)}(x_{\text{echo}}, t) = 2\pi(4\pi)^3 \frac{e^3 \rho_1 \rho_2}{m^2 k_0^2}\, e^{-i\omega_3 t}\, \frac{(\omega_1 - \omega_2)(L_1 - L_2)}{\omega_1 \omega_2 \omega_3}$$

$$\times \int_{-\infty}^{\infty} \frac{dv}{v} \frac{\frac{df_0}{dv}}{\varepsilon(\omega_1, \frac{\omega_1}{v})\varepsilon(\omega_2, \frac{\omega_2}{v})\varepsilon(\omega_3, \frac{\omega_3}{v})} \tag{13.44}$$

Also note that there is a 'black spot' where $E^{(2)}(x,t) = 0$:

$$x = \frac{L_1 + L_2}{2} \tag{13.45}$$

13.2 Echoes in a Semi-Bounded Plasma by Two Stationary Oscillating Charges

We consider a plasma consisting of electrons and immobile ions, the latter forming the uniform background. The plasma occupies the half-space $x \geq 0$ interfaced with a vacuum region $x < 0$. In this section, we investigate echoes arising by two oscillating charges placed on two vertical locations to the interface, as given by (13.26):

$$\rho_0(x,t) = \rho_1 e^{-i\omega_1 t}\delta[k_0(x - L_1)] + \rho_2 e^{-i\omega_2 t}\delta[k_0(x - L_2)] \tag{13.46}$$

Then, the meaningful spatial coordinate is the vertical coordinate x, which is non-ignorable. The other two coordinates, y and z, are of infinite directions, and therefore can be treated as ignorable coordinates that have the translational invariance ($\partial/\partial y = \partial/\partial z = 0$) or as having the periodic boundary condition (in this case, $\sim e^{ik_z z}$, say). In this section, we ignore y and z coordinates. Then the electron distribution function $f(x,v,t)$ and the electric field $E(x,t)$ will depend on x-coordinate only. In this specification, we are restricting to the echoes that are not associated with surface waves. We are interested in the effects caused by the presence of the bounding surface $x = 0$ when the plasma is impressed by the external charges as prescribed in (13.46).

We have the nonlinear Vlasov equation and the Poisson equation to describe the electrostatic perturbation:

$$\frac{\partial}{\partial t} f(v, x, t) + v \frac{\partial f}{\partial x} - \frac{e}{m} E(x, t) \frac{\partial f}{\partial v} = 0 \tag{13.47}$$

$$\frac{\partial E}{\partial x} = 4\pi \left(-e \int dv f + \rho_0(x, t) \right) \tag{13.48}$$

In order to have the kinematic boundary condition $f(v, 0, t) = f(-v, 0, t)$ fulfilled, the electric field is oddly continued at $x = 0$:

$$E(-x) = -E(x) \tag{13.49}$$

In addition, we need the electric boundary condition: the electric displacement $D(x)$ is continuous across the interface. This can be seen from

$$\frac{\partial D(x, \omega)}{\partial x} = \frac{\partial}{\partial x} \left(E + \frac{4\pi i}{\omega} J \right) = 4\pi \rho_0(x, \omega)$$

Since ρ_0 does not contribute to the surface current, the infinitesimal integration gives

$$D|_{x=0^+} = D|_{x=0^-} : \ (\varepsilon E)_{0+} = E_0 \tag{13.50}$$

where E_0 is the vacuum electric field.

The governing equations (13.46)–(13.48) are identical with the foregoing equations (13.26)–(13.28) except that the electric field has a discontinuity at $x = 0$ due to the oddly continuation in (13.49). The discontinuity will give rise to a new term in the Fourier k space.

Let us Fourier transform the above equations with respect to t and x to write

$$\rho_0(k, \omega) = \frac{2\pi}{k_0} \left[\rho_1 \delta(\omega - \omega_1)\, e^{-ikL_1} + \rho_2 \delta(\omega - \omega_2)\, e^{-ikL_2} \right] \tag{13.51}$$

$$\begin{aligned} &-i(\omega - kv) f(v, k, \omega) - \frac{e}{m}\, E(k, \omega) \frac{\partial f_0}{\partial v} \\ &- \frac{e}{4\pi^2 m} \int_{-\infty}^{\infty} dk' \int_{-\infty}^{\infty} d\omega' E(k - k', \omega - \omega') \frac{\partial}{\partial v} f(k', \omega', v) = 0 \end{aligned} \tag{13.52}$$

where f is perturbation. The Poisson equation is transformed in the form

$$ikE(k, \omega) + N(\omega) = 4\pi \left[-e \int dv f(k, \omega, v) + \rho_0(k, \omega) \right] \tag{13.53}$$

where $N(\omega) = E^{(1)}(0^-, \omega) - E^{(1)}(0^+, \omega)$.

The first order equation of (13.52) gives

$$f^{(1)}(k,v,\omega) = \frac{i\,\frac{e}{m} E^{(1)}(k,\omega) f_0'(v)}{\omega - kv} \tag{13.54}$$

Using (13.54) in (13.53) gives

$$E^{(1)}(k,\omega) = \frac{i}{k\ \varepsilon(k,\omega)}(N - 4\pi\rho_0(k,\omega)) \tag{13.55}$$

Using (13.55) in (13.54) yields

$$f^{(1)}(k,v,\omega) = -\frac{e}{m}\frac{f_0'(v)}{k\ \varepsilon(k,\omega)(\omega - kv)}(N - 4\pi\rho_0) \tag{13.56}$$

The unknown constant N should be determined by satisfying the electric boundary condition at the interface $x = 0$ (13.50). Then the first order quantities $f^{(1)}$ and $E^{(1)}$ will be used for the second order quantities $f^{(2)}$ and $E^{(2)}$ by iteration.

• **Determination of** N**.** This can be done by connecting the plasma field with the vacuum field by the boundary condition. In the plasma side ($x > 0$), we have from (13.55), per $D = \varepsilon E$,

$$D^{(1)}(k,\omega) = \frac{i}{k}\Big(N(\omega) - 4\pi\rho_0(k,\omega)\Big) \tag{13.57}$$

Inverting (13.57) gives in the region $x > 0$

$$\begin{aligned} D^{(1)}(x,\omega) = -\frac{1}{2}\ NH(x) + \frac{4\pi^2}{k_0}[&\rho_1\delta(\omega-\omega_1)H(x-L_1) \\ &+ \rho_2\delta(\omega-\omega_2)H(x-L_2)] \end{aligned} \tag{13.58}$$

where we used the formula

$$\int_{-\infty}^{\infty} dk\ \frac{e^{ikx}}{k} = i\pi H(x)$$

where $H(x)$ is the step function: $H(x) = 1$ if $x > 0$ or -1 if $x < 0$. Therefore, we obtain

$$D^{(1)}(0^+,\omega) = -\frac{N}{2} - \frac{4\pi^2}{k_0}[\rho_1\delta(\omega-\omega_1) + \rho_2\delta(\omega-\omega_2)] \tag{13.59}$$

Setting this quantity to D_0, the vacuum side quantity, we have

$$N = -2D_0 - \frac{8\pi^2}{k_0}[\rho_1\delta(\omega-\omega_1) + \rho_2\delta(\omega-\omega_2)] \tag{13.60}$$

Using (13.60) in (13.55) gives

$$E^{(1)}(k,\omega) = \frac{-2i}{k\ \varepsilon(k,\omega)}\left[D_0 + \frac{4\pi^2}{k_0}\left(\rho_1\delta(\omega-\omega_1)(1+e^{-ikL_1}) + \rho_2\delta(\omega-\omega_2)(1+e^{-ikL_2})\right)\right] \tag{13.61}$$

In (13.61), we determined $E^{(1)}$ in terms of the vacuum field D_0 and the external charges. D_0 and '1' in the factors $(1+e^{-ikL_1})$ and $(1+e^{-ikL_2})$ arise due to the presence of the boundary surface $x = 0$, which will be absent from the infinite plasma. Check this equation against (13.34).

• **Second order solution and echo occurrence.** To investigate the echo occurrence condition, we can easily derive the second order electric field equation in the form

$$E^{(2)}(k,\omega) = \frac{ie^3}{m^2}\frac{1}{\varepsilon(k,\omega)}\int_{-\infty}^{\infty} dv \frac{\frac{df_0}{dv}}{(\omega-kv)^2}\int_{-\infty}^{\infty} d\omega' \int_{-\infty}^{\infty} dk' \ \frac{E^{(1)}(\omega',k',v)E^{(1)}(\omega-\omega',k-k')}{\omega'-k'v} \tag{13.62}$$

where (13.61) will be used. We write the product terms which survive the phase-mixing if a suitable condition is met:

$$\frac{E^{(1)}(\omega',k')E^{(1)}(\omega-\omega',k-k')}{\omega'-k'v}$$

$$= \frac{-\alpha}{k'(k-k')(\omega'-k'v)\varepsilon(k',\omega')\varepsilon(k-k',\omega-\omega')}$$

$$\times\Big[\rho_1\rho_2\delta(\omega'-\omega_2)\delta(\omega-\omega'-\omega_1)e^{-ik'L_2}e^{-i(k-k')L_1} \tag{13.63}$$

$$+\rho_1\rho_2\delta(\omega'-\omega_1)\delta(\omega-\omega'-\omega_2)e^{-ik'L_1}e^{-i(k-k')L_2} \tag{13.64}$$

$$+b\ \rho_1^2\delta(\omega'-\omega_1)\delta(\omega-\omega'-\omega_1)e^{-i(k-k')L_1} \tag{13.65}$$

$$+b\ \rho_2^2\delta(\omega'-\omega_2)\delta(\omega-\omega'-\omega_2)e^{-i(k-k')L_2} \tag{13.66}$$

$$+b\ \rho_1\rho_2\delta(\omega'-\omega_2)\delta(\omega-\omega'-\omega_1)e^{-i(k-k')L_1} \tag{13.67}$$

$$+b\ \rho_1\rho_2\delta(\omega'-\omega_1)\delta(\omega-\omega'-\omega_2)e^{-i(k-k')L_2}\Big] \tag{13.68}$$

where $\alpha = (8\pi^2/k_0)^2$. Contributions from the two terms in (13.63) and (13.64) yield the results obtained in (13.43) and (13.44)—for an infinite plasma. The terms multiplied by b are the boundary terms which are irrelevant to an infinite plasma. The factor 'b' will be put to 1 eventually. We remind that two algebraic approximations in the Fourier inversion are used: (i) pick up only the free streaming term singularity $\omega = kv$ (ii) in evaluating the double pole differentiate only the exponential function.

Repeating similar algebra performed in the foregoing analysis, we obtain the following echo locations and the corresponding electric fields. These echoes do not occur in an infinite plasma.

At $x_{\text{echo}} = \frac{L_1}{2}$:

$$E^{(2)}(x_{\text{echo}}, t) = -\alpha\rho_1^2 \frac{eL_1}{8\pi m} \frac{\omega_p^2}{\omega_1^2} e^{-2i\omega_1 t} \int_{-\infty}^{\infty} dv \frac{\frac{1}{v}\frac{df_0}{dv}}{\varepsilon\left(\frac{2\omega_1}{v}, 2\omega_1\right)\varepsilon\left(\frac{\omega_1}{v}, \omega_1\right), \varepsilon\left(\frac{\omega_1}{v}, \omega_1\right)}$$

At $x_{\text{echo}} = \frac{L_2}{2}$:

$$E^{(2)}(x_{\text{echo}}, t) = -\alpha\rho_2^2 \frac{eL_2}{8\pi m} \frac{\omega_p^2}{\omega_2^2} e^{-2i\omega_2 t} \int_{-\infty}^{\infty} dv \frac{\frac{1}{v}\frac{df_0}{dv}}{\varepsilon\left(\frac{2\omega_2}{v}, 2\omega_2\right)\varepsilon\left(\frac{\omega_2}{v}, \omega_2\right), \varepsilon\left(\frac{\omega_2}{v}, \omega_2\right)}$$

At $x_{\text{echo}} = \frac{L_1\omega_1}{\omega_3}$ $(\omega_3 = \omega_2 + \omega_1)$:

$$E^{(2)}(x_{\text{echo}}, t) = -\alpha\rho_1\rho_2 \frac{eL_1}{4\pi m} \frac{\omega_p^2}{\omega_1\omega_3} e^{-i\omega_3 t} \int_{-\infty}^{\infty} dv \frac{\frac{1}{v}\frac{df_0}{dv}}{\varepsilon\left(\frac{\omega_1}{v}, \omega_1\right)\varepsilon\left(\frac{\omega_2}{v}, \omega_2\right)\varepsilon\left(\frac{\omega_3}{v}, \omega_3\right)}$$

At $x_{\text{echo}} = \frac{L_2\omega_2}{\omega_3}$:

$$E^{(2)}(x_{\text{echo}}, t) = -\alpha\rho_1\rho_2 \frac{eL_2}{4\pi m} \frac{\omega_p^2}{\omega_2\omega_3} e^{-i\omega_3 t} \int_{-\infty}^{\infty} dv \frac{\frac{1}{v}\frac{df_0}{dv}}{\varepsilon\left(\frac{\omega_1}{v}, \omega_1\right)\varepsilon\left(\frac{\omega_2}{v}, \omega_2\right)\varepsilon\left(\frac{\omega_3}{v}, \omega_3\right)}$$

The above equations predict possible echo spots and the corresponding electric fields. We mention that *surface wave echoes* may be dealt with by including the z-coordinate in the governing equations.

Reference:

H. J. Lee and M. J. Lee, *Echo in a semi-bounded plasma*, The Open Plasma Phys. Journal **8** (Bentham), (2015), pp. 1–7.

H. J. Lee and M. J. Lee, *Surface wave echo in a semi-bounded plasma*, J. of Modern Phys. **7** (2016), pp. 1400–1412.

13.3 Surface Wave Echo in a Semi-Bounded Plasma

In Section 13.2, we considered the one-dimensional electric field given rise to by two oscillating charges in a semi-bounded plasma; the echoes reflected by the plasma-vacuum boundary. In this section, we investigate electrostatic *echoes associated with a surface wave propagating in a semi-infinite plasma.* In contrast with the last section, the z-direction is not ignorable; the electric field is two-dimensional. We start with the nonlinear Vlasov equation for electrons and the Poisson equation

$$\frac{\partial}{\partial t} f(\mathbf{v}, \mathbf{r}, t) + \mathbf{v} \cdot \frac{\partial f}{\partial \mathbf{r}} - \frac{e}{m} \mathbf{E}(\mathbf{r}, t) \cdot \frac{\partial f}{\partial \mathbf{v}} = 0 \tag{13.69}$$

$$\text{with } \mathbf{r} = \hat{\mathbf{x}} x + \hat{\mathbf{z}} z, \quad \mathbf{v} = \hat{\mathbf{x}} v_x + \hat{\mathbf{z}} v_z, \quad \mathbf{E} = \hat{\mathbf{x}} E_x + \hat{\mathbf{z}} E_z$$

$$\nabla \cdot \mathbf{E} = \frac{\partial E_x}{\partial x} + \frac{\partial E_z}{\partial z} = -4\pi e \int f \, d^3 v + 4\pi \rho_0(x, z, t) \tag{13.70}$$

where ρ_0 represents the external charges as given by (13.46).

Breaking down (13.69) and (13.70) order by order and Fourier transforming, we have

$$-i(\omega - \mathbf{k} \cdot \mathbf{v}) f^{(1)}(\mathbf{k}, \mathbf{v}, \omega) = \frac{e}{m} \, \mathbf{E}^{(1)}(\mathbf{k}, \omega) \cdot \frac{df_0}{d\mathbf{v}} \tag{13.71}$$

$$i\mathbf{k} \cdot \mathbf{E}^{(1)}(\mathbf{k}, \omega) = -4\pi e \int f^{(1)}(\mathbf{k}, \mathbf{v}, \omega) d^3 v + \rho_0(\mathbf{k}, \omega) + S(k_z, \omega) \tag{13.72}$$

where $S(k_z, \omega)$ is due to the jump in E_x at $x = 0$.

$$\begin{aligned} -i(\omega - \mathbf{k} \cdot \mathbf{v}) f^{(2)}(\mathbf{k}, \mathbf{v}, \omega) &= \frac{e}{m} \, \mathbf{E}^{(2)}(\mathbf{k}, \omega) \cdot \frac{df_0}{d\mathbf{v}} + \frac{e}{m} \int_{-\infty}^{\infty} \frac{d\omega'}{2\pi} \\ &\times \int_{-\infty}^{\infty} \frac{dk'_x}{2\pi} \int_{-\infty}^{\infty} \frac{dk'_z}{2\pi} \mathbf{E}^{(1)}(\mathbf{k} - \mathbf{k}', \omega - \omega') \cdot \frac{\partial}{\partial \mathbf{v}} f^{(1)}(\mathbf{k}', \omega', \mathbf{v}) \end{aligned} \tag{13.73}$$

$$i\,\mathbf{k} \cdot \mathbf{E}^{(2)}(\mathbf{k}, \omega) = -4\pi e \int f^{(2)}(\mathbf{k}, \mathbf{v}, \omega) d^3 v \tag{13.74}$$

Equations (13.71) and (13.72), and $\nabla \times \mathbf{E} = 0$ give

$$\mathbf{E}^{(1)}(\mathbf{k}, \omega) = \frac{i\,\mathbf{k}}{k^2 \, \varepsilon(\mathbf{k}, \omega)} (S(k_z, \omega) - 4\pi \rho_0(\mathbf{k}, \omega)) \tag{13.75}$$

where

$$\varepsilon(\mathbf{k}, \omega) = 1 + \frac{\omega_p^2}{k^2} \int d^3 v \, \frac{\mathbf{k} \cdot \frac{df_0}{d\mathbf{v}}}{\omega - \mathbf{k} \cdot \mathbf{v}} \tag{13.76}$$

is the dielectric function.

• **Determination of** $S(k_z, \omega)$**.** This can be done by connecting the plasma field with the vacuum field by the boundary conditions: $[\varepsilon E_x] = 0]$ and

$[E_z] = 0$. On plasma side,

$$\begin{aligned} D_x(0^+) &= \varepsilon(\mathbf{k},\omega)E_x^{(1)}(0^+, k_z, \omega) \\ &= \lim_{x\to 0}\int_{-\infty}^{\infty}\frac{dk_x}{2\pi}\, e^{ik_x x}\,\frac{ik_x}{k_x^2+k_z^2}(S(k_z,\omega) - 4\pi\rho_0(\mathbf{k},\omega)) \end{aligned} \tag{13.77}$$

Fourier transforming (13.46) gives

$$\rho_0(\mathbf{k},\omega) = \frac{4\pi^2}{k_0}\,(\rho_1\delta(\omega-\omega_1)\,\delta(k_z)\,e^{-ik_x L_1} + 1 \to 2) \tag{13.78}$$

Using (13.78) in (13.77) and carrying out the integral yield

$$D_x(0^+) = -\frac{1}{2}S(k_z,\omega) - \frac{8\pi^3}{k_0}\,\rho_1\,\delta(\omega-\omega_1)\,\delta(k_z) + 1 \to 2 \tag{13.79}$$

where we used

$$\int_{-\infty}^{\infty}\frac{k_x dk_x}{k_x^2+k_z^2} = i\pi \tag{13.80}$$

$$\int_{-\infty}^{\infty} dk\,\frac{e^{ikx}}{k} = i\pi H(x) \tag{13.81}$$

where $H(x)$ is the step function: $H(x) = 1$ for $x > 0$ and $H(x) = -1$ for $x < 0$.

We also need

$$E_z^{(1)}(0^+, k_z, \omega) = \lim_{x\to 0}\int_{-\infty}^{\infty}\frac{dk_x}{2\pi}e^{ik_x x}\frac{ik_z}{\varepsilon(\mathbf{k},\omega)(k_x^2+k_z^2)}(S(k_z,\omega) - 4\pi\rho_0(\mathbf{k},\omega)) \tag{13.82}$$

The second integral vanishes due to $\delta(k_z)$. Hence, we can write

$$E_z^{(1)}(0^+, k_z, \omega) = \frac{ik_z}{2\pi} S(k_z, \omega) I(k_z, \omega) \tag{13.83}$$

where

$$I(k_z, \omega) = \int_{-\infty}^{\infty} \frac{dk_x}{\varepsilon(\mathbf{k}, \omega)(k_x^2 + k_z^2)} \tag{13.84}$$

• **Vacuum solution.** In the vacuum, the potential ϕ ($\mathbf{E} = -\nabla\phi$) satisfies $\frac{\partial^2 \phi}{\partial x^2} + \frac{\partial^2 \phi}{\partial z^2} = 0$, whose solution appropriate for the surface wave can be assumed of the form

$$\phi(x, z) = \tilde{\phi}\, e^{ik_z z + k_z x} \tag{13.85}$$

Hence

$$E_x(k_z, x) = -k_z \tilde{\phi}\, e^{ik_z z + k_z x} \tag{13.86}$$

$$E_z(k_z, x) = -i\, k_z \tilde{\phi}\, e^{ik_z z + k_z x} \tag{13.87}$$

Equations (13.79), (13.83), (13.86), and (13.87) give the boundary equations

$$\frac{S}{2} + \frac{8\pi^3 \rho_1}{k_0} \delta(\omega - \omega_1)\delta(k_z) = \xi \tag{13.88}$$

$$\frac{k_z}{2\pi} SI = -\xi \tag{13.89}$$

where $\xi = k_z \tilde{\phi}\, e^{ik_z z}$. The above two equations determine S in the form

$$S = \frac{-16\pi^4}{k_0(\pi + k_z I(k_z, \omega))} (\rho_1 \delta(\omega - \omega_1)\delta(k_z) + 1 \to 2) \tag{13.90}$$

When the external charges are nonexistent, we obtain the electrostatic surface wave dispersion relation as

$$1 + \frac{k_z}{\pi} \int_{-\infty}^{\infty} \frac{dk_x}{\varepsilon(\mathbf{k}, \omega) k^2} = 0 \tag{13.91}$$

• Second order electric field is arranged in the form

$$E_i^{(2)}(\mathbf{k}, \omega) = -\frac{\omega_p^2 k_i}{8\pi^3 \varepsilon_L(\mathbf{k}, \omega) k^2} \int \frac{d^3 v}{\omega - \mathbf{k} \cdot \mathbf{v}} \int_{-\infty}^{\infty} d\omega' \int_{-\infty}^{\infty} dk_x'$$
$$\times \int_{-\infty}^{\infty} dk_z'\, \mathbf{E}^{(1)}(\mathbf{k} - \mathbf{k}', \omega - \omega') \cdot \frac{\partial f^{(1)}}{\partial \mathbf{v}}(\mathbf{k}', \omega', \mathbf{v}) \tag{13.92}$$

where $i = x, z$ and ε_L is the dielectric permittivity

$$\varepsilon_L = 1 + \frac{\omega_p^2}{k^2}\int d^3v\, \frac{\mathbf{k}\cdot\frac{\partial f_0}{\partial \mathbf{v}}}{\omega - \mathbf{k}\cdot\mathbf{v}}$$

After integrating by parts, (13.92) becomes

$$E_i^{(2)}(\mathbf{k},\omega) = \frac{\omega_p^2 k_i}{8\pi^3\varepsilon_L(\mathbf{k},\omega)k^2}\int \frac{d^3v}{(\omega-\mathbf{k}\cdot\mathbf{v})^2}\int_{-\infty}^{\infty} d\omega' \int_{-\infty}^{\infty} dk_x' \times \int_{-\infty}^{\infty} dk_z'\; \mathbf{k}\cdot\mathbf{E}^{(1)}(\mathbf{k}-\mathbf{k}',\omega-\omega')f^{(1)}(\mathbf{k}',\omega',\mathbf{v}) \quad (13.93)$$

which becomes upon using (13.71)

$$E_i^{(2)}(\mathbf{k},\omega) = \frac{ie\omega_p^2 k_i}{8\pi^3 m\; \varepsilon_L(\mathbf{k},\omega)k^2}\int d^3v \frac{\frac{\partial f_0}{\partial v_j}}{(\omega-\mathbf{k}\cdot\mathbf{v})^2}\int_{-\infty}^{\infty} d\omega' \int_{-\infty}^{\infty} dk_x' \times \int_{-\infty}^{\infty} dk_z'\; \frac{E_j^{(1)}(\mathbf{k}',\omega')\;\mathbf{k}\cdot\mathbf{E}^{(1)}(\mathbf{k}-\mathbf{k}',\omega-\omega')}{\omega'-\mathbf{k}'\cdot\mathbf{v}} \quad (13.94)$$

where the beat term reads

$$\frac{E_j^{(1)}(\mathbf{k}',\omega')\mathbf{k}\cdot\mathbf{E}^{(1)}(\mathbf{k}-\mathbf{k}',\omega-\omega')}{\omega'-\mathbf{k}'\cdot\mathbf{v}} = \frac{-k_j'(\mathbf{k}\cdot\mathbf{k}'')(S'-4\pi\rho_0')(S''-4\pi\rho_0'')}{\varepsilon'\varepsilon'' k'^2 k''^2(\omega'-\mathbf{k}'\cdot\mathbf{v})} \quad (13.95)$$

where $\mathbf{k}'' = \mathbf{k}-\mathbf{k}'$ and other notations have the obvious meaning, and

$$S(\mathbf{k},\omega) - 4\pi\rho_0(\mathbf{k},\omega) = -\frac{16\pi^3}{k_0}\left[\rho_1\delta(\omega-\omega_1)\delta(k_z)\left(\frac{\pi}{\pi+k_z I(k_z,\omega)} + e^{-ik_x L_1}\right) + 1 \to 2\right] \quad (13.96)$$

The product $(S'-4\pi\rho_0')(S''-4\pi\rho_0'')$ has 8 beat terms, half of them corresponds to $1 \leftrightarrow 2$. We shall consider the beat term

$$\rho_1\rho_2\delta(\omega'-\omega_1)\delta(\omega''-\omega_2)\delta(k_z')\delta(k_z'')\; e^{-ik_x'' L_2}\; \frac{\pi}{\pi+k_z' I'} + 1 \leftrightarrow 2 \quad (13.97)$$

Then after $\int dk_z'$-integral, we have

$$E_i^{(2)}(\mathbf{k},\omega) \sim \frac{k_i\;\delta(k_z)e^{-ik_x L_2}}{k^2\varepsilon(\mathbf{k},\omega)}\int d^3v\; \frac{\frac{\partial f_0}{\partial v_x}}{(\omega-\mathbf{k}\cdot\mathbf{v})^2}\int d\omega' \int dk_x' \times \frac{(k^2-k_x k_x')k_x'\delta(\omega'-\omega_1)\delta(\omega''-\omega_2)\; e^{-ik_x'' L_2}}{\varepsilon'(k_x',\omega')\varepsilon''(k_x-k_x',\omega-\omega')k_x'^2(k_x-k_x')^2(\omega'-k_x' v_x)} \quad (13.98)$$

where we suppressed the constant factor. Due to $\delta(k_z)$, $E_z^{(2)}(\mathbf{k},\omega)=0$. After $\int dk'_x$-integral, we get

$$E_x^{(2)}(\mathbf{k},\omega) \sim \frac{\delta(k_z)e^{-2ik_xL_2}}{\varepsilon(k_x,\omega)} \int d^3v\, \frac{\frac{\partial f_0}{\partial v_x}}{(\omega-k_xv_x)^2} \times \int d\omega' \frac{\delta(\omega'-\omega_1)\delta(\omega-\omega'-\omega_2)\; e^{i\frac{\omega'}{v_x}L_2}}{\omega'\varepsilon'\left(\frac{\omega'}{v_x},\omega'\right)\varepsilon''(k_x-\frac{\omega'}{v_x},\omega-\omega')(k_x-\frac{\omega'}{v_x})} \tag{13.99}$$

After $\int d\omega'$-integral, we have

$$E_x^{(2)}(\mathbf{k},\omega) \sim \frac{\delta(k_z)e^{-2ik_xL_2}}{\varepsilon(k_x,\omega)\omega_1} \times \int d^3v \frac{\frac{\partial f_0}{\partial v_x}}{(\omega-k_xv_x)^2} \frac{\delta(\omega-\omega_1-\omega_2)\; e^{i\frac{\omega_1}{v_x}L_2}}{\varepsilon'\left(\frac{\omega_1}{v_x},\omega_1\right)\varepsilon''(k_x-\frac{\omega_1}{v_x},\omega-\omega_1)(k_x-\frac{\omega_1}{v_x})} \tag{13.100}$$

In Fourier inverting (13.100), $\int dk_z \int d\omega$ can be done easily due to the delta functions, giving

$$E_x^{(2)}(k_x,t) \sim \frac{e^{-i\omega_3 t}e^{-2ik_xL_2}}{\varepsilon(k_x,\omega_3)\omega_1} \times \int d^3v \frac{\frac{\partial f_0}{\partial v_x}}{(\omega_3-k_xv_x)^2} \frac{e^{i\frac{\omega_1}{v_x}L_2}}{\varepsilon(\frac{\omega_1}{v_x},\omega_1)\varepsilon(k_x-\frac{\omega_1}{v_x},\omega_2)(k_x-\frac{\omega_1}{v_x})} \tag{13.101}$$

Inverting (13.101) gives

$$E_x^{(2)}(x,t) \sim \frac{e^{-i\omega_3 t}(x-2L_2)}{\omega_1\omega_2} \int d^3v \frac{\frac{1}{v_x}\frac{\partial f_0}{\partial v_x}\exp(\frac{i\omega_3}{v_x}(x-2L_2)+\frac{i\omega_1}{v_x}L_2)}{\varepsilon(\frac{\omega_3}{v_x},\omega_3)\varepsilon(\frac{\omega_2}{v_x},\omega_2)\varepsilon(\frac{\omega_1}{v_x},\omega_1)} \tag{13.102}$$

We obtain the echo spots:

$$x_{\text{echo}} = \left(2-\frac{\omega_1}{\omega_3}\right)L_2,\ \left(2-\frac{\omega_2}{\omega_3}\right)L_1 \tag{13.103}$$

References

Gould, R. W., T. M. O'Neil, and J. H. Malmberg, *Plasma wave echo*, Phys. Rev. Lett. **19** (1967), pp. 219–222.

Lee, H. J. and M. J. Lee, *Echo in a semi-bounded plasma*, Open Plasma Phys. J. **8** (Bentham), (2015), pp. 1–7.

Lee, H. J. and M. J. Lee, *Surface wave echo in a semi-bounded plasma*, J. Modern Phys. **7** (2016), pp. 1400–1412.

Further interesting subjects which are not dealt with in this Chapter include parametric instabilities in a bounded plasma, and fluctuation of electric field in a bounded plasma, and spatial echoes in a plasma slab. For these topics, refer to:

Stenflo, L. *Nonlinear interaction of electrostatic waves in a plasma*, Ann. Physik. **7**. Folge, Bd. 27, Heft 3, S. 289–295 (1971)–this is a kinetic theory.

Sitenko, A. G. *Fluctuations and Non-linear Wave Interactions in Plasmas*, Pergamon, Oxford, (1982) Chapter 11.

Lee, H. J. and Y. K. Lim, *Surface wave echo in a plasma slab*, Phys. Sci. Int. J. **12**, No. 2, (2016), pp. 1–16 PSIJ.28647.

Chapter 14

Surface Waves in Cylindrical Plasma

14.1 Electrostatic Surface Waves in Cylindrical Plasma

Let us consider a cylindrical plasma which has two boundaries interfaced with vacuum: $z = 0$ and $r = a$, where z is the axial coordinate of a cylindrical plasma and r is the radial coordinate. The equilibrium density is assumed to be inhomogeneous, denoted by $N(z, r)$, which has a step function profile:

$$\begin{aligned} N(z,r) &= N_0 \qquad z > 0; \qquad 0 < r < a \qquad \text{plasma} \\ &= 0 \qquad z < 0; \qquad r > a \qquad \text{vacuum} \end{aligned} \tag{14.1}$$

We begin with the following linearized fluid equations for electrons, written in cylindrical coordinates, with ions being immobile.

$$-i\omega\mathbf{v}(r,z,\theta) = \frac{e}{m}\nabla\phi(r,z,\theta) - \frac{\beta^2}{N(r,z)}\nabla n(r,z,\theta) \tag{14.2}$$

$$-i\omega n + N(r,z)\nabla\cdot\mathbf{v} + v_r\frac{\partial N}{\partial r} + v_z\frac{\partial N}{\partial z} = 0 \tag{14.3}$$

$$\nabla^2\phi = 4\pi e n \tag{14.4}$$

where $\mathbf{v} = \hat{\mathbf{r}}v_r + \hat{\theta}v_\theta + \hat{\mathbf{z}}v_z$, $\nabla = \hat{\mathbf{r}}\frac{\partial}{\partial r} + \frac{\hat{\theta}}{r}\frac{\partial}{\partial\theta} + \hat{\mathbf{z}}\frac{\partial}{\partial z}$, ϕ is the electric potential, n is the perturbation of number density, and β is the thermal velocity. The electric field components are obtained by

$$E_r = -\frac{\partial\phi}{\partial r}, \quad E_\theta = -\frac{1}{r}\frac{\partial\phi}{\partial\theta}, \quad E_z = -\frac{\partial\phi}{\partial z} \tag{14.5}$$

In components, (14.2) gives

$$v_r = \frac{i}{\omega}\left(\frac{e}{m}\frac{\partial\phi}{\partial r} - \frac{\beta^2}{N}\frac{\partial n}{\partial r}\right) \tag{14.6}$$

$$v_\theta = \frac{i}{\omega}\frac{1}{r}\left(\frac{e}{m}\frac{\partial\phi}{\partial\theta} - \frac{\beta^2}{N}\frac{\partial n}{\partial\theta}\right) \tag{14.7}$$

$$v_z = \frac{i}{\omega}\left(\frac{e}{m}\frac{\partial\phi}{\partial z} - \frac{\beta^2}{N}\frac{\partial n}{\partial z}\right) \tag{14.8}$$

Taking divergence of (14.2) gives

$$\nabla\cdot\mathbf{v} = \frac{i}{\omega}\left[\frac{e}{m}\nabla^2\phi - \frac{\beta^2}{N}\nabla^2 n + \frac{\beta^2}{N^2}\left(\frac{\partial n}{\partial r}\frac{\partial N}{\partial r} + \frac{\partial n}{\partial z}\frac{\partial N}{\partial z}\right)\right] \tag{14.9}$$

Eliminating $\nabla\cdot\mathbf{v}$ in (14.3) and (14.9) yields

$$\omega^2 n - \frac{e}{m}N\nabla^2\phi + \beta^2\nabla^2 n - \frac{e}{m}\frac{\partial N}{\partial r}\frac{\partial\phi}{\partial r} - \frac{e}{m}\frac{\partial N}{\partial z}\frac{\partial\phi}{\partial z} = 0 \tag{14.10}$$

which upon using (14.4) takes the form

$$(\omega^2 - \omega_p^2(r,z))\nabla^2\phi + \beta^2\nabla^2\nabla^2\phi - \frac{\partial\omega_p^2}{\partial r}\frac{\partial\phi}{\partial r} - \frac{\partial\omega_p^2}{\partial z}\frac{\partial\phi}{\partial z} = 0 \tag{14.11}$$

where $\omega_p^2(r,z) = 4\pi N(r,z)e^2/m$, the plasma frequency squared. In cylindrical coordinates, (14.11) is written as

$$\begin{aligned}&\omega^2\left(\frac{\partial^2\phi}{\partial r^2} + \frac{1}{r}\frac{\partial\phi}{\partial r} + \frac{1}{r^2}\frac{\partial^2\phi}{\partial\theta^2} + \frac{\partial^2\phi}{\partial z^2}\right) - \omega_p^2\left(\frac{1}{r}\frac{\partial\phi}{\partial r} + \frac{1}{r^2}\frac{\partial^2\phi}{\partial\theta^2}\right) - \frac{\partial}{\partial r}\left(\omega_p^2\frac{\partial\phi}{\partial r}\right) - \frac{\partial}{\partial z}\left(\omega_p^2\frac{\partial\phi}{\partial z}\right)\\ &+\beta^2\left(\frac{\partial^2}{\partial r^2} + \frac{1}{r}\frac{\partial}{\partial r} + \frac{1}{r^2}\frac{\partial^2}{\partial\theta^2} + \frac{\partial^2}{\partial z^2}\right)\left(\frac{\partial^2\phi}{\partial r^2} + \frac{1}{r}\frac{\partial\phi}{\partial r} + \frac{1}{r^2}\frac{\partial^2\phi}{\partial\theta^2} + \frac{\partial^2\phi}{\partial z^2}\right) = 0\end{aligned} \tag{14.12}$$

This equation is useful for deriving the boundary conditions at the interfaces $r = a$ and $z = 0$. In the above equation, $\omega_p(r,z)$ has a step function profile at $r = a$ and $z = 0$ whose derivatives are delta function-like. These terms are important to derive the jump relations across the interfaces of relevant physical quantities. Boundary conditions at the interface $r = a$ or $z = 0$ can be obtained by taking the infinitesimal integration of the relevant equations across the interface:

$$\lim_{\delta\to 0}\int_{a-\delta}^{a+\delta}(\cdots)dr \quad \text{or} \quad \lim_{\delta\to 0}\int_{-\delta}^{\delta}(\cdots)dz \tag{14.13}$$

- **Cold plasma boundary conditions.** The infinitesimal integration of (14.6) and (14.8) gives the boundary conditions

$$[\phi(r)]_{a^-}^{a+} = 0 \quad \text{at } r = a \tag{14.14}$$

$$[\phi(z)]_{0^-}^{0+} = 0 \quad \text{at } z = 0 \tag{14.15}$$

Boundary conditions in (14.14) and (14.15) are valid both in cold and warm plasma. As to $E_\theta = -\frac{1}{r}\frac{\partial \phi}{\partial \theta}$, we conclude that E_θ has no jump at $r = a$ or $z = 0$ if ϕ is a continuous function of θ:

$$[E_\theta] = 0 \quad \text{at } r = a \quad \text{or} \quad z = 0 \tag{14.16}$$

Let us consider the cold plasma version of (14.12) ($\beta = 0$):

$$\omega^2\left(\frac{\partial^2\phi}{\partial r^2} + \frac{1}{r}\frac{\partial \phi}{\partial r} + \frac{1}{r^2}\frac{\partial^2\phi}{\partial \theta^2} + \frac{\partial^2\phi}{\partial z^2}\right) - \omega_p^2\left(\frac{1}{r}\frac{\partial \phi}{\partial r} + \frac{1}{r^2}\frac{\partial^2\phi}{\partial \theta^2}\right) - \frac{\partial}{\partial r}\left(\omega_p^2\frac{\partial \phi}{\partial r}\right) - \frac{\partial}{\partial z}\left(\omega_p^2\frac{\partial \phi}{\partial z}\right) = 0 \tag{14.17}$$

If we apply the infinitesimal integral operation across $r = a$ to (14.17), we are left with

$$\left[\frac{\partial \phi}{\partial r}\right]_{a^-}^{a^+} = \left[\frac{\omega_p^2(r)}{\omega^2}\frac{\partial \phi}{\partial r}\right]_{a^-}^{a^+} \tag{14.18}$$

Hence, (14.18) yields the result

$$E_r(a^+) = \varepsilon E_r(a^-) \quad \text{for cold plasma} \tag{14.19}$$

where $\varepsilon = 1 - \omega_p^2/\omega^2$, the dielectric constant. In words, the normal component of the electric displacement is continuous in a cold plasma. Similarly, infinitesimal integration across $z = 0$ yields

$$\left[\frac{\partial \phi}{\partial z}\right]_{0^-}^{0^+} = \left[\frac{\omega_p^2}{\omega^2}\frac{\partial \phi}{\partial z}\right]_{0^-}^{0^+}$$

which is

$$E_z(0^-) = \varepsilon\ E_z(0^+) \quad \text{for cold plasma} \tag{14.20}$$

Mathematically, a surface term is derived from a delta function-like density distribution over the interface which may be traced to the derivative of the step function density profile.

- **Warm plasma boundary condition.** Equation (14.6), upon being integrated over the infinitesimal interval across $r = a$ yields $[\phi]_{a^-}^{a^+} = 0$. Note that $\frac{1}{N(r)}\frac{\partial n}{\partial r}$ is not a perfect differential, and thus vanishes upon integration. Hence,

$$[E_\theta] = 0, \quad [E_z] = 0 \quad \text{at } r = a \quad \text{(warm plasma)} \tag{14.21}$$

The tangential components of the electric field are continuous both in cold and warm plasmas.

Let us write (14.7) in the form

$$-i\omega r N(r,z) v_\theta = \frac{e}{m}\, N(r,z)\, \frac{\partial \phi}{\partial \theta} - \beta^2\, \frac{\partial n}{\partial \theta} \tag{14.22}$$

The only perfect differential term with respect to the coordinate r is contained in n through $\nabla^2\phi$ in the last term (See Eq. (14.4)). Therefore we have

$$\left[\frac{\partial \phi}{\partial r}\right]_{a^-}^{a^+} = 0 \tag{14.23}$$

$$\text{or} \quad [E_r]_{r=a} = 0 \quad \text{(warm plasma)} \tag{14.24}$$

The normal component of the electric field is continuous in a warm plasma. This contrasts with the corresponding cold plasma boundary condition in (14.19). If we integrate (14.22) across $z = 0$, we obtain analogously

$$\left[\frac{\partial \phi}{\partial z}\right]_{z=0} = 0$$

$$\text{or} \quad [E_z]_{z=0} = 0 \quad \text{(warm plasma)} \tag{14.25}$$

We have more boundary conditions, which can be extracted from (14.12). Performing the infinitesimal integration of (14.12) across $r = a$ yields

$$\left[\frac{\partial \phi}{\partial r}\right]_{a^-}^{a^+} - \left[\frac{\omega_p^2(r)}{\omega^2}\frac{\partial \phi}{\partial r}\right]_{a^-}^{a^+} + \frac{\beta^2}{\omega^2}\left[\frac{\partial}{\partial r}\nabla^2\phi\right]_{a^-}^{a^+} + \frac{\beta^2}{\omega^2}\frac{\partial^2}{\partial z^2}\left[\frac{\partial \phi}{\partial r}\right]_{a^-}^{a^+} = 0 \tag{14.26}$$

Introducing the explicit notations ϕ_1 for vacuum and ϕ_2 for plasma and noting $\nabla^2\phi_1 = 0$, the above equation gives

$$\beta^2\, \frac{\partial}{\partial r}\nabla^2\phi_2 - \omega_p^2\, \frac{\partial \phi_2}{\partial r} = 0 \quad \text{at } r = a \tag{14.27}$$

where we used (14.23). Consulting (14.6) affirms that (14.27) is equivalent to

$$v_r\Big|_{r=a} = 0 \quad \text{(warm plasma)} \tag{14.28}$$

which is the specular reflection boundary condition.

Performing integration $\int_{-\delta}^{\delta}(\cdots)dz$ on (14.12), we have

$$\omega^2 \left[E_z\right]_{0^-}^{0^+} - \left[\omega_p^2 E_z\right]_{0^-}^{0^+} + \beta^2 \left[\frac{\partial}{\partial z}\nabla \cdot \mathbf{E}\right]_{0^-}^{0^+} + \beta^2 \left[\frac{\partial^2 E_z}{\partial r^2} + \frac{1}{r}\frac{\partial E_z}{\partial r} + \frac{1}{r^2}\frac{\partial^2 E_z}{\partial \theta^2}\right]_{0^-}^{0^+} = 0 \tag{14.29}$$

The last term takes the form as shown above due to the commutativeness of $\frac{\partial}{\partial z}$ and the other operators but it vanishes due to (14.25). Also the first term vanishes as well. Analogously to (14.27), (14.29) can be put into the form

$$\beta^2 \frac{\partial}{\partial z}\nabla^2\phi_2 - \omega_p^2 \frac{\partial \phi_2}{\partial z} = 0 \quad \text{at } z = 0 \tag{14.30}$$

Consulting (14.8) affirms that (14.27) is equivalent to

$$v_z\big|_{z=0} = 0 \quad \text{(warm plasma)} \tag{14.31}$$

which is the specular reflection boundary condition. Another alternative form of the specular reflection boundary condition is

$$\frac{\partial}{\partial z}\left(\phi_2 - \lambda_D^2 \nabla^2 \phi_2\right) = 0 \quad \text{at} \quad z = 0 \tag{14.32}$$

where $\lambda_D = \beta/\omega_p$ is the Debye length.

• **Boundary conditions derived from Maxwell equations.** $\nabla \times \mathbf{E} = 0$, the electrostatic relation, gives the following three equations in cylindrical coordinates (here only, we use ρ instead of r as the radial coordinate):

$$\frac{1}{\rho}\left[\frac{\partial E_z}{\partial \theta} - \frac{\partial}{\partial z}(\rho E_\theta)\right] = 0, \quad \frac{\partial E_\rho}{\partial z} - \frac{\partial E_z}{\partial \rho} = 0, \quad \frac{1}{\rho}\left[\frac{\partial}{\partial \rho}(\rho E_\theta) - \frac{\partial E_\rho}{\partial \theta}\right] = 0$$

By inspection, we draw the following relations for a cylindrical boundary:

$$[E_\theta]_{z=0} = 0, \quad [E_\rho]_{z=0} = 0, \quad [E_z]_{\rho=a} = 0, \quad [E_\theta]_{\rho=a} = 0 \tag{14.33}$$

In dealing with electromagnetic waves, $\nabla \cdot \mathbf{B} = 0$ for the magnetic induction may give useful relations. In cylindrical coordinates we have

$$\frac{\partial B_\rho}{\partial \rho} + \frac{B_\rho}{\rho} + \frac{1}{\rho}\frac{\partial B_\theta}{\partial \theta} + \frac{\partial B_z}{\partial z} = 0 \tag{14.34}$$

which gives

$$[B_\rho]_{\rho=a} = 0, \qquad [B_z]_{z=0} = 0 \tag{14.35}$$

which state that the normal to the boundary components of **B** are continuous. Boundary conditions in (14.33) and (14.35) are seen to be independent of the previous-obtained boundary equations.

• **Electrostatic waves in infinite warm cylindrical plasma column surrounded by vacuum.** Here we assume that the wavelength of the standing wave excited in the plasma column is infinite (much greater than the length of the column), and ignore the z-direction by putting $\partial/\partial z = 0$.
Vacuum equation:

$$\nabla^2\phi_1(r,\theta) = 0: \quad a < r < b \tag{14.36}$$

Plasma equation: it can be written down from (14.11) with $\omega_p = \text{const.}$

$$(\nabla^2 + \kappa^2)\nabla^2\phi_2(r,\theta) = 0: \quad 0 \le r < a \tag{14.37}$$

$$\kappa^2 = \frac{\omega^2 - \omega_p^2}{\beta^2}, \quad (\beta^2 = 3T_e/m_e) \tag{14.38}$$

Equation (14.38) is called the Bohm–Gross dispersion relation of a warm plasma wave:

$$\omega^2 = \omega_p^2 + \frac{3T_e}{m_e}\kappa^2.$$

The Laplace equation takes the form

$$\nabla^2\phi(r,\theta) = \frac{\partial^2\phi}{\partial r^2} + \frac{1}{r}\frac{\partial\phi}{\partial r} + \frac{1}{r^2}\frac{\partial^2\phi}{\partial\theta^2} = 0 \tag{14.39}$$

Now we solve (14.39) by the method of separation of variable.

$$\phi(r,\theta) = R(r)\Theta(\theta) \tag{14.40}$$

$$\frac{d^2\Theta}{d\theta^2} + n^2\Theta = 0 \tag{14.41}$$

$$\frac{d^2R}{dr^2} + \frac{1}{r}\frac{dR}{dr} - \frac{n^2}{r^2}R = 0 \tag{14.42}$$

where n is integer, separation constant. Equation (14.41) gives $\Theta \sim e^{\pm in\theta}$. Equation (14.42) is solved by $R \sim r^{\pm n}$. Hence, the vacuum solution can be written as

$$\phi_1(r,\theta) = \sum_{n=-\infty}^{\infty} \left(A_n r^n + B_n r^{-n}\right) e^{in\theta} \quad a < r < b \tag{14.43}$$

The plasma equation (14.37) is solved by $\phi_2 = \phi_a + \phi_b$ where

$$\nabla^2\phi_a = 0 \tag{14.44}$$

$$(\nabla^2 + \kappa^2)\phi_b = 0 \tag{14.45}$$

$$\phi_a(r,\theta) = \sum_{n=-\infty}^{\infty} C_n\, r^n\, e^{in\theta} \quad 0 \le r < b \tag{14.46}$$

$$\frac{d^2\phi_b}{dr^2} + \frac{1}{r}\,\frac{d\phi_b}{dr} + \left(\kappa^2 - \frac{n^2}{r^2}\right)\phi_b = 0 \tag{14.47}$$

The solution can be written as $\phi_b(r) = \alpha_n J_n(\kappa r) + \beta_n Y_n(\kappa r)$, where Y_n is *Weber's Bessel functions of the second kind* which goes to $-\infty$ as $r \to 0$, so it should be rejected. Hence

$$\phi_b(r,\theta) = \sum_{n=-\infty}^{\infty} D_n J_n(\kappa r)\; e^{in\theta} \tag{14.48}$$

$$\phi_2(r,\theta) = \sum_{n=-\infty}^{\infty} \left[C_n r^n + D_n J_n(\kappa r)\right]\; e^{in\theta} \tag{14.49}$$

In regard to the Bessel functions, refer to Hildebrand (new) pp. 147–153.

To apply the specular reflection boundary condition, $v_r|_{r=a} = 0$, we use (14.27) and write it as

$$\frac{\partial}{\partial r}\sum_{n=-\infty}^{\infty}\left[\omega_p^2 C_n r^n + \omega^2 D_n J_n(\kappa r)\right]\; e^{in\theta} = 0 \quad \text{at } r = a \tag{14.50}$$

Matching conditions:

$$[E_\theta]_a = 0: \;\; A_n a^n + B_n a^{-n} = C_n a^n + D_n J_n(\kappa a) \tag{14.51}$$

$$[E_r]_a = 0: \;\; n\; A_n a^{n-1} - nB_n a^{-n-1} = nC_n a^{n-1} + \kappa D_n J_n'(\kappa a) \tag{14.52}$$

$$v_r|_{r=a} = 0: \;\; \omega_p^2 n C_n a^{n-1} + \omega^2 \kappa D_n J_n'(\kappa a) = 0 \tag{14.53}$$

$$\phi_1|_b = 0: \;\; A_n b^n + B_n b^{-n} = 0 \tag{14.54}$$

where prime denotes derivative with respect to the argument. After some algebra, one obtains the dispersion relation

$$\frac{\left(\frac{a}{b}\right)^{2n} - 1}{\left(\frac{a}{b}\right)^{2n} + 1} = \frac{1}{\varepsilon}\left[1 - \frac{\omega_p^2}{\omega^2}\,\frac{nJ_n(\kappa a)}{\kappa a J_n'(\kappa a)}\right] \tag{14.55}$$

If $b \to \infty$, we have

$$\frac{2\omega^2}{\omega_p^2} - 1 - \frac{n}{\kappa a}\,\frac{J_n(\kappa a)}{J_n'(\kappa a)} \tag{14.56}$$

We can write from (14.38)

$$\kappa a = \frac{1}{\sqrt{3}}\frac{a}{\lambda_D}\sqrt{\frac{\omega^2}{\omega_p^2}-1}, \quad \lambda_D = \sqrt{\frac{T_e}{4\pi n_0 e^2}}, \quad \text{Debye length} \tag{14.57}$$

There are a series of resonance frequencies $\omega_n (n = 0, 1, 2, \ldots)$ that satisfy (14.56) in a warm plasma column.

14.2 Electromagnetic Waves in Infinite Cold Cylindrical Plasma Column

The governing equations are:

$$\frac{\partial \mathbf{v}}{\partial t} = -\frac{e}{m}\mathbf{E} \tag{14.58}$$

$$\frac{\partial n}{\partial t} + N\nabla \cdot \mathbf{v} = 0 \tag{14.59}$$

$$\nabla \cdot \mathbf{E} = -4\pi e n \tag{14.60}$$

$$\nabla \times \mathbf{E} = -\frac{1}{c}\frac{\partial \mathbf{B}}{\partial t} \tag{14.61}$$

$$\nabla \times \mathbf{B} = \frac{1}{c}\frac{\partial \mathbf{E}}{\partial t} - \frac{4\pi e N}{c}\mathbf{v} \tag{14.62}$$

Note that (14.58)–(14.62) are combined into the form $\varepsilon \nabla \cdot \mathbf{E} = 0$ with $\varepsilon = 1 - \omega_p^2/\omega^2$, dielectric function. Hence, we have both in plasma and vacuum surrounding it,

$$\nabla \cdot \mathbf{B} = 0, \quad \nabla \cdot \mathbf{E} = 0 \tag{14.63}$$

The waves are propagating in the axial direction with phasor, $\sim e^{ikz-i\omega t}$. Put $\partial/\partial t = -i\omega$. Eliminate $\mathbf{E}$ or $\mathbf{B}$ in the above equations to get

$$\left[\nabla^2 + \frac{1}{c^2}(\omega^2 - \omega_p^2)\right]\begin{pmatrix}\mathbf{B}\\ \mathbf{E}\end{pmatrix} = 0 \tag{14.64}$$

Since $\left[\nabla^2 \mathbf{B}\right]_z = \nabla^2 B_z$ in cylindrical coordinates, (14.64) for $\hat{z}$-component, becomes

$$\left[\frac{1}{r}\frac{\partial}{\partial r} r \frac{\partial}{\partial r} + \frac{1}{r^2}\frac{\partial^2}{\partial \theta^2} + \frac{1}{c^2}(\omega^2 - \omega_p^2) - k^2\right]\begin{pmatrix}B_z\\ E_z\end{pmatrix} = 0 \tag{14.65}$$

It is useful to write (14.61) and (14.62) in the form

$$\nabla \times \mathbf{E} = \frac{i\omega}{c}\,\mathbf{B} \tag{14.66}$$

$$\nabla \times \mathbf{B} = -\frac{i\omega}{c}\,\varepsilon\,\mathbf{E} \tag{14.67}$$

If B_z and E_z are obtained by solving (14.65), other field components can be determined in terms of B_z and E_z. To show this, let us write (14.66) and (14.67) in cylindrical components

$$\frac{i\omega}{c}\,B_r = \frac{1}{r}\,\frac{\partial E_z}{\partial \theta} - \frac{\partial E_\theta}{\partial z} \tag{14.68}$$

$$-\frac{i\omega}{c}\,B_\theta = \frac{\partial E_z}{\partial r} - \frac{\partial E_r}{\partial z} \tag{14.69}$$

$$\frac{i\omega}{c}\,\varepsilon E_\theta = \frac{\partial B_z}{\partial r} - \frac{\partial B_r}{\partial z} \tag{14.70}$$

$$-\frac{i\omega}{c}\,\varepsilon E_r = \frac{1}{r}\,\frac{\partial B_z}{\partial \theta} - \frac{\partial B_\theta}{\partial z} \tag{14.71}$$

The above equations can be arranged in the following form by eliminating $B_r, B_\theta, E_r, E_\theta$ from the appropriate pair of equations

$$B_r = \frac{-i\,c^2}{c^2k^2 - \varepsilon\omega^2}\left[k\,\frac{\partial B_z}{\partial r} - \frac{\varepsilon\omega}{c}\,\frac{1}{r}\,\frac{\partial E_z}{\partial \theta}\right] \tag{14.72}$$

$$B_\theta = \frac{-i\,c^2}{c^2k^2 - \varepsilon\omega^2}\left[\frac{k}{r}\,\frac{\partial B_z}{\partial \theta} + \frac{\varepsilon\omega}{c}\,\frac{\partial E_z}{\partial r}\right] \tag{14.73}$$

$$E_r = \frac{-i\,c^2}{c^2k^2 - \varepsilon\omega^2}\left[k\,\frac{\partial E_z}{\partial r} + \frac{\omega}{c}\,\frac{1}{r}\,\frac{\partial B_z}{\partial \theta}\right] \tag{14.74}$$

$$E_\theta = \frac{-i\,c^2}{c^2k^2 - \varepsilon\omega^2}\left[\frac{k}{r}\,\frac{\partial E_z}{\partial \theta} - \frac{\omega}{c}\,\frac{\partial B_z}{\partial r}\right] \tag{14.75}$$

Remark. For an azimuthally symmetric wave ($\partial/\partial\theta = 0$), the above 4 equations are decoupled into two independent sets: E-wave, (E_z, E_r, B_θ) and B-wave, (B_z, B_r, E_θ). The E-wave set ((14.73) and (14.74)) is completed with $\nabla\cdot\mathbf{E} = 0$. The B-wave set ((14.72) and (14.75)) is completed with $\nabla\cdot\mathbf{B} = 0$.

- **Solution of Eq. (14.65).** Equation (14.65) is solved by

$$B_z(r,\theta,z,t) = e^{ikz-i\omega t}\sum_{n=-\infty}^{\infty} e^{in\theta}\phi_n(\alpha r) \tag{14.76}$$

where $\phi_n(r)$ should solve the modified Bessel equation

$$x^2 \frac{d^2\phi_n(x)}{dx^2} + x \frac{d\phi_n}{dx} - \phi_n(n^2 + x^2) = 0 \tag{14.77}$$

with $x = r\alpha$ and

$$\alpha = \left(k^2 + \frac{\omega_p^2 - \omega^2}{c^2}\right)^{\frac{1}{2}} \tag{14.78}$$

The form of (14.76) holds for $E_z(r, \theta, z, t)$, too. The plasma solution $\phi_n(\alpha r)$ is a linear combination of *modified Bessel functions* $I_n(\alpha r)$ and $K_n(\alpha r)$, but K_n should be rejected because it blows up at $r = 0$.

The vacuum solution should be written as $\phi_n(\lambda r)$ with

$$\lambda = \left(k^2 - \frac{\omega^2}{c^2}\right)^{\frac{1}{2}} \tag{14.79}$$

and $\phi_n(\lambda r)$ is a linear combination of modified Bessel functions $I_n(\lambda r)$ and $K_n(\lambda r)$. If the vacuum is extended to $r \to \infty$, the function I_n function should be rejected. We distinguish the vacuum and plasma solutions by subscripts 1 and 2, respectively.

Vacuum solution

$$E_{z1}(r, \theta, k, \omega) = \sum_{-\infty}^{\infty} e^{in\theta} \left[A_n K_n(\lambda r) + C_n I_n(\lambda r)\right] \quad a < r < b \tag{14.80}$$

$$B_{z1}(r, \theta, k, \omega) = \sum_{-\infty}^{\infty} e^{in\theta} \left[a_n K_n(\lambda r) + c_n I_n(\lambda r)\right] \quad a < r < b \tag{14.81}$$

Plasma solution

$$E_{z2}(r, \theta, k, \omega) = \sum_{-\infty}^{\infty} e^{in\theta} F_n I_n(\alpha r) \quad 0 \le r < a \tag{14.82}$$

$$B_{z2}(r, \theta, k, \omega) = \sum_{-\infty}^{\infty} e^{in\theta} f_n I_n(\alpha r) \quad 0 \le r < a \tag{14.83}$$

We consider azimuthally symmetric waves: $\partial/\partial\theta = n = 0$.

Vacuum solution

$$E_{z1}(r) = A_0 K_0(\lambda r) + C_0 I_0(\lambda r) \quad a < r < b \tag{14.84}$$

$$B_{z1}(r) = a_0 K_0(\lambda r) + c_0 I_0(\lambda r) \quad a < r < b \tag{14.85}$$

Plasma solution

$$E_{z2}(r) = F_0 I_0(\alpha r) \quad 0 < r < a \tag{14.86}$$

$$B_{z2}(r) = f_0 I_0(\alpha r) \quad 0 < r < a \tag{14.87}$$

• **E-wave.** Boundary conditions are: $[E_z]_a = 0$, $[B_\theta]_a = 0$, and $E_z|_b = 0$. The condition $[B_\theta]_a = 0$ can be replaced by $[\varepsilon E_r]_a = 0$. We have the following relations which are valid only in the case of θ-symmetry.

$$E_{r1} = \frac{-ik}{\lambda^2} \frac{\partial E_{z1}}{\partial r} \tag{14.88}$$

$$E_{r2} = \frac{-ik}{\alpha^2} \frac{\partial E_{z2}}{\partial r} \tag{14.89}$$

$$B_{\theta 1} = \frac{-i\omega}{c\lambda^2} \frac{\partial E_{z1}}{\partial r} \tag{14.90}$$

$$B_{\theta 2} = \frac{-i\omega}{c\alpha^2} \varepsilon \frac{\partial E_{z2}}{\partial r} \tag{14.91}$$

Let us enforce the boundary conditions.

$$[B_\theta] = 0: \quad A_0 K_0'(\lambda a) + C_0 I_0'(\lambda a) = \frac{\varepsilon\lambda}{\alpha} F_0 I_0'(\alpha a) \tag{14.92}$$

$[\varepsilon E_r]_a = 0$ yields identical relation.

$$[E_z]_a = 0: \quad A_0 K_0(\lambda a) + C_0 I_0(\lambda a) = F_0 I_0(\alpha a) \tag{14.93}$$

$$E_z|_b = 0: \quad A_0 K_0(\lambda b) + C_0 I_0(\lambda b) = 0 \tag{14.94}$$

The above three equations give the dispersion relation

$$\frac{\lambda\varepsilon}{\alpha} \frac{I_0'(\alpha a)}{I_0(\alpha a)} = \frac{K_0(\lambda b) I_0'(\lambda a) - I_0(\lambda b) K_0'(\lambda a)}{K_0(\lambda b) I_0(\lambda a) - I_0(\lambda b) K_0(\lambda a)} \tag{14.95}$$

When $b \to \infty$, $K_0(\lambda b) = 0$, and one obtains dispersion relation

$$\frac{\lambda\varepsilon}{\alpha} \frac{I_0'(\alpha a)}{I_0(\alpha a)} = \frac{K_0'(\lambda a)}{K_0(\lambda a)}$$

When $a \to \infty$, we have the electromagnetic wave dispersion relation in semi-infinite plasma.

$$\omega^2 = \frac{1}{2}\left(\omega_p^2 + 2c^2k^2 \pm \sqrt{\omega_p^4 + 4c^4k^4}\right)$$

• **B-wave.** To investigate the B-wave, we use B_{z1} and B_{z2} in (14.81) and (14.83).

In vacuum,

$$B_{z1}(r) = a_0 K_0(\lambda r) + c_0 I_0(\lambda r) \quad a < r < b \tag{14.96}$$

In plasma,

$$B_{z2}(r) = d_0\ I_0(\alpha r) \quad 0 \leq r < a \tag{14.97}$$

Boundary conditions: $[E_\theta]_a = 0$, that is the tangential components are continuous on the cylindrical surface $r = a$. We also have $[B_r]_a = 0$. But these two conditions, $[E_\theta] = 0$ and $[B_r] = 0$, are redundant as will be shown below. At $r = b$, the plasma-vacuum boundary, we use $\frac{dB_z}{dr}|_{\ r=b} = 0$. We need one more condition, which is $[B_z] = 0$ at $r = a$ as can be obtained by integrating (14.70) over the infinitesimal segment, $r = a^-$ to $r = a^+$. To apply these boundary conditions, we use (14.72) and (14.75) which derives the following relations:

$$B_{r1} = \frac{-ik}{\lambda^2}\frac{\partial B_{z1}}{\partial r} \tag{14.98}$$

$$B_{r2} = \frac{-ik}{\alpha^2}\frac{\partial B_{z2}}{\partial r} \tag{14.99}$$

$$E_{\theta 1} = \frac{i}{\lambda^2}\frac{\omega}{c}\frac{\partial B_{z1}}{\partial r} \tag{14.100}$$

$$E_{\theta 2} = \frac{i}{\alpha^2}\frac{\omega}{c}\frac{\partial B_{z2}}{\partial r} \tag{14.101}$$

Using (14.96) and (14.97) in the above equations yield

$$E_{\theta 1} = \frac{i\,\omega}{c\lambda}(a_0 K_0'(\lambda r) + c_0 I_0'(\lambda r)) \tag{14.102}$$

$$E_{\theta 2} = \frac{i\,\omega}{c\alpha} d_0 I_0'(\alpha r) \tag{14.103}$$

$$B_{r1} = -\frac{ik}{\lambda}(a_0 K_0'(\lambda r) + c_0 I_0'(\lambda r)) \tag{14.104}$$

$$B_{r2} = -\frac{ik}{\alpha} d_0 I_0'(\alpha r) \tag{14.105}$$

Inspection of the above four equations shows that $[E_\theta] = 0$ and $[B_r] = 0$ are redundant. The matching conditions are explicitly written down as:

$$\left.\frac{\partial B_z}{\partial r}\right|_b = 0{:}a_0 K_0'(\lambda b) + c_0 I_0'(\lambda b) = 0 \tag{14.106}$$

$$[B_z]_a = 0{:}a_0 K_0(\lambda a) + c_0 I_0(\lambda a) = d_0 I_0(\alpha a) \tag{14.107}$$

$$[E_\theta]_a = 0{:}\frac{\alpha}{\lambda}\ \left[a_0 K_0'(\lambda a) + c_0 I_0'(\lambda a)\right] = d_0 I_0'(\alpha a) \tag{14.108}$$

The solvability condition for the above three equations gives the dispersion relation of the B-wave in the form

$$\frac{\lambda}{\alpha}\,\frac{I_0'(\alpha a)}{I_0(\alpha a)} = \frac{I_0'(\lambda b)K_0'(\lambda a) - I_0'(\lambda a)K_0'(\lambda b)}{I_0'(\lambda b)K_0(\lambda a) - I_0(\lambda a)K_0'(\lambda b)} \tag{14.109}$$

When $b \to \infty$, $K_0'(\lambda b) = 0$, and one obtains the dispersion relation

$$\frac{\lambda}{\alpha}\,\frac{I_0'(\alpha a)}{I_0(\alpha a)} = \frac{K_0'(\lambda a)}{K_0(\lambda a)} \tag{14.110}$$

Check that B-wave in semi-infinite medium is forbidden.

14.3 Electromagnetic Surface Waves in a Relativistic Cold Plasma in Infinite Cylindrical Column

A relativistic calculation is desirable to display more consistent wave equations in the analysis of the electromagnetic waves for a beam plasma. Non-relativistic analysis derives non-symmetrical equations between the electric and the magnetic wave equations. The latter disparity disappears if one calculates the relativistic current by using the relativistic equation of motion for a moving plasma. Use of cylindrical coordinates in the dynamical equations and Maxwell equations can fit more closely some laboratory and astrophysical conditions. In a cylindrical plasma, the separation of TM and TE modes is not possible since they are inherently coupled through the boundary conditions. However, for an azimuthally symmetric wave, the separation of two uncoupled sets of the electric and magnetic field components (E- and B-waves) is possible. The dielectric tensor of a streaming plasma is not isotropic.

We begin with the relativistic Euler equation

$$\frac{\partial \mathbf{p}}{\partial t} + (\mathbf{v}\cdot\nabla)\mathbf{p} = -e\left(\mathbf{E} + \frac{1}{c}\,\mathbf{v}\times\mathbf{B}\right) \tag{14.111}$$

$$\mathbf{p} = m\gamma\mathbf{v} \tag{14.112}$$

$$\gamma = \frac{1}{\sqrt{1-\frac{v^2}{c^2}}} \tag{14.113}$$

$$\frac{\partial n}{\partial t} + \nabla\cdot(n\mathbf{v}) = 0 \tag{14.114}$$

We have the Maxwell equations

$$\nabla\times\mathbf{E} = -\frac{1}{c}\,\frac{\partial \mathbf{B}}{\partial t} \tag{14.115}$$

$$\nabla \times \mathbf{B} = \frac{1}{c}\frac{\partial \mathbf{E}}{\partial t} + \frac{4\pi}{c}\mathbf{J} \tag{14.116}$$

$$\nabla \cdot \mathbf{E} = -4\pi e n \tag{14.117}$$

In (14.111), we see that the electric and the magnetic forces enter the equation of motion as equal status; this contrasts with the non-relativistic equation of motion in which the magnetic force is assumed to be much smaller than the electric force. This may be the reason why the symmetry of the electric and the magnetic wave equations is not conserved in the non-relativistic theory.

- **Linearization.** We linearize the dynamical equations by setting

$$n(r,\theta,z) = N(r) + n_1(r,\theta)\ e^{ik_z z - i\omega t} \tag{14.118}$$

where $N(r)$ is the zero order number density and the subscript 1 denotes the perturbation.

$$\mathbf{v} = \hat{\mathbf{z}} v_0 + \mathbf{v}_1(r,\theta)\ e^{ik_z z - i\omega t} \tag{14.119}$$

where v_0 is the constant zero order velocity. Then, we calculate per binomial expansion

$$\gamma_0 = \frac{1}{\sqrt{1 - \frac{v_0^2}{c^2}}}, \qquad \gamma_1 = \gamma_0^3\ \frac{v_0 v_{z1}}{c^2} \tag{14.120}$$

Then, (14.111) gives

$$i\omega' \left(\gamma_0 \mathbf{v}_1 + \hat{\mathbf{z}} v_{z1} \gamma_0^3\ \frac{v_0^2}{c^2}\right) = \frac{e}{m}\left(\mathbf{E} + \frac{v_0}{c}\ \hat{\mathbf{z}} \times \mathbf{B}\right)$$

which gives

$$\mathbf{v}_1 = \hat{\mathbf{z}}\ \frac{iev_0^2}{mc^2\gamma_0\omega'}\ E_z - \frac{ie}{\gamma_0 m\omega'}\left(\mathbf{E} + \frac{v_0}{c}\ \hat{\mathbf{z}} \times \mathbf{B}\right) \tag{14.121}$$

where $\omega' = \omega - k_z v_0$ is the Doppler-shifted frequency. In components, we have

$$i\omega'\gamma_0 v_r = \frac{e}{m}\left(E_r - \frac{v_0}{c}B_\theta\right) \tag{14.122}$$

$$i\omega'\gamma_0 v_\theta = \frac{e}{m}\left(E_\theta + \frac{v_0}{c}B_r\right) \tag{14.123}$$

$$i\omega'\gamma_0^3 v_z = \frac{e}{m}E_z \tag{14.124}$$

Equation (14.114) gives

$$-i\omega' n_1 + \nabla \cdot (N\mathbf{v}_1) = 0 \tag{14.125}$$

From (14.121) we obtain

$$\nabla \cdot \mathbf{v}_1 = \frac{-i}{\omega' \gamma_0} \frac{e}{m} \left[\nabla \cdot \mathbf{E} - i\frac{v_0^2}{c^2} k_z E_z - \frac{v_0}{c} (\nabla \times \mathbf{B})_z \right] \tag{14.126}$$

Using (14.117) and (14.122)–(14.126) in (14.125) yields in cylindrical coordinates

$$\frac{\partial}{\partial r}\left[\frac{\omega_p^2}{\gamma_0 \omega'^2}\left(E_r - \frac{v_0}{c}B_\theta\right)\right] + \frac{\omega_p^2}{\gamma_0 \omega'^2}\frac{1}{r}\left(E_r - \frac{v_0}{c}B_\theta\right) + \frac{\omega_p^2}{\gamma_0 \omega'^2}\frac{1}{r}\frac{\partial}{\partial \theta}\left(E_\theta + \frac{v_0}{c}B_r\right)$$
$$+ \frac{\omega_p^2}{\gamma_0^3 \omega'^2} i k_z E_z - \left(\frac{\partial E_r}{\partial r} + \frac{E_r}{r} + \frac{1}{r}\frac{\partial E_\theta}{\partial \theta} + i k_z E_z\right) = 0 \tag{14.127}$$

where $\omega_p(r) = \sqrt{4\pi e^2 N(r)/m}$ is the plasma frequency. This is the wave equation for the electromagnetic field propagating in inhomogeneous plasma with a varying density $N(r)$. This equation is useful in deriving a connection formula matching a plasma quantity and the corresponding vacuum quantity across the interface $r = a$ (say), by performing the infinitesimal integration $\int_{a-}^{a+} (\cdots) dr$, where $N(r)$ has a step function-like profile as in (14.1). Picking up the perfect differential terms, we obtain the jump condition at $r = a$.

$$\left[\frac{\omega_p^2}{\gamma_0 \omega'^2}\left(E_r - \frac{v_0}{c}B_\theta\right)\right] - [E_r] = 0 \tag{14.128}$$

where the large bracket denotes the jump at $r = a$. If the plasma is stationary, putting $v_0 = 0$ and $\gamma_0 = 1$ in the above equation gives

$$(1 - \frac{\omega_p^2}{\omega^2}) E_r(a^-) = E_r(a^+) \tag{14.129}$$

with a^+ and a^- denoting the vacuum and the plasma side of the interface $r = a$, respectively. Equation (14.129) is the familiar jump condition that the normal component of the electric displacement is continuous across the interface. Equation (14.128) is written more explicitly as

$$E_r(a^+) - \left(1 - \frac{\omega_p^2}{\gamma_0 \omega'^2}\right) E_r(a^-) = \frac{\omega_p^2}{\gamma_0 \omega'^2} \frac{v_0}{c} B_\theta(a^-) \tag{14.130}$$

In a beam plasma, the normal component of the electric displacement is *not* continuous across the interface; it is discontinuous by the amount of the surface current represented by the right side of (14.130).

Next, we turn to the $\nabla \times \mathbf{B}$ equation in (14.116), where the current is linearized in the form

$$\mathbf{J}_1 = -eN\mathbf{v}_1 - \hat{\mathbf{z}}ev_0 n_1 = -eN\mathbf{v}_1 + \hat{\mathbf{z}}\frac{v_0}{4\pi}\ \nabla \cdot \mathbf{E} \tag{14.131}$$

Using (14.121) for $\mathbf{v}_1$, (116) takes the form

$$\nabla \times \mathbf{B} = -\frac{i\omega}{c}\left(1 - \frac{\omega_p^2(r)}{\gamma_0\omega\omega'}\right)\mathbf{E} + \frac{i\ \omega_p^2(r)}{\gamma_0\omega'}\ \frac{v_0}{c^2}\hat{\mathbf{z}} \times \mathbf{B} + \hat{\mathbf{z}}\frac{v_0}{c}\left[\nabla \cdot \mathbf{E} - i\frac{v_0}{c^2}\ \frac{\omega_p^2(r)}{\gamma_0\omega'}E_z\right] \tag{14.132}$$

Specializing as ω_p = const. for $r < a$ (plasma region) and $\omega_p = 0$ for $r > a$ (vacuum region), the wave equation (14.127) is written separately

$$\nabla \cdot \mathbf{E} = 0 \quad \text{(vacuum)} \tag{14.133}$$

$$\left(1 - \frac{\omega_p^2}{\gamma_0\omega'^2}\right)\nabla \cdot \mathbf{E} + \frac{\omega_p^2}{\gamma_0\omega'^2}\left(\frac{v_0^2}{c^2}ik_zE_z + \frac{v_0}{c}(\nabla \times \mathbf{B})_z\right) = 0 \quad \text{(plasma)} \tag{14.134}$$

Taking the z-component, (14.132) gives in the plasma region

$$(\nabla \times \mathbf{B})_z = -i\ \frac{\omega}{c}E_z\left(1 - \frac{\omega_p^2}{\omega\omega'\gamma_0^3}\right) + \frac{v_0}{c}\nabla \cdot \mathbf{E} \quad \text{(plasma)} \tag{14.135}$$

The z-component of the curl of (14.115) is

$$ik_z\nabla \cdot \mathbf{E} - \nabla^2 E_z = \frac{i\omega}{c}(\nabla \times \mathbf{B})_z \tag{14.136}$$

We have 3 equations (14.134)–(14.136) for the 3 quantities $\nabla^2 E_z$, $\nabla \cdot \mathbf{E}$, and $(\nabla \times \mathbf{B})_z$, which can be determined in terms of E_z. Eliminating $(\nabla \times \mathbf{B})_z$ in (14.134) and (14.136) by means of (14.135) yields, after some algebra,

$$\nabla^2 E_z + \frac{\omega^2 - \omega_p^2/\gamma_0}{c^2}\ E_z = 0 \tag{14.137}$$

In later development, we shall derive the similar equation for B_z:

$$\nabla^2 B_z + \frac{\omega^2 - \omega_p^2/\gamma_0}{c^2}\ B_z = 0 \tag{14.138}$$

The complete symmetry between E_z and B_z equations is the triumph of the relativistic analysis; in the parallel non-relativistic calculation we obtain

$$\left(\omega'^2 - \omega_p^2/\gamma_0^2\right) \nabla^2 E_z + \frac{(\omega_p^2 - \omega\omega')^2}{c^2} E_z = 0$$

whose expression is quite annoying and unsatisfactory.

Equations (14.137) and (14.138) are important since E_z and B_z satisfy their own equations, uncoupled with other field variables, and consequently, each can be solved independently of other field components.

Next, the z-component of the curl of (14.132) reads

$$\nabla^2 B_z + \frac{\omega^2}{c^2} B_z \left(1 - \frac{\omega_p^2}{\gamma_0 \omega \omega'}\right) + \frac{\omega_p^2}{\gamma_0 \omega'} \frac{v_0}{c^2} k_z B_z = 0$$

which becomes (14.138) after further algebra.

Equation (14.132) gives in cylindrical components,

$$\frac{1}{r}\frac{\partial B_z}{\partial \theta} - \frac{\partial B_\theta}{\partial z} = \frac{-i\omega}{c}\left(1 - \frac{\omega_p^2(r)}{\gamma_0 \omega \omega'}\right) E_r - \frac{i\,\omega_p^2(r)}{\gamma_0 \omega'} \frac{v_0}{c^2} B_\theta \tag{14.139}$$

$$\frac{\partial B_r}{\partial z} - \frac{\partial B_z}{\partial r} = \frac{-i\omega}{c}\left(1 - \frac{\omega_p^2(r)}{\gamma_0 \omega \omega'}\right) E_\theta + \frac{i\,\omega_p^2(r)}{\gamma_0 \omega'} \frac{v_0}{c^2} B_r \tag{14.140}$$

$$\frac{B_\theta}{r} + \frac{\partial B_\theta}{\partial r} - \frac{1}{r}\frac{\partial B_r}{\partial \theta} = \frac{-i\omega'}{c}\left(1 - \frac{\omega_p^2(r)}{\omega'^2 \gamma_0^3}\right) E_z + \frac{v_0}{c}\left(\frac{\partial E_r}{\partial r} + \frac{E_r}{r} + \frac{1}{r}\frac{\partial E_\theta}{\partial \theta}\right) \tag{14.141}$$

Writing (14.115) in cylindrical components, we have

$$\frac{1}{r}\frac{\partial E_z}{\partial \theta} - \frac{\partial E_\theta}{\partial z} = \frac{i\omega}{c} B_r \tag{14.142}$$

$$\frac{\partial E_r}{\partial z} - \frac{\partial E_z}{\partial r} = \frac{i\omega}{c} B_\theta \tag{14.143}$$

$$\frac{E_\theta}{r} + \frac{\partial E_\theta}{\partial r} - \frac{1}{r}\frac{\partial E_r}{\partial \theta} = \frac{i\omega}{c} B_z \tag{14.144}$$

From (14.139)–(14.144), useful boundary conditions at $r = a$ can be derived. Equation (14.140) gives by picking up the perfect differential term in dr

$$[B_z] = 0 \tag{14.145}$$

Likewise, (14.141) yields

$$[B_\theta] = \frac{v_0}{c}\,[E_r] \tag{14.146}$$

Equation (14.146) indicates that the tangential component of magnetic field in a beam plasma is *not* continuous. We also have, owing to $\nabla \cdot \mathbf{B} = 0$,

$$[B_r] = 0 \tag{14.147}$$

Equations (14.143) and (14.144) yield, respectively,

$$[E_z] = 0 \tag{14.148}$$

$$[E_\theta] = 0 \tag{14.149}$$

The boundary conditions obtained above are not all independent; some of them are redundant.

The cylindrical coordinate components of the electric and magnetic fields in (14.139)–(14.144) are all coupled. The couplings can be delineated in a more orderly fashion by mixing the equations in different ways.

Eliminating B_θ from (14.139) and (14.143) gives

$$E_r = \frac{-i\omega}{c\alpha^2}\left[\frac{1}{r}\frac{\partial B_z}{\partial \theta} + \left(\frac{ck_z}{\omega} - \frac{v_0}{c}\frac{\omega_p^2/\gamma_0}{\omega\omega'}\right)\frac{\partial E_z}{\partial r}\right] \tag{14.150}$$

Eliminating E_r from (14.139) and (14.143) yields

$$B_\theta = \frac{-i\omega}{c\alpha^2}\left[\left(1 - \frac{\omega_p^2/\gamma_0}{\omega\omega'}\right)\frac{\partial E_z}{\partial r} + \frac{ck_z}{\omega}\frac{1}{r}\frac{\partial B_z}{\partial \theta}\right] \tag{14.151}$$

Eliminating E_θ from (14.140) and (14.142) gives

$$B_r = \frac{i\omega}{c\alpha^2}\left[\left(1 - \frac{\omega_p^2/\gamma_0}{\omega\omega'}\right)\frac{1}{r}\frac{\partial E_z}{\partial \theta} - \frac{ck_z}{\omega}\frac{\partial B_z}{\partial r}\right] \tag{14.152}$$

Eliminate B_r from (14.140) and (14.142) yields

$$E_\theta = \frac{i\omega}{c\alpha^2}\left[\frac{\partial B_z}{\partial r} - \left(\frac{ck_z}{\omega} - \frac{v_0}{c}\frac{\omega_p^2/\gamma_0}{\omega\omega'}\right)\frac{1}{r}\frac{\partial E_z}{\partial \theta}\right] \tag{14.153}$$

where

$$\alpha = \left(k_z^2 + \frac{\omega_p^2/\gamma_0 - \omega^2}{c^2}\right)^{\frac{1}{2}} \tag{14.154}$$

If E_z and B_z are known, other field components can be determined.

• **Cylindrically symmetric case:** $\frac{\partial}{\partial\theta} = 0$. For future reference, we write down the θ-symmetric case equations.

$$E_r = \frac{-i\omega}{c\alpha^2}\left(\frac{ck_z}{\omega} - \frac{v_0}{c}\frac{\omega_p^2/\gamma_0}{\omega\omega'}\right)\frac{\partial E_z}{\partial r} \tag{14.155}$$

$$B_\theta = \frac{-i\omega}{c\alpha^2}\left(1 - \frac{\omega_p^2/\gamma_0}{\omega\omega'}\right)\frac{\partial E_z}{\partial r} \tag{14.156}$$

$$B_r = -\frac{i\omega}{c\alpha^2}\left(\frac{ck_z}{\omega}\right)\frac{\partial B_z}{\partial r} \tag{14.157}$$

$$E_\theta = \frac{i\omega}{c\alpha^2}\frac{\partial B_z}{\partial r} \tag{14.158}$$

The above equations indicate that E_r and B_θ are determined by E_z only—constituting the *E-wave* set. B_r and E_θ are determined by B_z only—constituting the *B-wave* set.

• **Solution of the Helmholtz Equation: E_z and B_z wave equation.** Equation (137) is written in cylindrical coordinates

$$\frac{1}{r}\frac{\partial}{\partial r}\left(r\,\frac{\partial E_z}{\partial r}\right) + \frac{1}{r^2}\frac{\partial^2 E_z}{\partial\theta^2} - \alpha^2 E_z = 0 \tag{14.159}$$

which is solved by

$$E_z(r,\theta,z,t) = e^{ik_z z - i\omega t}\sum_{n=-\infty}^{\infty} e^{in\theta}\phi_n(\alpha r) \tag{14.160}$$

where $\phi_n(r)$ should solve the modified Bessel equation

$$x^2\,\frac{d^2\phi_n(x)}{dx^2} + x\,\frac{d\phi_n}{dx} - \phi_n(n^2 + x^2) = 0 \quad (x = \alpha r) \tag{14.161}$$

The plasma solution $\phi_n(\alpha r)$ is a linear combination of modified Bessel functions $I_n(\alpha r)$ and $K_n(\alpha r)$, but K_n should be rejected because it blows up at $r = 0$. The vacuum solution should be written as $\phi_n(\lambda r)$ with

$$\lambda = \left(k_z^2 - \frac{\omega^2}{c^2}\right)^{\frac{1}{2}} \tag{14.162}$$

and $\phi_n(\lambda r)$ is a linear combination of modified Bessel functions $I_n(\lambda r)$ and $K_n(\lambda r)$. If the vacuum is extended to $r \to \infty$, the function I_n should be rejected.

Vacuum solution $a < r < b$:

$$E_{z1}(r,\theta,k_z,\omega) = \sum_{n=-\infty}^{\infty} e^{in\theta}\left[A_n K_n(\lambda r) + C_n I_n(\lambda r)\right] \tag{14.163}$$

$$B_{z1}(r,\theta,k_z,\omega) = \sum_{n=-\infty}^{\infty} e^{in\theta}\left[F_n K_n(\lambda r) + G_n I_n(\lambda r)\right] \tag{14.164}$$

Plasma solution $0 \leq r < a$:

$$E_{z2}(r,\theta,k_z,\omega) = \sum_{n=-\infty}^{\infty} L_n\ e^{in\theta} I_n(\alpha r) \tag{14.165}$$

$$B_{z2}(r,\theta,k_z,\omega) = \sum_{n=-\infty}^{\infty} D_n\ e^{in\theta} I_n(\alpha r) \tag{14.166}$$

where $A_n,\ C_n,\ D_n,\ F_n,\ G_n,\ L_n$ are all constants.

- **θ-symmetric wave. Vacuum solution**

$$E_{z1} = A_0 K_0(\lambda r) + C_0 I_0(\lambda r) \tag{14.167}$$

$$E_{r1} = \frac{-ik_z}{\lambda}(A_0 K_0'(\lambda r) + C_0 I_0'(\lambda r)) \tag{14.168}$$

$$B_{\theta 1} = \frac{-i}{\lambda}\frac{\omega}{c}(A_0 K_0'(\lambda r) + C_0 I_0'(\lambda r)) \tag{14.169}$$

$$B_{z1} = F_0 K_0(\lambda r) + G_0 I_0(\lambda r) \tag{14.170}$$

$$B_{r1} = \frac{-ik_z}{\lambda}(F_0 K_0'(\lambda r) + G_0 I_0'(\lambda r)) \tag{14.171}$$

$$E_{\theta 1} = \frac{i}{\lambda}\frac{\omega}{c}(F_0 K_0'(\lambda r) + G_0 I_0'(\lambda r)) \tag{14.172}$$

where $'$ is the derivative with respect to the argument, and we used (14.155)–(14.158).

Plasma solution.

$$E_{z2} = L_0 I_0(\alpha r) \tag{14.173}$$

$$E_{r2} = -\frac{i}{c\alpha}\left(ck_z - \frac{v_0}{c}\,\frac{\omega_p^2/\gamma_0}{\omega'}\right) L_0 I_0'(\alpha r) \tag{14.174}$$

$$B_{\theta 2} = -\frac{i\omega}{c\alpha}\left(1 - \frac{\omega_p^2/\gamma_0}{\omega\omega'}\right) L_0 I_0'(r\alpha) \tag{14.175}$$

$$B_{z2} = D_0 I_0(\alpha r) \tag{14.176}$$

$$B_{r2} = -\frac{ik_z}{\alpha} D_0 I_0'(r\alpha) \tag{14.177}$$

$$E_{\theta 2} = \frac{i\omega}{c\alpha} D_0 I_0'(r\alpha) \tag{14.178}$$

Boundary Equations.

E-wave boundary conditions: $[E_z] = 0$, Eq. (14.130), Eq. (14.146), $E_z|_b = 0$
B-wave boundary conditions: $[B_z] = 0$, $[B_r] = 0$, $[E_\theta] = 0$, $\frac{\partial B_z}{\partial r}|_b = 0$

- **E-wave.**

$$[E_z] = 0: A_0 K_0(\lambda a) + C_0 I_0(\lambda a) = L_0 I_0(\alpha a) \tag{14.179}$$

$$(14.130): A_0 K_0'(\lambda a) + C_0 I_0'(\lambda a) = \frac{\lambda}{\alpha}\left(1 - \frac{\omega_p^2}{\omega'^2\gamma_0^3}\right) L_0 I_0'(\alpha a) \tag{14.180}$$

Equation (14.146) is redundant with the above equation.

$$E_z|_b = 0: \quad A_0 k_0(\lambda b) + C_0 I_0(\lambda b) = 0 \tag{14.181}$$

The above three equations yield the E-wave dispersion relation:

$$\frac{\lambda}{\alpha}\left(1 - \frac{\omega_p^2}{\omega'^2\gamma_0^3}\right)\frac{I_0'(a\alpha)}{I_0(a\alpha)} = \frac{K_0(b\lambda)I_0'(a\lambda) - I_0(b\lambda)K_0'(a\lambda)}{K_0(b\lambda)I_0(a\lambda) - I_0(b\lambda)K_0(a\lambda)} \tag{14.182}$$

If $b \to \infty$ (infinite radial extension of the vacuum region), $K_0(\lambda b) \to 0$, giving

$$\frac{\lambda}{\alpha}\left(1 - \frac{\omega_p^2}{\omega'^2\gamma_0^3}\right)\frac{I_0'(a\alpha)}{I_0(a\alpha)} = \frac{K_0'(a\lambda)}{K_0(a\lambda)}$$

which becomes per $I_0' = I_1$ and $K_0' = -K_1$,

$$\frac{\lambda}{\alpha}\left(1 - \frac{\omega_p^2}{\omega'^2\gamma_0^3}\right)\frac{I_1(a\alpha)}{I_0(a\alpha)} = -\frac{K_1(a\lambda)}{K_0(a\lambda)} \tag{14.183}$$

If $a \to \infty$, $I_0 \simeq I_1$, $K_0 \simeq K_1$, and therefore, we have

$$\alpha + \lambda\left(1 - \frac{\omega_p^2}{\omega'^2\gamma_0^3}\right) = 0 \tag{14.184}$$

This is the dispersion relation of the electromagnetic wave in semi-infinite plasma.

- **B-wave.**

$$[B_z] = 0{:}F_0K_0(a\lambda) + G_0I_0(\lambda a) = D_0I_0(a\alpha) \tag{14.185}$$

$$[B_r] = 0{:}\frac{-ik_z}{\lambda}\left(F_0K_0'(a\lambda) + G_0I_0'(a\lambda)\right) + \frac{ik_z}{\alpha}D_0I_0'(a\alpha) = 0 \tag{14.186}$$

$[E_\theta] = 0$ is redundant with $[B_r] = 0$.

$$\left.\frac{\partial B_z}{\partial r}\right|_b = 0: \quad F_0K_0'(b\lambda) + G_0I_0'(b\lambda) = 0 \tag{14.187}$$

The above three equations yield the B-wave dispersion relation:

$$\frac{\lambda}{\alpha}\frac{I_0'(a\alpha)}{I_0(a\alpha)} = \frac{I_0'(b\lambda)K_0'(a\lambda) - I_0'(a\lambda)K_0'(b\lambda)}{I_0'(b\lambda)K_0(a\lambda) - I_0(a\lambda)K_0'(b\lambda)} \tag{14.188}$$

If $b \to \infty$ (infinite radial extension of the vacuum region), $K_0'(\lambda b) \to 0$, giving

$$\frac{\lambda}{\alpha}\frac{I_0'(a\alpha)}{I_0(a\alpha)} = \frac{K_0'(a\lambda)}{K_0(a\lambda)} \tag{14.189}$$

In the limit of $a \to \infty$, (14.189) becomes

$$\alpha = -\lambda \tag{14.190}$$

Clearly (14.190) is impossible, and we conjecture that (14.188) would have no solutions. This corresponds to the fact that the transverse electric mode of surface wave is impossible.

14.4 Electrostatic Surface Wave in Magnetized Cylindrical Plasma

In this section, we derive the surface wave dispersion relations in a cold magnetized plasma contained in infinite-length cylinder of radius "a", interfaced with vacuum. We use the following fluid equations written in cylindrical coordinates:

$$-i\omega\mathbf{v}(r,\theta,z) = \frac{e}{m}\nabla\phi - \omega_c\mathbf{v}\times\hat{\mathbf{z}} \tag{14.191}$$

$$-i\omega n(r,\theta,z) + N(r)\nabla\cdot\mathbf{v} + v_r\frac{\partial N}{\partial r} = 0 \tag{14.192}$$

$$\nabla^2\phi = 4\pi e n \tag{14.193}$$

Here we assume that the equilibrium density $N(r)$ has density gradients in the radial direction. At the interface $r = a$, N changes abruptly with infinite gradient, otherwise homogeneous. This artificiality is designed solely to derive the boundary conditions at the interface by 'infinitesimal integration' across the interface. Equation (14.191) yields

$$v_r = \frac{e}{m}\,\frac{\omega}{\omega^2-\omega_c^2}\left(i\,\frac{\partial\phi}{\partial r}+\frac{\omega_c}{\omega}\,\frac{1}{r}\,\frac{\partial\phi}{\partial\theta}\right) \tag{14.194}$$

$$v_\theta = \frac{e}{m}\,\frac{\omega}{\omega^2-\omega_c^2}\left(-\frac{\omega_c}{\omega}\,\frac{\partial\phi}{\partial r}+i\,\frac{1}{r}\,\frac{\partial\phi}{\partial\theta}\right) \tag{14.195}$$

$$v_z = \frac{i}{\omega}\,\frac{e}{m}\,\frac{\partial\phi}{\partial z} \tag{14.196}$$

Equations (14.192)–(14.196) give the wave equation written in terms of ϕ:

$$\begin{aligned}&\frac{\partial}{\partial r}\left[\left(1-\frac{\omega_p^2}{\omega^2-\omega_c^2}\right)\frac{\partial\phi}{\partial r}\right]+\left(1-\frac{\omega_p^2}{\omega^2-\omega_c^2}\right)\frac{1}{r}\,\frac{\partial\phi}{\partial r}+\frac{\partial}{\partial z}\left[\left(1-\frac{\omega_p^2}{\omega^2}\right)\frac{\partial\phi}{\partial z}\right]\\&+\frac{i\omega_c}{\omega}\,\frac{1}{r}\,\frac{\partial}{\partial r}\left(\frac{\omega_p^2}{\omega^2-\omega_c^2}\right)\frac{\partial\phi}{\partial\theta}+\left(1-\frac{\omega_p^2}{\omega^2-\omega_c^2}\right)\frac{1}{r^2}\,\frac{\partial^2\phi}{\partial\theta^2}=0\end{aligned} \tag{14.197}$$

where $\omega_p^2(r) = 4\pi e^2 N(r)/m$. If we take the infinitesimal integration across the interface $r = a$ in the manner $\int_{-\epsilon}^{\epsilon}(\cdots)dr$, we can extract useful boundary conditions. In this operation, we pick up only the perfect differentials that survive the limit $\epsilon \to 0$. Eq. (14.194) yields,

$$[\phi]_{r=a} = 0 \tag{14.198}$$

Integrating (14.197) yields

$$\left[\left(1-\frac{\omega_p^2}{\omega^2-\omega_c^2}\right)\frac{\partial\phi}{\partial r}\right]_{r=a} = 0 \tag{14.199}$$

where the square bracket denotes the jump across the interface. Putting ω_p = const. or zero in (14.197) gives the wave equation in the plasma or in the vacuum. We also restrict to an azimuthally symmetric field ($\partial/\partial\theta = 0$). We have

$$\frac{\partial^2\phi}{\partial r^2}+\frac{1}{r}\,\frac{\partial\phi}{\partial r}+\frac{\partial^2\phi}{\partial z^2} = 0, \text{ in vacuum} \tag{14.200}$$

$$\frac{\partial^2\phi}{\partial r^2}+\frac{1}{r}\,\frac{\partial\phi}{\partial r}+\xi^2\,\frac{\partial^2\phi}{\partial z^2} = 0, \text{ in plasma} \tag{14.201}$$

where ξ is defined by

$$\xi = \sqrt{\left(1 - \frac{\omega_p^2}{\omega^2}\right) \Big/ \left(1 - \frac{\omega_p^2}{\omega^2 - \omega_c^2}\right)}$$

Now let us solve the above equations for an infinite cylindrical plasma. Since the z-direction is infinite, we can assume $\phi(z) \sim e^{ik_z z}$ and the plasma equation in (14.201) becomes

$$\frac{\partial^2 \phi(r)}{\partial r^2} + \frac{1}{r}\frac{\partial \phi}{\partial r} + \xi^2 k_z^2 \phi(r) = 0 \tag{14.202}$$

where we suppressed the z-dependent phase. Equation (14.202)(78) is solved by the functions I_0 and K_0, the modified Bessel functions of first and second kind of order zero, respectively. K_0 is rejected since it blows up at $r = 0$, so the plasma solution takes the form

$$\phi_p(r) = AI_0(k_z \xi r) \quad 0 < r < a \tag{14.203}$$

In vacuum, the function I_0 should be rejected since it blows up at $r = \infty$. We have

$$\phi_v(r) = BK_0(k_z r) \quad r > a \tag{14.204}$$

where A, B are constants to be determined by boundary conditions. Equations (14.198) and (14.199) gives

$$AI_0(\xi k_z a) = BK_0(k_z a) \tag{14.205}$$

$$\nu \xi A I_0'(\xi k_z a) = BK_0'(k_z a) \tag{14.206}$$

where prime denotes derivative with respect to the argument, and ν is defined by

$$\nu = 1 - \frac{\omega_p^2}{\omega^2 - \omega_c^2}$$

The above two equations yield the dispersion relation

$$\nu \xi \frac{I_0'(\xi k_z a)}{I_0(\xi k_z a)} = \frac{K_0'(k_z a)}{K_0(k_z a)} \tag{14.207}$$

Using the formula for Bessel functions

$$\frac{d}{dx} I_0(\alpha x) = \alpha I_1(\alpha x), \quad \frac{d}{dx} K_0(\alpha x) = -\alpha K_1(\alpha x) \tag{14.208}$$

(14.207) becomes

$$\nu\xi \frac{I_1(\xi k_z a)}{I_0(\xi k_z a)} = -\frac{K_1(k_z a)}{K_0(k_z a)} \tag{14.209}$$

If $k_z a \to \infty$, $I_0 \approx I_1$, $K_0 \approx K_1$, and therefore (14.210) becomes

$$\nu\xi = -1 \quad \text{or} \quad \left(1 - \frac{\omega_p^2}{\omega^2 - \omega_c^2}\right)\sqrt{\frac{1 - \frac{\omega_p^2}{\omega^2}}{1 - \frac{\omega_p^2}{\omega^2 - \omega_c^2}}} = -1 \tag{14.210}$$

Scrutinizing (14.211) indicates that ω^2 should be in the range $\omega_c^2 < \omega^2 < \omega_p^2$. Solving (14.211) for ω^2 gives

$$\omega^2 = \frac{1}{2}(\omega_p^2 + \omega_c^2) \tag{14.211}$$

which is the upper-hybrid frequency.

By taking $a \to \infty$, we removed the cylindrical boundary, and (14.212) is the eigen mode of the upper-hybrid frequency oscillating in the semi-infinite plasma.

14.5 Electromagnetic Surface Waves of a Relativistic Beam in a Magnetized Plasma Propagating in a Cylindrical Column

This section is derived from

Reference: Hee J. Lee, *Electromagnetic surface waves in a relativistic magnetized plasma propagating in a cylindrical column*, Applied Phys. Res. **16**, No. 2 (2024), pp. 1–11.

In this section, dispersion relations of electromagnetic surface waves in a magnetized relativistic plasma propagating in a cylindrical column surrounded by vacuum are derived by means of the relativistic fluid equations. The zero order relativistic beam is assumed to stream along the direction of the axial magnetic field. We solve the E- and B-wave modes analytically in terms of the cylindrical coordinates in *weak or intermediate values of the magnetic field* in which the cyclotron frequency is much smaller than the relativistic Doppler-shifted frequency. The entangled field components of the E- and B-waves are decoupled by assuming that *the flow is incompressible.* The static magnetic field splits the E- and B-waves, thus doubling the modes in unmagnetized plasma. By taking the radius of the cylinder infinite, the dispersion relations in a semi-infinite plasma are derived. The beam-generated surface waves are interesting in the context of both astrophysical and laboratory plasma researches.

We begin with the relativistic Euler equation

$$\frac{\partial \mathbf{p}}{\partial t} + (\mathbf{v} \cdot \nabla)\mathbf{p} = -e\left(\mathbf{E} + \frac{1}{c}\,\mathbf{v} \times (\mathbf{B} + \mathbf{B}_0)\right) \tag{14.212}$$

where $\mathbf{B}$ is the wave magnetic field and $\mathbf{B}_0 = \hat{\mathbf{z}}B_0$ is the static magnetic field imbedded in the plasma, and

$$\mathbf{p} = m\gamma\mathbf{v} \tag{14.213}$$

$$\gamma = \frac{1}{\sqrt{1 - \frac{v^2}{c^2}}} \tag{14.214}$$

We have the continuity equation for the electron number density

$$\frac{\partial n}{\partial t} + \nabla \cdot (n\mathbf{v}) = 0 \tag{14.215}$$

We have the Maxwell equations

$$\nabla \times \mathbf{E} = -\frac{1}{c}\,\frac{\partial \mathbf{B}}{\partial t} \tag{14.216}$$

$$\nabla \times \mathbf{B} = \frac{1}{c}\,\frac{\partial \mathbf{E}}{\partial t} + \frac{4\pi}{c}\mathbf{J} \tag{14.217}$$

$$\nabla \cdot \mathbf{E} = -4\pi e n_1 \tag{14.218}$$

In the development, we shall use the subscript "1" for the perturbed quantities as already used in the above equation.

We linearize the dynamical equations by setting

$$n(r, \theta, z) = N(r) + n_1(r, \theta)\ e^{ik_z z - i\omega t} \tag{14.219}$$

where $N(r)$ is the zero order number density that has a step function profile: $N(r) = N_0,$ const. for $0 < r < a$ in plasma; $N = 0,\ r > a$ in vacuum. Also we have

$$\mathbf{v} = \hat{\mathbf{z}}v_0 + \mathbf{v}_1(r, \theta)\ e^{ik_z z - i\omega t} \tag{14.220}$$

where v_0 is the constant zero order velocity. Then, we calculate per binomial expansion

$$\gamma_0 = \frac{1}{\sqrt{1 - \frac{v_0^2}{c^2}}}, \qquad \gamma_1 = \gamma_0^3\ \frac{v_0 v_{z1}}{c^2} \tag{14.221}$$

Equation (14.212) gives

$$i\omega'\left(\gamma_0\mathbf{v}_1 + \hat{\mathbf{z}}v_{z1}\gamma_0^3\ \frac{v_0^2}{c^2}\right) = \frac{e}{m}\left(\mathbf{E} + \frac{v_0}{c}\ \hat{\mathbf{z}} \times \mathbf{B}\right) + \omega_c\mathbf{v}_1 \times \hat{\mathbf{z}} \tag{14.222}$$

where $\omega' = \omega - k_z v_0$ is the Doppler-shifted frequency, $\omega_c = eB_0/mc$ is the cyclotron frequency, and $\hat{\mathbf{z}} = \mathbf{B}_0/B_0$ is the unit vector.

Solving (14.222) for $\mathbf{v}_1$ gives

$$\mathbf{v}_1 = \frac{1}{i\bar{\omega}\xi}\left[\mathbf{F} - \frac{i\omega_c}{\bar{\omega}}\mathbf{F}\times\hat{\mathbf{z}} - \hat{\mathbf{z}}\frac{e}{m}E_z\left(1 - \frac{\xi}{\gamma_0^2}\right)\right] \tag{14.223}$$

where

$$\mathbf{F} = \frac{e}{m}\left(\mathbf{E} + \frac{v_0}{c}\hat{\mathbf{z}}\times\mathbf{B}\right) \tag{14.224}$$

$$\xi = 1 - \frac{\omega_c^2}{\bar{\omega}^2}, \quad \bar{\omega} = \gamma_0\omega' \tag{14.225}$$

Equation (14.223) gives the z-component of the velocity immediately

$$v_{z1} = -\frac{ie}{m\bar{\omega}\gamma_0^2}E_z \tag{14.226}$$

From the continuity equation, (14.215), we obtain

$$n_1 = -\frac{i}{\omega'}\nabla\cdot(N\mathbf{v}_1) \tag{14.227}$$

The current is calculated by

$$\mathbf{J} = -eN\mathbf{v}_1 - ev_0 n_1\hat{\mathbf{z}} \tag{14.228}$$

Equation (14.223) gives in cylindrical coordinates

$$v_r = \frac{-i}{\xi\bar{\omega}}\left(E_r - \frac{v_0}{c}B_\theta\right) - \frac{\omega_c}{\xi\bar{\omega}^2}\left(E_\theta + \frac{v_0}{c}B_r\right) \tag{14.229}$$

$$v_\theta = \frac{-i}{\xi\bar{\omega}}\left(E_\theta + \frac{v_0}{c}B_r\right) + \frac{\omega_c}{\xi\bar{\omega}^2}\left(E_r - \frac{v_0}{c}B_\theta\right) \tag{14.230}$$

$$v_z = \frac{-iE_z}{\bar{\omega}\gamma_0^2} \tag{14.231}$$

where the common factor e/m on the field components has been suppressed. Using the above equations in the Maxwell equations, we can write the cylindrical components of the Maxwell equations (5) and (6) with azimuthal symmetry ($\partial/\partial\theta = 0$) in the form

$$E_\theta = -\frac{\omega}{ck_z}B_r \tag{14.232}$$

$$ik_z E_r - \frac{\partial E_z}{\partial r} = \frac{i\omega}{c}B_\theta \tag{14.233}$$

$$\frac{\partial E_\theta}{\partial r} + \frac{E_\theta}{r} = \frac{i\omega}{c} B_z \tag{14.234}$$

$$-k_z B_\theta = -\frac{\omega}{c} E_r + \frac{\omega_p^2}{c\bar{\omega}\xi} \left[E_r - \frac{v_0}{c} B_\theta - \frac{i\omega_c}{\bar{\omega}} \left(E_\theta + \frac{v_0}{c} B_r \right) \right] \tag{14.235}$$

$$ik_z B_r - \frac{\partial B_z}{\partial r} = \frac{-i\omega}{c} E_\theta$$

$$-\frac{\omega_p^2}{c\bar{\omega}\xi} \left[-i \left(E_\theta + \frac{v_0}{c} B_r \right) + \frac{\omega_c}{\bar{\omega}} \left(E_r - \frac{v_0}{c} B_\theta \right) \right] \tag{14.236}$$

$$\frac{\partial B_\theta}{\partial r} + \frac{B_\theta}{r} = -\frac{i\omega q}{c} E_z + \frac{v_0}{c} \nabla \cdot \mathbf{E} \tag{14.237}$$

In addition, we have

$$\frac{\partial B_r}{\partial r} + \frac{B_r}{r} + ik_z B_z = 0 \quad (\nabla \cdot \mathbf{B} = 0) \tag{14.238}$$

Equations (14.223), (14.227), and (14.228) determine the dielectric function of the beam plasma, but the actual calculation would be enormously complex.

Let us consider the case of infinite axial magnetic field. Taking $\xi \to -\infty$ in (14.223), we have $v_{x1} = v_{y1} = 0$, and only v_{z1} is nonzero as given by (14.226). This means that the particle motion is only along the magnetic field while the perpendicular motion is entirely suppressed, which is a great simplifying factor. Then, we obtain

$$n_1 = -\frac{ieNk_z}{m\gamma_0\bar{\omega}^2} E_z \tag{14.239}$$

$$\mathbf{J} = \hat{\mathbf{z}} \, \frac{i\, Ne^2\omega}{m\gamma_0\bar{\omega}^2} E_z \tag{14.240}$$

From the relations for conductivity and dielectric permittivity, $J_i = \sigma_{ij} E_j$, $\varepsilon_{ij} = \delta_{ij} + \frac{4\pi i}{\omega}\sigma_{ij}$, we obtain the relativistic dielectric tensor in the form

$$\varepsilon_{ij} = \begin{pmatrix} 1 & 0 & 0 \\ 0 & 1 & 0 \\ 0 & 0 & \varepsilon_{zz} \end{pmatrix} \qquad \varepsilon_{zz} = 1 - \frac{\omega_p^2}{\gamma_0\bar{\omega}^2}$$

where $\omega_p = \sqrt{\frac{4\pi Ne^2}{m}}$ is the plasma frequency. In the non-relativistic limit, the above dielectric tensor agrees with the well-known form (Krall and Trivelpiece, 1973).

• **Wave equations for E_z and B_z.** Now we shall investigate the case of a weak or intermediate value of B_0 such that $\omega_c^2/\bar{\omega}^2 \ll 1$, and neglect the terms of higher powers of $(\frac{\omega_c}{\bar{\omega}})^2$ in the subsequent analysis. Now $\xi = 1$ in (14.229) and (14.230). The wave equations of the primary importance are the equations for E_z and B_z. Taking curl of (14.216) and (14.217) and taking the z-component of the resultant equations give

$$\nabla^2 E_z + \frac{\omega^2 q}{c^2} E_z = i\left(k_z - \frac{v_0\omega}{c^2}\right)\nabla\cdot\mathbf{E} \tag{14.241}$$

$$\nabla^2 B_z + \frac{\omega^2}{c^2} B_z = \frac{4\pi e N}{c}(\nabla\times\mathbf{v}_1)_z \tag{14.242}$$

where

$$q = 1 - \frac{\omega_p^2}{\omega\bar{\omega}\gamma_0^2}$$

Equations (14.241) and (14.242) are valid in cartesian and cylindrical coordinate systems (not spherical). The right hand sides of (14.241) and (14.242) can be expressed in terms of B_z and E_z, respectively, after some algebra, yielding the following coupled equations:

$$\nabla^2 B_z + \frac{\omega^2 - \omega_p^2/\gamma_0}{c^2}\, B_z = -i\,\beta E_z \tag{14.243}$$

where

$$\beta = \frac{\omega_c}{c}\,\left(k_z - \frac{v_0\omega}{c^2}\right)\,\frac{\omega_p^2}{\bar{\omega}^2} \tag{14.244}$$

$$\nabla^2 E_z + \frac{\omega^2 - \omega_p^2/\gamma_0}{c^2}\, E_z = i\,\beta' B_z \tag{14.245}$$

where

$$\beta' = \frac{\omega_c}{c}\left(k_z - \frac{v_0\omega}{c^2}\right)\frac{\omega_p^2}{\bar{\omega}^2 - \omega_p^2/\gamma_0} \tag{14.246}$$

Equations (14.243) and (14.245) are the two coupled wave equations to be solved, which look quite symmetrical. If $\omega_c = 0$, they are decoupled.

Now we proceed to solve the wave equations in cylindrical coordinates (r, θ, z) in the plasma region $0 \le r < a$. We shall look for only the azimuthally symmetric solution $(\partial/\partial\theta = 0)$. Introducing a non-dimensional variable

$$x \equiv \alpha r, \quad \alpha = \sqrt{k_z^2 - \frac{\omega^2 - \omega_p^2/\gamma_0}{c^2}} \tag{14.247}$$

Equations (14.243) and (14.245) are written as

$$\frac{\partial^2 B_z}{\partial x^2} + \frac{1}{x}\frac{\partial B_z}{\partial x} - B_z(x) = -i\nu E_z(x) \tag{14.248}$$

$$\frac{\partial^2 E_z}{\partial x^2} + \frac{1}{x}\frac{\partial E_z}{\partial x} - E_z(x) = i\nu' B_z(x) \tag{14.249}$$

where

$$\nu = \beta/\alpha^2, \quad \nu' = \beta'/\alpha^2 \tag{14.250}$$

The complete solution of the nonhomogeneous Bessel equation in Eq. (14.248), is composed of the homogeneous solution and the particular solution. The homogeneous solution is a linear combination of *modified Bessel functions of order zero*, $I_0(x)$ and $K_0(x)$, but $K_0(x)$ blows up at $x = 0$ and should be rejected. The particular solution can be obtained by employing the method of variation of parameter in the following form

$$B_z(x) = K_0(x)\int_0^x h(\zeta)I_0(\zeta)\zeta d\zeta - I_0(x)\int_0^x h(\zeta)K_0(\zeta)\zeta d\zeta + C_2 I_0(x) \tag{14.251}$$

where C_2 is the integration constant and $h(x)$ is the right side function in (14.248). This solution also can be proven by a direct substitution with noting the Wronskian identity

$$K_0'(x)I_0(x) - I_0'(x)K_0(x) = \frac{1}{x} \quad \left(' = \frac{d}{dx}\right) \tag{14.252}$$

Clearly, the second term in the right side of (14.251) is regular. The first term in the right side is also regular at $x = 0$ since the singularity of $K_0(x)$ at $x = 0$, which varies as $\sim \ln x - \gamma + \ln 2 + \textit{positive powers}$, is obliterated by the following multiplication of higher order terms. Hence, we can write the solutions of (14.248) and (14.249) as follows:

$$E_z(x) = C_1 I_0(x) + i\nu' \int_0^x d\zeta\ \zeta\ B_z(\zeta)\left[K_0(x)I_0(\zeta) - I_0(x)K_0(\zeta)\right] \tag{14.253}$$

$$B_z(x) = C_2 I_0(x) - i\nu \int_0^x d\zeta\ \zeta\ E_z(\zeta)\left[K_0(x)I_0(\zeta) - I_0(x)K_0(\zeta)\right] \tag{14.254}$$

In substituting (14.253) into (14.254) or vice versa, we neglect the term $\nu\nu' \ll 1$, which is of order $\omega_c^2/\bar{\omega}^2$. Thus, we obtain

$$E_z(x) = C_1 I_0(x) + C_2 i\nu' \Phi_0(x) \tag{14.255}$$

$$B_z(x) = -C_1 i\nu \Phi_0(x) + C_2 I_0(x) \tag{14.256}$$

where

$$\Phi_0(x) = \int_0^x d\zeta\; \zeta\; I_0(\zeta)\,(K_0(x)I_0(\zeta) - I_0(x)K_0(\zeta)) \tag{14.257}$$

The integration constants C_1 and C_2 in (14.255) and (14.256) are not independent. If the solutions in (14.255) and (14.256) are substituted into (14.248) and (14.249), we have a linear homogeneous algebraic system of equations for the constants C_1 and C_2. The latter constants are not uniquely determined but only their ratio is determined. Let us find it. If the said substitutions are done, we have

$$C_1(L\Phi_0(x) - I_0(x)) = C_2 i\nu'\Phi_0(x) \tag{14.258}$$

$$C_2(L\Phi_0(x) - I_0(x)) = -C_1 i\nu\Phi_0(x) \tag{14.259}$$

where $L = \frac{\partial^2}{\partial x^2} + \frac{1}{x}\frac{\partial}{\partial x} - 1$ and we used $LI_0(x) = 0$. The above two equations give the relation

$$C_2 = \pm i\sqrt{\frac{\nu}{\nu'}}\; C_1 \tag{14.260}$$

Using (14.260) in (14.255) and (14.256) gives

$$E_z(x) = C_1(I_0(x) \mp \sqrt{\nu\nu'}\;\Phi_0(x)) \tag{14.261}$$

$$B_z(x) = C_2(I_0(x) \mp \sqrt{\nu\nu'}\;\Phi_0(x)) \tag{14.262}$$

Hence, it follows that

$$\frac{E_z(x)}{B_z(x)} = \frac{C_1}{C_2} \tag{14.263}$$

which in turn gives after substituting into (14.248) and (14.249)

$$\frac{\partial^2 E_z}{\partial x^2} + \frac{1}{x}\,\frac{\partial E_z}{\partial x} - (1 \pm \sqrt{\nu\nu'})E_z = 0 \tag{14.264}$$

$$\frac{\partial^2 B_z}{\partial x^2} + \frac{1}{x}\,\frac{\partial B_z}{\partial x} - (1 \pm \sqrt{\nu\nu'})B_z = 0 \tag{14.265}$$

where $x = \alpha r$. This is an important development: the coupled equations (14.248) and (14.249) are separated to become uncoupled in two identical Bessel equations. Solving (14.264) and (14.265) in terms of the original cylindrical variable r in the region $0 \leq r < a$, we write

$$E_z(r) = A_1 I_0(\Lambda r) \tag{14.266}$$

$$B_z(r) = A_2 I_0(\Lambda r) \tag{14.267}$$

where $A_1,\ A_2$ are constants and

$$\Lambda = \left[(1 \pm \sqrt{\nu\nu'})\left(k_z^2 - \frac{\omega^2 - \omega_p^2/\gamma_0}{c^2}\right)\right]^{\frac{1}{2}} \tag{14.268}$$

Equations (14.266) and (14.267) should read as representing two different modes of $\pm$ signs contained in Λ. Physically, the unmagnetized mode is split into two modes in magnetized plasma as indicated by the $\pm$ signs.

• **Radial and circumferential field components.** In (14.232)–(14.238), E_z and B_z are known. Integrating (14.234) yields E_θ, and B_r follows from (14.232). Or integrating (14.238) gives B_r. The results will be identical. Hence, the B-wave components are determined without much difficulty. However, in the E-wave equations, the E-wave components are entangled with the B-wave components, and the complexity is unwieldy. To make further progress, we assume that the plasma is *incompressible*: $\nabla \cdot \mathbf{v_1} = 0$. The incompressibility removes the sound wave component in the Maxwell equation, thus projecting the electromagnetic characteristics. It mathematically amounts to $\nabla \cdot \mathbf{E} = 0$ per the continuity equation and the Poisson equation. Thus, we solve, in lieu of (14.27),

$$\frac{\partial B_\theta}{\partial r} + \frac{B_\theta}{r} = -\frac{i\omega q}{c} E_z \tag{14.269}$$

Knowing B_θ, E_r follows from (14.233), or equivalently it can be determined by solving

$$\frac{\partial E_r}{\partial r} + \frac{E_r}{r} = -ik_z E_z \tag{14.270}$$

This way, the B-wave components are all solved.

Using (14.267) and letting $\Lambda r \equiv x$ in (14.238) give

$$\frac{\partial B_r(x)}{\partial x} + \frac{B_r}{x} = -\frac{ik_z A_2}{\Lambda} I_0(x) \tag{14.271}$$

or

$$\frac{d}{dx}(x B_r) = -\frac{ik_z A_2}{\Lambda} x I_0(x) \tag{14.272}$$

Using $xI_0(x) = \frac{d}{dx}(xI_1(x))$ in the above equation yields

$$B_r(x) = -\frac{ik_z A_2}{\Lambda} I_1(x) \tag{14.273}$$

Then, we have from (14.232)

$$E_\theta(x) = \frac{i\omega}{c\Lambda} A_2 I_1(x) \tag{14.274}$$

Analogously, it follows from (14.269) and (14.270)

$$B_\theta(x) = -\frac{i\omega q}{c\Lambda} A_1 I_1(x) \tag{14.275}$$

$$E_r(x) = -\frac{ik_z}{\Lambda} A_1 I_1(x) \tag{14.276}$$

• **Dispersion relation** is obtained by connecting the plasma and the vacuum fields by boundary conditions at the interface. The longitudinal vacuum fields, E_z^v and B_z^v, satisfy

$$\nabla^2 E_z^v + \frac{\omega^2}{c^2} E_z^v = 0 \tag{14.277}$$

$$\nabla^2 B_z^v + \frac{\omega^2}{c^2} B_z^v = 0 \tag{14.278}$$

whose solutions, after discarding $I_0(\lambda r)$, which blows up as $r \to \infty$, are

$$E_z^v(r) = GK_0(\lambda r) \tag{14.279}$$

$$B_z^v(r) = HK_0(\lambda r) \tag{14.280}$$

where G and H are constants and

$$\lambda = \sqrt{k_z^2 - \frac{\omega^2}{c^2}} \tag{14.281}$$

Then, all the transverse components are determined by E_z^v and B_z^v. Solving $\nabla \cdot \mathbf{E}^v = 0$ yields

$$E_r^v(r) = \frac{ik_z}{\lambda} GK_1(\lambda r) \tag{14.282}$$

Solving $\nabla \cdot \mathbf{B}^v = 0$ yields

$$B_r^v(r) = \frac{ik_z}{\lambda} HK_1(\lambda r) \tag{14.283}$$

which in turn gives

$$E_\theta^v(r) = -\frac{i\omega}{c\lambda} HK_1(\lambda r) \tag{14.284}$$

We have the vacuum equation corresponding to (14.237):

$$\frac{\partial B_\theta^v}{\partial r} + \frac{B_\theta^v}{r} = -\frac{i\omega}{c} E_z^v \tag{14.285}$$

which is solved by

$$B_\theta^v(r) = \frac{i\omega}{c\lambda} G K_1(\lambda r) \tag{14.286}$$

The boundary equations that connect the plasma and the vacuum fields can be obtained *self-consistently* by the infinitesimal integration over the interface $\lim_{\delta\to 0}\int_{a-\delta}^{a+\delta}(\cdots)dr$. Equations (14.233) and (14.236) give, respectively, by picking up the perfect differential terms, the connection formulas $[E_z] = 0$ and $[B_z] = 0$ at $r = a$.

Equation (14.237) gives the well-known formula

$$[B_\theta] = \frac{v_0}{c}[E_r] \tag{14.287}$$

Applying the boundary conditions gives the following equations:

$$[E_z] = 0: \quad A_1 I_0(a\Lambda) = G K_0(a\lambda) \tag{14.288}$$

Using the above equations gives

$$(14.287): \frac{I_1(a\Lambda)}{\Lambda}\left(1 - \frac{\omega_p^2}{\gamma_0\bar{\omega}^2}\right) A_1 = -\frac{G}{\lambda} K_1(a\lambda) \tag{14.289}$$

$$[B_z] = 0: A_2 I_0(a\Lambda) = H K_0(a\lambda) \tag{14.290}$$

$$[B_r] = 0: -\frac{A_2}{\Lambda} I_1(a\Lambda) = \frac{H}{\lambda} K_1(a\lambda) \tag{14.291}$$

$$[E_\theta] = 0: \frac{A_2}{\Lambda} I_1(a\Lambda) = -\frac{H}{\lambda} K_1(a\lambda) \tag{14.292}$$

The last two equations are redundant.

Equations (14.288) and (14.289) yield the *E-wave dispersion relation*

$$\left(1 - \frac{\omega_p^2}{\gamma_0\bar{\omega}^2}\right)\frac{I_1(a\Lambda)}{I_0(a\Lambda)} = -\frac{\Lambda}{\lambda}\frac{K_1(a\lambda)}{K_0(a\lambda)} \tag{14.293}$$

Equations (14.290) and (14.291) yields the *B-wave dispersion relation*

$$-\frac{1}{\Lambda}\frac{I_1(\Lambda a)}{I_0(\Lambda a)} = \frac{1}{\lambda}\frac{K_1(\lambda a)}{K_0(\lambda a)} \tag{14.294}$$

A noteworthy characteristics of magnetized surface wave, as compared with unmagnetized surface wave, is that the modes are doubled as expressed by the $\pm$ signs in the dispersion relations (see Eq. (14.268)). The static magnetic field B_0 splits the mode in unmagnetized plasma.

By taking the limit of $a \to \infty$ in (14.293) and (14.294), we can derive the corresponding surface wave dispersion relations in the semi-infinite plasma.

If $a\Lambda$ or $a\lambda \to \infty$, $I_0 \approx I_1$, $K_0 \approx K_1$, and therefore, (14.293) and (14.294) become, respectively,

$$\Lambda + \lambda \left(1 - \frac{\omega_p^2}{\gamma_0 \bar{\omega}^2} \right) = 0 \tag{14.295}$$

$$\Lambda = -\lambda \tag{14.296}$$

Clearly, (14.296) is impossible, and this appears to be corresponding to the fact that the transverse electric mode of surface wave is impossible. Equation (14.294) would have no solutions. Equation (14.295) is of the same form as and corresponds to the tranverse magnetic mode of the surface wave (surface polariton) in a semi-infinite plasma.

References

Alexandrov, A. F., L. S. Bogdankevich, and A. A. Rukhadze, *Principles of Plasma Electrodynamics*, Chapter 6 Springer (1984), p. 159.

Barr, H. C. and T. J. M. Boyd, *Surface waves in hot plasmas.* J. Phys. A: Gen. Phys. **5** No. 7 (1972), pp. 1108–1118.

Boardman, A. D. *Electromagnetic Surface Modes* Wiley (1982).

Clemmow, P. C. and J. P. Dougherty, *Electrodynamics of Particles and Plasmas*, Addison-Wesley (1969).

Gradov, O. M. and L. Stenflo, *Linear theory of cold bounded plasma.* Phys. Rep. **94** (1983), pp. 111–137.

Krall, N. A. and A. W. Trivelpiece, *Principles of Plasma Physics*, McGraw-Hill (1973).

Lee, H. J. *Electrostatic surface waves in a magnetized two-fluid plasma*, Plasma Phys. Control. Fusion **37** (1995), pp. 755–762.

Lee, H. J. and S. H. Cho, *Boundary conditions for surface waves propagating along the interface of plasma flow and free space.* J. Plasma Phys. **58** Pt 3 (1997), pp. 409–419.

Lee, H. J. and S. H. Cho, *Response to 'Comment on "Boundary conditions for surface waves propagating along the interface of plasma flow and free space"'* J. Plasma Phys. **61** (1999), p. 173.

Lee, H. J. "*Comment on 'Kinetic theory of surface waves in plasma jets"*', Phys. Plasmas **12** (2005), 094701.

Lee, H. J. and Y. K. Lim, *Kinetic theory of surface waves in a plasma slab*, J. Korean Phys. Soc. **50**, (2007), pp. 1056–1061.

Lee, H. J. and S. H. Cho, *Surface waves in a streaming plasma propagating in a duct: kinetic theory*, J. Plasma Phys. **88** (2022), p. 1.

Lee, H. J. and Y. K. Lim, *Surface waves in a relativistic plasma stream propagating in duct: Kinetic theory*, J. Modern Phys. **13** (2022), p. 1065.

Lee, H. J. *Electromagnetic surface waves in a relativistic magnetized plasma propagating in a cylindrical column*, Applied Phys. Res. **16**, No. 2 (2024), pp. 1–11.

Swanson, D. G. *Plasma Waves*, Academic Press (1989).

Trivelpiece, A. W. and R. W. Gould, *Space charge waves in cylindrical plasma columns*, J. Appl. Phys. **30** (1959), p. 1784.

Chapter 15

Parametric Instability

Parametric instabilities are nonlinear interactions involving waves (external and normal modes) and particles (*wave-particle interaction*) and can be explained in terms of the nonlinearities represented by the quadratic terms in the governing equations. In the fluid equations, we have the nonlinear term $\mathbf{v} \cdot \nabla \mathbf{v}$ in the equation of motion and the nonlinear current $en\mathbf{v}$ in the continuity equation. In the Vlasov equation, we have $\mathbf{E} \cdot \frac{\partial f}{\partial \mathbf{v}}$. Parametric instabilities basically involve three waves and are conveniently regarded as scattering event in which the energy and momentum conservations among the waves involved are satisfied. Then the conservation relations become the matching conditions, which we may call the resonance conditions to enhance the interactive oscillations. Although parametric instabilities start from the nonlinear interactions, the actual analysis becomes a linear one, once the resonance conditions are satisfied with the driving wave (pump) being regarded as the zero order quantity. Thus, the literature on this subject compiled a vast amount of papers. We begin with the simple case (Section 15.1) where the electrostatic plasma normal modes (Langmuir wave and the ion acoustic wave) are coupled via a driving electric field in the form of a sinusoidal time variation. This case of a sinusoidal driving electric field is a special one in which a propagating wave is approximated by assuming an infinite wave length ($k \simeq 0$) *(dipole approximation)*, and a two-time scale perturbational analysis can be employed since the physical resonance condition eliminates the secular terms in the perturbation series. In the case that a dipole approximation is not appropriate, the propagating nature of the three waves involved should be fully respected, with the resonance conditions:

$$\omega_1 = \omega_2 + \omega_3, \quad \mathbf{k}_1 = \mathbf{k}_2 + \mathbf{k}_3$$

where the frequency (wave vector) relation signifies the energy (momentum) conservation.

In the field of parametric instability, we have another *nonlinear Landau damping* that should be regarded as unconnected with the nonlinear Landau damping discussed in earlier chapter. The latter signifies the repetitive damping over multiple cycles of a plasma wave due to Landau damping. The nonlinear Landau damping in parametric instability involves a beat wave with the resonance conditions in the form

$$\omega_1 - \omega_2 = (\mathbf{k}_1 - \mathbf{k}_2) \cdot \mathbf{v}$$

where subscript 1(2) denotes the primary (scattered) wave; thus the beat wave damps on the particle distribution. The beat wave nonlinear Landau damping is an important mechanism representing nonlinear wave-particle interaction and expands the scope of the *parametric decay instability* by including a quasi-mode.

15.1 Coupled Mode Parametric Instability

If a plasma is irradiated by a high frequency oscillating electric field (pump), the pump of frequency ω_0 can couple two plasma normal modes and can give rise to a parametric instability as will be shown below. We consider two electrostatic plasma waves which are the high frequency electron wave (Langmuir wave) and the low frequency ion acoustic wave. We easily see that, by mixing of two waves through the nonlinear quadratic terms, the pump coupled with the low (high) frequency wave produces a high (low) frequency daughter wave. Therefore we have a coupled system of wave equations that may give rise to an instability. This kind of interaction is called the *parametric decay instability.* This terminology is understood from the resonance condition satisfied among the three waves, $\omega_0 = \omega_{\text{low}} + \omega_{\text{high}}$, which suggests that the external electromagnetic wave (pump) of frequency ω_0 is split into two plasma normal modes via the nonlinear coupling. We shall investigate this instability by using the two descriptions of plasma: fluid and kinetic equations. First, we use the fluid equations and perform a perturbational approach by assuming a 'weak coupling'. We shall see that the dominant coupling terms are derived from the resonant terms. We have

$$\frac{\partial n_\alpha}{\partial t} + \nabla \cdot (n_\alpha \mathbf{v}_\alpha) = 0 \quad (\alpha = e,\ i) \tag{15.1}$$

$$m_\alpha n_\alpha \left(\frac{\partial \mathbf{v}_\alpha}{\partial t} + \mathbf{v}_\alpha \cdot \nabla \mathbf{v}_\alpha \right) = -\nabla p_\alpha + e_\alpha n_\alpha (\mathbf{E} + \mathbf{E}_0 \sin \omega_0 t) - m_\alpha n_\alpha \nu_\alpha \mathbf{v}_\alpha \tag{15.2}$$

$$\nabla \cdot \mathbf{E} = 4\pi \sum_\alpha e_\alpha n_\alpha \tag{15.3}$$

We assume that the equilibrium state is represented by uniform density N, and the zero order velocity is the particle quiver velocity $\mathbf{v}_{0\alpha}$ as given by

$$\mathbf{v}_{0\alpha}(t) = -\frac{e_\alpha}{m_\alpha \omega_0}\mathbf{E}_0 \cos\omega_0 t \tag{15.4}$$

If the above equations are linearized with respect to the perturbed quantities $n'_\alpha(\mathbf{r},t) = n_\alpha - N$ and $\mathbf{v}'_\alpha = \mathbf{v}_\alpha - \mathbf{v}_{0\alpha}(t)$, then we obtain

$$\left(\frac{\partial}{\partial t} + \mathbf{v}_{0\alpha}\cdot\nabla\right)^2 n'_\alpha + \nu_\alpha\left(\frac{\partial}{\partial t} + \mathbf{v}_{0\alpha}\cdot\nabla\right) n'_\alpha - \frac{\gamma_\alpha T_\alpha}{m_\alpha}\nabla^2 n'_\alpha + \omega_{p\alpha}^2(n'_\alpha - n'_\beta) = 0 \tag{15.5}$$

where $\omega_{p\alpha} = (4\pi N e^2/m_\alpha)^{1/2}$ and the subscript β becomes e (i) if α is i (e). In obtaining the above equation we used the equation of state $\nabla p_\alpha = \gamma_\alpha T_\alpha \nabla n_\alpha$.

Exercise. Derive (15.5). In the absence of the pump ($\mathbf{v}_{\alpha\mathbf{0}} = 0$), the electron and the ion equations are only weakly coupled through the Poisson equation: the coupling disappears if we put the mass ratio $m_e/m_i \to 0$. Note that the zero order velocity $\mathbf{v}_{\alpha\mathbf{0}}$ changes the differential operator $\partial/\partial t \to \partial/\partial t + \mathbf{v}_{\alpha\mathbf{0}}\cdot\nabla$, as is evident from the linearized continuity equation, $(\partial/\partial t + \mathbf{v}_{\alpha\mathbf{0}}\cdot\nabla)n_\alpha + N_\alpha\nabla\cdot\mathbf{v}_\alpha = 0$.

The above equation is Fourier transformed with respect to the spatial coordinate: $n'_\alpha(\mathbf{r},t) = n'_\alpha(t)e^{i\mathbf{k}\cdot\mathbf{r}}$. Also we transform as

$$n'_\alpha(t) = f_\alpha(t)e^{-ia_\alpha \sin\omega_0 t};$$

where

$$a_\alpha = \frac{-e_\alpha \mathbf{k}\cdot\mathbf{E}_0}{m_\alpha \omega_0^2}$$

is the ratio of particle excursion length in the external field to the wavelength of the oscillation. This parameter will be assumed to be $\ll 1$ (weak pump). Then, (15.5) takes the form

$$D_e f_e \equiv \left(\frac{d^2}{dt^2} + \nu_e\frac{d}{dt} + \omega_e^2\right) f_e = \omega_{\mathrm{pe}}^2 f_i e^{ia\sin\omega_0 t} \tag{15.6}$$

$$D_i f_i \equiv \left(\frac{d^2}{dt^2} + \nu_i\frac{d}{dt} + \omega_i^2\right) f_i = \omega_{pi}^2 f_e e^{-ia\sin\omega_0 t} \tag{15.7}$$

$$\text{where } a = a_e - a_i \simeq e\mathbf{k}\cdot\mathbf{E}_0/m_e\omega_0^2, \quad \omega_\alpha^2 = \omega_{p\alpha}^2 + k^2\gamma_\alpha T_\alpha/m_\alpha \tag{15.8}$$

Equations (15.6) and (15.7) are the basic equations for parametric decay instability induced by homogeneous oscillating electric field (here the pump is assumed to have infinite wavelength: dipole approximation).

Case of no applied external field: $a = 0$. Eliminating f_e or f_i in (15.6) and (15.7) gives

$$\left(\frac{d^2}{dt^2} + \nu_i \frac{d}{dt} + \omega_i^2\right)\left(\frac{d^2}{dt^2} + \nu_e \frac{d}{dt} + \omega_e^2\right) f_e = \omega_{\text{pe}}^2 \omega_{pi}^2 f_e$$

Neglecting the damping terms, the dispersion relation of natural oscillations is

$$(\omega^2 - \omega_e^2)(\omega^2 - \omega_i^2) = \omega_{\text{pe}}^2 \omega_{pi}^2$$

Approximate solutions of the above quadratic equation (in ω^2) can be obtained, keeping only the first order terms in $\omega_i^2/\omega_e^2 (\simeq m_e/m_i)$:

$$\omega^2 = \omega_e^2 = \omega_{\text{pe}}^2 + k^2 \frac{\gamma_e T_e}{m_e} \quad \text{(high frequency wave)} \tag{15.9}$$

$$\omega^2 = \omega_i^2 - \omega_{\text{pe}}^2 \omega_{pi}^2/\omega_e^2 = \omega_{pi}^2 \left[1 + \frac{k^2 \gamma_i T_i}{m_i \omega_{pi}^2} - \frac{\omega_{\text{pe}}^2}{\omega_{\text{pe}}^2 + k^2 \gamma_e T_e/m_e}\right] \tag{15.10}$$

The above expression becomes

$$\omega^2 = k^2 \, \frac{\gamma_i T_i + \gamma_e T_e}{m_i} \quad \text{(low frequency wave)} \tag{15.11}$$

when the electron Debye length (λ_e) is such that $k^2 \lambda_e^2 \ll 1$. This analysis shows that the natural oscillations of plasma in electrostatic approximation are the two waves, Langmuir wave (electron plasma wave) and ion acoustic wave. The ion acoustic wave is severely Landau-damped unless $T_e \gg T_i$, so as a propagating mode of plasma wave it is characterized by the dispersion relation

$$\omega^2 = k^2 \, \frac{\gamma_e T_e}{m_i} \tag{15.12}$$

We have two methods to solve the coupled equations (15.6) and (15.7): (1) directly solve the differential equations and (2) use Fourier transform.

15.1.1 *Multiple time perturbation analysis*

This perturbation method is used to solve directly the coupled equations (15.6) and (15.7). Eliminating f_i gives

$$L \, f_e = 0 \tag{15.13}$$

where

$$L = D_i D_e - \omega_{pi}^2\omega_{\rm pe}^2 - ia\omega_0\left(-\omega_0 \sin\omega_0 t + \nu_i \cos\omega_0 t + 2\cos\omega_0 t\ \frac{d}{dt}\right)D_e - a^2\omega_0^2\cos^2\omega_0 t D_e$$

$$a = a_e - a_i \approx e\mathbf{k}\cdot\mathbf{E}_0/m_e\omega_0^2 \tag{15.14}$$

Let $a = \epsilon \ll 1$ be the expansion parameter in the multiple time perturbation series for the solution of (15.13). Then we write

$$f_e = f_0 + \epsilon f_1 + \epsilon^2 f_2 + \cdots$$
$$t = t_0 + \epsilon t_1 + \epsilon^2 t_2 + \cdots$$
$$\frac{\partial}{\partial t} = \frac{\partial}{\partial t_0} + \epsilon\frac{\partial}{\partial t_1} + \epsilon^2\frac{\partial}{\partial t_2} + \cdots \tag{15.15}$$

The zeroth order equation of (15.13) is

$$(D_i^{(0)} D_e^{(0)} - \omega_{pi}^2\omega_{pi}^2) f_0 = 0 \tag{15.16}$$

$$\text{with } D_\alpha^{(0)} = \frac{\partial^2}{\partial t_0^2} + \omega_\alpha^2$$

where we assumed that the collision frequency ν_α is a quantity of $O(\epsilon)$. Equation (15.16) determines two normal modes of electrostatic perturbation in the absence of the pump:

$$f_0 = A_2(t_1)e^{i\omega_2 t_0} + A_{-2}(t_1)e^{-i\omega_2 t_0} + A_3(t_1)e^{i\omega_3 t_0} + A_{-3}(t_1)e^{-i\omega_3 t_0} \tag{15.17}$$

where ω_2 is given by (15.9) (high frequency) and ω_3 is given by (15.11) (low frequency). The first order equation of (15.13) is

$$(D_i^{(0)} D_e^{(0)} - \omega_{pi}^2\omega_{pi}^2) f_1 = -L_1 f_0 \tag{15.18}$$

$$L_1 = 4\frac{\partial^4}{\partial t_0^3 t_1} + (\nu_i + \nu_e)\frac{\partial^3}{\partial t_0^3} + 2(\omega_e^2 + \omega_i^2)\frac{\partial^2}{\partial t_0 \partial t_1} + (\nu_i\omega_e^2 + \nu_e\omega_i^2)\frac{\partial}{\partial t_0}$$
$$+ ia\omega_0\left(\omega_0 \sin\ \omega_0 t - 2\cos\omega_0 t\ \frac{\partial}{\partial t_0}\right)\left(\frac{\partial^2}{\partial t_0^2} + \omega_e^2\right) \tag{15.19}$$

To facilitate calculation of $L_1 f_0$, we consider a typical term, $A(t_1)\ e^{i\omega t_0}$, operated by L_1. We obtain

$$L_1 A(t_1)e^{i\omega t_0} = \left(F_0(\omega)A + F_1(\omega)\ \frac{\partial A}{\partial t_1}\right)e^{i\omega t_0} + F_+(\omega)Ae^{i(\omega+\omega_0)t_0} + F_-(\omega)Ae^{i(\omega-\omega_0)t_0} \tag{15.20}$$

$$F_0(\omega) = i\omega\left(\nu_i\omega_e^2 + \nu_e\omega_i^2 - (\nu_e+\nu_i)\omega^2\right) \tag{15.21}$$

$$F_1(\omega) = 2i\omega(\omega_e^2 + \omega_i^2 - 2\omega^2) \tag{15.22}$$

$$F_\pm(\omega) = a\omega_0(\omega_e^2 - \omega^2)\left(\omega \pm \frac{\omega_0}{2}\right) \tag{15.23}$$

Then (15.18) can be written as

$$(D_i^{(0)}D_e^{(0)} - \omega_{pi}^2\omega_{pi}^2)f_1 = -\sum_j\left[\left(F_0(\omega_j)A_j + F_1(\omega_j)\frac{\partial A_j}{\partial t_1}\right)e^{i\omega_j t_0}\right.$$
$$\left. + F_+(\omega_j)A_j e^{i(\omega_j+\omega_0)t_0} + F_-(\omega_j)A_j e^{i(\omega_j-\omega_0)t_0}\right] \tag{15.24}$$

where the summation index j runs over ± 2 and ± 3. Also note $\omega_{-j} = -\omega_j$ and $F_\pm$ are the mixing terms of the pump and a natural mode.

The secularity causing terms in (15.24) are those terms oscillating with frequencies $\pm\omega_2$ or $\pm\omega_3$, which are the frequencies of the homogeneous solution. In addition to these terms, there are also resonant terms which cause the secularity by satisfying the following resonant conditions:

$$-\omega_0 + \omega_2 + \omega_3 = \Delta' : \text{ Stokes coupling}$$
$$-\omega_0 + \omega_2 - \omega_3 = \Delta'' : \text{ anti-Stokes coupling}$$

Strict satisfaction of resonant conditions means $\Delta' = \Delta'' = 0$, but we allowed for a small frequency mismatch. All these secularity causing terms in (15.24) should be removed for each frequency. For example: if we take $j = 2$, the terms $F_+(\pm\omega_3)$ are the resonant terms. So we write

$$\left[F_1(\omega_2)\frac{\partial A_2}{\partial t_1} + F_0(\omega_2)A_2\right]e^{i\omega_2 t_0} + F_+(-\omega_3)A_{-3}e^{i(\omega_0-\omega_3)t_0} + F_+(\omega_3)A_3 e^{i(\omega_0+\omega_3)t_0} = 0$$

Canceling the exponential factors yields (15.25), and the other equations are likewise obtained as follows:

$$F_1(\omega_2)\frac{\partial A_2}{\partial t_1} + F_0(\omega_2)A_2 + F_+(-\omega_3)A_{-3}e^{-i\Delta' t_1} + F_+(\omega_3)A_3 e^{-i\Delta'' t_1} = 0 \tag{15.25}$$

$$F_1(-\omega_2)\frac{\partial A_{-2}}{\partial t_1} + F_0(-\omega_2)A_{-2} + F_-(\omega_3)A_3 e^{i\Delta' t_1} + F_-(-\omega_3)A_{-3}e^{i\Delta'' t_1} = 0 \tag{15.26}$$

$$F_1(\omega_3)\,\frac{\partial A_3}{\partial t_1} + F_0(\omega_3)A_3 + F_+(-\omega_2)A_{-2}e^{-i\Delta' t_1} + F_-(\omega_2)A_2 e^{i\Delta'' t_1} = 0 \tag{15.27}$$

$$F_1(-\omega_3)\,\frac{\partial A_{-3}}{\partial t_1} + F_0(-\omega_3)A_{-3} + F_-(\omega_2)A_2 e^{i\Delta' t_1} + F_+(-\omega_2)A_{-2}e^{-i\Delta'' t_1} = 0 \tag{15.28}$$

It would be useful to know that $\omega_2^2 \approx \omega_e^2 \approx \omega_{\mathrm{pe}}^2, \omega_i^2 \ll \omega_e^2$. Using the definitions for F's, the above expressions are simplified to yield

$$\frac{\partial A_2}{\partial t_1} + \Gamma_e A_2 + \alpha(A_{-3}e^{-i\Delta' t_1} + A_3 e^{-i\Delta'' t_1}) = 0 \tag{15.29}$$

$$\frac{\partial A_{-2}}{\partial t_1} + \Gamma_e A_{-2} + \alpha(A_3 e^{i\Delta' t_1} + A_{-3}e^{i\Delta'' t_1}) = 0 \tag{15.30}$$

$$\frac{\partial A_3}{\partial t_1} + \Gamma_i A_3 + \beta(A_{-2}e^{-i\Delta' t_1} - A_2 e^{i\Delta'' t_1}) = 0 \tag{15.31}$$

$$\frac{\partial A_{-3}}{\partial t_1} + \Gamma_i A_{-3} + \beta(A_2 e^{i\Delta' t_1} - A_{-2}e^{-i\Delta'' t_1}) = 0 \tag{15.32}$$

$$\Gamma_{e,i} = \nu_{e,i}/2, \quad \alpha = ia\omega_0^2\omega_e^2/4\omega_2^3,$$
$$\beta = a\omega_0(\omega_2 - \omega_0/2)\omega_{pi}^2\omega_{\mathrm{pe}}^2/(2i\omega_3\omega_2^2\omega_e^2)$$

where α and β are the coupling constants.

Summary. (+)-subscripted wave and (−)-subscripted wave propagate in opposite directions. Equation (15.29) describes slow change (first order) of ω_2-wave due to the coupling of the pump and the low frequency waves $\omega_{\pm 3}$. Equation (15.31) describes slow change of ω_3-wave due to the coupling of the pump and the high frequency waves $\omega_{\pm 2}$. We have the four equations (15.29)–(15.32) for the 4 unknown functions $A_{\pm 2}$ and $A_{\pm 3}$ which describe the resonant interactions among the 4 four waves when the Stokes and anti-Stokes resonant conditions are met; they include all the possible couplings among the waves propagating both of ± directions, with both resonance conditions of Stokes and anti-Stokes being met. Now we consider more restricted couplings.

- If we exclude the low frequency wave propagating in the negative direction, put $A_3 = 0$ to write

$$\frac{\partial A_2}{\partial t_1} + \Gamma_e A_2 + \alpha A_{-3}e^{-i\Delta' t_1} = 0 \tag{15.33}$$

$$\frac{\partial A_{-2}}{\partial t_1} + \Gamma_e A_{-2} + \alpha A_{-3} e^{i\Delta'' t_1} = 0 \tag{15.34}$$

$$\frac{\partial A_{-3}}{\partial t_1} + \Gamma_i A_{-3} + \beta \left(A_2 e^{i\Delta' t_1} - A_{-2} e^{-i\Delta'' t_1} \right) = 0 \tag{15.35}$$

- When the anti-Stokes coupling is excluded, we have

$$\frac{\partial A_2}{\partial t_1} + \Gamma_e A_2 + \alpha A_{-3} e^{-i\Delta' t_1} = 0 \tag{15.36}$$

$$\frac{\partial A_{-3}}{\partial t_1} + \Gamma_i A_{-3} + \beta A_2 e^{i\Delta' t_1} = 0 \tag{15.37}$$

Equations (15.36) and (15.37) are a subset of (15.33)–(15.35) as they are obtained by excluding the anti-Stokes coupling terms in the latter equations. Equations (15.33)–(15.35) yields two kinds of instability: *parametric decay instability* and *oscillating two-stream instability* (OTS). The parametric decay instability can be also obtained from (15.36) and (15.37) in exactly identical form. Equation (15.35) is suggestive of the characteristics of the OTS instability. Putting $\Delta' = \Delta'' = 0$ in (15.35) gives the β-term as sin $\omega_3 t$, a standing wave form which grows in the fashion of *purely growing*. This non-propagating instability is also called the *modulational instability*. It is a result of a four wave interaction involving the pump, two side band waves ($\omega_{\pm 2}$), and the low frequency wave ω_{-3}, while the parametric decay instability is a three-wave interaction.

For future reference, we derive the dispersion relations from the above two sets of equations.

- Let us eliminate $A_{\pm 2}$ in (15.33)–(15.35):

$$\frac{\partial^3 A'_{-3}}{\partial t_1^3} + \frac{\partial^2 A'_{-3}}{\partial t_1^2}[2\gamma + i(\Delta'' - \Delta')] + (\gamma - i\Delta')(\gamma + i\Delta'')\frac{\partial A'_{-3}}{\partial t_1} - i\alpha\beta(\Delta' + \Delta'')A'_{-3} = 0 \tag{15.38}$$

where $A'_{-3} = A_{-3} e^{\Gamma_i t}$, $\gamma = \Gamma_e - \Gamma_i$. Since $\frac{\partial}{\partial t_1} = \frac{\partial}{\partial t} - \frac{\partial}{\partial t_0} = -i\omega + i\omega_3$, putting $\frac{\partial}{\partial t_1} = i(\omega_3 - \omega) + \Gamma_i$ gives the dispersion relation

$$[i(\omega - \omega_3) - \Gamma_i][-\omega^2 + \Gamma_e^2 - 2i\omega\Gamma_e + \delta^2] - 2i\delta\alpha\beta = 0 \tag{15.39}$$

where $\delta = \omega_0 - \omega_2$, and we used $\Delta' - \Delta'' = 2\omega_3$, $\Delta' + \Delta'' = -2\delta$, $\Delta'\Delta'' = \delta^2 - \omega_3^2$. Equation (15.39) can be cast into a slightly different form by multiplying both sides by $i(\omega + \omega_3 + i\Gamma_i)$:

$$(\omega^2 + 2i\omega\Gamma_i - \Gamma_i^2 - \omega_3^2)(\omega^2 + 2i\omega\Gamma_e - \Gamma_e^2 - \delta^2)$$
$$= -2\alpha\beta\delta(\omega + \omega_3 + i\Gamma_i) \approx -4\alpha\beta\delta\ \omega_3 \tag{15.40}$$

where we used $\omega + \omega_3 \approx 2\omega_3$. The coupling coefficient takes the value

$$\alpha\beta = \frac{a^2\omega_{pi}^2\omega_{\rm pe}^2}{16\omega_2\omega_3} = \frac{e^2k^2E_0^2}{16m_im_e\omega_2\omega_3} \tag{15.41}$$

If Γ_i is neglected as compared to ω_3, (15.40) is the dispersion relation for the OTS instability derived in Nishikawa I [Nishikawa, K. (1968), *Parametric excitation of coupled waves I. General formulation*, J. Phys. Soc. Japan, **24** (1968), Eq. (31); our ω_3 is ω_1 therein, and ω_2 stands for the same high frequency].

• Next, we eliminate A_2 from (15.36) and (15.37) to derive the dispersion relation of the parametric decay instability:

$$\frac{\partial^2 A_{-3}}{\partial t_1^2} + (\Gamma_e + \Gamma_i - i\Delta')\frac{\partial A_{-3}}{\partial t_1} + \Gamma_i(\Gamma_e - i\Delta')A_{-3} = \alpha\ \beta\ A_{-3} \tag{15.42}$$

Putting $\frac{\partial}{\partial t_1} = i(\omega_3 - \omega)$ in the above equation gives the dispersion relation in the form

$$(\omega - \omega_3)^2 + (\omega - \omega_3)\ (i\Gamma_i + i\Gamma_e + \Delta') - \Gamma_i\Gamma_e + i\Gamma_i\Delta' + \alpha\ \beta$$
$$= (\omega - \omega_3 + i\Gamma_i)(\omega - \omega_3 + \Delta' + i\Gamma_e) + \alpha\ \beta$$
$$= (\omega - \omega_3 + i\Gamma_i)(\omega - \omega_0 + \omega_2 + i\Gamma_e) + \alpha\ \beta = 0 \tag{15.43}$$

Equation (15.43) agrees with Nishkawa's dispersion relation [Nishikawa, (1968a), Eq. (16); note that Nishikawa's $\Delta = -\Delta'$].

Exercise. Explain the physical context of the relation $\frac{\partial}{\partial t_1} = i(\omega_3 - \omega)$ contained in (15.39) and (15.43).

15.1.2 *Fourier transform derivation of the parametric dispersion relation*

Instead of using a direct perturbational method, we analyze (15.6)–(15.7) by Fourier transform to derive (15.40) and (15.43). We use the Bessel function identity in (15.6) and (15.7)

$$e^{ia\sin\omega_0 t} = \sum_p J_p(a)e^{ip\omega_0 t}$$

and obtain Fourier transformed equations

$$D_e(\omega)f_e(\omega) + \omega_{\rm pe}^2 \sum_{p=-\infty}^{\infty} J_p(a)f_i(\omega + p\omega_0) = 0 \tag{15.44}$$

$$D_i(\omega)f_i(\omega) + \omega_{pi}^2 \sum_{p=-\infty}^{\infty} J_p(-a)f_e(\omega + p\omega_0) = 0 \tag{15.45}$$

$$D_\alpha(\omega) = \omega^2 + i\ \nu_\alpha\omega - \omega_\alpha^2 \tag{15.46}$$

We consider $a \ll 1$ (weak pump), so we retain only $p = 0,\ \pm 1$ terms in the Bessel series:

$$D_e(\omega)f_e(\omega) + \omega_{\rm pe}^2[f_i(\omega) + J_1(a)f_i(\omega + \omega_0) + J_{-1}(a)f_i(\omega - \omega_0)] = 0 \tag{15.47}$$

$$D_i(\omega)f_i(\omega) + \omega_{pi}^2[f_e(\omega) + J_1(-a)f_e(\omega + \omega_0) + J_{-1}(-a)f_e(\omega - \omega_0)] = 0 \tag{15.48}$$

where we used $J_0(\pm a) = 1$. By shifting, we can write

$$D_e(\omega + \omega_0)f_e(\omega + \omega_0) + \omega_{\rm pe}^2[f_i(\omega + \omega_0) + J_1(a)f_i(\omega + 2\omega_0) + J_{-1}(a)f_i(\omega)] = 0 \tag{15.49}$$

$$D_e(\omega - \omega_0)f_e(\omega - \omega_0) + \omega_{\rm pe}^2[f_i(\omega - \omega_0) + J_1(a)f_i(\omega) + J_{-1}(a)f_i(\omega - 2\omega_0)] = 0 \tag{15.50}$$

In the above two equations, we consider the frequency ω is in the region of the low frequency near the ion acoustic wave frequency. In the absence of the pump, $f_i(\omega)$ is the ion density perturbation. Thus, $f_i(\omega{\pm}\omega_0)$, $f_i(\omega{\pm}2\omega_0)$, etc. which are perturbations in the high frequency field, are small as compared to $f_i(\omega)$. Neglecting those small terms, the preceding equations yield the dispersion relation

$$D_i(\omega) - \frac{\omega_{\rm pe}^2\omega_{pi}^2}{D_e(\omega)} = \omega_{\rm pe}^2\omega_{pi}^2 J_1^2(a)\left[\frac{1}{D_e(\omega - \omega_0)} + \frac{1}{D_e(\omega + \omega_0)}\right] \tag{15.51}$$

Since ω is of low frequency, $D_e(\omega) \approx -\omega_e^2$. And we have the relation

$$\begin{aligned}
D_i(\omega) - \frac{\omega_{\rm pe}^2\omega_{pi}^2}{D_e(\omega)} &\simeq \omega^2 + i\nu_i\omega - \omega_i^2 + \frac{\omega_{\rm pe}^2\omega_{pi}^2}{\omega_e^2} \\
&\simeq \omega^2 + i\nu_i\omega - \omega_{pi}^2 + \frac{\omega_{\rm pe}^2\omega_{pi}^2}{\omega_e^2} \\
&\simeq \omega^2 + i\nu_i\omega - \frac{k^2\gamma_e T_e}{m_i},
\end{aligned}$$

Then, the dispersion relation takes the form

$$\omega^2 + i\omega\nu_i - \frac{k^2\gamma_e T_e}{m_i} = \frac{1}{4}\, a^2\, \omega_{\rm pe}^2\omega_{pi}^2 \left[\frac{1}{D_e(\omega-\omega_0)} + \frac{1}{D_e(\omega+\omega_0)}\right] \tag{15.52}$$

where we used $J_1(a) = a/2$. In (15.52), $D_e(\omega \mp \omega_0)$ corresponds to, respectively, the Stokes and anti-Stokes coupling. Retaining only $D_e(\omega-\omega_0)$ term recovers (15.43). Keeping both terms recovers (15.40). To show this, we need further simplification of (15.52) to filter out small terms by employing the *resonance approximation.* The approximation is based on $\omega \approx \omega_3 \ll \omega_0 \approx \omega_e = \omega_2$. The LHS of (15.52) is, putting $\omega_3^2 = k^2\gamma_e T_e/m_i$,

$$\text{LHS} = \omega^2 - \omega_3^2 + i\omega\nu_i = (\omega+\omega_3)(\omega-\omega_3) + 2i\omega\Gamma_i \approx 2\omega_3(\omega - \omega_3 + i\Gamma_i) \tag{15.53}$$

$$D_e(\omega-\omega_0) \approx -2\omega_2(\omega - \omega_0 + \omega_2 + i\Gamma_e) \tag{15.54}$$

$$D_e(\omega+\omega_0) \approx 2\omega_2(\omega + \omega_0 - \omega_2 + i\Gamma_e) \tag{15.55}$$

Using the above relations in (15.52) recovers (15.40) and (15.43). It is noteworthy that the multiple time perturbation scheme automatically takes care of the resonance approximations.

So far we derived the dispersion relation of the coupled mode parametric instability in two ways with the identical results. Next, the growth rates and thresholds of the instabilities will be calculated. Let us write (15.40) or (15.52) in the form

$$\left(\omega^2 + i\omega\nu_i - k^2\frac{T_e}{m_i}\right)\left[\left(\omega + \frac{i\nu_e}{2}\right)^2 - \delta^2\right] + \frac{\delta}{\omega_0}\omega_{pi}^2|\mathbf{k}\cdot\mathbf{v}_0|^2 = 0 \tag{15.56}$$

where $\delta = \omega_0 - \omega_e$, $\mathbf{v}_0 = \frac{e\mathbf{E}_0}{m_e\omega_0}$ is the electron quiver velocity in the pump field. The form of the dispersion relation (15.56) is similar to the dispersion relation of two stream instability (hence the name *oscillating two-stream instability*). When v_0 is small, (15.56) relates with the growth of the ion acoustic wave. The solution can be sought by writing $\omega = kv_s + i\gamma$ ($v_s = \sqrt{\gamma_e T_e/m_i}$, the speed of the ion acoustic wave, γ is the growth rate). In this case, the instability is called a *decay instability.* However, when v_0 is large, ω can become as large as kv_s. Then (15.56) becomes

$$\omega^2(\omega^2 - \delta^2) + \frac{\delta}{\omega_0}\omega_{pi}^2|\mathbf{k}\cdot\mathbf{v}_0|^2 = 0$$

This mode is entirely independent of the ion acoustic wave. Furthermore it disappears when $v_0 = 0$ (while the ion acoustic wave doesn't). Such a mode which disappears if the pump is turned off is called a *quasi mode*. Returning to (15.56), it yields two different solutions: see Sections 15.1.3 and 15.1.4.

Reference: Nishikawa, K. *Parametric excitation of coupled waves II. Parametric plasmon-photon interaction*, J. Phys. Soc. Japan **24**, (1968), p. 1152–1158.

15.1.3 *Purely growing instability (oscillating two stream instability)*

When $\delta = \omega_0 - \omega_e < 0$, (15.56) permits $\omega = i\gamma$ solution, as can be seen from

$$(\gamma^2 + k^2 v_s^2)\left[\left(\gamma + \frac{\nu_e}{2}\right)^2 + \delta^2\right] + \frac{\delta}{\omega_0}\omega_{pi}^2 |\mathbf{k}\cdot\mathbf{v}_0|^2 = 0 \tag{15.57}$$

Exercise. Show that in the regime of strong growth ($\gamma \gg kv_s$) the maximum growth rate is

$$\gamma_{\max} = \left[\frac{m_e\omega_0|\mathbf{k}\cdot\mathbf{v}_0|^2}{2m_i}\right]^{1/3}$$

which is obtained at $\gamma = -\delta$. Neglect $k^2 v_s^2 \ll \gamma^2$ and obtain the optimum γ per $\frac{\partial}{\partial\delta}(15.57) = 0$.

Exercise. Show that in the regime of weak growth the maximum growth rate is

$$\gamma_{\max} = \frac{\omega_0}{2}\frac{v_0^2}{u^2} - \frac{\nu_e}{2} \quad (u = \sqrt{\gamma_e T_e/m_e})$$

which is obtained at $\delta = -\frac{\omega_0}{2}\frac{v_0^2}{u^2}$. The threshold is obtained from setting $\gamma = 0$ which gives

$$\left(\frac{v_0}{u}\right)^2 \geq \frac{\nu_e}{\omega_0} \tag{15.58}$$

15.1.4 *Parametric decay instability when $\delta > 0$*

Put $\omega = kv_s + i\gamma$ and $\delta = kv_s (>0)$ and assume $kv_s \gg \gamma$ in (15.56) to obtain

$$4\gamma^2 + 2\gamma(\nu_i + \nu_e) + \nu_i\nu_e - \frac{\omega_{pi}^2 k^2 v_0^2}{\omega_0 k v_s} = 0 \tag{15.59}$$

The growth rate is found by solving the above quadratic equation. The threshold for the instability is simply obtained by setting $\gamma = 0$:

$$\frac{v_0^2}{u^2} = \frac{\nu_e\nu_i}{\omega_0 k v_s} \tag{15.60}$$

Note that the threshold for the decay instability depends on the product of the two collision frequencies.

The parametric decay instability can be analyzed by retaining only the Stokes coupling in (15.52),

$$\omega^2 + i\omega\nu_i - k^2 v_s^2 = \frac{\omega_{pi}^2 k^2 v_0^2}{D(\omega - \omega_0)} \tag{15.61}$$

with the result in (15.59). It is interesting that (15.56) which have contributions from both of the Stokes and anti-stokes components contains the solution of (15.61) as a subset. One can verify that (15.61) yields the growth rate as given by (15.59) [use (15.54), put $\delta = kv_s$, and neglect $\gamma \ll kv_s$].

15.2 Physical Derivation of Coupled Wave Equations

The coupled equations in the preceding section were derived mathematically. For the sake of more transparent physical presentation of the parametric interactions, we explicitly exhibit the parametric coupling terms in the governing equations; this helps to identify the important nonlinear mechanism involved in the parametric instability.

Parametric decay instability occurs when a high frequency electromagnetic wave (represented by the quiver velocity (15.4)) decays into a high frequency electron wave (Langmuir wave) and a low frequency ion acoustic wave. We use:

- High frequency electron equation of motion:

$$\frac{\partial \mathbf{v}_e}{\partial t} = -\frac{e}{m_e}\mathbf{E} - \frac{\gamma_e T_e}{N m_e}\nabla n_e - \nu_e \mathbf{v}_e \tag{15.62}$$

- Low frequency electron equation of motion: (the low frequency quantities will be denoted by δ)

$$-\frac{e}{m_e}\delta\mathbf{E} - \frac{T_e}{N m_e}\nabla \delta n_e = \mathbf{v}_0 \cdot \nabla \mathbf{v}_e \tag{15.63}$$

where $\mathbf{v}_0$ is given by (15.4). In (15.63), the electron inertia is neglected, and the last term, the nonlinear term, is the electron ponderomotive force which will give rise to a low frequency beat by mixing two high frequency waves.

- Low frequency ion equation of motion:

$$\frac{\partial}{\partial t}\delta\mathbf{v}_i = \frac{e}{m_i}\delta\mathbf{E} - \nu_i \delta\mathbf{v}_i \tag{15.64}$$

- High frequency electron continuity equation:

$$\frac{\partial n_e}{\partial t} + N\nabla \cdot \mathbf{v}_e = 0 \tag{15.65}$$

- Low frequency ion continuity equation:

$$\frac{\partial\,\delta n_i}{\partial t} + N\nabla\cdot\delta\mathbf{v}_i = 0 \tag{15.66}$$

Equations (15.62)–(15.66) should be supplemented by Maxwell equation.

$$\frac{\partial \mathbf{E}}{\partial t} = 4\pi e(N\mathbf{v}_e + \delta n_e \mathbf{v}_0): \quad \text{high frequency} \tag{15.67}$$

In (15.67) $\mathbf{v}_0\delta n_e$, the nonlinear electron current, is the important nonlinearity, together with the ponderomotive force in (15.63), which excites the decay instability. Both of the nonlinearities are electron's. Equations (15.62), (15.65), and (15.67) are arranged in the form

$$\left(\frac{\partial^2}{\partial t^2} + \nu_e\frac{\partial}{\partial t} + \omega_{\rm pe}^2 - \frac{\gamma_e T_e}{m_e}\nabla\nabla\cdot\right)\mathbf{v}_e = -\omega_{\rm pe}^2\frac{\delta n_e}{N}\mathbf{v}_0 \tag{15.68}$$

The low frequency equations in (15.63) and (15.64) are combined to give

$$-\frac{m_i}{m_e}\left(\frac{\partial}{\partial t}\,\delta\mathbf{v}_i + \nu_i\delta\mathbf{v}_i\right) - \frac{T_e}{Nm_e}\nabla\delta n_e = \mathbf{v_0}\cdot\nabla\mathbf{v_e} \tag{15.69}$$

Perform $\nabla\cdot$ on the above equation and use (15.66) in which we put $\delta n_i \simeq \delta n_e$ (this charge neutrality condition is justified since we are considering the low frequency wave) to get

$$\left(\frac{\partial^2}{\partial t^2} + \nu_i\frac{\partial}{\partial t} - \frac{T_e}{m_i}\nabla^2\right)\frac{\delta n_e}{N} = \frac{m_e}{m_i}\nabla\cdot(\mathbf{v}_0\cdot\nabla\mathbf{v}_e) \tag{15.70}$$

Equations (15.68) and (15.70) are the two coupled equations describing parametric interactions. The RHS of (15.68) represents the nonlinear current which excites the scattered field, and the RHS of (15.70) is the ponderomotive force which gives rise to, in turn, the electron density perturbation δn_e.

Dipole approximation is not used. In (15.68) and (15.70), the electron quiver velocity $\mathbf{v}_0$ (given in (15.4)) is written in the dipole approximation. In this section we remove this restriction to derive the parametric equations. Consider a large amplitude plane polarized electromagnetic pump wave is propagated in the plasma:

$$\mathbf{E}_p(\mathbf{r},t) = \mathbf{E}_0 e^{i\mathbf{k}_0\cdot\mathbf{r}-i\omega_0 t} + \mathbf{E}_0^* e^{-i\mathbf{k}_0\cdot\mathbf{r}+i\omega_0 t}, \quad (* = \text{ c.c. complex conjugate}) \tag{15.71}$$

The particle quiver velocity

$$\mathbf{v}_{0\alpha}(\mathbf{r},t) = \frac{ie_\alpha}{m_\alpha\omega_0}\mathbf{E}_0\, e^{i(\mathbf{k}_0\cdot\mathbf{r}-\omega_0 t)} + c.c. \tag{15.72}$$

represents the equilibrium electron velocity in the plasma under consideration. The frequency ω_0 is so high that only the electrons oscillate with the high quiver velocity in the pump field with the ions forming a stationary background (also note that the ion nonlinearities are much smaller than the electron's by the mass ratio m_e/m_i). The electric vector is polarized so as to be $\mathbf{k}_0 \cdot \mathbf{E}_0 = 0$. And the pump wave satisfies the usual electromagnetic dispersion relation,

$$\omega_0^2 = \omega_{\rm pe}^2 + c^2 k_0^2 \tag{15.73}$$

When the pump frequency is near to the plasma frequency, the dipole approximation ($\mathbf{k}_0 \simeq 0$) is valid. When the pump is a propagating wave as given by (15.72), (15.70) should be modified to include another coupling term:

$$\left(\frac{\partial^2}{\partial t^2} + \nu_i \frac{\partial}{\partial t} - \frac{T_e}{m_i}\nabla^2\right)\frac{\delta n_e}{N} = \frac{m_e}{m_i}\nabla \cdot [\mathbf{v}_0 \cdot \nabla \mathbf{v}_e + \mathbf{v}_e \cdot \nabla \mathbf{v}_0] \tag{15.74}$$

Fourier-transforming (15.68) yields

$$\begin{aligned}(-\omega^2 - i\omega\nu_e + \omega_{\rm pe}^2)\mathbf{v}_e(\mathbf{k},\omega) &+ \frac{\gamma_e T_e}{m_e}\mathbf{k}\cdot\mathbf{v}_e(\mathbf{k},\omega)\mathbf{k} \\ &= -\frac{\omega_{\rm pe}^2}{N}[\mathbf{v}_0 \delta n_e(\omega-\omega_0,\mathbf{k}-\mathbf{k}_0) + \mathbf{v}_0^* \delta n_e(\omega+\omega_0,\mathbf{k}+\mathbf{k}_0)]\end{aligned} \tag{15.75}$$

where

$$\mathbf{v}_0 = \frac{-ie\mathbf{E}_0}{m_e\omega_0} \tag{15.76}$$

Fourier-transforming (15.74) yields

$$\begin{aligned}&\left(-\omega^2 - i\omega\nu_i + \frac{T_e}{m_i}k^2\right)\frac{\delta n_e(\omega,\mathbf{k})}{N} \\ &\quad = \frac{m_e}{m_i}[-\mathbf{v}_0\cdot(\mathbf{k}-\mathbf{k}_0)\ \ \mathbf{k}\cdot\mathbf{v}_e(\omega-\omega_0,\mathbf{k}-\mathbf{k}_0) - \mathbf{v}_0^*\ \cdot(\mathbf{k}+\mathbf{k}_0)\mathbf{k} \\ &\quad\quad \cdot\mathbf{v}_e(\omega+\omega_0,\mathbf{k}+\mathbf{k}_0) - \mathbf{v}_0\cdot\mathbf{k}\ \ \mathbf{k}_0\cdot\mathbf{v}_e(\omega-\omega_0,\mathbf{k}-\mathbf{k}_0) \\ &\quad\quad + \mathbf{v}_0^*\ \cdot\mathbf{k}\ \ \mathbf{k}_0\cdot\mathbf{v}_e(\omega+\omega_0,\mathbf{k}+\mathbf{k}_0)]\end{aligned} \tag{15.77}$$

Putting $\mathbf{k}_0 \cdot \mathbf{v}_0 = 0$ simplifies the above equation,

$$\begin{aligned}\left(\omega^2 + i\omega\nu_i - \frac{T_e}{m_i}k^2\right)\frac{\delta n_e(\omega,\mathbf{k})}{N} &= \frac{m_e}{m_i}[(\mathbf{k}\cdot\mathbf{v}_0)\ (\mathbf{k}+\mathbf{k}_0)\cdot\mathbf{v}_e(\omega-\omega_0,\mathbf{k}-\mathbf{k}_0) \\ &\quad + (\mathbf{k}\cdot\mathbf{v}_0^*)\ (\mathbf{k}-\mathbf{k}_0)\cdot\mathbf{v}_e(\omega+\omega_0,\mathbf{k}+\mathbf{k}_0)]\end{aligned} \tag{15.78}$$

In (15.75), we can set $\mathbf{v}_e \| \mathbf{k}$ since we are considering electrostatic waves. Then we have

$$\mathbf{v}_e(\omega - \omega_0, \mathbf{k} - \mathbf{k}_0) = \frac{\omega_{\text{pe}}^2}{N} \frac{\mathbf{v}_0^* \, \delta n_e(\omega, \mathbf{k})}{D(\omega - \omega_0, \mathbf{k} - \mathbf{k}_0)} \tag{15.79}$$

$$\mathbf{v}_e(\omega + \omega_0, \mathbf{k} + \mathbf{k}_0) = \frac{\omega_{\text{pe}}^2}{N} \frac{\mathbf{v}_0 \, \delta n_e(\omega, \mathbf{k})}{D(\omega + \omega_0, \mathbf{k} + \mathbf{k}_0)} \tag{15.80}$$

$$D(\omega, \mathbf{k}) = \omega^2 + i\omega\nu_e - \omega_{\text{pe}}^2 - \frac{\gamma_e T_e}{m_e} k^2 \tag{15.81}$$

where we neglected the off-resonance terms, $\delta n_e(\omega \pm 2\omega_0, \mathbf{k} \pm 2\mathbf{k}_0)$. Then (15.75) and (15.78) give, with the aid of above equations, the following dispersion relation,

$$\omega^2 + i\omega\nu_i - \frac{T_e k^2}{m_i} = \omega_{pi}^2 (\mathbf{k} \cdot \mathbf{v}_0)(\mathbf{k} \cdot \mathbf{v}_0^*) \left[\frac{1}{D(\omega - \omega_0, \mathbf{k} - \mathbf{k}_0)} + \frac{1}{D(\omega + \omega_0, \mathbf{k} + \mathbf{k}_0)} \right] \tag{15.82}$$

• **Wave vector resonance condition.** The foregoing steps leading to the dispersion relation (15.82) requires an intensive caution in the algebra of Fourier transform, so an easier and straight way is desirable in hindsight. For this purpose, we stipulate from the beginning the resonance conditions among the central wave (k_0, ω_0) and the two sidebands:

$$k' = k_0 - k, \quad \omega' = \omega_0 - \omega : \text{Stokes} \tag{15.83}$$

$$k'' = k_0 + k, \quad \omega'' = \omega_0 + \omega : \text{anti-Stokes} \tag{15.84}$$

where we understand that ω and k belong to the low frequency wave. For simplicity, we shall assume $\nabla = \hat{\mathbf{x}} \frac{\partial}{\partial x}$, $\mathbf{v}_e = \hat{\mathbf{x}} v_e$, and $\mathbf{v}_0 = \hat{\mathbf{x}} v_0$ to write (15.68) and (15.70) in the form

$$\left(\frac{\partial^2}{\partial t^2} + \nu_e \frac{\partial}{\partial t} + \omega_{\text{pe}}^2 - \frac{\gamma_e T_e}{m_e} \frac{\partial^2}{\partial x^2} \right) v_e = -\omega_{\text{pe}}^2 \frac{\delta n_e}{N} v_0 \tag{15.85}$$

$$\left(\frac{\partial^2}{\partial t^2} + \nu_i \frac{\partial}{\partial t} - \frac{T_e}{m_i} \frac{\partial^2}{\partial x^2} \right) \frac{\delta n_e}{N} = \frac{m_e}{m_i} \frac{\partial}{\partial x} \left(v_0 \frac{\partial}{\partial x} v_e \right) \tag{15.86}$$

We write

$$v_0 = \tilde{v}_0 e^{ik_0 x - i\omega_0 t} + \tilde{v}_0^* e^{-ik_0 x + i\omega_0 t} \tag{15.87}$$

$$v_e = v_e' e^{ik' x - i\omega' t} + v_e'^* e^{-ik' x + i\omega' t} + v_e'' e^{ik'' x - i\omega'' t} + v_e''^* e^{-ik'' x + i\omega'' t} \tag{15.88}$$

$$\frac{\delta n_e}{N} = n(x, t) = \tilde{n} e^{ikx - i\omega t} + \tilde{n}^* e^{-ikx + i\omega t} \tag{15.89}$$

Substituting (15.88) and (15.89) into (15.85) yields after discarding all the terms which does not satisfy the resonance conditions

$$(\omega'^2 + i\omega'\nu_e - \omega_{\rm pe}^2 - k'^2u^2)v_e' = \omega_{\rm pe}^2\tilde{v}_0\tilde{n}^* \quad (u^2 = \gamma_e T_e/m_e) \tag{15.90}$$

$$(\omega''^2 + i\omega''\nu_e - \omega_{\rm pe}^2 - k''^2u^2)v_e'' = \omega_{\rm pe}^2\tilde{v}_0\tilde{n} \tag{15.91}$$

$$(\omega^2 + i\omega\nu_i - v_s^2k^2)\tilde{n} = \frac{m_e}{m_i}k^2(\tilde{v}_0 v_e'^* + \tilde{v}_0^* v_e'') \quad (v_s^2 = T_e/m_i) \tag{15.92}$$

We can show that the above three equations agree with (15.82).

15.3 Parametric Decay Instability and Nonlinear Landau Damping

The threshold for the parametric decay instability can be derived not only from the dispersion relation (15.56) but also from consideration of power flow among the two interacting waves: the low frequency wave $n(\omega)$ and the scattered wave (lower side band wave) $v'(\omega')$. We begin with (15.92) where we neglect the anti-Stokes component:

$$\tilde{n} = \frac{m_e}{m_i}\frac{k^2\tilde{v}_0 v_e'^*}{\omega^2 + i\nu_i\omega - v_s^2k^2} \tag{15.93}$$

This is the low frequency ion density perturbation determined as a beat mode of the two high frequency waves with frequencies ω_0 and ω'. In the low frequency region, the quasi-neutrality holds, and therefore, this can be put equal to the electron density perturbation, which couples the pump to drive a high frequency nonlinear *electron current*:

$$J_e' = -eN\tilde{v}_0\tilde{n}^* = -\frac{m_e}{m_i}\frac{eNk^2|\tilde{v}_0|^2 v_e'}{\omega^2 - i\nu_i\omega - v_s^2k^2} \tag{15.94}$$

The dissipation rate of the ω'-wave energy is obtained by

$$\begin{aligned}\frac{\partial}{\partial t}|E'|^2 &= -4\pi(J_e'E'^* + J'^*E') \\ &= 4\pi eNk^2|\tilde{v}_0|^2\frac{m_e}{m_i}\left[\frac{v_e'E'^*}{\omega^2 - i\nu_i\omega - v_s^2k^2} + \frac{v_e'^*E'}{\omega^2 + i\nu_i\omega - v_s^2k^2}\right]\end{aligned}$$

Here we assume $\nu_i \gg \omega,\ kv_s$, and keep only $\nu_i\omega$-term in the denominator and use $v_e' = \frac{-ie}{m_e}\frac{E'}{\omega'}$ to obtain

$$\frac{\partial}{\partial t}|E'|^2 = 4\pi eNk^2|\tilde{v}_0|^2\frac{m_e}{m_i}|E'|^2\frac{e}{m_e\omega'}\frac{2}{\nu_i\omega}$$

Equating the above expression to the collisional dissipation rate of the ω'-wave, i.e. $\nu_e|E'|^2$, we obtain the threshold of the parametric decay instability as

$$\frac{\tilde{v}_0^2}{u^2} = \frac{1}{2}\frac{\nu_e\nu_i}{kv_s\omega_0} \tag{15.95}$$

where we put $\omega \simeq kv_s$, $\omega' \simeq \omega_{\text{pe}} \simeq \omega_0$. Except the factor $\frac{1}{2}$, (15.95) is the same as (15.60).

When the frequency ν_i dominates over ω and kv_s, the ion acoustic wave loses its identity as a normal mode; it is a degenerated virtual mode due to heavy damping. When the plasma has $T_e \sim T_i$, the ion wave is heavily Landau-damped on ions if the wave-particle resonant condition,

$$\omega_0 - \omega' = (k_0 - k') \times \text{ ion thermal velocity}$$

is satisfied. When we identify ν_i as the Landau damping rate of the ion acoustic wave, the parametric decay process described above is called *nonlinear Landau damping.* The beat wave energy is channeled to the scattered wave to give rise to a parametric decay instability.

15.4 More on Oscillating Two Stream Instability (OTSI)

The purely growing wave is a zero-frequency wave. This is an extended parametric decay instability which arises when both Stokes and anti-Stokes couplings are included. At the onset of the OTSI, the 'dispersion relation' is obtained by putting $\omega = 0$ in (15.52):

$$-\frac{T_e}{m_i}k^2 = \omega_{pi}^2 k^2 v_0^2\left[\frac{1}{\omega_0^2 - i\omega_0\nu_e - \omega_{ek}^2} + \frac{1}{\omega_0^2 + i\omega_0\nu_e - \omega_{ek}^2}\right] = \frac{\omega_{pi}^2 k^2 v_0^2}{\omega_0}\frac{\delta}{\delta^2 + \frac{\nu_e^2}{4}}$$

which gives

$$v_0^2 = \frac{T_e}{m_i}\frac{\omega_0}{\omega_{pi}^2}\frac{\delta^2 + \frac{\nu_e^2}{4}}{\delta}$$

The right hand side has the minimum value at $\delta = \nu_e/2$, and the corresponding v_0 is

$$\frac{v_0^2}{u^2} = \frac{\nu_e}{\omega_0} \quad \text{(see (15.58))}$$

At the onset of OTSI, the scattered waves (ω', ω'') form a standing wave. In (15.88) we substitute the following expressions (obtained per $\omega = k_0 = 0$), introducing a phase α,

$$v_e' = |v_e|e^{-i\alpha}e^{-ikx-i\omega_0 t} + \text{ c.c.} = 2|v_e|\cos(kx+\omega_0 t+\alpha)$$
$$v_e'' = |v_e|e^{-i\alpha}e^{ikx-i\omega_0 t} + \text{ c.c.} = 2|v_e|\cos(kx-\omega_0 t-\alpha)$$

So we can represent the scattered field as a standing wave

$$v_e = |v_e|\cos kx\cos(\omega_0 t+\alpha) \tag{15.96}$$

The corresponding electric field is

$$E_s = -\frac{m_e}{e}\frac{\partial v_e}{\partial t} = |E_s|\cos kx\ \sin(\omega_0 t+\alpha) \tag{15.97}$$

with

$$|E_s| = \frac{m_e}{e}\omega_0|v_e|$$

The low frequency equations are

$$-\frac{e}{m_e}\delta E - \frac{T_e}{Nm_e}\frac{\partial\delta n_e}{\partial x} = v_0\frac{\partial v_e}{\partial x} \qquad \text{(cf. (15.63))} \tag{15.98}$$
$$\frac{\partial}{\partial t}\delta v_i = \frac{e}{m_i}\delta E - \frac{T_i}{Nm_i}\frac{\partial}{\partial x}\delta n_i \qquad \text{(cf. (15.64))} \tag{15.99}$$

In (15.99) we neglected the ion collision frequency ν_i which is not essential in OTSI and included T_i although it is not essential. Eliminating δE in the above two equations, we get

$$\frac{\partial}{\partial t}\delta v_i + \frac{T_i+T_e}{Nm_i}\frac{\partial}{\partial x}\delta n_e = -\frac{m_e}{m_i}\frac{\partial}{\partial x}(v_0 v_e) \tag{15.100}$$

We put $\partial/\partial t = 0$ in (15.100) because we consider a zero frequency wave to obtain

$$\delta n_e = -\frac{Nm_e}{T_e+T_i}\langle v_0 v_e\rangle \tag{15.101}$$

where $< \ldots >$ is the time average and $\langle v_0 v_e\rangle$ is the ponderomotive force. Note that to pick up the zero frequency component from the RHS we have to take time average. For a pump of the form

$$E_0 = |E_0|\sin\ \omega_0 t, \quad \text{or } v_0 = \xi\ \omega_0\cos\omega_0 t$$

where $\xi = \frac{e|E_0|}{m_e\omega_0^2}$ is the amplitude of the electron oscillation, the time average is

$$<v_0 v_e> = \frac{\xi|E_s|}{2}\frac{e}{m_e}\cos kx\cos\alpha \tag{15.102}$$

where we used (15.96). The current responsible for the excitation of the scattered wave is

$$J = -e\delta n_e\, v_0 = -e\omega_0\xi\cos\omega_0 t\;\delta n_e$$

The power per unit volume delivered by this current to the standing wave is given by

$$\begin{aligned}<E_s\; J> &= <E_s\;\cos kx\;\sin(\omega_0 t+\alpha)\; J> \\ &= \frac{1}{2}\;|E_s|^2\;\xi^2\frac{e^2 N\omega_0}{T_e+T_i}<\cos^2 kx\cos^2\omega_0 t>\sin\alpha\cos\alpha\end{aligned}$$

where the time and space average give factor (1/2), respectively, and we put $\alpha = \pi/4$, corresponding to the maximum power, to get

$$<E_s J> = \frac{1}{16}|E_s|^2\;\xi^2\;\frac{Ne^2\omega_0}{T_e+T_i}$$

At the threshold, this power must be equal to the power of the scattered wave dissipated by collision, $\frac{\nu_e}{16\pi}|E_s|^2$, giving in terms of the Debye lengths,

$$\xi^2 = 4\frac{\nu_e}{\omega_{\rm pe}}\;(\lambda_e^2+\lambda_i^2)\quad\text{or}\quad |E_0|^2 = 16\pi N\;\frac{\nu_2}{\omega_{\rm pe}}(T_e+T_i) \tag{15.103}$$

which can be put into the form (use $|v_0| = \xi\omega_0$)

$$\frac{|v_0|^2}{u^2} = 4\;\frac{\nu_e}{\omega_{\rm pe}}\quad\text{(cf. (15.58))}$$

The factor 4 discrepancy from (15.58) arises because of our use of sin and cos representation instead of the exponential functions.

• **Two-time scale analysis of parametric interaction.** The parametric interaction between the electron and ion waves is basically a phenomenon involving two time scales of great difference; slow and fast oscillations. Mode couplings among the plasma normal modes make the amplitudes of the participating waves change slowly while the phases of the waves oscillate rapidly. Without the coupled mode interaction, the Fourier amplitudes of the normal modes remain constant in time and space. Naturally, analysis of parametric instability is facilitated by employing two time perturbation method which clearly delineate the different time scales characterizing the dynamics of mode coupling. This section complements the outright perturbation scheme presented in the earlier section. Here we shall assume $k_0 = 0$ (dipole approximation) for the pump,

$$v_0(t) = \tilde{v}_0 e^{-i\omega_0 t} + \tilde{v}_0^* e^{i\omega_0 t}$$

Then the x coordinate in (15.85) and (15.86) can be Fourier transformed away by replacing $\partial/\partial x = ik$ with the understanding that k now is the wavenumber of the low frequency wave. Then the two equations read

$$\left(\frac{\partial^2}{\partial t^2} + \nu_e \frac{\partial}{\partial t} + \omega_{\text{pe}}^2 + u^2 k^2\right) v_e(k,t) = -\omega_{\text{pe}}^2 n(k,t) v_0(t) \tag{15.104}$$

$$\left(\frac{\partial^2}{\partial t^2} + \nu_i \frac{\partial}{\partial t} + v_s^2 k^2\right) n(k,t) = -\frac{m_e}{m_i} k^2 v_0(t) v_e(k,t) \tag{15.105}$$

Let us write $t = t_0 + \epsilon t_1$ where t_0 is the fast time which scales the rapidly oscillating phases and t_1 is the slow time which scales the slowly changing amplitudes. Accordingly the time derivative is written

$$\frac{\partial}{\partial t} = \frac{\partial}{\partial t_0} + \epsilon \frac{\partial}{\partial t_1} \tag{15.106}$$

Also we write

$$n(k,t) = \tilde{n}(k,t_1) e^{-i\omega t_0} + \tilde{n}^*(k,t_1) e^{+i\omega t_0} \tag{15.107}$$

Because of the coupling in (15.104), there will appear quasi plasma waves whose frequencies are up and down shifted:

$$v_e' = A'(t_1) e^{-i\omega' t_0} + A'^*(t_1) e^{i\omega' t_0} \tag{15.108}$$

$$v_e'' = A''(t_1) e^{-i\omega'' t_0} + A''^*(t_1) e^{i\omega'' t_0} \tag{15.109}$$

$$\text{where } \omega' = \omega_0 - \omega, \qquad \omega'' = \omega_0 + \omega \tag{15.110}$$

Using (15.106) in (15.104), the quasi waves should satisfy

$$\left[\frac{\partial^2}{\partial t_0^2} + \omega_{ek}^2 + \epsilon\left(2\frac{\partial}{\partial t_0}\frac{\partial}{\partial t_1} + \nu_e \frac{\partial}{\partial t_0}\right)\right] A'(t_1) e^{-i\omega' t_0} = -\omega_{\text{pe}}^2 \tilde{n}^* v_0 e^{-i(\omega_0 - \omega))t_0} \tag{15.111}$$

$$\left[\frac{\partial^2}{\partial t_0^2} + \omega_{ek}^2 + \epsilon\left(2\frac{\partial}{\partial t_0}\frac{\partial}{\partial t_1} + \nu_e \frac{\partial}{\partial t_0}\right)\right] A''(t_1) e^{-i\omega'' t_0} = -\omega_{\text{pe}}^2 \tilde{n} v_0 e^{-i(\omega_0 + \omega)t_0} \tag{15.112}$$

$$\text{where} \quad \omega_{ek}^2 = \omega_{\text{pe}}^2 + u^2 k^2 \tag{15.113}$$

and all the non-resonant terms are discarded in the RHS. In the ϵ^0 order we have

$$\omega'^2 = \omega_{ek}^2, \qquad \omega''^2 = \omega_{ek}^2 \tag{15.114}$$

In view of (15.110), the two equations in (15.114) can't be both right. So we relax the resonance conditions by introducing the frequency mismatches,

$$-\omega_0 + \omega + \omega' = \Delta' \tag{15.115}$$

$$-\omega_0 - \omega + \omega'' = \Delta'' \tag{15.116}$$

Then in ϵ^1 order, we obtain

$$\left(\frac{\partial}{\partial t_1} + \Gamma_e\right) A' = -\frac{i}{2}\frac{\omega_{\text{pe}}^2}{\omega'}\tilde{v}_0 \tilde{n}^* e^{i\Delta' t_1} \tag{15.117}$$

$$\left(\frac{\partial}{\partial t_1} + \Gamma_e\right) A'' = -\frac{i}{2}\frac{\omega_{\text{pe}}^2}{\omega''}\tilde{v}_0 \tilde{n} e^{i\Delta'' t_1} \tag{15.118}$$

where $\Gamma_e = \nu_e/2$. Likewise, (15.105) gives the linear dispersion relation

$$\omega^2 = v_s^2 k^2 \tag{15.119}$$

and the equation for the amplitude

$$\left(\frac{\partial}{\partial t_1} + \Gamma_i\right) \tilde{n} = -\frac{i}{2}\frac{m_e}{m_i}\frac{k^2}{\omega}(\tilde{v}_0 A'^* e^{i\Delta' t_1} + \tilde{v}_0^* A'' e^{-i\Delta'' t_1}) \quad (\Gamma_i = \nu_i/2) \tag{15.120}$$

Equations (15.117)–(15.120) completely agree with the earlier mode coupling equations (15.33)–(15.35); but the coupling coefficients are more explicit.

Exercise. Verify that $A_{-3} \equiv \tilde{n}\, e^{\Gamma_i t}$ satisfies (15.38).

Exercise. Verify that A_{-3} satisfies (15.42) if the anti-Stokes coupling is discarded.

15.5 Various Parametric Processes

Reference: W. L. Kruer, *The physics of laser plasma interactions*, Addison-Wesley (1988).

15.5.1 *Stimulated Raman scattering*

A large amplitude electromagnetic wave (light wave) disturbs a plasma to give rise to an electron density perturbation which in turn scatters off the incident wave into another electromagnetic wave of a different frequency. To investigate this process, we use the following equations.

$$\frac{\partial \mathbf{v}}{\partial t} + \mathbf{v} \cdot \nabla \mathbf{v} = -\frac{e}{m_e}\left[\mathbf{E} + \frac{1}{c}\mathbf{v} \times \mathbf{B}\right] - \frac{\nabla p}{N m_e} - \nu_e \mathbf{v}_e \tag{15.121}$$

The electron velocity $\mathbf{v}$ can be separated into two components: a component in the light field ($\mathbf{v}_L$) and a component in the plasma wave field ($\mathbf{v}_e$). In the light field, the magnetic force can be neglected for non-relativistic motion and the second term in (15.121) can be neglected as compared to the first term.

$$\frac{|\partial v/\partial t|}{|\mathbf{v} \cdot \nabla \mathbf{v}|} = \frac{\omega v}{k v^2} = \frac{c}{v} \gg 1$$

So we use

$$\frac{\partial \mathbf{v}_L}{\partial t} = \frac{-e}{m_e}\mathbf{E}_L \quad \text{(light wave)} \tag{15.122}$$

In the electrostatic plasma wave equation, the ponderomotive force, the beat of two light waves (the pump wave and the scattered light wave) should be included:

$$\frac{\partial \mathbf{v}_e}{\partial t} + \mathbf{v}_L \cdot \nabla \mathbf{v}_L = \frac{e}{m_e}\nabla\phi - \frac{\nabla p}{N m_e} - \nu_e \mathbf{v}_e - \frac{e}{c m_e}\mathbf{v}_L \times \mathbf{B} \tag{15.123}$$

Use the vector identity, $\mathbf{v}_L \cdot \nabla \mathbf{v}_L = (\nabla \times \mathbf{v}_L) \times \mathbf{v}_L + \nabla v_L^2/2$. Note that (15.122) gives

$$\frac{\partial}{\partial t}\nabla \times \mathbf{v}_L = -\frac{e}{m_e}\nabla \times \mathbf{E}_L \tag{15.124}$$

Here Maxwell equation

$$\nabla \times \mathbf{E}_L = -\frac{1}{c}\frac{\partial \mathbf{B}}{\partial t} \tag{15.125}$$

gives $\nabla \times \mathbf{v}_L = \frac{e\mathbf{B}}{m_e c}$. So we have a cancellation in (15.123):

$$\frac{\partial \mathbf{v}_e}{\partial t} = -\nabla\frac{v_L^2}{2} + \frac{e}{m_e}\nabla\phi - \frac{\nabla p}{N m_e} - \nu_e \mathbf{v}_e \quad \text{(plasma wave)} \tag{15.126}$$

where $\mathbf{v}_L = \mathbf{v}_0 + \mathbf{v}_L'$ ($'$ denotes the scattered light wave). The other necessary equations are

$$\frac{\partial n_e}{\partial t} + N\nabla \cdot \mathbf{v}_e = 0 \tag{15.127}$$

$$\nabla^2\phi = 4\pi e n_e \tag{15.128}$$

Taking divergence of (15.126) and putting $p = \gamma_e T_e n_e$ yield

$$\left(\frac{\partial^2}{\partial t^2} + \nu_e \frac{\partial}{\partial t} + \omega_{\text{pe}}^2 - \frac{\gamma_e T_e}{m_e}\nabla^2\right) n_e = N\nabla^2(\mathbf{v}_0 \cdot \mathbf{v}'_L) \tag{15.129}$$

The equations for the scattered light wave are (15.122), (15.125), and

$$\nabla \times \mathbf{B} = \frac{1}{c}\frac{\partial \mathbf{E}}{\partial t} + \frac{4\pi}{c}\mathbf{J} \tag{15.130}$$

$$\mathbf{J} = -e(N\mathbf{v}'_L + n_e\mathbf{v}_0) \tag{15.131}$$

Taking curl of (15.125) yields ($\nabla \cdot \mathbf{E}'_L = 0$)

$$-\nabla^2\mathbf{E}'_L = -\frac{1}{c^2}\frac{\partial^2 \mathbf{E}'_L}{\partial t^2} - \frac{4\pi e}{c^2}\left[N\frac{\partial \mathbf{v}'_L}{\partial t} + \frac{\partial}{\partial t}(n_e\mathbf{v}_0)\right]$$

Use of (15.122) in the above equation gives

$$\left(c^2\nabla^2 - \frac{\partial^2}{\partial t^2} - \omega_{\text{pe}}^2\right)\mathbf{v}'_L = \frac{4\pi e^2}{m_e}(n_e\mathbf{v}_0) \tag{15.132}$$

Note that the important nonlinearities are the ponderomotive force in the electron equation in (15.123) and the nonlinear current in (15.131) for the scattered field. Equations (15.129) and (15.132) are the coupled equations for the Raman scattering.

15.5.2 *Stimulated Brillouin scattering*

Similarly to Raman scattering, a large amplitude light wave perturbs a plasma to excite a density fluctuation associated with an ion acoustic wave, which in turn scatters off the incident wave into another light wave of different frequency. Since the low frequency ion motion is interconnected with the low frequency electron motion through the Poisson equation, the low frequency equations of both electrons and ions should be used. Electron equation is given by (15.126) in which we can neglect the electron inertia:

$$-\frac{m_e}{2}\nabla v_L^2 + e\nabla\phi - \frac{\gamma_e T_e}{N}\nabla n_e = 0 \tag{15.133}$$

Ion equations are:

$$\left(\frac{\partial}{\partial t} + \nu_i\right)\mathbf{v}_i = -\frac{e}{m_i}\nabla\phi \tag{15.134}$$

$$\frac{\partial n_i}{\partial t} + N\nabla \cdot \mathbf{v}_i = 0 \tag{15.135}$$

The above two equations give

$$\left(\frac{\partial}{\partial t}+\nu_i\right)\frac{\partial n_i}{\partial t}=\frac{eN}{m_i}\nabla^2\phi \tag{15.136}$$

Instead of the Poisson equation, one can use the charge neutrality which is valid in a low frequency region, $n_e \simeq n_i$. So n_i in the above equation can be replaced by n_e. Taking divergence of (15.133) and eliminating $\nabla^2\phi$ yield

$$\frac{\partial^2 n_e}{\partial t^2}+\nu_i\frac{\partial n_e}{\partial t}-\frac{\gamma_e T_e}{m_i}\nabla^2 n_e=N\ \frac{m_e}{m_i}\nabla^2(\mathbf{v}_0\cdot\mathbf{v}'_L) \tag{15.137}$$

Equations (15.132) and (15.137) are the coupled equations to describe the Brillouin process.

15.5.3 *2 $\omega_{\mathbf{pe}}$ instability*

Here we consider an instability in which a light wave decays into two electron plasma waves. The frequency of the light wave is approximately equal to twice of the plasma frequency, hence the name. We need only the electron equation which we write in the form

$$\frac{\partial\mathbf{v}}{\partial t}+\mathbf{v}\cdot\nabla\mathbf{v}=-\frac{e}{m_e}(\mathbf{E}_0-\nabla\phi)-\frac{c}{m_e}\mathbf{v}\times\mathbf{B}-\frac{\gamma_e T_e}{Nm_e}\nabla n_e-\nu_e\mathbf{v}'_e \tag{15.138}$$

where

$$\mathbf{v}=\mathbf{v}_0+\mathbf{v}'_e$$

with $\mathbf{v}_0$ the light wave field and $\mathbf{v}'_e$ the plasma wave field ($\nabla\times\mathbf{v}'_e$=0). As discussed earlier, we have

$$\frac{\partial\mathbf{v}_0}{\partial t}=-\frac{e}{m_e}\mathbf{E}_0 \tag{15.139}$$

$$\frac{\partial\mathbf{v}'_e}{\partial t}+\frac{1}{2}\nabla v^2=\frac{e}{m_e}\nabla\phi-\frac{\gamma_e T_e}{Nm_e}\nabla n_e-\nu_e\mathbf{v}'_e \tag{15.140}$$

Maxwell equation in the electrostatic form is

$$\frac{\partial}{\partial t}\nabla\phi+4\pi e(n_e\mathbf{v}_0+N\mathbf{v}'_e)=0 \tag{15.141}$$

$$\nabla^2\phi=4\pi e n_e \tag{15.142}$$

The above two equations give

$$\frac{\partial n_e}{\partial t}+N\nabla\cdot\mathbf{v}'_e+\mathbf{v}_0\cdot\nabla n_e=0 \tag{15.143}$$

which is the continuity equation [in this way we can clearly see that the nonlinear term stems from the nonlinear current in (15.141)]. Taking divergence of (15.140) and using the preceding equations yield

$$-\left(\frac{\partial}{\partial t}+\nu_e\right)\left(\frac{\partial n_e}{\partial t}+\mathbf{v}_0\cdot\nabla n_e\right)+N\nabla^2\mathbf{v}_0\cdot\mathbf{v}_e'=\omega_{\rm pe}^2 n_e-\frac{\gamma_e T_e}{m_e}\nabla^2 n_e \tag{15.144}$$

Equations (15.143) and (15.144) are the two coupled equations for $2\omega_{\rm pe}$ instability.

15.6 Unified Theory of Parametric Instability

In the preceding sections, parametric instabilities were considered case by case. We attempt here to integrate them to form a unified picture, keeping in mind that a large amplitude pump wave beat with low frequency wave to excite Stokes and anti-Stokes daughter waves (15.146). The latter beat with the pump again to give rise to a low frequency ponderomotive force that enhances the low frequency wave (15.156). This input and feed back mechanism of parametric instabilities can be integrated to establish general formula, comprising the various cases. When the pump is a large amplitude electromagnetic wave, the parametric interaction is often called *stimulated scattering.* The nonlinearities responsible for the instabilities are the ponderomotive force in electron equation of motion (ion ponderomotive force is negligible by the mass ratio) and the nonlinear electron current (ion nonlinear current is also negligible by the mass ratio). For the scattered field (side band), Maxwell equations read

$$\nabla\times\mathbf{E}_s=-\frac{1}{c}\frac{\partial\mathbf{B}_s}{\partial t} \tag{15.145}$$

$$\nabla\times\mathbf{B}_s=\frac{1}{c}\frac{\partial\mathbf{E}_s}{\partial t}+\frac{4\pi}{c}\mathbf{J}_s-\frac{4\pi e}{c}\mathbf{v}_0\delta n_e \tag{15.146}$$

where $\mathbf{J}_s(=-eN\mathbf{v}_s)$ is the linear current contributing to the scattered field and the last term is the electron nonlinear current responsible for the scattered field, with δn_e the electron density fluctuation. Eliminating $\mathbf{B}_s$ in (15.145) and (15.146) and Fourier transforming them give

$$\mathbf{k}\times(\mathbf{k}\times\mathbf{E}_s)+\frac{\omega^2}{c^2}\mathbf{E}_s(\mathbf{k},\omega)+\frac{4\pi i\omega}{c^2}\mathbf{J}_s(\mathbf{k},\omega)=\mathbf{Q} \tag{15.147}$$

$$\text{where } \mathbf{Q}=\frac{4\pi e i\omega}{c^2}\int \mathbf{v}_0(\mathbf{r},t)\delta n_e(\mathbf{r},t)\,e^{-i\mathbf{k}\cdot\mathbf{r}+i\omega t}d^3r\,dt \tag{15.148}$$

The pump obeys the equation

$$\frac{\partial \mathbf{v}_0}{\partial t} = -\frac{e}{m_e}\mathbf{E}_0\ e^{i\mathbf{k}_0\cdot\mathbf{r}-i\omega_0 t} + c.c$$

$$\text{therefore } \mathbf{v}_0(\mathbf{r},t) = \frac{-ie}{m_e\omega_0}\mathbf{E}_0 e^{i\mathbf{k}_0\cdot\mathbf{r}-i\omega_0 t} + \frac{ie}{m_e\omega_0}\mathbf{E}_0^* e^{-i\mathbf{k}_0\cdot\mathbf{r}+i\omega_0 t} \tag{15.149}$$

The linear current $\mathbf{J}_s$ can be written in terms of $\mathbf{E}_s$: $J_{\text{si}} = \sigma_{ij}E_{sj}$. We here consider only an isotropic plasma (no magnetic field, no beam). Then (15.147) is written as, using $\varepsilon_{ij} = \delta_{ij} + \frac{4\pi i}{\omega}\sigma_{ij} = \delta_{ij}\varepsilon_T + (\varepsilon_L - \varepsilon_T)\frac{k_ik_j}{k^2}$,

$$\left[\left(\frac{\omega^2}{c^2}\varepsilon_T - k^2\right)\delta_{ij} + \left(k^2 + (\varepsilon_L - \varepsilon_T)\frac{\omega^2}{c^2}\right)\frac{k_ik_j}{k^2}\right] E_{sj} = Q_i \tag{15.150}$$

The inverse matrix of (15.150) is obtained as

$$\text{Inverse of } [\quad]_{ij} = \frac{\delta_{ij}}{\beta} - \frac{\alpha}{\beta(\alpha+\beta)}\frac{k_ik_j}{k^2} = \frac{\delta_{ij}}{\beta} - \frac{k_ik_j}{k^2}\left(\frac{1}{\beta} - \frac{1}{\alpha+\beta}\right)$$

$$\alpha = k^2 + (\varepsilon_L - \varepsilon_T)\frac{\omega^2}{c^2}, \quad \beta = \frac{\omega^2}{c^2}\varepsilon_T - k^2,$$

Thus, we write

$$E_{\text{si}}(\mathbf{k},\omega) = c^2\ \Lambda_{ij}Q_j \tag{15.151}$$

$$\text{where } \Lambda_{ij} = \frac{\delta_{ij} - \frac{k_ik_j}{k^2}}{\omega^2\varepsilon_T - c^2k^2} + \frac{k_ik_j}{k^2\omega^2\varepsilon_L} \tag{15.152}$$

Exercise. Show that (15.151) satisfies (15.150).

Carrying out the Fourier integral in (15.148) we have

$$\mathbf{Q} = \frac{4\pi i e\omega}{c^2}\ [\mathbf{v}_0\delta n_e(\omega-\omega_0, \mathbf{k}-\mathbf{k}_0) + \mathbf{v}_0^*\delta n_e(\omega+\omega_0, \mathbf{k}+\mathbf{k}_0)] \tag{15.153}$$

$$\text{where } \mathbf{v}_0 = \frac{-ie}{m_e\omega_0}\mathbf{E}_0$$

From (15.151) and (15.153) we can write

$$E_{\text{si}}(\mathbf{k},\omega) = \Lambda_{ij}(\mathbf{k},\omega)4\pi i e\omega[v_{0j}\delta n_e(\omega-\omega_0) + v_{0j}^*\delta n_e(\omega+\omega_0)]$$

Shifting gives

$$\mathbf{E}^- \equiv \mathbf{E}_s(\omega-\omega_0, \mathbf{k}-\mathbf{k}_0) = 4\pi i e(\omega-\omega_0)\ \Lambda_{ij}(\omega-\omega_0, \mathbf{k}-\mathbf{k}_0)v_{0j}^*\delta n_e(\omega,\mathbf{k}) \tag{15.154}$$

$$\mathbf{E}^+ \equiv \mathbf{E}_s(\omega+\omega_0, \mathbf{k}+\mathbf{k}_0) = 4\pi i e(\omega+\omega_0)\ \Lambda_{ij}(\omega+\omega_0, \mathbf{k}+\mathbf{k}_0)v_{0j}\delta n_e(\omega,\mathbf{k}) \tag{15.155}$$

where we neglected $\delta n_e(\omega \pm 2\omega_0, \mathbf{k} \pm 2\mathbf{k}_0)$. In the above $\mathbf{E}^{\pm}$ are, respectively, anti-Stokes and Stokes component of the scattered field. So far we obtained the scattered field components in terms of the electron density perturbation δn_e. We calculate δn_e by means of fluid equations,

$$\left(\frac{\partial}{\partial t} + \nu_e\right)\mathbf{v}_e = \frac{e}{m_e}\nabla\phi - \frac{\gamma_e T_e}{N m_e}\nabla\delta n_e - \nabla(\mathbf{v}_0 \cdot \mathbf{v}_s) \tag{15.156}$$

$$\frac{\partial \delta n_e}{\partial t} + N\nabla \cdot \mathbf{v}_e = 0 \tag{15.157}$$

$$-\nabla^2\phi = 4\pi e(\delta n_i - \delta n_e) \tag{15.158}$$

$$\frac{\partial \mathbf{v}_i}{\partial t} = -\frac{e}{m_i}\nabla\phi \tag{15.159}$$

$$\frac{\partial \delta n_i}{\partial t} + N\nabla \cdot \mathbf{v}_i = 0 \tag{15.160}$$

If δn_e is a high frequency perturbation ($\omega \gg \omega_{pi}$), the ion perturbation can be neglected ($\delta n_i = 0$). If the frequency ω is sufficiently small, the ion response to the electric force $\nabla\phi$ should be retained, and the Poisson equation should be used. Equations (15.158)–(15.160) are Fourier-transformed and combined to give

$$-k^2\phi = \frac{4\pi e}{1 - \frac{\omega_{pi}^2}{\omega^2}}\delta n_e(\mathbf{k}, \omega)$$

Then (15.156) and (15.157) yield

$$\left[\omega(\omega + i\nu_e) - \frac{\omega_{\rm pe}^2}{1 - \frac{\omega_{pi}^2}{\omega^2}} - \frac{\gamma_e T_e k^2}{m_e}\right]\delta n_e(\mathbf{k}, \omega)$$

$$= -N\int e^{-i\mathbf{k}\cdot\mathbf{r} + i\omega t}\nabla^2(\mathbf{v}_s \cdot \mathbf{v}_0(\mathbf{r}, t))\, d^3r\, dt$$

$$= Nk^2[\mathbf{v}_0 \cdot \mathbf{v}_s(\omega - \omega_0, \mathbf{k} - \mathbf{k}_0) + \mathbf{v}_0^* \cdot \mathbf{v}_s(\omega + \omega_0, \mathbf{k} + \mathbf{k}_0)] \tag{15.161}$$

$$\text{where } \mathbf{v}_s = -\frac{ie}{m_e\omega}\mathbf{E}_s \tag{15.162}$$

Equation (15.162) is valid because the scattered wave has the wave velocity much greater than the electron thermal velocity.

Exercise. Show that the linear part of (15.161) gives the Langmuir wave and the ion acoustic wave, respectively, according as $\omega^2 \gg \omega_{pi}^2$ or $\omega^2 \ll \omega_{pi}^2$.

From (15.154), (15.155), and (15.161), we obtain the dispersion relation

$$\omega(\omega + i\nu_e) - \frac{\omega_{\rm pe}^2}{1 - \frac{\omega_{pi}^2}{\omega^2}} - \frac{\gamma_e T_e k^2}{m_e} = k^2 \omega_{\rm pe}^2 \left[\frac{|\mathbf{k}_+ \times \mathbf{v}_0|^2}{k_+^2 (\omega_+^2 \varepsilon_T^+ - c^2 k_+^2)} + \frac{|\mathbf{k}_+ \cdot \mathbf{v}_0|^2}{k_+^2 \omega_+^2 \varepsilon_L^+} \right.$$
$$\left. + \frac{|\mathbf{k}_- \times \mathbf{v}_0|^2}{k_-^2 (\omega_-^2 \varepsilon_T^- - c^2 k_-^2)} + \frac{|\mathbf{k}_- \cdot \mathbf{v}_0|^2}{k_-^2 \omega_-^2 \varepsilon_L^-} \right] \tag{15.163}$$

where $\omega_\pm = \omega \pm \omega_0$, $\mathbf{k}_\pm = \mathbf{k} \pm \mathbf{k}_0$, $\varepsilon_{L,T}^\pm = \varepsilon_{L,T}(\omega_\pm, \mathbf{k}_\pm)$, $|\mathbf{k} \times \mathbf{v}|^2 = k^2 |\mathbf{v}|^2 - (\mathbf{k} \cdot \mathbf{v})(\mathbf{k} \cdot \mathbf{v}^*)$.

• **Electrostatic dispersion relation.** When the scattered wave can be approximated as an electrostatic mode, (15.147) reduces to, per $\mathbf{k} \times \mathbf{E}_s = 0$,

$$\varepsilon_{ij} E_{sj} = \frac{c^2}{\omega^2} Q_i$$

Using

$$\varepsilon_{ij} = \varepsilon_L \frac{k_i k_j}{k^2} + \varepsilon_T \left(\delta_{ij} - \frac{k_i k_j}{k^2} \right), \quad E_{\rm si} = \frac{k_i}{k} E_s$$

we have

$$\varepsilon_{ij} E_{sj} = \varepsilon_L \frac{k_i k_j}{k^2} E_{sj} = \frac{c^2}{\omega^2} Q_i$$

Performing $\mathbf{k} \cdot$ into both sides gives

$$E_{sj} = \frac{c^2}{\varepsilon_L \omega^2 k^2} k_j k_i Q_i$$

Comparing this equation with (15.151), we identify as

$$\Lambda_{ij} = \frac{c^2}{\varepsilon_L \omega^2 k^2} k_i k_j$$

in this electrostatic case. This identification is directly obtained by taking the limit $c \to \infty$ in (15.152). The corresponding electrostatic parametric dispersion relation takes the form of (15.163) with $|\mathbf{k}_\pm \times \mathbf{v}_0|^2$-terms discarded.

References: Drake, J. F., Kaw, P. K., Lee, Y. C., Schmidt, G., Liu, C. S., and Rosenbluth, M. N. *Parametric instabilities of electromagnetic waves in plasmas*, Phys. Fluids **17** (1974), p. 778; L. Stenflo, *Stimulated scattering of large amplitude waves in the ionosphere*, Physica Scripta **T30** (1990), pp. 166–169; L. Stenflo and P. K. Shukla, *Theory of stimulated scattering of large-amplitude waves*, J. Plasma Phys. **64** part 4 (2000), pp. 353–357.

15.7 Kinetic Theory of Parametric Instability

15.7.1 *Time-oscillating pump*

In this section, we derive the kinetic theory version of the parametric dispersion relation, (15.52). This is a kinetic theory of the parametric decay instability by sinusoidal high frequency electric field (dipole approximation for the pump).

We use the Vlasov equation in the form

$$\frac{\partial f_\alpha}{\partial t} + \mathbf{v}\cdot\frac{\partial f_\alpha}{\partial \mathbf{r}} + \frac{e_\alpha}{m_\alpha}\left(\mathbf{E}_0 \sin\omega_0 t - \frac{\partial \phi(\mathbf{r},t)}{\partial \mathbf{r}}\right)\cdot\frac{\partial f_\alpha}{\partial \mathbf{v}} = 0 \quad (15.164)$$

Linearizing this by putting $f_\alpha = F_\alpha(\mathbf{v},t) + f'_\alpha(\mathbf{r},\mathbf{v},t)$, we have

$$\frac{\partial F_\alpha}{\partial t} + \frac{e_\alpha}{m_\alpha}\sin\omega_0 t\ \mathbf{E}_0\cdot\frac{\partial F_\alpha}{\partial \mathbf{v}} = 0 \quad (15.165)$$

$$\frac{\partial f'_\alpha}{\partial t} + \mathbf{v}\cdot\frac{\partial f'_\alpha}{\partial \mathbf{r}} + \frac{e_\alpha}{m_\alpha}\mathbf{E}_0\sin\omega_0 t\cdot\frac{\partial f'_\alpha}{\partial \mathbf{v}} - \frac{e_\alpha}{m_\alpha}\frac{\partial \phi(\mathbf{r},t)}{\partial \mathbf{r}}\cdot\frac{\partial F_\alpha}{\partial \mathbf{v}} = 0 \quad (15.166)$$

$$\nabla^2\phi = -4\pi\sum_\alpha e_\alpha \int f'_\alpha d^3v \quad (15.167)$$

In the following, we shall omit the prime on f'. The zero order equation (15.165) is solved in the following form by characteristic method.

$$F_\alpha(\mathbf{v},t) = F_\alpha(\mathbf{u}_\alpha), \qquad \mathbf{u}_\alpha = \mathbf{v} + \frac{e_\alpha \mathbf{E}_0}{m_\alpha\omega_0}\cos\omega_0 t \quad (15.168)$$

Note that $\mathbf{u}_\alpha$ is the constant of the motion of the characteristic equation of (15.165).

Exercise. Verify that a Maxwellian

$$F(v,t) = \sqrt{\frac{m}{2\pi T}}\ \exp\left[-\frac{m}{2T}\ (v + \frac{eE_0}{m\omega_0}\cos\omega_0 t)^2\right]$$

satisfies $\frac{\partial F}{\partial t} + \frac{e}{m}E_0\ \sin\omega_0 t\ \frac{\partial F}{\partial v} = 0$.

Therefore, it is advantageous to transform old variable set $(t,\mathbf{r},\mathbf{v}) \rightarrow$ new variable set $(t',\mathbf{r},\mathbf{u}(\mathbf{v},t))$ with $t = t'$ in (15.166): we have

$$\frac{\partial}{\partial t} = \frac{\partial}{\partial t'} + \frac{\partial u_j}{\partial t}\frac{\partial}{\partial u_j} = \frac{\partial}{\partial t'} - \frac{e_\alpha}{m_\alpha}\sin\omega_0 t\ E_{0j}\frac{\partial}{\partial u_j}, \qquad \frac{\partial}{\partial v_j} = \frac{\partial}{\partial u_j} \quad (15.169)$$

Also Fourier-transforming with respect to $\mathbf{r}$, one obtains

$$\frac{\partial}{\partial t} f_\alpha(\mathbf{k},\mathbf{u},t) + i\mathbf{k}\cdot\mathbf{u} f_\alpha + i\omega_0 a_\alpha \cos\omega_0 t\ f_\alpha - i\phi \frac{e_\alpha}{m_\alpha}\mathbf{k}\cdot\frac{\partial F_\alpha}{\partial \mathbf{u}} = 0, \tag{15.170}$$

$$\text{where } a_\alpha = -\frac{e_\alpha}{m_\alpha \omega_0^2}\mathbf{k}\cdot\mathbf{E}_0, \tag{15.171}$$

We further transform as $f_\alpha = g_\alpha e^{-ia_\alpha \sin\omega_0 t}$ to obtain

$$\frac{\partial}{\partial t} g_\alpha(\mathbf{k},\mathbf{u},t) + i\mathbf{k}\cdot\mathbf{u}\ g_\alpha - i\frac{e_\alpha}{m_\alpha}\ \phi\ \mathbf{k}\cdot\frac{\partial F_\alpha}{\partial \mathbf{u}} e^{ia_\alpha \sin\omega_0 t} = 0 \tag{15.172}$$

$$k^2\phi = 4\pi \sum_\alpha e_\alpha e^{-ia_\alpha \sin\omega_0 t} \int g_\alpha\ d^3u \tag{15.173}$$

Substituting (15.173) into (15.172) for ϕ, we obtain

$$\frac{\partial g_e}{\partial t} + i\mathbf{k}\cdot\mathbf{u}\ g_e - i\frac{\omega_{\rm pe}^2}{k^2}\mathbf{k}\cdot\frac{\partial F_e}{\partial \mathbf{u}}\int g_e d^3u = -i\frac{\omega_{\rm pe}^2}{k^2}\mathbf{k}\cdot\frac{\partial F_e}{\partial \mathbf{u}} e^{ia\sin\omega_0 t}\int g_i d^3u \tag{15.174}$$

$$\frac{\partial g_i}{\partial t} + i\mathbf{k}\cdot\mathbf{u}\ g_i - i\frac{\omega_{pi}^2}{k^2}\mathbf{k}\cdot\frac{\partial F_i}{\partial \mathbf{u}}\int g_i d^3u = -i\frac{\omega_{pi}^2}{k^2}\mathbf{k}\cdot\frac{\partial F_i}{\partial \mathbf{u}} e^{-ia\sin\omega_0 t}\int g_e d^3u \tag{15.175}$$

where 'a' is defined in (15.8). Kinetic theory yields integro-differential equations. The above coupled equations should be compared with (15.6) and (15.7). Fourier transform the above equations with respect to t to obtain

$$\begin{aligned}(\omega - \mathbf{k}\cdot\mathbf{u})g_e(\omega) &+ \frac{\omega_{\rm pe}^2}{k^2}\mathbf{k}\cdot\frac{\partial F_e}{\partial \mathbf{u}}\int g_e(\omega)d^3u \\ &= \frac{\omega_{\rm pe}^2}{k^2}\mathbf{k}\cdot\frac{\partial F_e}{\partial \mathbf{u}}\Sigma_p J_p(a)\int g_i(\omega + p\omega_0)d^3u\end{aligned} \tag{15.176}$$

$$\begin{aligned}(\omega - \mathbf{k}\cdot\mathbf{u})g_i(\omega) &+ \frac{\omega_{pi}^2}{k^2}\mathbf{k}\cdot\frac{\partial F_i}{\partial \mathbf{u}}\int g_i(\omega)d^3u \\ &= \frac{\omega_{pi}^2}{k^2}\mathbf{k}\cdot\frac{\partial F_i}{\partial \mathbf{u}}\Sigma_p J_p(-a)\int g_e(\omega + p\omega_0)d^3u\end{aligned} \tag{15.177}$$

$$\text{where } g_\alpha(\omega) = g_\alpha(\omega,\mathbf{k},\mathbf{u})$$

Divide the above equations by $(\omega - \mathbf{k}\cdot\mathbf{u})$ and integrate over $\int d^3u$ to write

$$\left[1+\frac{1}{\chi_e(\omega)}\right] G_e(\omega) = \Sigma_p J_p(a)\; G_i(\omega + p\omega_0) \tag{15.178}$$

$$\left[1+\frac{1}{\chi_i(\omega)}\right] G_i(\omega) = \Sigma_p J_p(-a)\; G_e(\omega + p\omega_0) \tag{15.179}$$

$$\chi_\alpha(\omega) = \frac{\omega_{p\alpha}^2}{k^2}\int d^3u \frac{\mathbf{k}\cdot\frac{\partial F_\alpha}{\partial \mathbf{u}}}{\omega - \mathbf{k}\cdot\mathbf{u}}, \quad \text{susceptibility,}$$

$$G_\alpha(\omega) = \int d^3u\; g_\alpha(\mathbf{k}, \mathbf{u}, \omega)$$

and the $\mathbf{k}$-dependence was suppressed. For a weak pump, we can take three terms in the Bessel series: $p = 0,\ \pm 1$. We write

$$\left[1+\frac{1}{\chi_e(\omega)}\right] G_e(\omega) = G_i(\omega) + J_1(a)G_i(\omega+\omega_0) + J_{-1}(a)G_i(\omega-\omega_0) \tag{15.180}$$

$$\left[1+\frac{1}{\chi_i(\omega)}\right] G_i(\omega) = G_e(\omega) + J_1(-a)G_e(\omega+\omega_0) + J_{-1}(-a)G_e(\omega-\omega_0) \tag{15.181}$$

The above two equations correspond to (15.47) and (15.48). The parametric dispersion relation can be obtained in analogous way. Shifting in (15.180) gives

$$\left[1+\frac{1}{\chi_e(\omega+\omega_0)}\right] G_e(\omega+\omega_0)$$
$$= G_i(\omega+\omega_0) + J_1(a)G_i(\omega+2\omega_0) + J_{-1}(a)G_i(\omega) \tag{15.182}$$

$$\left[1+\frac{1}{\chi_e(\omega-\omega_0)}\right] G_e(\omega-\omega_0)$$
$$= G_i(\omega-\omega_0) + J_1(a)G_i(\omega) + J_{-1}(a)G_i(\omega-2\omega_0) \tag{15.183}$$

Eliminate $G_e(\omega \pm \omega_0)$ in (15.181) by using (15.182) and (15.183). In doing so, neglect $G_i(\omega \pm \omega_0),\ \ G_i(\omega \pm 2\omega_0)$ as off-resonant to obtain

$$1+\frac{1}{\chi_i(\omega)} - \frac{\chi_e(\omega)}{1+\chi_e(\omega)} = J_1^2(a)\left[\frac{\chi_e(\omega+\omega_0)}{1+\chi_e(\omega+\omega_0)} + \frac{\chi_e(\omega-\omega_0)}{1+\chi_e(\omega-\omega_0)}\right] \tag{15.184}$$

The linear dispersion relation is obtained by putting $a = 0$ in the above:

$$1 + \chi_i(\mathbf{k}, \omega) + \chi_e(\mathbf{k}, \omega) = 0 \tag{15.185}$$

In (15.184), we approximate:

$$\chi_e(\omega) = \frac{-\omega_{\rm pe}^2}{\omega^2 - \gamma_e T_e k^2/m_e}, \quad \chi_i = -\frac{\omega_{pi}^2}{\omega^2}$$

$$\frac{1}{\chi_i(\omega, \mathbf{k})} + \frac{1}{1 + \chi_e(\omega, \mathbf{k})} = J_1^2(a)\omega_{\rm pe}^2\omega_{pi}^2 \left[\frac{1}{D_e(\omega + \omega_0, \mathbf{k})} + \frac{1}{D_e(\omega - \omega_0, \mathbf{k})}\right] \tag{15.186}$$

where D_e is defined in (15.46). In the LHS, neglect ω^2 as compared to $\omega_{\rm pe}^2$, $T_e k^2/m_e$ and $k^2\lambda_e^2 \ll 1$, then the LHS of the above equation becomes $\omega^2 - k^2\gamma_e T_e/m_i$, yielding the identical form as (15.52). (15.186) agrees with Eq. (17) in Sanmartin (1970). Equation (15.184) was obtained by Silin (1965).

Reference: Sanmartin, J. R. *Electrostatic plasma instabilities excited by a high frequency electric field*, Phys. Fluids **13** (1970), pp. 1533–1542. V. P. Silin, *Parametric resonance in a plasma*, Sov. Phys. JETP **21** (1965), p. 1127: JETP **24** (1967), p. 1242.

15.7.2 *Kinetic theory of parametric decay instability: Nonlinear Landau damping*

The parametric decay instability is given rise to by wave-particle interaction when the low frequency beat wave (denoted by subscript 3 below) of two high frequency waves satisfy the Landau resonance conditions. See Section 15.3 for its dipole approximation version. In this section, we study the kinetic process of the nonlinear Landau damping by using the Vlasov equation. This treatment will make the physical process of the decay instability more clear. We consider three waves with frequencies and wave vectors $(\mathbf{k}_j, \omega_j)$ $(j = 1, 2, 3)$ in which the high frequency waves are designated by $j = 1, 2$ and the low frequency wave is subscripted by '3'. We assume that the three waves satisfy the resonance conditions,

$$\omega_1 - \omega_2 = \omega_3, \quad \mathbf{k}_1 - \mathbf{k}_2 = \mathbf{k}_3 \tag{15.187}$$

The Vlasov equations corresponding to each frequency are

$$\frac{\partial f_1}{\partial t} + \mathbf{v} \cdot \frac{\partial f_1}{\partial \mathbf{r}} - \frac{e}{m_e}\mathbf{E}_1 \cdot \frac{\partial F_{0e}}{\partial \mathbf{v}} = 0 \tag{15.188}$$

$$\frac{\partial f_2}{\partial t} + \mathbf{v} \cdot \frac{\partial f_2}{\partial \mathbf{r}} - \frac{e}{m_e}\mathbf{E}_2 \cdot \frac{\partial F_{0e}}{\partial \mathbf{v}} = 0 \tag{15.189}$$

$$\frac{\partial f_{3e}}{\partial t} + \mathbf{v} \cdot \frac{\partial f_{3e}}{\partial \mathbf{r}} - \frac{e}{m_e}\mathbf{E}_3 \cdot \frac{\partial F_{0e}}{\partial \mathbf{v}} - \frac{e}{m_e}\left(\mathbf{E}_1 \cdot \frac{\partial f_2}{\partial \mathbf{v}} + \mathbf{E}_2 \cdot \frac{\partial f_1}{\partial \mathbf{v}}\right) = 0 \tag{15.190}$$

$$\frac{\partial f_{3i}}{\partial t} + \mathbf{v} \cdot \frac{\partial f_{3i}}{\partial \mathbf{r}} + \frac{e}{m_i}\mathbf{E}_3 \cdot \frac{\partial F_{0i}}{\partial \mathbf{v}} = 0 \tag{15.191}$$

where the subscript $e(i)$ denote electron (ion). $F_{0e,i}$ are the respective zero order distribution functions. Note that the high frequency ion equation is not used (ions are stationary in a high frequency field); all the important nonlinearities are contained in the electron equations; two high frequency waves beat to produce the low frequency perturbation f_{3e}; ion nonlinearities are small by the mass ratio and can be neglected in (15.191). We also have Maxwell equations

$$\frac{\partial \mathbf{E}_j}{\partial t} = -4\pi \mathbf{J}_j \quad (j = 1, 2, 3) \tag{15.192}$$

$$\mathbf{J}_1 = -en_0 \int \mathbf{v} f_1 d^3v - en_3 \int \mathbf{v} f_2 d^3v \tag{15.193}$$

$$\mathbf{J}_2 = -en_0 \int \mathbf{v} f_2 d^3v - en_3 \int \mathbf{v} f_1 d^3v \tag{15.194}$$

$$\mathbf{J}_3 = en_0 \int \mathbf{v}(f_{3i} - f_{3e}) d^3v \tag{15.195}$$

where

$$n_3 = \int f_{3i}\, d^3v \tag{15.196}$$

is the low frequency number density fluctuation; by the charge neutrality in the low frequency region, we have $n_e \simeq n_i \equiv n_3$. In the above governing equations, the nonlinear terms are: (1) $\mathbf{E} \cdot \frac{\partial f}{\partial \mathbf{v}}$ in the low frequency electron equation in (15.190), which corresponds to the low frequency ponderomotive force in fluid equation; (2) the nonlinear currents in the two high frequency electron currents $\mathbf{J}_1$ and $\mathbf{J}_2$, the last terms of (15.193) and (15.194). Let us introduce the Fourier coefficients by putting

$$f_j(\mathbf{r}, \mathbf{v}, t) = f_j(\mathbf{v}) e^{i\mathbf{k}\cdot\mathbf{r} - i\omega_j t} + f_j^*(\mathbf{v}) e^{-i\mathbf{k}\cdot\mathbf{r} + i\omega_j t} \quad (j = 1, 2, 3)$$

where

$$f_j(\mathbf{v}) = \frac{ie}{m_e} \frac{\mathbf{E}_j \cdot \frac{d}{d\mathbf{v}} F_{0e}}{\omega_j - \mathbf{k}_j \cdot \mathbf{v}} \quad (j = 1, 2) \tag{15.197}$$

$$f_{3e}(\mathbf{v}) = \frac{ie}{m_e} \frac{\mathbf{E}_3 \cdot \frac{d}{d\mathbf{v}} F_{0e} + \mathbf{E}_1 \cdot \frac{d}{d\mathbf{v}} f_2^* + \mathbf{E}_2^* \cdot \frac{d}{d\mathbf{v}} f_1}{\omega_3 - \mathbf{k}_3 \cdot \mathbf{v}} \tag{15.198}$$

$$f_{3i}(\mathbf{v}) = -\frac{ie}{m_i} \frac{\mathbf{E}_3 \cdot \frac{d}{d\mathbf{v}} F_{0i}}{\omega_3 - \mathbf{k}_3 \cdot \mathbf{v}} \tag{15.199}$$

Equating $\int f_{3e} d^3v = \int f_{3i} d^3v$ (charge neutrality) gives $\mathbf{E}_3$ in terms of the beat terms. The result is substituted into (15.199) to obtain

$$\begin{aligned} n_3 = &-\frac{e^2}{m_i m_e} \int d^3v \frac{\hat{\mathbf{k}}_3 \cdot \frac{d}{d\mathbf{v}} F_{0i}}{\omega_3 - \mathbf{k}_3 \cdot \mathbf{v}} \int d^3v \frac{\mathbf{E}_1 \cdot \frac{d}{d\mathbf{v}} f_2^* + \mathbf{E}_2^* \cdot \frac{d}{d\mathbf{v}} f_1}{\omega_3 - \mathbf{k}_3 \cdot \mathbf{v}} \\ &\div \left[\frac{ie}{m_e} \int d^3v \frac{\hat{\mathbf{k}}_3 \cdot \frac{d}{d\mathbf{v}} F_{0e}}{\omega_3 - \mathbf{k}_3 \cdot \mathbf{v}} + \frac{ie}{m_i} \int d^3v \frac{\hat{\mathbf{k}}_3 \cdot \frac{d}{d\mathbf{v}} F_{0i}}{\omega_3 - \mathbf{k}_3 \cdot \mathbf{v}} \right] \end{aligned} \tag{15.200}$$

where $\hat{\mathbf{k}}_3 = \mathbf{k}_3/k_3$, and we used $\mathbf{E}_3 = \hat{\mathbf{k}}_3 E_3$. Equation (15.200) gives the low frequency number density perturbation generated as the two high frequency waves beat. Using the one-dimensional Maxwellian

$$F_{0\alpha} = \sqrt{\frac{m_\alpha}{2\pi T_\alpha}} e^{-\frac{m_\alpha v^2}{2T_\alpha}} \qquad (\alpha = e,\ i)$$

the velocity integrals in (15.200) can be calculated. Assume $\frac{\omega_3}{k_3} \ll v_{eth}$ (electron thermal velocity) and expand $1/(\omega_3 - k_3 v)$ in the powers of $\omega_3/k_3 v$ to obtain

$$\int dv \frac{\hat{\mathbf{k}}_3 \cdot \frac{d}{d\mathbf{v}} F_{0e}}{\omega_3 - \mathbf{k}_3 \cdot \mathbf{v}} = \frac{m_e}{T_e k_3} \tag{15.201}$$

The ion integral in the denominator of (15.200) should be integrated along the Landau contour since the wave-particle resonance condition can be satisfied. So we write

$$\int dv \frac{\hat{\mathbf{k}}_3 \cdot \frac{d}{d\mathbf{v}} F_{0i}}{\omega_3 - \mathbf{k}_3 \cdot \mathbf{v}} = -\frac{m_i}{T_i k_3} W \tag{15.202}$$

$$\text{with} \quad W = \int_{-\infty}^{\infty} \frac{v F_{oi}}{\frac{\omega_3}{k_3} - v}\, dv$$

Note that $\omega_3/k_3 \sim v_{thi}$ (ion thermal velocity), which is the wave-particle resonance condition. The latter may be rewritten in view of (15.187) as

$$\omega_1 - \omega_2 = (k_1 - k_2) \cdot v_{\text{thi}}$$

Now, in (15.200), we have, per (15.197) and (15.198) (assume all the $\mathbf{k}$'s are parallel), the integral

$$\int \frac{E_1 \frac{\partial f_2^*}{\partial v} + E_2^* \frac{\partial f_1}{\partial v}}{\omega_3 - k_3 v}\, dv = \frac{-ie}{m_e} E_1 E_2^* k_3 \int dv\, \frac{F_{0e}'}{(\omega_3 - k_3 v)^2} \left(\frac{1}{\omega_1 - k_1 v} - \frac{1}{\omega_2 - k_2 v} \right)$$
$$= \frac{ie}{T_e \omega_{\text{pe}}^2} E_1 E_2^* \tag{15.203}$$

where we put $\omega_1 \simeq \omega_2 \simeq \omega_{\text{pe}}$, and the prime denotes v-differentiation.

Exercise. Prove (15.203). Assume $\frac{\omega_3}{k_3} \ll v_{eth} \ll \frac{\omega_1}{k_1}, \frac{\omega_2}{k_2}$ and expand the velocity-dependent functions in appropriate power series.

Collecting the foregoing results, we obtain

$$n_3 = \frac{e^2}{m_e T_i \omega_{\text{pe}}^2} \frac{W}{1 - \frac{T_e}{T_i} W} E_1 E_2^* \tag{15.204}$$

The dispersion relation of the linear low frequency wave is obtained by putting the denominator equal to zero:

$$1 - \frac{T_e}{T_i} W \equiv \varepsilon(k_3, \omega_3) = 0$$

from which we can write

$$\varepsilon_r = 1 - \frac{T_e}{T_i} W_r \quad \text{(real part)} \tag{15.205}$$

$$\varepsilon_i = -\frac{T_e}{T_i} W_i \quad \text{(imaginary part)} \tag{15.206}$$

$$W_r = \frac{k_3}{\omega_3} \int dv\, F_{0i}\, v \left(1 + \frac{k_3 v}{\omega_3} + \cdots \right) = \frac{T_i}{m_i} \frac{k_3^2}{\omega_3^2} \tag{15.207}$$

$$\text{Therefore, } \varepsilon_r = 1 - \frac{T_e}{m_i} \frac{k_3^2}{\omega_3^2} \tag{15.208}$$

To calculate W_i, we write

$$W = -\sqrt{\frac{m_i}{2\pi T_i}} \int \frac{v e^{-\frac{m_i v^2}{2T_i}}}{v - \frac{\omega_3}{k_3}} dv \tag{15.209}$$

from which we obtain by picking up the half-residue at the Landau pole

$$W_i = \pi i (-) \sqrt{\frac{m_i}{2\pi T_i}} \left[v e^{-\frac{m_i v^2}{2T_i}} \right]_{v = \frac{\omega_3}{k_3}} = -\pi i \sqrt{\frac{m_i}{2\pi T_i}} \frac{\omega_3}{k_3} e^{-T_e/2T_i} \tag{15.210}$$

The Landau damping rate (γ) of the low frequency ion acoustic wave is obtained from the well known formula

$$\gamma = -\frac{\varepsilon_i}{\partial\varepsilon_r/\partial\omega_3} = i\sqrt{\frac{\pi}{8}}\,\omega_3\left(\frac{T_e}{T_i}\right)^{3/2} e^{-T_e/2T_i} \tag{15.211}$$

In terms of γ, we have

$$W_i = \frac{T_i}{T_e}\frac{2\gamma}{\omega_3} \tag{15.212}$$

Using (15.207) and (15.212) in (15.204) gives

$$n_3 = \frac{(k_3^2 + i2\gamma\omega_3\frac{m_i}{T_e})e^2E_iE_2^*}{m_e m_i\omega_{\rm pe}^2\,(\omega_3^2 - v_s^2k_3^2 - 2i\gamma\omega_3)} \tag{15.213}$$

Equation (15.213) should be compared with (15.93). It can be ascertained that the imaginary part in the numerator cancels out in the calculation of the dissipation rate of the ω_2-wave energy. The result for the threshold is same as (15.95) with the replacement of $\nu_i \to 2\gamma$.

Exercise. Show that

$$n_3 = \frac{i\omega_3 - 2\gamma}{2\gamma}\cdot\frac{e^2E_1E_2^*}{T_e m_e\omega_{\rm pe}^2}$$

$$\frac{\partial}{\partial t}|E_2|^2 = \frac{\omega_3}{\gamma}\cdot\frac{e^2|E_1|^2|E_2|^2}{T_e\; m_e\;\omega_1}$$

$$\text{Threshold: } |E_1|^2 = \frac{4\pi n_0\; T_e\;\gamma\;\nu_e}{\omega_3\;\omega_{\rm pe}} \quad\text{or}\quad \frac{|v_1|^2}{u^2} = \frac{\gamma\nu_e}{\omega_1k_3v_s} \quad \text{(see (15.95))} \tag{15.214}$$

Reference: Y.-Y. Kuo and J. A. Fejer, *Spectral-line structures of saturated parametric instabilities*, Phys. Rev. Lett. **29** No. 25 (1972), pp. 1667–1670; H. J. Lee, *Kinetic theory of coherent resonant interactions of electrostatic plasma waves*, J. Korean Phys. Soc. **7** (1984), pp. 256–265.

15.7.3 *Nonlinear dispersion relation for parametric decay instability*

The content of the last section can be reformulated in terms of the nonlinear dispersion relations, which make the parametric couplings more transparent. The nonlinear dispersion relation can be converted to differential equations for the wave amplitude. It is a two-time analysis. Fourier transforming

(15.192) and dotting $\mathbf{k}_1$ into the result give

$$-i\omega_1\mathbf{k}_1\cdot\mathbf{E}_1 = i\omega_{\rm pe}^2\left[\int d^3v\,\frac{\mathbf{k}_1\cdot\mathbf{v}\,\mathbf{E}_1\cdot\frac{d}{d\mathbf{v}}F_{0e}}{\omega_1-\mathbf{k}_1\cdot\mathbf{v}} + \frac{n_3}{n_0}\int d^3v\,\frac{\mathbf{k}_1\cdot\mathbf{v}\,\mathbf{E}_2\cdot\frac{d}{d\mathbf{v}}F_{0e}}{\omega_2-\mathbf{k}_2\cdot\mathbf{v}}\right]$$

Let us use

$$\frac{\mathbf{k}_1\cdot\mathbf{v}}{\omega_1-\mathbf{k}_1\cdot\mathbf{v}} = -1+\frac{\omega_1}{\omega_1-\mathbf{k}_1\cdot\mathbf{v}}$$

and integrate by parts the first integral to obtain

$$\left(1-\omega_{\rm pe}^2\int\frac{F_{0e}\,d^3v}{(\omega_1-\mathbf{k}_1\cdot\mathbf{v})^2}\right)\mathbf{k}_1\cdot\mathbf{E}_1 = -\frac{\omega_{\rm pe}^2}{\omega_1}\frac{n_3}{n_0}\int d^3v\,\frac{\mathbf{k}_1\cdot\mathbf{v}\,\mathbf{E}_2\cdot\frac{d}{d\mathbf{v}}F_{0e}}{\omega_2-\mathbf{k}_2\cdot\mathbf{v}}$$

The integral in the right-hand side is equal to

$$\begin{aligned}
-\frac{m_e}{T_e}\frac{E_2}{k_2}&\int d^3v\,\frac{\mathbf{k}_1\cdot\mathbf{v}\,\mathbf{k}_2\cdot\mathbf{v}\,F_{0e}}{\omega_2-\mathbf{k}_2\cdot\mathbf{v}}\\
&= -\frac{m_e}{T_e}\frac{\omega_2E_2}{k_2}\int d^3v\,F_{0e}\frac{\mathbf{k}_1\cdot\mathbf{v}}{\omega_2-\mathbf{k}_2\cdot\mathbf{v}}\\
&= -\frac{m_e}{T_e}\frac{E_2}{k_2}\int d^3v\,F_{0e}\,\mathbf{k}_1\cdot\mathbf{v}\left(1+\frac{\mathbf{k}_2\cdot\mathbf{v}}{\omega_2}+\cdots\right)\\
&= -\frac{m_e}{T_e\omega_2}\frac{E_2}{k_2}\int d^3v\,F_{0e}\,(k_{1x}k_{2x}v_x^2+k_{1y}k_{2y}v_y^2+k_{1z}k_{2z}v_z^2)\\
&= -\frac{m_e}{T_e\omega_2}\frac{E_2}{k_2}\mathbf{k}_1\cdot\mathbf{k}_2\int d^3v\,F_{0e}\,v_x^2 = -\frac{1}{\omega_2}\mathbf{k}_1\cdot\mathbf{E}_2
\end{aligned}$$

We also have

$$n_3 = \int f_{3i}d^3v = \frac{ie\,W}{T_ik_3}E_3 \tag{15.215}$$

Collecting the above results, we finally can write

$$\varepsilon(k_1,\omega_1)\mathbf{k}_1\cdot\mathbf{E}_1 = \frac{ieW}{k_3T_i}\,(\mathbf{k}_1\cdot\mathbf{E}_2)\,E_3 \tag{15.216}$$

$$\text{where }\varepsilon(k_1,\omega_1) = 1-\omega_{\rm pe}^2\int\frac{F_{0e}d^3v}{(\omega_1-\mathbf{k}_1\cdot\mathbf{v})^2} = 1-\frac{\omega_{\rm pe}^2}{\omega_1^2}(1+3k_1^2\lambda_e^2) \tag{15.217}$$

is the dielectric function of ω_1-wave. Likewise we obtain

$$\varepsilon(k_2,\omega_2)\mathbf{k}_2\cdot\mathbf{E}_2 = -\frac{ieW^*}{k_3T_i}\,(\mathbf{k}_2\cdot\mathbf{E}_1)\,E_3^* \tag{15.218}$$

$$\text{where }\varepsilon(k_2,\omega_2) = 1-\omega_{\rm pe}^2\int\frac{F_{0e}d^3v}{(\omega_2-\mathbf{k}_2\cdot\mathbf{v})^2} = 1-\frac{\omega_{\rm pe}^2}{\omega_2^2}(1+3k_2^2\lambda_e^2) \tag{15.219}$$

is the dielectric function of ω_2-wave. To obtain the ω_3-equation, it is more convenient to use the Poisson equation

$$i\,\mathbf{k}_3 \cdot \mathbf{E}_3 = 4\pi e \int (f_{3i} - f_{3e}) d^3v$$

Using (15.201) and (15.202) and carrying out the integral of (15.198), one obtains

$$\varepsilon(k_3, \omega_3) E_3 = \frac{i}{4\pi n_0 e} \left[(\mathbf{k}_2 \cdot \mathbf{E}_1)\mathbf{E}_2^* \cdot \hat{\mathbf{k}_3} - (\mathbf{k}_1 \cdot \mathbf{E}_2^*)\mathbf{E}_1 \cdot \hat{\mathbf{k}_3} \right] \tag{15.220}$$

$$\text{where} \quad \varepsilon(k_3, \omega_3) = 1 - \frac{T_e}{T_i} W \quad (\lambda_e^2 k_3^2 \ll 1) \tag{15.221}$$

This is the low frequency dielectric function as we defined earlier (see the equation below (15.204)). Equations (15.216), (15.218), and (15.220) are the three coupled equations for the three waves connected by the resonance condition (15.187). It is immediately seen that no couplings take place if $\mathbf{k}_1 \perp \mathbf{k}_2$, so we assume all the $\mathbf{k}'s$ to be parallel. Then we have

$$\varepsilon(k_1, \omega_1) E_1 = \frac{ieW}{k_3 T_i} E_2 E_3 \tag{15.222}$$

$$\varepsilon(k_2, \omega_2) E_2 = -\frac{ieW^*}{k_3 T_i} E_1 E_3^* \tag{15.223}$$

$$\varepsilon(k_3, \omega_3) E_3 = -\frac{ik_3}{4\pi n_0 e} E_1 E_2^* \tag{15.224}$$

Eliminating E_2 gives

$$1 - \frac{T_e}{T_i} W = \frac{W}{4\pi n_0 T_i} \frac{|E_1|^2}{\varepsilon(k_2, \omega_2)} \tag{15.225}$$

This is the desired nonlinear dispersion relation.

Exercise. Use (15.212) and show that the nonlinear dispersion relation (15.225) can be cast into the form

$$\omega_3^2 + 2i\gamma\omega_3 - \frac{T_e}{m_i} k_3^2 = \frac{k_3^2 \omega_{\text{pe}}^2}{4\pi n_0 m_i} \frac{|E_1|^2}{\omega_2^2 - \omega_{\text{pe}}^2 - 3k_2^2 \frac{T_e}{m_e}} \tag{15.226}$$

which is equivalent to (15.61).

References: H. J. Lee, *Kinetic theory of coherent resonant interactions of electrostatic plasma waves*, J. Korean Phys. Soc. **17** (1984), pp. 256–265; H. J. Lee, *Time evolution of coherently resonant waves in a collisionless homogeneous plasma*, J. Korean Phys. Soc. **7** (1982), pp. 71–80.

15.7.4 *Parametric instability due to an oscillating electric field in a magnetized plasma: Kinetic theory*

One method will be presented to derive the dispersion relation of the electrostatic parametric instability due to an oscillating electric field in a magnetized plasma by using the Vlasov equation. We only attempt to derive the results corresponding to (15.172) and (15.173). Then the remaining job can be carried out analogously to the analysis in Section 15.7.1. The Vlasov equation in a magnetized plasma reads ($\mathbf{B}_0 = \hat{\mathbf{z}} B_0$)

$$\frac{\partial f_\alpha}{\partial t} + \mathbf{v} \cdot \frac{\partial f_\alpha}{\partial \mathbf{r}} + \frac{e_\alpha}{m_\alpha} \left(\mathbf{E}_0 \sin \omega_0 t + \frac{1}{c} \mathbf{v} \times \mathbf{B}_0 - \frac{\partial \phi(\mathbf{r}, t)}{\partial \mathbf{r}} \right) \cdot \frac{\partial f_\alpha}{\partial \mathbf{v}} = 0 \tag{15.227}$$

Linearizing, after putting $f_\alpha = F_\alpha(\mathbf{v}, t) + f'_\alpha(\mathbf{r}, \mathbf{v}, t)$, gives

$$\frac{\partial F_\alpha}{\partial t} + \frac{e_\alpha}{m_\alpha} \left(\mathbf{E}_0 \sin \omega_0 t + \frac{1}{c} \mathbf{v} \times \mathbf{B}_0 \right) \cdot \frac{\partial F_\alpha}{\partial \mathbf{v}} = 0 \tag{15.228}$$

$$\frac{\partial f'_\alpha}{\partial t} + \mathbf{v} \cdot \frac{\partial f'_\alpha}{\partial \mathbf{r}} + \frac{e_\alpha}{m_\alpha} \left(\mathbf{E}_0 \sin \omega_0 t + \frac{1}{c} \mathbf{v} \times \mathbf{B}_0 \right) \frac{\partial f'_\alpha}{\partial \mathbf{v}} - \frac{e_\alpha}{m_\alpha} \frac{\partial \phi(\mathbf{r}, t)}{\partial \mathbf{r}} \cdot \frac{\partial F_\alpha}{\partial \mathbf{v}} = 0 \tag{15.229}$$

The characteristic equation of (15.228) is

$$\begin{aligned} \frac{dt}{1} &= \frac{dv_x}{\frac{e_\alpha}{m_\alpha} \sin \omega_0 t \; E_{0x} + \omega_{c\alpha} v_y} = \frac{dv_y}{\frac{e_\alpha}{m_\alpha} \sin \omega_0 t \; E_{0y} - \omega_{c\alpha} v_x} \\ &= \frac{dv_z}{\frac{e_\alpha}{m_\alpha} \; E_{0z} \sin \omega_0 t} = \frac{dF_\alpha}{0} \end{aligned}$$

Or, writing separately, we have

$$\frac{dv_x}{dt} = \frac{e_\alpha}{m_\alpha} E_{0x} \sin \omega_0 t + \omega_{c\alpha} v_y \tag{15.230}$$

$$\frac{dv_y}{dt} = \frac{e_\alpha}{m_\alpha} E_{0y} \sin \omega_0 t - \omega_{c\alpha} v_x \tag{15.231}$$

$$\frac{dv_z}{dt} = \frac{e_\alpha}{m_\alpha} E_{0z} \sin \omega_0 t, \tag{15.232}$$

Then F_α is a function of the constants of motion of the above equations.

Now (15.230), (15.231), and (15.232) will be solved. Eliminating v_y gives

$$\frac{d^2 v_x}{dt^2} + \omega_{c\alpha}^2 v_x = \frac{e_\alpha}{m_\alpha} (\omega_0 E_{0x} \cos \omega_0 t + \omega_{c\alpha} E_{0y} \sin \omega_0 t), \quad \left(\omega_{c\alpha} = \frac{e_\alpha B_0}{c \, m_\alpha} \right) \tag{15.233}$$

whose solution consists of the particular solution and the homogeneous solution, as is obtained below. Also the rest of the equations (15.231) and (15.232) can be solved straightly, giving

$$v_x = u_x \sin\ \omega_{c\alpha}t + u_y \cos\omega_{c\alpha}t - \frac{e_\alpha/m_\alpha}{\omega_0^2 - \omega_{c\alpha}^2}(\omega_0 E_{0x}\cos\omega_0 t + \omega_{c\alpha}E_{0y}\sin\ \omega_0 t) \tag{15.234}$$

$$v_y = u_x \cos\omega_{c\alpha}t - u_y \sin\ \omega_{c\alpha}t - \frac{e_\alpha/m_\alpha}{\omega_0^2 - \omega_{c\alpha}^2}(\omega_0 E_{0y}\cos\omega_0 t - \omega_{c\alpha}E_{0x}\sin\ \omega_0 t) \tag{15.235}$$

$$v_z = -\frac{e_\alpha}{m_\alpha \omega_0}E_{0z}\cos\omega_0 t + u_z \tag{15.236}$$

where u_x, u_y, u_z are the constants of motion of integration. Then, the general solution of (15.228) can be written as

$$F_\alpha(\mathbf{v}, t) = F_\alpha(u_x, u_y, u_z) \tag{15.237}$$

Solving for the constants of integration, u_x, u_y, u_z, we have

$$\begin{aligned} u_x = v_x \sin\ \omega_{c\alpha}t + v_y \cos\omega_{c\alpha}t + \frac{e_\alpha/m_\alpha}{\omega_0^2 - \omega_{c\alpha}^2}\Big[\omega_0\cos\omega_0 t(E_{0x}\sin\omega_{c\alpha}t \\ + E_{0y}\cos\omega_{c\alpha}t) - \omega_{c\alpha}\sin\omega_0 t(E_{0x}\cos\omega_{c\alpha}t - E_{0y}\sin\omega_{c\alpha}t)\Big] \end{aligned} \tag{15.238}$$

$$\begin{aligned} u_y = v_x \cos\omega_{c\alpha}t - v_y \sin\ \omega_{c\alpha}t + \frac{e_\alpha/m_\alpha}{\omega_0^2 - \omega_{c\alpha}^2}\Big[\omega_0\cos\omega_0 t(E_{0x}\cos\omega_{c\alpha}t \\ - E_{0y}\sin\omega_{c\alpha}t) + \omega_{c\alpha}\sin\omega_0 t(E_{0x}\sin\omega_{c\alpha}t + E_{0y}\cos\omega_{c\alpha}t)\Big] \end{aligned} \tag{15.239}$$

$$u_z = v_z + \frac{e_\alpha}{m_\alpha\omega_0}E_{0z}\cos\omega_0 t \tag{15.240}$$

Exercise. Prove that $F_\alpha(\mathbf{v}, t) = F_\alpha(u_x(t), u_y(t), u_z(t))$ satisfies (15.228).

Next, we make a change of variable in (15.229): $f(\mathbf{r}, \mathbf{v}, t) \to f(\mathbf{r}, \mathbf{u}(\mathbf{v}, t), t)$ by introducing $\mathbf{u}$ via (15.238)–(15.240). In (15.229), we change

$$\frac{\partial f}{\partial t} \to \frac{\partial f}{\partial t} + \frac{\partial f}{\partial \mathbf{u}}\cdot\frac{\partial \mathbf{u}}{\partial t}, \quad \frac{\partial f}{\partial \mathbf{v}} = \frac{\partial f}{\partial \mathbf{u}}$$

Then, among the terms in (15.229), we have

$$\frac{\partial f_\alpha}{\partial \mathbf{u}}\cdot\frac{\partial \mathbf{u}}{\partial t} + \frac{e_\alpha}{m_\alpha}\left(\mathbf{E}_0\sin\omega_0 t + \frac{1}{c}\mathbf{v}\times\mathbf{B}_0\right)\cdot\frac{\partial f_\alpha}{\partial \mathbf{v}} = 0$$

This equation holds because it is nothing but (15.228) with F being replaced by f. Therefore, we are left with

$$\frac{\partial f_\alpha(\mathbf{u})}{\partial t} + i\mathbf{k}\cdot\mathbf{v}f_\alpha - i\frac{e_\alpha}{m_\alpha}\,\phi(\mathbf{k},t)\mathbf{k}\cdot\frac{\partial F_\alpha}{\partial \mathbf{v}} = 0 \tag{15.241}$$

where we omitted prime on f_α, and $\mathbf{v}$ should be written in terms of $\mathbf{u}$. It can be easily shown by direct calculation that

$$\mathbf{k}\cdot\frac{\partial F_\alpha}{\partial \mathbf{v}} = \mathbf{q}_\alpha(t)\cdot\frac{\partial F_\alpha}{\partial \mathbf{u}}$$

$$\text{where } \mathbf{q}_\alpha = (k_x \sin\omega_{c\alpha}t + k_y\cos\omega_{c\alpha}t, k_x\cos\omega_{c\alpha}t - k_y\sin\omega_{c\alpha}t, k_z) \tag{15.242}$$

which is the rotation of the wave vector $\mathbf{k}$ by the rotation matrix

$$\mathbf{R}_\alpha = \begin{pmatrix} \sin\omega_{c\alpha}t & \cos\omega_{c\alpha}t & 0 \\ \cos\omega_{c\alpha}t & -\sin\omega_{c\alpha}t & 0 \\ 0 & 0 & 1 \end{pmatrix}$$

Using (15.234)–(15.236) in (15.241) gives

$$\begin{aligned} &\frac{\partial}{\partial t}f_\alpha(\mathbf{q}_\alpha,t,\mathbf{u}) + i\mathbf{q}_\alpha\cdot\mathbf{u}\;f_\alpha - \frac{ie_\alpha}{m_\alpha}\phi(\mathbf{k},t)\mathbf{q}_\alpha\cdot\frac{\partial F_\alpha}{\partial \mathbf{u}} \\ &-i\;f_\alpha\left[\beta_\alpha\omega_0\cos\omega_0 t\;\mathbf{k}_\perp\cdot\mathbf{E}_{0\perp} + \beta_\alpha\omega_{c\alpha}\sin\;\omega_0 t\;\hat{\mathbf{z}}\cdot(\mathbf{k}\times\mathbf{E}_0)\right. \\ &\left.+\frac{e_\alpha}{m_\alpha\omega_0}k_z E_{0z}\cos\omega_0 t\right] = 0 \end{aligned} \tag{15.243}$$

$$\beta_\alpha = \frac{e_\alpha}{m_\alpha}\,\frac{1}{\omega_0^2 - \omega_{c\alpha}^2}$$

The symbol $\perp$ is referred to the direction of $\mathbf{B}_0$. We further transform

$$f_\alpha = g_\alpha e^{i\xi_\alpha(t)}$$

with

$$\xi_\alpha(t) = \left(\beta_\alpha\mathbf{k}_\perp\cdot\mathbf{E}_{0\perp} + \frac{e_\alpha}{m_\alpha\omega_0^2}k_z E_{oz}\right)\sin\omega_0 t - \beta_\alpha\frac{\omega_{c\alpha}}{\omega_0}\hat{\mathbf{z}}\cdot(\mathbf{k}\times\mathbf{E}_0)\cos\omega_0 t \tag{15.244}$$

Then we have

$$\frac{\partial}{\partial t}g_\alpha(\mathbf{k},\mathbf{u},t) + i\mathbf{q}_\alpha(\mathbf{k},t)\cdot\mathbf{u}g_\alpha - \frac{ie_\alpha}{m_\alpha}\phi\;\mathbf{q}_\alpha\cdot\frac{\partial F_\alpha}{\partial \mathbf{u}}e^{-i\xi_\alpha(t)} = 0 \tag{15.245}$$

The Poisson equation takes the form (the Jacobian is 1)

$$k^2 \ \phi(\mathbf{k}, t) = 4\pi \sum_\alpha e_\alpha \ e^{i\xi_\alpha(t)} \int g_\alpha d^3u \tag{15.246}$$

Equations (15.245) and (15.246) are the main results of this section and their further analysis is left uncarried. Equations (15.245) and (15.246) are identical in the formal structure with (15.172) and (15.173). The analysis of (15.245) and (15.246) may proceed as is done in Section 9.7.1 but the details would be considerably more complex because we have still the time-dependent coefficient $\mathbf{q}_\alpha$, and ξ_α is composed of three terms as compared with single term in the case of unmagnetized plasma. The three terms in ξ_α represent the relative strength of the pump with respect to the direction of the magnetic field. Note that in the case of unmagnetized plasma ($\omega_{c\alpha} = 0$) we have $\xi_\alpha = -a_\alpha$, and (15.245) and (15.246) reduce to (15.172) and (15.173), respectively. Tzoar (1969) dealt with the same problem by entirely different transformation.

Reference: Tzoar, N. *Parametric excitations in plasma in a magnetic field*, Phys. Rev. **178** (1969), pp. 356–362. P. K. Kaw, *Parametric excitation of electromagnetic waves in magnetized plasmas*, Advances in Plasma Physics, Vol. 6, Ed. by A. Simon and W. B. Thompson, Wiley, New York (1976).

15.8 Compact Form of the Dispersion Relation of Parametric Instability in a Magnetized Plasma

This section extends Section 15.6 by considering a magnetized plasma with a static magnetic field $\mathbf{B}_0 = \hat{\mathbf{z}}B_0$. The mathematical steps are analogous to Section 15.6, and a compact form of the dispersion relation of parametric instability in a magnetized plasma is derived by using fluid equations. We consider a high frequency electromagnetic pump wave propagating in the plasma, coupled with electrostatic plasma normal modes. The latter can be the high frequency Langmuir wave or the low frequency ion acoustic wave. We hope to integrate the various parametric processes in a magnetized plasma in a compact form. Let us include the magnetic force terms in the two fluid equations in (15.156)–(15.160).

$$\frac{\partial \mathbf{v}_e}{\partial t} = \frac{e}{m_e}\nabla\phi - \frac{e}{m_e c}\mathbf{v}_e \times \mathbf{B}_0 - \frac{\gamma_e T_e}{N m_e}\nabla\delta n_e \ - \nabla(\mathbf{v}_0 \cdot \mathbf{v}_s) \tag{15.247}$$

$$\frac{\partial \mathbf{v}_i}{\partial t} = -\frac{e}{m_i}\nabla\phi + \frac{e}{m_i c}\mathbf{v}_i \times \mathbf{B}_0 \tag{15.248}$$

$$\frac{\partial \delta n_e}{\partial t} + N\nabla \cdot \mathbf{v}_e = 0 \tag{15.249}$$

$$\frac{\partial \delta n_i}{\partial t} + N\nabla \cdot \mathbf{v}_i = 0 \tag{15.250}$$

$$-\nabla^2 \phi = 4\pi e(\delta n_i - \delta n_e) \tag{15.251}$$

The last term in (15.247) is the nonlinear force (ponderomotive force), which is important for the parametric interaction under consideration. Another important nonlinearity is the nonlinear current that must be included in the equation for $\mathbf{v}_s$, the scattered field, which is (15.262) below. These two nonlinear terms are responsible for the parametric instabilities that we intend to integrate in a compact form. Using (15.247) and (15.249), δn_e can be determined in terms of $(\phi - \frac{m_e}{e}\mathbf{v}_0 \cdot \mathbf{v}_s)$. General formal solution can be written in terms of the electron and ion susceptibilities, $\chi_{e,i}$, which give the linear relations in terms of Fourier variables

$$\chi_e \phi = \frac{4\pi e}{k^2}\delta n_e \tag{15.252}$$

$$\chi_i \phi = -\frac{4\pi e}{k^2}\delta n_i \tag{15.253}$$

Then δn_e is obtained from (15.247) and (15.249) in the form

$$\delta n_e = \frac{\chi_e\, k^2}{4\pi e}\left(\phi - \frac{m_e}{e}\mathbf{v}_0 \cdot \mathbf{v}_s\right) \tag{15.254}$$

Equations (15.251), (15.252), and (15.253) give $(1 + \chi_e + \chi_i)\phi = 0$, so the linear dispersion relation reads

$$1 + \chi_e + \chi_i \equiv \varepsilon_L = 0 \tag{15.255}$$

where ε_L is the dielectric constant. From (15.252) and (15.255), we can write

$$\phi = -\frac{4\pi e}{k^2}\,\frac{\delta n_e}{1 + \chi_i} \tag{15.256}$$

Using (15.256) in (15.254) yields

$$\delta n_e(\mathbf{k}, \omega) = -\frac{N k^2}{\omega_{\rm pe}^2}\,\frac{\chi_e(1 + \chi_i)}{\varepsilon_L}\,[\mathbf{v}_0 \cdot \mathbf{v}_s^*]_{(\mathbf{k},\omega)} \tag{15.257}$$

Equation (15.257) is general in the sense that $\chi_{e,i}$ can be calculated kinetically as well as fluidly.

• **Susceptibility.** We have

$$\frac{\partial \mathbf{P}}{\partial t} = \mathbf{J}, \quad J_i = \sigma_{ij} E_j = -i\omega P_i$$

where $\mathbf{P}$ is polarization. The susceptibility is defined by $\chi_{ij} = \frac{4\pi i}{\omega}\sigma_{ij}$. Therefore, we have the relation in one-dimension $\mathbf{P}_\alpha = \frac{\chi_\alpha}{4\pi}\,\mathbf{E}$, and therefore

$$-i\omega \nabla \cdot \mathbf{P}_\alpha = \nabla \cdot \mathbf{J}_\alpha = i\omega\ \rho_\alpha, \quad \frac{\chi_\alpha}{4\pi}\nabla \cdot \mathbf{E} = -e_\alpha \delta n_\alpha \quad (\alpha = e.\ i)$$

from which (15.252)–(15.253) follow.

When one uses the fluid equations above, we get

$$\chi_i = \frac{-\ \eta_i \omega_{pi}^2}{\omega^2 - \omega_{ci}^2} \tag{15.258}$$

$$\chi_e = \frac{-\eta_e \omega_{\rm pe}^2}{\omega^2 - \omega_{\rm ce}^2 - \eta_e \ \frac{\gamma_e T_e k^2}{m_e}} \tag{15.259}$$

$$\text{where} \ \ \eta_\alpha = 1 - \frac{\omega_{c\alpha}^2}{\omega^2}\frac{k_z^2}{k^2} \ \ (\alpha = e,\ i) \tag{15.260}$$

with $\omega_{p\alpha}$ and $\omega_{c\alpha}$ plasma and cyclotron frequency, respectively. The correctness of (15.257) can be directly checked from the governing equations (15.247)–(15.251). One can solve the vector equation (15.247) by Fourier transforming and resolving the resulting equation into the three directions: $\mathbf{k}$, $\hat{\mathbf{z}}$, and $\mathbf{k} \times \hat{\mathbf{z}}$. Then we obtain the expression

$$\delta n_e = -\frac{N}{\omega_{\rm pe}^2 \eta_e}\, \frac{\chi_e(1+\chi_i)}{\varepsilon_L}\, \mathbf{k} \cdot \left(\mathbf{k} + \frac{\omega_{\rm ce}}{\omega}\mathbf{k} \times \hat{\mathbf{z}} - k_z \frac{\omega_{\rm ce}^2}{\omega^2}\hat{\mathbf{z}} \right) [\mathbf{v}_0 \cdot \mathbf{v}_s^*]_{(\mathbf{k},\omega)} \tag{15.261}$$

where we used (15.258)–(15.260). It is immediately seen that (15.261) reduces to (15.257).

The electrostatic normal modes in a magnetized plasma are obtained from the linear dispersion relation $\varepsilon_L = 1 + \chi_i + \chi_e = 0$ with (15.258)–(15.260). Also the linear dispersion relation can be put into the form

$$0 = \frac{\varepsilon_L}{(1+\chi_i)\chi_e} = \frac{1}{1+\chi_i} + \frac{1}{\chi_e} \quad \text{(cf. (15.257))}$$

Exercise. Verify the following relations in unmagnetized plasma.

$$\frac{1}{1+\chi_i} + \frac{1}{\chi_e} = \begin{cases} -\dfrac{1}{\omega_{pi}^2}\left(\omega^2 - \dfrac{\gamma_e T_e k^2}{m_i}\right) & (\omega \ll \omega_{pi}) \\[2ex] -\dfrac{1}{\omega_{\rm pe}^2}\left(\omega^2 - \omega_{\rm pe}^2 - \dfrac{\gamma_e T_e k^2}{m_e}\right) & (\omega \gg \omega_{pi}) \end{cases}$$

Next we consider the scattered wave (side band) which we represent by primed quantities, $\mathbf{k}', \omega'$. We first deal with the Stokes coupling with the resonance conditions

$$\mathbf{k}' = \mathbf{k}_0 - \mathbf{k}, \quad \omega' = \omega_0 - \omega$$

The Maxwell equations in Fourier variables are arranged in the form (cf. (15.146))

$$\mathbf{k}' \times (\mathbf{k}' \times \mathbf{E}') + \frac{\omega'^2}{c^2}\mathbf{E}' + \frac{4\pi i\omega'}{c^2}\mathbf{J}' = \frac{4\pi i e\omega'}{c^2}\,[\tilde{\mathbf{v}}_0 \delta n_e^*]_{(\mathbf{k}',\omega')} \tag{15.262}$$

where the last term is the electron nonlinear current (tilded quantity is a complex Fourier amplitude and * denotes the complex conjugate of a Fourier amplitude). The linear current

$$\mathbf{J}' = -eN\mathbf{v}' \tag{15.263}$$

is calculated from the cold electron equation of motion

$$-i\,\omega'\mathbf{v}' = -\omega_{\text{ce}}\mathbf{v}' \times \hat{\mathbf{z}} - \frac{e}{m_e}\mathbf{E}' \tag{15.264}$$

where we neglected the wave magnetic force $\mathbf{v}' \times \mathbf{B}'$ assuming $v'^2/c^2 \ll 1$. Using (15.263) for $\mathbf{J}'$ and (15.264) for $\mathbf{E}'$, (15.262) is written as

$$\left(\frac{\omega'^2 - \omega_{\text{pe}}^2}{c^2} - k'^2\right)\mathbf{v}' + \mathbf{k}'\,\mathbf{k}' \cdot \mathbf{v}' + i\,\frac{\omega_{\text{ce}}}{\omega'}\left(\frac{\omega'^2}{c^2} - k'^2\right)\mathbf{v}' \times \hat{\mathbf{z}}$$

$$+\,i\,\frac{\omega_{\text{ce}}}{\omega'}\mathbf{k}'\,\mathbf{v}' \cdot \hat{\mathbf{z}} \times \mathbf{k}' = \frac{\omega_{\text{pe}}^2}{Nc^2}\,\tilde{\mathbf{v}}_0 \delta n_e^* \equiv \mathbf{Q} \tag{15.265}$$

Exercise. Consider unmagnetized plasma ($\omega_{\text{ce}} = 0$). Solve the above equation for $\mathbf{v}'$ in terms of $\mathbf{Q}$. Substitute the result into (15.257) and obtain the parametric dispersion relation

$$\frac{1}{k^2}\left(\frac{1}{\chi_e(\mathbf{k},\omega)} + \frac{1}{1 + \chi_i(\mathbf{k},\omega)}\right) = \frac{1}{k'^2}\left[\frac{|\tilde{\mathbf{v}}_0 \cdot \mathbf{k}'|^2}{\omega_{\text{pe}}^2 - \omega'^2} + \frac{|\mathbf{k}' \times \tilde{\mathbf{v}}_0|^2}{c^2k'^2 + \omega_{\text{pe}}^2 - \omega'^2}\right] \tag{15.266}$$

The contribution from the anti-Stokes component can be easily included. Upon inclusion of that component in the above equation, we recover Eq. (13) in Drake *et al.* (1974). Compare (15.266) with (15.163).

Equation (15.265) can be solved for $\mathbf{v}$ in terms of the source term $\mathbf{Q}$ by resolving the equation into the three directions, $\mathbf{k}$, $\hat{\mathbf{z}}$, and $\mathbf{k} \times \hat{\mathbf{z}}$. [Primes will be omitted for a while.] After a long algebra, we obtain

$$\mathbf{v} = \frac{1}{\lambda^2(1-A_3^2)}\left(\mathbf{F}(\mathbf{Q}) - \ \mathbf{F}(\mathbf{k})\ \frac{\frac{\omega_{\rm ce}}{\omega}\,G(\mathbf{Q}) + \mathbf{k}\cdot\mathbf{F}(\mathbf{Q})}{A_2 + \mathbf{k}\cdot\mathbf{F}(\mathbf{k})}\right) \tag{15.267}$$

$$\text{where } \mathbf{F}(\mathbf{k}) = \mathbf{k} - iA_3\ \mathbf{k}\times\hat{\mathbf{z}} - A_3^2(\hat{\mathbf{z}}\cdot\mathbf{k})\ \hat{\mathbf{z}} \tag{15.268}$$

$$\mathbf{F}(\mathbf{Q}) = \mathbf{Q} - iA_3\ \mathbf{Q}\times\hat{\mathbf{z}} - A_3^2(\hat{\mathbf{z}}\cdot\mathbf{Q})\ \hat{\mathbf{z}} \tag{15.269}$$

$$G(\mathbf{Q}) = i\ \mathbf{Q}\cdot\hat{\mathbf{z}}\times\mathbf{k} - A_3\ (\hat{\mathbf{z}}\times\mathbf{k})\cdot(\hat{\mathbf{z}}\times\mathbf{Q}) \tag{15.270}$$

$$\lambda^2 = \frac{\omega^2 - \omega_{\rm pe}^2}{c^2} - k^2 \tag{15.271}$$

$$A_2 = \lambda^2 - \frac{\omega_{\rm ce}}{\omega}\left(\frac{\omega^2}{c^2} - k_z^2\right) A_3 \tag{15.272}$$

$$A_3 = \frac{\omega_{\rm ce}}{\omega}\ \frac{1}{\lambda^2}\ \left(\frac{\omega^2}{c^2} - k^2\right) \tag{15.273}$$

One can check the correctness of (15.267) by substituting it into (15.265) and ascertaining many cancellations. After combining (15.257) and (15.267) we obtain the desired dispersion relation of the parametric instability in the intermediate form

$$\frac{1}{k^2}\left(\frac{1}{\chi_e(\omega,\mathbf{k})} + \frac{1}{1+\chi_i(\omega,\mathbf{k})}\right) = -\frac{1}{\omega_{\rm pe}^2}\tilde{\mathbf{v}}_0\cdot\tilde{\mathbf{v}}_s^* \tag{15.274}$$

where (15.267) should be substituted for $\mathbf{v}_s$ with $\mathbf{Q}$ now standing for $\frac{\omega_{\rm pe}^2}{c^2}$. [$\delta n_e$ has been canceled on both sides.] On the RHS of (15.274), the scattered wave quantities depend on $\mathbf{k}'$ and ω'—the Stokes component. Using Eq. (15.267) gives

$$\begin{aligned}\tilde{\mathbf{v}}_0\cdot\tilde{\mathbf{v}}_s^* = \frac{1}{\lambda^2(1-A_3^2)}\Bigg[&|\tilde{\mathbf{v}}_0|^2 - A_3^2|\tilde{\mathbf{v}}_0\cdot\hat{\mathbf{z}}|^2 - \frac{\tilde{\mathbf{v}}_0\cdot\mathbf{F}^*(\mathbf{k})}{A_2+\mathbf{k}\cdot\mathbf{F}(\mathbf{k})} \\ &\times\left(\mathbf{k}\cdot\mathbf{F}^*(\tilde{\mathbf{v}}_0) + \frac{\omega_{\rm ce}}{\omega}G^*(\tilde{\mathbf{v}}_0)\right)\Bigg]\end{aligned} \tag{15.275}$$

to which the factor $\omega_{\rm pe}^2/c^2$ should be attached. Note that $\mathbf{k}\cdot\mathbf{F}(\mathbf{k})$ is real. We calculate:

$$\begin{aligned}\mathbf{k}\cdot\mathbf{F}^*(\tilde{\mathbf{v}}_0)\tilde{\mathbf{v}}_0\cdot\mathbf{F}^*(\mathbf{k}) &= \left(\mathbf{k}\cdot(\tilde{\mathbf{v}}_0^* + iA_3\tilde{\mathbf{v}}_0^*\times\hat{\mathbf{z}} - A_3^2\hat{\mathbf{z}}\cdot\tilde{\mathbf{v}}_0^*\hat{\mathbf{z}})\right) \\ &\quad\times\left(\tilde{\mathbf{v}}_0\cdot(\mathbf{k} + iA_3\mathbf{k}\times\hat{\mathbf{z}} - A_3^2\hat{\mathbf{z}}\cdot\mathbf{k}\ \hat{\mathbf{z}})\right) \\ &= \left(\tilde{\mathbf{v}}_0^*\cdot(\mathbf{k} - iA_3\mathbf{k}\times\hat{\mathbf{z}} - A_3^2\hat{\mathbf{z}}\cdot\mathbf{k}\ \hat{\mathbf{z}})\right) \\ &\quad\times\left(\tilde{\mathbf{v}}_0\cdot(\mathbf{k} + iA_3\mathbf{k}\times\hat{\mathbf{z}} - A_3^2\hat{\mathbf{z}}\cdot\mathbf{k}\ \hat{\mathbf{z}})\right) \\ &= |\tilde{\mathbf{v}}_0\cdot(\mathbf{k} + iA_3\mathbf{k}\times\hat{\mathbf{z}} - A_3^2\hat{\mathbf{z}}\cdot\mathbf{k}\ \hat{\mathbf{z}})|^2\end{aligned}$$

$$G^*(\tilde{\mathbf{v}}_0)\tilde{\mathbf{v}}_0 \cdot \mathbf{F}^*(\mathbf{k}) = (-\tilde{\mathbf{v}}_0^* \cdot (A_3\mathbf{k} + i\hat{\mathbf{z}} \times \mathbf{k} - A_3\hat{\mathbf{z}} \cdot \mathbf{k}\,\hat{\mathbf{z}})) \times \left(\tilde{\mathbf{v}}_0 \cdot (\mathbf{k} - iA_3\hat{\mathbf{z}} \times \mathbf{k} - A_3^2\,\hat{\mathbf{z}} \cdot \mathbf{k}\,\hat{\mathbf{z}})\right) \quad (15.276)$$

In (15.274) and (15.275), we see that the denominator $\lambda^2(1 - A_3^2) = 0$ gives the linear dispersion relation of the side band wave. $\lambda^2 = 0$ yields the usual electromagnetic wave $\omega^2 = \omega_{\rm pe}^2 + c^2k^2$, and the relation $A_3 = \pm 1$ can be rearranged in the form

$$\frac{c^2k^2}{\omega^2} = 1 - \frac{\omega_{\rm pe}^2}{\omega(\omega \mp \omega_{\rm ce})} \quad (15.277)$$

where +(−) sign corresponds to the L (R, whistler) wave. We put $A_3 = \pm 1$ in (15.276). This approximation may be permissible for a weak pump. Thus we have

$$G^*(\mathbf{v}_0)\tilde{\mathbf{v}}_0 \cdot \mathbf{F}^*(\mathbf{k}) = \mp|\tilde{\mathbf{v}}_0 \cdot (\mathbf{k} \pm i\mathbf{k} \times \hat{\mathbf{z}} - \hat{\mathbf{z}} \cdot \mathbf{k}\,\hat{\mathbf{z}})|^2 \quad (15.278)$$

where the upper (lower) signs correspond to the R(L) wave. Collecting the foregoing results, the dispersion relation can be written as

$$\frac{1}{k^2}\left(\frac{1}{\chi_e(\mathbf{k},\omega)} + \frac{1}{1+\chi_i(\mathbf{k},\omega)}\right) = -\frac{1}{c^2}\,\frac{1}{\lambda^2(1-A_3^2)}\left[|\tilde{\mathbf{v}}_0|^2 - A_3|\hat{\mathbf{z}} \cdot \tilde{\mathbf{v}}_0|^2 - \frac{|\tilde{\mathbf{v}}_0 \cdot (\mathbf{k}' + iA_3\mathbf{k}' \times \hat{\mathbf{z}} - A_3^2\hat{\mathbf{z}} \cdot \mathbf{k}'\,\hat{\mathbf{z}})|^2 \mp \frac{\omega_{\rm ce}}{\omega'}|\tilde{\mathbf{v}}_0 \cdot (\mathbf{k}' \pm i\mathbf{k}' \times \hat{\mathbf{z}} - \hat{\mathbf{z}} \cdot \mathbf{k}'\,\hat{\mathbf{z}})|^2}{A_2 - A_3^2k_z'^2 + k'^2}\right] \quad (15.279)$$

where $\lambda,\ A_2,\ A_3$ are functions of primed variables, $\omega',\ \mathbf{k}'$. Equation (15.279) is the final result of the electromagnetic parametric dispersion relation when the waves are coupled through the Stokes coupling. The contribution from the anti-Stokes coupling is easily added to the RHS by writing additional terms with the replacement of $(')$ → $('')$ in the Stokes terms, which implies the resonance conditions $\mathbf{k}'' = \mathbf{k}_0 + \mathbf{k}$ and $\omega'' = \omega_0 + \omega$.

Next we show that (15.279) recovers the earlier result by Drake *et al.* for unmagnetized plasma in the limit of $\omega_{\rm ce} = 0$. The right-hand side of (15.279) reduces, upon putting $\omega_{\rm ce} = 0$, to

$$-\frac{1}{\omega'^2 - \omega_{\rm pe}^2 - c^2k'^2}\left[|\tilde{\mathbf{v}}_0|^2 - \frac{c^2|\tilde{\mathbf{v}}_0 \cdot \mathbf{k}'|^2}{\omega'^2 - \omega_{\rm pe}^2}\right]$$

We use the identity in the above

$$|\tilde{\mathbf{v}}_0|^2 = \frac{|\tilde{\mathbf{v}}_0 \cdot \mathbf{k}'|^2 + |\mathbf{k}' \times \tilde{\mathbf{v}}_0|^2}{k'^2} \quad (15.280)$$

to obtain the dispersion relation

$$\frac{1}{k^2}\left(\frac{1}{\chi_e(\mathbf{k},\omega)}+\frac{1}{1+\chi_i(\mathbf{k},\omega)}\right)=\frac{1}{k'^2}\left[\frac{|\tilde{\mathbf{v}}_0\cdot\mathbf{k}'|^2}{\omega_{\text{pe}}^2-\omega'^2}+\frac{|\mathbf{k}'\times\tilde{\mathbf{v}}_0|^2}{c^2k'^2+\omega_{\text{pe}}^2-\omega'^2}\right] \tag{15.281}$$

for unmagnetized plasma, which is Eq. (13) in Drake *et al.*

Next we consider the electrostatic limit ($c\to\infty$) of (15.279). From (15.271)–(15.273), we obtain

$$\lambda^2(1-A_3^2)\to -k'^2\left(1-\frac{\omega_{\text{ce}}^2}{\omega'^2}\right)$$

$$A_3\to\frac{\omega_{\text{ce}}}{\omega'}$$

$$\begin{aligned}A_2-A_3^2k_z'^2+k'^2 &\to \frac{\omega'^2-\omega_{\text{pe}}^2}{c^2}-\frac{\omega_{\text{ce}}}{\omega'}\left(\frac{\omega'^2}{c^2}-k_z'^2\right)A_3-A_3^2k_z'^2\\ &=\frac{\omega'^2-\omega_{\text{pe}}^2-\omega_{\text{ce}}^2}{c^2}\end{aligned}$$

We see that c^2 cancels and the electrostatic dispersion relation is found to be

$$\begin{aligned}\frac{1}{k^2}\left(\frac{1}{\chi_e(\mathbf{k},\omega)}+\frac{1}{1+\chi_i(\mathbf{k},\omega)}\right)&=\frac{(1-\frac{\omega_{\text{ce}}^2}{\omega'^2})^{-1}}{k'^2(\omega_{\text{pe}}^2+\omega_{\text{ce}}^2-\omega'^2)}\\ &\times\left[\left|\tilde{\mathbf{v}}_0\cdot\left(\mathbf{k}'+i\frac{\omega_{\text{ce}}}{\omega'}\mathbf{k}'\times\hat{\mathbf{z}}-\frac{\omega_{\text{ce}}^2}{\omega'^2}\,\hat{\mathbf{z}}\cdot\mathbf{k}'\hat{\mathbf{z}}\right)\right|^2\right.\\ &\left.\mp\frac{\omega_{\text{ce}}}{\omega'}\left|\tilde{\mathbf{v}}_0\cdot\left(\mathbf{k}'\pm i\mathbf{k}'\times\hat{\mathbf{z}}-\hat{\mathbf{z}}\cdot\mathbf{k}'\,\hat{\mathbf{z}}\right)\right|^2\right]\end{aligned} \tag{15.282}$$

where the upper and the lower signs corresponds to two cyclotron waves of different helicity. If $\omega_{\text{ce}}=0$, the above equation reduces to the electrostatic limit of (15.281). The singularity at $\omega'=\omega_{\text{ce}}$ (upper signs) or $\omega'=-\omega_{\text{ce}}$ (lower signs) is only apparent since the numerator also vanishes in this case.

References

Drake, J. F., P. K. Kaw, Y. C. Lee, G. Schmidt, C. S. Liu, and M. N. Rosenbluth, *Parametric instabilities of electromagnetic waves in plasmas*, Phys. Fluids **17** (1974), p. 778.

Kaw, P. K. *Parametric Excitation of Electromagnetic Waves in Magnetized Plasmas*, Advances in Plasma Physics Vol. 6, Ed. by A Simon and W. B. Thompson, Wiley, New York (1976).

Kruer, W. L. *The Physics of Laser Plasma Interactions*, Addison-Wesley (1988).

Kuo, Y.-Y. and J. A. Fejer, *Spectral-line structures of saturated parametric instabilities*, Phys. Rev. Lett. **29** No. 25 (1972), pp. 1667–1670.

Lee, H. J. *An electrostatic parametric instability due to thermal coupling* J. Korean Phys. Soc. **13** (1980), pp. 64–73.

Lee, H. J. *Kinetic theory of coherent resonant interactions of electrostatic plasma waves* J. Korean Phys. Soc. **7** pp. 256–265 (1984)

Lee, H. J. *Kinetic theory of coherent resonant interactions of electrostatic plasma waves*, J. Korean Phys. Soc. **17** pp. 256–265 (1984)

Lee, H. J. *Time evolution of coherently resonant waves in a collisionless homogeneous plasma*, J. Korean Phys. Soc. **7** (1982), pp. 71–80.

Nishikawa, K. *Parametric excitation of coupled waves I. General formulation,* J. Phys. Soc. Japan **24** (1968), pp. 916–922.

Nishikawa, K. *Parametric excitation of coupled waves II. Parametric plasmon-photon interaction,* J. Phys. Soc. Japan **24** (1968), pp. 1152–1158.

Sanmartin, J. R. *Electrostatic plasma instabilities excited by a high frequency electric field*, Phys. Fluids **13** (1970), pp. 1533–1542.

Silin, V. P. *Parametric resonance in a plasma*, Sov. Phys. JETP **21** (1965), p. 1127: JETP **24** (1967), p. 1242.

Stenflo, L. J. Plasma Phys. **61** p.129;

Stenflo, L. *Resonant three-wave interactions in plasmas*, Physica Scripta **T50** (1994), pp. 15–19.

Stenflo, L. and P. K. Shukla, *Theory of stimulated scattering of large-amplitude waves* J. Plasma Phys. **64** part 4 (2000), pp. 353–357.

Tzoar, N. *Parametric excitations in plasma in a magnetic field*, Phys. Rev. **178** (1969), pp. 356–362.

Chapter 16

Nonlinear Theory of Vlasov Equation

The Vlasov equation contains the nonlinear term $\mathbf{E} \cdot \frac{\partial f}{\partial \mathbf{v}}$ from which various nonlinear effects are derived. One step higher development over the linearized Vlasov equation is the *quasilinear theory* which replaces the Vlasov–Poisson (or Vlasov–Maxwell) system of equations by a set of two linear equations. The quasilinear theory is in essence two-time or two-space scale analysis of the distribution function. The nonlinear term in the Vlasov equation has been studied under the classification called *wave-particle interaction* and *wave-wave interaction*. Basically the mathematical scheme to deal with the nonlinearity is iteration with linear solutions.

16.1 Quasilinear Theory of Nonlinear Vlasov Equation

In the linear theory, the zero order distribution function which is assumed to be independent of $\mathbf{r}$ and t determine the growth or damping rates of the excited wave. As the wave grows, the zero order distribution function will be affected by the wave. This was discussed in conjunction with the trapped particle dynamics in earlier chapter. However, the linear theory itself doesn't take account of the influence of the growing wave onto the zero order distribution function. The quasilinear theory remedies this shortcoming of the linear Vlasov equation by allowing a slow time variation for the "zero order" distribution function. The zero order distribution function f_0 is the spatially homogeneous part extracted from $f(\mathbf{r}, \mathbf{v}, t)$ (the extraction can be done by picking up the $\mathbf{k} = 0$ component from the Fourier series) and furthermore, f_0 is endowed with slow time variation.

We start with the Vlasov equation for the electron distribution function $f(\mathbf{r}, \mathbf{v}, t)$ (ions are assumed to form a uniform background of immobile positive charges)

$$\frac{\partial}{\partial t} f(\mathbf{r}, \mathbf{v}, t) + \mathbf{v} \cdot \frac{\partial f}{\partial \mathbf{r}} - \frac{e}{m} \mathbf{E}(\mathbf{r}, t) \cdot \frac{\partial f}{\partial \mathbf{v}} = 0 \tag{16.1}$$

and the Poisson equation

$$\nabla \cdot \mathbf{E} = 4\pi e N \left(1 - \int d^3v\ f(\mathbf{r}, \mathbf{v}, t)\right) \tag{16.2}$$

Let us Fourier transform (16.1) and (16.2) with respect to $\mathbf{r}$. Transforming the product term proceeds as follows.

$$I \equiv \int d^3 r e^{-i\mathbf{k}\cdot\mathbf{r}} \mathbf{E}(\mathbf{r}) f(\mathbf{r})$$

We substitute into the integral I the expression of the inverse transform

$$\mathbf{E}(\mathbf{r}) = \left(\frac{1}{2\pi}\right)^3 \int d^3k' e^{i\mathbf{k}'\cdot\mathbf{r}} \mathbf{E}(\mathbf{k}')$$

Then we can write

$$\begin{aligned} I &= \int d^3 r e^{-i\mathbf{k}\cdot\mathbf{r}} f(\mathbf{r}) \left(\frac{1}{2\pi}\right)^3 \int d^3k' e^{i\mathbf{k}'\cdot\mathbf{r}} \mathbf{E}(\mathbf{k}') \\ &= \left(\frac{1}{2\pi}\right)^3 \int d^3k' \mathbf{E}(\mathbf{k}') \int d^3 r e^{-i(\mathbf{k}-\mathbf{k}')\cdot\mathbf{r}} f(\mathbf{r}) \\ &= \left(\frac{1}{2\pi}\right)^3 \int d^3k' \mathbf{E}(\mathbf{k}') f(\mathbf{k}-\mathbf{k}') = \left(\frac{1}{2\pi}\right)^3 \int d^3k' \mathbf{E}(\mathbf{k}-\mathbf{k}') f(\mathbf{k}') \end{aligned}$$

The last equality is due to the property of the convolution integral. If one wishes to change the above Fourier integral to Fourier series, put $\left(\frac{1}{2\pi}\right)^3 \int d^3k' \to \sum_{\mathbf{k}'}$. Then (16.1) yields

$$\frac{\partial}{\partial t} f(\mathbf{k}, \mathbf{v}, t) + i\mathbf{k}\cdot\mathbf{v} f - \frac{e}{m} \sum_{\mathbf{k}'} \mathbf{E}(\mathbf{k}', t) \cdot \frac{\partial}{\partial \mathbf{v}} f(\mathbf{k}-\mathbf{k}', \mathbf{v}, t) = 0 \tag{16.3}$$

In the above equation, we separate the equation corresponding to $\mathbf{k} = 0$ and $\mathbf{k} \neq 0$:

$$\mathbf{k} = 0 : \frac{\partial}{\partial t} f(\mathbf{k}=0, \mathbf{v}, t) = \frac{e}{m} \sum_{\mathbf{k}'} \mathbf{E}(-\mathbf{k}', t) \cdot \frac{\partial}{\partial \mathbf{v}} f(\mathbf{k}', \mathbf{v}, t) \quad \text{(A)}$$

$$\begin{aligned} \mathbf{k} \neq 0 : \frac{\partial}{\partial t} f(\mathbf{k}, \mathbf{v}, t) + i\mathbf{k}\cdot\mathbf{v} f - \frac{e}{m} \mathbf{E}(\mathbf{k}, t) \cdot \frac{\partial f(\mathbf{k}=0, \mathbf{v}, t)}{\partial \mathbf{v}} \\ = \frac{e}{m} \sum_{\mathbf{k}'}{}' \mathbf{E}(\mathbf{k}', t) \cdot \frac{\partial}{\partial \mathbf{v}} f(\mathbf{k}-\mathbf{k}', \mathbf{v}, t) \quad \text{(B)} \end{aligned}$$

In (A) we changed $\mathbf{k}' \to -\mathbf{k}'$. In (B) the prime on $\sum$ signifies that the $\mathbf{k}' = \mathbf{k}$ term is excluded in the $\sum$.

Let us assume in Eq. (B) $f(\mathbf{k}=0,\mathbf{v},t) \gg f(\mathbf{k},\mathbf{v},t)$ and neglect the nonlinear term in the RHS:

$$\frac{\partial}{\partial t}f(\mathbf{k},\mathbf{v},t)+i\mathbf{k}\cdot\mathbf{v}f-\frac{e}{m}\mathbf{E}(\mathbf{k},t)\cdot\frac{\partial f(\mathbf{k}=0,\mathbf{v},t)}{\partial\mathbf{v}}=0 \quad \text{(C)}$$

Neglect of the RHS of (B), the 'collision term', appears to be an analogous approximation to the Vlasov equation which was also made 'collisionless' by neglecting the higher order correlation term. In the higher order theory, RHS of (B) should be included to develop the coupled mode equations. This subject will be dealt with later. Equation (C) is refinement over the Vlasov equation in as much as we take account of the time dependence of $f(\mathbf{k}=0,\mathbf{v},t)$ which we assume to be slowly varying. To express this idea, we shall write the function $f(\mathbf{k}=0,\mathbf{v},t)$ as $f(\mathbf{k}=0,\mathbf{v},\epsilon t)$, with ϵt meaning the slow time. Thus we have two quasilinear equations (C) and (A) which we write in the form

$$\frac{\partial}{\partial t}f(\mathbf{k},\mathbf{v},t)+i\mathbf{k}\cdot\mathbf{v}f-\frac{e}{m}\mathbf{E}(\mathbf{k},t)\cdot\frac{\partial f(\mathbf{k}=0,\mathbf{v},\epsilon t)}{\partial\mathbf{v}}=0 \tag{16.4}$$

$$\frac{\partial}{\partial t}f(\mathbf{k}=0,\mathbf{v},\epsilon t)=\frac{e}{m}\sum_{\mathbf{k}}\mathbf{E}(-\mathbf{k},t)\cdot\frac{\partial}{\partial\mathbf{v}}f(\mathbf{k},\mathbf{v},t) \tag{16.5}$$

Together with (16.4) and (16.5), we have the Poisson equation expanded in Fourier series:

$$\nabla\cdot\sum_{k}\mathbf{E}(\mathbf{k},t)e^{i\mathbf{k}\cdot\mathbf{r}}=4\pi eN\left[1-\int d^3v\sum_{k}f(\mathbf{k},\mathbf{v},t)e^{i\mathbf{k}\cdot\mathbf{r}}\right]$$

The $\mathbf{k}=0$ term in the RHS integral cancels 1, and we have

$$i\mathbf{k}\cdot\mathbf{E}(\mathbf{k},t)=-4\pi Ne\int f(\mathbf{k},\mathbf{v},t)d^3v \tag{16.6}$$

Equations (16.4)–(16.6) are the self-consistent quasilinear equations. Clearly (16.4) and (16.5) are improvement over the linearized Vlasov equation. The slow function $f(\mathbf{k}=0,\mathbf{v},\epsilon t)$ corresponds to the zero order equilibrium distribution function F_0 in the linearized equation, which we assumed to be unvarying in time. In the quasilinear theory, the corresponding quantity changes slowly of which time variation is determined according to the evolution of the system, $\mathbf{E}(\mathbf{k},t)$ and f($\mathbf{k}$,v,t), as expressed by (16.5).

For simplicity we go to the one-dimensional case: take $\mathbf{k}=\hat{\mathbf{x}}k$, say, and integrate (16.4) with respect to the two velocity components perpendicular to **k**, to obtain (note $\mathbf{E_k} \parallel \mathbf{k}$)

$$\frac{\partial f_k}{\partial t} + ikvf_k + \frac{e}{m} ik\phi_k \frac{\partial f_0(v, \epsilon t)}{\partial v} = 0 \tag{16.7}$$

$$k^2 \phi_k(t) = -4\pi eN \int dv f_k(v, t) \tag{16.8}$$

$$\frac{\partial}{\partial t} f_0(v, \epsilon t) = \frac{e}{m} \sum_k ik\phi_{-k} \frac{\partial f_k}{\partial v} \tag{16.9}$$

where we changed notations slightly in obvious way and introduced potential, $E_k = -ik\phi_k$. The first order linear equation (16.7) can be solved to obtain

$$f_k(v, t) = -\frac{e}{m} ik \int_0^t dt' e^{ikv(t'-t)} \phi_k(t') \frac{\partial}{\partial v} f_0(v, \epsilon t') \tag{16.10}$$

Using (16.10) in (16.8) gives

$$\begin{aligned} \phi_k(t) &= ik \frac{\omega_{p0}^2}{k^2} \int_0^t dt' \phi_k(t') \int_{-\infty}^{\infty} dv e^{ikv(t'-t)} \frac{\partial}{\partial v} f_0(v, \epsilon t') \\ &= \omega_{p0}^2 \int_0^t dt' \phi_k(t')(t'-t) \int_{-\infty}^{\infty} dv e^{ikv(t'-t)} f_0(v, \epsilon t') \end{aligned}$$

from which we obtain

$$\frac{d^2\phi_k}{dt^2} = -\left(\omega_{p0}^2 \int_{-\infty}^{\infty} f_0(\epsilon t, v) dv\right) \phi_k(t) - ik\omega_{p0}^2 \int_0^t dt' \phi_k(t') \int_{-\infty}^{\infty} dv e^{ikv(t'-t)} v^2 \frac{\partial f_0}{\partial v} \tag{16.11}$$

where $\omega_{p0}^2 = 4\pi Ne^2/m$. The last term vanishes asymptotically as $t \to \infty$ since the exponential factor oscillates rapidly to yield almost complete cancellation in the integral. The quantity in the round parenthesis defines a slowly changing plasma frequency $\omega_p^2(\epsilon t)$. We write

$$\frac{d^2\phi_k}{dt^2} + \omega_p^2(\epsilon t)\phi_k(t) = 0 \tag{16.12}$$

The asymptotic result, (16.12), clearly reveals the two-time nature inherited from (16.7) and suggests to use the method of two-time perturbation analysis or WKB approximation to deal with (16.7)–(16.9). Here we adopt the latter approach in continuing development.

We review the relevant physics implied in (16.7). The linear theory indicates that the plasma is stable if $f_0(v, \epsilon t)$ is a single-humped function of v. If there is a small second hump in the distribution function at velocities somewhat larger than the thermal velocity, a weak instability occurs with

$\omega_k = \omega_k^r + i\gamma_k$ with $\gamma_k \ll \omega_k^r$, and as given by (7.96). Clearly the wave will not keep on growing without limit. The physical mechanism of the instability is understood by the role of the resonant particles. Particles with velocities a little less (greater) than the phase velocity of a wave are accelerated (decelerated) on the average. The sign of $\partial f_0/\partial v$ at the wave velocity determines whether the wave would grow or damp, and the double humped distribution in Fig. 18.2 predicts instability. This physical process, which accelerates the slower particles and decelerates the faster particles changes the shape of the zero order distribution $f_0(v)$ to form a plateau, and $f_0(v)$ should not be considered as time-independent.

Equations (16.7) and (16.8) are solved by introducing the WKB phasor in analogous manner as the linearized Vlasov equation is solved:

$$f_k(v,t) = \widetilde{f}_k(v)e^{\int_0^t dt'[-i\omega_k(t')+\gamma_k(t')]} \tag{16.13}$$

$$\phi_k(t) = \widetilde{\phi}_k e^{\int_0^t dt'[-i\omega_k(t')+\gamma_k(t')]} \tag{16.14}$$

Using (16.13) in (16.7) gives

$$f_k(v,t) = \frac{e}{m}(-ik)\phi_k \frac{\frac{\partial}{\partial v} f_0(v,t)}{ikv - i\omega_k(t) + \gamma_k(t)} \tag{16.15}$$

Using (16.15) in (16.8) yields

$$\frac{\omega_{p0}^2}{k}\int dv \, \frac{\frac{\partial}{\partial v} f_0(v,t)}{kv - \omega_k(t) - i\gamma_k(t)} = 1 \tag{16.16}$$

Equation (16.16) is solved for γ_k in the form

$$\gamma_k = \frac{\pi}{2}\frac{\omega_k \omega_{\text{pe}}^2}{k^2}\left[\frac{\partial f_0}{\partial v}\right]_{v=\omega_k/k} \qquad \text{(cf. (7.96))} \tag{16.17}$$

On the other hand, substituting (16.15) into (16.9) gives

$$\frac{\partial f_0}{\partial t} = \frac{e^2}{m^2}\sum_k |E_k|^2 \frac{\partial}{\partial v}\left[\frac{\gamma_k}{(kv-\omega_k)^2+\gamma_k^2}\frac{\partial}{\partial v} f_0(v)\right] \tag{16.18}$$

where we used the reality condition $\omega_{-k} = -\omega_k$. When $\omega_k/\gamma_k \gg 1$, we have

$$\frac{\gamma_k}{(kv-\omega_k)^2+\gamma_k^2} \simeq \pi\delta(kv-\omega_k)$$

The Fourier series in (16.18) can be converted to Fourier integral by changing $\sum_k \to \frac{1}{2\pi}\int dk$:

$$\frac{\partial f_0}{\partial t} = \frac{e^2}{2m^2}\frac{\partial}{\partial v}\int_{-\infty}^{\infty} dk|E_k|^2\delta(kv-\omega_k)\frac{\partial f_0}{\partial v}$$

Carrying out the integration, we obtain

$$\frac{\partial}{\partial t}f_0(v,t) = \frac{e^2}{2m^2}\frac{\partial}{\partial v}\left[\frac{1}{|v|}|E_k|^2\frac{\partial f_0(v,t)}{\partial v}\right] \tag{16.19}$$

with $v = \omega_k/k$.

Equation (16.14) gives the relation, $\frac{\partial}{\partial t}|E_k|^2 = 2\gamma_k|E_k|^2$, which can be written in the form

$$\frac{\partial}{\partial t}|E_k|^2 = \frac{4\pi^2 e^2}{m\omega_{\mathrm{pe}}}v^2\frac{\partial f_0(v,t)}{\partial v}|E_k|^2 \tag{16.20}$$

Equations (16.19) and (16.20) are the main results of the quasilinear theory.

Alternative derivation. Use (16.10) to derive the quasilinear equation (16.19).

Let us write the potential in the WKB form

$$\phi_k(t) = \phi_k(0)\exp\left(\int_0^t[-i\omega_k+\gamma_k(\epsilon t')]dt'\right) \tag{16.21}$$

Use (16.10) and (16.21) in (16.9) and the reality condition $\omega_{-k} = -\omega_k, \gamma_{-k} = \gamma_k$ to get

$$\begin{aligned}\frac{\partial}{\partial t}f_0(v,\epsilon t) &= \frac{e}{m}\sum_k ik\phi_{-k}(0)e^{\int_0^t[i\omega_k+\gamma_k(\epsilon t')]dt'}\\ &\quad\times\left(-\frac{e}{m}ik\int_0^t dt' e^{ikv(t'-t)}\phi_k(0)e^{\int_0^{t'}[-i\omega_k+\gamma_k(\epsilon t'')]dt''}\frac{\partial}{\partial v}f_0(v,\epsilon t')\right)\end{aligned} \tag{16.22}$$

Putting $|\phi_k(t)|^2 = \phi_k(0)\phi_{-k}(0)e^{2\int_0^t\gamma_k(\epsilon t'')dt''}$ in (16.22), we obtain

$$\frac{\partial}{\partial t}f_0(v,\epsilon t) = \frac{\partial}{\partial v}\int_0^t dt'\left(\frac{e}{m}\right)^2\sum_k k^2|\phi_k(t)|^2 e^{i(kv-\omega_k)(t'-t)}e^{\int_t^{t'}\gamma_k(\epsilon t'')dt''}\frac{\partial}{\partial v}f_0(v,\epsilon t')$$

In the above integral, the functions of slow time variable ϵt can be taken out of the integral, assigning them the values at time t:

$$\frac{\partial}{\partial t}f_0(v,\epsilon t) = \left(\frac{e}{m}\right)^2 \frac{\partial}{\partial v}\left[\sum_k k^2|\phi_k(t)|^2 \left(\int_0^t dt' e^{[i(kv-\omega_k)+\gamma_k(t)](t'-t)}\right)\frac{\partial}{\partial v}f_0(v,\epsilon t)\right]$$

Carrying out the $\int dt'$-integral, we obtain

$$\frac{\partial}{\partial t}f_0(v,\epsilon t) = \left(\frac{e}{m}\right)^2 \frac{\partial}{\partial v}\left[\sum_k k^2|\phi_k(t)|^2 \frac{1-e^{-i(kv-\omega_k)t-\gamma_k t}}{i(kv-\omega_k)+\gamma_k}\frac{\partial}{\partial v}f_0(v,\epsilon t)\right] \tag{16.23}$$

The exponential term vanishes asymptotically. Using the familiar relation

$$\frac{1}{i(kv-\omega_k)+\gamma_k} = \frac{P}{i(kv-\omega_k)+\gamma_k} + \pi\ \delta(kv-\omega_k)$$

Equation (16.23) can be written as

$$\frac{\partial f_0}{\partial t} = \frac{\partial}{\partial v}D(v)\frac{\partial f_0}{\partial v} \tag{16.24}$$

$$D(v) = \left(\frac{e}{m}\right)^2 \frac{\partial}{\partial v}\sum_k k^2|\phi_k(t)|^2\left[\frac{\gamma_k}{(kv-\omega_k)^2+\gamma_k^2} + \pi\ \delta(kv-\omega_k)\right] \tag{16.25}$$

where neglecting the non-resonant part yields (16.19) and we used the reality condition $\omega_{-k} = -\omega_k$ and γ_k is given by (16.17). To complement (16.25) we have

$$\frac{\partial}{\partial t}|\phi_k|^2 = 2\gamma_k(t)|\phi_k|^2 \tag{16.26}$$

• **Extraction of f_0 from f by spatial averaging.** In the preceding section, we focused on the mathematical formulation of the quasilinear theory only with the physical motivation as to how we can remedy the limitation of the linearized Vlasov equation in view of the trapped particle dynamics. The physical motivation can be strengthened by the development presented in this section. The same quasilinear equations can be derived more physically by the process of spatially averaging the interacting terms over the plasma volume. We start from the nonlinear equation (16.1) and split up $f(\mathbf{r},\mathbf{v},t)$ into two parts:

$$f(\mathbf{r},\mathbf{v},t) = f_0(\mathbf{v},t) + f_1(\mathbf{r},\mathbf{v},t) \tag{16.27}$$

$$\text{where } f_0(\mathbf{v},t) = \langle f(\mathbf{r},\mathbf{v},t)\rangle_r \tag{16.28}$$

is the spatially averaged distribution function, regarded as a slowly varying function of t. The waves are represented by $f_1(\mathbf{r},\mathbf{v},t)$, and we have

$<f_1(\mathbf{r}, \mathbf{v}, t)>_r = 0$. With this separation, we assume $f_1 \ll f_0$. Averaging (16.1) over space, we can write

$$\frac{\partial}{\partial t}<f(\mathbf{r}, \mathbf{v}, t)>_r + \mathbf{v} \cdot \left\langle \frac{\partial f}{\partial \mathbf{r}} \right\rangle_r - \frac{e}{m} \left\langle \mathbf{E} \cdot \frac{\partial f}{\partial \mathbf{v}} \right\rangle_r = 0 \tag{16.29}$$

In the second term, we have typically

$$\frac{v_x}{V} \int \frac{\partial f}{\partial x} dx dy dz = \frac{v_x}{V} \int dy\ dz\ f(x, y, z) \bigg|_{x=\text{boundary}}$$

where V is the volume of the plasma, and we put the boundary at ∞. Since there are no particles at the boundary this integral vanishes. The third term in (16.29) is simplified by using the assumption $<\mathbf{E}>_r = 0$, which is true for the ensemble of randomly phased plasma waves.

$$\begin{aligned} \left\langle \mathbf{E} \cdot \frac{\partial f}{\partial \mathbf{v}} \right\rangle_r &= \left\langle \mathbf{E} \cdot \frac{\partial f_0}{\partial \mathbf{v}} \right\rangle_r + \left\langle \mathbf{E} \cdot \frac{\partial f_1}{\partial \mathbf{v}} \right\rangle_r \\ &= <\mathbf{E}>_r \cdot \frac{\partial f_0}{\partial \mathbf{v}} + \left\langle \mathbf{E} \cdot \frac{\partial f_1}{\partial \mathbf{v}} \right\rangle_r = \left\langle \mathbf{E} \cdot \frac{\partial f_1}{\partial \mathbf{v}} \right\rangle_r \end{aligned}$$

Then (16.29) becomes

$$\frac{\partial f_0}{\partial t} = \frac{e}{m} \left\langle \mathbf{E} \cdot \frac{\partial f_1}{\partial \mathbf{v}} \right\rangle_r \tag{16.30}$$

Equation for f_1 is obtained by linearizing (16.1) ($f_1 \ll f_0$) in the form

$$\frac{\partial f_1}{\partial t} + \mathbf{v} \cdot \frac{\partial f_1}{\partial \mathbf{r}} - \frac{e}{m} \mathbf{E}(\mathbf{r}, t) \cdot \frac{\partial f_0(\mathbf{v}, \epsilon t)}{\partial \mathbf{v}} = 0 \tag{16.31}$$

Together with the Poisson equation

$$\nabla \cdot \mathbf{E} = -4\pi e \int d^3 v f_1(\mathbf{r}, \mathbf{v}, t) \tag{16.32}$$

Equations (16.31) and (16.32) are formally identical with the linearized Vlasov and Poisson equations. We obtain the normal mode Langmuir wave frequencies

$$\omega(k, \epsilon t) = \omega_{\text{pe}}(\epsilon t) \left(1 + \frac{3}{2} k^2 \lambda_{\text{e}}^2\right) + i \frac{\pi}{2} \frac{\omega_{\text{pe}}^3}{k^2} \left[\frac{\partial g(u, \epsilon t)}{\partial u}\right]_{\frac{\omega}{k}} \tag{16.33}$$

where ϵt is the slow time, expressing the slow change of $f_0(v, \epsilon t)$. Note that the Langmuir frequency, the real part (ω_r) as well as the imaginary part (ω_i), is now slowly changing in t. For simplicity, we assume the one-dimensional propagation; $\mathbf{k} = k\hat{\mathbf{x}}$. We consider that the linear waves form a spectrum of waves with phasor $\exp[ikx - i\omega(k, \epsilon t)t]$ with $\omega(k, \epsilon t)$ as given by (16.33). The first order electric field can be expressed as spatial Fourier representation (in the following we suppress the slow time indicator ϵt):

$$E(x,t) = \int_{-\infty}^{\infty} \frac{dk}{2\pi} E(k) e^{ikx - i\omega(k)t} \tag{16.34}$$

$$f_1(v,x,t) = \int_{-\infty}^{\infty} \frac{dk}{2\pi} f_1(v,k) e^{ikx - i\omega(k)t} \tag{16.35}$$

$$\begin{aligned} <Ef_1>_x &= \frac{1}{L}\int dx \int \frac{dk'}{2\pi} E(k') e^{ik'x - i\omega(k')t} \int \frac{dk}{2\pi} f_1(v,k) e^{ikx - i\omega(k)t} \\ &= \frac{1}{2\pi L}\int dk \int dk' E(k') f_1(v,k) e^{-i[\omega(k)+\omega(k')]t} \delta(k+k') \\ &= \frac{1}{2\pi L}\int dk \int dk' E(-k) f_1(v,k) e^{-i[\omega(k)+\omega(-k)]t} \end{aligned} \tag{16.36}$$

Substituting the linear relation

$$f_1(v,k) = i\frac{e}{m} E(k) \frac{\partial f_0/\partial v}{\omega - kv} \tag{16.37}$$

into (16.36) gives

$$<Ef_1>_x = \frac{i}{2\pi L}\frac{e}{m}\int dk E(-k) E(k) \frac{\partial f_0/\partial v}{\omega - kv} e^{-i[\omega(k)+\omega(-k)]t} \tag{16.38}$$

Since $E(x,t)$ in (16.34) is a real quantity, we have the following parity relations:

$$E(-k) = E^*(k), \quad \omega_r(-k) = -\omega_r(k), \quad \omega(-k) = \omega_i(k) \tag{16.39}$$

where the asterisk denotes the complex conjugate. The real (imaginary) part of ω is an odd (even) function of k. Consequently, we have

$$E(k)E(-k) = |E(k)|^2, \quad \omega(k) + \omega(-k) = 2i\omega_i(k)$$

and (16.38) reduces to

$$<Ef_1>_x = \frac{i}{2\pi L}\frac{e}{m}\int dk |E(k)|^2 \frac{\partial f_0/\partial v}{\omega - kv} e^{2i\omega_i(k)t} \tag{16.40}$$

Therefore the time evolution of $f_0(v,t)$, (16.30), takes the form

$$\frac{\partial f_0}{\partial t} = \frac{i}{2\pi L}\frac{e^2}{m^2}\frac{\partial}{\partial v}\int dk|E(k)|^2\frac{\partial f_0/\partial v}{\omega - kv}e^{2i\omega_i(k)t} \tag{16.41}$$

In the preceding presentation of the quasilinear equations, we saw that the zero order distribution function f_0 (see Eq. (16.28)) as defined by spatial average of the exact distribution function allows a complete separation of the rapidly varying distribution function $f_1(\mathbf{r},\mathbf{v},t)$ from the slowly changing zero order $f_0(\mathbf{v},\epsilon t)$. The quasilinear equations were derived by calculating the spatial average of the product terms, each of which is rapidly varying. [see (16.30)].

References: D. R. Nicholson, *Introduction to plasma theory*, Wiley, New York (1983); P. M. Bellan, *Fundamentals of plasma physics*, Cambridge Univ. Press, Cambridge (2006).

• **Extraction of f_0 from f by averaging over the random phases.** In this section we show that the separation of f_1 and f_0 can be achieved alternatively by taking average over the random phases of the fluctuating waves. We start from (16.1) and (16.2) where we write the electric field $\mathbf{E}$ as a wave packet:

$$\mathbf{E}(\mathbf{r},t) = \sum_k \mathbf{E}_k(t)e^{i\mathbf{k}\cdot\mathbf{r} - i\omega(\mathbf{k})t} \tag{16.42}$$

where $\omega(\mathbf{k})$ is determined by the linear dispersion relation. Equation (16.42) is a two-time representation of $\mathbf{E}(\mathbf{r},t)$, in which the complex amplitude is slowly varying while the phase is rapidly oscillating. The slow change of the amplitude is caused by Landau damping or the growth (γ), depending upon the shape of the initial distribution function. The slow change of the amplitude implies $\gamma(\mathbf{k}) \ll \omega(\mathbf{k})$. Thus the slow amplitude can be represented as

$$\mathbf{E}_k(t) = \hat{\mathbf{a}}(\mathbf{k})E_k(t)e^{i\theta(\mathbf{k})} \quad (\hat{\mathbf{a}}(\mathbf{k})\text{ polarization vector; } E_k(t)\text{ real})$$

where we assume that $\theta(\mathbf{k})$ corresponding to different $\mathbf{k}$ are random and independent. This means that the amplitude vanishes when averaged over the phase:

$$<\mathbf{E}_k(t)> = \frac{\int_0^{2\pi} d\theta(\mathbf{k})\mathbf{E}_k(t)}{\int_0^{2\pi} d\theta(\mathbf{k})} = 0$$

To show this, it is sufficient to calculate

$$\int_0^{2\pi} d\theta(\mathbf{k})[\cos\theta(\mathbf{k}) \quad \text{or} \quad \sin\theta(\mathbf{k})] = 0$$

It follows from this assumption of random phases that

$$<\mathbf{E}_k(t)\cdot\mathbf{E}^*_{k'}(t)> = |\mathbf{E}_k(t)|^2\delta(\mathbf{k}-\mathbf{k}')$$

We plan to work on the quantity which is the deviation from the mean value:

$$f_1(\mathbf{r},\mathbf{v},t) = f(\mathbf{r},\mathbf{v},t) - f_0(\mathbf{v},t) \tag{16.43}$$

where $f_0 \equiv <f>$ is the average of $f(\mathbf{r},\mathbf{v},t)$ over the phases. Accordingly, we have $<f_1(\mathbf{r},\mathbf{v},t)> = 0$. Substituting $f = f_0 + f_1$ into (16.1) and taking average over the ensemble of phases, we obtain

$$\frac{\partial f_0}{\partial t} + \mathbf{v}\cdot\frac{\partial f_0}{\partial \mathbf{r}} - \frac{e}{m}\left\langle \mathbf{E}(\mathbf{r},t)\cdot\frac{\partial f_1}{\partial \mathbf{v}}\right\rangle = 0 \tag{16.44}$$

We consider the plasma to be uniform in the original state, and the averaged distribution function f_0 will be assumed to be independent of spatial coordinate. Then The above equation becomes

$$\frac{\partial f_0}{\partial t} - \frac{e}{m}\left\langle \mathbf{E}(\mathbf{r},t)\cdot\frac{\partial f_1}{\partial \mathbf{v}}\right\rangle = 0 \tag{16.45}$$

Subtracting (16.44) from (16.1) gives

$$\begin{aligned}\frac{\partial}{\partial t}f_1(\mathbf{r},\mathbf{v},t) + \mathbf{v}\cdot\frac{\partial f_1}{\partial \mathbf{r}} - \frac{e}{m}\mathbf{E}(\mathbf{r},t)\cdot\frac{\partial f_0}{\partial \mathbf{v}}\\ - \frac{e}{m}\mathbf{E}(\mathbf{r},t)\cdot\frac{\partial f_1}{\partial \mathbf{v}} - \frac{e}{m}\left\langle \mathbf{E}(\mathbf{r},t)\cdot\frac{\partial f_1}{\partial \mathbf{v}}\right\rangle = 0\end{aligned}$$

In the *quasilinear approximation*, the last two terms in the above are neglected which account for nonlinear wave-wave interactions. Then, we have

$$\frac{\partial}{\partial t}f_1(\mathbf{r},\mathbf{v},t) + \mathbf{v}\cdot\frac{\partial f_1}{\partial \mathbf{r}} - \frac{e}{m}\mathbf{E}(\mathbf{r},t)\cdot\frac{\partial f_0}{\partial \mathbf{v}} = 0 \tag{16.46}$$

In the two quasilinear equations, (16.45) and (16.46), which are completely identical to (16.30) and (16.31), the average is taken over the random phases in contrast with the spatial average as indicated in (16.30). However, in view of the correlation relationship between the random phased electric fields (the equation above (16.43)), it is reasonably clear that the spatial average (see (16.36)) and the average over the phases would produce the same result. The subsequent development of (16.45) and (16.46) is entirely parallel to the foregoing presentation. Our further development of (16.45) and (16.46) contains complexity due to multidimensional analysis. In addition, we shall consider, in accordance with the wave packet form of electric field (16.42), wave packet form of f_1:

$$f_1(\mathbf{r}, \mathbf{v}, t) = \sum_k f_1(\mathbf{k}, \mathbf{v}, \epsilon t) e^{i\mathbf{k}\cdot\mathbf{r} - i\omega(\mathbf{k})t} \tag{16.47}$$

where $f_1(\mathbf{k}, \mathbf{v}, \epsilon t)$ is a slowly varying function of t. Substituting (16.47) into (16.46) gives for the Fourier amplitude

$$\frac{\partial f_k(\epsilon t)}{\partial t} + i(\mathbf{k}\cdot\mathbf{v} - \omega_k) f_k + i\frac{e}{m}\phi_k(t)\mathbf{k}\cdot\frac{\partial f_0(\epsilon t, \mathbf{v})}{\partial \mathbf{v}} = 0 \quad \text{(cf. (16.7))} \tag{16.48}$$

where and in the following the subscript '1' is omitted. Keeping in mind that the first term in (16.48) is due to Landau damping or growth, we make transformation $f_k(\epsilon t, \mathbf{v}) = g_k(\mathbf{v}) e^{\gamma_k t}$. Then (16.48) becomes

$$(i\mathbf{k}\cdot\mathbf{v} - i\omega_k + \gamma_k) g_k + i\frac{e}{m}\phi_k(t)\ e^{-\gamma_k t}\mathbf{k}\cdot\frac{\partial f_0(\epsilon t, \mathbf{v})}{\partial \mathbf{v}} = 0$$

$$\text{or}\quad (i\mathbf{k}\cdot\mathbf{v} - i\omega_k + \gamma_k) f_k(\mathbf{v}, \epsilon t) + i\frac{e}{m}\phi_k(t)\ \mathbf{k}\cdot\frac{\partial f_0(\epsilon t, \mathbf{v})}{\partial \mathbf{v}} = 0 \tag{16.49}$$

We instated the slow time derivative in (16.48) as imaginary part of frequency ω_k in (16.49). We have the Poisson equation in the form

$$k^2\phi_k(t) = -4\pi e N \int d^3v f_k(\mathbf{v}, t) \tag{16.50}$$

Using (16.49) in (16.50) gives the linear dispersion relation

$$1 + \frac{\omega_{\text{pe}}^2}{k^2}\int d^3v\ \frac{\mathbf{k}\cdot\frac{\partial f_0}{\partial \mathbf{v}}}{\omega_k - \mathbf{k}\cdot\mathbf{v} + i\gamma_k} = 0$$

Next, let us develop (16.45). Substituting $\mathbf{E}(\mathbf{r}, t)$ and $f(\mathbf{r}, \mathbf{v}, t)$ in their Fourier representation and using (16.49), we can write (16.45) in the form

$$\begin{aligned}\frac{\partial f_0}{\partial t} &= \frac{e}{m}\left\langle \sum_{k'} \mathbf{E}_{k'}(t) e^{i\mathbf{k}'\cdot\mathbf{r} - i\omega_{k'}t}\cdot\frac{\partial}{\partial \mathbf{v}}\sum_k f_k(t) e^{i\mathbf{k}\cdot\mathbf{r} - i\omega_k t}\right\rangle \\ &= \frac{e}{m}\left\langle \sum_{k'} -i\mathbf{k}'\phi_{k'}(t) e^{i\mathbf{k}'\cdot\mathbf{r} - i\omega_{k'}t}\cdot\frac{\partial}{\partial \mathbf{v}}\sum_k e^{i\mathbf{k}\cdot\mathbf{r} - i\omega_k t}\frac{-i\frac{e}{m}\phi_k\mathbf{k}\cdot\frac{\partial f_0}{\partial \mathbf{v}}}{i\mathbf{k}\cdot\mathbf{v} - i\omega_k + \gamma_k}\right\rangle\end{aligned}$$

The average taken over random phases gives $\delta(\mathbf{k} + \mathbf{k}')$. Note that $\omega_k + \omega_{-k} = 0, \phi_k\phi_{-k} = |\phi_k|^2$. Thus the above equation reduces to

$$\frac{\partial f_0}{\partial t} = \left(\frac{e}{m}\right)^2 \sum_k |\phi_k|^2 \mathbf{k} \cdot \frac{\partial}{\partial \mathbf{v}} \frac{\mathbf{k} \cdot \frac{\partial f_0}{\partial \mathbf{v}}}{i\mathbf{k} \cdot \mathbf{v} - i\omega_k + \gamma_k} \quad \text{(cf. (16.23))} \tag{16.51}$$

$\gamma_k = \gamma_k(\epsilon t)$, a slowly varying function of t. In linear theory where the distribution function is constant, the dependence of E_k^2 on t is simply exponential, $E_k^2 \sim e^{2\gamma_k t}$, but in quasilinear theory in which the distribution has a weak time dependence, we have

$$\frac{dE_k^2}{dt} = 2\gamma_k(t) E_k^2.$$

[The original author of the material presented in this subsection is anonymous to the author of this book, who transported the analysis with gratitude.]

16.2 Nonlinear Theory of Vlasov–Poisson Plasma

In this section, the nonlinear term that we discarded in the linear and quasilinear theory of the Vlasov equation is accounted for. The Vlasov equation is expanded in a perturbation series with the nonlinear term iterated by the linear solution. In terms of Fourier coefficients, the nonlinear term produces quadratic and triadic terms, and even higher order terms. The series solution is investigated under certain resonance conditions. Then, the Vlasov–Poisson system (or Vlasov–Maxwell system) can be reduced to a form from which important physical consequence can be elicited. The resonance conditions are two kinds of which products are: nonlinear Landau damping or growth (*wave–particle interaction*) and mode coupling (*wave–wave-interaction*).

Reference: R. Z. Sagdeev and A. A. Galeev, *Nonlinear plasma theory*, Benjamin, New York (1969).

16.2.1 *Straight expansion of Vlasov–Poisson equation in powers of wave amplitude*

We start from the Vlasov equation for the distribution function $f_\alpha(\mathbf{r}, \mathbf{v}, t)$

$$\frac{\partial}{\partial t} f_\alpha(\mathbf{r}, \mathbf{v}, t) + \mathbf{v} \cdot \frac{\partial f_\alpha}{\partial \mathbf{r}} + \frac{e_\alpha}{m_\alpha} \mathbf{E}(\mathbf{r}, t) \cdot \frac{\partial f_\alpha}{\partial \mathbf{v}} = 0 \tag{16.52}$$

and the Poisson equation

$$\nabla \cdot \mathbf{E} = 4\pi \sum_\alpha \int d^3v \; f_\alpha(\mathbf{r}, \mathbf{v}, t), \quad (\mathbf{E} = -\nabla\phi) \tag{16.53}$$

Expand f in the series

$$f_\alpha = f_{\alpha 0}(\mathbf{v}) + f_\alpha^{(1)}(\mathbf{r}, \mathbf{v}, t) + f_\alpha^{(2)}(\mathbf{r}, \mathbf{v}, t) + f_\alpha^{(3)}(\mathbf{r}, \mathbf{v}, t) + \cdots$$

and solve (16.52) by successive approximation.

$$\frac{\partial}{\partial t} f_\alpha^{(n)}(\mathbf{r}, \mathbf{v}, t) + \mathbf{v} \cdot \frac{\partial f_\alpha^{(n)}}{\partial \mathbf{r}} + \frac{e_\alpha}{m_\alpha} \mathbf{E}(\mathbf{r}, t) \cdot \frac{\partial f_\alpha^{(n-1)}}{\partial \mathbf{v}} = 0$$
$$(n = 1, 2, 3, \ldots, f^{(0)} = f_0) \tag{16.54}$$

The first order equation gives

$$f_\alpha^{(1)}(\mathbf{k}, \mathbf{v}, \omega) = -\frac{e_\alpha}{m_\alpha} \frac{\mathbf{k} \cdot \frac{\partial f_{\alpha 0}}{\partial \mathbf{v}}}{\omega - \mathbf{k} \cdot \mathbf{v}} \phi(\mathbf{k}, \omega) \tag{16.55}$$

The second order equation gives

$$f_\alpha^{(2)}(\mathbf{k}, \mathbf{v}, \omega) = -\frac{e_\alpha}{m_\alpha} \sum_{k', \omega'} \frac{\mathbf{k}' \cdot \frac{\partial f_\alpha^{(1)}(\mathbf{k}'', \omega'', \mathbf{v})}{\partial \mathbf{v}}}{\omega - \mathbf{k} \cdot \mathbf{v}} \phi(\mathbf{k}', \omega')$$
$$(\mathbf{k}'' = \mathbf{k} - \mathbf{k}', \omega'' = \omega - \omega')$$

Using (16.55) in the above gives (for algebraic details, see Problem 16.4)

$$f_\alpha^{(2)}(\mathbf{k}, \mathbf{v}, \omega) = -\frac{(e_\alpha/m_\alpha)^2}{\omega - \mathbf{k} \cdot \mathbf{v}} \frac{1}{2} \sum_{k', \omega'} \left[\mathbf{k}' \cdot \frac{\partial}{\partial \mathbf{v}} \frac{\mathbf{k}'' \cdot \frac{\partial f_{\alpha 0}}{\partial \mathbf{v}}}{\omega'' - \mathbf{k}'' \cdot \mathbf{v}} + term \; (' \leftrightarrow '') \right] \phi(\mathbf{k}', \omega') \phi(\mathbf{k}'', \omega'') \tag{16.56}$$

Substituting the above result into (16.53) yields the equation for ϕ in the form

$$\varepsilon^{(1)}(\mathbf{k}, \omega)\phi(\mathbf{k}, \omega) + \sum_{k = k' + k'', \omega = \omega' + \omega''} \varepsilon^{(2)}(\mathbf{k}', \omega', \mathbf{k}'', \omega'')\phi(\mathbf{k}', \omega')\phi(\mathbf{k}'', \omega'')$$
$$+ \sum_{k = k' + k'' + k''', \; \omega = \omega' + \omega'' + \omega'''} \varepsilon^{(3)}(\mathbf{k}', \omega', \mathbf{k}'', \omega'', \mathbf{k}''', \omega''')\phi(\mathbf{k}', \omega')\phi(\mathbf{k}'', \omega'')\phi(\mathbf{k}''', \omega''') + \cdots = 0 \tag{16.57}$$

where

$$\varepsilon^{(1)}(\mathbf{k},\omega) = 1 + \Sigma_\alpha \frac{\omega_{p\alpha}^2}{k^2} \int d^3v \, \frac{\mathbf{k}\cdot\frac{\partial f_{\alpha 0}}{\partial \mathbf{v}}}{\omega - \mathbf{k}\cdot\mathbf{v}} \tag{16.58}$$

$$\varepsilon^{(2)}(\mathbf{k}',\omega',\mathbf{k}'',\omega'') = -\frac{1}{2}\sum_\alpha \frac{e_\alpha}{m_\alpha}\frac{\omega_{p\alpha}^2}{k^2}\int \frac{d^3v}{\omega - \mathbf{k}\cdot\mathbf{v}} \times \left[\mathbf{k}'\cdot\frac{\partial}{\partial \mathbf{v}}\frac{\mathbf{k}''\cdot\frac{\partial f_{\alpha 0}}{\partial \mathbf{v}}}{\omega'' - \mathbf{k}''\cdot\mathbf{v}} + \mathbf{k}''\cdot\frac{\partial}{\partial \mathbf{v}}\frac{\mathbf{k}'\cdot\frac{\partial f_{\alpha 0}}{\partial \mathbf{v}}}{\omega' - \mathbf{k}'\cdot\mathbf{v}}\right] \tag{16.59}$$

Equation (16.57) is solved by expanding in powers of the wave amplitude. The lowest order solution is obtained from

$$\varepsilon^{(1)}(\mathbf{k},\omega)\phi^{(1)}(\mathbf{k},\omega) = 0 \tag{16.60}$$

whose solution we write in the form

$$\phi^{(1)}(\mathbf{k},\omega) = \phi_k\delta(\omega - \omega(\mathbf{k})) \tag{16.61}$$

where $\omega(\mathbf{k})$ is the root of $\varepsilon^{(1)}(\mathbf{k},\omega) = 0$; the normal mode frequency of the linear solution. Note in (16.60) that $\phi^{(1)}(\mathbf{k},\omega) \sim \delta(\varepsilon^{(1)}(\mathbf{k},\omega))$. The second order equation for $\phi(\mathbf{k},\omega)$ in (16.57) reads

$$\varepsilon^{(1)}(\mathbf{k},\omega)\phi^{(2)}(\mathbf{k},\omega) + \sum_{k=k'+k'',\ \omega=\omega'+\omega''} \varepsilon^{(2)}(\mathbf{k}',\omega',\mathbf{k}'',\omega'')\phi^{(1)}(\mathbf{k}',\omega')\phi^{(1)}(\mathbf{k}'',\omega'') = 0 \tag{16.62}$$

Using (16.61) in (16.62) gives

$$\phi^{(2)}(\mathbf{k},\omega) = -\sum_{k=k'+k''} \frac{\varepsilon^{(2)}(\mathbf{k}',\omega',\mathbf{k}'',\omega'')}{\varepsilon^{(1)}(\mathbf{k},\omega)}\phi_{k'}\phi_{k''}\delta(\omega - \omega(\mathbf{k}') - \omega(\mathbf{k}'')) \tag{16.63}$$

To develop the equation for $|\phi(\mathbf{k},t)|^2$, the wave kinetic equation, we multiply (16.57) by $\phi^*(\mathbf{k},\Omega)e^{-i\omega t}e^{i\Omega t}$ and integrate over $\int d\Omega d\omega$:

$$\int d\omega \int d\Omega \text{ Eq. (16.57)} \times \phi^*(\mathbf{k},\Omega)e^{-i\omega t}e^{i\Omega t} \tag{16.64}$$

We take in (16.64) the imaginary part only. It makes sense because it yields the desired kinetic equation; the real part produces its own relation but is of no concern. Also note

$$|\phi(\mathbf{k},t)|^2 = \left(\frac{1}{2\pi}\right)^2 \int e^{-i\omega t}\phi(\mathbf{k},\omega)d\omega \int e^{i\Omega t}\phi^*(\mathbf{k},\Omega)d\Omega$$

The first term in (16.57) gives the integral

$$I = \int d\omega \int d\Omega\ \varepsilon^{(1)}(\mathbf{k},\omega)\phi(\mathbf{k},\omega)\phi^*(\mathbf{k},\Omega)e^{-i\omega t}e^{i\Omega t} = \phi^*(\mathbf{k},t)J \tag{16.65}$$

where

$$J = \int d\omega e^{-i\omega t}\varepsilon^{(1)}(\mathbf{k},\omega)\phi(\mathbf{k},\omega) = \int d\omega e^{-i\omega t}[\varepsilon_k'(\omega) + i\varepsilon_k''(\omega)]\phi(\mathbf{k},\omega)$$

where we separated $\varepsilon^{(1)} = \varepsilon_k'(\omega) + i\varepsilon_k''(\omega)$ into real and imaginary parts. Our purpose is to express J explicitly in terms of $\phi(\mathbf{k},t)$. Since $\phi(\mathbf{k},\omega)$ is peaked around $\omega(\mathbf{k}) \equiv \omega_k$, we can write

$$\phi(\mathbf{k},t) = \int d\omega e^{-i\omega t}\phi(\mathbf{k},\omega) \simeq \Delta\omega e^{-i\omega_k t}\ \phi_k(\omega_k)$$

$$\int d\omega \varepsilon_k''(\omega)e^{-i\omega t}\phi(\mathbf{k},\omega) \simeq \Delta\omega\varepsilon_k''(\omega_k)e^{-i\omega_k t}\ \phi_k(\omega_k) = \varepsilon_k''(\omega_k)\phi(\mathbf{k},t)$$

Therefore

$$J = i\ \varepsilon_k''(\omega_k)\phi(\mathbf{k},t) + \int d\omega\left(\varepsilon'(\mathbf{k},\omega_k) + (\omega-\omega_k)\left[\frac{\partial\varepsilon'}{\partial\omega}\right]_{\omega_k}\right)\phi(\mathbf{k},\omega)e^{-i\omega t}$$

where we Taylor-expanded $\varepsilon'(\omega)$ in the neighborhood of $\omega = \omega_k$. Note $\varepsilon'(\mathbf{k},\omega_k) = 0$.

Hence

$$\begin{aligned}
J &= i\ \varepsilon_k''(\omega_k)\phi(\mathbf{k},t) + \left[\frac{\partial\varepsilon'}{\partial\omega}\right]_{\omega_k}\int d\omega(\omega-\omega_k)\phi(\mathbf{k},\omega)e^{-i\omega t} \\
&= i\ \varepsilon_k''(\omega_k)\phi(\mathbf{k},t) + \left[\frac{\partial\varepsilon'}{\partial\omega}\right]_{\omega_k}\left(-\omega_k\phi(\mathbf{k},t) + \int d\omega\omega\phi(\mathbf{k},\omega)e^{-i\omega t}\right) \\
&= i\ \varepsilon_k''(\omega_k)\phi(\mathbf{k},t) + \left[\frac{\partial\varepsilon'}{\partial\omega}\right]_{\omega_k}\left(-\omega_k\phi(\mathbf{k},t) + i\ \frac{\partial\phi(\mathbf{k},t)}{\partial t}\right)
\end{aligned}$$

Using the above in (16.65) gives for the imaginary part

$$\text{Im}\ I = \varepsilon_k''(\omega_k)|\phi(\mathbf{k},t)|^2 + \left[\frac{\partial\varepsilon'}{\partial\omega}\right]_{\omega_k}\frac{1}{2}\frac{\partial}{\partial t}|\phi(\mathbf{k},t)|^2 \tag{16.66}$$

Next, we consider the quadratic term of (16.57) in the integral (16.64). The quadratic term is expressed as follows by putting it in terms of the products of ϕ's:

$$\sum \varepsilon^{(2)}(\mathbf{k}',\omega',\mathbf{k}'',\omega'')[\phi^{(1)}(\mathbf{k}',\omega')\phi^{(1)}(\mathbf{k}'',\omega'') + \phi^{(2)}(\mathbf{k}',\omega')\phi^{(2)}(\mathbf{k}'',\omega'') + \phi^{(1)}(\mathbf{k}',\omega')\phi^{(2)}(\mathbf{k}'',\omega'') + \phi^{(2)}(\mathbf{k}',\omega')\phi^{(1)}(\mathbf{k}'',\omega'') + \cdots] \tag{16.67}$$

The above expression is multiplied by $\phi^{(1)*}(\mathbf{k})\delta(\omega - \Omega)e^{i\Omega t}$ and is integrated over $d\Omega$. Here we invoke the random phase approximation which states that the ensemble average takes the form: $<\phi^{(1)}(\mathbf{k})\phi^{(1)*}(\mathbf{k}')> = |\phi^{(1)}(\mathbf{k})|^2\delta_{k,k'}$. Clearly the odd power terms of $\phi^{(1)}(\mathbf{k})$ will be ensemble averaged to zero, and thus the first and the second terms in the RHS of (16.67) will vanish. We also neglect terms that are higher than fourth order in $\phi^{(1)}(\mathbf{k})$. Using (16.63) for the remaining last two terms in (16.67), we obtain the kinetic equation in the following form (Sagdeev and Galeev, 1969, III-9)

$$\begin{aligned} &\left[\frac{\partial\varepsilon'}{\partial\omega}\right]_{\omega_k} \frac{1}{2}\frac{\partial}{\partial t}|\phi(\mathbf{k},t)|^2 \\ &= -\varepsilon''_k(\omega_k)|\phi(\mathbf{k},t)|^2 + 2\mathrm{Im}\sum_{\mathbf{k}=\mathbf{k}'+\mathbf{k}''} \frac{|\varepsilon^{(2)}(\mathbf{k}',\mathbf{k}'',\omega_{k'},\omega_{k''})|^2}{\varepsilon^{(1)*}(\mathbf{k}'+\mathbf{k}'',\ \omega_{k'}+\omega_{k''})}|\phi_{k'}|^2|\phi_{k''}|^2 \\ &\quad + \mathrm{Im}\sum_{\mathbf{k}'}\left[4\,\frac{\varepsilon^{(2)}(\mathbf{k}',\mathbf{k}-\mathbf{k}',\omega_{k'},\omega_k-\omega_{k'})\varepsilon^{(2)}(\mathbf{k},-\mathbf{k}',\omega_k,-\omega_{k'})}{\varepsilon^{(1)}(\mathbf{k}-\mathbf{k}',\omega_k-\omega_{k'})}\right. \\ &\qquad \left. -\,3\,\varepsilon^{(3)}(\mathbf{k}',-\mathbf{k}',\mathbf{k},\omega_{k'},-\omega_{k'},\omega_k)\right]|\phi_{k'}|^2|\phi_k|^2 \end{aligned} \tag{16.68}$$

The first term of the RHS of (16.68) is the quasilinear growth rate since it is

$$\omega_i(\mathbf{k}) = -\frac{\varepsilon''_k(\omega_k)}{\partial\varepsilon'_k/\partial\omega_k}$$

The second term describes the rate at which $\phi_{k'}$ and $\phi_{k''}$ merge to ϕ_k through wave-wave interaction. The third term determines the rate at which ϕ_k and $\phi_{k'}$ generate $\phi_{k''}$. Furthermore, this term contains a contribution to wave-particle interaction (nonlinear Landau damping) as will be shown in the following. The resonance condition for a wave-wave interaction to occur (in the second term) depends on the type of the dispersion curve. In the case of electron plasma oscillation, The dispersion curve is non-decay type.

• **Decay type and non-decay type.** Let $\overrightarrow{\kappa} \equiv (\mathbf{k}, \omega)$ represent the four-vector as indicated. For $\overrightarrow{\kappa_1}$ and $\overrightarrow{\kappa_2}$ which satisfy $\varepsilon(\overrightarrow{\kappa_1}) = 0$ and $\varepsilon(\overrightarrow{\kappa_2}) = 0$, respectively, the dispersion curve is of non-decay type, if $\varepsilon(\overrightarrow{\kappa_1} + \overrightarrow{\kappa_2}) = 0$ cannot be satisfied. If $\varepsilon(\overrightarrow{\kappa_1} + \overrightarrow{\kappa_2}) = 0$ can be satisfied, the wave is decay type. The phase velocity decreases (increases) with k in the non-decay(decay) type wave. The resonance condition $\mathbf{k} = \mathbf{k}' + \mathbf{k}''$ can be satisfied among three waves belonging to a branch of the dispersion curve, if the dispersion curve is of decay type. If the dispersion curve is of non-decay type, the resonance condition can be satisfied by involving a third component belonging to different branch (for instance, two Langmuir waves and an ion acoustic wave). In the linear portion of the ion acoustic wave spectrum, the wave can decay. The change of wave energy (LHS of (16.68)) of a Langmuir wave is determined by the quasilinear growth rate and nonlinear Landau damping in the third term, as is discussed below.

16.2.2 *Nonlinear Landau damping*

To evaluate the nonlinear Landau damping term in (16.68), we identify as follows. $\mathbf{k} \equiv \mathbf{k}_1$: a high frequency Langmuir wave, $\mathbf{k}' \equiv \mathbf{k}_2$: another high frequency Langmuir wave, $\mathbf{k}'' = \mathbf{k} - \mathbf{k}' = \mathbf{k}_1 - \mathbf{k}_2 \equiv \mathbf{k}_3$: low frequency wave, which should be an ion acoustic wave because Langmuir wave is not decayable. Therefore the third term for a single triad is

$$\frac{\varepsilon^{(2)}(\mathbf{k}_2, \omega_2, \mathbf{k}_3, \omega_3)\varepsilon^{(2)}(\mathbf{k}_1, \omega_1, -\mathbf{k}_2, -\omega_2)}{\varepsilon^{(1)}(\mathbf{k}_3, \omega_3)}|\phi_{k_1}|^2|\phi_{k_2}|^2 \tag{16.69}$$

where $\varepsilon^{(2)}$ are electron's and $\varepsilon^{(1)}$ consists of both electron and ion contributions. To see how the third term came about more clearly, let us go back to the nonlinear dispersion relation in (16.57), and write the three-wave interaction in the form

$$\varepsilon^{(1)}(\mathbf{k}_1, \omega_1)\phi(\mathbf{k}_1, \omega_1) = -\varepsilon^{(2)}(\mathbf{k}_2, \omega_2, \mathbf{k}_3, \omega_3)[\phi^{(1)}(\mathbf{k}_2, \omega_2)\phi^{(2)}(\mathbf{k}_3, \omega_3) + \phi^{(2)}(\mathbf{k}_2, \omega_2)\phi^{(1)}(\mathbf{k}_3, \omega_3)] \tag{16.70}$$

The second term in the above is nonresonant and need not be considered. Using

$$\phi^{(2)}(\mathbf{k}_3, \omega_3) = -\frac{\varepsilon^{(2)}(\mathbf{k}_1, \omega_1, -\mathbf{k}_2, -\omega_2)}{\varepsilon^{(1)}(\mathbf{k}_3, \omega_3)}\phi_{k_1}\phi_{-k_2}$$

in (16.70) gives (16.69).

Now let us calculate the dielectric functions, i.e. the integrals in (16.58) and (16.59). Let us use the inequalities

$$v_i \ll \frac{\omega_3}{k_3} \ll v_e \ll \frac{\omega_1}{k_1}, \frac{\omega_2}{k_2} \quad (v_{e,i} : \text{thermal velocity})$$

and neglect small terms. Then we obtain

$$\begin{aligned}
&\varepsilon^{(2)}(\mathbf{k}_1, \omega_1, -\mathbf{k}_2, -\omega_2) \\
&= \frac{1}{2} \frac{\omega_{\text{pe}}^2}{k_3^2} \frac{e}{m_e} \int \frac{d^3 v}{\omega_3 - \mathbf{k}_3 \cdot \mathbf{v}} \left[\mathbf{k}_1 \cdot \frac{\partial}{\partial \mathbf{v}} \frac{-\mathbf{k}_2 \cdot \frac{\partial f_{e0}}{\partial \mathbf{v}}}{-\omega_2 + \mathbf{k}_2 \cdot \mathbf{v}} - \mathbf{k}_2 \cdot \frac{\partial}{\partial \mathbf{v}} \frac{\mathbf{k}_1 \cdot \frac{\partial f_{e0}}{\partial \mathbf{v}}}{\omega_1 - \mathbf{k}_1 \cdot \mathbf{v}} \right] \\
&= \frac{1}{2} \frac{\omega_{\text{pe}}^2}{k_3^2} \frac{e}{m_e} \int \frac{d^3 v}{\omega_3 - \mathbf{k}_3 \cdot \mathbf{v}} \left[\frac{\mathbf{k}_1 \cdot \mathbf{k}_2 \mathbf{k}_2 \cdot \frac{\partial f_{e0}}{\partial \mathbf{v}}}{(\omega_2 - \mathbf{k}_2 \cdot \mathbf{v})^2} + \frac{\mathbf{k}_1 \cdot \frac{\partial}{\partial \mathbf{v}} \mathbf{k}_2 \cdot \frac{\partial f_{e0}}{\partial \mathbf{v}}}{\omega_2 - \mathbf{k}_2 \cdot \mathbf{v}} \right. \\
&\quad \left. - \frac{\mathbf{k}_1 \cdot \mathbf{k}_2 \mathbf{k}_1 \cdot \frac{\partial f_{e0}}{\partial \mathbf{v}}}{(\omega_1 - \mathbf{k}_1 \cdot \mathbf{v})^2} - \frac{\mathbf{k}_2 \cdot \frac{\partial}{\partial \mathbf{v}} \mathbf{k}_1 \cdot \frac{\partial f_{e0}}{\partial \mathbf{v}}}{\omega_1 - \mathbf{k}_1 \cdot \mathbf{v}} \right]
\end{aligned} \tag{16.71}$$

where we used $\mathbf{k}_1 \cdot \frac{\partial}{\partial \mathbf{v}}(\mathbf{k}_2 \cdot \mathbf{v}) = \mathbf{k}_1 \cdot \mathbf{k}_2$. The second and the fourth terms in the above cancel since $(\mathbf{k}_2 \cdot \mathbf{v})$ and $(\mathbf{k}_1 \cdot \mathbf{v})$ in the denominators can be neglected as compared to $\omega_1 \simeq \omega_2 \simeq \omega_{\text{pe}}$. Also neglect ω_3 as compared to $(\mathbf{k}_3 \cdot \mathbf{v})$, then the remaining terms give

$$\varepsilon^{(2)}(\mathbf{k}_1, \omega_1, -\mathbf{k}_2, -\omega_2) \simeq -\frac{1}{2} \frac{e}{T_e} \frac{\mathbf{k}_1 \cdot \mathbf{k}_2}{k_3^2} \tag{16.72}$$

$$\begin{aligned}
&\varepsilon^{(2)}(\mathbf{k}_2, \omega_2, \mathbf{k}_3, \omega_3) \\
&= \frac{1}{2} \frac{\omega_{\text{pe}}^2}{k_1^2} \frac{e}{m_e} \int \frac{d^3 v}{\omega_1 - \mathbf{k}_1 \cdot \mathbf{v}} \left[\mathbf{k}_2 \cdot \frac{\partial}{\partial \mathbf{v}} \frac{\mathbf{k}_3 \cdot \frac{\partial f_{e0}}{\partial \mathbf{v}}}{\omega_3 - \mathbf{k}_3 \cdot \mathbf{v}} + \mathbf{k}_3 \cdot \frac{\partial}{\partial \mathbf{v}} \frac{\mathbf{k}_2 \cdot \frac{\partial f_{e0}}{\partial \mathbf{v}}}{\omega_2 - \mathbf{k}_2 \cdot \mathbf{v}} \right] \\
&= \frac{1}{2} \frac{\omega_{\text{pe}}^2}{k_1^2} \frac{e}{m_e} \int d^3 v \left[\frac{\mathbf{k}_1 \cdot \mathbf{k}_2}{(\omega_1 - \mathbf{k}_1 \cdot \mathbf{v})^2} \frac{\mathbf{k}_3 \cdot \frac{\partial f_{e0}}{\partial \mathbf{v}}}{\omega_3 - \mathbf{k}_3 \cdot \mathbf{v}} + \frac{\mathbf{k}_1 \cdot \mathbf{k}_3}{(\omega_1 - \mathbf{k}_1 \cdot \mathbf{v})^2} \frac{\mathbf{k}_2 \cdot \frac{\partial f_{e0}}{\partial \mathbf{v}}}{\omega_2 - \mathbf{k}_2 \cdot \mathbf{v}} \right]
\end{aligned}$$

where we integrated by parts. In the above, $(\mathbf{k}_2 \cdot \mathbf{v})$ and $(\mathbf{k}_1 \cdot \mathbf{v})$ in the denominators can be neglected as compared to $\omega_1 \simeq \omega_2 \simeq \omega_{\text{pe}}$. The second term vanishes because it is an odd function integral. Also neglect ω_3 as compared to $\mathbf{k}_3 \cdot \mathbf{v}$. Then we obtain

$$\varepsilon^{(2)}(\mathbf{k}_2, \omega_2, \mathbf{k}_3, \omega_3) \simeq -\frac{1}{2} \frac{e}{T_e} \frac{\mathbf{k}_1 \cdot \mathbf{k}_2}{k_1^2} \tag{16.73}$$

$$\varepsilon^{(1)}(\mathbf{k}_3, \omega_3) = 1 + \frac{\omega_{pi}^2}{k_3^2} \int d^3v \frac{\mathbf{k}_3 \cdot \frac{\partial f_{i0}}{\partial \mathbf{v}}}{\omega_3 - \mathbf{k}_3 \cdot \mathbf{v}} + \frac{\omega_{\text{pe}}^2}{k_3^2} \int d^3v \frac{\mathbf{k}_3 \cdot \frac{\partial f_{e0}}{\partial \mathbf{v}}}{\omega_3 - \mathbf{k}_3 \cdot \mathbf{v}} \tag{16.74}$$

Equation (16.74) agrees with the definition of the longitudinal dielectric permeability. In the last term (electron term), ω_3 can be neglected, and it becomes $1/\lambda_e^2 k_3^2$. The ion term integral is calculated along the Landau contour because the ions can resonate with the low frequency ion wave. We write

$$\varepsilon^{(1)}(\mathbf{k}_3, \omega_3) = 1 + \frac{1}{\lambda_e^2 k_3^2} - \frac{W}{k_3^2 \lambda_i^2} \tag{16.75}$$

$$W = \int d^3v \, \frac{v_z f_{io}(v^2)}{\frac{\omega_3}{k_3} - v_z} = \int dv_z \, \frac{v_z f_{io}(v_z)}{\frac{\omega_3}{k_3} - v_z} \tag{16.76}$$

where $f_{i0}(v_z) = \sqrt{\frac{m_i}{2\pi T_i}} \exp[-\frac{m_i v_z^2}{2T_i}]$, one-dimensional Maxwellian, and we take $\mathbf{k}_3$ to be parallel to the z-direction without loss of generality. The function W is related with the plasma dispersion function $Z(\zeta)$:

$$W = -1 - \zeta Z(\zeta); \quad Z(\zeta) = \frac{1}{\sqrt{\pi}} \int_{-\infty}^{\infty} \frac{e^{-x^2}}{x - \zeta} \, dx, \quad \zeta = \frac{\omega_3/k_3}{\sqrt{2T_i/m_i}}$$

[Sagdeev and Galeev, 1969, introduced a similar function $W_S(\zeta) = -\frac{i}{\sqrt{\pi}} Z(\zeta)$. Our function W in (16.76) is equal to $-1 - i\sqrt{\pi} \, \zeta \, W_S(\zeta)$.]

The real part of $W (\equiv W_r)$ is the principal value of the integral, and can be evaluated by expanding the denominator in the powers of the small quantity $k_3 v_z/\omega_3$. One obtains

$$W_r = \frac{T_i}{m_i} \frac{k_3^2}{\omega_3^2} \tag{16.77}$$

We also have the useful expression for the imaginary part of W:

$$W_i = -\sqrt{\frac{\pi}{2}} \left(\frac{T_e}{T_i} \right)^{\frac{1}{2}} e^{-\frac{T_e}{2T_i}} \tag{16.78}$$

Thus, finally, the kinetic equation (16.68) in which the dominant nonlinear interaction is due to the nonlinear Landau damping is obtained by using

(16.72), (16.73), and (16.75) in the form

$$\frac{\partial}{\partial t}\left[k^2|\phi(\mathbf{k},t)|^2\right] = \mathrm{Im}\sum_{k'} \frac{(\mathbf{k}\cdot\mathbf{k}')^2\omega_{\mathrm{pe}}}{4\pi N T_i}\left(\frac{T_i}{T_e}\right)^2 |\phi(\mathbf{k},t)|^2|\phi(\mathbf{k}',t)|^2\frac{1}{\frac{T_i}{T_e}-W} \tag{16.79}$$

$$\zeta = \frac{\omega_k - \omega_{k'}}{|\mathbf{k}-\mathbf{k}'|\,\sqrt{2T_i/m_i}}, \qquad N\text{: plasma density}$$

where we used $\partial\varepsilon'/\partial\omega_k = 2/\omega_{\mathrm{pe}}$.

Using (16.77) in (16.75) gives the real and imaginary parts of $\varepsilon^{(1)}(\omega_3, k_3)$:

$$\varepsilon'(\omega_3,k_3) = 1 + \frac{1}{\lambda_e^2 k_3^2} - \frac{\omega_{pi}^2}{\omega_3^2}; \quad \varepsilon'' = -\frac{W_i}{k_3^2\lambda_i^2} \tag{16.80}$$

We obtain the low frequency ion acoustic wave dispersion relation from $\varepsilon' = 0$:

$$\omega_3^2 = k_3^2\,\frac{T_e}{m_i} \quad \text{or} \quad W_r = \frac{T_i}{T_e} \tag{16.81}$$

where we neglected $\lambda_e^2 k_3^2 \ll 1$. Ion Landau damping rate is calculated by

$$\gamma_3 = -\frac{\varepsilon''}{\partial\varepsilon'/\partial\omega_3} = \frac{\omega_3}{2}\,\frac{T_e}{T_i}W_i = -\sqrt{\frac{\pi}{8}}\,\omega_3\left(\frac{T_e}{T_i}\right)^{\frac{3}{2}} e^{-\frac{T_e}{2T_i}} \tag{16.82}$$

One can show

$$\mathrm{Im}\left[\frac{1}{\frac{T_i}{T_e}-W}\right] = \frac{1}{W_i} = \frac{1}{2}\,\frac{\omega_3}{\gamma_3}\,\frac{T_e}{T_i} \tag{16.83}$$

Using the foregoing relations, (16.2.2) takes the form for single triad of interacting waves

$$\frac{\partial}{\partial t}\left[k_1^2|\phi(\mathbf{k}_1,t)|^2\right] = \frac{1}{2}\,\frac{(\mathbf{k}_1\cdot\mathbf{k}_2)^2\omega_{\mathrm{pe}}}{4\pi N T_e}\frac{\omega_3}{\gamma_3}|\phi(\mathbf{k}_1,t)|^2|\phi(\mathbf{k}_2,t)|^2 \tag{16.84}$$

This is the rate of energy transfer to the wave k_1 by resonant interaction of k_2 and k_3 waves through nonlinear Landau damping.

16.2.3 *Derivative expansion*

We derived differential equation (16.68) from the algebraic equation (16.57) by Fourier inversion. In the course of Fourier inversion, ω acted like the operator $i\frac{\partial}{\partial t}$. Converting nonlinear dispersion relation to a differential equation with respect to t can be accomplished more directly by the correspondence $\omega \to \omega_k + i\frac{\partial}{\partial t}$. Let us consider the equation

$$\varepsilon(\omega, k)\phi(k, \omega) = R \tag{16.85}$$

where R represents the nonlinear term and $\varepsilon(\omega, k)$ is regarded as an operator. The Fourier representation $\phi(k, \omega)$ is considered to be a slow function of t due to the nonlinear term in R : $\phi(k, \omega) \to \phi(k, \omega, \epsilon t)$. Taylor expansion gives

$$\varepsilon = \varepsilon'\left(\omega_k + i\frac{\partial}{\partial t}\right) + i\varepsilon'' = \varepsilon'(\omega_k) + \frac{\partial \varepsilon'}{\partial \omega_k} i\frac{\partial}{\partial t} + i\varepsilon''; \quad \varepsilon'(\omega_k) = 0$$

Therefore

$$\frac{\partial \phi(k, \omega, \epsilon t)}{\partial t} + \frac{\varepsilon''}{\partial \varepsilon'/\partial \omega_k}\phi(k, \omega, \epsilon t) = \frac{R}{i\partial \varepsilon'/\partial \omega_k} \tag{16.86}$$

Let us elaborate this. The linear solution can be written as $\phi(k, \omega, \epsilon t) = A(\epsilon t)\delta(\omega - \omega_k)$ with ω_k the normal mode frequency satisfying $\varepsilon(\omega_k) = 0$. Expanding $\varepsilon(\omega)$ near ω_k as before, we have

$$\left(-\omega_k\frac{\partial \varepsilon'}{\partial \omega_k} + i\varepsilon''(\omega_k)\right)\phi(k, \omega, \epsilon t) + \frac{\partial \varepsilon'}{\partial \omega_k}\omega\int \phi(k, t)e^{i\omega t}dt = R$$

where we put $\varepsilon'(\omega_k) = 0$. The following integral is evaluated as

$$\begin{aligned}
\omega\int \phi(k, t)e^{i\omega t}dt &= \omega\int dte^{i\omega t}\frac{1}{2\pi}\int d\omega'\phi(k, \omega')e^{-i\omega' t} \\
&= -i\int dt\frac{\partial}{\partial t}e^{i\omega t}\frac{1}{2\pi}\int d\omega'\phi(k, \omega')e^{-i\omega' t} \\
&= i\int dte^{i\omega t}\frac{1}{2\pi}\int d\omega'\frac{\partial}{\partial t}\left(\phi(k, \omega'\epsilon t)e^{-i\omega' t}\right) \\
&= i\int dte^{i\omega t}\frac{1}{2\pi}\int d\omega'\left(\frac{\partial}{\partial t}\phi(k, \omega', \epsilon t) - i\omega'\phi(k, \omega', \epsilon t)\right)e^{-i\omega' t} \\
&= i\int d\omega'\delta(\omega - \omega')\left(\frac{\partial}{\partial t}\phi(k, \omega', \epsilon t) - i\omega'\phi(k, \omega', \epsilon t)\right) \\
&= i\left(\frac{\partial}{\partial t}\phi(k, \omega, \epsilon t) - i\omega\phi(k, \omega, \epsilon t)\right)
\end{aligned}$$

Therefore, we have

$$\left(-\omega_k\frac{\partial \varepsilon'}{\partial \omega_k} + i\varepsilon''(\omega_k)\right)\phi(k, \omega, \epsilon t) + i\frac{\partial \varepsilon'}{\partial \omega_k}\left(\frac{\partial}{\partial t}\phi(k, \omega, \epsilon t) - i\omega\phi(k, \omega, \epsilon t)\right) = R$$

In the above, the first term and the last term cancel because $\phi(k,\omega) \sim \delta(\omega - \omega_k)$. So we finally obtain (16.86).

16.2.4 *Coupled mode equations: Wave–wave interaction*

A plasma can be populated by various collective oscillations excited by instability and spontaneous noise, in addition to particles. In such a *turbulent plasma*, the interaction between the various components of waves plays an important role. Anomalous diffusion of a plasma across a magnetic field can take place in such a turbulent state of plasma. *Langmuir paradox* is a phenomenon that electrons are Maxwell–Boltzmann distributed even in a very rarefied plasma. This can be explained by strong scattering of electrons by collective oscillations in turbulent plasma. Dynamical equation for describing wave-wave interactions in such an aggregate of interacting oscillations is useful to study turbulent plasma. This term is applied to a plasma which has a large number of collective degrees of wave oscillation, given rise to spontaneously or by instability. In (16.68), the second term on the RHS represents mathematically the dynamical effect of the interaction between the waves. Instead of exploring the second term, we study this category of wave–wave interaction by means of the basic fluid equations. Fluid equations are advantageous if we focus on the wave–wave interaction because the resonant particle dynamics are automatically skimmed.

In this section, coupled mode equations are derived for a plasma whose wave constituents are the high frequency Langmuir waves and the low frequency ion acoustic waves. The constituent waves beat each other to produce dynamical effects for the turbulent plasma. The fluid equations will be used for each component of the waves with superscript 'l' for Langmuir wave and 's' for ion acoustic wave (subscript 'e' and 'i' denote electron and ion quantities). Readers should also consult Chapter 15.

$$\mathbf{v}_e = \mathbf{v}_e^l + \mathbf{v}_e^s, \quad \mathbf{v}_i = \mathbf{v}_i^s + \mathbf{v}_i^l$$

Here the superscript 'l' and 's' can be also read as high and low frequency, respectively. The ions in the high frequency field are stationary ($\mathbf{v}_i^l = 0$), so $\mathbf{v}_i = \mathbf{v}_i^s$. The electrons in the low frequency ion acoustic wave field will be assumed to form Boltzmann distribution which is static (electron inertia is neglected), so we put $\mathbf{v}_e^s = 0$.

$$\mathbf{v}_e = \mathbf{v}_e^l, \quad \mathbf{v}_i = \mathbf{v}_i^s \tag{16.87}$$

$$\mathbf{E} = -\nabla\phi, \quad \phi = \phi^l + \phi^s, \quad n_i = n_0 + n_i^s \tag{16.88}$$

$$n_e = n_0 + n_e^l + n_e^s, \quad n_e^s = \frac{en_o}{T_e}\phi^s \tag{16.89}$$

$$\frac{\partial \mathbf{v}_e^l}{\partial t} = -\frac{e}{m_e}\mathbf{E}^l - \frac{3T_e}{n_0 m_e}\nabla n_e^l - \mathbf{v}_e^l \cdot \nabla \mathbf{v}_e^l \tag{16.90}$$

The last term of (16.90) is written considering the possibility that two Langmuir wave components beat to produce the third Langmuir wave component. But Langmuir wave spectrum is not of decay type, and the last term is put to zero. The high (Langmuir) and low (ion acoustic), electron and ion equations are summarized as follows:

$$\frac{\partial \mathbf{v}_e^l}{\partial t} = -\frac{e}{m_e}\mathbf{E}^l - \frac{3T_e}{n_0 m_e}\nabla n_e^l \tag{16.91}$$

$$\frac{\partial n_e^l}{\partial t} + n_0 \nabla \cdot \mathbf{v}_e^l + \nabla \cdot (n_e^s \mathbf{v}_e^l) = 0 \tag{16.92}$$

$$\nabla \cdot \mathbf{E}^l = -\nabla^2 \phi^l = -4\pi e n_e^l \tag{16.93}$$

$$-\frac{e}{m_e}\mathbf{E}^s - \frac{T_e}{n_0 m_e}\nabla n_e^s = \mathbf{v}_e^l \cdot \nabla \mathbf{v}_e^l \tag{16.94}$$

$$\frac{\partial n_e^s}{\partial t} + n_0 \nabla \cdot \mathbf{v}_e^s = 0 \tag{16.95}$$

$$\nabla \cdot \mathbf{E}^s = 4\pi e(n_i^s - n_e^s) \tag{16.96}$$

$$\frac{\partial \mathbf{v}_i^s}{\partial t} = \frac{e}{m_i}\mathbf{E}^s \tag{16.97}$$

$$\frac{\partial n_i^s}{\partial t} + n_0 \nabla \cdot \mathbf{v}_i^s = 0 \tag{16.98}$$

$$\frac{\partial \mathbf{E}^l}{\partial t} = 4\pi e \mathbf{v}_e^l (n_0 + n_e^s) \tag{16.99}$$

$$\frac{\partial \mathbf{E}^s}{\partial t} = 4\pi e n_0 (\mathbf{v}_e^s - \mathbf{v}_i^s) \tag{16.100}$$

$$n_e^s = \frac{en_0}{T_e}\phi^s \tag{16.101}$$

Not all the equations are independent. Equation (16.99) is seen to be equivalent to (16.93) upon using (16.92). Equation (16.100) is equivalent to (16.96), upon using (16.95) and (16.98). Equation (16.101) will be used in lieu of (16.95). One can use (16.99) and (16.100) in lieu of the Poisson equations (16.93) and (16.96). Combining (16.91) and (16.92) yields

$$\left(\frac{\partial^2}{\partial t^2} + \omega_{\text{pe}}^2 - \frac{3T_e}{m_e}\nabla^2\right) n_e^l = -\nabla \cdot \frac{\partial}{\partial t}(n_e^s \mathbf{v}_e^l) = -\nabla \cdot \left(n_e^s \frac{\partial \mathbf{v}_e^l}{\partial t}\right)$$

where we neglected the time derivative of the low frequency quantity. In the above, use (16.91) with neglect of the thermal correction term and (16.101) to write

$$\left(\frac{\partial^2}{\partial t^2}+\omega_{\rm pe}^2-\frac{3T_e}{m_e}\nabla^2\right)n_e^l=\frac{n_0e^2}{T_em_e}\nabla\cdot(\mathbf{E}^l\phi^s)$$

Performing Fourier transform, $\times\int e^{-i\mathbf{k}\cdot\mathbf{r}}d^3r$, gives

$$\begin{aligned}\left(\frac{\partial^2}{\partial t^2}+\omega_{\rm pe}^2+\frac{3T_e}{m_e}k^2\right)&n_e^l(\mathbf{k},t)\\ &=\frac{n_0e^2}{T_em_e}i\mathbf{k}\cdot\int e^{-i\mathbf{k}\cdot\mathbf{r}}\mathbf{E}^l(\mathbf{r},t)\phi^s(\mathbf{r},t)d^3r\\ &=\frac{n_0e^2}{T_em_e}i\mathbf{k}\cdot\int d^3re^{-i\mathbf{k}\cdot\mathbf{r}}\left[\int\mathbf{E}^l(\mathbf{k}',t)e^{i\mathbf{k}'\cdot\mathbf{r}}\frac{d^3k'}{(2\pi)^3}\right]\left[\int\phi^s(\mathbf{k}'',t)e^{i\mathbf{k}''\cdot\mathbf{r}}\frac{d^3k''}{(2\pi)^3}\right]\end{aligned}$$

Integral $\int d^3r$ gives $(2\pi)^3\delta(\mathbf{k}-\mathbf{k}'-\mathbf{k}'')$. Using in the above the following relation

$$n_e^l(\mathbf{k},t)=-\frac{k^2}{4\pi e}\phi^l(\mathbf{k}.t)$$

we obtain

$$\begin{aligned}\left(\frac{\partial^2}{\partial t^2}+\omega_{\rm pe}^2+\frac{3T_e}{m_e}k^2\right)&\phi^l(\mathbf{k},t)\\ &=-\frac{e\,\omega_{\rm pe}^2}{T_e}\int\frac{d^3k'}{(2\pi)^3}\int d^3k''\frac{\mathbf{k}\cdot\mathbf{k}'}{k^2}\phi^l(\mathbf{k}',t)\phi^s(\mathbf{k}'',t)\delta(\mathbf{k}-\mathbf{k}'-\mathbf{k}'')\end{aligned}\tag{16.102}$$

Note the appearance of the wave vector matching condition, $\delta(\mathbf{k}-\mathbf{k}'-\mathbf{k}'')$. Taking divergence of (16.94) and using (16.97) and (16.98) and the charge neutrality $n_i^s\simeq n_e^s$ (acceptable in the low frequency region) yield

$$\left(\frac{\partial^2}{\partial t^2}-\frac{T_e}{m_i}\nabla^2\right)n_e^s=\frac{m_e}{m_i}n_0\nabla\cdot(\mathbf{v}_e^l\cdot\nabla\mathbf{v}_e^l)\tag{16.103}$$

Fourier transforming (16.103) yields

$$\left(\frac{\partial^2}{\partial t^2}+\frac{T_e}{m_i}k^2\right)n_e^s(\mathbf{k},t)=\frac{m_e}{m_i}n_0i\mathbf{k}\cdot\int d^3re^{-i\mathbf{k}\cdot\mathbf{r}}\mathbf{v}_e^l(\mathbf{r},t)\cdot\nabla\mathbf{v}_e^l(\mathbf{r},t)$$

$$= \frac{m_e}{m_i} n_0 i\mathbf{k} \cdot \int d^3 r e^{-i\mathbf{k}\cdot\mathbf{r}} \int \mathbf{v}_e^l(\mathbf{k}', t) e^{i\mathbf{k}'\mathbf{r}} \frac{d^3k'}{(2\pi)^3} \int \frac{d^3k''}{(2\pi)^3} \cdot i\mathbf{k}'' \mathbf{v}_e^l(\mathbf{k}'', t) e^{i\mathbf{k}''\mathbf{r}}$$

$$\left(\frac{\partial^2}{\partial t^2} + \frac{T_e}{m_i} k^2\right) \phi^s(\mathbf{k}, t) = -\frac{m_e}{m_i} \frac{T_e}{e} \int \frac{d^3k'}{(2\pi)^3} \int d^3k'' \delta(\mathbf{k} - \mathbf{k}' - \mathbf{k}'') \mathbf{k}'' \cdot \mathbf{v}_e^l(\mathbf{k}', t) \mathbf{k} \cdot \mathbf{v}_e^l(\mathbf{k}'', t) \tag{16.104}$$

where we used (16.101). Equations (16.102) and (16.104) are the set of the mode coupling equations in a turbulent plasma whose wave constituents are Langmuir and ion acoustic waves. Equations (16.102) and (16.104) will be further reduced to standard forms of coupled mode equations. Let us first consider the linear solutions of (16.102) and (16.104), with neglect of the RHS. We write from (16.102)

$$\left(\frac{\partial^2}{\partial t^2} + \omega_e^2(\mathbf{k})\right) \phi^l(\mathbf{k}, t) = 0, \quad \omega_e^2(\mathbf{k}) = \omega_{\mathrm{pe}}^2 + \frac{3T_e}{m_e} k^2$$

whose solution can be written as

$$\phi^l(\mathbf{k}, t) = A(\mathbf{k}) e^{-i\omega_e^+(\mathbf{k})t} + B(\mathbf{k}) e^{-i\omega_e^-(\mathbf{k})t},$$

$$\omega_e^{\pm}(\mathbf{k}) = \pm\sqrt{\omega_{\mathrm{pe}}^2 + \frac{3T_e}{m_e} k^2} \tag{16.105}$$

Because of the nonlinear driving term on the RHS, the amplitudes $A(\mathbf{k})$ and $B(\mathbf{k})$ become slowly varying, so we write the nonlinear solution in the form

$$\phi^l(\mathbf{k}, t) = A(\mathbf{k}, t) e^{-i\omega_e^+(\mathbf{k})t} + B(\mathbf{k}, t) e^{-i\omega_e^-(\mathbf{k})t} \tag{16.106}$$

Implicitly, (16.106) is the two-time scale solution: the amplitude is slowly changing while the phase changes rapidly.

Since $\phi(\mathbf{r}, t)$ is a real quantity, we have

$$\phi(\mathbf{r}, t) = \int_{-\infty}^{\infty} d^3k e^{i\mathbf{k}\cdot\mathbf{r}} \phi(\mathbf{k}, t) = \int_{-\infty}^{\infty} d^3k e^{-i\mathbf{k}\cdot\mathbf{r}} \phi^*(\mathbf{k}, t)$$

$$= \int_{-\infty}^{\infty} d^3k e^{i\mathbf{k}\cdot\mathbf{r}} \phi^*(-\mathbf{k}, t)$$

$$\phi(\mathbf{k}, t) = \phi^*(-\mathbf{k}, t), \quad \phi(\mathbf{k}, 0) = \phi^*(-\mathbf{k}, 0)$$

Therefore

$$A(\mathbf{k}) = A^*(-\mathbf{k}), \quad B(\mathbf{k}) = B^*(-\mathbf{k})$$

Likewise, the ion sound wave solution is written as

$$\phi^s(\mathbf{k},t) = S^+(\mathbf{k},t)e^{-i\omega_s^+(\mathbf{k})t} + S^-(\mathbf{k},t)e^{-i\omega_s^-(\mathbf{k})t}, \quad \omega_s^\pm(k) = \pm\sqrt{\frac{T_e}{m_i}}\, k \tag{16.107}$$

Neglecting the thermal correction ($k^2\lambda_e^2 \ll 1$), we have

$$\frac{\partial}{\partial t}\mathbf{v}_e^l(\mathbf{k},t) = i\,\frac{e}{m_e}\mathbf{k}\phi^l(\mathbf{k},t)$$

Using (16.106) in the above equation gives

$$\mathbf{v}_e^l(\mathbf{k},t) = -\frac{e}{m_e}\mathbf{k}\left[\frac{A(\mathbf{k},t)}{\omega_e^+(\mathbf{k})}e^{-i\omega_e^+(\mathbf{k})t} + \frac{B(\mathbf{k},t)}{\omega_e^-(\mathbf{k})}e^{-i\omega_e^-(\mathbf{k})t}\right] \tag{16.108}$$

Using (16.106), LHS of (16.102) is obtained after neglecting the second derivative, $\partial^2 A/\partial t^2$,

$$\text{LHS of } (16.102) = -2i\omega_e^+(\mathbf{k})\,\frac{\partial A}{\partial t}e^{-i\omega_e^+(\mathbf{k})t} - 2i\omega_e^-(\mathbf{k})\,\frac{\partial B}{\partial t}e^{-i\omega_e^-(\mathbf{k})t}$$

We shall consider the Langmuir and ion sound waves to propagate toward + direction only:

$$\phi^s(\mathbf{k},t) = S^+(\mathbf{k},t)e^{-i\omega_s^+(\mathbf{k})t} \equiv S(\mathbf{k},t)e^{-i\omega_s(\mathbf{k})t} \tag{16.109}$$

$$\phi^l(\mathbf{k},t) = A(\mathbf{k},t)e^{-i\omega_e^+(\mathbf{k})t} \equiv A(\mathbf{k},t)e^{-i\omega_e(\mathbf{k})t} \tag{16.110}$$

This simplification makes it easier to see the conspicuous feature of the mode coupling equations. Then (16.102) takes the form

$$\begin{aligned} i\frac{\partial A(\mathbf{k},t)}{\partial t} &= \frac{e\,\omega_{\text{pe}}^2}{2T_e}\frac{1}{\omega_e(k)}\int\frac{d^3k'}{(2\pi)^3}\frac{\mathbf{k}\cdot\mathbf{k}'}{k^2} \\ &\times\int d^3k''A(\mathbf{k}',t)S(\mathbf{k}'',t)e^{i[\omega_e(k)-\omega_e(k')-\omega_s(k'')]t}\delta(\mathbf{k}-\mathbf{k}'-\mathbf{k}'') \end{aligned} \tag{16.111}$$

The important point of observation is the emergence of the exponential phasor $e^{i[\omega_e(k)-\omega_e(k')-\omega_s(k'')]t}$ (frequency matching condition) and the δ-function, $\delta(\mathbf{k}-\mathbf{k}'-\mathbf{k}'')$ which is the wave vector matching condition.

Next we work on the ion acoustic wave equation. Using (16.108)–(16.110), Eq. (16.104) takes the form in the parallel way

$$
i\frac{\partial S(\mathbf{k},t)}{\partial t} = \frac{eT_e}{2\omega_s(k)m_i m_e}\int \frac{d^3k'}{(2\pi)^3}\int d^3k''\,(\mathbf{k}'\cdot\mathbf{k}'')(\mathbf{k}\cdot\mathbf{k}'')
\times \frac{A(\mathbf{k}',t)A(\mathbf{k}'',t)}{\omega_e(k')\omega_e(k'')} e^{i[\omega_s(k)-\omega_e(k')-\omega_e(k'')]t}\delta(\mathbf{k}-\mathbf{k}'-\mathbf{k}'') \quad (16.112)
$$

Here we again see the frequency and wave vector matching conditions. Equations (16.111) and (16.112) are the desired mode coupling equations in a *turbulent plasma* populated with Langmuir waves and ion acoustic waves. The coupling is solely due to the wave vector-matching resonances as represented by the δ functions. In contrast, (15.222)–(15.224) in Chapter 15, the coupled equations due to the wave-particle interaction, have the coefficients that are related with Landau damping rate.

Remark. Let us Fourier transform in time (16.102) by performing $\times \int(\cdots)e^{i\omega t}dt$ to get

$$
(-\omega^2+\omega_e^2)\phi^l(\mathbf{k},\omega) = -\frac{e\,\omega_{\rm pe}^2}{T_e}\int\frac{d^3k'}{(2\pi)^3}\int d^3k''\frac{\mathbf{k}\cdot\mathbf{k}'}{k^2}\delta(\mathbf{k}-\mathbf{k}'-\mathbf{k}'')\int_{-\infty}^{\infty}dte^{i\omega t}\phi^l(\mathbf{k}',t)\phi^s(\mathbf{k}'',t)
$$
$$
= -\frac{e\,\omega_{\rm pe}^2}{T_e}\int\frac{d^3k'}{(2\pi)^3}\int d^3k''\frac{\mathbf{k}\cdot\mathbf{k}'}{k^2}\delta(\mathbf{k}-\mathbf{k}'-\mathbf{k}'')
\times\int_{-\infty}^{\infty}dte^{i\omega t}\int_{-\infty}^{\infty}\frac{d\omega'}{2\pi}e^{-i\omega' t}\phi^l(\mathbf{k}',\omega')\int_{-\infty}^{\infty}\frac{d\omega''}{2\pi}e^{-i\omega'' t}\phi^s(\mathbf{k}'',\omega'')
$$

Integrating $\int dt$ gives $\delta(\omega-\omega'-\omega'')$. So the above equation can be written in the form

$$
(\omega^2-\omega_e^2)\phi^l(\mathbf{k},\omega) = \frac{e\,\omega_{\rm pe}^2}{T_e}\int\frac{d^3k'}{(2\pi)^3}\frac{\mathbf{k}\cdot\mathbf{k}'}{k^2}\int\frac{d\omega'}{2\pi}\,\phi^l(\mathbf{k}',\omega')\phi^s(\mathbf{k}-\mathbf{k}',\omega-\omega') \quad (16.113)
$$

where we have $\omega^2 - \omega_e^2 \simeq 2\omega_e(\omega - \omega_e)$, because $\omega \simeq \omega_e$ per assuming that the nonlinear frequency shift is small. If we include damping of the wave ϕ^l, we put $\omega \to \omega - i\gamma_k$, and (16.113) can be put into the form

$$\omega - \omega_e(k) - i\gamma_k = \frac{e\,\omega_{\rm pe}^2}{2\omega_e(k)T_e}\int \frac{d^3k'}{(2\pi)^3}\frac{\mathbf{k}\cdot\mathbf{k}'}{k^2}\int \frac{d\omega'}{2\pi}\phi^l(\mathbf{k}',\omega')\phi^s(\mathbf{k}-\mathbf{k}',\omega-\omega') \tag{16.114}$$

Equation (16.114) is equivalent to (16.111) (see Problem 3).

References:

- R. Z. Sagdeev and A. A. Galeev, *Nonlinear plasma theory*, Benjamin, New York (1969).
- B. B. Kadomtsev, *Plasma turbulence* Academic Press, London (1965), Chap. II
- Al'tshul L. M. and Karpman V. I. Sov Phys. JETP **20** (1965), p. 1043; Galeev, A. A., Karpman, V. I. and Sagdeev R. Z. Nucl. Fusion **5** (1965), p. 20, for original derivation of Equation (10.68).
- Camac, M., Kantrowitz, A. R., Litvak, M. M., Patrick, R. M., and Petschek, H. E. *Shock waves in collision-free plasmas*, Nuc. Fusion Supplement **Pt. 2** (1962), pp. 423–444, where the authors derived Boltzmann-like kinetic equation for the wave action of a wave under consideration in the distribution of other waves with the collision term provided by the wave-wave merging and wave splitting. The wave-wave interaction term in (16.68) can be further developed to a kinetic equation for the wave action along this line. See also the comment at the end of Section 11.5.

Problems

16.1 Prove

$$W = -1 - \zeta Z(\zeta)$$

where W is the function in (16.76), $Z(\zeta)$ is the plasma dispersion function.

16.2 Obtain (16.78), the expression for the imaginary part of W:

$$W_i = -\sqrt{\frac{\pi}{2}}\left(\frac{T_e}{T_i}\right)^{\frac{1}{2}} e^{-\frac{T_e}{2T_i}}$$

Solution. W_i is the negative of the imaginary part of the integral

$$\int dv\, \frac{v\ f_{i0}(v)}{v - \frac{\omega + i\nu}{k}} : \quad \nu \to 0^+$$

[ω has a small imaginary part in accord with the causality.] Use

$$\frac{1}{v - u \mp i\nu} = P\,\frac{1}{v-u} \pm i\pi\delta(v-u)$$

$$W_i = -\pi\delta\left(v - \frac{\omega_3}{k_3}\right) v\ f_{i0}(v); \quad \frac{\omega_3}{k_3} = \sqrt{\frac{T_e}{m_i}}$$

which gives (16.78).

16.3 Explain that (16.114) is equivalent to (16.111).

Solution. Equation (16.126) involves implicitly two different time scales. To be clear, it is written explicitly by introducing the slow time ϵt:

$$i\frac{\partial A(\mathbf{k}, \epsilon t)}{\partial t} = \frac{e\,\omega_{\text{pe}}^2}{2T_e}\frac{1}{\omega_e(k)}\int \frac{d^3k'}{(2\pi)^3}\frac{\mathbf{k}\cdot\mathbf{k}'}{k^2}$$
$$\times \int d^3k'' A(\mathbf{k}', \epsilon t) S(\mathbf{k}'', \epsilon t) e^{i[\omega_e(k) - \omega_e(k') - \omega_s(k'')]t}\delta(\mathbf{k} - \mathbf{k}' - \mathbf{k}'')$$

where we put in the notations used in (16.114) the following relations

$$A(\mathbf{k}', \epsilon t)e^{-i\omega_e(k')t} = A(\mathbf{k}', t)$$
$$S(\mathbf{k}'', \epsilon t)e^{-i\omega_e(k'')t} = S(\mathbf{k}'', t)$$

Then we obtain

$$i\frac{\partial A(\mathbf{k}, \epsilon t)}{\partial t}e^{-i\omega_e(k)t} = \frac{e\,\omega_{\text{pe}}^2}{2T_e}\frac{1}{\omega_e(k)}\int \frac{d^3k'}{(2\pi)^3}\frac{\mathbf{k}\cdot\mathbf{k}'}{k^2}$$
$$\times \int d^3k''\delta(\mathbf{k} - \mathbf{k}' - \mathbf{k}'')A(\mathbf{k}', t)S(\mathbf{k}'', t)$$

Multiply both sides by $e^{i\omega t}$ and perform $\int dt$ integral. Then LHS becomes by derivative expansion $[\omega - \omega_e(k)]A(\mathbf{k}, \omega)$. RHS is a convolution integral, giving the form of (16.114).

16.4 Carry out the algebra leading to (16.56).

$$\phi = \phi^{(1)}(\mathbf{r}, t) + \phi^{(2)}(\mathbf{r}, t) + \phi^{(3)}(\mathbf{r}, t) + \cdots$$

$$\nabla^2(\phi^{(1)} + \phi^{(2)} + \phi^{(3)} + \cdots) + 4\pi\sum_\alpha e_\alpha \int d^3v(f_\alpha^{(1)} + f_\alpha^{(2)} + f_\alpha^{(3)} + \cdots) = 0 \tag{16.115}$$

We break down (16.52) and (16.53) order by order:

$$\frac{\partial}{\partial t} f_\alpha^{(1)} + \mathbf{v} \cdot \frac{\partial f_\alpha^{(1)}}{\partial \mathbf{r}} - \frac{e_\alpha}{m_\alpha} \nabla \phi^{(1)}(\mathbf{r}, t) \cdot \frac{\partial f_{\alpha 0}}{\partial \mathbf{v}} = 0 \tag{16.116}$$

$$\nabla^2 \phi^{(1)} + 4\pi \Sigma_\alpha \, e_\alpha \int d^3v \; f_\alpha^{(1)} = 0 \tag{16.117}$$

$$\frac{\partial}{\partial t} f_\alpha^{(2)} + \mathbf{v} \cdot \frac{\partial f_\alpha^{(2)}}{\partial \mathbf{r}} - \frac{e_\alpha}{m_\alpha} \left[\nabla \phi^{(2)} \cdot \frac{\partial f_{\alpha 0}}{\partial \mathbf{v}} + \nabla \phi^{(1)} \cdot \frac{\partial f_\alpha^{(1)}}{\partial \mathbf{v}} \right] = 0 \tag{16.118}$$

$$\nabla^2 \phi^{(2)} + 4\pi \Sigma_\alpha \, e_\alpha \int d^3v \; f_\alpha^{(2)} = 0 \tag{16.119}$$

$$\frac{\partial}{\partial t} f_\alpha^{(3)} + \mathbf{v} \cdot \frac{\partial f_\alpha^{(3)}}{\partial \mathbf{r}} - \frac{e_\alpha}{m_\alpha} \left[\nabla \phi^{(3)} \cdot \frac{\partial f_{\alpha 0}}{\partial \mathbf{v}} + \nabla \phi^{(1)} \cdot \frac{\partial f_\alpha^{(2)}}{\partial \mathbf{v}} + \nabla \phi^{(2)} \cdot \frac{\partial f_\alpha^{(1)}}{\partial \mathbf{v}} \right] = 0 \tag{16.120}$$

$$\nabla^2 \phi^{(3)} + 4\pi \Sigma_\alpha \, e_\alpha \int d^3v \; f_\alpha^{(3)} = 0 \tag{16.121}$$

$$(16.116): \quad f_\alpha^{(1)}(\mathbf{k}, \mathbf{v}, \omega) = -\frac{e_\alpha}{m_\alpha} \frac{\mathbf{k} \cdot \frac{\partial f_{\alpha 0}}{\partial \mathbf{v}}}{\omega - \mathbf{k} \cdot \mathbf{v}} \, \phi^{(1)}(\mathbf{k}, \omega) \tag{16.122}$$

$$(16.117): \quad \varepsilon^{(1)}(\mathbf{k}, \omega) \; \phi^{(1)}(\mathbf{k}, \omega) = 0 \tag{16.123}$$

$$\varepsilon^{(1)}(\mathbf{k}, \omega) = 1 + \Sigma_\alpha \, \frac{\omega_{p\alpha}^2}{k^2} \int d^3v \; \frac{\mathbf{k} \cdot \frac{\partial f_{\alpha 0}}{\partial \mathbf{v}}}{\omega - \mathbf{k} \cdot \mathbf{v}} \tag{16.124}$$

$$(16.118): \; f_\alpha^{(2)}(\mathbf{k}, \mathbf{v}, \omega) = -\frac{e_\alpha}{m_\alpha} \frac{\mathbf{k} \cdot \frac{\partial f_{\alpha 0}}{\partial \mathbf{v}}}{\omega - \mathbf{k} \cdot \mathbf{v}} \, \phi^{(2)}(\mathbf{k}, \omega)$$

$$- \frac{e_\alpha}{m_\alpha} \sum_{k', \omega'} \frac{\mathbf{k}' \cdot \frac{\partial f_\alpha^{(1)}(\mathbf{k}'', \omega'', \mathbf{v})}{\partial \mathbf{v}}}{\omega - \mathbf{k} \cdot \mathbf{v}} \, \phi^{(1)}(\mathbf{k}', \omega') \tag{16.125}$$

where $\mathbf{k}'' = \mathbf{k} - \mathbf{k}', \quad \omega'' = \omega - \omega'$

$$(16.119): \; \varepsilon^{(1)}(\mathbf{k}, \omega) \phi^{(2)}(\mathbf{k}, \omega)$$

$$= -\sum_{k', \omega'} \int d^3v \frac{\mathbf{k}' \cdot \frac{\partial f_\alpha^{(1)}(\mathbf{k}'', \omega'', \mathbf{v})}{\partial \mathbf{v}}}{\omega - \mathbf{k} \cdot \mathbf{v}} \, \phi^{(1)}(\mathbf{k}', \omega') \tag{16.126}$$

Use (16.122) in (16.126) to obtain

$$\varepsilon^{(1)}(\mathbf{k},\omega)\phi^{(2)}(\mathbf{k},\omega)$$
$$= -\sum_{k',\omega'} \varepsilon^{(2)}(\mathbf{k}',\omega',\mathbf{k}'',\omega'')\phi^{(1)}(\mathbf{k}',\omega')\phi^{(1)}(\mathbf{k}'',\omega'') \tag{16.127}$$

$$\varepsilon^{(2)}(\mathbf{k}',\omega',\mathbf{k}'',\omega'')$$
$$= -\sum_{\alpha} \frac{e_\alpha}{m_\alpha}\frac{\omega_{p\alpha}^2}{k^2}\int \frac{d^3v}{\omega - \mathbf{k}\cdot\mathbf{v}}\,\mathbf{k}'\cdot\frac{\partial}{\partial \mathbf{v}}\frac{\mathbf{k}''\cdot\frac{\partial f_{\alpha 0}}{\partial \mathbf{v}}}{\omega'' - \mathbf{k}''\cdot\mathbf{v}} \tag{16.128}$$

The expression for $\varepsilon^{(2)}$ can be symmetrized by adding the term obtained by interchanging ($'$) and ($''$) and dividing by 2 (the term obtained by interchanging is the same as the original term):

$$\varepsilon^{(2)}(\mathbf{k}',\omega',\mathbf{k}'',\omega'') = -\frac{1}{2}\sum_{\alpha} \frac{e_\alpha}{m_\alpha}\frac{\omega_{p\alpha}^2}{k^2}\int \frac{d^3v}{\omega - \mathbf{k}\cdot\mathbf{v}}$$
$$\times\left[\mathbf{k}'\cdot\frac{\partial}{\partial \mathbf{v}}\frac{\mathbf{k}''\cdot\frac{\partial f_{\alpha 0}}{\partial \mathbf{v}}}{\omega'' - \mathbf{k}''\cdot\mathbf{v}} + \mathbf{k}''\cdot\frac{\partial}{\partial \mathbf{v}}\frac{\mathbf{k}'\cdot\frac{\partial f_{\alpha 0}}{\partial \mathbf{v}}}{\omega' - \mathbf{k}'\cdot\mathbf{v}}\right] \tag{16.129}$$

Now (16.127) can be written in the form (16.62). The perturbation series in (16.115) can be written, upon using (16.123) and (16.62), in the form of (16.56).

References

Al'tshul, L. M. and V. I. Karpman, Sov Phys. JETP **20** (1965), p. 1043.

Bellan, P. M. *Fundamentals of Plasma Physics*, Cambridge University Press, Cambridge (2006).

Camac, M., A. R. Kantrowitz, M. M. Litvak, R. M. Patrick, and H. E. Petschek, *Shock waves in collision-free plasmas*, Nuc. Fusion Supplement **Pt. 2** (1962).

Galeev, A. A., V. I. Karpman, and R. Z. Sagdeev, Nucl. Fusion **5** (1965), p. 20.

Kadomtsev, B. B. *Plasma Turbulence*, Academic Press London (1965) Chap. II.

Nicholson, D. R. *Introduction to Plasma Theory*, John Wiley, New York (1983).

Sagdeev, R. Z. and A. A. Galeev, *Nonlinear Plasma Theory*, Benjamin, New York (1969).

Chapter 17

Surface Waves in Magnetized Plasmas

This chapter is the continuation of Chapter 9. Dispersion relations of the electrostatic waves in *bounded magnetized plasmas* are derived kinetically. Dispersion relation of the Alfven surface wave in a semi-bounded plasma is derived kinetically by using the low frequency dielectric tensor, Eq. (9.240). Recall that (9.240) is obtained by solving the Vlasov equation with the magnetic force by expanding the distribution function in the harmonic series of $e^{in\phi}$ (ϕ is the azimuthal angle in the cylindrical velocity space).

17.1 Magnetohydrodynamic Surface Wave

17.1.1 *Fluid theory*

We write the linearized ideal MHD equations that we introduced in Section 3.10, Eqs. (3.81)–(3.86):

$$\frac{\partial \rho}{\partial t} + \rho_0 \nabla \cdot \mathbf{v} = 0 \tag{17.1}$$

$$\rho_0 \frac{\partial \mathbf{v}}{\partial t} = -\nabla p + \frac{1}{c}\mathbf{J} \times \mathbf{B}_0 \tag{17.2}$$

$$\mathbf{E} + \frac{1}{c}\mathbf{v} \times \mathbf{B}_0 = 0 \tag{17.3}$$

$$\nabla \times \mathbf{E} = -\frac{1}{c}\frac{\partial \mathbf{B}}{\partial t} \tag{17.4}$$

$$\nabla \times \mathbf{B} = \frac{4\pi}{c}\mathbf{J} \tag{17.5}$$

$$p = v_s^2\, \rho \tag{17.6}$$

where $v_s = \sqrt{\frac{\gamma p_0}{\rho_0}}$ is the speed of sound. Equation (17.6) is obtained by linearizing the adiabatic relation $p\rho^{-\gamma} = \text{const}$.

By eliminating ρ and $\mathbf{B}$, these equations can be arranged in the form

$$\frac{\partial^2 \mathbf{v}}{\partial t^2} - v_s^2 \nabla\nabla \cdot \mathbf{v} + \mathbf{v}_A \times \nabla \times [\nabla \times (\mathbf{v} \times \mathbf{v}_A)] = 0 \tag{17.7}$$

where we introduced the Alfven velocity

$$\mathbf{v}_A = \frac{\mathbf{B}_0}{\sqrt{4\pi\rho_0}} \tag{17.8}$$

Let us take $\mathbf{B}_0 = \hat{\mathbf{z}}B_0$. Without loss of generality, we can assume that the y-direction is ignorable ($\partial/\partial y = 0$). Then, in components, (17.7) gives

$$\frac{\partial^2 v_x}{\partial t^2} - v_s^2 \frac{\partial}{\partial x}\left(\frac{\partial v_x}{\partial x} + \frac{\partial v_z}{\partial z}\right) - v_A^2\left(\frac{\partial^2 v_x}{\partial x^2} + \frac{\partial^2 v_x}{\partial z^2}\right) = 0 \tag{17.9}$$

$$\frac{\partial^2 v_y}{\partial t^2} = v_A^2 \frac{\partial^2 v_y}{\partial z^2} \tag{17.10}$$

$$\frac{\partial^2 v_z}{\partial t^2} - v_s^2 \frac{\partial}{\partial z}\left(\frac{\partial v_x}{\partial x} + \frac{\partial v_z}{\partial z}\right) = 0 \tag{17.11}$$

We wish to solve (17.9)–(17.11) by Fourier transform. Since the x-region is bounded by $0 \le x < \infty$, the above equations must be extended so that they are valid in the entire region $-\infty < x < \infty$. This can be done by extending the velocity components in the fashion

$$v_x(-x) = -v_x(x), \quad v_y(-x) = v_y(x), \quad v_z(-x) = v_z(x) \tag{17.12}$$

We can ascertain by inspection that (17.9)–(17.11) are invariant under the reflection $x \to -x$, provided the velocity components are extended in the fashion of (17.12). Now we can Fourier transform (17.9)–(17.11) in the usual way, but the jump of v_x at $x = 0$ given rise to by the oddly continuation should be included in the transformed equation.

Eliminating v_z between (17.9) and (17.11) and assuming $\mathbf{v}(x, z, t) = \mathbf{v}(x)\, e^{ik_z z - i\omega t}$ yield

$$\frac{\partial^2 v_x}{\partial x^2} - q^2 v_x = 0 \tag{17.13}$$

where

$$q = \sqrt{\frac{(\omega^2 - v_s^2 k_z^2)(\omega^2 - k_z^2 v_A^2)}{k_z^2 v_s^2 v_A^2 - \omega^2(v_A^2 + v_s^2)}} \tag{17.14}$$

Let us Fourier transform (17.13) by carrying out

$$\int_{-\infty}^{\infty} e^{-ik_x x} \frac{\partial^2 v_x}{\partial x^2} dx = -k_x^2 v_x(k_x) + k_x M(k_z, \omega) \tag{17.15}$$

where $M(k_z,\omega) = iv_x(x=0^-,k_z,\omega) - iv_x(x=0^+,k_z,\omega)$ is derived from the jump of $v_x(x)$ at $x=0$. The constant M will be determined by satisfying the electromagnetic boundary condition at $x=0$. The detailed form of M is unimportant. Equation (17.13) gives

$$v_x(k_x,k_z,\omega) = \frac{k_x}{k_x^2+q^2}\,M \tag{17.16}$$

Fourier inverting this equation yields the boundary value

$$v_x(0^+,k_z,\omega) = M\lim_{x\to 0}\int_{-\infty}^{\infty} e^{ik_x x}\,\frac{k_x}{k_x^2+q^2}dk_x \tag{17.17}$$

The integral in (17.17) can be done by using the contour integral and the residue theorem. To ensure $v_x \to 0$ as $x\to +\infty$, the contour must encircle the upper-half k_x-plane. We obtain

$$v_x(0^+,k_z,\omega) = i\,\pi M \tag{17.18}$$

Next, (17.11) gives

$$v_z(x,k_z,\omega) = \frac{ik_z v_s^2}{v_s^2k_z^2-\omega^2}\,\frac{\partial v_x(x,k_z,\omega)}{\partial x} \tag{17.19}$$

Fourier transforming this equation yields

$$v_z(k_x,k_z,\omega) = \frac{k_z v_s^2 M}{v_s^2k_z^2-\omega^2}\,\frac{q^2}{k_x^2+q^2} \tag{17.20}$$

Inverting this equation, we obtain the boundary value

$$v_z(0^+,k_z,\omega) = \frac{\pi k_z v_s^2 q M}{v_s^2k_z^2-\omega^2} \tag{17.21}$$

In order to have the connection between the velocity components and the electric and magnetic field components, we write the basic equations (17.1)–(17.6) in components:

$$-i\omega\rho_0 v_x + \frac{\partial p}{\partial x} = \frac{B_0}{c}J_y \tag{17.22}$$

$$ic\omega\rho_0 v_y = B_0 J_x \tag{17.23}$$

$$\rho_0 v_z = \frac{k_z}{\omega}p \tag{17.24}$$

$$E_x = -\frac{1}{c}B_0\ v_y \tag{17.25}$$

$$E_y = \frac{1}{c}B_0\ v_x \tag{17.26}$$

$$E_z = 0 \tag{17.27}$$

$$-k_z E_y = \frac{\omega}{c} B_x \tag{17.28}$$

$$ik_z E_x - \frac{\partial E_z}{\partial x} = \frac{i\omega}{c} B_y \tag{17.29}$$

$$\frac{\partial E_y}{\partial x} = \frac{i\omega}{c} B_z \tag{17.30}$$

$$-ik_z B_y = \frac{4\pi}{c} J_x \tag{17.31}$$

$$ik_z B_x - \frac{\partial B_z}{\partial x} = \frac{4\pi}{c} J_y \tag{17.32}$$

$$\frac{\partial B_y}{\partial x} = \frac{4\pi}{c} J_z \tag{17.33}$$

Using (17.32) in (17.22) gives

$$-i\omega\rho_0 v_x + \frac{\partial p}{\partial x} = \frac{B_0}{4\pi}\left(ik_z B_x - \frac{\partial B_z}{\partial x}\right) \tag{17.34}$$

which yields upon infinitesimal integration

$$[p] = -\frac{B_0}{4\pi}[B_z] \tag{17.35}$$

where the bracket $[\cdots]$ denotes the jump at $x = 0$.

Equation (17.30) gives $[E_y] = 0$, hence

$$\frac{B_0}{c} v_x(0^+) = E_y(0^-) \tag{17.36}$$

where we used (17.26). Equations (17.35) and (17.36) are two boundary equations that we may use.

Equation (17.35) needs to be further elaborated. Using (17.24) for p and (17.30) and (17.26) for B_z, (17.35) is written as

$$-i\,\frac{B_0}{\omega}\,\frac{\partial v_x}{\partial x}\bigg|_{0+} + \frac{4\pi\rho_0\omega}{k_z B_0}\,v_z(0^+) = B_z(0^-) \tag{17.37}$$

Alternatively, using (17.6) and (17.1) in (17.35) gives

$$-i\,\frac{B_0}{\omega}\,\frac{\partial v_x}{\partial x}\bigg|_{0+} - \frac{4\pi i\rho_0 v_s^2}{\omega B_0}\,(\nabla\cdot\mathbf{v})|_{0+} = B_z(0^-) \tag{17.38}$$

Equivalence of (17.37) and (17.38) can be easily proven, per (17.11).

• **Vacuum solution.** In the vacuum ($x > 0$), (17.4) and (17.5) are valid with $\mathbf{J} = 0$. Also we have $\nabla \cdot \mathbf{B} = 0$. Thus we have

$$\frac{\partial B_z}{\partial x} - ik_z B_x = 0 \tag{17.39}$$

$$\frac{\partial B_x}{\partial x} + ik_z B_z = 0 \tag{17.40}$$

$$\frac{\partial E_y}{\partial x} = \frac{i\omega}{c} B_z \tag{17.41}$$

The solution that is finite at $x \to -\infty$ is

$$B_z(x) = A\ e^{k_z x} \tag{17.42}$$

$$E_y(x) = \frac{i\omega A}{ck_z}\ e^{k_z x} \tag{17.43}$$

where A is constant and the phasor $e^{ik_z z - i\omega t}$ is omitted.

• **Dispersion relation.** The boundary conditions in (17.36) and (17.37) now are written

$$A = -i\ \frac{B_0}{\omega}\ \frac{\partial v_x}{\partial x}\bigg|_{0+} + \frac{4\pi\rho_0\omega}{k_z B_0}\ v_z(0^+) \tag{17.44}$$

$$\frac{i\omega A}{k_z} = B_0\ v_x(0^+) \tag{17.45}$$

Eliminating A gives

$$v_x(0^+) - \frac{1}{k_z}\ \frac{\partial v_x}{\partial x}\bigg|_{0+} = \frac{i\ \omega^2}{k_z^2 v_A^2}\ v_z(0^+) \tag{17.46}$$

We know the quantities in (17.46) except the middle term. Inverting (17.16) gives

$$v_x(x, k_z, \omega) = M \int_{-\infty}^{\infty} e^{ik_x x}\ \frac{k_x}{k_x^2 + q^2} dk_x$$

Hence

$$\frac{\partial v_x}{\partial x}\bigg|_{0+} = i\ M \int_{-\infty}^{\infty} \frac{k_x^2}{k_x^2 + q^2} dk_x = -i\ \pi M q \tag{17.47}$$

Using (17.18), (17.21), and (17.47) in (17.46) yields the Alfven surface wave dispersion relation in the form

$$1 + \frac{q}{k_z}\left(1 - \frac{v_s^2}{v_A^2}\ \frac{\omega^2}{k_z^2 v_s^2 - \omega^2}\right) = 0 \tag{17.48}$$

where q is given in (17.14).

Equation (17.48), after rearrangement, takes the form

$$\frac{q}{k_z} = \frac{v_A^2(k_z^2 v_s^2 - \omega^2)}{\omega^2(v_s^2 + v_A^2) - k_z^2 v_s^2 v_A^2} \tag{17.49}$$

Squaring (17.49) and referring to (17.14) yield

$$k_z^2 v_s^2 v_A^2 - \omega^2(v_s^2 + v_A^2) = k_z^2 v_A^2 \ \frac{\omega^2 - k_z^2 v_s^2}{\omega^2 - k_z^2 v_A^2} \tag{17.50}$$

which is solved for the phase velocity in the form

$$\frac{\omega}{k_z} = \pm\sqrt{2} \ \frac{v_A v_s}{\sqrt{v_A^2 + v_s^2}} \tag{17.51}$$

This is the surface wave dispersion relation of the magnetosonic wave in MHD plasma.

• **Alfven surface wave in incompressible plasma.** Taking divergence of (17.2) and using (17.5) give

$$\rho_0 \ \frac{\partial}{\partial t}\nabla \cdot \mathbf{v} = -\nabla^2 \ \left(p + \frac{1}{4\pi}\mathbf{B}_0 \cdot \mathbf{B}\right) \tag{17.52}$$

Therefore we have

$$\nabla^2 \ \left(p + \frac{1}{4\pi}\mathbf{B}_0 \cdot \mathbf{B}\right) = 0 \tag{17.53}$$

in incompressible plasma ($\nabla \cdot \mathbf{v} = 0$). Hence, P and B_z should solve the Laplace equation, varying in the form $\sim e^{ik_z z - k_z x}$. Since the variables are linearly related with each other, all the quantities should vary likewise in the form $\sim e^{ik_z z - k_z x}$. Thus we write

$$v_z = C \ e^{ik_z z - k_z x} \tag{17.54}$$

$$v_x = i \ C \ e^{ik_z z - k_z x} \tag{17.55}$$

$$E_y = i \ \frac{B_0}{c} C \ e^{ik_z z - k_z x} \tag{17.56}$$

$$B_x = -i \ \frac{k_z}{\omega} B_0 C \ e^{ik_z z - k_z x} \tag{17.57}$$

$$B_z = -\frac{k_z}{\omega} B_0 C \ e^{ik_z z - k_z x} \tag{17.58}$$

$$p = -i \ \frac{\omega}{k_z} \rho_0 C \ e^{ik_z z - k_z x} \tag{17.59}$$

where C is constant. The boundary condition in (17.36) gives

$$\frac{B_0}{c}\,C = \frac{\omega A}{ck_z} \tag{17.60}$$

The boundary condition in (17.37) gives

$$-\frac{k_z}{\omega}B_0\,C + \frac{4\pi\rho_0\omega}{k_z B_0}C = A \tag{17.61}$$

These two equations yield the dispersion relation of the Alfven surface wave in incompressible plasma

$$\frac{\omega}{k_z} = \sqrt{2}\,v_A \tag{17.62}$$

17.1.2 *Kinetic theory*

In this subsection, we use the simple dielectric tensor of MHD plasma derived via the harmonic expansion of the perturbed distribution function in Chapter 9—Equation (9.240).

We assume that the electric and magnetic field components propagate with phasor $e^{ik_z z - i\omega t}$ with the amplitude varying in the x-coordinate. In components, (17.4) and (17.5) are written

$$-k_z E_y = \frac{\omega}{c}B_x \tag{17.63}$$

$$ik_z E_x - \frac{\partial E_z}{\partial x} = \frac{i\omega}{c}B_y \tag{17.64}$$

$$\frac{\partial E_y}{\partial x} = \frac{i\omega}{c}B_z \tag{17.65}$$

$$-ik_z B_y = \frac{4\pi}{c}J_x \tag{17.66}$$

$$ik_z B_x - \frac{\partial B_z}{\partial x} = \frac{4\pi}{c}J_y \tag{17.67}$$

$$\frac{\partial B_y}{\partial x} = \frac{4\pi}{c}J_z \tag{17.68}$$

Using (9.240), the currents are expressed as

$$J_x = \frac{\omega}{4\pi i}\varepsilon_{xx}E_x \tag{17.69}$$

$$J_y = \frac{\omega}{4\pi i}(\varepsilon_{yy}E_y + \varepsilon_{yz}E_z) \tag{17.70}$$

$$J_z = \frac{\omega}{4\pi i}(-\varepsilon_{yz}E_y + \varepsilon_{zz}E_z) \tag{17.71}$$

The above equations will be Fourier transformed with respect to the x-coordinate in the entire region $-\infty < x < \infty$ with the extension

$$E_x(-x) = -E_x(x), \quad E_y(-x) = E_y(x), \quad E_z(-x) = E_z(x)$$
$$B_x(-x) = B_x(x), \quad B_y(-x) = -B_y(x), \quad B_z(-x) = -B_z(x) \tag{17.72}$$

For the oddly continued function, the jump arises:

$$\int_{-\infty}^{\infty} \frac{\partial B_y(x)}{\partial x} \, e^{-ik_x x} \, dx = C_1 + ik_x B_y(k_x) \tag{17.73}$$

Then we obtain

$$k_z E_y = -\frac{\omega}{c} B_x \tag{17.74}$$

$$k_z E_x - k_x E_z = \frac{\omega}{c} B_y \tag{17.75}$$

$$k_x E_y = \frac{\omega}{c} B_z \tag{17.76}$$

$$k_z B_y = \frac{\omega}{c} \varepsilon_{xx} E_x \tag{17.77}$$

$$-k_z B_x - iC_1 + k_x B_z = \frac{\omega}{c}(\varepsilon_{yy} E_y + \varepsilon_{yz} E_z) \tag{17.78}$$

$$iC_2 - k_x B_y = \frac{\omega}{c}(-\varepsilon_{yz} E_y + \varepsilon_{zz} E_z) \tag{17.79}$$

$$k_x B_x + k_z B_z = 0 \quad (\nabla \cdot \mathbf{B} = 0) \tag{17.80}$$

where C_1 and C_2 are the constants derived from the jump of $B_y(x)$ and $B_z(x)$, respectively.

Eliminating the magnetic field components, the above equations yield

$$E_x \left(k_z^2 - \frac{\omega^2}{c^2} \, \varepsilon_{xx} \right) - k_x k_z E_z = 0 \tag{17.81}$$

$$E_y \left(k^2 - \frac{\omega^2}{c^2} \varepsilon_{yy} \right) - \frac{\omega^2}{c^2} \varepsilon_{yz} E_z = C_1 \tag{17.82}$$

$$k_x k_z E_x - \frac{\omega^2}{c^2} \varepsilon_{yz} E_y - \left(k_x^2 - \frac{\omega^2}{c^2} \varepsilon_{zz} \right) E_z = C_2 \tag{17.83}$$

where $k^2 = k_x^2 + k_z^2$, C_1 and C_2 are redefined constants (old $C_{1,2}$ multiplied by $i\omega/c$). Solving the above simultaneous equations yield

$$E_x = \frac{k_x k_z}{\Delta} \left[C_1 \, \frac{\omega^2}{c^2} \varepsilon_{yz} + C_2 \left(k^2 - \frac{\omega^2}{c^2} \varepsilon_{yy} \right) \right] \tag{17.84}$$

$$E_y = \frac{1}{\Delta}\frac{\omega^2}{c^2}\left[C_1\left(k_x^2\varepsilon_{xx} - \left(\varepsilon_{xx}\frac{\omega^2}{c^2} - k_z^2\right)\varepsilon_{zz}\right) + C_2\varepsilon_{yz}\left(k_z^2 - \frac{\omega^2}{c^2}\varepsilon_{xx}\right)\right] \tag{17.85}$$

$$E_z = \frac{1}{\Delta}\left(k_z^2 - \frac{\omega^2}{c^2}\varepsilon_{xx}\right)\left[C_1\frac{\omega^2}{c^2}\varepsilon_{yz} + C_2\left(k^2 - \frac{\omega^2}{c^2}\varepsilon_{yy}\right)\right] \tag{17.86}$$

where Δ is the determinant consisting of the coefficients as calculated to be

$$\Delta = \left(k_z^2 - \frac{\omega^2}{c^2}\varepsilon_{xx}\right)\left[\left(k^2 - \frac{\omega^2}{c^2}\varepsilon_{yy}\right)\left(k_x^2 - \frac{\omega^2}{c^2}\varepsilon_{zz}\right) + \frac{\omega^4}{c^4}\varepsilon_{yz}^2\right] - k_x^2k_z^2\left(k^2 - \frac{\omega^2}{c^2}\varepsilon_{yy}\right) \tag{17.87}$$

Note $\varepsilon_{yy} = \varepsilon_{xx}$. We neglect $k_x^4/k_z^4 \to 0$ by assuming that the wave propagates predominantly along the direction of the static magnetic field; keep only k_x^2 term. Furthermore we shall assume $\varepsilon_{zz} \gg \varepsilon_{xx}$, which is valid in the low frequency $\omega \ll \omega_{ci}$ and for $k_z\lambda_e \ll 1$. Then (17.87) takes the form

$$\Delta = \left(k_z^2 - \frac{\omega^2}{c^2}\varepsilon_{xx}\right)\left[\left(\varepsilon_{xx} - \frac{c^2k^2}{\omega^2}\right)\varepsilon_{zz} + \varepsilon_{yz}^2\right] \tag{17.88}$$

In writing the expression for Δ, we suppressed constant factor, which amounts to redefining the constants $C_{1,2}$. Using (9.233), (9.238), and (9.239), and the relations

$$\frac{\omega_{pi}^2}{\omega_{ci}^2} = \frac{c^2}{v_A^2}, \quad \lambda_e^2\omega_{pi}^2 = v_s^2$$

Δ in (17.88) becomes

$$\Delta = \frac{1}{\Omega}(k_z^2v_A^2 - \omega^2)[(\omega^2 - k^2v_A^2)(\omega^2 - v_s^2k_z^2) - k_x^2\omega^2v_s^2] \tag{17.89}$$

where $\Omega = \lambda_e^2k_z^2\omega^4v_A^4/c^2$. We can let the factor $(k_z^2v_A^2 - \omega^2)/\Omega$ be absorbed into $C_{1,2}$, assuming that $\omega^2 \neq k_z^2v_A^2$. In the approximation made in the transition from (17.87) to (17.88), the surface wave associated with the pure Alfven wave appears eliminated. We conjecture that keeping k_x^4-term may be necessary to investigate the pure Alfvenic surface wave.

- **Zeros of Δ.** The zeros of

$$\Delta = (\omega^2 - k^2v_A^2)(\omega^2 - v_s^2k_z^2) - k_x^2\omega^2v_s^2 \tag{17.90}$$

are the poles of the integrand in the inversion integral $\int E_x(k_x)\exp(ik_xx)dk_x$. After a straightforward algebra, we can write

$$\Delta = k_x^2 + q^2 \tag{17.91}$$

with

$$q = \sqrt{\frac{(\omega^2 - v_s^2 k_z^2)(\omega^2 - k_z^2 v_A^2)}{v_A^2 v_s^2 k_z^2 - \omega^2 (v_s^2 + v_A^2)}} \tag{17.92}$$

where a 'constant' factor has been suppressed, and note (17.92) is identical with (17.14).

Inverting (17.84)–(17.86) gives

$$E_x(x) = i\, e^{-qx} \left[C_1 \, \frac{\omega^2}{v_A^2} \, \frac{\omega_{ci}}{\omega} q + C_2 k_z (\alpha^2 - q^2) \right] \tag{17.93}$$

$$E_y(x) = \alpha^2 \, \frac{\omega^2}{c^2} \, e^{-qx} \left[C_1 \, \frac{\varepsilon_{zz}}{q} + C_2 \, \frac{1}{k_z} \, \frac{\omega_{ci}}{\omega} \, \frac{c^2}{v_A^2} \right] \tag{17.94}$$

$$E_z(x) = e^{-qx} \alpha^2 \left[C_1 \, \frac{1}{k_z} \, \frac{\omega_{ci}}{\omega} \, \frac{\omega^2}{v_A^2} + C_2 \, \frac{1}{q} (\alpha^2 - q^2) \right] \tag{17.95}$$

$$B_x(x) = -k_z \alpha^2 \, \frac{\omega}{c} \, e^{-qx} \left[C_1 \, \frac{\varepsilon_{zz}}{q} + C_2 \, \frac{1}{k_z} \, \frac{\omega_{ci}}{\omega} \, \frac{c^2}{v_A^2} \right] \tag{17.96}$$

$$B_y(x) = \frac{ic}{\omega} \, \frac{\omega^2}{v_A^2} \, e^{-qx} \left[C_1 \frac{\omega^2}{v_A^2} \, \frac{\omega_{ci}}{\omega} \, \frac{q}{k_z} + C_2 (\alpha^2 - q^2) \right] \tag{17.97}$$

$$B_z(x) = iq\alpha^2 \, \frac{\omega}{c} \, e^{-qx} \left[C_1 \, \frac{\varepsilon_{zz}}{q} + C_2 \, \frac{1}{k_z} \, \frac{\omega_{ci}}{\omega} \, \frac{c^2}{v_A^2} \right] \tag{17.98}$$

where

$$\alpha = \sqrt{k_z^2 - \frac{\omega^2}{v_A^2}} \tag{17.99}$$

- **Vacuum solution.** In the vacuum region ($x < 0$), $\nabla \times \mathbf{B} = 0$ and $\nabla \cdot \mathbf{B} = 0$, $\nabla \times \mathbf{E} = \frac{i\omega}{c}\mathbf{B}$, $\nabla \cdot \mathbf{E} = 0$.

Hence, we write the solution

$$B_x(x) = -iA_1 \, e^{k_z x} \tag{17.100}$$

$$B_z(x) = A_1 \, e^{k_z x} \tag{17.101}$$

$$E_y(x) = \frac{i\omega}{ck_z} A_1 \, e^{k_z x} \tag{17.102}$$

$$E_z(x) = A_2 \, e^{k_z x} \tag{17.103}$$

$$E_x(x) = -iA_2 \, e^{k_z x} \tag{17.104}$$

$$B_y(x) = 0 \tag{17.105}$$

where A_1, A_2 are constants, and we suppressed the phasor $e^{ik_z z - i\omega t}$.

• **Boundary condition.** The infinitesimal integration of (17.64) and (17.65) gives

$$[E_z] = 0, \quad [E_y] = 0 \tag{17.106}$$

Equation (17.67) is written in full as

$$ik_z B_x(x) - \frac{\partial B_z(x)}{\partial x} = \frac{4\pi}{c}\int_{-\infty}^{\infty} J_y(k_x)e^{ik_x x}dk_x \tag{17.107}$$

In the inversion integral, k_x changes to the differential operator $k_x \to -i\frac{\partial}{\partial x}$. Hence

$$ik_z B_x(x) - \frac{\partial B_z(x)}{\partial x} = \frac{-i\omega}{c}\left(\varepsilon_{yy}(k_z,\omega)E_y(x,k_z,\omega) - \frac{\omega_{pi}^2}{k_z\omega\omega_{ci}}\frac{\partial E_z(x,k_z,\omega)}{\partial x}\right) \tag{17.108}$$

where we used (9.238). The infinitesimal integration of (17.108) gives the jump relation

$$[B_z] = -\frac{i}{ck_z}\left[\frac{\omega_{pi}^2}{\omega_{ci}}E_z\right] \tag{17.109}$$

where we integrated by parts to pick up the perfect differential, assuming that ω_{pi} has a step function profile. Analogously, (17.68) yields the jump relation

$$[B_y] = -\frac{i}{ck_z}\left[\frac{\omega_{pi}^2}{\omega_{ci}}E_y\right] \tag{17.110}$$

Equation (17.109) gives

$$\begin{aligned} &iq\alpha^2\frac{\omega}{c}\left[C_1\frac{\varepsilon_{zz}}{q} + C_2\frac{1}{k_z}\frac{\omega_{ci}}{\omega}\frac{c^2}{v_A^2}\right] - A_1 \\ &= -\frac{i}{ck_z}\frac{\omega_{pi}^2}{\omega_{ci}}\alpha^2\left[C_1\frac{1}{k_z}\frac{\omega_{ci}}{\omega}\frac{\omega^2}{v_A^2} + C_2\frac{1}{q}(\alpha^2 - q^2)\right] \end{aligned} \tag{17.111}$$

Equation (17.110) gives

$$\begin{aligned} &\frac{ic}{\omega}\frac{\omega^2}{v_A^2}\left[C_1\frac{\omega^2}{v_A^2}\frac{\omega_{ci}}{\omega}\frac{q}{k_z} + C_2(\alpha^2 - q^2)\right] \\ &= -\frac{i\alpha^2}{ck_z}\frac{\omega_{pi}^2}{\omega_{ci}}\frac{\omega^2}{c^2}\left[C_1\frac{\varepsilon_{zz}}{q} + C_2\frac{1}{k_z}\frac{\omega_{ci}}{\omega}\frac{c^2}{v_A^2}\right] \end{aligned} \tag{17.112}$$

$[E_y] = 0$ gives

$$\alpha^2 \frac{\omega}{c}\left[C_1 \frac{\varepsilon_{zz}}{q} + C_2 \frac{1}{k_z} \frac{\omega_{ci}}{\omega} \frac{c^2}{v_A^2}\right] = \frac{i}{k_z} A_1 \tag{17.113}$$

Eliminating A_1 between (17.111)) and (17.113) gives

$$\begin{aligned} i\alpha^2 \frac{\omega}{c}(q+k_z)&\left[C_1 \frac{\varepsilon_{zz}}{q} + C_2 \frac{1}{k_z} \frac{\omega_{ci}}{\omega} \frac{c^2}{v_A^2}\right] \\ &= -\frac{i}{ck_z} \frac{\omega_{pi}^2}{\omega_{ci}} \alpha^2 \left[C_1 \frac{1}{k_z} \frac{\omega_{ci}}{\omega} \frac{\omega^2}{v_A^2} + C_2 \frac{1}{q}(\alpha^2 - q^2)\right] \end{aligned} \tag{17.114}$$

Equations (17.112) and (17.114) yield the dispersion relation in the form

$$c^2 k_z^2 \lambda_e^2 q(q+k_z) = \alpha^2 v_s^2 \tag{17.115}$$

Equation (17.115) is still very complex. We restrict the region of interest to the region $q \gg k_z$. From (17.92), we can derive the relation

$$\frac{q^2}{k_z^2} = \frac{(v_p^2 - v_s^2)(v_p^2 - v_A^2)}{v_A^2 v_s^2 - v_p^2(v_s^2 + v_A^2)}, v_p^2 = \frac{\omega^2}{k_z^2} \tag{17.116}$$

From this relation, we see that the approximation $q \gg k_z$ appears to be valid in the region $v_p < v_s v_A / \sqrt{(v_s^2 + v_A^2)}$. Thus, in this region, neglect of k_z as compared to q reduces (17.115) to a quadratic equation for ω^2, giving

$$\frac{\omega^2}{k_z^2} = \frac{v_A^2 v_s^2 (v_s^2 - c^2 k_z^2 \lambda_e^2)}{v_s^2(v_s^2 + v_A^2) - c^2 v_A^2 k_z^2 \lambda_e^2} \tag{17.117}$$

When $k_z \lambda_e \to 0$, we have the dispersion relation

$$\frac{\omega^2}{k_z^2} = \frac{v_A^2 v_s^2}{v_s^2 + v_A^2} \tag{17.118}$$

which agrees with the result of the fluid theory analysis. When $c \to \infty$, $\omega/k_z \to v_s$; the electrostatic limit.

17.2 Surface Waves in Magnetized Vlasov Plasma

In this section, we investigate surface waves in magnetized semi-infinite plasmas by means of the Vlasov equation. We restrict consideration to the electrostatic waves. First, we note that there are three orientations involved in the problem: (i) the direction of the static magnetic field $\mathbf{B}_0$, which we will take as $\mathbf{B}_0 = \hat{\mathbf{z}} B_0$ (ii) the interface normal $\hat{\mathbf{n}}$ (iii) the wave vector $\mathbf{k} \perp \hat{\mathbf{n}}$. So we will investigate the following two different cases:

$$(1)\ \mathbf{B}_0 = \hat{\mathbf{z}} B_0; \quad \hat{\mathbf{n}} = \hat{\mathbf{z}}; \mathbf{k} \parallel \hat{\mathbf{x}} \text{ i.e wave} \sim e^{ik_x x - i\omega t}; \ x\text{-wave}; \ \frac{\partial}{\partial y} = 0$$

$$(2)\ \mathbf{B}_0 = \hat{\mathbf{z}} B_0; \quad \hat{\mathbf{n}} = \hat{\mathbf{x}}; \mathbf{k} \parallel \hat{\mathbf{z}} \text{ i.e wave} \sim e^{ik_z z - i\omega t}; \ z\text{-wave}; \ \frac{\partial}{\partial y} = 0$$

For the cases (1) and (2), the y-coordinate is ignorable. The names, x-, z-waves are referred to earlier work by Lee (Plasma Phys. Control. Fusion (1995)). The recipe to make the 'magnetized Vlasov equation' satisfy the kinetic boundary condition of the specular reflection is the same as that in 'unmagnetized Vlasov equation'. The reason is simply that no derivative $\frac{d\mathbf{B}_0}{dn}$ (the normal derivative) enters in the subsequent analysis. It is reminded that the normal component of the electric field is oddly-continued, which gives rise to the extra term due to the jump at the interface, when it is Fourier-transformed.

17.2.1 *Semi-infinite plasma:* $\mathbf{B}_0 = \hat{\mathbf{z}} B_0; \hat{\mathbf{n}} = \hat{\mathbf{z}}; \mathbf{k} \parallel \hat{\mathbf{x}}$*: x-wave*

The plasma occupying the region ($z > 0$) is bounded by vacuum ($z < 0$). The surface wave propagates along the x-direction ($\propto e^{ik_x x - i\omega t}$). We consider the electrostatic wave. The Vlasov equation is rewritten in the form

$$\frac{\partial}{\partial t} f_\alpha(v_x, v_y, v_z, x, z, t) + v_x \frac{\partial f_\alpha}{\partial x} + v_z \frac{\partial f_\alpha}{\partial z} + \omega_{c\alpha} v_y \frac{\partial f_\alpha}{\partial v_x} - \omega_{c\alpha} v_x \frac{\partial f_\alpha}{\partial v_y} + \frac{e_\alpha}{m_\alpha}\left(E_x(x,z)\frac{\partial f_{0\alpha}}{\partial v_x} + E_z(x,z)\frac{\partial f_{0\alpha}}{\partial v_z}\right) = 0 \quad \left(\omega_{c\alpha} = \frac{e_\alpha B_0}{m_\alpha c}\right) \tag{17.119}$$

where we put $E_y = -\partial\phi/\partial y = 0$. Equation (17.119) can be solved by introducing cylindrical coordinates in the velocity space: $v_x = v_\perp \cos\varphi$, $v_y = v_\perp \sin\varphi$. Also (17.119) is Fourier-transformed with respect to the variables, x, z, and t in the infinite region $-\infty < z < \infty$ with the extended field. Consulting Chapter 9, one obtains

$$f_\alpha(\mathbf{k}, \mathbf{v}, \omega) = \frac{ie_\alpha}{m_\alpha \omega_{c\alpha}} \sum_{n=-\infty}^{\infty} \sum_{l=-\infty}^{\infty} e^{i(n-l)\varphi} \frac{J_n(a_\alpha) J_l(a_\alpha)}{b_\alpha + l} \left[E_x \frac{l}{a_\alpha}\frac{\partial f_{\alpha 0}}{\partial v_\perp} + E_z \frac{\partial f_{\alpha 0}}{\partial v_z}\right] \tag{17.120}$$

$$a_\alpha = \frac{k_x v_\perp}{\omega_{c\alpha}}, \quad b_\alpha = \frac{k_z v_z - \omega}{\omega_{c\alpha}},$$

Fourier-transforming the Poisson equation

$$\nabla \cdot \mathbf{E}(\mathbf{r}, t) = 4\pi \sum_{\alpha} e_{\alpha} \int f_{\alpha} d^3 v \tag{17.121}$$

one obtains

$$\begin{aligned} &ik_x E_x(k_x, k_z, \omega) + ik_z E_z(k_x, k_z, \omega) \\ &\quad = 4\pi \sum_{\alpha} e_{\alpha} \int f_{\alpha}(\mathbf{k}, \mathbf{v}, \omega) d^3 v + M(k_x, \omega) \end{aligned} \tag{17.122}$$

where $M(k_x, \omega) = 2E_z(k_x, \omega, z = 0^+)$ comes from the discontinuity of $E_z(z)$ at $z = 0$. Equations (17.120) and (17.122) yield

$$E_j(k_x, k_z, \omega) = -i \frac{k_j}{k^2 \varepsilon_L} M(k_x, \omega)(j = x,\ z) \tag{17.123}$$

where $k_x E_z = k_z E_x$ was used and

$$\varepsilon_L = 1 + \sum_{\alpha} \frac{\omega_{p\alpha}^2}{k^2} \sum_{n=-\infty}^{\infty} \int d^3 v \frac{J_n^2(a_\alpha)}{\omega - k_z v_z - n\omega_{c\alpha}} \left[k_z \frac{\partial f_{\alpha 0}}{\partial v_z} + \frac{n\omega_{c\alpha}}{v_\perp} \frac{\partial f_{\alpha 0}}{\partial v_\perp} \right] \tag{17.124}$$

where $\omega_{p\alpha}$ is the plasma frequency. The quantity ε_L is the longitudinal dielectric constant. In infinite plasma, the electrostatic dispersion relation is written $\varepsilon_L = 0$. We need the normal component of the electric displacement,

$$D_z(k_x, k_z, \omega) = E_z(k_x, k_z, \omega) + \frac{4\pi i}{\omega} \sum_{\alpha} e_{\alpha} \int v_z f_{\alpha}(\mathbf{k}, \mathbf{v}, \omega) d^3 v \tag{17.125}$$

Using (17.120) in the above equation gives

$$D_z(k_x, k_z, \omega) = \varepsilon_z(k_x, k_z, \omega) E_z(k_x, k_z, \omega) = -i \frac{k_z \varepsilon_z}{k^2 \varepsilon_L} M(k_x, \omega) \tag{17.126}$$

where

$$\varepsilon_z = 1 + \sum_{\alpha} \frac{\omega_{p\alpha}^2}{\omega k_z} \sum_{n=-\infty}^{\infty} \int d^3 v \frac{v_z\ J_n^2(a_\alpha)}{\omega - k_z v_z - n\omega_{c\alpha}} \left[k_z \frac{\partial f_{\alpha 0}}{\partial v_z} + \frac{n\omega_{c\alpha}}{v_\perp} \frac{\partial f_{\alpha 0}}{\partial v_\perp} \right] \tag{17.127}$$

The vacuum equations, $\nabla \cdot \mathbf{E} = 0$ and $\nabla \times \mathbf{E} = 0$, are solved by

$$E_x(k_x, k_z, z) = F(k_x, \omega) e^{k_x z} \tag{17.128}$$

$$E_z(k_x, k_z, z) = -iF(k_x, \omega) e^{k_x z} \tag{17.129}$$

where F is an arbitrary function of (k_x, ω). The two boundary conditions at the interface $z = 0$ are the continuity of E_x and D_z. To reinstate the z-dependance, the Fourier inversion integral should be performed on (17.123) and (17.126). Then the boundary conditions at $z = 0$ give

$$M\,\frac{(-i)}{2\pi}\int_{-\infty}^{\infty} dk_z\,\frac{k_x}{k^2\varepsilon_L} = F \tag{17.130}$$

$$M\,\frac{(-i)}{2\pi}\int_{-\infty}^{\infty} dk_z\,\frac{k_z\varepsilon_z}{k^2\varepsilon_L} = -iF \tag{17.131}$$

from which we obtain the kinetic surface wave dispersion relation,

$$\int_{-\infty}^{\infty}\frac{ik_x + k_z\varepsilon_z}{k^2\varepsilon_L}\,dk_z = 0 \tag{17.132}$$

• **Cold plasma limit.** To check the correctness of (17.132), its cold plasma limit will be considered. We put

$$f_{\alpha 0}(\mathbf{v}) = \frac{\delta(v_\perp)}{v_\perp}\delta(v_z)\delta(\varphi)$$

into (12.302) for ε_L. In the limit of $a \to 0$, $J_n(a) \to \delta_{n,0}$, and the first term integral involving $\partial f_{\alpha 0}/\partial v_z$ yields, after integration by parts,

$$-\sum_\alpha \frac{\omega_{p\alpha}^2}{\omega^2}\frac{k_z^2}{k^2}$$

For the second term involving $\partial f_{\alpha 0}/\partial v_\perp$, we integrate by parts with respect to $v_\perp$, and use $\frac{2n}{a}J_n(a)J_n'(a) \to \frac{1}{2}(\delta_{n,1} - \delta_{n,-1})$ to obtain

$$-\sum_\alpha \frac{\omega_{p\alpha}^2}{\omega^2 - \omega_{c\alpha}^2}\frac{k_x^2}{k^2}$$

Adding the above two terms give

$$\varepsilon_L^{\text{cold}} = 1 - \sum_\alpha \frac{\omega_{p\alpha}^2}{k^2}\left(\frac{k_z^2}{\omega^2} + \frac{k_x^2}{\omega^2 - \omega_{c\alpha}^2}\right) \tag{17.133}$$

A similar calculation for (17.127) yields

$$\varepsilon_z^{\text{cold}} = 1 - \sum_\alpha \frac{\omega_{p\alpha}^2}{\omega^2} \tag{17.134}$$

Putting (17.133) and (17.134) into (17.132), the cold plasma dispersion relation is obtained

$$\int_{-\infty}^{\infty} dk_z \frac{k_z + \frac{ik_x}{1-\sum \omega_{p\alpha}^2/\omega^2}}{k_z^2 + \lambda_z^2} = 0 \tag{17.135}$$

where

$$\lambda_z = k_x \left[\frac{1 - \sum \frac{\omega_{p\alpha}^2}{\omega^2 - \omega_{c\alpha}^2}}{1 - \sum \frac{\omega_{p\alpha}^2}{\omega^2}} \right]^{1/2} \tag{17.136}$$

Equation (17.136) is the attenuation constant of the surface wave, as will be seen if one inverts (17.123). Carrying out the contour integral in (17.135), the final cold plasma dispersion relation is written

$$\frac{k_x}{\lambda_z} + 1 - \sum_\alpha \frac{\omega_{p\alpha}^2}{\omega^2} = 0 \tag{17.137}$$

This agrees with the earlier fluid dispersion relation by H. J. Lee [Plasma Phys. Control. Fusion (1995)].

17.2.2 *Semi-infinite plasma:* $\mathbf{B}_0 = \hat{\mathbf{z}} B_0; \hat{\mathbf{n}} = \hat{\mathbf{x}}; \mathbf{k} \parallel \hat{\mathbf{z}}$*: z-wave*

The plasma occupying the region ($x > 0$) is bounded by vacuum ($x < 0$). The surface wave propagates along the z-direction ($\propto e^{ik_z z - i\omega t}$). We also expect that the wave variable k_x and the Fourier-inversion variable k_z in x-wave interchange the roles for z-wave. Fourier transform of (17.121) gives

$$\begin{aligned} &ik_x E_x(k_x, k_z, \omega) + ik_z E_z(k_x, k_z, \omega) \\ &\quad = 4\pi \sum_\alpha e_\alpha \int f_\alpha(\mathbf{k}, \mathbf{v}, \omega) d^3v + N(k_z, \omega) \end{aligned} \tag{17.138}$$

where $N(k_z, \omega) = 2E_x(k_z, \omega, x = 0^+)$ comes from the discontinuity of $E_x(x)$ at $x = 0$. Equations (17.120) and (17.138) give

$$E_j(k_x, k_z, \omega) = -i \frac{k_j}{k^2 \varepsilon_L} N(k_z, \omega) (j = x,\ z) \tag{17.139}$$

where ε_L is given by (17.124). We need the normal component of the electric displacement,

$$D_x(k_x, k_z, \omega) = E_x(k_x, k_z, \omega) + \frac{4\pi i}{\omega} \sum_\alpha e_\alpha \int v_x f_\alpha(\mathbf{k}, \mathbf{v}, \omega) d^3v \tag{17.140}$$

Using (17.120) in the above equation gives

$$D_x(k_x, k_z, \omega) = \varepsilon_x(k_x, k_z, \omega) E_x(k_x, k_z, \omega) = -i\, \frac{k_x \varepsilon_x}{k^2 \varepsilon_L} N(k_z, \omega) \tag{17.141}$$

$$\varepsilon_x = 1 + \sum_\alpha \frac{\omega_{p\alpha}^2}{\omega k_x} \sum_{n=-\infty}^{\infty} \sum_{l=-\infty}^{\infty} \int d^3v \frac{e^{i(n-l)\varphi} v_\perp \cos\varphi J_n(a_\alpha) J_l(a_\alpha)}{\omega - k_z v_z - l\omega_{c\alpha}} \left[k_z \frac{\partial f_{\alpha 0}}{\partial v_z} + \frac{l\omega_{c\alpha}}{v_\perp} \frac{\partial f_{\alpha 0}}{\partial v_\perp} \right]$$

The integral is non-vanishing only when $n = l \pm 1$. Using the formula $J_{l-1}(a) + J_{l+1}(a) = \frac{2l}{a} J_l(a)$, one obtains

$$\varepsilon_x = 1 + \sum_\alpha \frac{\omega_{p\alpha}^2}{\omega k_x} \int d^3v \sum_{n=-\infty}^{\infty} \frac{\frac{n}{a_\alpha} J_n^2(a_\alpha)}{\omega - k_z v_z - n\omega_{c\alpha}} \left[n\omega_{c\alpha} \frac{\partial f_{\alpha 0}}{\partial v_\perp} + k_z v_\perp \frac{\partial f_{\alpha 0}}{\partial v_z} \right] \tag{17.142}$$

The vacuum solutions are slightly different from (17.128) and (17.129):

$$E_x(k_z, \omega, x) = -iG(k_z, \omega) e^{k_z x} \tag{17.143}$$

$$E_z(k_z, \omega, x) = G(k_z, \omega) e^{k_z x} \tag{17.144}$$

where G is an arbitrary function of (k_z, ω). The two boundary conditions on the interface $x = 0$ are the continuity of E_z and D_x. Omitting the intermediate steps, we write the dispersion relation,

$$\int_{-\infty}^{\infty} \frac{ik_z + k_x \varepsilon_x}{k^2 \varepsilon_L}\, dk_x = 0 \tag{17.145}$$

Equation (17.145) should be compared with (17.132). They have an identical structure, but the difference between ε_x and ε_z should be noted; compare (17.127) and (17.142).

• **Cold plasma limit.** We need to calculate the cold plasma limit of ε_x, (17.142), as has been done earlier. Owing to the relation, $\lim_{a \to 0} \frac{n}{a} J_n^2(a) = 0$, the second integral involving $\partial f_{\alpha 0}/\partial v_z$ vanishes. The first integral is integrated by parts and use is made of the relation, $\frac{2n}{a} J_n(a) J_n'(a) \to \frac{1}{2}(\delta_{n,1} - \delta_{n,-1})$ to obtain

$$\varepsilon_x^{\text{cold}} = 1 - \sum_\alpha \frac{\omega_{p\alpha}^2}{\omega^2 - \omega_{c\alpha}^2} \tag{17.146}$$

Using (17.133) and (17.146) in (17.145) yield the cold plasma dispersion relation,

$$\int_{-\infty}^{\infty} dk_x \frac{k_x + \frac{ik_z}{1-\sum \frac{\omega_{p\alpha}^2}{\omega^2-\omega_{c\alpha}^2}}}{k_x^2 + \lambda_x^2} = 0 \tag{17.147}$$

where

$$\lambda_x = k_z \left[\frac{1 - \sum \frac{\omega_{p\alpha}^2}{\omega^2}}{1 - \sum \frac{\omega_{p\alpha}^2}{\omega^2-\omega_{c\alpha}^2}} \right]^{1/2} \tag{17.148}$$

agrees with the attenuation constant obtained in [H. J. Lee Plasma Phys. Control. Fusion (1995), Eq. (30)]. Carrying out the contour integral in (17.147) gives the dispersion relation

$$\frac{k_z}{\lambda_x} + 1 - \sum_\alpha \frac{\omega_{p\alpha}^2}{\omega^2 - \omega_{c\alpha}^2} = 0 \tag{17.149}$$

which agrees with the dispersion relation obtained in [H. J. Lee, Plasma Phys. Control. Fusion (1995), Eq. (33)]. Finally we notice from (17.136) and (17.148) that the ratio of the attenuation constant to the wave number (λ/k) of x- and z-waves is reciprocal to each other. This reciprocity is discussed in [H. J. Lee Plasma Phys. Control. Fusion (1995)].

Problems

17.1 In his famous paper, Landau [L. Landau, *On the vibrations of the electronic plasma*, J. Physics (USSR) **10** (1946), p. 25] solved a boundary value problem in a semi-bounded plasma interfaced with a vacuum, under the condition of the specular reflection condition. His approach is useful for considering the surface wave echo. We restrict ourselves to electrostatic waves. Assume the wave has the phasor $e^{ik_z z - i\omega t}$ and the wave amplitude varies along the perpendicular direction x, while the y-coordinate is ignorable. Fourier transform the electron Vlasov equation to write in the form

$$\frac{\partial}{\partial x} f(\omega, x, k_z, v_x, v_z) - i\frac{\psi}{v_x} f - \frac{e}{mv_x}\mathbf{E}(x, k_z, \omega) \cdot \frac{\partial f_0}{\partial \mathbf{v}} = 0,$$
$$\psi = \omega - k_z v_z \tag{17.150}$$

Show that the above first order differential equation for x is solved in the form

$$f(x, v_x) = e^{i\frac{\psi}{v_x}x}\left[f(x_0, v_x)e^{-i\frac{\psi}{v_x}x_0} + \int_{x_0}^{x} dx' \frac{e}{mv_x}\mathbf{E}(x') \cdot \frac{\partial f_0}{\partial \mathbf{v}} e^{-i\frac{\psi}{v_x}x'}\right] \tag{17.151}$$

where x_0 is the reference point of the boundary condition. Note that the first term inside the bracket [ ] goes to zero as $x_0 \longrightarrow \infty$ provided $v_x < 0$. So we write

$$f^N(x, v_x) = \int_{\infty}^{x} dx' \frac{e}{mv_x}\mathbf{E}(x') \cdot \frac{\partial f_0}{\partial \mathbf{v}} e^{i\frac{\psi}{v_x}(x-x')} \quad \text{for } v_x < 0 \tag{17.152}$$

Thus, we have succeeded in constructing the solution that satisfies the boundary condition $f(v_x, \infty) = 0$ provided $v_x < 0$.

Taking $x_0 = -\infty$, the solution of (17.150) can be written as, provided $v_x > 0$:

$$f^P(x, v_x) = \int_{-\infty}^{x} dx' \frac{e}{mv_x}\mathbf{E}(x') \cdot \frac{\partial f_0}{\partial \mathbf{v}} e^{i\frac{\psi}{v_x}(x-x')} \quad \text{for } v_x > 0 \tag{17.153}$$

To satisfy the boundary condition at $x = \pm\infty$, the solution of the Vlasov equation should be written in separate forms, which we denote by f^N, f^P.

These equations for f^N, f^P can be derived from the Fourier expression for $f(k_x, v_x)$:

$$f(k_x, v_x) = \frac{-ie}{mv_x}\frac{1}{k_x - \frac{\psi}{v_x}}\left(E_x(k_x)\frac{\partial f_0}{\partial v_x} + E_z(k_x)\frac{\partial f_0}{\partial v_z}\right) \tag{17.154}$$

$$f(x, v_x) = \frac{-ie}{2\pi mv_x}\int_{-\infty}^{\infty} dk_x \frac{e^{ik_x x}}{k_x - \psi/v_x}\left(E_x(k_x)\frac{\partial f_0}{\partial v_x} + E_z(k_x)\frac{\partial f_0}{\partial v_z}\right) \tag{17.155}$$

Using the expression $E_x(k_x) = \int_{-\infty}^{\infty} d\xi\ e^{-ik_x\xi}E_x(\xi)$ in the above gives

$$f(x, v_x) = \frac{-ie}{2\pi mv_x}\int_{-\infty}^{\infty} d\xi \left(E_x(\xi)\frac{\partial f_0}{\partial v_x} + E_z(\xi)\frac{\partial f_0}{\partial v_z}\right)\int_{-\infty}^{\infty} dk_x \frac{e^{ik_x(x-\xi)}}{k_x - \psi/v_x} \tag{17.156}$$

Let us consider the integral

$$I = \int_{-\infty}^{\infty} dk \, \frac{e^{ik(x-\xi)}}{k - \frac{\omega}{v}} \tag{17.157}$$

where ω has a small imaginary part. So the singularity on the real k-axis is slightly below (above) the real k-axis when $v < 0$ ($v > 0$). When $v < 0$, the $\int dk$-integral should follow the convex-up path going around the singularity $k = \omega/v$ by being pushed upward as $\omega_i \to 0$. When $v > 0$, the $\int dk$-integral should follow the concave-down path going around the singularity $k = \omega/v$ by being pushed downward as $\omega_i \to 0$. So the $\int dk$-integral is sum of the principal part plus the contribution from the infinitesimal semi-circle:

$$\int_{-\infty}^{\infty} dk \, \frac{e^{ik(x-\xi)}}{k - \frac{\omega}{v}} = P \int_{-\infty}^{\infty} dk \, \frac{e^{ik(x-\xi)}}{k - \frac{\omega}{v}} \mp i\pi \exp\left[i \, \frac{\omega}{v}(x - \xi)\right] \tag{17.158}$$

where $-(+)$ sign corresponds to $v < 0$ ($v > 0$). The principal value integral is

$$P \int_{-\infty}^{\infty} dk \, \frac{e^{ik(x-\xi)}}{k - \frac{\omega}{v}} = i\pi \, \exp\left[i \, \frac{\omega}{v}(x - \xi)\right] S(x - \xi) \tag{17.159}$$

where S is the step function defined as $S(x) = 1$ for $x > 0$ and $S(x) = -1$ for $x < 0$. Thus, we finally obtain f^N and f^P as given by (17.152) and (17.153), respectively.

17.2 Show that, when $\omega_i > 0$ is finite, inversion of (17.154) yields (17.152) and (17.153).

References

Braginskii, S. I. and A. P. Kazantsev, *Magnetohydrodynamic waves in a rarefied plasma* (work done in 1957; the Journal of this article is unknown to the author of this book.)

Lee, H. J. and C.-G. Kim, *Kinetic theory of electrostatic surface waves in magnetized plasmas*, J. Korean Phys. Soc. **54** (2009), pp. 85–93.

Lee, H. J. *Electrostatic surface waves in a magnetized two-fluid plasma*, Plasma Phys. Control. Fusion **37** (1995), pp. 755–762.

Chapter 18

Two-Stream Instability

A plasma with components of different streaming velocity (beams) tends to be unstable if a certain condition is met. In comparison, a moving plasma with a uniform velocity as a whole would not be different from a stationary plasma, as is evident from the Galilean point of view. A plasma consisting of two groups of electrons streaming with different speeds can be unstable because of the growing Langmuir oscillation (in the background of immobile ions). Similarly, a plasma with two ion beams in the background of hot Maxwellian electrons can be unstable because of the growing ion acoustic wave perturbation (because in this case ion wave Landau damping can be weak)—*ion acoustic instability.*

Plasmas with beams draw attention since such plasmas can be used to make a traveling wave tube or an amplifier in laboratory. Solar wind is also a streaming plasma appearing in nature. Two-stream instabilities can be used to heat plasma by injecting a particle beam into the plasma. Streaming of charged particles is also given rise to by currents induced in a plasma by external electromagnetic fields. In this case, two-stream instabilities arise as the result of a beam-plasma interaction. Analytically, the streaming velocity $\mathbf{V}_0$ is the zero order quantity upon which the small perturbation $\mathbf{v}$ should be superposed. The term $\mathbf{k} \cdot \mathbf{V}_0$ that appears in the dispersion relation significantly modifies the dispersion relation of stationary plasma wave.

The study of the dispersion relations of beam plasma wave has established the criterion of two different kinds of instability: convective vs absolute instability.

18.1 Analysis of Two-Stream Instability

A plasma can be regarded as an assembly of beams individualized by different initial velocities. The plasma susceptibility can be obtained by summing up all the *cold* individual beam susceptibilities. Combining this result with

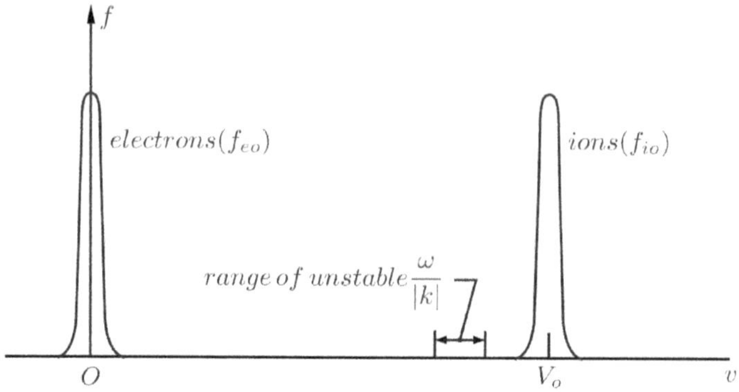

Fig. 18.1. Distributions for a two-stream instability.

the Poison equation yields the usual dispersion relation, which agrees with that obtained from the Vlasov equation. To obtain a beam instability, we need at least two beams of different streaming speed. This requires a hump-like component superposed on a smoothly varying velocity distribution function; for example, *a bump-in-tail instability* [N. A. Krall and A. W. Trivelpiece, *Principles of plasma physics*, McGraw-Hill, New York (1973)]. See Figs. 18.1 and 18.2. In terms of the distribution function, a cold beam can be represented by a delta function. Figure 18.1 shows the distribution function for a cold ion beam streaming at speed V_0 through a cold electron 'beam' (in fact, the figure represents a stationary group of electrons). For a warm beam with thermal spread, we can use the distribution function of the form (see Fig. 18.2)

$$f_0 = \frac{N_1}{N}\left(\frac{m}{2\pi T_1}\right)^{3/2} \exp\left(-\frac{mv^2}{2T_1}\right) + \frac{N_2}{N}\left(\frac{m}{2\pi T_2}\right)^{1/2}$$
$$\times \exp\left(-\frac{m(v_z - V_0)^2}{2T_2}\right) \delta(v_x)\delta(v_y) \quad (N = N_1 + N_2)$$

In the following, we suppose that a plasma consists of several *cold* streaming species characterized by zero order velocities V_{jo} and zero order constant number densities N_{j0}, where $j = 1, 2, \ldots$ is the index designating each streaming species. We assume that $\Sigma_j e_j N_{j0} = 0$ (charge neutrality), and there is no electric field in the unperturbed state. Each beam individualized by the index 'j' is governed by the cold fluid equations,

$$\frac{\partial \mathbf{v}_j}{\partial t} + \mathbf{v}_j \cdot \nabla \mathbf{v}_j = \frac{e_j}{m_j}\mathbf{E} \tag{18.1}$$

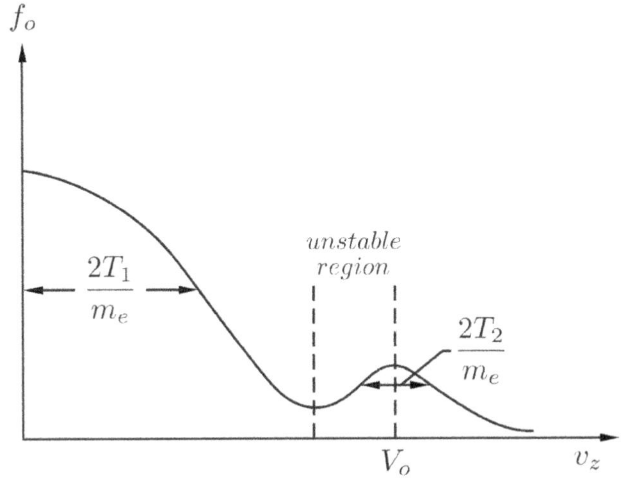

Fig. 18.2. Distribution for 'bump in tail' instability.

$$\frac{\partial n_j}{\partial t} + \nabla \cdot (n_j \mathbf{v}_j) = 0 \tag{18.2}$$

$$\nabla \cdot \mathbf{E} = 4\pi \sum_j e_j n_j \tag{18.3}$$

Linearizing the above equations by writing $\mathbf{v}_j = \mathbf{V}_{j0} + \mathbf{v}'_j$, $n_j = N_{j0} + n'_j$ where the primed quantities are the perturbations, we obtain

$$\frac{\partial \mathbf{v}'_j}{\partial t} + \mathbf{V}_{j0} \cdot \nabla \mathbf{v}'_j = \frac{e_j}{m_j} \mathbf{E} \tag{18.4}$$

$$\frac{\partial n'_j}{\partial t} + \mathbf{V}_{j0} \cdot \nabla n'_j + N_{j0} \nabla \cdot \mathbf{v}'_j = 0 \tag{18.5}$$

$$\nabla \cdot \mathbf{E} = 4\pi \sum_j e_j n'_j \tag{18.6}$$

If the first order quantities vary as $e^{i\mathbf{k}\cdot\mathbf{r} - i\omega t}$, the linearized equations take the form

$$-i(\omega - \mathbf{k} \cdot \mathbf{V}_{j0}) \mathbf{v}'_j = \frac{e_j}{m_j} \mathbf{E} \tag{18.7}$$

$$-i(\omega - \mathbf{k} \cdot \mathbf{V}_{j0}) n'_j + i N_{j0} \mathbf{k} \cdot \mathbf{v}'_j = 0 \tag{18.8}$$

$$i\mathbf{k} \cdot \mathbf{E} = 4\pi \sum_j e_j n'_j \tag{18.9}$$

Solving for n'_j and using (18.9) give the dispersion relation

$$\sum_j \frac{\omega_{pj}^2}{(\omega - \mathbf{k}\cdot\mathbf{V}_{j0})^2} = 1 \tag{18.10}$$

where ω_{pj} is the plasma frequency of the jth beam.

We consider only two streams of particles with $\mathbf{k}\|\mathbf{V}_{j0}$ ($j = 1, 2$). These may be a stream of electrons and a stream of ions with different velocities. Or these may be two streams of electrons, limiting to the waves of high frequencies in which the ions form stationary background. Then the dispersion relation (8.10) becomes

$$\frac{\omega_{p1}^2}{(\omega - kV_1)^2} + \frac{\omega_{p2}^2}{(\omega - kV_2)^2} = 1 \tag{18.11}$$

Since this is a quartic equation in ω, ω cannot be obtained as a simple function of k. If we introduce the phase velocity $v_\phi = \omega/k$, the above equation can be written as

$$k^2 = \frac{\omega_{p1}^2}{(v_\phi - V_1)^2} + \frac{\omega_{p2}^2}{(v_\phi - V_2)^2} \equiv F(v_\phi) \tag{18.12}$$

This equation has four roots for v_ϕ, possibly including two pairs of complex conjugate. The general form of $F(v_\phi)$ is shown in Figs. 18.3(a) and 18.3(b). In Fig. 18.3(a), there are four real roots of (18.12), corresponding to four propagating waves. In Fig. 18.3(b), there are only two real roots; the remaining two solutions are a pair of complex conjugate, one of which corresponding to a growing wave in time. The instability is called the two-stream instability. We can find the value k_c [for $k > k_c$ ($k < k_c$), we have

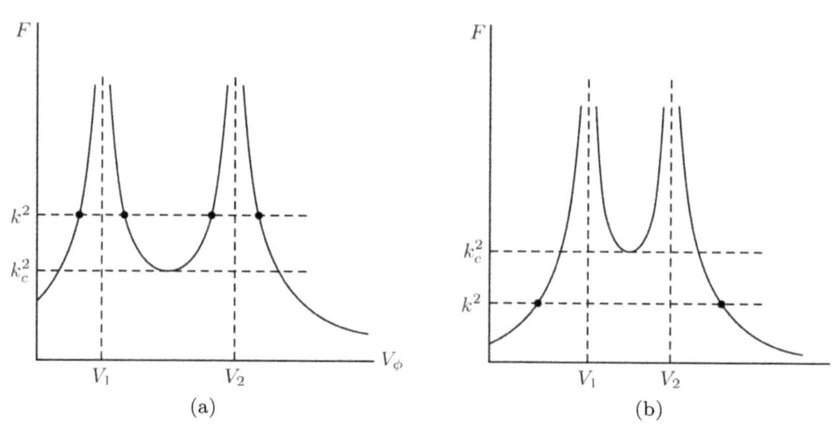

Fig. 18.3. Two-stream instability (a) stable (b) unstable.

stable (unstable) stream] from the condition $dF/dv_\phi=0$, which gives

$$v_\phi = \frac{V_2\omega_{p1}^{2/3} + V_1\omega_{p2}^{2/3}}{\omega_{p1}^{2/3} + \omega_{p2}^{2/3}} \tag{18.13}$$

Using (18.13) in (18.12) yields

$$k_c^2 = \frac{\omega_{p1}^{2/3} + \omega_{p2}^{2/3}}{(V_1 - V_2)^2} \tag{18.14}$$

We can see from (18.14) that large (small) k_c corresponds to the case that the two streaming velocities are narrowly (widely) separated. Analysis of (18.12) requires a numerical work, but there is a special case for which the dispersion relation can be solved analytically. The quartic equation (18.11) (in ω) becomes a quadratic equation in ω^2 if

$$V_1 = -V_2 \equiv V, \quad \omega_{p1}^2 = \omega_{p2}^2 \equiv \omega_p^2$$

This includes the case where two identical electron beams travel in opposite direction. In this case, (18.11) becomes

$$\frac{\omega_p^2}{(\omega - kV)^2} + \frac{\omega_p^2}{(\omega + kV)^2} = 1 \tag{18.15}$$

whose solution can be written as

$$\omega^2 = \omega_p^2 + k^2V^2 \pm \omega_p(\omega_p^2 + 4k^2V^2)^{1/2} \tag{18.16}$$

The plot ω^2 vs. k^2 is shown in Fig. 18.4. Note the resemblance with the surface polariton dispersion curve (see Eq. (12.88′)). $\oplus$ curve gives rise to convective instability since real ω corresponds to complex k. $\ominus$ curve shows that for $k < k_c$ we have 'non-convective' instability.

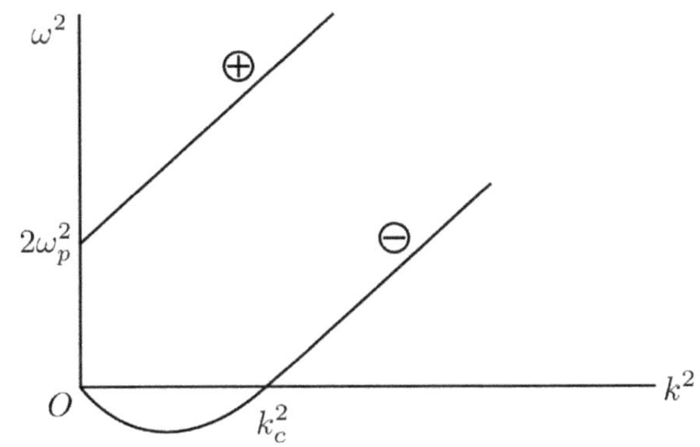

Fig. 18.4. Plot of Eq. (18.16).

18.2 Absolute Instability and Convective Instability

The dispersion relation $D(k,\omega) = 0$ is, in general, an algebraic equation connecting a complex $k = k_r + ik_i$ to a complex $\omega = \omega_r + i\omega_i$. With the phasor $e^{ikx-i\omega t}$, a wave is said to be unstable (damped) if we have $\omega_i > 0$ ($\omega_i < 0$) for some real wavenumber k. One also speaks of amplifying (evanescent) wave if $k_i < 0$ ($k_i > 0$) for some real ω. This classification does not apply when ω and k are both complex at the same time. In particular, when $\omega_i > 0$, there are two kinds of instability: absolute and convective instability. In convective instability, the instability propagates away from a fixed point of observation so that the disturbance at a fixed point in the medium may decay with time (as in a traveling wave). On the other hand, an absolute instability grows with time at every point in space (as in a standing wave). See Fig. 18.5.

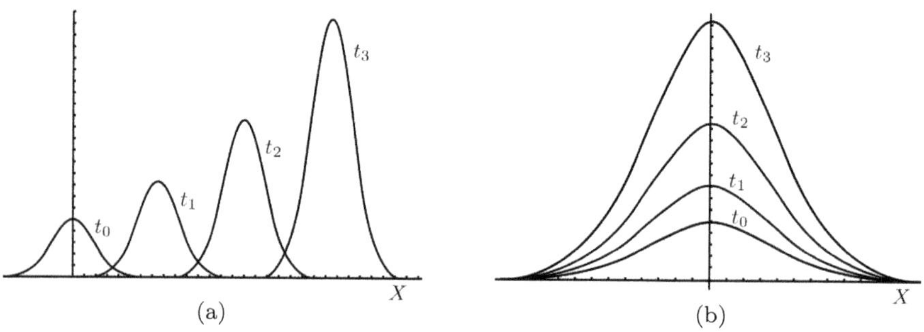

Fig. 18.5. Convective (a) and absolute (b) instability at four equidistant times (after M. Brambilla, *Kinetic Theory of Plasma Waves*, p. 43, Clarendon Oxford, 1998).

We approach this problem through an example of specific model equations

$$\frac{\partial \rho_1}{\partial t} + c_1 \frac{\partial \rho_1}{\partial x} = a\rho_2 \tag{18.17}$$

$$\frac{\partial \rho_2}{\partial t} - c_2 \frac{\partial \rho_2}{\partial x} = a\rho_1 \tag{18.18}$$

The above equations describe two oppositely propagating waves which are coupled (with 'a' the coupling constant). The corresponding dispersion relation is obtained from $(-i\omega + ikc_1)\rho_1 = a\rho_2, (-i\omega - ikc_2)\rho_2 = a\rho_1$, or

$$(kc_1 - \omega)(kc_2 + \omega) = a^2 \tag{18.19}$$

In uncoupled state, ρ_1 is constant along the characteristic line $\frac{dx}{dt} = c_1$, and ρ_2 is constant along the characteristic line $\frac{dx}{dt} = -c_2$. Let us recall that

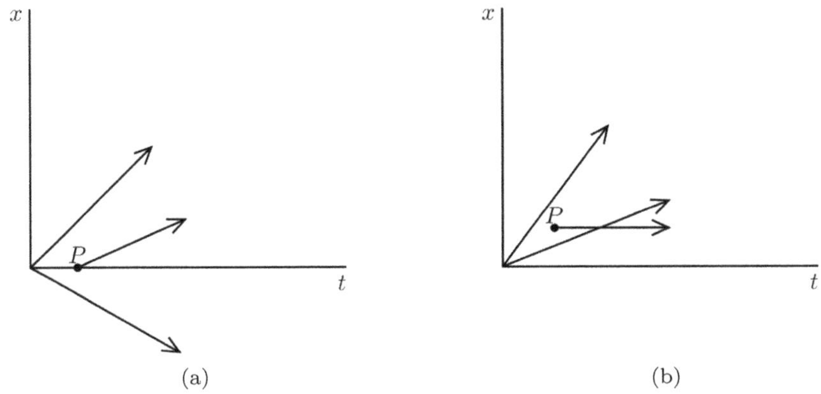

Fig. 18.6. Absolute (a) and convective (b) instability.

a characteristic line is a dividing line of two *different nature* of solutions which cannot be connected through Taylor expansion across it. This property of the characteristic line defines the *region of influence* for a perturbation propagating on (x, t)-plane. In Fig. 18.6(a), we drew two characteristic lines emanating from the origin on the (x, t)-plane. Note that the signs of the slope of the characteristic lines are opposite. In this case, the region of influence is the wedge-shaped region comprising the t-axis. A moving detector of the disturbance situated at P initially, which is subject to the perturbation, cannot escape from the region of disturbance since the t-axis is always inside the region of influence. If the disturbance at P grows, it is an *absolute instability*. In Fig. 18.6(b), we drew two characteristic lines emanating from the origin with the same sign of the slopes. A moving detector situated at the point P will be eventually out of the region of influence, in this case. The perturbation at the point P is convected away, and the instability is a *convective instability*.

In Fig. 18.7, we plotted two model dispersion relations,

$$(kc_1 - \omega)(kc_2 + \omega) = \pm a^2 \tag{18.20}$$

where the + sign corresponds (18.19). Firstly, it should be noted that the asymptotes of the hyperbolas are $\frac{\omega}{k} = c_1$ and $\frac{\omega}{k} = -c_2$ which coincides with the slopes of the characteristics, respectively. In general we have

$$\lim_{k \to \infty} \frac{\omega}{k} = V, \text{ the slope of characteristics} \tag{18.21}$$

Figure 18.7 corresponds to the case that the characteristics run with slopes of different signs, and therefore the unstable solutions of (18.19), if there is any, should indicate an absolute instability. The $\oplus$ curve shows that there

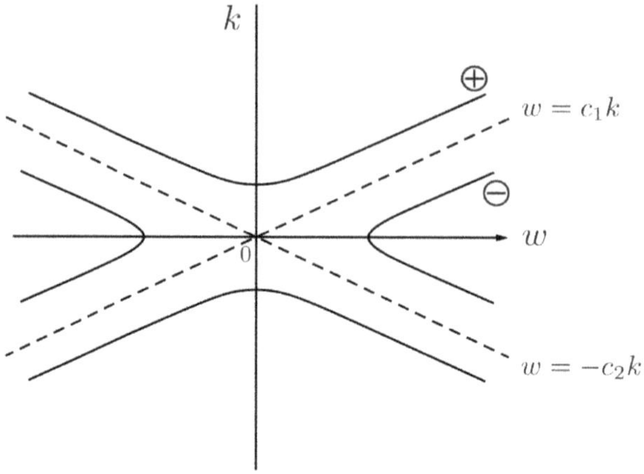

Fig. 18.7. Dispersion relation for two beams traveling in the opposite directions.

is a section of real k (on the k-axis in the diagram) to which no real ω is corresponding. Solving the quadrature, one finds that ω is complex for

$$-\frac{2a}{c_1+c_2} < k < \frac{2a}{c_1+c_2}.$$

Clearly we can have an absolute instability. The $\ominus$ curve shows that there is a section of real ω (on the ω-axis in the diagram) to which no real k is corresponding. Solving the quadrature, one finds that k is complex for

$$-\frac{2a\sqrt{c_1c_2}}{c_1+c_2} < \omega < \frac{2a\sqrt{c_1c_2}}{c_1+c_2}.$$

The complex k means that the wave can be amplifying or evanescent in this case.

In Fig. 18.8, we plotted another set of model dispersion relation

$$(kc_1-\omega)(kc_2-\omega) = \pm a^2.$$

The curves correspond to the characteristics running in the same direction. The $\oplus$ curve in Fig. 18.8 shows that ω is real for all real k ($-\infty < k < \infty$). Also k is real for all real ω ($-\infty < \omega < \infty$). There is no instability nor damping; no amplification nor evanescence in this case. The $\ominus$ curve indicates that there is a segment of real ω,

$$-\frac{2a\sqrt{c_1c_2}}{|c_1-c_2|} < \omega < \frac{2a\sqrt{c_1c_2}}{|c_1-c_2|}$$

which corresponds to complex k. Also there is a section of real k,

$$-\frac{2a}{|c_1-c_2|} < k < \frac{2a}{|c_1-c_2|}$$

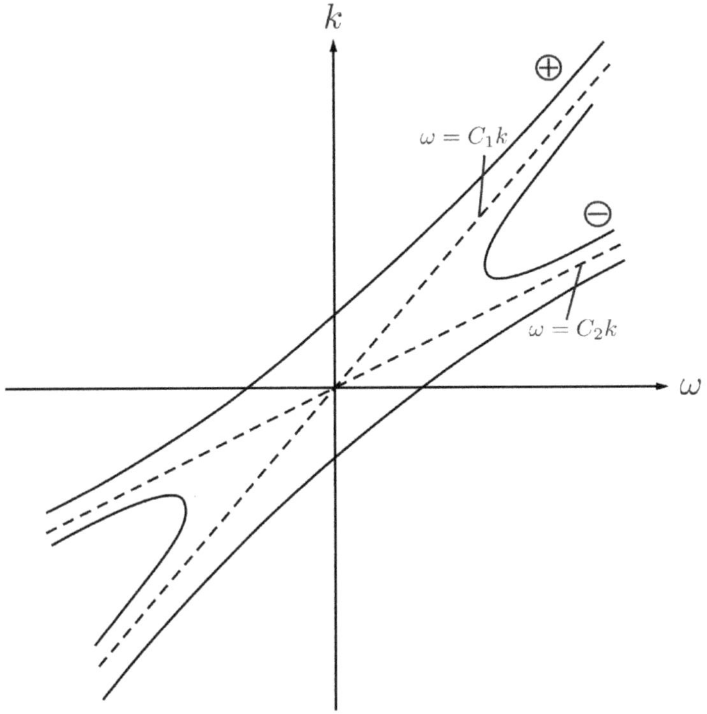

Fig. 18.8. Dispersion relation for two beams traveling in the same direction.

which corresponds to complex ω. In this range of k, we have a convective instability. In summary, a convective instability can be given rise to if real ω corresponds to complex k, and an absolute instability when real ω corresponds to real k.

Remark. To be more precise in finding the characteristics of the system of equations (18.17) and (18.18), we eliminate ρ_1 or ρ_2 to get

$$-c_1c_2\rho_{xx} + (c_1 - c_2)\rho_{xt} + \rho_{tt} - a^2\rho = 0 \tag{18.22}$$

where ρ stands for ρ_1 or ρ_2, and the subscript x or t denotes the partial derivative: $\rho_x = \frac{\partial \rho}{\partial x}$, $\rho_{xx} = \frac{\partial^2 \rho}{\partial x^2}$, etc. We seek the solution of the above equation in the form $\rho(x,t) = \rho(\xi)$ where $\xi = x - c_1 t$. Then (18.22) is solved by $\rho(\xi) = 0$: in words, $\rho = 0$ along the characteristics $\frac{dx}{dt} = c_1$ or $x - c_1 t = \xi =$ constant. Equation (18.22) also permits solution in the form $\rho(x,t) = \rho(\eta)$ where $\eta = x + c_2 t$. Then (18.22) is solved by $\rho(\eta) = 0$: in words, $\rho = 0$ along the characteristics $\frac{dx}{dt} = -c_2$ or $x + c_2 t = \eta =$ constant. Clearly $\frac{dx}{dt} = c_1$ and $\frac{dx}{dt} = -c_2$ are two families of the characteristic lines of (18.22).

18.3 Instability Analysis via Fourier–Laplace Inversion

Fourier–Laplace transform is useful for the investigation of the asymptotic behavior of the waves excited by some initial perturbation. By inverting the dispersion relation, we can distinguish the absolute and convective instabilities from the asymptotic behavior of the waves. Let us consider an initial value problem posed by Vlasov and Poisson equations:

$$\frac{\partial}{\partial t}g(x,v,t) + v\frac{\partial g}{\partial x} - \frac{e}{m}E(x,t)\frac{dg_0(v)}{dv} = 0 \tag{18.23}$$

$$\frac{\partial E}{\partial x} = -4\pi e\int_{-\infty}^{\infty} g(x,v,t)dv \tag{18.24}$$

with the initial condition $g(x,v,0) = \delta(x)I(v)$. Fourier-Laplace transform of the above two equations with the given initial condition yields (see Eq. (8.43))

$$E(x,t) = \frac{en_0}{\pi}\int_{-\infty+i\sigma}^{\infty+i\sigma} d\omega e^{-i\omega t}\int_{-\infty}^{\infty}\frac{dk}{k^2}e^{ikx}\frac{\int_{-\infty}^{\infty}\frac{I(v)}{v-\omega/k}dv}{\varepsilon(\omega,k)} \tag{18.25}$$

$$\text{where}\quad \varepsilon(\omega,k) = 1 - \frac{\omega_p^2}{k^2}\int_{-\infty}^{\infty}\frac{\frac{dg_0}{dv}}{v-w/k}dv \tag{18.26}$$

is the dielectric function.

If we are interested in only the normal modes of plasma, we should solve $\varepsilon(\omega,k) = 0$ to derive the dispersion relations. [This is seen from (18.25) by putting $I(v) = 0$ in the numerator.] Solutions of the algebraic equation $\varepsilon(\omega,k) = 0$ connect complex k to complex ω, generally. We talk about instability or damping for real k; amplifying or evanescent wave for real ω. However, convective and absolute instabilities can be distinguished only after examination of the asymptotic behavior of the response function $E(x,t)$.

Equation (18.25) contains all the information in regard to the asymptotic behavior of the waves excited by the initial perturbation. It is convenient to introduce the Green function in terms of which the response in (18.25) can be investigated:

$$G(x,t) = \int_{-\infty+i\sigma}^{\infty+i\sigma}\frac{d\omega}{2\pi}e^{-i\omega t}\int_{-\infty}^{\infty}\frac{dk}{2\pi}e^{ikx}\frac{1}{D(\omega,k)}\quad (D(\omega,k) = k^2\varepsilon(\omega,k)) \tag{18.27}$$

The integration paths in the above indicate that k is real and ω is complex, but we can or have to deform, if necessary, the contour as far as the contour remains inside the analytic region. Thus we are free from any such restrictions for k and ω being real. We have two-fold integral in (18.27).

We shall consider two model dispersion relations $D(k,\omega) = 0$ (Equations (18.19) and (18.20)):

$$D(k,\omega) = \left(k - \frac{\omega}{c}\right)\left(k - \frac{\omega}{c}\right) + k_0^2 = 0 \tag{18.28}$$

$$D(k,\omega) = \left(k - \frac{\omega}{c}\right)\left(k + \frac{\omega}{c}\right) - k_0^2 = 0 \tag{18.29}$$

Assume $x > 0$, then the k-contour should encircle the upper-half k-plane. The k-integration can be done by calculating the residues at the poles:

$$G(x,t) = \int_{-\infty+i\sigma}^{\infty+i\sigma} \frac{d\omega}{2\pi} e^{-i\omega t} i \sum_{k_s} \frac{e^{ik_s x}}{[\partial D/\partial k]_{k_s}} \tag{18.30}$$

where the k-poles (k_s) are obtained from $D(\omega,k) = 0$:

$$D(\omega, k_s) = 0 \quad (s = 1, 2, \ldots)$$

Let us first consider the wave in (18.28).

$$k_s = \frac{\omega}{c} \pm ik_0 \tag{18.31}$$

We have $\partial D/\partial k = 2(k - \omega/c) = \pm 2ik_0$. The imaginary part of $\omega = \omega_r + i\omega_i$ can be large, so that both signs should be included. Using these values in (18.30) gives

$$G(x,t) = \int_{-\infty+i\sigma}^{\infty+i\sigma} \frac{d\omega}{2\pi} e^{-i\omega t} e^{i\frac{\omega}{c}x} \frac{1}{2k_0}\left(e^{-k_0 x} - e^{k_0 x}\right) \tag{18.32}$$

which yields

$$G(x,t) = \frac{1}{2k_0}\left(e^{-k_0 x} - e^{k_0 x}\right)\delta\left(t - \frac{x}{c}\right) \tag{18.33}$$

This gives an amplifying wave along the trajectory $x = ct$. This is a convective instability.

Next, we consider the wave in (18.29). Let us expand $D(\omega,k)$ in (18.30) as a Taylor series in the vicinity of $\omega = \omega_s$ and $k = k_s$, which are the values satisfying $\partial D/\partial k = 0$ and $D(\omega,k) = 0$: we find $k_s = 0$ and $\omega_s = ick_0$.

$$D(\omega,k) = D(\omega_s,k_s) + \left.\frac{\partial D}{\partial k}\right|_s (k - k_s) + \left.\frac{\partial D}{\partial \omega}\right|_s (\omega - \omega_s) + \frac{1}{2}(k - k_s)^2 \left.\frac{\partial^2 D}{\partial k^2}\right|_s + \cdots . \tag{18.34}$$

The first two terms are zero. $\frac{\partial^2 D}{\partial k^2}|_s = 2$ per (18.29). $\frac{\partial D}{\partial \omega}\big|_s = -2\omega_s/c^2$. Then, (18.34) takes the form

$$D(\omega, k) = k^2 - \frac{2}{c^2}\omega_s(\omega - \omega_s) \tag{18.35}$$

which gives a simple pole at $k = \frac{\sqrt{2}}{c}\sqrt{\omega_s(\omega - \omega_s)}$. Then we have

$$\frac{\partial D}{\partial k} = 2k = \frac{2\sqrt{2}}{c}\sqrt{\omega_s}\sqrt{\omega - \omega_s} \tag{18.36}$$

Using this in (18.30) gives

$$G(x, t) = \int_{-\infty+i\sigma}^{\infty+i\sigma} \frac{d\omega}{2\pi} e^{-i\omega t} \frac{ic}{2\sqrt{2}\sqrt{\omega_s}} \frac{1}{\sqrt{\omega - \omega_s}} \tag{18.37}$$

$$= \frac{ice^{-i\omega_s t}}{2\sqrt{2}\sqrt{\omega_s}} \int_{-\infty+i\sigma}^{\infty+i\sigma} d\omega \frac{e^{-i(\omega-\omega_s)t}}{\sqrt{\omega - \omega_s}} \tag{18.38}$$

The $\int d\omega$ integral in the above has the value $i^{-\frac{1}{2}}2\sqrt{\pi}/\sqrt{t}$. This can be proven by a contour integration along the contour in the cut ω plane where ω_s is the branch point. Thus we finally obtain

$$G(x, t) = \frac{\sqrt{i}}{2} \frac{c}{\sqrt{2\pi}} \frac{e^{-i\omega_s t}}{\sqrt{\omega_s t}} \sim \frac{e^{ck_0 t}}{\sqrt{t}} \tag{18.39}$$

which indicates an absolute instability, independent of x.

The method presented above can be generalized for a dispersion relation $D(\omega, k)$, provided we can find ω_c and k_c which satisfy $D(\omega_c, k_c) = 0$ and $\frac{\partial D}{\partial k}|_c = 0$. The latter conditions are equivalent to that $D(\omega_c, k_c)$ has a *double root* at $\omega = \omega_c$ and $k = k_c$. It is seen that this method is not applicable to (18.28).

General formula. When a double root is formed in $D(k, \omega)$ at $\omega = \omega_c$ and $k = k_c$, we expand the dispersion function near $\omega = \omega_c$ and $k = k_c$:

$$D(\omega, k) \simeq (\omega - \omega_c)\frac{\partial D}{\partial \omega}\bigg|_{\omega_c, k_c} + \frac{1}{2}(k - k_c)^2 \frac{\partial^2 D}{\partial k^2}\bigg|_{\omega_c, k_c} = 0 \tag{18.40}$$

which yields in turn

$$k - k_c = \pm\sqrt{2}i \left[(\omega - \omega_c)\frac{\partial D}{\partial \omega}\bigg|_c \div \frac{\partial^2 D}{\partial k^2}\bigg|_c\right]^{\frac{1}{2}} \tag{18.41}$$

where the subscript 'c' stands for (ω_c, k_c). We also have

$$\frac{\partial D}{\partial k} \simeq \left.\frac{\partial^2 D}{\partial k^2}\right|_c (k - k_c) \tag{18.42}$$

Using (18.41) in (18.42) gives

$$\frac{\partial D}{\partial k} = \pm\sqrt{2}i\sqrt{\omega - \omega_c}\left[\frac{\partial^2 D}{\partial k^2}\frac{\partial D}{\partial \omega}\right]_c^{\frac{1}{2}} \tag{18.43}$$

We take a single term corresponding to the upper-most singularity in the ω-plane which dominates the time-asymptotic behavior of $G(x,t)$. Using (18.43) in (18.30) gives

$$G(t,x) = \left[2\frac{\partial^2 D}{\partial k^2}\frac{\partial D}{\partial \omega}\right]_c^{-\frac{1}{2}} e^{ik_c x - i\omega_c t}\int_{-\infty+i\sigma}^{\infty+i\sigma}\frac{d\omega}{2\pi}\frac{e^{-i(\omega-\omega_c)t}}{\sqrt{\omega - \omega_c}} \tag{18.44}$$

After the Laplace inversion integral, we obtain

$$G(t,x) = \left[2\pi i\frac{\partial^2 D}{\partial k^2}\frac{\partial D}{\partial \omega}\right]_c^{-\frac{1}{2}}\frac{e^{ik_c x - i\omega_c t}}{\sqrt{t}} \tag{18.45}$$

When the imaginary part of ω_c is > 0, $G(t,x)$ in (18.45) grows indefinitely in time regardless of x, which is an absolute instability.

Now we summarize the results obtained in the forgoing discussion. Let us suppose that the dispersion relation $D(\omega,k) = 0$ has unstable roots ($\mathrm{Im}\,\omega > 0$) for real k. Then the system is unstable. What determines the instability as convective or absolute? If the dispersion relation $D(\omega,k) = 0$ has a double root at $k = k_c$ and $\omega = \omega_c$, we have an absolute instability. Otherwise, the instability is convective. The condition $\partial D/\partial k_c = 0$ at $\omega = \omega_c$ is equivalent to the condition $\partial\omega_c/\partial k_c = 0$, i.e. the group velocity is zero. The wave energy grows in time at every point while the wave is spreading out without propagation.

Problems

18.1 Determine the nature of instability for

$$1 - \frac{\omega_p^2}{\omega^2} - \frac{\omega_b^2}{(\omega - kv_b)^2} = 0.$$

Hint: This equation can be cast into the form,

$$(\omega^2 - \omega_p^2)[(\omega - kv_b)^2 - \omega_b^2] = \omega_b^2\omega_p^2$$

Draw the asymptotes of the hyperbolas, which are the two characteristic lines. State why we conclude that we have a convective instability.

18.2 Evaluate the integral in (18.38). The integral in (18.38) is transformed via $\omega \to \omega_s + iz$ to

$$\int_{-\infty+i\sigma}^{\infty+i\sigma} d\omega \frac{e^{-i(\omega-\omega_s)t}}{\sqrt{\omega-\omega_s}} = -\sqrt{i}\int_{-i\infty+c}^{i\infty+c} dz \frac{e^{zt}}{\sqrt{z}} = i^{-\frac{1}{2}}\frac{2\sqrt{\pi}}{\sqrt{t}}$$

This is the standard form of the inverse Laplace transform of $1/\sqrt{z}$. [See F. B. Hildebrand, *Advanced Calculus for Applications*, Prentice-Hall (1976), p. 627].

18.3 Evaluate the integral in (8.27) with $D(k,\omega)$ given by (8.29) in the following way.

$$I = \int_{-\infty}^{\infty} \frac{dk}{2\pi}\frac{e^{ikx}}{D(k,\omega)} = \int_{-\infty}^{\infty} \frac{dk}{2\pi} e^{ikx}\frac{1}{2\kappa}\left(\frac{1}{k-\kappa} - \frac{1}{k+\kappa}\right)$$

where $\kappa = \sqrt{\omega^2/c^2 + k_0^2}$. When $x > 0$, we can assume that only the singularity $k = \kappa$ contributes to the integral by using the contour enclosing the upper k-plane. Hence,

$$I = \frac{i}{2}\frac{c}{\sqrt{\omega^2 + c^2k_0^2}} \exp\left(\frac{ix}{c}\sqrt{\omega^2 + c^2k_0^2}\right)$$

Hence,

$$G(x,t) = \int_{-\infty+i\sigma}^{\infty+i\sigma} \frac{d\omega}{2\pi} e^{-i\omega t}\frac{ic}{2}\frac{\exp\left(\frac{ix}{c}\sqrt{(\omega+\omega_s)(\omega-\omega_s)}\right)}{\sqrt{(\omega+\omega_s)(\omega-\omega_s)}}$$

where $\omega_s = ick_0$. Consider the pole at $\omega = \omega_s$, the branch point. Putting $\omega = \omega_s$ in the above, evaluate the integral to obtain

$$G(x,t) = \int_{-\infty+i\sigma}^{\infty+i\sigma} \frac{d\omega}{2\pi} e^{-i\omega t}\frac{ic}{2}\frac{1}{\sqrt{2\omega_s}\sqrt{\omega-\omega_s}}$$

$$= \frac{c\sqrt{i}}{2\sqrt{2\pi\omega_s}}\frac{e^{-i\omega_s t}}{\sqrt{t}}$$

which agrees with (18.39).

18.4 Evaluate the integral

$$G(x,t) = \int_{-\infty+i\sigma}^{\infty+i\sigma} \frac{d\omega}{2\pi} e^{-i\omega t}\int_{-\infty}^{\infty} \frac{dk}{2\pi}\frac{e^{ikx}}{\left(k - \frac{\omega}{c}\right)^2} = -x\delta\left(t - \frac{x}{c}\right)$$

Take the limit $k_0 \to 0$ in (18.33) to recover this result.

18.5 Obtain the kinetic dispersion relation when the zero order distribution functions $F_{i0} = \delta(\mathbf{v} - \mathbf{V}_0)$ and $F_{e0} = \delta(\mathbf{v})$ are perturbed electrostatically. Use (8.24), Chapter 8, and ascertain that the result agrees with (18.10).

References

Akhiezer, A., I. Akhiezer, R. Polovin, A. Sitenko, K. Stepanov, *Collective Oscillations in a Plasma*, Pergamon, Oxford, (1967).

Brambilla, M. *Kinetic Theory of Plasma Waves*, Clarendon, Oxford (1998).

Briggs, R. J. *Electron-Stream Interaction with Plasmas*, MIT Press, Cambridge, Massachusetts (1964).

Clemmow, P. C. and J. P. Daugherty, *Electrodynamics of particles and Plasmas*, Addison-Wesley, London (1969).

Krall, N. A. and A. W. Trivelpiece, *Principles of Plasma Physics*, San Francisco Press (1986).

Appendix A

Fourier Transform, Delta Function, and Useful Formulas

A.1 Fourier Transform, Delta Function

We define the Fourier transform pair as follows:

$$\tilde{f}(k) = \int_{-\infty}^{\infty} f(x)e^{-ikx}dx \tag{A.1}$$

$$f(x) = \frac{1}{2\pi}\int_{-\infty}^{\infty} \tilde{f}(k)e^{ikx}dk \tag{A.2}$$

We often drop the tilde on Fourier transforms when there is no danger of confusion. Instead, the argument is written explicitly. In case of a space-time function $f(x,t)$, its Fourier transform is defined as

$$f(k,\omega) = \int_{-\infty}^{\infty}\int_{-\infty}^{\infty} f(x,t)e^{-ikx+i\omega t}dxdt$$

Its inversion is

$$f(x,t) = \left(\frac{1}{2\pi}\right)^2 \int_{-\infty}^{\infty}\int_{-\infty}^{\infty} f(k,\omega)e^{ikx-i\omega t}dkd\omega$$

Our convention of Fourier transform is consistent with the form of wave expressed by the phasor $e^{ikx-i\omega t}$. With this phasor, it is often tacitly assumed that the imaginary part of the frequency ω_i ($\omega = \omega_r + i\omega_i$) is infinitesimally positive. This signifies the *causality* that the perturbation was non-existent in the infinitely remote past ($t = -\infty$).

The operational property of the delta function is

$$\int_{-\infty}^{\infty} f(x)\delta(x-a)dx = \int_{a-\epsilon}^{a+\epsilon} f(x)\delta(x-a)dx = f(a) \tag{A.3}$$

We take $f(x) = \delta(x)$ in (A.1). Then (A.1) gives $\tilde{f}(k) = 1$ due to (A.3). Then (A.2) gives

$$\delta(x) = \frac{1}{2\pi}\int_{-\infty}^{\infty} e^{ikx}dk = \frac{1}{2\pi}\lim_{n\to\infty}\int_{-n}^{n} e^{ikx}dk = \lim_{n\to\infty}\frac{\sin(nx)}{\pi x} \tag{A.4}$$

which is one of the well known expressions for the *Dirac delta function*. Clearly, the RHS of (A.4) is real. So we can write

$$\delta(x) = \frac{1}{2\pi}\text{Re}\int_{-\infty}^{\infty} e^{ikx}dk = \frac{1}{2\pi}\int_{-\infty}^{\infty} \cos\ kx\ dk \tag{A.5}$$

This is true because $\int_{-\infty}^{\infty} \sin kx\ dk = 0$. (A.5) can be written as

$$\delta(x) = \frac{1}{\pi}\int_{0}^{\infty} \cos kx\ dk = \frac{1}{\pi}\ \text{Re}\int_{0}^{\infty} e^{ikx}dk \tag{A.6}$$

Let us consider the integral

$$\begin{aligned}
&\lim_{t\to\infty}\int_0^t e^{ix(t-t')}dt' + \text{c.c.}\\
&= \lim_{t\to\infty}\int_0^t e^{ix\xi}d\xi + \text{c.c. per } (t-t'=\xi)\\
&= \int_0^{\infty} e^{ix\xi}d\xi + \ \text{c.c.} = 2\ \text{Re}\int_0^{\infty} e^{ix\xi}d\xi
\end{aligned}$$

Thus we have the identity

$$\delta(x) = \frac{1}{2\pi}\left[\lim_{t\to\infty}\int_0^t e^{ix(t-t')}dt' + \text{c.c.}\right] \tag{A.7}$$

One can verify that direct integration of (A.7) yields (A.4). Direct integration of (A.6) also recovers (A.4). It is left as an exercise to prove the asymptotic relation $(t > 0)$

$$\lim_{t\to\infty}\frac{1-e^{-ixt}}{x} = P\ \frac{1}{x} + i\pi\delta(x) \tag{A.8}$$

where P denotes the principal value. It is useful to know the following properties of δ-function.

$$\frac{\partial}{\partial x}\delta(x-a) = -\frac{\partial}{\partial a}\delta(x-a) \tag{A.9}$$

$$\frac{d}{dt}\delta[X(t)] = \frac{d}{dX}\delta(X)\ \frac{dX(t)}{dt} \tag{A.10}$$

We can also prove from the above equation by integration by parts

$$\int_{-\infty}^{\infty} f(x)\delta'(x-a)dx = -f'(a) \tag{A.11}$$

$$\int_{-\infty}^{\infty} f(x)\delta''(x-a)dx = f''(a) \tag{A.12}$$

An important property of the delta function is

$$\delta[g(x)] = \sum_{a_i} \frac{\delta(x-a_i)}{|g'(a_i)|} \tag{A.13}$$

where a_i $(i = 1, 2, 3, \ldots)$ is a root of $g(x) = 0$. In particular, we have

$$\delta(cx) = \frac{1}{|c|}\delta(x) \tag{A.14}$$

As an application of (A.13), a formula often used in this book is

$$\int_{-\infty}^{\infty} \frac{\partial g}{\partial v}\delta(\omega - kv)dv = \frac{1}{k}\left[\frac{\partial g}{\partial v}\right]_{v=\frac{\omega}{k}} \tag{A.15}$$

As another application of (A.13), we can write

$$\delta[(x-a)(x-b)] = \frac{\delta(x-a)+\delta(x-b)}{|a-b|} \tag{A.16}$$

The delta function is visualized as

$$\delta(x) = \begin{cases} 0 & \text{when } x \neq 0 \\ \infty & \text{when } x = 0 \end{cases} \tag{A.17}$$

so that it has unit area when integrated:

$$\int_{-\infty}^{\infty} \delta(x)dx = 1 \tag{A.18}$$

There are many functions (*distribution*) which satisfy the properties (A.17) and (A.18):

$$\delta(x) = \lim_{\beta\to\infty}\sqrt{\frac{\beta}{\pi}}e^{-\beta x^2} = \lim_{\epsilon\to 0}\sqrt{\frac{1}{\pi\epsilon}}e^{-\frac{x^2}{\varepsilon}} \tag{A.19}$$

A distribution of x is a delta function $\delta(x)$ when it is described by

$$\delta(x) = \begin{cases} \frac{1}{2\epsilon} & \text{when } |x| < \epsilon \\ 0 & \text{when } |x| > \epsilon \end{cases} \tag{A.20}$$

where $\epsilon \to 0$ is implied. Also we have

$$\delta(x) = \lim_{\epsilon \to 0} \frac{\epsilon}{\pi(\epsilon^2 + x^2)} \tag{A.21}$$

(A.21) is useful for deriving *Plemelj's formula* (A.22):

$$\frac{1}{\omega \pm i\epsilon} = \frac{\omega \mp i\epsilon}{\omega^2 + \epsilon^2} = \frac{\omega}{\omega^2 + \epsilon^2} \mp i\pi\delta(\omega) \ per \ (A.21)$$

The operational product with the first term is

$$\lim_{\epsilon \to 0} \int_{-\infty}^{\infty} \frac{f(\omega)\omega}{\omega^2 + \epsilon^2} d\omega = \lim_{\epsilon \to 0} \int_{-\infty}^{-\epsilon} \frac{f(\omega)\omega}{\omega^2 + \epsilon^2} d\omega$$

$$+ \lim_{\epsilon \to 0} \int_{\epsilon}^{\infty} \frac{f(\omega)\omega}{\omega^2 + \epsilon^2} d\omega + \lim_{\epsilon \to 0} \int_{-\epsilon}^{\epsilon} \frac{f(\omega)\omega}{\omega^2 + \epsilon^2} d\omega$$

$$= P \int_{-\infty}^{\infty} f(\omega) \frac{d\omega}{\omega} + f(0) \lim_{\epsilon \to 0} \int_{-\epsilon}^{\epsilon} \frac{\omega}{\omega^2 + \epsilon^2} d\omega$$

The last term vanishes because it is an odd function integral. So we obtain the Plemelj formula

$$\lim_{\epsilon \to 0} \frac{1}{\omega \pm i\epsilon} = P\frac{1}{\omega} \mp i\pi\delta(\omega) \tag{A.22}$$

An alternative derivation of (A.22) is given later.

In the following, we write useful relations.

$$\delta(-x) = \delta(x) \tag{A.23}$$

$$x\delta(x) = 0 \tag{A.24}$$

$$\delta(x) = -x\delta'(x) \tag{A.25}$$

$$\int_{-\infty}^{\infty} \delta(x-a)\delta(x-b)dx = \delta(a-b) \tag{A.26}$$

$$\sum_{n=-\infty}^{\infty} e^{inx} = 2\pi\delta(x) \tag{A.27}$$

The last formula is the summation version of (A.4): $\sum_n \to \int dn$.

The three-dimensional delta functions are

$$\delta(\mathbf{r} - \mathbf{r}') = \delta(x - x')\delta(y - y')\delta(z - z') \quad \text{(cartesian coordinates)} \tag{A.28}$$

$$\delta(\mathbf{r}-\mathbf{r}') = \frac{1}{\rho}\delta(\rho-\rho')\delta(\varphi-\varphi')\delta(z-z') \quad \text{(cylindrical coordinates)} \tag{A.29}$$

$$\begin{aligned}\delta(\mathbf{r}-\mathbf{r}') &= \frac{1}{r^2}\delta(r-r')\delta(\cos\theta-\cos\theta')\delta(\varphi-\varphi') \\ &= \frac{1}{r^2}\delta(r-r')\delta(\Omega-\Omega') \quad \text{(spherical coordinates)}\end{aligned} \tag{A.30}$$

where Ω is the solid angle.

From the Gauss theorem we can prove

$$\nabla_r^2 \frac{1}{|\mathbf{r}-\mathbf{r}'|} = -4\pi\delta(\mathbf{r}-\mathbf{r}') = -\nabla_{r'}^2 \frac{1}{|\mathbf{r}-\mathbf{r}'|} \tag{A.31}$$

A.2 Step Function

We deal with the Fourier transform of the Heaviside step function

$$H(t) = \begin{cases} 1 & t > 0 \\ 0 & t < 0 \end{cases} \tag{A.32}$$

The key equation for the development presented in the text is

$$P\int_{-\infty}^{\infty} \frac{e^{i\omega t}}{\omega} d\omega = \begin{cases} i\pi & t > 0 \\ -i\pi & t < 0 \end{cases} \tag{A.33}$$

This equation is a kind of symmetrical step function. (A.33) results from the contour integral in the complex z-plane

$$\oint_C \frac{e^{izt}}{z}\, dz \tag{A.34}$$

where the contour consists of the real axis and the infinite semi-circle winding the upper or the lower half plane, depending upon the sign of t. The details are referred to standard text books of Mathematical Physics. Here we present another derivation of (A.33). We have the well-known integral

$$\int_{-\infty}^{\infty} \frac{\sin\, x}{x}\, dx = \pi \tag{A.35}$$

Note that $x = 0$ is not singular. A change of variable in the above $x = ny$ with $n > 0$ gives

$$\int_{-\infty}^{\infty} \frac{\sin\, nx}{\pi x}\, dx = 1 \quad (n > 0) \tag{A.36}$$

Note that the integral is independent of the value of n. Multiply both sides of (A.36) by (-1) and put $-n = m$ to write

$$\int_{-\infty}^{\infty} \frac{\sin\, mx}{\pi x}\, dx = -1 \quad (m < 0) \tag{A.37}$$

Combining (A.36) and (A.37) yields (A.33), since $\int_{-\infty}^{\infty} \frac{\cos\, \omega t}{\omega}\, d\omega = 0$. Making a change of variable $\omega \to -\omega$ in (A.33) gives

$$P \int_{-\infty}^{\infty} \frac{e^{-i\omega t}}{\omega}\, d\omega = \begin{cases} -i\pi & t > 0 \\ i\pi & t < 0 \end{cases} \tag{A.38}$$

Next we Fourier transform $H(t)$ in (A.32).

$$\begin{aligned} H(\omega) &= \int_{-\infty}^{\infty} H(t)\, e^{i\omega t} dt = \int_{0}^{\infty} e^{i\omega t} dt \\ &= \frac{i}{\omega} = P\, \frac{i}{\omega} + \lambda\, \delta(\omega) \end{aligned} \tag{A.39}$$

where we used $\omega_i > 0$. To determine λ, we Fourier invert (A.39) and fit it to $H(t)$.

$$H(t) = \frac{1}{2\pi} \int_{-\infty}^{\infty} d\omega\, e^{-i\omega t} \left(P\, \frac{i}{\omega} + \lambda\, \delta(\omega) \right)$$

Using (A.38), one finds $\lambda = \pi$ and it follows:

$$H(\omega) = P\, \frac{i}{\omega} + \pi\, \delta(\omega) \tag{A.40}$$

We identify the real and imaginary parts of $H(\omega)$

$$\text{Re}\ H(\omega) = \pi\, \delta(\omega), \quad \text{Im}\ H(\omega) = P\frac{1}{\omega} \tag{A.41}$$

and see that they satisfy the Kramers–Kronig relations. Also one can show as a corollary of (A.38)

$$P \int_{-\infty}^{\infty} \frac{e^{-i\omega t}}{\omega - \omega_0}\, d\omega = \begin{cases} -i\pi\ e^{-i\omega_0 t} & t > 0 \\ i\pi\ e^{-i\omega_0 t} & t < 0 \end{cases} \tag{A.42}$$

$$P \int_{-\infty}^{\infty} \frac{e^{i\omega t}}{\omega - \omega_0}\, d\omega = \begin{cases} i\pi\ e^{i\omega_0 t} & t > 0 \\ -i\pi\ e^{i\omega_0 t} & t < 0 \end{cases} \tag{A.43}$$

Next let us consider integral

$$I = P \int_{-\infty}^{\infty} \frac{F(\omega)}{\omega - \omega_0}\, d\omega$$

The idea for evaluation of this integral is to express $F(\omega)$ by the Fourier integral and apply (A.42) and (A.43).

$$I = \int_{-\infty}^{\infty} dt\ F(t)\ P \int_{-\infty}^{\infty} \frac{e^{i\omega t}}{\omega - \omega_0} d\omega$$

$$= i\pi \int_{0}^{\infty} dt\ F(t)\ e^{i\omega_0 t} - i\pi \int_{-\infty}^{0} dt\ F(t)\ e^{i\omega_0 t} \tag{A.44}$$

Therefore we obtain

$$I = P \int_{-\infty}^{\infty} \frac{F(\omega)}{\omega - \omega_0}\ d\omega = i\pi\ (F^+(\omega_0) - F^-(\omega_0)) \tag{A.45}$$

where

$$F^+(\omega_0) = \int_{0}^{\infty} dt\ F(t)\ e^{i\omega_0 t}$$

$$F^-(\omega_0) = \int_{-\infty}^{0} dt\ F(t)\ e^{i\omega_0 t}$$

Clearly we have

$$F(\omega_0) = \int_{-\infty}^{\infty} F(t)\ e^{i\omega_0 t} dt = F^+(\omega_0) + F^-(\omega_0) \tag{A.46}$$

(A.45) and (A.46) give

$$F^+(\omega_0) = \frac{1}{2\pi i}\left[P \int_{-\infty}^{\infty} \frac{F(\omega) d\omega}{\omega - \omega_0} + i\pi F(\omega_0)\right] \tag{A.47}$$

$$F^-(\omega_0) = \frac{-1}{2\pi i}\left[P \int_{-\infty}^{\infty} \frac{F(\omega) d\omega}{\omega - \omega_0} - i\pi F(\omega_0)\right] \tag{A.48}$$

This result can be obtained alternatively by using Cauchy integral formula. (A.47) and (A.48) are expressions for the half-range Fourier integrals.

$H(x)$ has an infinite slope at $x = 0$, and elsewhere the slope is zero. Thus we have

$$\frac{dH}{dx} = \delta(x) \tag{A.49}$$

Integrating the above, we have

$$H(x) = \int_{-\infty}^{x} \delta(t) dt \tag{A.50}$$

Clearly, this integral is zero when $x < 0$, and 1 when $x > 0$.

Using (A.33), the step functions can be written in the form

$$H(x) = \frac{1}{2} + \frac{1}{2\pi i} P \int_{-\infty}^{\infty} \frac{e^{i\omega x}}{\omega} d\omega \tag{A.51}$$

Differentiating (A.51) gives $dH(x)/dx = \delta(x)$ as expected.

Example. We have the following two equations for step function $H(t)$:

$$H(at) = H(t) \quad \text{if } a > 0 \tag{A.52}$$

$$H(at) = 1 - H(t) \quad \text{if } a < 0 \tag{A.53}$$

Differentiating (A.52) gives

$$a \frac{dH(at)}{d(at)} = \frac{dH}{dt} \quad \text{or} \quad a\delta(at) = \delta(t) \ \text{ if } a > 0 \tag{A.54}$$

Differentiating (A.53) gives

$$a \frac{dH(at)}{d(at)} = -\frac{dH}{dt} \quad \text{or} \quad a\delta(at) = -\delta(t) \ \text{ if } a < 0 \tag{A.55}$$

Therefore we conclude $\delta(at) = \delta(t)/|a|$.

We have another kind of step function:

$$S(x) = \begin{cases} 1 & \text{if } x \geq 0 \\ -1 & \text{if } x < 0 \end{cases} \tag{A.56}$$

$S(x)$ has a parity: $S(-x) = -S(x)$; odd function of x. Clearly,

$$H(x) = \frac{1}{2}[S(x) + 1] \tag{A.57}$$

We also have

$$\frac{dS}{dx} = 2\frac{dH}{dx} = 2\delta(x) \tag{A.58}$$

$$S(x) = \frac{1}{i\pi} P \int_{-\infty}^{\infty} \frac{e^{ikx}}{k} dk = \frac{1}{\pi} P \int_{-\infty}^{\infty} \frac{\sin kx}{k} dk \tag{A.59}$$

Here $k = 0$ is not a singularity, thus we don't need the principal value symbol 'P'. Thus we simply write

$$S(x) = \frac{1}{i\pi} \int_{-\infty}^{\infty} \frac{e^{ikx}}{k} dk \tag{A.60}$$

which in turn gives

$$\frac{dS}{dx} = \frac{1}{\pi} \int_{-\infty}^{\infty} e^{ikx} dk = 2\ \delta(x) \tag{A.61}$$

A.3 Derivation of Plemelj Formula

Consider the integral

$$I = \lim_{\epsilon \to 0} \int_a^b \frac{f(x)dx}{x - x_0 + i\epsilon} \qquad (\epsilon > 0 \quad and \quad a < x_0 < b)$$

which can be written as

$$\begin{aligned} I &= \lim_{\epsilon \to 0} \int_a^b \frac{f(x) - f(x_0)}{x - x_0 + i\epsilon}\, dx \; + f(x_0) \lim_{\epsilon \to 0} \int_a^b \frac{dx}{x - x_0 + i\epsilon} \\ &= \int_a^b \frac{f(x) - f(x_0)}{x - x_0} dx + f(x_0)[\ln(b - x_0 + i\epsilon) - \ln(a - x_0 + i\epsilon)] \end{aligned}$$

Note that the complex numbers $b - x_0 + i\epsilon = (b - x_0)e^{i0^+}$ and $a - x_0 + i\epsilon = (x_0 - a)e^{i\pi}$. Therefore,

$$I = \int_a^b \frac{f(x) - f(x_0)}{x - x_0} dx + f(x_0) \left(\ln \frac{b - x_0}{x_0 - a} - i\pi \right) \tag{A.62}$$

On the other hand

$$\begin{aligned} P \int_a^b \frac{f(x)dx}{x - x_0} &= \int_a^b \frac{f(x) - f(x_0)}{x - x_0} dx + f(x_0) P \int_a^b \frac{dx}{x - x_0} \\ &= \int_a^b \frac{f(x) - f(x_0)}{x - x_0} dx + f(x_0) \lim_{\epsilon \to 0} \left[\int_a^{x_0 - \epsilon} \frac{dx}{x - x_0} + \int_{x_0 + \epsilon}^b \frac{dx}{x - x_0} \right] \end{aligned}$$

The second term $[\cdots]$ gives upon integration, $\ln(-\epsilon) - \ln(a - x_0) + \ln(b - x_0) - \ln(\epsilon)$. Note that $\ln(-\epsilon) = \ln \epsilon + i\pi$ and $\ln(a - x_0) = \ln(x_0 - a) + i\pi$. Thus the second term equals to $f(x_0) \ln \frac{b - x_0}{x_0 - a}$. Therefore, we have

$$P \int_a^b \frac{f(x)dx}{x - x_0} = \int_a^b \frac{f(x) - f(x_0)}{x - x_0} dx + f(x_0) \ln \frac{b - x_0}{x_0 - a} \tag{A.63}$$

Equations (A.62) and (A.63) give

$$I = \lim_{\epsilon \to 0} \int_a^b \frac{f(x)dx}{x - x_0 + i\epsilon} = P \int_a^b \frac{f(x)dx}{x - x_0} - i\pi f(x_0) \tag{A.64}$$

Likewise we can prove that for $\epsilon < 0$ the last term changes sign, i.e. $+i\pi f(x_0)$:

$$\lim_{\epsilon \to 0} \int_a^b \frac{f(x)dx}{x - x_0 - i\epsilon} = P \int_a^b \frac{f(x)dx}{x - x_0} + i\pi f(x_0) \tag{A.65}$$

A.4 Convolution Integral

Convolution of two functions is

$$(f * g)_{\mathbf{r}} = \int_{-\infty}^{\infty} g(\mathbf{r}')f(\mathbf{r}-\mathbf{r}')d^3r' = \int_{-\infty}^{\infty} f(\mathbf{r}')g(\mathbf{r}-\mathbf{r}')d^3r' \quad \text{(A.66)}$$

It can be easily proved that

$$I = \int e^{-i\mathbf{k}\cdot\mathbf{r}}(f * g)_{\mathbf{r}}\ d^3r = \tilde{f}(\mathbf{k})\ \tilde{g}(\mathbf{k}) \quad \text{(A.67)}$$

In words, the Fourier transform of a convolution integral is the product of each Fourier transform. The inverse of $\tilde{f}(\mathbf{k})\ \tilde{g}(\mathbf{k})$=

$$\text{Inverse of } \tilde{f}(\mathbf{k})\ \tilde{g}(\mathbf{k}) = \left(\frac{1}{2\pi}\right)^3 \int e^{i\mathbf{k}\cdot\mathbf{r}}\tilde{f}(\mathbf{k})\ \tilde{g}(\mathbf{k})d^3k = (f * g)_{\mathbf{r}}$$

In words, the Fourier transform (or the inverse transform) of the product of two functions is the convolution. We prove the identity (A.67) by means of the technique used repeatedly in the text. We have in (A.67)

$$f(\mathbf{k}) = \int f(\mathbf{r})e^{-i\mathbf{k}\cdot\mathbf{r}}d^3r$$

$$g(\mathbf{k}) = \int g(\mathbf{r})e^{-i\mathbf{k}\cdot\mathbf{r}}d^3r$$

where the tilde is omitted. Substitute the inverse relations

$$f(\mathbf{r}') = \left(\frac{1}{2\pi}\right)^3 \int f(\mathbf{k}')e^{i\mathbf{k}'\cdot\mathbf{r}'}d^3k'$$

$$g(\mathbf{r}-\mathbf{r}') = \left(\frac{1}{2\pi}\right)^3 \int g(\mathbf{k}'')e^{i\mathbf{k}''\cdot(\mathbf{r}-\mathbf{r}')}d^3k''$$

Therefore,

$$I = \int d^3r e^{-i\mathbf{k}\cdot\mathbf{r}} \int d^3r' \left[\left(\frac{1}{2\pi}\right)^3 \int f(\mathbf{k}')e^{i\mathbf{k}'\cdot\mathbf{r}'}d^3k'\right]\left[\left(\frac{1}{2\pi}\right)^3 \int g(\mathbf{k}'')e^{i\mathbf{k}''\cdot(\mathbf{r}-\mathbf{r}')}d^3k''\right]$$

In the above, use

$$\left(\frac{1}{2\pi}\right)^3 \int d^3r' e^{i(\mathbf{k}'-\mathbf{k}'')\cdot\mathbf{r}'} = \delta(\mathbf{k}'-\mathbf{k}'')$$

$$\left(\frac{1}{2\pi}\right)^3 \int d^3r e^{-i\mathbf{k}\cdot\mathbf{r}+i\mathbf{k}''\cdot\mathbf{r}} = \delta(\mathbf{k}-\mathbf{k}'')$$

Then we have

$$I = \int f(\mathbf{k}')d^3k'g(\mathbf{k}'')d^3k''\delta(\mathbf{k}' - \mathbf{k}'')\delta(\mathbf{k} - \mathbf{k}'') = f(\mathbf{k})g(\mathbf{k}) \quad \text{Q.E.D.}$$

• Fourier Transform of

$$\begin{aligned}
[A(x)B(x)] &= \int e^{-ikx}A(x)B(x)dx = \frac{1}{2\pi}\int dxe^{-ikx}A(x)\int e^{ik'x}B(k')dk' \\
&= \frac{1}{2\pi}\int dk'B(k')\int e^{-i(k-k')x}A(x)dx = \frac{1}{2\pi}\int dk'B(k')A(k-k') \\
&= \frac{1}{2\pi}\int dk'B(k')\int dk''A(k'')\delta(k''-k+k') \qquad (A.68)
\end{aligned}$$

A.5 First Order Linear Differential Equation

$$\frac{dy}{dx} + a(x)y + b(x) = 0 \qquad (A.69)$$

$$\text{Transform: } f(x) = y(x)e^{\int^x a(x')dx'} \qquad (A.70)$$

$$\frac{df}{dx} = \left(\frac{dy}{dx} + a(x)y\right)e^{\int^x a(x')dx'} \qquad (A.71)$$

$$\frac{df}{dx} + b(x)e^{\int^x a(x')dx'} = 0 \qquad (A.72)$$

Indefinite integral form:

$$y(x) = e^{-\int^x a(x')dx'}\left[C - \int^x dx'b(x')e^{\int^{x'} a(x'')dx''}\right] \quad (C: \text{ const.}) \qquad (A.73)$$

Definite integral form:

$$\int_{f(x_0)}^{f(x)} df(x') + \int_{x_0}^{x} dx'b(x')e^{\int^{x'} a(x'')dx''} = 0 \qquad (A.74)$$

$$f(x) - f(x_0) = -\int_{x_0}^{x} dx'b(x')e^{\int^{x'} a(x'')dx''} \qquad (A.75)$$

$$y(x)e^{\int^x a(x')dx'} - y(x_0)e^{\int^{x_0} a(x')dx'} = -\int_{x_0}^{x} dx'b(x')e^{\int^{x'} a(x'')dx''} \qquad (A.76)$$

$$y(x) = e^{-\int^x a(x')dx'}\left[y(x_0)e^{\int^{x_0} a(x')dx'} - \int_{x_0}^{x} dx'b(x')e^{\int^{x'} a(x'')dx''}\right] \qquad (A.77)$$

A.6 Differentiation of an Integral: Leibnitz's Rule

$$I(\alpha) = \int_{x_1(\alpha)}^{x_2(\alpha)} f(x,\alpha)dx \tag{A.78}$$

$$\frac{dI}{d\alpha} = f(x_2,\alpha)\,\frac{dx_2}{d\alpha} - f(x_1,\alpha)\,\frac{dx_1}{d\alpha} + \int_{x_1(\alpha)}^{x_2(\alpha)} \frac{\partial f}{\partial \alpha}\,dx \tag{A.79}$$

Appendix B

Multiple Scale Perturbation Analysis

In usual perturbation analysis, an equation to be solved is expanded straightforwardly in a power series of ϵ, the small parameter involved in the equation under consideration. In this case the higher order solution usually contains secular terms which diverge as the independent variable t becomes large. This means that the naive straight expansion is unable to predict correct approximate solution, and this naive perturbation scheme should be remedied. We consider the time t consists of times of different scales:

$$t_0 = t, \quad t_1 = \epsilon t, \quad t_2 = \epsilon^2 t, \quad t_3 = \epsilon^3 t, \dots$$

$$\frac{d}{dt} = \frac{\partial}{\partial t_0} + \epsilon \frac{\partial}{\partial t_1} + \epsilon^2 \frac{\partial}{\partial t_2} + \cdots \tag{B.1}$$

Representing a dynamical change in terms of different time scales can be compared with reading time on a watch (quoted from Nayfeh: A. H. Nayfeh (1981) *Introduction to Perturbation Techniques*, Wiley). '$\epsilon = \frac{1}{60}$, variations on the scale t_0 can be observed on the second arm of a watch, variations on the scale t_1 can be observed on the minute arm of a watch, and variations on the scale t_2 can be observed on the hour arm of a watch. Thus, t_0 represents a fast scale, t_1 represents a slower scale, t_2 represents an even slower scale, and so on'. By this scheme, uniform solutions can be obtained by removing the secularity-causing terms order by order. In this Appendix, this perturbation method is illustrated to present the essential features of the analysis through engaging with two examples. First we consider the *Van der Pol equation* (in dimensionless form)

$$\frac{d^2x}{dt^2} + x - \epsilon(1 - x^2)\frac{dx}{dt} = 0 \quad (\epsilon \ll 1) \tag{B.2}$$

While we expand t according to the scheme (B.1), we also expand the function x as

$$x = x_0(t_0, t_1, t_2, \ldots) + \epsilon x_1(t_0, t_1, t_2, \ldots) + \epsilon^2 x_2(t_0, t_1, t_2, \ldots) + \cdots \tag{B.3}$$

One easily obtain

$$\frac{d^2}{dt^2} = \frac{\partial^2}{\partial t_0^2} + \epsilon\, 2 \frac{\partial}{\partial t_0}\frac{\partial}{\partial t_1} + \epsilon^2 \left(2\frac{\partial}{\partial t_0}\frac{\partial}{\partial t_2} + \frac{\partial^2}{\partial t_1^2}\right) + \cdots \tag{B.4}$$

Putting (B.3) and (B.4) into (B.2) yields

$$\epsilon^0 : \ \frac{\partial^2 x_0}{\partial t_0^2} + x_0 = 0 \tag{B.5}$$

$$\epsilon^1 : \ \frac{\partial^2 x_1}{\partial t_0^2} + x_1 = -2\frac{\partial^2 x_0}{\partial t_0 \partial t_1} + (1 - x_0^2)\frac{\partial x_0}{\partial t_0} \tag{B.6}$$

(B.5) is solved by

$$x_0 = A(t_1, t_2, \ldots)e^{it_0} + A^*(t_1, t_2, \ldots)e^{-it_0} \tag{B.7}$$

where $*$ denotes the complex conjugate. If (B.7) is substituted into the RHS of (B.6), various harmonics of $e^{\pm it_0}$ are generated: terms of $e^{\pm it_0}$, $e^{\pm 2it_0}$, and $e^{\pm 3it_0}$. Among these, those terms which are oscillating with the phase $e^{\pm it_0}$ are the secularity causing terms which should be removed by putting the sum of them equal to zero. [Recall that the solution of $\frac{d^2y}{dt^2} + \omega^2 y = \cos\omega t$ is $y \sim t\sin(\omega t)$. This is the secular solution which diverges as t becomes large.] Thus removal of the secular terms in the RHS of (B.6) requires

$$2\frac{\partial A}{\partial t_1} - A + A^2 A^* = 0 \tag{B.8}$$

$$2\frac{\partial A^*}{\partial t_1} - A^* + A^{*2} A = 0 \tag{B.9}$$

Since (B.8) and (B.9) are complex conjugate to each other, it is sufficient to consider only one of them. In fact (B.8) is the condition to determine the undetermined coefficient A in the zero order solution (B.7). In order to solve (B.8), we write A in the form

$$A = \frac{1}{2}\alpha(t_1)e^{i\beta(t_1)} \tag{B.10}$$

with α and β real function of t_1. Then (B.7) becomes

$$x_0 = \alpha(t_1)\cos(t_0 + \beta(t_1)) \tag{B.11}$$

Substituting (B.10) into (B.8) gives

$$\frac{d\alpha}{dt_1} + i\alpha\frac{d\beta}{dt_1} - \frac{\alpha}{2} + \frac{\alpha^3}{8} = 0$$

Separating real and imaginary parts, we obtain

$$\frac{d\alpha}{dt_1} = \frac{\alpha}{2} - \frac{\alpha^3}{8} \tag{B.12}$$

$$\frac{d\beta}{dt_1} = 0 \quad \text{or} \quad \beta = \beta_0 \ = \ \text{const.} \tag{B.13}$$

We solve (B.12) by writing as

$$dt_1 = \frac{8d\alpha}{4\alpha - \alpha^3} = \frac{8d\alpha}{\alpha(2-\alpha)(2+\alpha)} = \frac{2d\alpha}{\alpha} + \frac{d\alpha}{2-\alpha} - \frac{d\alpha}{2+\alpha}$$

Integrating the above equation gives

$$t_1 + c = 2\ln\alpha - \ln|2-\alpha| - \ln(2+\alpha) \quad (c: \ \text{const.})$$

$$\text{or} \quad t_1 + c = \ln\frac{\alpha^2}{|4-\alpha^2|}$$

$$\text{or} \quad \frac{\alpha^2}{4-\alpha^2} = e^{t_1+c} = e^{\epsilon t+c} \tag{B.14}$$

Solving (B.14) for α^2 gives

$$\alpha^2 = \frac{4e^{\epsilon t+c}}{1+e^{\epsilon t+c}} = \frac{4}{1+e^{-\epsilon t-c}}$$

$$\text{or} \quad \alpha(t) = 2(1+e^{-\epsilon t-c})^{-\frac{1}{2}}$$

The constant c can be determined in terms of the 'initial' amplitude $\alpha_0 = \alpha(t_1 = 0)$, giving the final expression for $x(t)$:

$$x(t) = \frac{2}{1+\left(\frac{4}{\alpha_0^2}-1\right)e^{-\epsilon t}}\cos(t+\beta_0) \tag{B.15}$$

Equation (B.15) shows the secularity-free behavior that $x \to 2\cos(t+\beta_0)$ as $t \to \infty$, regardless of the value of α_0.

As another example we shall consider the *Duffing equation*:

$$\frac{d^2u}{dt^2} + f(u) = 0$$

where $f(u)$ is a polynomial of u. We consider a special case

$$\frac{d^2u}{dt^2} + u + \epsilon u^3 = 0 \quad (\epsilon \ll 1) \tag{B.16}$$

We point out that (B.16) allows for exact solutions by the technique of 'energy integral'. Here we use multiple time scale perturbation method. Expand u as

$$u(t) = u_0(t_0, t_1, \ldots) + \epsilon u_1(t_0, t_1, \ldots) + \cdots \tag{B.17}$$

Substituting (B.10) and (B.17) into (B.16) and breaking down the resultant equation order by order, we obtain

$$\epsilon^0 : \quad \frac{\partial^2 u_0}{\partial t_0^2} + u_0 = 0 \tag{B.18}$$

$$\epsilon^1 : \quad \frac{\partial^2 u_1}{\partial t_0^2} + u_1 = -2\frac{\partial^2 u_0}{\partial t_0 \partial t_1} - u_0^3 \tag{B.19}$$

The general solution of (B.18) can be written as

$$u_0 = \alpha(t_1, t_2, \ldots) \cos[t_0 + \beta(t_1, T_2, \ldots)] \tag{B.20}$$

[We can write the solution of (B.18) in terms of the exponential functions as in (B.7).] Substituting (B.20) into (B.19) yields

$$\frac{\partial^2 u_1}{\partial t_0^2} + u_1 = 2\frac{\partial \alpha}{\partial t_1} \sin(t_0 + \beta) + \left(2\alpha\frac{\partial \beta}{\partial t_1} - \frac{3}{4}\alpha^3\right) \cos(t_0 + \beta)$$
$$- \frac{1}{4}\alpha^3 \cos(3t_0 + 3\beta) \tag{B.21}$$

To eliminate the secular terms in the RHS of the above equation, we put

$$\frac{\partial \alpha}{\partial t_1} = 0$$

$$2\alpha \, \frac{\partial \beta}{\partial t_1} - \frac{3}{4}\alpha^3 = 0$$

Without the secularity-causing terms, (B.21) gives the particular solution

$$u_1 = \frac{1}{32}\alpha^3 \cos(3t_0 + 3\beta) \tag{B.22}$$

Appendix C

Strained Coordinate Perturbation Method

This is another perturbation method which is appropriate to extract the Korteweg–de Vries equation from the nonlinear ion acoustic equations (10.67)–(10.69) in Chapter 10. Here we write them again

$$\frac{\partial^2 \phi}{\partial x^2} = 4\pi e(n_0 e^{\frac{e\phi}{T_e}} - n_i) \tag{C.1}$$

$$\frac{\partial n_i}{\partial t} + \frac{\partial}{\partial x}(n_i v) = 0 \tag{C.2}$$

$$\frac{\partial v}{\partial t} + v\frac{\partial v}{\partial x} = -\frac{e}{m_i}\frac{\partial \phi}{\partial x} \tag{C.3}$$

It is more convenient to write the above equations in dimensionless form by introducing appropriate scaling factors:

$$n_i = n_0 n', \quad \phi = \frac{T_e}{e}\phi', \quad x = \lambda_e x', \quad t = \omega_{\rm pi}^{-1} t' \tag{C.4}$$

where n_0 is the equilibrium electron (or ion) number density, $\lambda_e = \sqrt{\frac{T_e}{4\pi n_0 e^2}}$ is the electron Debye length, $\omega_{\rm pi} = \sqrt{\frac{4\pi n_0 e^2}{m_i}}$ is ion plasma frequency, and the primed variables are dimensionless. Naturally the dimensionless velocity is $v' = \frac{v}{\lambda_e \omega_{\rm pi}}$ where $\lambda_e \omega_{\rm pi} = \sqrt{\frac{T_e}{m_i}}$ is the ion acoustic wave speed. Then (C.1)–(C.3) become

$$\frac{\partial^2 \phi'}{\partial x'^2} = e^{\phi'} - n' \tag{C.5}$$

$$\frac{\partial n'}{\partial t'} + \frac{\partial}{\partial x'}(n'v') = 0 \tag{C.6}$$

$$\frac{\partial v'}{\partial t'} + v'\frac{\partial v'}{\partial x'} = -\frac{\partial \phi'}{\partial x'} \tag{C.7}$$

The boundary conditions are now $n' = 1,\ v' = 0,\ \phi' = 0$ as $x' \to \pm\infty$. In the following, we omit the primes with the understanding that all the quantities are dimensionless. We expand

$$n = 1 + \epsilon n^{(1)} + \epsilon^2 n^{(2)} + \cdots$$

$$\phi = \epsilon\phi^{(1)} + \epsilon^2\phi^{(2)} + \cdots$$

$$v = \epsilon v^{(1)} + \epsilon^2 v^{(2)} + \cdots$$

We break down (C.5)–(C.7) order by order:

$$\frac{\partial n^{(1)}}{\partial t} + \frac{\partial v^{(1)}}{\partial x} = 0 \tag{C.8}$$

$$\frac{\partial v^{(1)}}{\partial t} + \frac{\partial \phi^{(1)}}{\partial x} = 0 \tag{C.9}$$

$$\frac{\partial^2 \phi^{(1)}}{\partial x^2} - \phi^{(1)} + n^{(1)} = 0 \tag{C.10}$$

$$\frac{\partial n^{(2)}}{\partial t} + \frac{\partial v^{(2)}}{\partial x} = -\frac{\partial}{\partial x}(n^{(1)}v^{(1)}) \tag{C.11}$$

$$\frac{\partial v^{(2)}}{\partial t} + \frac{\partial \phi^{(2)}}{\partial x} = -v^{(1)}\frac{v^{(1)}}{\partial x} \tag{C.12}$$

$$\frac{\partial^2 \phi^{(2)}}{\partial x^2} - \phi^{(2)} + n^{(2)} = \frac{1}{2}[\phi^{(1)}]^2 \tag{C.13}$$

The linear equations (C.8)–(C.10) are combined to yield the linear Boussinesq equation

$$\left(\frac{\partial^4}{\partial x^2 \partial t^2} + \frac{\partial^2}{\partial x^2} - \frac{\partial^2}{\partial t^2}\right)\phi^{(1)} = 0 \tag{C.14}$$

The nonlinear (ϵ^2 order) equations (C.11)–(C.13) yield nonlinear Boussinesq equation

$$\left(\frac{\partial^4}{\partial x^2 \partial t^2} + \frac{\partial^2}{\partial x^2} - \frac{\partial^2}{\partial t^2}\right)\phi^{(2)} = \frac{1}{2}\frac{\partial^2}{\partial t^2}[\phi^{(1)}]^2$$

$$-\frac{1}{2}\frac{\partial^2}{\partial x^2}[v^{(1)}]^2 + \frac{\partial^2}{\partial x \partial t}[n^{(1)}v^{(1)}] \tag{C.15}$$

The foregoing algebra shows that a straight expansion in the series of ϵ gives Boussinesq equation. In order to derive KdV equation, we should stretch the x and t coordinates *appropriately*. In order to guess the appropriate stretching, we consider the dispersion relation that is derived from the linear equation (C.14).

$$\omega^2 = \frac{k^2}{1+k^2} \quad \text{(dimensionless), or} \quad \omega^2 = \frac{\lambda_e^2 \omega_{pi}^2 k^2}{1+\lambda_e^2 k^2} \quad \text{(dimensional)} \tag{C.16}$$

Examination of the linear dispersion relation (C.16) can give a clue as to how we stretch x and t coordinates in (C.5)–(C.7) to isolate the KdV equation in the limit of long wavelength. The x and t dependence of the linear wave is through the phasor $\theta(x,t) = kx - \omega t$. Expanding ω in (C.16) in the powers of k (small in the long wave region), we obtain $\theta = k(x \pm t) \pm \frac{1}{2}k^3 + \cdots$. This suggests that the x and t coordinates are stretched as

$$\xi = \epsilon^{\frac{1}{2}}(x - \lambda t), \quad \tau = \epsilon^{\frac{3}{2}} t \tag{C.17}$$

As might be expected, the acceptable wave speed λ turns out to be ± 1. The fractional power on ϵ is introduced to terminate perturbation equation at the lowest order (guided by algebraic scrutiny). In accordance with (C.17), we transform as

$$\frac{\partial}{\partial x} = \epsilon^{\frac{1}{2}} \frac{\partial}{\partial \xi}, \quad \frac{\partial}{\partial t} = \epsilon^{\frac{3}{2}} \frac{\partial}{\partial \tau} - \epsilon^{\frac{1}{2}} \lambda \frac{\partial}{\partial \xi} \tag{C.18}$$

Then (C.5) gives in the lowest two orders of ϵ

$$\phi^{(1)} = n^{(1)} \tag{C.19}$$

$$\frac{\partial^2}{\partial \xi^2}\phi^{(1)} = -n^{(2)} + \phi^{(2)} + \frac{1}{2}[\phi^{(1)}]^2 \tag{C.20}$$

From (C.6) we obtain

$$\epsilon^{\frac{3}{2}}: \quad \frac{\partial v^{(1)}}{\partial \xi} - \lambda \frac{\partial n^{(1)}}{\partial \xi} = 0 \tag{C.21}$$

$$\epsilon^{\frac{5}{2}}: \quad \frac{\partial n^{(1)}}{\partial \tau} - \lambda \frac{\partial n^{(2)}}{\partial \xi} + \frac{\partial}{\partial \xi}[v^{(2)} + v^{(1)} n^{(1)}] = 0 \tag{C.22}$$

Next, (C.7) yields

$$\epsilon^{\frac{3}{2}}: \quad -\lambda \frac{\partial v^{(1)}}{\partial \xi} + \frac{\partial \phi^{(1)}}{\partial \xi} = 0 \tag{C.23}$$

$$\epsilon^{\frac{5}{2}}: \quad \frac{\partial v^{(1)}}{\partial \tau} - \lambda \frac{\partial v^{(2)}}{\partial \xi} + \frac{\partial \phi^{(2)}}{\partial \xi} + v^{(1)} \frac{\partial v^{(1)}}{\partial \xi} = 0 \tag{C.24}$$

Equations (C.19), (C.21), and (C.23) give $\lambda^2 n^{(1)} = n^{(1)}$ or $\lambda \pm 1$. We first consider the case $\lambda = 1$. Then we have

$$\phi^{(1)} = n^{(1)} = v^{(1)} \tag{C.25}$$

Eliminating $n^{(2)}$ and $v^{(2)}$ from (C.20), (C.22), and (C.24), and using (C.25), we obtain the evolution equation for $n^{(1)}$:

$$\frac{\partial n^{(1)}}{\partial \tau} + n^{(1)} \frac{\partial n^{(1)}}{\partial \xi} + \frac{1}{2} \frac{\partial^3 n^{(1)}}{\partial \xi^3} = 0 \tag{C.26}$$

which is the KdV equation. It is to be noted that the equations (C.19)–(C.24) are closed: no equations and no terms are left over.

Appendix D

Vector Algebra

D.1 e-Symbol

A vector is denoted by a boldface letter symbol. A vector may also be represented by its components using indexes. For example,

$$\mathbf{A} = (A_1, A_2, A_3) = A_i \quad (i = 1, 2, 3) \tag{D.1}$$

The indexes $i = 1, 2, 3$ corresponds to x, y, z component, respectively.

The scalar product of two vectors $\mathbf{A}$ and $\mathbf{B}$ is written in index notation as

$$\mathbf{A} \cdot \mathbf{B} = \sum_{i=1}^{3} A_i B_i = A_i B_i \quad \text{(summation convention)} \tag{D.2}$$

The summation convention is an agreement that repeated indexes are summed over.

Vector algebra is greatly facilitated by introducing the three dimensional *e-symbol* (Levi-Civita symbol) e_{ijk} defined by

$$e_{ijk} = \begin{cases} 1 & \text{for even permutation of } ijk \\ -1 & \text{for odd permutation of } ijk \\ 0 & \text{for two or more indexes equal} \end{cases} \tag{D.3}$$

Thus we have

$$e_{123} = e_{231} = e_{312} = 1$$

$$e_{132} = e_{213} = e_{321} = -1$$

$$\text{all other } e_{ijk} = 0$$

The e-symbol may be thought as three dimensional extension of the Kronecker delta,

$$\delta_{ij} = \begin{cases} 1 & \text{if } i = j \\ 0 & \text{if } i \neq j \end{cases} \tag{D.4}$$

Let us consider a vector product

$$\mathbf{A} \times \mathbf{B} = (A_2B_3 - A_3B_2, A_3B_1 - A_1B_3, A_1B_2 - A_2B_1) \tag{D.5}$$

Also consider

$$e_{ikm}A_kB_m \tag{D.6}$$

Expression (D.6) obviously represents a vector since there is only one free index (i) while the indexes k and m are to be summed over 1 to 3. (D.6) gives the following values:

$$\begin{aligned} &\text{when } i = 1, \quad e_{1km}A_kB_m = e_{123}A_2B_3 + e_{132}A_3B_2 = A_2B_3 - A_3B_2 \\ &\text{when } i = 2, \quad e_{2km}A_kB_m = e_{213}A_1B_3 + e_{231}A_3B_1 = A_3B_1 - A_1B_3 \\ &\text{when } i = 3, \quad e_{3km}A_kB_m = e_{312}A_1B_2 + e_{321}A_2B_1 = A_1B_2 - A_2B_1 \end{aligned}$$

Thus, in view of (D.5), we can write

$$(\mathbf{A} \times \mathbf{B})_i = e_{ikm}A_kB_m \tag{D.7}$$

Example. Triple scalar product.

$$\begin{aligned} \mathbf{C} \cdot \mathbf{A} \times \mathbf{B} &= C_i(\mathbf{A} \times \mathbf{B})_i = C_ie_{ikm}A_kB_m \\ &= B_me_{mik}C_iA_k \quad (\textit{per } e_{ikm} = e_{mik}) \\ &= B_m(\mathbf{C} \times \mathbf{A})_m \\ &= \mathbf{B} \cdot \mathbf{C} \times \mathbf{A} \end{aligned}$$

Problem. Prove the relation

$$e_{ikm}e_{psm} = \delta_{ip}\delta_{ks} - \delta_{is}\delta_{kp} \tag{D.8}$$

Equation (D.8) is an important relation often encountered in vector algebra. Its usefulness will be demonstrated in the following example.

Triple vector product.

$$\begin{aligned} \mathbf{A} \times (\mathbf{B} \times \mathbf{C}) &= e_{ijk}A_j(\mathbf{B} \times \mathbf{C})_k \\ &= e_{ijk}A_je_{klm}B_lC_m \\ &= e_{ijk}A_je_{lmk}B_lC_m \\ &= A_j(B_iC_j - B_jC_i) \\ &= \mathbf{B}(\mathbf{A} \cdot \mathbf{C}) - \mathbf{C}(\mathbf{A} \cdot \mathbf{B}) \end{aligned}$$

D.2 Vector Differentiation

The vector differential operator ∇ can be defined as

$$\nabla = \frac{\partial}{\partial x_i} \tag{D.9}$$

In view of (D.9), the three vector derivatives are written

$$\text{gradient: } \nabla\phi = \frac{\partial}{\partial x_i} \tag{D.10}$$

$$\text{divergence: } \nabla \cdot \mathbf{v} = \frac{\partial v_i}{\partial x_i} \tag{D.11}$$

$$\text{curl: } \nabla \times \mathbf{v} = e_{ijk}\,\frac{\partial v_k}{\partial x_j} \tag{D.12}$$

When the vector $\mathbf{v}$ in (D.11) is a position vector, $\mathbf{v} = x_i$ $(i = 1, 2, 3)$,

$$\nabla \cdot \mathbf{r} = \frac{\partial x_i}{\partial x_i} = 3 \tag{D.13}$$

Also

$$\nabla \cdot \nabla\phi = \frac{\partial}{\partial x_i}\,\frac{\partial \phi}{\partial x_i} = \frac{\partial^2 \phi}{\partial x_1^2} + \frac{\partial^2 \phi}{\partial x_2^2} + \frac{\partial^2 \phi}{\partial x_3^2} = \nabla^2\phi \tag{D.14}$$

$$\begin{aligned}
\nabla \times \nabla \times \mathbf{v} &= e_{ijk}\,\frac{\partial}{\partial x_j}(\nabla \times \mathbf{v})_k \\
&= e_{ijk}\,\frac{\partial}{\partial x_j} e_{klm}\,\frac{\partial v_m}{\partial v_l} \\
&= \frac{\partial}{\partial x_j}\left(\frac{\partial v_j}{\partial x_i} - \frac{\partial v_i}{\partial x_j}\right) \\
&= \nabla\nabla \cdot \mathbf{v} - \nabla^2 \mathbf{v}
\end{aligned} \tag{D.15}$$

$$\begin{aligned}
\mathbf{v} \times \nabla \times \mathbf{v} &= e_{ijk} v_j e_{klm}\,\frac{\partial v_m}{\partial x_l} \\
&= v_j\left(\frac{\partial v_j}{\partial x_i} - \frac{\partial v_i}{\partial x_j}\right) \\
&= \frac{1}{2}\,\frac{\partial}{\partial x_i}(v_j v_j) - v_j\,\frac{\partial v_i}{\partial x_j} \\
&= \frac{1}{2}\nabla v^2 - (\mathbf{v} \cdot \nabla)\mathbf{v}
\end{aligned} \tag{D.16}$$

D.3 Vector Integration

Without proof we state that

$$\int_S \mathbf{A} \cdot d\mathbf{S} = \int_V \nabla \cdot \mathbf{A} d\tau \tag{D.17}$$

where S is some closed surface and V is the volume bounded by S. The conversion of the surface integral into the volume integral is known as *Gauss theorem* or *divergence theorem*. Equation (D.17) is written in index notation

$$\int_S A_i dS_i = \int_V \frac{\partial A_i}{\partial x_i} d\tau \tag{D.18}$$

Consider

$$\begin{aligned}\int_S d\mathbf{S} \times \mathbf{A} &= \int_S e_{ijk} dS_j A_k d\tau \\ &= \int_V \frac{\partial}{\partial x_j} e_{ijk} A_k d\tau\end{aligned}$$

where we used (D.19). Hence,

$$\int_S d\mathbf{S} \times \mathbf{A} = \int_S \nabla \times \mathbf{A}\ d\tau \tag{D.19}$$

We have

$$\int_S \phi d\mathbf{S} = \int_S \phi dS_i = \int_V \frac{\partial \phi}{\partial x_i} d\tau = \int_V \nabla \phi\ d\tau \tag{D.20}$$

$$\int_S \underline{\sigma} d\mathbf{S} = \int_S \sigma_{ij} dS_j = \int_V \frac{\partial \sigma_{ij}}{\partial x_j}\ d\tau = \int_V \nabla \cdot \underline{\sigma}\ d\tau \tag{D.21}$$

We write without proof the Stokes theorem:

$$\int_C \mathbf{A} \cdot d\mathbf{x} = \int_S \nabla \times \mathbf{A} \cdot d\mathbf{S} \tag{D.22}$$

where **S** is some open surface and C is the rim of **S**. In index notation, (D.22) is written as

$$\int_C A_i dx_i = \int_S e_{ijk} \frac{\partial A_k}{\partial x_j} dS_i \tag{D.23}$$

Putting formally $A_i \to \phi$ (scalar) into (D.23), we have

$$\int_C \phi d\mathbf{x} = \int_S e_{ijk} \frac{\partial \phi}{\partial x_j} dS_i = \int_S e_{kij} dS_i \frac{\partial \phi}{\partial x_j} = \int_S d\mathbf{S} \times \nabla \phi \tag{D.24}$$

D.4 Vector Identities

(1) $\mathbf{A} \cdot \mathbf{B} \times \mathbf{C} = \mathbf{A} \times \mathbf{B} \cdot \mathbf{C}$
(2) $\mathbf{A} \times (\mathbf{B} \times \mathbf{C}) = (\mathbf{A} \cdot \mathbf{C})\mathbf{B} - (\mathbf{A} \cdot \mathbf{B})\mathbf{C}$
(3) $(\mathbf{A} \times \mathbf{B}) \cdot (\mathbf{C} \times \mathbf{D}) = (\mathbf{A} \cdot \mathbf{C})(\mathbf{B} \cdot \mathbf{D}) - (\mathbf{A} \cdot \mathbf{D})(\mathbf{B} \cdot \mathbf{C})$
(4) $(\mathbf{A} \times \mathbf{B}) \times (\mathbf{C} \times \mathbf{D}) = (\mathbf{A} \times \mathbf{B} \cdot \mathbf{D})\mathbf{C} - (\mathbf{A} \times \mathbf{B} \cdot \mathbf{C})\mathbf{D}$
(5) $\nabla(fg) = f\nabla g + g\nabla f$
(6) $\nabla \cdot (f\mathbf{A}) = f\nabla \cdot \mathbf{A} + \mathbf{A} \cdot \nabla f$
(7) $\nabla \times (f\mathbf{A}) = f\nabla \times \mathbf{A} + \nabla f \times \mathbf{A}$
(8) $\nabla \cdot (\mathbf{A} \times \mathbf{B}) = \mathbf{B} \cdot \nabla \times \mathbf{A} - \mathbf{A} \cdot \nabla \times \mathbf{B}$
(9) $\nabla \times (\mathbf{A} \times \mathbf{B}) = \mathbf{A}\nabla \cdot \mathbf{B} - \mathbf{B}\nabla \cdot \mathbf{A} + (\mathbf{B} \cdot \nabla)\mathbf{A} - (\mathbf{A} \cdot \nabla)\mathbf{B}$
(10) $\mathbf{A} \times (\nabla \times \mathbf{B}) = (\nabla\mathbf{B}) \cdot \mathbf{A} - (\mathbf{A} \cdot \nabla)\mathbf{B}$
(11) $\nabla(\mathbf{A} \cdot \mathbf{B}) = \mathbf{A} \times (\nabla \times \mathbf{B}) + \mathbf{B} \times (\nabla \times \mathbf{A}) + (\mathbf{A} \cdot \nabla)\mathbf{B} + (\mathbf{B} \cdot \nabla)\mathbf{A}$
(12) $\nabla^2 f = \nabla \cdot \nabla f$
(13) $\nabla^2 \mathbf{A} = \nabla(\nabla \cdot \mathbf{A}) - \nabla \times \nabla \times \mathbf{A}$
(14) $\nabla \times \nabla f = 0$
(15) $\nabla \cdot \nabla \times \mathbf{A} = 0$
(16) $\frac{1}{2}\nabla \mathbf{A}^2 = \mathbf{A} \times \nabla \times \mathbf{A} + (\mathbf{A} \cdot \nabla)\mathbf{A}$

If $\mathbf{r}$ is the position vector drawn from the origin to a general point (x, y, z), then

$$\nabla \times \mathbf{r} = 0 \tag{D.25}$$

$$\nabla r = \frac{\mathbf{r}}{r} \tag{D.26}$$

$$\nabla \frac{1}{r} = -\frac{\mathbf{r}}{r^3} \tag{D.27}$$

$$\nabla \cdot \frac{\mathbf{r}}{r^3} = -\nabla^2 \frac{1}{r} = 4\pi\delta(\mathbf{r}) \tag{D.28}$$

D.5 Cylindrical Coordinates

Cylindrical coordinates (r, ϕ, z).

$$\nabla \cdot \mathbf{A} = \frac{1}{r}\frac{\partial}{\partial r}(rA_r) + \frac{1}{r}\frac{\partial A_\phi}{\partial \phi} + \frac{\partial A_z}{\partial z}$$

$$\nabla f = \hat{\mathbf{r}}\frac{\partial f}{\partial r} + \hat{\phi}\frac{1}{r}\frac{\partial f}{\partial \phi} + \hat{\mathbf{z}}\frac{\partial f}{\partial z}$$

$$(\nabla \times \mathbf{A})_r = \frac{1}{r}\frac{\partial A_z}{\partial \phi} - \frac{\partial A_\phi}{\partial z}$$

$$(\nabla \times \mathbf{A})_\phi = \frac{\partial A_r}{\partial z} - \frac{\partial A_z}{\partial r}$$

$$(\nabla \times \mathbf{A})_z = \frac{1}{r}\frac{\partial}{\partial r}(rA_\phi) - \frac{1}{r}\frac{\partial A_r}{\partial \phi}$$

$$\nabla^2 f = \frac{1}{r}\frac{\partial}{\partial r}\left(r\frac{\partial f}{\partial r}\right) + \frac{1}{r^2}\frac{\partial^2 f}{\partial \phi^2} + \frac{\partial^2 f}{\partial z^2}$$

$$(\nabla^2 \mathbf{A})_r = \nabla^2 A_r - \frac{2}{r^2}\frac{\partial A_\phi}{\partial \phi} - \frac{A_r}{r^2}$$

$$(\nabla^2 \mathbf{A})_\phi = \nabla^2 A_\phi + \frac{2}{r^2}\frac{\partial A_r}{\partial \phi} - \frac{A_\phi}{r^2}$$

$$(\nabla^2 \mathbf{A})_z = \nabla^2 A_z$$

Unit vectors

$$\hat{\mathbf{r}} = \hat{\mathbf{x}}\ \cos\phi + \hat{\mathbf{y}}\ \sin\phi$$
$$\hat{\phi} = -\hat{\mathbf{x}}\ \sin\phi + \hat{\mathbf{y}}\ \cos\phi$$
$$\hat{\mathbf{z}} = \hat{\mathbf{z}}$$
$$\hat{\mathbf{x}} = \hat{\mathbf{r}}\ \cos\phi - \hat{\phi}\ \sin\phi$$
$$\hat{\mathbf{y}} = \hat{\mathbf{r}}\ \sin\phi + \hat{\phi}\ \cos\phi$$
$$\hat{\mathbf{z}} = \hat{\mathbf{z}}$$

Bibliography

The following list of references includes the papers and the books that are referred in the text as well as materials for general background and further reading. The relevant chapters are not indicated but the *titles* of the material are furnished.

Bibliography

Akhiezer, A., I. Akhiezer, R. Polovin, A. Sitenko, and K. Stepanov, *Collective Oscillations in a Plasma*, Pergamon, Oxford (1967).

Alfven, H. and C.-G. Fälthammar, *Cosmical electrodynamics*, Clarendon, Oxford (1963).

Al'tshul L. M. and V. I. Karpman, *The kinetics of waves in a weakly turbulent plasma*, Sov Phys. JETP **20** (1965) pp. 1043–1056.

Al'Tshul', L. M. and V. I. Karpman, *Theory of nonlinear oscillations in a collisionless plasma*, Sov. Phys. JETP, **22** (1966) pp. 361–369.

Alexandrov, A. F., L. S. Bogdankevich, and A. A. Rukhadze, *Principles of Plasma Electrodynamics*, Springer-Verlag, New York (1984).

Arfken, G. B. and H. J. Weber, *Mathematical Methods for Physicists*, 5th ed. Academic Press, Harcourt (2001).

Barr, H. C. and T. J. M. Boyd, *Surface waves in hot plasmas*, J. Phys. A: Gen. Phys. **5** pp. 1108–1118.

Bekefi, G. *Radiation Processes in Plasmas*, John Wiley, New York (1966).

Benny, D. J. *A general theory for interactions between long and short waves*, Studies in Appl. Math., **56** pp. 81–94.

Bernstein, I. B., J. M. Greene, and M. D. Kruskal, *Exact nonlinear plasma oscillations*, Phys. Rev. **108** (1957) p. 546.

Bernstein, I. B. *Waves in a plasma in a magnetic field*, Phys. Rev., **109** (1958) p. 10.

Bhatnagar, P. L., E. P. Gross, and M. Krook, Phys. Rev. **94** (1954) p. 511.

Bodner, S. E. and E. A. Frieman, *The quasi-linear theory of plasma oscillations and instability*, in the eighth Lockheed symposium of magnetohydrodynamics, Stanford University Press, Stanford, California (1964).

Bohm, D. and E. P. Gross, *Theory of plasma oscillations. A. Origin of medium-like behavior*, Phys. Rev. **75** (1949) p. 1851.

Bohm, D. and E. P. Gross, *Theory of plasma oscillations. B. Excitation and damping of oscillations*, Phys. Rev. **75** (1949) p. 1864.

Bohm, D. and E. P. Gross, *Effects of plasma boundaries in plasma oscillations*, Phys. Rev. **79** (1950) p. 992.

Braginskii, S. I. and A. P. Kazantsev, *MHD eaves in a rarefied plasma*–the Journal is unknown to the author. (Kinetic derivation of Alfven wave) (1957).

Braginskii, S. I. *Transport phenomena in a completely ionized Two-temperature plasma,* Soviet Phys. JETP, **6** (1958) p. 358.

Braginskii, S. I. *Transport Process in a Plasma*, in Reviews of plasma physics Vol. 1: Consultant's Bureau, New York (1965).

Brambilla, M. *Kinetic Theory of Plasma Waves*, Clarendon, Oxford (1998) p. 43.

Briggs, R. J. *Electron-Stream Interaction with Plasmas*, MIT Press, Cambridge, Massachusetts, (1964).

Camac, M., A. R. Kantrowitz, M. M. Litvak, R. M. Patrick, and H. E. Petschek, *Shock waves in collision-free plasmas*, Nuc. Fusion Supplement **Pt. 2** (1962) pp. 423–444.

Chandrasekhar, S. *Adiabatic invariants in the motions of charged particles: The plasma in a magnetic field*, A symposium on magnetohydrodynamics, Ed. Landshoff, R. K. M., Stanford University Press (1958).

Chandrasekhar, S., A. N. Kaufman, and K. M. Watson, *Properties of an ionized gas at low density in a magnetic field*, Annals Phys. **5** (1958) p. 1.

Chandrasekhar, S. *Plasma Physics*; Notes compiled by S. K. Trehan, Phoenix Books, University of Chicago Press (1960).

Chandrasekhar, S. *Hydrodynamic and Hydromagnetic Stability*, Dover, New York (1961) Chapter X, XI.

Chapman, S. and T. G. Cowling, *The Mathematical Theory of non-Uniform Gases*, 2nd ed. Cambridge University Press (1964).

Chew, G. F., M. L. Goldberger, and F. E. Low, *The Boltzmann equation and the one-fluid hydromagnetic equations in the absence of particle collisions*, Proc. Royal Soc. London **236. A** (1956) p. 112.

Clemmow, P. C. and J. P. Daugherty, *Electrodynamics of Particles and Plasmas*, Addison-Wesley, London (1969) p. 405.

Cohen, B. I. and C. E. Max, *Stimulated scattering of light by ion modes in a homogeneous plasma: Space-time evolution*, Phys. Fluids **22** (1979) p. 1115.

Cohen, B. I. *Compact dispersion relations for parametric instabilities of electromagnetic waves in magnetized plasmas*, Phys. Fluids **30** (1987) p. 2676.

Davidson, R. C. *Methods in Nonlinear Plasma Theory*, Academic Press, New York (1972).

Dawson, J. and C. Oberman, *High frequency conductivity and the emission and absorption coefficients of a fully ionized plasmas*, Phys. Fluids **5** p. 517

Dodd, R. K., J. C. Eilbeck, J. D. Gibbon, and H. C. Morris, *Solitons and Nonlinear Wave Equations*, Academic Press, New York (1982).

Drake, J. F., P. K. Kaw, Y. C. Lee, G. Schmidt, C. S. Liu, and M. N. Rosenbluth, *Parametric instabilities of electromagnetic waves in plasmas,* Phys. Fluids **17** (1974) p. 778.

Drummond, W. E. and D. Pines, *Nonlinear stability of plasma oscillations*, Nucl. Fusion: 1962 Supplement, **Pt 3** (1962) p. 1049.

Engel, R. D. *Nonlinear stability of of the extraordinary wave in a plasma*, Phys. Fluids **8** (1965) p. 939.

Fejer, J. A. and E. Leer, *Purely growing parametric instability in an inhomogeneous plasma,* J. Geophys. Res. **77** (1972) p. 700.

Fejer, J. A. and E. Leer, *Excitation of parametric instabilities by radio waves in the ionosphere*, Radio Science **7** (1972) p. 481.

Fejer, J. A. and Y.-Y. Kuo *Structure in the nonlinear saturation spectrum of parametric instabilities,* Phys. Fluids **16** No. 9 (1973) pp. 1490–1496.

Gould, R. W., T. M. O'Neil and J. H. Malmberg, *Plasma wave echo*, Phys. Rev. Lett. **19** (1967) pp. 219–222.

Gradov, O. M. and L. Stenflo, *Linear theory of a cold bounded plasma*, Phys. Rep. **94** (1983) p. 111–137.

Hall, L. S. and W. Heckrotte, *Instabilities: convective versus absolute*, Phys. Rev. **166** (1968) p. 120.

Hildebrand, F. B. *Advanced Calculus for Applications*, Prentice-Hall, Englewood Cliffs, N.J. (1976).

Hildebrand, F. B. *Methods of Applied Mathematics*, Prentice-Hall, Englewood Cliffs, N.J. (1965).

Infeld, E. *Nonlinear evolution of the Kruskal-Schwarzschild instability of a plasma above a vacuum*, Phys. Rev. A **39** (1989) pp. 723–727.

Jackson, J. D. *Longitudinal plasma oscillations*, J. Nucl. Energy Part C **1** (1960) p. 171.

Kadomtsev, B. B. *Plasma Turbulence*, Academic Press (1965), p. 18.

Kadomtsev, B. B. *Cooperative effects in plasmas*, Reviews of Plasma Physics Vol 22 Ed. V. D. Shafranov, Consultant Bureau, New York (2001).

Karpman, V. I. *Nonlinear Waves in Dispersive Media*, Pergamon (1975).

Kaw, P. K. and J. B. McBride, *Surface waves on a plasma,* Phys. Fluids **13** (1970) pp. 1784–1793.

Kaw, P. K. *Parametric excitation of electromagnetic waves in magnetized plasmas*, Advances in Plasma Physics Vol. 6, Ed. by A Simon and W. B. Thompson, Wiley, New York (1976).

Klimontovich, Y. L. *The Statistical Theory of non-Equilibrium Processes in a Plasma*, MIT Press Cambridge, Masschusetts (1967).

Korteweg, D. J. and G. de Vries, *On the change of form of long waves advancing in a rectangular canal, and on a new type of long stationary waves*, Phil. Mag., **39** (1895) pp. 422–443.

Krall, N. A. and A. W. Trivelpiece, *Principles of Plasma Physics*, San Francisco Press (1986).

Kruer, W. L. *The Physics of Laser Plasma Interactions*, Addison-Wesley (1988).

Kulsrud, R. M. *Adiabatic invariant of the harmonic oscillator*, Phys. Rev. **106** (1957) pp. 205–210.

Kurlsrud, R. M. *Adiabatic invariant of the harmonic oscillator*, Proceedings of the International School of Physics Enrico Fermi Course XXV Advanced Plasma Theory Academic Press (1964).

Kuo, Y.-Y. and J. A. Fejer *Spectral-line structures of saturated parametric instabilities*, Phys. Rev. Lett. **29** No. 25 (1972) pp. 1667–1670.

Landau, L. *On the vibrations of the electronic plasma*, J. Physics (USSR) **10** (1946) p. 25.

Landau, L. D. and E. M. Lifshitz, *Electrodynamics of Continuous Media*, Pergamon (1960) p. 288.

Landau, L. D. *Mechanics*, Pergamon, Oxford (1960).

Landau, L. D. *The transport equation in the case of Coulomb interactions*, in Collected papers of L. D. Landau, Ed. D. Ter Haar, Pergamon, Oxford (1965) p. 163.

Laval, G. and R. Pellat, *Quasilinear Theory*, Plasma physics, Universite de Grenoble, Summer school of theoretical phys. Les Houches Ed. C. Dewitt and J. Peyraud (1972).

Lee, H. J., J. F. McKenzie, and W. I. Axford, *Discontinuous transitions in a current carrying plasma* Astrophys. and Space Sci. **51** (1977) pp. 3–32.

Lee, H. J. *Surface waves in magnetized thermal plasma*, J. Korean Phys. Soc. **28** (1995) pp. 51–59.

Lee, H. J. *Electrostatic surface waves in a magnetized two-fluid plasma*, Plasma Phys. Control. Fusion **37** (1995) pp. 755–762.

Lee, H. J. and S. H. Cho, *Boundary conditions for surface waves propagating along the interface of plasma flow and free space*, J. Plasma Phys. **58** pt. 3 (1997) pp. 409–419.

Lee, H. J. and Y. K. Lim, *Modulation of a TM mode surface wave on a free boundary between plasma and vacuum*, Plasma Phys. and Control. Fusion, **40** (1998) pp 1619–1633.

Lee, H. J. and Y. K. Lim, *Kinetic theory of surface waves in a plasma slab*, J. Korean Phys. Soc. **50** (2007) pp. 1056–1061.

Lee, H. J. and C.-G. Kim, *Kinetic theory of electrostatic surface waves in magnetized plasmas,* J. Korean Phys. Soc. **54** (2009) pp. 85–93.

Lee, H. J. and M. Y. Song, *Causality and collisionless damping in plasma,* J. Modern Phys. **4** (2013) pp. 555–558.

Lee, H. J. *Fluctuation of the electric field in a plasma*, J. Korean Phys. Soc. **66** (2015) pp. 1167–1185.

Lee H. J. and M. J. Lee, *Surface wave echo Echo in a semi-bounded plasma*, J. of Modern Phys. **7** (2016) pp. 1400–1412.

Lee, H. J. *Nonlinear analysis of the cold fluid-Poisson plasma by using the characteristic method*, J. of the Korean Phys. Soc. **69** No. 7 (2016) pp. 1191–1211.

Lee, H. J. *Causality in plasma electrodynamics*, J. Korean Phys. Soc. **73** No. 1 (2018) pp. 65–85.

Leibovich, S. and A. R. Seebass, *Nonlinear Waves*, Examples of dissipative and dispersive systems leading to the Burgers and the Korteweg-deVries equations, Cornell University Press, Ithaca (1974).

Liboff, R. L. *Kinetic theory: Classical, Quantum, and Relativistic Descriptions, Third edition*, Springer (2003).

Lifshitz, E. M. and Pitaevskii, L. P., *Physical Kinetics*, Vol. 10, Landau and Lifshitz Course of Theoretical Physics, Butterworth-Heinemann, Oxford (1981).

Lighthill, E. J. *Waves in Fluids*, Cambridge University Press (1978).

Longmire, C. L. *Elementary Plasma Physics*, Interscience Publishers (1963).

Mathews, J. and R. L. Walker, *Mathematical Methods of Physics*, Benjamin, New York (1970).

McBride, J. B. and P. K. Kaw, *Low frequency surface waves on a warm plasma*, Phys. Lett. **33A** (1970) pp. 72–73.

Montgomeery, D. C. and D. A. Tidman, *Plasma Kinetic Theory*, McGraw-Hill, New York, (1964).

Montgomery, D. C. *Theory of the Unmagnetized Plasma*, Gordon and Beach, New York, (1971).

Nayfeh, A. H. *Introduction to Perturbation Techniques*, Wiley, New York (1981).

Nicholson, D. R. *Introduction to Plasma Theory*, John Wiley, New York (1983).

Nishikawa, K. *Parametric excitation of coupled waves I. General formulation,* J. Phys. Soc. Japan **24** (1968) pp. 916–922.

Nishikawa, K. *Parametric excitation of coupled waves II. Parametric plasmon-photon interaction,* J. Phys. Soc. Japan **24** (1968) pp. 1152–1158.

Nishikawa, K. and M. Wakatani, *Plasma Physics*, Springer-Verlag, Heidelberg, (1990).

O'Neil, T. *Collisionless damping of nonlinear plasma oscillations,* Phys. Fluids **8** (1965) p. 2255.

Porkolab, M. and R. P. H. Chang, *Instabilities and induced scattering due to nonlinear Landau damping of longitudinal plasma waves in a magnetic field,* Phys. Fluids **15** (1972) p. 283.

Porkolab, M. and R. P. H. Chang, *Nonlinear wave effects in laboratory plasmas: A comparison between theory and experiment,* Rev. Modern Phys. **50** No. 4 (1978) pp. 746–794.

Reif, F. *Fundamentals of Statistical and Thermal Physics*, McGraw-Hill International Ed., (1985).

Rostoker, N. Nucl. Fusion **1** (1960) p. 101.

Rostoker, N. and M. N. Rosenbluth, Phys. Fluids **3** (1960) p. 1.

Sagdeev, R. Z. and A. A. Galeev, *Nonlinear Plasma Theory*, Benjamin, New York (1969).

Sagdeev, R. Z. *The 1976 Oppenheimer lectures: Critical problems in plasma astrophysics I. Turbulence and nonlinear waves*, Rev. Mod. Phys. **51** No. 1 (1979) pp. 1–9.

Sanmartin, J. R. *Electrostatic plasma instabilities excited by a high frequency electric field,* Phys. Fluids **13** (1970) pp. 1533–1542.

Schram, P. P. J. M. *Kinetic Theory of Gases and Plasmas*, Kluwer Academic, Dordrecht (1991).

Shivarova, A. and I. Zhelyazkov, *Surface waves in a homogeneous plasma sharply bounded by a dielectric*, Plasma Physics **20** (1978) pp. 1049–1073.

Shkarofsky, I. P., T. W. Johnston, and M. P. Bachynski, *The Particle Kinetics of Plasmas*, Addison-Wesley, London (1966).

Silin, V. P. *Parametric resonance in a plasma*, Sov. Phys. JETP **21** (1965) p. 1127; JETP **24** (1967) p. 1242.

Sitenko, A. G., Nguyen-Van-Chong, Pavlenko, V. N. *Contribution to the theory of echo phenomena in a plasma,* Nucl. Fusion **10** (1970) pp. 259–267.

Sitenko, A. G., V. N. Pavlenko and V. I. Zasenko, *Echo in a half-space plasma,* Phys. Letters **53A** (1975) pp. 259–260.

Sitenko, A. G., V. N. Pavlenko and V. I. Zasenko, *Echo surface waves in plasmas,* Soviet J. Plasma Phys. **2** (1976) pp. 448–450.

Sitenko, A. G., *Fluctuations and Non-linear Wave Interactions in Plasmas*, Pergamon, Oxford (1982).

Sitenko, A. and V. Malnev. *Plasma Physics Theory*, Chapman and Hall, London (1995).

Stenflo, L. *Nonlinear interaction of electrostatic waves in a plasma,* Ann. Physik. **7**. Folge, Bd. 27, Heft 3, (1971) S. 289–295.

Stenflo, L. *Stimulated scattering of electromagnetic waves by magnetosonic modes in a plasma,* J. Plasma Phys. **34**, Pt. 1, (1985) p. 95.

Stenflo, L. *Theory of nonlinear plasma surface waves,* Physica Scripta **T63** (1996) pp. 59–62.

Stenflo, L. *Resonant three-wave interactions in plasmas*, Physica Scripta **T50** pp. 15–19 (1994).

Stenflo, L. and P. K. Shukla, *Theory of stimulated scattering of large-amplitude waves*, J. Plasma Phys. **64** part 4 (2000) pp. 353–357.

Stenflo, L. *Stimulated scattering of large amplitude waves in the ionosphere*, Physica Scripta **T30** (1990) pp. 166–169.

Stix, T. H. *Waves in Plasmas*, AIP, New York (1992).

Sturrock, P. A. *Plasma Physics*, Cambridge University Press (1994).

Swanson, D. G. *Plasma Waves*, Academic Press, New York (1989).

Thau, S. A. *Linear Dispersive Waves* in Nonlinear waves, Ed. S. Leibovich and A. R. Seebass, Cornell University Press (1970).

Tzoar, N. *Parametric excitations in plasma in a magnetic field*, Phys. Rev. **178** (1969) pp. 356–362.

Van Kampen, N. G. and B. U. Felderhof, *Theoretical Methods in Plasma Physics*, John Wiley (1967), Chapter XV.

Vedenov, A. A. *Quasi-linear plasma theory (theory of a weakly turbulent plasma)*, Plasma Phys. (J. Nucl. Energy Part C) **5** (1963) p. 169.

Weinstock, R. *Calculus of Variations* McGraw-Hill (1952).

Whitham, G. B. *Linear and Nonlinear Waves*, Wiley, New York (1974) Chapter 2, Fig. 2.2.

Yourgrau, W. and S. Mandelstam, *Variational Principles in Dynamics and Quantum Theory*, Dover (1968).

Index

G

H

I

M

N

S

T

www.ingramcontent.com/pod-product-compliance
Lightning Source LLC
LaVergne TN
LVHW071411210926
848486LV00004B/242
9789819822737